M000312619

KENT'S TECHNOLOGY
OF CEREALS

Related titles

Kent's Technology of Cereals, 4th Edition; Woodhead Publishing
(ISBN 978-1-85573-660-3)

Cereal Biotechnology; Woodhead Publishing
(ISBN 978-1-85573-498-2)

Woodhead Publishing Series in Food Science,
Technology and Nutrition

KENT'S TECHNOLOGY OF CEREALS

AN INTRODUCTION FOR STUDENTS OF FOOD SCIENCE AND AGRICULTURE

FIFTH EDITION

KURT A. ROSENTRATER

A.D. EVERS

WP
WOODHEAD
PUBLISHING
An imprint of Elsevier

Woodhead Publishing is an imprint of Elsevier
The Officers' Mess Business Centre, Royston Road, Duxford, CB22 4QH, United Kingdom
50 Hampshire Street, 5th Floor, Cambridge, MA 02139, United States
The Boulevard, Langford Lane, Kidlington, OX5 1GB, United Kingdom

Notices
Knowledge and best practice in this field are constantly changing. As new research and experience broaden our understanding, changes in research methods, professional practices, or medical treatment may become necessary.

Practitioners and researchers must always rely on their own experience and knowledge in evaluating and using any information, methods, compounds, or experiments described herein. In using such information or methods they should be mindful of their own safety and the safety of others, including parties for whom they have a professional responsibility.

To the fullest extent of the law, neither the Publisher nor the authors, contributors, or editors, assume any liability for any injury and/or damage to persons or property as a matter of products liability, negligence or otherwise, or from any use or operation of any methods, products, instructions, or ideas contained in the material herein.

Library of Congress Cataloging-in-Publication Data
A catalog record for this book is available from the Library of Congress

British Library Cataloguing-in-Publication Data
A catalogue record for this book is available from the British Library

ISBN: 978-0-08-100529-3 (print)
ISBN: 978-0-08-100532-3 (online)

For information on all Woodhead Publishing publications
visit our website at https://www.elsevier.com/books-and-journals

Working together
to grow libraries in
developing countries

www.elsevier.com • www.bookaid.org

Publisher: Andre G. Wolff
Acquisition Editor: Nina Rosa Bandeira
Editorial Project Manager: Karen R. Miller
Production Project Manager: Debasish Ghosh
Cover Designer: Matthew Limbert

Typeset by TNQ Books and Journals

K. Rosentrater would like to dedicate this book to his family, Kari, Emma and Alec, and to thank them for their constant support, especially as he has pursued grain and food issues in various far-flung locales.

A. Evers would like to dedicate this book to his wife, Tessa, in recognition of the support she has provided, and tolerance she has shown, through 53 years of married life.

Tony Blakeney (1948–2015) was to have been a coauthor to this edition. Sadly, after a period of ill health he died before contracts were signed. Tony did much to promote education and exchange of knowledge about cereals, both in his native Australia and worldwide. It is appropriate and a pleasure that we recall here the generous scientist that Tony was.

Contents

Biography

Norman L. Kent

My father, Norman Kent, was born and brought up in South London, attending Battersea Grammar School. He won a scholarship to Emmanuel College, Cambridge, where he took a first in botany, and went on to carry out research into the diseases of celery, which led to a lifelong reluctance to eat it. He had gone up to Cambridge in 1934 and continued there, after being awarded his PhD, under the team led by Professor Robert McCance and Dr. Elsie Widdowson, on their work that was devoted to nutrition and in particular to the provision of adequate nutrients to the British nation during a time of shortage in World War II.

Sometime after the war, the United States government provided grants to distinguished scientists outside the United States, and my father received one of them. At the completion of the work he was invited to present his results in America, and this enabled him to visit a number of labs in that country.

My parents married in 1941 and moved to St. Albans, Hertfordshire, when my father took up a post at the British Flour Millers laboratory there, where he remained for the rest of his working life.

The same nutritional priority existed at the St. Albans labs and my father made a significant contribution to the collaborative work there. His own publications were on the occurrence of individual mineral elements in wheat and its milling products. As a botanical microscopist, he also studied the fine structure of the wheat grain and the fates of different cells during milling.

Besides being a meticulous scientist, he was a fine writer, and while at the St. Albans lab produced the *Bulletin* describing the work there. When the labs merged with the British Baking Industries Research Association at Chorleywood to form the Flour Milling and Baking Research Association, my father took charge of the information department, and as well as producing some of the publications himself, he oversaw the production of many of the periodicals that emanated from Chorleywood at that time. He continued in this role until he retired.

His writing skills were internationally valued for his participation in the drafting of many Standards for the International Standards Organisation, and he worked, too, for the Food and Agriculture Organisation.

While still at St. Albans, my father received a call from the University of Nottingham, which realized that it had inadequate expertise on its permanent staff to cover cereals and cereals processing. They asked for a series of lectures that he delivered to food science students to fill this gap. These lectures formed the basis for this book when another call came, this time from a publisher, for a student textbook. It was an immediate success, combining as it did, readability with concentrated information and insight. He produced the first three editions and then collaborated with former colleague Tony Evers on the fourth edition of the book after retirement. There are now few mills and cereals laboratories in which a copy of this book cannot be found.

For a scientist, my father was a surprisingly artistic man – an inspired gardener who created delightful gardens at the houses we lived in, an excellent photographer who delighted in the photographic opportunities travel afforded, and very musical. He played the piano well and was an accomplished singer, for many years a member of the Royal Choral Society in London.

He was also very active, rowing for Emmanuel College while at Cambridge, taking climbing holidays in North Wales with a group of friends every year, and cycling to and from the labs as long as they were in St. Albans. Walking in the Hertfordshire countryside with my mother became a regular activity after retirement and kept him fit well into his eighties.

I know he would have been gratified that his book continues to be valued by food scientists and to learn of this further revised edition. My mother and I are very grateful to Tony Evers and Kurt Rosentrater for undertaking to bring this seminal book up to date.

My father died in 2006.

Celia Kent

Preface to the fifth edition

With the publication of the first edition of this book in 1966, Dr. Norman Kent initiated what would become a classic text in the cereal grains literature. His intent was to provide a comprehensive introduction to all major aspects of cereal grains, their production and their processing, with the express intent to educate students. Since that time, the industry has seen many changes and new technologies. Even so, on a fundamental level, cereal grains, their chemistries and their processing operations, have changed only a little over the decades. The fourth edition was published in 1994, and while that version of the book has been a useful text for many students over the years, an updated version is necessary for a new generation of students.

During a beautiful April day a few years back we met with Celia Kent, daughter of the late Norman Kent, in the Royal Horticultural Society's garden at Wisley in Surrey, England, to discuss this undertaking and to plan a new edition of this work.

The new version of this classic book has been thoroughly updated and revised throughout, both in terms of trends and statistics and also processing operations. We have also rearranged some of the topics for what we hope is a better flow. It should serve as a timely and expansive resource that will be useful for students, researchers and industrial practitioners alike, covering the full spectrum of cereal grain production, chemistry, processing and uses in foods, feeds, fuels, industrial materials and other applications.

Kurt A. Rosentrater   A.D. Evers
May, 2017

Maize and soybean artwork by E. Rosentrater.

Preface to the fourth edition

The principal purpose of the fourth edition is to update the material – including the statistics – of the third edition, while maintaining an emphasis on nutrition and, in particular, the effects of processing on the nutritive value of the products as compared with that of the raw cereals.

However, some new material has been introduced, notably sections dealing with extrusion cooking and the use of cereals for animal feed, and the section on industrial uses for cereals has been considerably enlarged.

A change in the fourth edition, which readers of earlier editions will notice, is the order in which the material is presented. Instead of devoting a separate chapter to each of the cereals, other than wheat, chapters in the fourth edition are devoted to distinct subjects, e.g., dry milling, wet milling, malting, brewing and distilling, pasta, domestic and small-scale processing, feed and industrial uses, in each of which all the cereals, as may be appropriate, are considered. Besides avoiding a certain degree of repetition, we feel that this method of presentation may give a better understanding of the subject, particularly as to how the various cereals compare with each other.

Preface to the third edition

The general plan of the third edition of this textbook, intended for the use of students of food science and of agriculture, closely follows that of the previous edition.

An attempt has been made to show the importance of the various cereals and cereal products as staple items in the diet, relating their nutritive value to their chemical composition. The structure of cereal grains is described in order to provide a basis for understanding their processing, and consideration is given to the effects of processing on the nutritive value of the processed products.

By-products and their uses are also mentioned, and reference is made to relevant food legislation and standards. At the time of writing, new regulations to replace the (UK) Bread and Flour Regulation 1963 are still awaited.

The sections dealing with sorghum and the millets have been further expanded, with the introduction of some information about the domestic processing of these cereals. A new section on wet milling processes to produce gluten and starch from wheat or wheat flour is included in Chapter 7.

The valued assistance I have received from many of my erstwhile colleagues at the Flour Milling and Baking Research Association is gratefully acknowledged. My particular thanks are due to Mrs. Connie French, Librarian, who went to considerable trouble in searching out information; to Dr. Norman Chamberlain, whose crisp comments on Chapter 10 were most valuable; and to Mr. Brian Stewart, who provided hitherto unpublished data from which Fig. 32 was constructed.

I would like to thank the firms who most generously provided the data for Tables 59 and 60; the firms who have willingly supplied pictures; and the editors of journals, other individuals and publishers who have kindly allowed me to reproduce pictures and data from their publications.

The picture of ergotized rye (Fig. 39) is Crown Copyright and is reproduced with the permission of the Controller of Her Majesty's Stationery Office. The Controller of H.M.S.O. has also kindly given permission for data from Crown Copyright publications to be quoted.

St. Albans, Herts   N. L. Kent
June 1982

Preface to the second edition

The second edition of *Technology of Cereals* closely follows the plan of the first edition, with the addition of some new material, the deletion of sections that are no longer relevant, and a general updating of information to take account of the changes that have occurred during the past 10 years.

Sorghum has been covered more comprehensively, and material on triticale and millet has been introduced. Chapter 10, Bread-baking Technology, has been largely rewritten, and includes new sections on chemical development processes and microwave baking. I am greatly indebted to my colleague Dr. N. Chamberlain, who reviewed Chapter 10 and offered valuable suggestions.

Since the publication of the first edition, the United Kingdom has joined the "Common Market", and opportunity has been taken to mention EEC Regulations and Directives that concern the flour-milling and baking industries and the EEC standards for intervention and denaturation. The new French and the new Canadian wheat-grading systems are also mentioned.

It is again a pleasure to express my thanks to those firms and individuals who supplied pictures or data, and to the authors, editors and publishers who have kindly allowed reproduction of illustrations, including the Controller of H.M.S.O. for permission to reproduce Crown copyright material.

Flour Milling and Baking Research Association,   N. L. Kent
Chorleywood
Rickmansworth, Herts
January 1974

Preface to the first edition

This introduction to the technology of the principal cereals is intended, in the first place, for the use of students of food science. A nutritional approach has been chosen, and the effects of processing treatments on the nutritive value of the products have been emphasized. Throughout, both the merits and the limitations of individual cereals as sources of food products have been considered in a comparative way.

I am greatly indebted to Dr. T. Moran, C.B.E., Director of Research, for his encouragement and advice, and to all my senior colleagues in the Research Association of British Flour Millers for their considerable help in the writing of this book. My thanks are also due to Miss R. Bennett of the British Baking Industries Research Association and Mr. M. Butler of the Ryvita Co. Ltd who have read individual chapters and offered valuable criticism, and particularly to Professor J. A. Johnson of Kansas State University and Professor J. Hawthorn of the University of Strathclyde, Glasgow, who have read and criticized the whole of the text.

I wish to thank the firms that have supplied pictures or data, viz. Henry Simon Ltd, Kellogg Co. of Great Britain Ltd, and also the authors, editors and publishers who have allowed reproduction of illustrations, including the Controller of H.M.S.O. for permission to reproduce Crown copyright material (Fig. 38, and data in Tables 1, 22, 26, 55 and 72).

Research Association of British Flour Millers   N. L. Kent
Cereals Research Station
St Albans, Herts
July 1964

Acknowledgements

We would like to extend our appreciation to Celia Kent and to the estate of Dr. Norman Kent for allowing us to proceed with this new edition. Many thanks to Nina Bandeira at Elsevier and Heather Dean at Iowa State University for helping to initiate this project and to breathe life back into this book. We could not have undertaken this endeavour without their efforts and support. We also would like to thank Karen Miller and Debasish Ghosh at Elsevier for working closely with us as we developed the text and for coordinating the translation of our manuscripts into the finished product.

During the preparation of this edition the authors have called upon many people and organizations, seeking their cooperation. We wish to acknowledge the generous responses to our requests for advice and permission to reproduce published and in some cases unpublished findings and illustrations. Special thanks are due to Drs. Sinead Drea, Ulla Holopainen-Mantila, John King, Mark Nesbitt, Mary Parker, and John Taylor. Among the companies and organizations consulted, we would like to thank Buhler AG, CPM-Roskamp Champion, FluidQuip, Grain Milling Federation (RSA), Prosea Foundation, and Satake (Europe Ltd.), who were particularly helpful.

Introduction to cereals and pseudocereals and their production

Kent's Technology of Cereals, Fifth Edition
http://dx.doi.org/10.1016/B978-0-08-100529-3.00001-3

1

1.1 CEREALS AND PSEUDOCEREALS

Cereals are cultivated grasses that grow throughout the temperate and tropical regions of the world.

Broad characterization into two categories serves to distinguish most types of true cereals. Warm-season cereals are produced and consumed in tropical lowlands year-round and in temperate climates during the frost-free season. Warm-season cereals vary in their water requirements, for example, the majority of rice is grown in flooded fields for at least part of its growth period while sorghum and millets can be grown in arid conditions.

Cool-season cereals are well-adapted to temperate climates. Most varieties of a particular species are either winter or spring types. Winter varieties are sown in the autumn. Some types may require a period of low temperature at a suitable moisture content before germination can occur; this condition is known as seed dormancy. When germination occurs seedlings grow vegetatively until they enter a resting phase induced by low temperatures. Although growth is suspended, the young plants need to be capable of withstanding low temperatures. They resume growing in the spring and mature to bear fruit in late spring or early summer. Dormancy is a condition that is declining in abundance through selective breeding, and it can be overcome artificially by a process of 'stratification' whereby seeds are subjected to the required conditions of moisture and cold temperature without being in the soil. Winter varieties do not flower until spring because they require vernalization (exposure to low temperatures for a genetically determined length of time).

Spring cereals do not require vernalization; they are planted in early spring and mature later in the same year. They are selected for cultivation where winters are either too warm to meet the vernalization requirements of winter varieties or too cold for the winter varieties to survive. Spring cereals typically yield less than winter cereals.

There are some varieties, described as 'facultative', that can be sown either in autumn or spring because they exhibit good cold tolerance but no dormancy. Dormancy is also an important factor in determining grain quality since absence of dormancy can lead to premature germination known as preharvest sprouting (see Chapter 3).

For flowering to occur in some varieties it is also necessary for growing plants to experience an appropriate period of daylight. Such varieties are described as photoperiod-sensitive, and among the cereals both long-day and short-day species exist. Wheat is an example of a long-day plant (Dubcovsky et al., 2006) and African rice is an example of a short-day type (Brink and Belay, 2006). Even in less-sensitive types, day-length and temperature exert an influence on the time of flowering.

Some plants, including buckwheat, quinoa and species of amaranth, may sometimes be referred to as cereals but, as they are not grasses or even monocotyledons, they cannot be classed as true cereals. Their fruits are capable of replacing those of cereals for some purposes but they are more correctly described as pseudocereals. Although they represent an important food source in some countries, world production of individual pseudocereals is less than a quarter of that of the least grown true cereal.

The range of latitudes over which some cereals and pseudocereals can be grown are shown in Fig. 1.1.

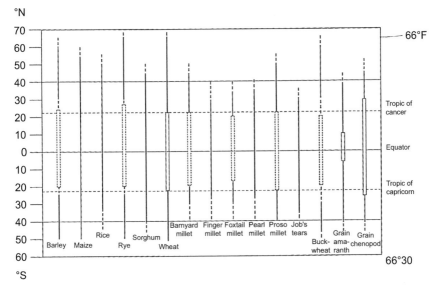

FIGURE 1.1   Bar chart of the global distribution of the main cereals. ▬▬▬▬ suitable; ▭▭▭▭ suitable at high elevations; ------ marginal; ⌐------⌐ marginal even at high elevations. *Reproduced from Grubben, G.J.H., Partohardjono, S. (Eds.), 1996. Plant Resources of South-East Asia No. 10 Cereals. Backhuys Publishers, Leiden, with permission.*

1.1.1   General characteristics of grasses

As members of the monocotyledonous family Poaceae (formally Graminiae) (Black et al., 2006) grasses share the following characteristics, but these are developed to different degrees in the various members.

1.1.1.1   Vegetative features

- Erect stems (culms) with nodes in the stem, which is otherwise known as a 'culm'.
- A single leaf at each node.
- Leaves in two opposite ranks.
- Leaves consist of sheath and blade.
- Tendency to form branches at nodes and adventitious roots at the bases of nodes.
- Lower branches may take root and develop into tillers and behave ultimately like true stems. While all tillers have the potential to develop inflorescences capable of bearing fruit, those that arise late in the season may not do so.

When the moisture and temperature conditions for germination are met, seminal roots arise from the germinating embryo but are very soon replaced by an adventitious nodal root system that arises from the culm

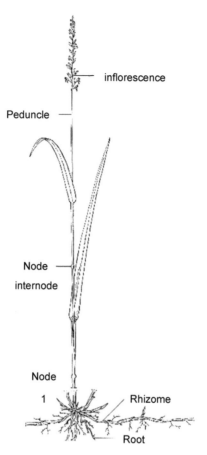

FIGURE 1.2   Characteristic elements of a typical grass plant. *Reproduced from Grubben, G.J.H., Partohardjono, S. (Eds.), 1996. Plant Resources of South-East Asia No. 10 Cereals. Backhuys Publishers, Leiden, with permission.*

and tillers. The roots are fibrous and usually penetrate 1 to 2 m into the soil. Maize also has stout prop roots arising from the lower nodes above the soil surface.

The stem is cylindrical, with elongated, usually hollow internodes (but pith-filled in maize and sorghum), connected by short, harder, disc-shaped solid nodes that look very different from the internodes and from which the buds and leaves originate. The lower nodes usually remain very short (Fig. 1.2).

Each leaf is comprises a sheath, a ligule and a blade. The sheath clasps the stem tightly and at its upper end passes into a parallel-veined, typically long and narrow blade. The ligule, a short membranous or ciliate rim, is present at the junction of sheath and blade. The base of the blade or the top of the sheath often bears auricles (ear- or teeth-like appendages) (Fig. 1.3).

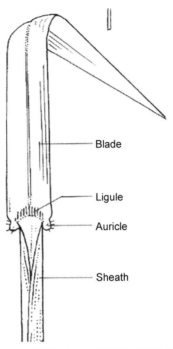

- Blade
- Ligule
- Auricle
- Sheath

FIGURE 1.3 Details of a typical grass leaf. *Reproduced from Grubben, G.J.H., Partohardjono, S. (Eds.), 1996. Plant Resources of South-East Asia No. 10 Cereals. Backhuys Publishers, Leiden, with permission.*

1.1.1.2 Reproductive features (inflorescence)

The inflorescence is a specialized leafless branch system, which usually terminates the stem. Morphologically, its basic units, called spikelets, are partial inflorescences, but functionally they can be compared with the flowers of petaloid plants. On the outside, two opposite rows of scales are arranged alternately along an axis (rachilla); the two lower scales, the glumes, are empty, but the remainder form part of a floret, whose floral parts are enclosed by the lemma on the outside and the delicate membranous scale, the palea, on the inside; glumes and lemmas often terminate in one or more long, stiff bristles termed awns; the floral parts consist of two or three tiny scales, the lodicules. At an appropriate time, the lodicules swell as their water content increases, and as a result of the pressure that they apply to the inside of the lemma and palea this causes them to separate, allowing the anthers to protrude and shed pollen that can then be dispersed. Three stamens (six in rice) each with a delicate filament and a two-celled versatile anther and a pistil consisting of a superior single-loculed ovary with a single ovule and two styles, each ending in a feathery stigma. Bisexual spikelets are the rule, although some of their florets are often unisexual or barren. Separate male and female spikelets are occasionally borne on the same plant (e.g., on maize), rarely on separate plants (Figs 1.4–1.6).

The spikelets are arranged in various ways, ranging, in different species, from a single spike or raceme through an intermediate stage of several spikes, arranged digitately or along an axis, to a many-branched panicle.

Both types of inflorescence are widely found in grasses. The diagram in Fig. 1.7 shows the differences between them.

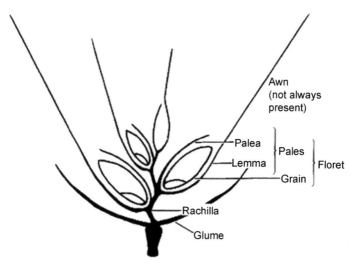

FIGURE 1.4   Diagram of a spikelet.

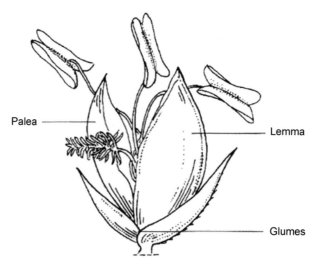

FIGURE 1.5   Structures of a floret subtended by glumes and enclosed within a palea and lemma.

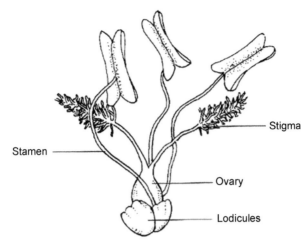

FIGURE 1.6 Structures of a floret. *Reproduced from Grubben, G.J.H., Partohardjono, S. (Eds.), 1996. Plant Resources of South-East Asia No. 10 Cereals. Backhuys Publishers, Leiden, with permission.*

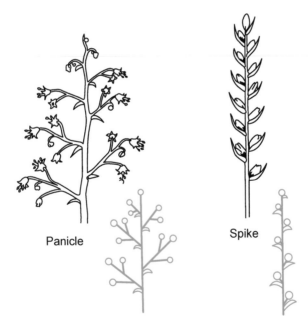

FIGURE 1.7 Diagrammatic representation of grass inflorescence types. *Reproduced from Raven, P.H., Evert, R.F., Eichhorn, S.E., 1992. Biology of Plants, fifth ed. Worth Publishers Inc., New York.*

Wheat barley, rye and triticale exhibit only the spike morphology. A spike may be defined as an indeterminate inflorescence bearing sessile flowers on an unbranched axis.

Most of the other cereals have only the panicle as its floral form. The panicle is defined as an inflorescence in which the flowers are borne on branches of the main axis or on further branches of these. Rice, oats, sorghum and the millets are examples of paniculate species.

Maize is unique among cereals in that it has a male inflorescence borne at the terminus of the culm and consisting of panicle of staminate racemes, and one or more female inflorescences born as a lateral branch or branches described as pistillate spikes.

Grass florets open for only a few hours to expose the sexual organs to wind pollination. Cross-pollination may be ensured by protandry (pollen becoming mature and dispersed before the stigma of the same floret becomes receptive); the pollen is viable for less than a day. However, several cereal species are self-pollinated as the anthers shed their pollen before emerging from between the lemma and palea. Thus, barley is about 95% self-pollinated and rice, oats and wheat are 97%–98% self-pollinated. Finger millet also shows a high degree of self-pollination. On the other hand, rye is almost entirely cross-pollinated and may even be self-sterile; maize engages in about 95% crossing and pearl millet about 80% crossing.

Differences in inflorescence morphology among cereal species are accompanied by differences in the number of grains borne in each inflorescence. While there are wide variations even within a species, Table 1.1 gives an indication of the inherent differences among types. Grain number

TABLE 1.1 Numbers of grains borne on an inflorescence of some cereals

Cereal type	No. of grains per inflorescence
Maize	400–600
Rice	100–150
Wheat	40–75
Sorghum	800–3000
Millet	Mean 1600 for pearl millet
Barley	
2 row	18–30
6 row	25–60
Rye	75–80
Triticale	31–50

Various sources.

per inflorescence is not a parameter that is frequently measured and consequently little information is available. These values should be seen only in the context of a comparison among cereal inflorescences.

1.1.1.3 Grain

Following pollination, the fruit (or grain) develops. Systematically, it is described as a caryopsis. As is general for fruits the caryopsis comprises tissues from two generations because the outer layers are derived from the carpel, which is part of the parent plant, while only the products of fertilization (the embryo and endosperm, which includes the aleurone) are of the filial generation. The outer layers of the caryopsis are the outer and inner pericarp. Within, lies the seed comprising seed coats (testa and remains of the nucellus), the starchy endosperm and the embryo at the base of the abaxial face. It has a flat haustorial cotyledon (the scutellum) and a special outer sheath (the coleoptile) protecting the plumule during soil penetration (Fig. 1.8).

The fruit is surrounded by the husk, which consists of the dried and hardened palea and lemma. The husk of wheat, maize, millet and some sorghum types is easily removed by threshing, but the husk of barley rice and oats is tenacious and is traditionally removed by pounding, impact or abrasion and winnowing. All these processes are now conducted by machine in commercial settings.

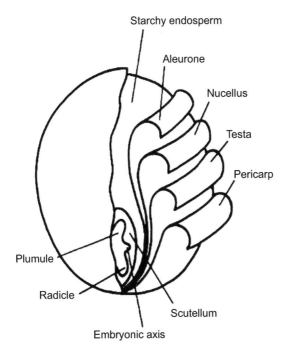

FIGURE 1.8 Gyeneralized cereal grain showing the relationships among the tissues.

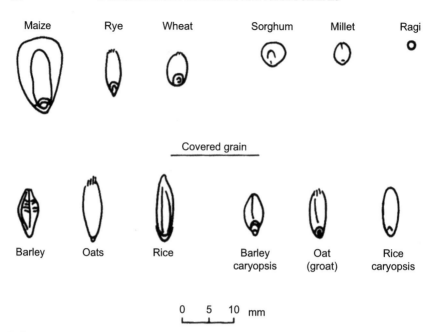

FIGURE 1.9   Grains of cereals showing comparative sizes and shapes. The caryopses of the three husked grains (barley, oats, rice) are shown with and without their surrounding palea and lemma.

Large differences exist among the sizes of grains, a selection of which are shown in Fig. 1.9.

1.1.2 Yield of grain

For cereals, yield may be defined as the weight (e.g., hg or tonne) or volume (bu) of cleaned grain harvested from a unit cultivated area (e.g., ha). To some degree yield is determined by the prevailing conditions, such as soil and climate, but growers can influence yield by controlling seed rate and thus the number of plants per unit area and by applying appropriate amounts and types of fertilizer. Control of weeds, diseases and pests is also an important yield determinant. There are also several plant characteristics that can be manipulated by breeders in order to improve yield, such as the number and weights of grains on an ear, the number of tillers on a plant and the harvest index.

For cereals, harvest index (HI) is the proportion of the total dry matter of the plant (usually the above-ground parts) that is grain. Definitive values are difficult to establish as not all values quoted are expressed on a consistent moisture content basis (Unkovich et al., 2010). HI is usually expressed as a fraction but sometimes as a percentage. In plant breeding it has been the most important yield component involved in

TABLE 1.2   Some agronomic characteristics of cereals

Cereal type	Plant height	Harvest index
Maize	4–4.8 m max recorded >8.5 m (dent)	0.25–0.56
Rice	0.96 m	0.34–0.55
Wheat	0.5–1.2 m	0.35–0.55
Sorghum	0.6–1.0 m	0.25–0.56
Millet (pearl)	1.3–3 m	0.16–0.40
Barley	0.25–0.9 m	0.25–0.42
Oats	0.8–1.2	0.31–0.62
Rye	1.2 m	0.28–0.38
Triticale	0.9–1.1 m	0.24–0.48

Various sources.

improving crop yields. One of the major contributions to the increase in harvest index is the decrease in culm length that breeding has achieved. Reduction in culm length has also reduced the susceptibility of aerial parts of the plants to become bent close to soil level as a result of rain or wind and possibly disease; when this condition becomes permanent it is known as lodging.

Because of the wide range of growing systems under which cereals are grown, there are large variations in plant and grain characteristics, but Table 1.2 shows some values reported for plant height and harvest index for cereal species.

1.1.3 Growing cereals

Growing cereals involves a range of agricultural operations. All involve a physical element but some also involve a chemical input. The manner in which both are applied varies according to the level of sophistication involved. For example, all cereals require a seedbed to be prepared when or before the seed is sown in order to achieve good contact between seed and surrounding soil, but there are many ways in which this is achieved. The traditional method is to plough and harrow, and this method persists in many instances, although the types of ploughs and the means of drawing them vary considerably. Rice farming in Asia is dominated by millions of small farmers, with an average landholding of 1 ha, cultivating small paddies, which may be waterlogged. Hence the use of draught animals such as water buffalo is still the favoured practice. At the other extreme, wheat is grown on the North American prairies on 'quarters' of 160 acres (65 ha) and cultivation involves use of powerful tractors. Clearly there is a considerable difference in the amount of person-time involved in preparing a unit area, and the cost of labour in different locations is a major factor in determining the chosen method. The possibilities of driverless vehicles for agricultural operations are being explored and indeed exploited to some extent. In the interests of saving fuel and labour, seedbed preparation and sowing are sometimes combined by adopting the technique of 'direct drilling' into unploughed soil, with the remains of the previous crop still in place.

While most crops are sown by hand or by animal- or tractor-drawn implements, rice seed can also be successfully broadcast from aircraft into flooded fields (Akesson and Yates, 1976). Distribution of fertilizer and pesticide from aircraft can also be advantageous, especially where fields are flooded as may be the case for rice. For many cereal crops, but particularly where flooding is practiced at any stage in the growth of rice, advantage can arise from levelling the field. When flooding or irrigating this ensures more uniform application and retention of water, and in the case of flooding the volume of water required to achieve sufficient depth throughout is reduced. Hence, levelling is a long-established practice but the means of levelling is very variable, ranging from movement of soil by a log dragged behind a draught animal to the employment of large earth-moving machines (Jat et al., 2009). In recent years, such machines have been guided by laser technology, and the use of smaller laser-guided machines is beginning to be introduced into the more traditional rice-producing areas of the world.

With ever-increasing costs of inputs, growers are welcoming innovative ways of achieving more precise placements of seed, through precision drilling, whereby row and seed spacing are precisely controlled, and applications of pesticide and fertilizer respond to the calculated needs of the crop. The needs are assessed by soil analysis, to quantify nutrient deficiencies, and visual examinations to assess the risk of infection or parasite attack. To assist with visual assessments, camera-carrying drones have proved useful as problem areas can thereby be identified from above. By combining GPS-related micromapping techniques with information on soil condition and composition, onboard computers are capable of controlling implements so that appropriate, metered applications, are delivered, leading to considerable savings through treatments being applied only where and when needed. Such reduced use of chemicals is also beneficial to the environment as pesticides can kill beneficial insects and fertilizer above and beyond the needs of the crops, can lead to contamination of water courses, adversely affecting the health of animals and wild pants.

As with other operations, the degree of mechanization involved with harvesting varies widely, from traditional methods using handheld knives to combine harvesting with cutter bars as wide as 36 ft (11 m).

While the importance of the efficient and economical production of cereals is universally recognized, the agricultural practices of growers is increasingly scrutinized by the wider public in relation to the effects on the environment. This and the increasing costs of all agricultural inputs has led to the evolution of systems described as 'sustainable'. While this is difficult to define in detail, the principles involved are clear. They require that the production of food, fibre or other plant or animal products should be achieved using techniques that protect the environment, public health, human communities and animal welfare.

With the increase in knowledge that has been achieved through research and observation, it has been established that timing of treatments in relation to the stage of development of a growing crop is critical. Although there are considerable differences among cereal species in their development, it has been possible to define developmental stages that are common to several species. Scales of development have been devised to define the stages in the plant life cycle; the one shown in Fig. 1.10 is appropriate to all the cool-season cereals plus rice. Although some aspects of it are relevant, it is less readily applied to maize, sorghum and the millets. The system was devised to enable accurate communication among scientists, agricultural chemical suppliers and growers.

1.1.3.1 Growth degree days

The time interval between different growth stages varies according to species and variety, but even for a single variety the interval can vary as a result of ambient conditions. Scientists and growers, striving for a consistent basis on which to make comparisons and predictions regarding the growth of both plants and invertebrates, have evolved a system whereby time and temperature (the main ambient influence on growth rate) are included in a calculation that provides the required consistency. The expression that has been accepted is known as growth degree days (GDDs). It takes account of the number of 24-h periods and the maximum and minimum temperatures, at which growth is possible, experienced during the chosen growth interval. For some scientific purposes it may be desirable to monitor temperature more frequently, thus providing, for example, a 'growth degree hours' record. Because numerically different temperature scales are adopted in different settings, it is not possible to make universal comparisons without knowledge of whether the Fahrenheit or Celsius convention has been used. Further, there are different ways in which the temperature element of the expression is determined; it might be a simple mean of minimum and maximum temperatures experienced during the relevant 24-h period or it may be more complex. A review of methods has been published online by the University of California's Integrated Pest Program, at http://ipm.ucanr.edu/WEATHER/ddconcepts.html.

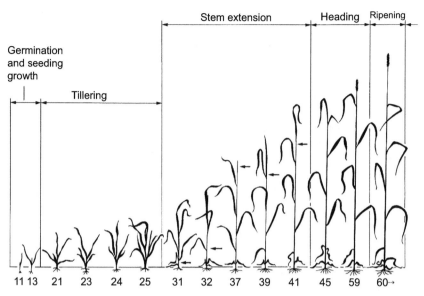

FIGURE 1.10   Growth stages of a grass. *Seedling growth*: GS10 First leaf through coleoptile, GS11 First leaf unfolded (ligule visible), GS13 3 leaves unfolded, GS15 5 leaves unfolded, GS19 9 or more leaves unfolded; *Tillering*: GS20 Main shoot only, GS21 Main shoot and 1 tiller, GS23 Main shoot and 3 tillers, GS25 Main shoot and 5 tillers, GS29 Main shoot and 9 or more tillers; *Stem elongation*: GS30 Ear at 1 cm (pseudostem erect), GS31 First node detectable, GS32 Second node detectable, GS33 Third node detectable, GS37 Flag leaf just visible, GS39 Flag leaf blade all visible; *Booting*: GS41 Flag leaf sheath extending, GS43 Flag leaf sheath just visibly swollen, GS45 Flag leaf sheath swollen, GS47 Flag leaf sheath opening, GS49 First awns visible (if an awned variety) Barley only; *Ear emergence*: GS51 First spikelet of ear just visible above flag leaf ligule, GS55 Half of ear emerged above flag leaf ligule, GS59 Ear completely emerged above flag leaf ligule; *Flowering*: GS61 Start of flowering, GS65 Flowering half-way, GS69 Flowering complete; *Milk development*: GS71 Grain watery ripe, GS73 Early milk, GS75 Medium milk, GS77 Late milk; *Dough development*: GS83 Early dough, GS85 Soft dough, GS87 Hard dough (thumbnail impression held); *Ripening*: GS91 Grain hard (difficult to divide), GS92 Grain hard (not dented by thumbnail), GS93 Grain loosening in daytime. *Based on data from Tottman, D.R., Makepeace, R.J., Broad, H., 1979. An explanation of the decimal code for the growth stages of cereals, with illustrations. Annals of Applied Biology 93 (2), 221–234.*

1.1.4 The importance of cereals

Domestication of poaceous cereal crops such as maize (corn), wheat, rice, barley and millet lies at the foundation of sedentary living and civilization around the world, and the Poaceae still constitute the most economically important plant family in modern times, providing forage, building materials (bamboo, thatch) and fuel (ethanol), as well as food.

The majority of cereal grains are utilized as dry or dried grains and this confers an important advantage: they can be stored under ambient conditions for long periods. Sweetcorn, a type of maize, is an exception as it may be consumed while still retaining a higher proportion of moisture.

Cereal grains are important principally as a source of nutrients for both humans and livestock. Their main contribution to diets is starch, an energy source that is present as a storage product in the endosperm, but there are other major and minor nutrients such as protein, oil and vitamins that are also important. Starch is capable of being converted to other organic compounds such as alcohol, for consumption in cereal-based drinks or for use as a biofuel.

Inevitably, some cereal species are more extensively grown and consumed than others but together they contribute more food energy worldwide than any other type of crop, they are therefore staple crops. Largely as a result of their distribution, consumption of the different cereals also varies geographically.

As well as producing grains, cereal crops can be harvested and may be preserved green for forage purposes. The extent to which this occurs varies among species and locations, and it is only the ripened and harvested grain that is included in the statistical results reproduced here.

The importance of cereals (grains) is illustrated by the fact that in 2012 three cereal species were listed in the world's top 11 food commodities (by value), and the production value of rice alone is only marginally less than that of milk, the most valuable (Fig. 1.11). Sugar cane is also a grass but it is not included among the cereals because its value does not lie in its fruit.

In terms of quantity, the major cereals were listed in the top four commodities (Fig. 1.12) and barley was ranked 12 with 133 mt grown.

The global significance of the three major cereals in relation to the others is shown in Fig. 1.13.

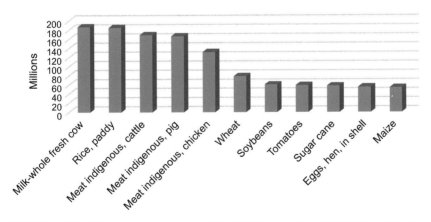

FIGURE 1.11   Values of top food commodities, 2012 (International $1000). http://faostat. fao.org/site/339/default.aspx. *Based on data from FAOSTAT.*

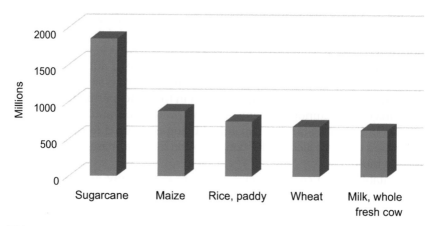

FIGURE 1.12   Quantities of top food commodities, 2012 (tonnes). http://faostat.fao.org/ site/339/default.aspx. *Based on data from FAOSTAT.*

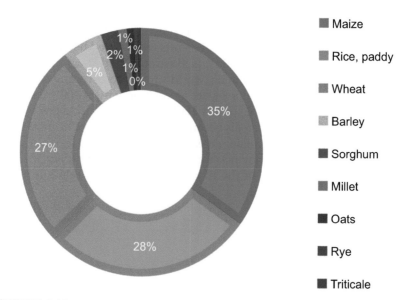

FIGURE 1.13   Relative contributions to world cereal grain production (by weight) of different cereals (mean of 2009–13). (value for triticale unavailable but similar to rye). *Based on data from FAOSTAT.*

It should be noted, when comparing yields, that in the case of rice, production values are reported as 'paddy'. These include the contribution of the adherent husk, which is around 20% of the grain weight. Similarly, barley hull contributes 13% and oat husk 25% of grain weight.

Together the three major cereals contribute 88% of total cereals production. While all of them have shown the same trend over the period from

1961 to 2012, it is maize that has shown the greatest increase (Fig. 1.14) and this is a reflection of increases both in area harvested (Fig. 1.15) and yield (Fig. 1.16), both of which have exhibited the same trend for 50 years.

The increase in rice production is also a result of an increase in both components, while in the case of wheat, increase in production has resulted almost entirely from a tripling of yield during the period.

Of the three major cereals, rice has the greatest monetary value as can be seen from Fig. 1.17.

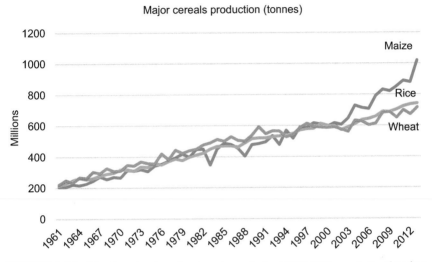

FIGURE 1.14 World production of maize, rice and wheat 2003–13 (tonnes). *Based on data from FAOSTAT.*

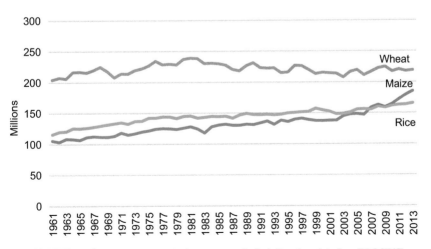

FIGURE 1.15 Area harvested of major cereals (ha). *Based on data from FAOSTAT.*

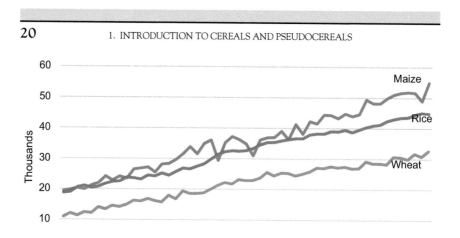

FIGURE 1.16 Variation in yields of major cereal types (hg/ha). *Based on data from FAOSTAT.*

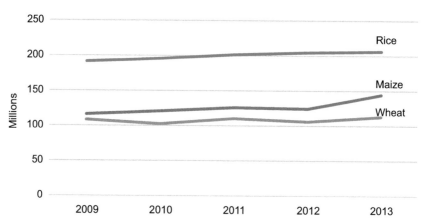

FIGURE 1.17 Variation in values of major cereals (International $1000). *Based on data from FAOSTAT.*

Prices paid to producers have fluctuated, but in spite of inflation, increases have been small over a period of more than 20 years as shown by the records for two European wheat-producing countries (Fig. 1.18).

Among the lesser cereals, barley production in 2013 was double that of 1961. However, it declined from a peak in the 1970–90s when it was more than twice the 1961 value (Fig. 1.19). During the period there was a marginal decline in the area harvested (Fig. 1.21) so all of the increase in production resulted from improvement in yield (Fig. 1.20).

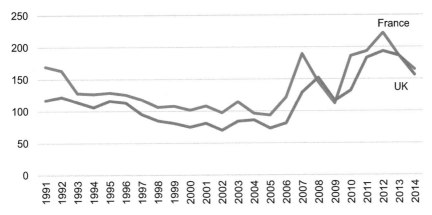

FIGURE 1.18 Variation in wheat prices paid to producers (standard local currency per tonne). *Based on data from FAOSTAT.*

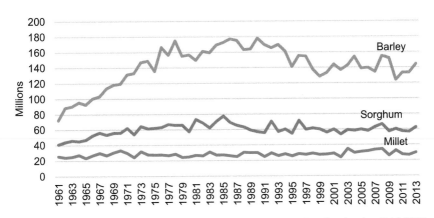

FIGURE 1.19 Production of lesser cereals (tonnes) 1961–2013. *Based on data from FAOSTAT.*

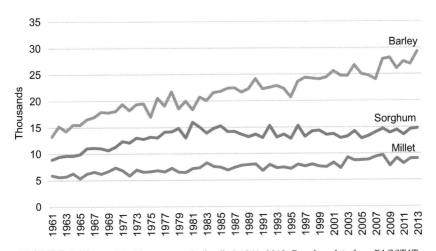

FIGURE 1.20 Yield of lesser cereals (hg/ha) 1961–2013. *Based on data from FAOSTAT.*

During the same period sorghum and millet production showed a comparatively small increase in both yield (Fig. 1.20) and total production (Fig. 1.19), but the area harvested declined slightly (Fig. 1.21).

Variations in total values of lesser cereals are shown in Fig. 1.22.

Among the minor cereals production of both oats and rye declined significantly between 1961 and 2013 (Fig. 1.23). This is entirely due to an approximately fourfold reduction in area harvested (Fig. 1.24) as yields of both species showed an increase in yield during the period (Fig. 1.25).

As triticale species were newly introduced, the relevant United Nations Food and Agricultural Organisation (FAO) records were not compiled until 1975, and world triticale production did not exceed 1 million tonnes

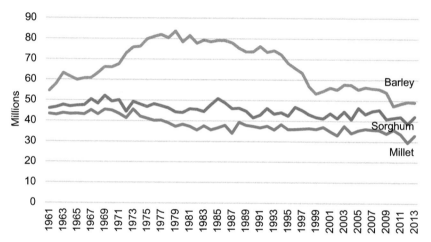

FIGURE 1.21   Area harvested of lesser cereals (ha). *Based on data from FAOSTAT.*

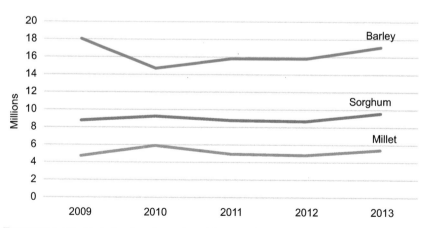

FIGURE 1.22   Variation in values of world crops of minor cereals (International $1000). *Based on data from FAOSTAT.*

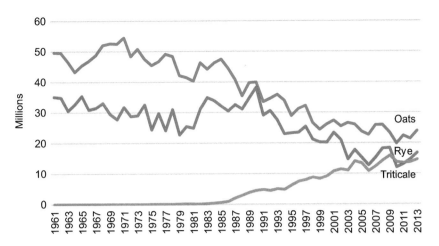

FIGURE 1.23 Variation in world production of minor cereals (tonnes). *Based on data from FAOSTAT.*

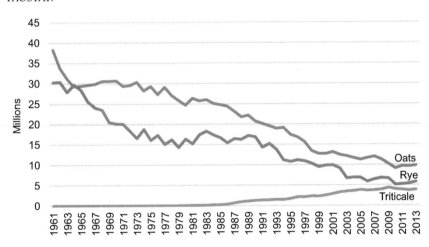

FIGURE 1.24 Variation in area harvested of minor cereals (ha). *Based on data from FAOSTAT.*

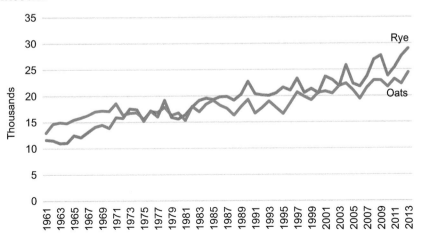

FIGURE 1.25 Variation in yield of oats and rye (hg/ha) 1961–2013. *Based on data from FAOSTAT.*

until 1987. Triticale production has increased progressively since that date to a level close to that of rye (Fig. 1.23). The majority of this increase has resulted from more planting (Fig. 1.24) as yield has not increased significantly since 1987 (Fig. 1.26).

In recent years the value of the triticale crop has been higher than that of the rye crop (Fig. 1.27).

It should be remembered that although FAO records show each cereal as a single species, they actually comprise several subspecies or races, many varieties (see Chapter 2) and in some cases even different species. This is particularly true of the cereals that are grouped as millet wherein different species are associated with different regions.

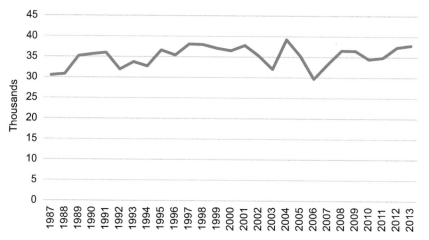

FIGURE 1.26   Variation in the yield of triticale (hg/ha). *Based on data from FAOSTAT.*

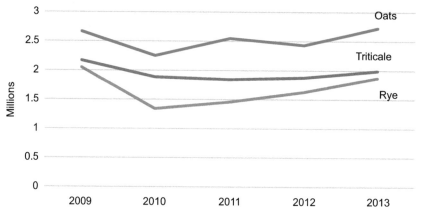

FIGURE 1.27   Variation in the values of minor cereals (International $1000). *Based on data from FAOSTAT.*

1.1.5 Multicropping

Maize, rice, sorghum and the millets are warm-season cereals, grown under conditions in which seasons are less distinct than those experienced by cool-season cereals typically grown in temperate climates. Growers of some warm-season cereals are able to benefit from the production of more than one crop each year, either involving crops of the same cereal (double- or even triple-cropping) or crops of different species, one of which may not be a cereal (intercropping). In regions experiencing monsoons, rice can be grown in the rainy season and another crop in the dry season.

In the double-cropping system the second crop may result from reseeding or from 'ratooning'. Ratooning is the practice of removing the majority of the above-ground vegetative parts of the plant following harvest and by watering and adding fertilizer, encouraging growth of more stems (tillers) from the base, capable of bearing grains for the second harvest. The ratoon crop takes around half the time of the main crop to reach maturity and normally produces around half the grain yield of the first crop. This method is widely used for both sorghum and rice in Texas and other southern states of the United States and to a lesser extent elsewhere in the world.

1.1.6 Efficiency of production

Although many cereal species can be grown over a wide range of geographical conditions and agronomic practices, their yield and quality vary considerably. Even within Europe a difference greater than threefold exists between the yields of wheat in the two major producing countries, France and the Russian Federation (Fig. 1.28).

Even the French average is less than half of the highest yield achieved in a single field (16.52 tonnes/ha achieved in the United Kingdom in 2015) (Anon., 2015).

Climate is not the only factor affecting yield as the type and variety are also significant. Soil type and amount of added fertilizer are also influential. In general, where spring and winter sowing is possible, winter varieties produce higher yields.

A major contributory factor in the case of rice yields lies in the method of cultivation adopted as the entries in Table 1.3 illustrate. The table also shows the relative contributions of the different cultivation systems.

These FAO values are for 1985 and average yields have increased since then. However, more recent figures show that among different Asian countries a twofold difference exists between the two main producers (Fig. 1.29).

If all producing Asian countries are taken into account, there is a tenfold difference, as Turkey's 2009–13 mean is around 82,000 hg/ha and Brunei Darussalam's is around 8000 hg/ha.

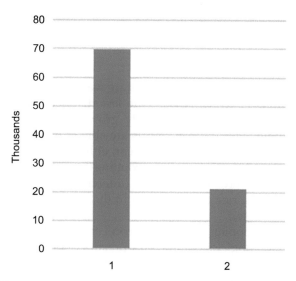

FIGURE 1.28 Comparison of wheat yield in France (1) and Russian Federation (2) hg/ha. *Based on data from FAOSTAT.*

TABLE 1.3 Global statistics relating to rice grown under different conditions (1985)

Ecosystem	Area (million ha)	(%)*	Yield (tonne/ha)	Production (million tonne)	(%)*
Irrigated	67	49	4.7	313	72
Rain-fed lowland	40	29	2.1	84	19
Upland	18	13	11	21	5
Deep-water/ tidal wetland	13	9	1.5	19	4
Total	138	100	3.2a	437	100

Based on data from FAOSTAT.
*It is the percentage of the total rice crop.

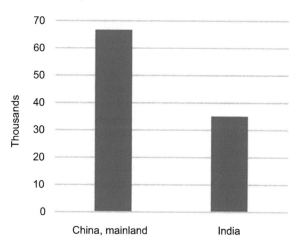

FIGURE 1.29 Rice (paddy) yield in the two main producing countries (hg/ha). Average of 2009–13. *Based on data from FAOSTAT.*

1.1.7 Seed certification

The Organisation for Economic Co-operation and Development (OECD) has established criteria for the certification of seed in international trade, and these principles have become the basis of many national schemes. The schemes cover many types of seed as well as cereals. The OECD 'List of Varieties Eligible for Certification' is updated annually, based on submissions of acceptable varieties by designated authorities in the contributing countries (OECD, 2015).

1.1.8 Pests, diseases and weeds

Cereal growers seek to maximize yields to the point that the cost of the agricultural practices (e.g., tillage and fertilizer application) can be justified by the returns. There are also factors that are both unpredictable and beyond their control such as weather conditions. The extent of some other factors are also unpredictable but, as the results can be disastrous, growers take a cautious approach to them, ensuring precautions are taken to minimize their effects. Such hazards include pests, diseases and weeds.

Pests, in the context of cereal crops, are normally considered to be predatory vertebrate or invertebrate animals that reduce crops either by consuming or damaging grains or other parts of the plant that contribute to its ability to produce maximum yield. As a result of their mobility, pests are also capable of spreading disease organisms from plant to plant and crop to crop. Pest control methods include the use of pesticide sprays in the case of insect pests and preventing of settling in the case of birds.

Tillage weeds are annual or perennial plants that have adapted to surviving in cultivated ground in a number of ways. Many annual weeds thrive because of their ability to germinate, flower and set seed quickly, all year round. Others can remain dormant in the ground for years and germinate when the ground is disturbed. Perennial plants can also cause problems for the arable farmer especially those that can establish from root fragments. Crops themselves can become weeds if they reoccur in subsequent crops – these are known as volunteers. Weeds compete with crop plants for space, light and nutrients, and some produce seeds that are capable of being harvested with and hence contaminating grain of the crop.

Reduction of weed populations is achievable by several methods, some of which attempt to control the number of weed seeds introduced into the crop, some that reduce the weeds' ability to compete and some that seek to kill or remove the weeds at some stage in their development. Ensuring clean crop seed, uncontaminated by weed seeds, is sown is one obvious strategy, but the tillage regime employed is also important, as is the adoption of a beneficial crop rotation. In terms of competitiveness, a healthy, well-nourished crop with plants appropriately spaced, that establishes rapidly, can reduce the space for weeds to intrude. To combat weeds that do manage to establish it is possible, in small-scale or organic production settings, to remove them manually by 'rogueing', or, in larger-scale or less-restricted enterprises, by selective herbicide spraying. The ultimate extension of this last strategy is the production, by genetic modification, of crop seed that is resistant to a broad-spectrum herbicide so that spraying the crop at a later date with that herbicide kills all plants present except the crop plants.

Diseases may be defined as factors that interfere with normal function or structure of a plant. While they are mostly caused by pathogenic organisms such as fungal, bacterial or viral species, they can also result from adverse environmental factors such as nutrient deficiency. A further hazard associated with some fungal diseases is the production by the pathogen of toxins (mycotoxins) that remain with the grain and its products throughout processing.

Many pathogenic diseases are seed-borne or soil-borne so they become more prevalent as seeds are resown in the same locality or where a crop is grown repeatedly in the same field. Other diseases arise as a result of an infection of the aerial parts of the crop plant. Such diseases thrive and spread successfully where monoculture is practised over a large area.

Some pests and diseases are specific to the host cereal species but it is nevertheless possible to make valid generalizations about types of infestation in relation to stages of development of plants.

Infective stages of disease-causing organisms may be present in (infection) or on (infestation) seed before it is planted; infective stages of other pathogens can enter the below-ground or near-ground parts of the plant from the soil. Yet others can, as the plant develops, be carried onto aerial parts by air and water movements (e.g., wind, rain and splashing) or introduction by parasitic invertebrates.

An example of a soil-borne pathogen is take-all (*Gaeumannomyces* spp.), which attacks most cool-season cereals when it is present in the soil. The amount of fungus in the soil increases rapidly as successive crops of the same type are grown on the same plot, up to a maximum of 3 years, after which decline occurs albeit at a slower rate. Symptoms are patches of stunted growth within the field, empty ears or even bare ground; these are evident in the early growth stages of plants and they persist to maturity. Losses of 20% are not unusual in wheat.

Ergot (*Claviceps* spp.) is a common disease of grasses including cereals. It is not strictly a seed-borne disease as it is spread by sclerotia (singular sclerotium), the nutrient-containing fungal mycelial mass that develops in a floret, on one generation of cereal plant, and behaves as a resting stage until infecting plants of a later generation. Sclerotia are generally larger than the grain that they replace but they can be harvested with the grains and, being brittle, they can break into grain-sized fragments that are difficult to separate from grain samples. Infection occurs through spores, generated from a reactivated sclerotium in the ground, falling on opening florets.

Further examples of pathogens infecting at a relatively late stage of plant growth and infecting mainly the leaves and husks of the crop are the rusts (*Puccinia* spp.), powdery mildews (*Blumeria, Erysiphe* spp.) and leaf blotches (*Septoria, Mycosphaerella, Sphaeosphaeria, Stagonospora* spp.). Under appropriate conditions of temperature and humidity, spores of these fungi can spread rapidly through a crop, dispersed by wind or rain, to form spots or blotches that reduce leaf area and debilitate photosynthetic activity causing losses that can exceed 20%. Rusts produce coloured spots or stripes (pustules) that release asexually produced spores; they persist between crops by infecting alternate hosts, usually broad-leaved plants such as *Berberis, Mahonia* or *Thalictrum* when the sexual phase of their life-cycle occurs. Mildews appear as white clusters of fungus that release spores into the wind; toward the end of the season the white pustules contain dark bodies (cleistothecia) in which the sexual phase of the life cycle occurs. The leaf blotches generally kill off areas of leaf tissue to produce lesions containing microscopic structures (pycnidia) from which spores are exuded to be dispersed by rain splash; the sexual phase occurs within older lesions. Leaf blotches can persist on stubble and trash of previous host crops.

The time in the life cycle of a plant when disease symptoms become obvious may not be an indication of the time when infection occurs, for example, most cereal species can be infected by smuts or bunts (e.g., *Tilletia* spp.). Spores of the pathogen are present on the seed before it is sown but awareness of the symptoms does not arise until the grains develop.

There are some fungal pathogens that change the appearance of grains, making them unattractive to processors but not necessarily adversely affecting yield. An example of this phenomenon is black point. It is most prevalent on wheat, particularly durum wheat, but it also arises on barley and oats. Identifying the causal organism has proved difficult but species frequently considered to be involved are *Alternaria, Cladosporium* and *Helminthosporium*.

Bacterial diseases are less numerous than their fungal equivalents. They include leaf blight of rice (*Xanthomonas oryzae*), which leads to wilting of young plants and can lead to huge losses (up to 70%), and bacterial stripe of wheat (*Xanthomonas campestris*), which can occur on all small grain cereals worldwide, although its affects are not usually significant (www.wheatdoctor.org). Other bacteria genera associated with crop spoilage include *Erwinia* and *Pseudomonas*.

Viruses are the smallest pathogens. The infectious viral particle is called a virion, which is a stable, nonmultiplying stage. Viruses multiply in the host plant, and transmission may occur via several means: insects and mites (especially sucking insects, such as aphids), nematodes, seeds, pollen, fungi, soil and mechanically. Viral diseases are often difficult to detect because infected hosts may not display visible symptoms, or symptoms may closely resemble those of various physiological disorders or genetic abnormalities. Identification can be facilitated by determining which vectors are present and the host range; in many cases, positive identification requires the use of an electron microscope and serological techniques (CIMMYT, 1987). In many cases the name of a pathogenic virus includes the name of the crop species on which it was first identified, but other crops may also be susceptible. Viruses are frequently referred to by initial letters of their designation. Viruses causing Tungro, meaning 'degenerated growth', may be caused by Rice Tungro Bacilliform Virus (RTBV), but symptoms can be enhanced by the spherical form, RTSV. Leafhoppers are responsible for spreading both types.

The most widespread and least host-specific cereal virus is probably barley yellow dwarf virus (BYDV). It can be distributed by more than 20 species of aphid. It has been recorded wherever cereals are grown. Soil-borne Wheat Mosaic Virus (SBWMV) is an example of a virus with a fungal vector (*Polymyxa graminis*) (CIMMYT, 1987).

Disease resistance can be improved by selective breeding, whereby resistance recognized in one variety or species can be introduced to others. However, resistance is frequently overcome by natural generation of new strains of the pathogen. This is particularly likely to occur where monoculture involves not just the same crop species but the same variety of it.

1.1.8.1 Invertebrate pests

Slugs, of which there are more than 100 species, are pests to most crops, and they are most prevalent in damp heavy soils. Snails are most damaging to rice plants growing in flooded paddies. The golden apple snail (there are more than 100 species including *Pomacea canaliculata*) are distributed via irrigation canals, and they damage seedlings by cutting young shoots (www.knowledgebank.irri.org).

Among the many insect field pests are stalk borers such as the African stalk borer (*Brusseola fusca*) and spotted stem borer (*Chilo portellus*). These are the most destructive pests of maize and sorghum in southern Africa. In Europe and North America, the caterpillar of the European corn borer (*Ostrinia nubilalis*) is capable of causing much damage to stems and cobs of maize, although it was originally associated with millet. As with herbicide resistance, genetic modification has been carried out whereby a bacterial gene is introduced into the crop seed, allowing the plant to synthesize a protein that is toxic to specific insects. Resistance to corn rootworm, corn earworm and corn borer has now been introduced into maize in this way.

Nematodes are not as destructive in cereals as they are in other crop types, however, they are associated with introduction of diseases caused by other infective agencies.

1.1.8.2 Vertebrate pests

In regions of the world where rabbits (European rabbit *Oryctolagus* spp.) thrive – and they are found in all continents except Asia and Antarctica – they can do serious harm to any cereal crop and indeed any crop. Rats are usually associated with stored grain but the rice field rat (*Rattus argentiventer*) is capable of decimating fields of rice or maize. The rats damage the crop at all stages of growth from the seed, of which they consume all or part, to the grain filling stage, where they gnaw at the base of the tillers and consume the developing grains on the ears that are thus made accessible. Levees around rice paddies provide the favoured nesting sites for the rats.

In Ethiopia bands of the graminivorous gelada monkey (*Theropithecus gelada*) can cause serious damage to all cereal crops (Kifle et al., 2013).

Birds of many species can also be pests as ears develop. In Africa weaver birds are particularly destructive.

Some diseases, pests and weeds are listed later in this chapter in relation to the respective cereals crops but more detailed information about the hazards and their treatments is readily available on websites, including those of the FAO, and texts including those from the Centre for Agriculture and Bioscience International.

1.1.8.3 Genetic modification

Reference was made before to introduction of genes from other organisms into crop plant seed that confer resistance to both herbicides and insects. The adoption of this technology among maize producers in United States has been phenomenal as shown in Fig. 1.30.

Such rapid take-up arises only when innovation leads to increased profitability, and in the case of genetically engineered maize this has indeed been experienced. However, the strategy is not without its critics whose main objection derives from the fact that natural genetic responses have conferred resistance to the herbicide (glyphosate) in previously susceptible weed species. Products of genetically modified crops are not accepted by some importing countries.

1.1.8.4 Integrated pest management

Pests, diseases and weeds are controllable by chemical means and this can take the form of seed dressing or crop spraying. Such methods are both expensive and controversial as a result of possible residues of

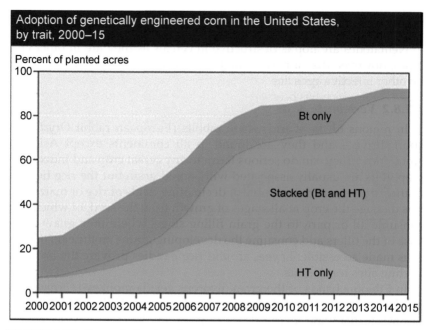

FIGURE 1.30   Rates of adoption of genetically engineered maize in the United States by trait. *Bt*, seed with insect resistance from *Bacillus thuringiensis*; *HT*, seed with herbicide resistance. *USDA, Economic Research Service using data from USDA, National Agricultural Statistics Service, June Agricultural Survey.*

undesirable chemicals in grains used for food or feed, and inadequate targeting of the treatments. Thus, where spraying is involved, some drift of the spray to outside the crop area is almost inevitable. Also, where insecticide is concerned, species that are valuable as predators of the pest species may be adversely affected as well as the target pest species. Insects that are important pollinators in the wider context of crops to which their services are invaluable may also be killed or damaged by insecticide sprays; such a risk was considered to be sufficient grounds for the ban introduced in the European Union (EU) of some neonicotinoids in 2013 due to their alleged effects on honeybee populations. Awareness and concern are increasing that any measure that changes the established balance among species in the environment of the crop may bring unexpected results. In many countries, with the encouragement of ecologists, since the 1970s, strategies have been developed to take advantage of all available pest and weed control methods. Such strategies are described as 'Integrated Pest Control' (IPC) or 'Integrated Pest Management' (IPM).

The UN FAO defines IPM as 'the careful consideration of all available pest control techniques and subsequent integration of appropriate measures that discourage the development of pest populations and keep pesticides and other interventions to levels that are economically justified and reduce or minimize risks to human health and the environment'. IPM emphasizes the growth of a healthy crop with the least possible disruption to agro-ecosystems and encourages natural pest control mechanisms. This includes managing insects, plant pathogens and weeds. A related concept is that of 'sustainable agriculture', which has been defined as 'both highly productive and protective of the natural resources on which future productivity depends'.

In parts of Africa where, for many producers, the use of chemical control is unacceptable or unaffordable, an alternative strategy dubbed 'push-pull' has met with some success. The description arises from the fact that pests are repelled or deterred from the target crop (push) by including plants that repel the pests, which are simultaneously attracted (pull) to a trap crop where they are concentrated, leaving the target crop protected. It is claimed that the push-pull strategy is effective against both stem borer and a parasitic weed (*Striga* spp.) (Khan et al., 2008).

In Europe interest has recently been shown in the effects of cultivating noncrop plants on field margins. Reduction in aphids in the crop area have been reported as a result of increased predator populations in the marginal strips. Conversely, molluscs have been recorded as increasing. Although the environmental benefits of this practice are widely acknowledged, much remains to be done in researching the agricultural benefits, not least because of the wide variations in planting and management strategies involved (Marshall and Moonen, 2002).

1.1.8.5 *Effects of pests and diseases on grains*

As well as the number of harvestable grains in cereal crops being reduced by pest and disease attack, those grains that survive to be harvested from all species thus afflicted can be adversely affected. Harmless fungal mycelium is present in most grains but in some cases the fungal residues are more damaging and may even be toxic; reference has already been made to ergot and other fungi that produce mycotoxins.

Organisms that cause a reduction in capacity to photosynthesize lead to a reduction in the supply of nutrient to their fruits and hence can lead to a reduction in synthesis of the starch component of endosperm. As a result, grains are less well filled and they become shrivelled, shrunken or thin when mature. Some invertebrate pests feed on the grains, thus creating cavities and making the grains lighter, disfigured and as a result unacceptable.

1.1.9 Test weight

This is a long-established characteristic considered in the grain trade to be an indicator of flour yield and feed quality potential, through reflecting the proportion of well-filled grains in a sample. Its validity is based on the fact that, as endosperm has a higher specific gravity than bran and embryo, a high proportion of shrivelled grains, which contain little endosperm, is reflected by a low test weight. Test weight (otherwise known as bushel weight, hectolitre weight or specific weight) can be measured in a chondrometer, an instrument that allows a container of known volume to be filled with grain under repeatable conditions, so that the contents can be weighed and related to unit volume. Units in which test weight is expressed reflect the volume and gravimetric conventions in the relevant country but simple conversions can be made or found (e.g., Canadian Grain Commission online conversion tables). Test weight is dependent on a number of factors, not the least moisture content, which, as well as contributing to the sample weight, influences the manner in which grains pack. Packing is also affected by the grain shape, a characteristic that varies among species, types and varieties. Test weight features in many grading systems. In an extensive study Owens et al. (2007) found test weight to be of little value in predicting feeding value of wheat fed to broiler fowl.

1.2 CHARACTERISTICS OF INDIVIDUAL CEREAL TYPES

In this section cereals are considered in order of their world production values (see Fig. 1.13).

1.2.1 Maize (corn)

The term 'corn' is frequently applied to the predominant local cereal, viz. wheat in England, oats in Scotland. In the United States 'corn' means specifically maize.

1.2.1.1 Origin and types

Teosinte, the wild ancestor of maize, was domesticated around 9000 years ago in southern Mexico. Cobs about 25 mm long were found in caves near Puebla. Three thousand years ago maize spread to the Andes and to eastern North America. From 1492 it spread to Asia, Africa and Europe.

Classification of maize and related species is complex (see Iltis and Doebley, 1980). The species of commercial production is *Zea mays* L. (2n = 20), but in current classification, wild and domesticated forms are recognized as subspecies, with the domesticated form as *Zea mays* L. ssp. *mays*. Different types of maize have evolved through selection, differing mainly in the nature of the endosperm. They are shown with the principal characteristics in Table 1.4.

TABLE 1.4   Characteristics of commonly grown maize types

Type		Grain/endosperm characteristics	% of crop
Dent	Indentata Group	Floury core and crown. These contract disproportionately during ripening giving rise to a small indentation at the distal end of the grain. The main type grown in United States of America.	73
Flint	Indurata Group	Smaller than dent but bigger than popcorn, with hard endosperm. Grown in Latin America and Europe for food.	14
Floury	Amylacea Group	Soft and mealy endosperm; easy to grind. Grown in Andean region.	12
Sweet	Saccharata Group	Contains a higher proportion of sugar in relation to starch as conversion blocked by one or more recessive mutations. Harvested before mature.	<1
Waxy	Ceratina Group	Starch is composed entirely of amylopectin (normal is 70% amylopectin and 30% amylose). Preferred as food in E. Asia.	<1
Pop	Everta Group	Floury core surrounded by hard flinty shell. Used for preparing snack food.	<1

Maize grains may be white, yellow or reddish in colour. White is preferred in east and southern Africa; even in the United States, where the predominant colour is yellow, white maize commands a premium. In world production, white maize constitutes less than 20%. Yellow maize is preferred for poultry feeds; the reddish type is favoured in Japan.

While conventional maize is a valuable ingredient of diets of both humans and livestock, it suffers from a limitation in the nutritional quality of its protein. The main deficiencies are in the proportions present of the essential amino acids lysine and tryptophan. However, in the early 1960s, a mutant maize cultivar with similar total protein content but twice the amount of lysine and tryptophan and 90% bioavailable protein was discovered (Bressani, 1991). This nutritionally superior maize was named 'opaque-2' maize because of the chalky (i.e., not vitreous) appearance of its endosperm when cut. Subsequent research demonstrated that the mechanism for the improvement was the suppression of zein synthesis. Zein is the prolamine (see Chapter 4) protein of maize, and like the prolamines of many cereals it is deficient in some essential amino acids. Opaque-2 had lower yields and its soft, chalky grain made it more susceptible to ear rot and insect damage. Moreover, the taste and grain appearance dissatisfied consumers, who ultimately rejected the enhanced-protein varieties in the market. Subsequent conventional breeding efforts at CIMMYT generated numerous cultivars with improved agronomic characteristics, and these were referred to as quality protein maize (QPM). Further work, using the CIMMYT approach in a number of countries, is continuing as large-scale production of QPM maize promises to offer significant benefits (Moura Duarte et al., 2004).

1.2.1.2 Cultivation

The crop is grown in climates ranging from temperate to tropical, during the period when mean daily temperatures are above 15°C and frost-free. The photosynthetic pathway found in maize is C4, a characteristic found in tropical grasses, in which less water is utilized in the production of biomass.

For germination the lowest mean daily temperature is about 10°C, with 18–20°C being optimum. When mean daily temperatures during the growing season are greater than 20°C, early grain varieties take 80–110 days and medium varieties 110–140 days to mature. When mean daily temperatures are below 20°C, there is an extension in days to maturity of 10–20 days for each 0.5°C decrease depending on variety.

Maize tolerates hot and dry atmospheric conditions as long as sufficient water is available and temperatures are below 45°C. It is considered to be either a day-neutral or a short-day plant. The growth of maize is very responsive to radiation, however, five or six leaves near and above the cob are the source of assimilation for grain filling and light must penetrate to these leaves. Plant population varies from 20,000 to 30,000 plants per hectare for the large late varieties to 50,000 to 80,000 for small early varieties. Sowing depth is 5–7 cm with one or more seeds per sowing point.

The plant does well on most soils but less so on very heavy dense clay and very sandy soils. The soil should preferably be well aerated and well drained as the crop is susceptible to waterlogging. The fertility demands for grain maize are relatively high and amount, for high-producing varieties, up to about 200 kg/ha N, 50–80 kg/ha P and 60–100 kg/ha K. In general, the crop can be grown continuously as long as soil fertility is maintained (FAO water/crops website).

1.2.1.3 Diseases and pests

Diseases affecting maize, growing in different parts of the world, include downy mildew (*Sclerospora* and *Peronosclerospora* spp.), leaf blight (*Helminthosporium turcicum* and *Helminthosporium maydis*), rusts (*Puccinia* spp.), maize smut (*Ustilago maydis*), anthracnose (*Colletotrichum graminicola*), bacterial leaf blight (*Pseudomonas avanae*), stalk and ear rots (several pathogens e.g., bacterial ear rot *Erwinia chrysanthemi*).

Pests include stem borers (*Brusseola fusca*), stalk borers (e.g., *Papaipema nebris*) corn ear worm (*Helicoverpa zea*) and army worm (e.g., *Mythimna unipuncta*).

In Africa, particularly, species of the semiparasitic weed *Striga* are intrusive and damaging weeds. see also Section 1.1.8.4 (ref to info on pests and diseases earlier in chapter).

1.2.1.4 Uses

Maize is traditionally a feed grain and this continues to be an important use. However, some types, such as sweetcorn and popcorn are used primarily as human food. Food use of the main crop is also important in some countries, such as several in Africa. Elsewhere, corn grits are used in the manufacture of breakfast cereals, and cornstarch is used as a thickener in food products. Corn syrups are used as sweeteners in processed food and drinks. Recently maize has been used as a feedstock for production of biofuels including ethanol and diesel and in 2014 in the United States this use exceeded feed use (www.ers.usda.gov/topics/crops/corn/background.aspx).

The largest volume industrial process applied to maize grains is wet milling to produce starch as a main product; dry milling, to produce grits is applied to a smaller volume. In both processes the embryos, which are rich in lipid, are isolated and a product known as maize (corn) germ oil is produced by extraction. By-products of maize milling and germ oil extraction are suitable for inclusion in feed.

Maize is the largest component of global coarse grain (corn, sorghum, barley, oats, rye, millet and mixed grains) trade, generally accounting for about two-thirds of the volume over the past decade.

1.2.1.5 Global production

Around 80% of the world's maize crop production is shared between two continental regions: the Americas and Asia (Fig. 1.31).

While there has been an increase in production in other continents, the most consistent increase has been demonstrated for Asia where, over a 5-year period, production has been raised by about 30% (Fig. 1.32). The marginally greater contribution to this increase has been improved yield, which rose by 15% (Fig. 1.34), and area harvested expanded by 13% (Fig. 1.33).

Although the contribution by Oceania is less than 1% of world maize production, yields in that region are among the highest in the world. For more details of production see Chapter 2.

1.2.2 Rice

Although world production of maize exceeds that of rice, rice is the world's largest human-food crop. Rice contributes approximately 21% of

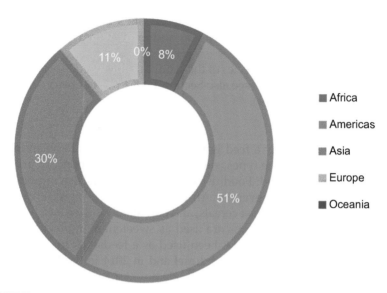

FIGURE 1.31   Proportional contributions to world maize production by continental regions (mean of 2009–13 values). *Based on data from FAOSTAT.*

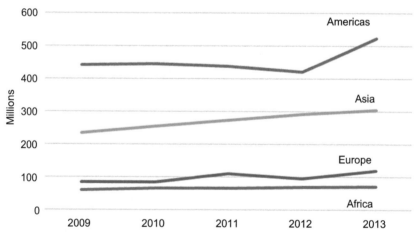

FIGURE 1.32   Variation in annual production of maize by continental regions (tonnes). *Based on data from FAOSTAT.*

world per capita caloric intake, and 27% of per capita calories in the developing countries. In highest-consumption countries, Vietnam, Cambodia and Myanmar, up to 80% of caloric intake is derived from rice. Of the 440 million tonnes of polished rice produced in the world in 2010, 85% went into direct human food supply. By contrast, 70% of wheat and only 15% of maize production was directly consumed by humans.

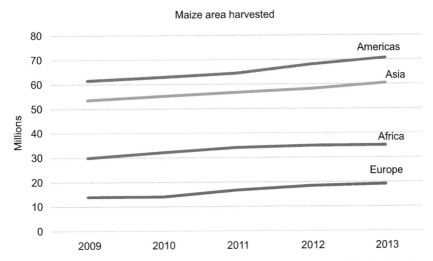

FIGURE 1.33 Variation in area harvested in continental regions (ha). *Based on data from FAOSTAT.*

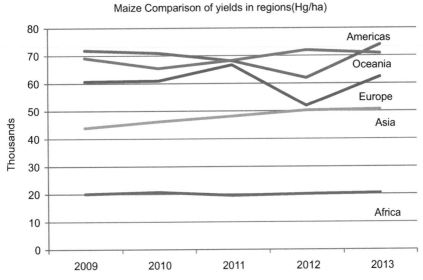

FIGURE 1.34 Variation in yields of maize in continental regions (hg/ha). *Based on data from FAOSTAT.*

1.2.2.1 Origin and types

Rice was domesticated 7000–8000 years ago in China in the Yangtze River valley. It is thought to have reached India 4000 years ago, possibly hybridizing with local wild rices, and Japan 2400 years ago. Rice culture gradually spread westward from northeastern Asia and was introduced to southern Europe in medieval times.

Rice belongs to the genus *Oryza*, which includes 20 wild species and the two cultivated annual species: *Oryza sativa*, known as Asian rice and *Oryza glaberrima* known as African rice. *O. sativa* whose wild ancestor, the perennial *Oryza rufipogon*, grows throughout South and Southeast Asia, is now grown in 112 countries on all continents except Antarctica (Black et al., 2006). *O. glaberrima*, originating from *Oryza barthii*, is confined to West Africa, and even here it is progressively being replaced by *O. sativa* (but see NERICA below). The 'wild rice' of the Great Lakes region of North America is a different species and indeed a different genus, though in the same tribe as rice. It is *Zizania palustris*.

O. sativa is characterized by subspecies (or races), the primary ones being *indica* and *japonica*. Whether these were independently domesticated is controversial. All members of the *Oryza* genus have 24 (2n) chromosomes, and while interspecific crossing is possible within each complex, it is difficult to recover fertile offspring from crosses across complexes.

Grains of *indica* type are long, narrow and slightly flattened. They are fluffy and remain separate when cooked, whereas grains of *japonica* are short, broad and thick with a rounded cross-section, they become moist and sticky on cooking.

Within the *indica* long-grain rice family are the fragrant rices. Because of their distinctive perfumes they command a price premium. The principal varieties defined as aromatic are the fragrant 'Hom Mali' rice produced in Thailand and the various types of basmati exclusively grown on the Himalayan foothills in India (in the states of Haryana and Punjab) and Pakistan (in the state of Punjab).

Japonica rice is mainly cultivated (and consumed) in temperate and tropical-upland climatic zones, in eastern and northern China, Taiwan, North and South Korea, the EU, Japan, Russia and Turkey. It is also grown in Australia and the United States (California).

Many of the rice hybrids available to producers derive from the basic *indica* varieties, including the semidwarf rice varieties, the introduction of which led to record yield increases throughout Asia in the 1960s and 1970s. Although research into genetically modified (GM) rice has intensified since the decoding of the rice genome in 2002, no GM rice has yet been officially released for commercial production.

Another type, known as glutinous, sweet or waxy rice (*Oryza sativa* L. var. *glutinosa*) and grown mainly in Southeast and East Asia, has a chalky, opaque endosperm, the cut surface of which has the appearance of paraffin wax. It differs from nonglutinous strains of japonica rice which also become sticky to some degree when cooked. The starch from glutinous rice contains 0.8%–1.3% of amylose. The amylose content of milled rice may be classified as waxy, 1%–2%; intermediate, 20%–25%; and high, >25%

In 1971 the African Rice Center (WARDA) was created by 11 African countries to contribute to poverty alleviation and food security in Africa, through research, development and partnership activities aimed at increasing the productivity and profitability of the rice sector in ways that ensure the sustainability of the farming environment. Today its membership comprises 25 countries. In the early 1990s a new rice known as NERICA (New rice for Africa) was developed, by conventional breeding techniques, at the Center by crossing African rice (*O. glaberrima*) with Asian rice (*O. sativa*) (Jones et al., 1997). NERICA varieties have unique characteristics such as higher yields of 1.5 tons per hectare (tonne/ha) without fertilizer application. Other qualities include early maturity, tolerance to major stresses, higher protein content and good taste compared with the traditional rice varieties.

1.2.2.2 Cultivation

Rice is a warm-season crop grown predominantly in tropical and subtropical climates as a semiaquatic annual grass although, in the tropics, it can survive as a perennial producing new tillers from nodes (see ratooning earlier) after harvest. Cultivation methods may be classified as paddy, upland and flood prone.

Paddy rice is grown in enclosures bounded by bunds or levees that become flooded so that plants are grown in standing water, the level of which may be controlled by irrigation to depths appropriate to growth stage. At times these will be greater than 15 cm. Upland (or dry land) rice is grown with (often sparse) natural rainfall alone.

Flood prone (deep water or 'floating') rice is grown in river basins or deltas that become uncontrollably flooded, often to 1.5 m deep, during the rainy season.

Alternative designations of the rice growing ecosystems, based on soil-water conditions, include irrigated lowland, irrigated upland, rain-fed lowland, rain-fed upland and deep-water/floating. In the case of rain-fed, lowland rice, where water levels are dependent entirely on rainfall, depths of 50–100 cm can occur.

The ability of rice to grow on saline soils has been proved in many parts of the world. Many researchers consider rice to be a crop with medium salt resistance and, since water reduces salt concentration, plant growth is not inhibited. Investigations carried out by physiologists have shown that rice is most susceptible to saline soils at the germination, shooting and flowering stages.

The environmental and socioeconomic conditions of rice production vary greatly from country to country as well as from location to location. The diverse environmental and socioeconomic conditions have affected the performance of rice production in the past. They also influence the opportunities for increasing rice production in the future. Rice farming in Asia is dominated by millions of small farmers with an average landholding of 1 ha.

As with some other cereals, in certain climates double cropping is practiced, whereby two crops of the same or different cereals can be produced in a season. Also under certain conditions, mainly in lowland systems, a second crop of rice can be obtained during the same season as the first, without sowing more seed. It is produced by the process of 'ratooning', whereby the parts of plants that remain in the ground after harvest produce more tillers, which bear grains after a period of maturation.

1.2.2.3 Diseases and pests

Rice blast (*Megaporthe grisea*) is a serious fungal disease but bacterial leaf blight (*Xanthomonas campestris* pv. *oryzae*) can also cause serious damage. The important tungro virus is spread by several species of leafhopper including *Nephotettix* spp. and *Recilia dorsalis* and can be devastating in south and southeast Asia.

Pests capable of serious direct plant damage include the brown planthopper (*Nilaparvata lugens*) and stem borers including the Asiatic rice borer (*Chilo suppressalis*) otherwise known as the striped rice stem borer and the yellow stem borer (*Scirpophaga incertulas*).

Snails such as *Pomacea canaliculata* and rats are also a problem.

The widespread barnyard grass (*Echinochloa crus-galli*), also grown as a millet, competes with crop plants for nitrogen and hosts pathogens capable of infecting rice.

1.2.2.4 Uses

Rice, more than any other cereal, is a food crop. It is mainly eaten cooked as whole grains after removal of hulls (as brown rice) and possibly after further milling, as white, or polished, rice, to remove pericarp. About 50% of the world's paddy is parboiled before milling; that is, it is subjected to a prescribed sequence of soaking, steaming and drying. Parboiling confers a number of advantages in that it leads to an increase in the yield of polished rice that has an enhanced nutritional value. Its storage potential is improved and it requires less cooking (Kaddus et al., 2002).

Rice flour, produced mainly from broken grains, has many food and industrial uses. Some rice grain products are fed to animals. Rice bran oil (extracted essentially from the embryos removed from the grain with the bran during special milling) has value as a cooking oil.

1.2.2.5 Global production

The degree of dominance of Asia in the production of rice is demonstrated in Fig. 1.35, which shows also that in global terms the contributions of Europe and Oceania are barely significant. However, it was in Oceania (essentially Australia as this is the only significant producer in the region) where not only the highest yields but also the greatest improvements in yields occurred during the period 2009–13 (See Chapter 2 for more details on regional production.) (Fig. 1.36).

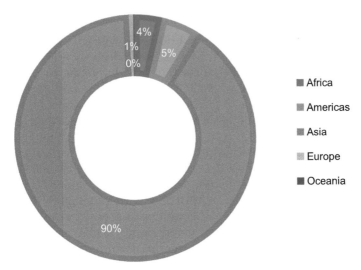

FIGURE 1.35 Proportional contributions to world rice production by continental regions (mean of 2009–13 values). *Based on data from FAOSTAT.*

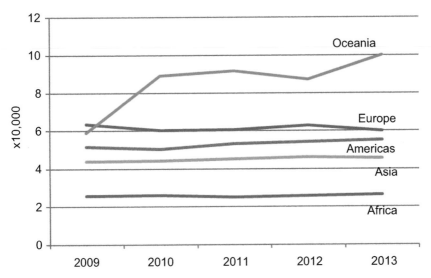

FIGURE 1.36 Variation in rice yields in continental regions (hg/ha). *Based on data from FAOSTAT.*

1.2.3 Wheat

Wheat is a crop of temperate regions but it is also cultivated in the higher lands of the subtropics and even the tropics; it is currently ahead of maize and second to rice as the main human food crop.

1.2.3.1 Origin and types

Both wheat and barley were first domesticated 10,000 to 11,000 years ago in the Fertile Crescent of the Middle East, which was described as surrounding the northern reaches of the Tigris and Euphrates rivers. Together with pulses such as lentils, wheat and barley were the earliest plants to be cultivated in the region.

Numerous examples of ancient wheat have been unearthed in archaeological investigations; the grains usually carbonized, although in some cases the anatomical structure is well preserved. The earliest cultivated wheats were both hulled species: einkorn *Triticum monococcum* ssp. *monococcum* (diploid, chromosome complement AA) and emmer *Triticum turgidum* ssp. *dicoccum* (tetraploid, AABB) both with grain tightly enclosed by tough husks (glumes: lemma and palea, see Chapter 3). These two species are cultivated as a staple only in mountainous regions including central and eastern Europe, but they are increasingly cultivated as 'health foods' elsewhere.

Naked or free-threshing forms (i.e., without adherent glumes) including the hexaploid bread, or common, wheat *Triticum aestivum* ssp. *aestivum* and tetraploid durum, or macaroni, wheat *T. turgidum* ssp. *durum*, evolved about 9000 years ago. Common or bread wheat (AABBDD) is a hexaploid allopolyploid; with three genomes, each corresponding to a normal diploid set of chromosomes, it resulted from accidental hybridization of emmer or durum wheat with the wild grass *Aegilops tauschii*.

There are about 16 cultivated species but the two most important are bread wheat and durum wheat. Each of these species has thousands of cultivars and landraces.

Wheat species and their thousands of varieties can be grouped into three groups based on chromosome numbers. A widely used classification of cultivated wheats (traditional names in brackets), and their probable wild ancestors and chromosome numbers (2n), is shown in Table 1.5.

By far the most important form of common or bread wheat is *T. aestivum* ssp. *aestivum*, but *T. aestivum* ssp. *compactum*, known as club wheat, is also grown commercially in a limited area in the western states of North America, and *T. aestivum* ssp. *sphaerococcum*, known as Indian dwarf or shot wheat, is found in northwest India and Iran.

Protein content

The proportion of protein in wheat grains is an important quality factor for both nutrition and processing purposes. While proteins are also present in other parts of the grain, it is the proteins in the starchy endosperm that are of most interest in the context of processing, as it is these proteins that end up in white flour. However, as the proportion of protein in other parts is fairly consistent, the whole grain protein content serves as a useful basis of comparison.

TABLE 1.5   Genetic constitution of wheats

Species	2n	Genomic constitution	Common name
DIPLOID WHEATS			
Triticum monococcum ssp. *aegilopoides* (*T. boeoticum*)	14	AA	Einkorn (wild)
T. monococcum ssp. *monococcum* (*T. monococcum*)	14	AA	Einkorn (cultivated)
TETRAPLOID WHEATS			
Triticum turgidum ssp. *dicoccoides* (*T. dicoccoides*)	28	AABB	Emmer (wild)
T. turgidum ssp. *dicoccum* (*T. dicoccum*)	28	AABB	Emmer (cultivated)
T. turgidum ssp. *durum* (*T. durum*)	28	AABB	Durum or Macaroni (cultivated)
T. turgidum ssp. *turanicum* (*T. turanicum*)	28	AABB	Kamut (cultivated)
HEXAPLOID WHEATS			
Triticum aestivum ssp. *spelta* (*T. spelta*)	42	AABBDD	Spelt (cultivated)
T. aestivum ssp. *aestivum* (*T. aestivum*)	42	AABBDD	Common or Bread (cultivated)

Among the main wheats of commerce protein content varies considerably, and a guide to some is given in Table 1.6. In the case of countries with grading systems, only samples conforming to specified protein content limits are permitted for inclusion in the premium grades and classes. Protein content is based on analytical quantification of nitrogen present in a ground sample and multiplication by an appropriate factor (see Chapter 4).

While wheats grown for stock feeding have requirements relating only to their agronomic and nutritional properties, the larger proportion of the crop that is destined for human consumption has important additional requirements related to the manner in which the endosperm behaves during processing. Much effort has been expended on understanding these characteristics and identifying means by which the desirable ones can be introduced into new varieties. It has not yet been possible to define some of the important characteristics in fundamental terms but there are long-established technological concepts that are widely used and these are explained next.

Endosperm texture

This is a property the importance of which lies in the way the starchy endosperm behaves during milling. The two extremes of texture

TABLE 1.6   Ranges of protein contents in wheat types

Wheat type	Approximate protein range (%)
United States Hard Red Spring (HRS)	11.5–18.0
Durum	10–16.5
Plate (Argentina)	10–16.0
Canada Western Red Spring (CWRS)	9–18.0
United States Hard Red Winter (HRW)	9–14.5
Russian	9–14.5
Australian	8–13.5
English	8–13.0
Other European	8–11.5
United States Soft Red Winter (SRW)	8–11.0
United States White	8–10.5

Based on data from Schruben, L.W., 1979. Principles of wheat protein pricing. In: Wheat Protein Conference, Agriculture Research Manual, ARM9 Science and Education Administration. USDA, Washington, DC and Kent-Jones, D.W., Amos, A.J., 1947. Modern Cereal Chemistry, fourth ed. Northern Publishing Co. Ltd., Liverpool.

description are hard and soft but a range exists between the extremes. Starchy endosperm in the mature grain is a dry cellular tissue with mostly weak cell walls, within which lie starch granules in a matrix of protein. When subjected to physical pressure and shear, such as is experienced during roller milling, the endosperm breaks into small particles, and indeed this is the purpose of flour milling as it is the mass of small particles produced that is known as flour.

The endosperm of softer wheats breaks down more readily than that of harder wheats, and thus, when similarly processed, softer wheats produce particles of lesser size than those from harder wheats. Hard wheats yield coarse, gritty flour, free-flowing and easily sifted, consisting of regular-shaped particles, many of which are whole endosperm cells, singly or in groups. Endosperm particles in flour from soft wheat are irregularly shaped, with some flattened particles, which become entangled and adhere together, sift with difficulty, and tend to clog the apertures of sieves. A proportion of quite small cellular fragments and free starch granules are present. The degree of mechanical damage to starch granules produced during milling is greater for hard wheats than for soft.

Hardness affects the ease of detachment of the endosperm from the bran. In hard wheats the endosperm cells come away more cleanly and tend to remain intact, whereas in soft wheats the subaleurone endosperm cells tend to fragment, part coming away while part is left attached to the bran.

It is now understood that the property that underlies the variation in hardness among types is the degree of adhesion between the starch granules and their surrounding protein matrix, and it is further understood that in soft wheats chemical agents are present that prevent adhesion. These agents have now been characterized as protein-lipid complexes, and the proteins concerned have been given the name puroindolines (see Bechtel et al., 2009).

Because of the strong bond between starch and protein in the endosperm of hard wheats, a cut face of a grain has a glassy or 'vitreous' appearance (sometimes described as steely, flinty or horny) and its natural dark amber colour is evident. In contrast, soft wheat, in which the protein matrix is discontinuous, has air spaces surrounding the starch granules, giving rise to light scattering at the starch/air interfaces. This gives cut endosperm a white, mealy appearance, sometimes described as starchy or chalky.

Texture *per se* is not a property that determines baking potential but it is often the case that hard wheats also have higher protein content and a vitreous appearance, as well as having good bread-baking properties. Vitreousness has thus become to some buyers, a visible indication of quality and it has been incorporated into some marketing standards.

Vitreousness is also a desirable character of durum wheats although here it is less likely that bread-making properties are of interest. Nevertheless, hardness is needed for milling into semolina, with minimum reduction to flour size particles, and high protein is required for pasta making, the use to which most durum wheat is put. As its name suggests, durum wheat has particularly hard endosperm. Club wheats on the other hand have very soft endosperm.

Strength

This property refers not to the physical strength of the endosperm but to the characteristics of the protein present. In fact, it refers to 'baking' strength, or the degree of suitability to provide flour capable of making good leavened bread.

A 'strong' wheat variety produces a flour with a protein complement that, on hydration and mixing, produces a gluten that can form films capable of containing expanding gases. Leavened bread made from strong wheat flours has a relatively large volume and a soft texture. Strong wheats have a relatively high-protein content but the qualities of the protein are also essential for the criterion of 'strength' to be met.

Spring wheats have a shorter growing season than winter wheats and accumulate less starch. They do however accumulate similar quantities of protein to winter varieties and consequently protein contributes a higher proportion. This, coupled with protein strength conferred by selective breeding, has established many spring wheats, including those of North

America, as the best wheats for bread-flour production. Because of their high quality many countries whose climates are less suited to production of strong spring wheats have imported such grains. The United Kingdom is an example of a country with a long history of importing bread wheat, but in the 1970s, in pursuit of self-sufficiency and to avoid levies on third-country imports imposed by the European Economic Community (now the EU), a focused breeding programme led to the introduction of (mainly winter) varieties with sufficient strength to replace imported types. At the same time bread-making techniques were developed that allowed acceptable quality to be achieved from lower protein flours (Chamberlain et al., 1962). A further facilitating factor was the coproduction of starch and vital gluten from wheat, making the latter available for inclusion in flour produced for bread making. The United Kingdom is now around 90% self-sufficient in bread-making wheat. Even where high-quality spring wheats are available, they may not be milled alone. Instead a 'grist' is made in which they are supplemented by 'filler' wheats with lower bread-making properties.

Although strength is a desirable characteristic of flours destined for bread production, there are many applications of flours for which 'weakness' is the desirable property. In these uses, such as biscuit (cookie) and confectionery applications, starch is the more important structural component. Because neither hardness nor strength can be measured in authentic fundamental units, empirical methods of measurement have been devised. In the case of hardness, several methods exist, but they do not necessarily rank all samples in the same order. A popular method is particle size index, whereby a sample ground by a standard method is sieved. Values may be expressed as the proportion passing through the sieve. Measuring the energy required to grind a sample is an alternative approach.

The relationship between protein content and particle size index has been explored and suitability of types based on these characteristics has been determined. Results are shown in Fig. 1.37.

Many countries sort their wheats and other cereals into grades depending on quality, purity and cleanliness before sale to buyers at home or abroad. In almost all systems the properties of varieties are established and only those varieties that have been shown to conform to certain standards are permitted within the more demanding classes. In the United Kingdom, where no official grading system exists, a Wheat Guide is published by the National Association of British and Irish Millers following annual assessment. In it, varieties are grouped according to suitability for end use. Group 1 varieties are most suited to bread making and therefore attract a premium. At the other extreme, Group 4 varieties are suitable mainly for feed, with a possibility of inclusion as 'fillers' in other grists. Most feed wheats are excluded from the grading system (http://www.nabim.org.uk/).

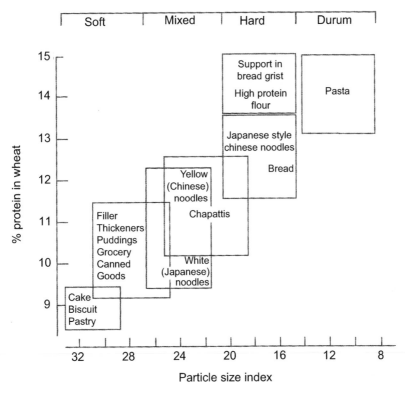

FIGURE 1.37 Indication of protein content and grain hardness requirements for a range of baked products. *Reproduced from O'Brien, L., Blakeney, A.B., 1985. A Census of Methodology Used in Wheat Variety Development in Australia. Cereal Chemistry Division, Royal Australian Chemical Institute, Melbourne. with permission.*

1.2.3.2 Cultivation

T. aestivum and Triticum durum wheats exist as winter and spring varieties. Cultivated varieties, which are of widely differing pedigree and are grown under varied conditions of soil and climate, show wide variations in characteristics. The climatic features in countries where spring wheat is grown – maximum rainfall in spring and early summer, and maximum temperature in the mid- and late-summer – favour production of rapidly maturing grain with endosperm of vitreous texture and high-protein content, traditionally suitable for bread making. Winter wheat, grown in a climate of relatively even temperature and rainfall, matures more slowly, producing a crop of higher yield and lower nitrogen content, better suited for biscuit and cake making than for bread, although in the United Kingdom, where winter wheat comprises about 96% of the total, winter wheat is used for bread making.

The yield of durum wheat, which is grown in drier areas, is lower than that of bread wheat.

1.2.3.3 Diseases and pests

Several rusts infect wheat. They include stem rust (*Puccinia graminis*), leaf rust (*Puccinia recondita*) and stripe or yellow rust (*Puccinia striiformis*). There are also several blotch diseases such as spot blotch (*Bipolaris soroboniana*), speckled leaf blotch (*Septoria tritici*) and glume blotch (*Septoria nodorum*). Other fungal diseases are head scab and foot/root rots (*Fusarium* spp.) Rhizoctonia root rot (*Rhizoctonia* spp.), loose smut (*Ustilago tritici*), common bunt or stinking smut (*Tilletia tritici* and other *Tilletia* spp.), karnal bunt (*Tilletia indica*) and powdery mildew (*Erysiphe graminis*).

Bacterial leaf streak (*Xanthomonas campestris*) is one of the bacterial diseases, and barley yellow dwarf virus BYDV is one of the viruses infecting wheat crops.

Aphids, termites, grasshoppers and leafhoppers, bugs, thrips, and sawflies are among the pests. In some years, orange wheat blossom midge (*Sitodiplosis mosellana*) can cause damage to yields and quality. Wheat bulb fly (*Delia coarctata*) is a pest of mainly winter crops in England.

There are many grasses and broad-leaved species that compete with wheat crops; a widespread weed *Phalaris minor* (littleseed canarygrass) is widely distributed but it is particularly troublesome where rice-wheat cropping systems are employed. Blackgrass (*Alopecurus myosuroides*) is also widespread and difficult to control but it is considered invasive only in western Europe.

1.2.3.4 Uses

The main food use of bread wheat (*T. aestivum*) is as a source of flour for production of baked products. Different types are suited to different products and the flours with highest values are milled from wheats with high levels of protein with appropriate qualities for making leavened and unleavened breads. Flours milled from types with lower protein contents and different characteristics are used for noodles, confectionery, biscuit (cookie) production and for household uses. Types with protein contents and quality unsuited for milling into flour are used for animal feed. Industrial uses similar those applicable to maize are also increasing. The main use of durum wheats (*T. durum*) is production of semolina or flour for pasta manufacture.

By-products of milling of all types of wheat are suitable for use as feed.

1.2.3.5 Global production

As with rice it is Asia which dominates wheat production although the degree of dominance is far less than in the case of rice (Fig. 1.38). Nevertheless, it is in Asia that the most consistent increase in production has occurred (Fig. 1.39).

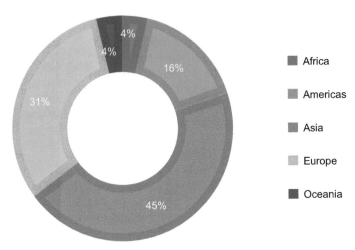

FIGURE 1.38 Proportional contributions to world wheat production by continental regions (mean of 2009–13 values). *Based on data from FAOSTAT.*

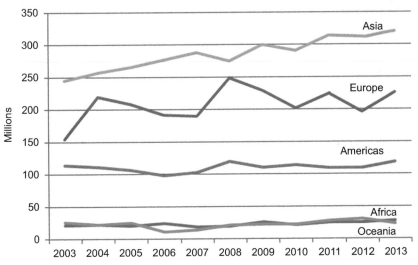

FIGURE 1.39 Variation in wheat production in continental regions (tonnes). *Based on data from FAOSTAT.*

1.2.4 Barley

Barley (*Hordeum vulgare*) is a cool-season grass with worldwide distribution. In the league of cereals production, it ranks fourth but quantities are less than a quarter of each of the major cereals, maize, rice and wheat. As a feed grain barley is ranked second to maize.

1.2.4.1 Origin and types

H. vulgare originated, like wheat, in the 'Fertile Crescent' (see Chapter 3) about 10,000–11,000 years ago, initially as a two-rowed type. The six-row form is thought to have resulted from a mutation in a cultivated crop about 9000 years ago There are about 30 species of wild *Hordeum*, most of which are diploid but some tetraploid and hexaploid types exist but they are low yielding and have high levels of sterility. The diploid chromosome complement is 2n = 14.

The cultivation of barley (*Hordeum* spp. was known to the ancient Egyptians, and grains of six-row barley) have been discovered in Egypt dating from pre-dynastic and early dynastic periods. Barley is mentioned in the Book of Exodus in connection with the 10 plagues.

Barley was used as a bread grain by the ancient Greeks and Romans. Hunter (1928) illustrates Greek coins dating from 413 to 50 BC which incorporate ears or grains of barley into their design. Barley was the general food of the Roman gladiators who were known as the *hordearii* (Percival, 1921). Calcined remains of cakes made from coarsely ground grain of barley and *Triticum monococcum*, dating from the Stone Age, have been found in Switzerland.

Bread made from barley and rye flour formed the staple diet of the country peasants and the poorer people of England in the 15th century while nobles ate wheaten bread. As wheat and oats became more generally available, and with the cultivation of potatoes, barley ceased to be used for bread making.

Commercially, barley exists in both two-rowed and six-rowed forms. The distinction refers to the form, of the ear; more specifically it is an indication of the number of fertile spikelets at each node (see Chapter 3). In six rowed types there are three fertile spikelets, each of which contains one grain. As spikelets are arranged alternately on opposite sides of the main axis of the ear, it can be seen to have six rows of grains when viewed from the top. In two rowed forms only the grain in the central spikelet bears a grain and hence only two rows of grains can be seen from above. As in rice, each grain is surrounded by hulls comprising adherent lemma and palea, except in the case of naked grained (hull-less) varieties which have been selected specifically to exhibit nonadherent hulls. As the hulls are unpalatable and indigestible, and hence have to be mechanically removed, the naked varieties are advantageous when the crop is valued as a human food source. Hull-less barley is a staple food crop in Tibet; it grew on nearly 300,000 acres in 1989. All naked barleys are six-rowed.

As with most major cereals types of barley possessing starch with various proportions of amylose and amylopectin exist.

1.2.4.2 Cultivation

Cultivated barleys occur as spring sown and winter sown varieties. A wide range of varieties exists with individual varieties having been bred to suit local conditions. Agricultural practices appropriate to barley are similar to those used for wheat growing. Seeds are sown to a depth of 2–5 cm about 15–18 cm apart.

1.2.4.3 Diseases and pests

Viruses are important pathogens in the case of barley and these include BYDV, which infects most cereals and barley stripe mosaic virus (BSMV), which infects mainly barley and wheat. BYDV is spread largely by aphids while BSMV is spread mainly by infected seed. Control measures for the two types thus need to be different.

Fungal pathogens of barley include powdery mildew (*Blumeria graminis*), downy mildew (*Sclerophthora rayssiae*), spot blotch (*Helminthosporium sativum*), barley stripe (*Pyrenophora graminea*), net blotch (*Pyrenophora teres*), scald (*Rhynchosporium secalis*), scab (*Gibberella zeae*), rusts (*Puccinia* spp.) and smuts (*Ustilago* spp.).

Several nematode species including the widely distributed cyst nematodes (*Heterodera* spp.) can be pests in nonresistant varieties.

As are all cereals, barley is attractive to aphids and they also introduce BYDV. In regions where they occur, various species of army worm are capable of devastating barley crops.

1.2.4.4 Uses

Barley is used principally as feed but it is also the cereal grain most used in production of malt. Malting barleys carry the highest value but only those that meet stringent requirements are selected for this purpose. Criteria include both genetic and compositional requirements, and a high proportion of grains capable of germinating is essential as the process requires high levels of enzymes associated with germination. Malt can be produced from all types of barley (and indeed all cereals) but two-rowed varieties are generally preferred and hulled varieties have the advantage over naked-grained varieties because the hulls serve as a filter medium useful to some stages in the malting process. Protein content is an important factor but, whereas for many cereals a high-protein level is desirable, for malting the requirement is for a relatively low-protein content. The majority of barleys are spring sown and these tend to accumulate less starch, and consequently have higher protein contents in their grains than winter barleys. Hence winter-sown crops are more likely to meet the malting requirement. For feed purposes, a higher protein content is more nutritious.

The most important constituents involved in malting are starch and the enzymes produced on germination. Starch from barley and other cereals may also be included as an adjunct.

1.2.4.5 Global production

Fig. 1.40 shows that Europe accounts for approximately 60% of the world's barley production, followed by Asia and the Americas.

Over the period 2010–14 there has been apparent increase in European production. However, this does not necessarily indicate a sustained trend as there is a high level of periodic variation in world production to which Europe's contribution is significant (see Figs 1.40 and 1.41).

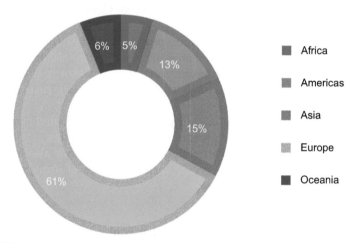

FIGURE 1.40 Proportional contributions to world barley production by continental regions (mean of 2010–14 values). *Based on data from FAOSTAT.*

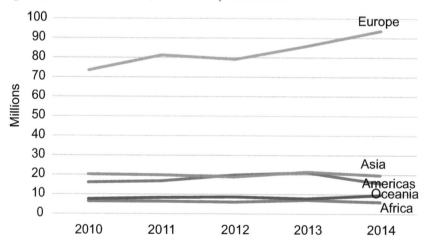

FIGURE 1.41 Variation in barley production by continental regions (tonnes). *Based on data from FAOSTAT.*

1.2.5 Sorghum

Sorghum is a coarse grass that bears loose panicles containing up to 200 seeds per panicle. It is an important crop and the chief food grain in parts of Africa, Asia, India/Pakistan and China, where it forms a large part of the human diet, but it is grown to a greater or lesser extent in all continents. In fact, the Americas produce similar quantities to Africa. It has many local names, for example in United States it is known as milo (Black et al., 2006).

1.2.5.1 Origin and types

Sorghum (*Sorghum bicolor*) is a warm-season cereal of African origin, which was first cultivated in the region of Ethiopia or Chad over 5000 years ago. It spread to India by 4000 years ago and later to China and to southern Africa by about 1500 years ago. Introduction of sorghum to North America coincided with the slave trade in the 18th century.

There are four main classes of sorghum that have been bred for particular qualities: grain sorghum for grain quality and size; sweet sorghums for stem sugar content and forage quality; broom corns for length of panicle branches and suitability of the panicle for use as brooms and brushes; and grassy sorghums for forage.

S. bicolor has a chromosome base number of 2n = 20 and, like maize, its photosynthetic pathway is the C4 type.

So called 'bird-resistant, bird-proof or bird-repellent' sorghums contain condensed tannins, in the nucellar layer and pericarp, that are distasteful to birds and give the crops some protection. Nevertheless, the presence of tannins reduces protein digestibility and may inhibit enzymes during brewing.

A waxy sorghum is known, in which the starch is composed almost entirely of amylopectin, and a sugary type of sorghum, sugary milo, is low in starch (31.5%) but contains 28.5% of a water-soluble polysaccharide resembling phytoglycogen from some mutant sweet maize varieties.

Other types of speciality sorghums are listed on www.nap.edu/catalog/2305/lost-crops-of-africa-volume-i-grains.

1.2.5.2 Cultivation

Sorghum is grown mainly in drought-prone low-rainfall areas, though it will tolerate more than 1 m/year rainfall, and in mountainous regions (above 1000 m). Plant height varies from under 1 m to more than 4 m, and most crops can take between 65 and 150 days to reach maturity. Unlike many other cereals sorghum can become dormant under adverse conditions, such as early drought, and can revive when wet conditions return. Seeds are planted 25–50 mm deep at rates that can be less than 5 kg/ha when sown or around 15 kg/ha when broadcast. Another practice is to transplant seedlings at 3–4 weeks old. Plants are spaced 20–100 cm apart within rows that are 50–100 cm apart, on the flat or on hills or ridges (Black et al., 2006). As with rice, ratooning is possible, leading to a second grain crop from the same plants. Ratooning requires the field to be relatively weed-free following harvest. Shortening the remaining stems, for example, by flailing, to a few inches above the ground, leads to tillering and eventual fruiting where sufficient water, temperature and nutrients are present. Fertilizer is usually applied at the time of flailing.

1.2.5.3 Diseases and pests

Sorghum is susceptible to grain moulds caused by a complex of fungal pathogens including *Curvularia lunata*, *Fusarium* spp. and *Phoma sorghina*. Among the leaf diseases anthracnose caused by *Colletotrichum sublineolum* is capable of causing significant damage and even plant death. Leaf blight (*Exserohilum turcicum*), a very widespread disease, more familiar as one of maize, also affects sorghum. Zonate leaf spot (*Gloeocercospora sorghi*) and tar spot (*Phyllacora sorghi*) can significantly reduce yield. Sorghum is also one of the species infected by charcoal spot (*Macrophomina phaseolina*) and downy mildew (*Peronosclerospora sorghi*), rust (*Puccinia purpurea*) and ergot (*Claviceps sorghi*).

Shoot fly (*Athorigona soccata*), stem borers (*Brusseola fusca* and *Chilo portellus*), sorghum midge (*Contarinia sorghicola*) and head bugs (*Calocoris angustatus*) are among relevant insect pests.

1.2.5.4 Uses

Uses of sorghum are heavily dependent on the location of production, which, because the proportion traded internationally is low, is also the location of use. In United States most is used as feed whereas in Africa its use as food predominates. Considerable versatility is associated with the ways in which sorghum is prepared for human consumption so that, as well as being made into a porridge or gruel, sorghum meal or flour can be used in noodles, breads and other fermented products. It is also consumed as milled whole grains. It is also the main or only constituent of the feedstock for malting for production of alcoholic and other beverages.

1.2.5.5 Global production

As with barley, the global area devoted to growing sorghum has declined over a period of 50 years but the slight increase in yield over the 10 years up to 2012 has led to a relatively stable production level over that period (Figs 1.19–1.21).

Among the continental regions, Africa and the Americas evenly shared around 80% of sorghum production during between 2010 and 2014, with Asia contributing most of the remaining (Fig. 1.42). Continental regional trends in production over this period are shown in Fig. 1.43.

In spite of being one of the major producers, African yields are three to four times lower than the Americas, Europe and Oceania (Fig. 1.44).

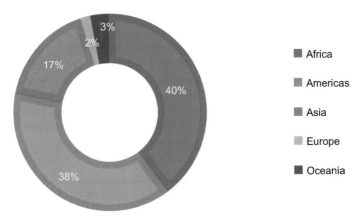

FIGURE 1.42   Proportional contributions to world sorghum production by continental regions (mean of 2010–14 values). *Based on data from FAOSTAT.*

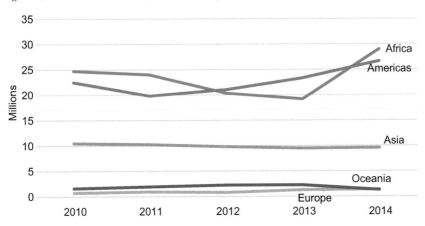

FIGURE 1.43   Variation in sorghum production by continental regions (tonnes). *Based on data from FAOSTAT.*

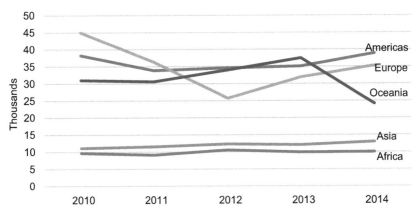

FIGURE 1.44   Variation in yield of sorghum in continental regions (hg/ha). *Based on data from FAOSTAT.*

1.2.6 Millets

Millet is a broad category of warm-season, small-seeded grasses including a number of genera that are not closely related. They are mainly, though not exclusively, cultivated without irrigation by pastoralists and small-scale producers, in semiarid conditions and in poor soils. Climatic regions in which they are grown include temperate, subtropical and tropical.

Some species of millet, with their chromosome number, are listed in Table 1.7 in descending order of importance where known.

TABLE 1.7   Millets and their properties

Common names	Botanical names	Chromosome number	World grain production (tonnes)	Where grown
Pearl millet (bulrush millet)	*Pennisetum glaucum* (*P. americanum, P. typhoides*)	2n = 14	~15 million	Africa (mainly west), India
Foxtail millet (Italian millet, German m. Hungarian m. Siberian m.)	*Setaria italica*	2n = 18	~5 million	Asia (mainly China), Europe
Proso millet (common millet, hog m. broomcorn m., Russian m.)	*Panicum miliaceum*	2n = 36 or 72	~4 million	E. Asia, India, Kazakhstan, Ukraine, Middle East, United States
Finger millet	*Eleusine coracana*	2n = 36	<4 million	India
Fonio (black fonio)	*Digitaria iburua*	2n = 54	<3 million	W. Africa
White fonio (acha, hungry rice)	*Digitaria exilis*		<3 million	W. Africa
Barnyard millet, (barnyard grass)	*Echinochloa crus-galli*	2n = 36 to 72	<3 million	Asia, Egypt, California
Japanese barnyard millet	*Echinochloa esculenta*	2n = 54		
Indian barnyard millet	*Echinochloa frumentacea*	2n = 36,54,56	<3 million	S. Asia
Teff	*Eragrostis tef*	2n = 40	<3 million	Ethiopia, Eritrea
Little millet	*Panicum sumatrense* (*P. miliare*)	2n = 36	<3 million	India
Kodo millet (creeping paspalum, dronkgras)	*Paspalum scrobiculatum*	2n = 40	<3 million	India

Coix lacryma-jobi Job's tears or adlay is used as food in India, but it is not included in the table as its main use is of a distinct form with glassy inflorescences, used in making seed jewellery.

1.2.6.1 Origin

The earliest evidence for domesticated common millet (*Panicum milliaceum*) and foxtail millet (*Setaria italica*) is in northern China, dating to 8000–7000 years ago. There is abundant evidence for cultivation of pearl millet (*Pennisetum glaucum*) and finger millet (*Eleusine coracana*) in India starting about 4000 years ago. Botanical evidence is that both were domesticated in Africa and later spread to Asia. Archaeological finds of these crops in Africa are still sparse but suggest that pearl millet originated in the West African savannah, while finger millet originated in the East African highlands.

1.2.6.2 Cultivation

Pearl millet is a hardy plant capable of yielding a crop where most other grain cereals would fail. In West Africa, pearl millet is grown in the north, where rainfall is less than 76 cm per annum, while sorghum replaces millet in the wetter south. Intercropping of pearl millet and sorghum (or other crop species) is sometime practiced. Pearl millet is usually sown in pockets on 'hills' or ridges, but broadcasting (particularly in India) or drilling in rows are also adopted.

Foxtail millet can be grown on marginal soils on which it yields reasonably but it thrives better on fertile soils. It can be grown in semiarid conditions but is susceptible to long periods of drought. It cannot tolerate waterlogging.

Foxtail millet is usually drilled in rows or broadcast. In India it may be grown in mixtures with finger millet, cotton and sorghum (Grubben and Partohardjono, 1996).

As well as being cultivated as crops, some of the species are regarded as intrusive weeds, causing a nuisance in other cereal crops. For example, barnyard millet is often encountered undesirably in rice crops.

1.2.6.3 Diseases and pests

Downy mildew (*Sclerospora graminicola*) affects most types of millet and this is also true of smuts (e.g., *Tolyposporium penicillariae*), ergot (e.g., *Claviceps microcephala*) and rusts (e.g., *Puccinia penniseti*) (the examples are relevant to pearl millet). Blast (*Pyricularia seriae*) affects foxtail millet, finger millet and pearl millets as well as oats and maize, although symptoms can vary according to crop species. Up to 50% losses due to blast have been reported.

Birds, particularly *Quelea* spp. in Africa, and parakeets, crows and migrating rosy starlings in India, can cause major problems with millets.

Insect pests are a greater hazard in West Africa than in India; they include stem borer (*Coniestaignefusalis*), millet midge (*Geiromiya penniseti*), grasshoppers, locusts and various Lepidoptera. In Africa finger millet and teff may receive sporadic attacks from the phyllophagous ladybird (*Chnootriba similis*). In Ethiopia bands of the graminivorous gelada monkey (*Theropithecus gelada*) can cause serious damage to all cereal crops.

The most damaging weeds are the semiparasitic *Striga* spp.

1.2.6.4 Uses

Almost all millet grown is consumed close to its site of production, as food. As there are many types of millet it is to be expected that a considerable range of products are consumed but the manners of preparing foods are similar. Much of the millet 'processing' occurs in the domestic environment. The grains are pounded using simple devices, so that endosperm can be separated as the main component for consumption. Fermentation may be carried out before separation. Grits or flours produced may be made into similar products to those produced from sorghum: porridge, flatbreads or boiled broken grains. It is also possible to produce alcoholic beverages from millet.

1.2.6.5 Global production

More than 90% of the millet crop is grown in Africa and Asia, but small quantities come from Europe and central parts of the Americas (Fig. 1.45).

Over a 50-year period the reduction in area of millets harvested (Fig. 1.21) has probably been more than compensated by improvements in yield (Fig. 1.21), leading to a slight increase in production. However, harvests (Fig. 1.46) and yields (Fig. 1.47) have been subject to considerable year-on-year fluctuation, so any short-term predictions on the basis of past trends must be regarded as hazardous.

1.2.7 Oats

Several species of the oat genus *Avena* are cultivated but the common or white oat (*Avena sativa*) is by far the most widely grown. It is a cool-season crop grown for both animal feed and human food. Oat grains are borne on a loose panicle rather than a spike, which is more usual among the cool-season cereals wheat, barley and rye.

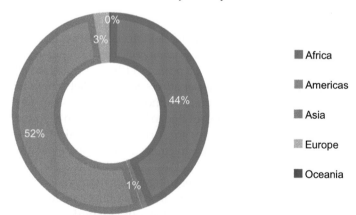

FIGURE 1.45 Proportional contributions to world millet production by continental regions (mean of 2010–14 values). *Based on data from FAOSTAT.*

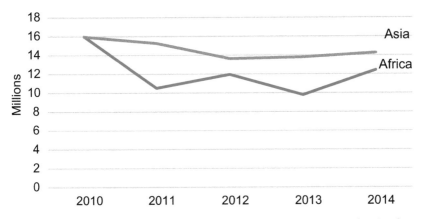

FIGURE 1.46 Variation in production of millet by the two main continental regional producers (tonnes). *Based on data from FAOSTAT.*

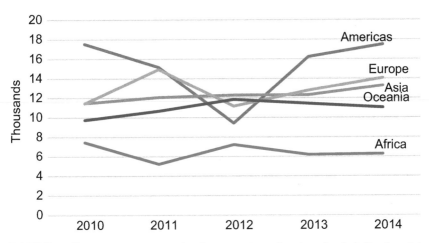

FIGURE 1.47 Variation in yield of millet in continental regions (hg/ha). *Based on data from FAOSTAT.*

1.2.7.1 Origin and types

Archaeological evidence of wild oats has been found in the Fertile Crescent as long as 7000–12,000 years ago. It is thought that the wild oat *Avena sterilis* spread as a weed in domesticated wheat and barley crops, eastwards from the Fertile Crescent, becoming domesticated in northern Europe 3000–4000 years ago. Oats as a crop reached America in 1602. Red- or Algerian-oat (*Avena byzantina*) is adapted to warmer and more arid climates. It is grown in the Near and Middle East and varieties grown south of 40 degrees in the United States are likely to have been derived from *A. byzantina*. Oat grains, as harvested, retain the tough husk comprising the dried lemma and palea, but the naked oat, *Avena nuda*, is a species that readily loses its husk during threshing. It has considerable potential for food and feed use, since the grains require no dehulling and have high-protein, oil and energy values. In spite of this it represents only a small proportion of the total world oat crop. *A. nuda* has been grown in China for 2000 years and is now grown principally in India, Tibet and China. *A. sativa*, *A byzantina* and *A. nuda* are hexaploid, with a chromosome number of 2n = 42.

A tetraploid species *Avena abyssinica* is cultivated but only in and around Ethiopia. There are two widespread hexaploid species that are not cultivated for grain, in fact they are troublesome weeds that compete well in other cereal crops; *Avena fatua* is abundant in cooler regions, and *A. sterilis* is found in warmer climates where it may be grown for forage.

1.2.7.2 Cultivation

Spring and winter varieties of *A. sativa* exist but the vernalization requirement (i.e., the period of exposure to cold to ensure flowering in the following summer) is about half that of wheat. Oats are generally sown 2–6 cm deep depending on soil type. Emergence is rapid and 4–5 cm of growth can be achieved within 5–7 days. Fast growing and vigorous tillering lead to good weed suppression, and because of a low requirement for weed and pest control and fertilizer application oats is regarded as a low-input crop. Long days are necessary for flowering to occur.

1.2.7.3 Diseases and pests

Oats are relatively free from diseases and pests with the exception of rusts; leaf and stem rusts and particularly crown rust (*Puccinia coronata* var. avenae) can greatly reduce crop yields. The rustic shoulder-knot moth (*Apamea sordens*), which feeds on many other cereals and grasses, is a minor pest.

1.2.7.4 Uses

The majority of oats are used for feed, but in some countries, breakfast cereals, including porridge, made from oats are popular. In recent years the health-promoting benefits of some of the complex carbohydrates that are more abundant in oats than in other cereal grains have led to the promotion of oats as a 'health food'. Antioxidants present in oats are potentially valuable constituents of some pharmaceutical products.

1.2.7.5 Global production

Of the continental regions, Europe is the main producer of oats, with the Americas producing less than half as much and other continental regions growing much less (Fig. 1.48). Little change in production has occurred in recent years (Fig. 1.49). Yields vary among continental regions, and in most locations there are relatively large annual fluctuations (Fig. 1.50).

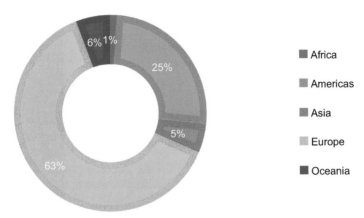

FIGURE 1.48 Proportional contributions to world oats production by continental regions (mean of 2010–13 values). *Based on data from FAOSTAT.*

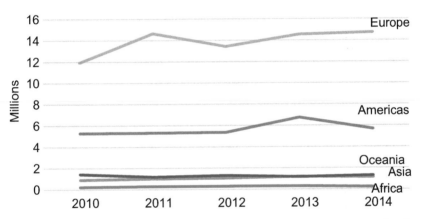

FIGURE 1.49 Variation in oats production by continental regions (tonnes). *Based on data from FAOSTAT.*

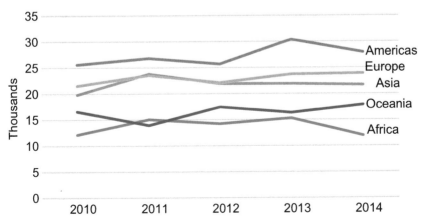

FIGURE 1.50 Variation in yield of oats in continental regions (hg/ha). *Based on data from FAOSTAT.*

1.2.8 Rye

Rye (*Secale cereale*) is a cool-season cereal grown almost exclusively in parts of Europe where dry summers and very cold dry winters are experienced.

1.2.8.1 Origin and types

In spite of its current distribution, rye originated in the Near East, where there is scattered evidence of domesticated forms from about 9000 years ago. However, it first became a well-established crop in Bronze Age Europe, about 3500 years ago. Like oats, the wild form may have spread from the Fertile Crescent to Europe as a weed of cereal crops.

Rye is a diploid with 2n = 14 chromosomes, but some tetraploid varieties have been produced. Unlike other temperate cereals rye exhibits a high degree of cross-pollination. This characteristic offers greater possibility for production of hybrids than in the self-fertile cereals but it also increases the difficulty of keeping the strain pure.

In Roman times the chief cereal crop in the south of Britain was probably wheat, but rye was introduced by Teutonic invaders who used it for making bread, and it was grown in East Anglia.

During the Middle Ages the poorer people in England ate bread made from rye, or from a rye/wheat mixture, known as 'maslin' or 'meslin', or from barley and rye. In 1764, according to Ashley (1928), bread made in the north of England contained 30% rye and that in Wales 40%. At this time, rye was still an important crop in the north of the country and, in addition, was regularly imported from Germany and Poland, where it was plentifully grown.

1.2.8.2 Cultivation

Rye is extremely hardy. It is cultivated right up to the Arctic Circle and 4000 m above sea level. It can grow in sandy soils of low fertility. Because of its lower value, less fertilizer may be applied to rye crops than other temperate cereals. Rye is more resistant to disease than wheat, although one fungal disease, ergot, is particularly associated with rye. Compared with wheat, rye plants are generally taller (>1 m) and in consequence are more prone to lodging during adverse weather conditions, as the plants mature.

1.2.8.3 Diseases and pests

Although ergot (*Claviceps purpurea*) arises in other cereals and grasses, it is associated mainly with rye. Brown rust (*P. recondita*) can damage stem and leaves, and snow mould (*Fusarium nivale*) can cause losses after germination, particularly in long spells of snow cover. Rye is included among many cereals affected by sharp eyespot (*Ceratobasidium cereale*) and like many cereals rye can be damaged by aphids. However, rye is relatively free from disease.

1.2.8.4 Uses

Food and feed uses of rye are almost equally balanced. Food uses are virtually all as bread, most popular in Northern and Eastern Europe. It is also much used in production of crispbreads. Rye is also malted to provide the basis of rye whiskeys (mainly in North America) and rye beers.

1.2.8.5 Global production

Europe produces nearly 90% of all rye, and in recent years some increase in production has occurred in that region (Figs 1.51 and 1.52) and yield increase has contributed to this (Fig. 1.53). However, of all the world's cereals rye has shown the greatest decline in production over a 50-year period (Fig. 1.23). Area harvested has been reduced by 80% (Fig. 1.24) and in spite of yield more than doubling over the period (Fig. 1.25) the effect on production has been to halve it Fig. 1.51.

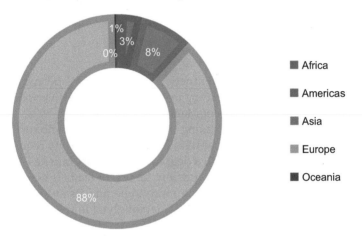

FIGURE 1.51 Proportional contributions to world rye production by continental regions (mean of 2010–14 values). *Based on data from FAOSTAT.*

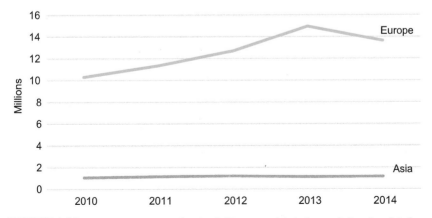

FIGURE 1.52 Variation in rye production in Europe and Asia (tonnes). *Based on data from FAOSTAT.*

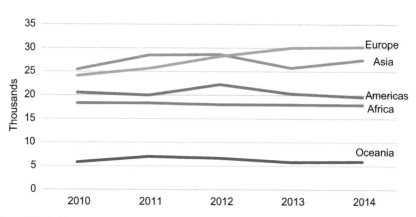

FIGURE 1.53   Variation in yield of rye in continental regions (hg/ha). *Based on data from FAOSTAT.*

1.2.9 Triticale

1.2.9.1 Origin and types

Triticale (*x Triticosecale*) is a synthetic hybrid of wheat and rye. It had its beginnings in the late 19th century in Europe, but intensive work began only in 1954, at the University of Manitoba, Canada.

The breeders' objective in making the crosses was to combine the grain quality of wheat with the disease resistance and low-fertility requirements of rye. This has been achieved to a degree but triticale grain has not found a place in mainstream bread production. Its use in feeds is generally as a replacement for maize. Both its common name and its systematic name reflect the two genera (*Triticum* and *Secale*) involved in the cross. Two distinct lines of hybridization have been adopted: one with *T. aestivum* ssp. *aestivum* as the female parent, producing octoploid lines (2n = 56); and one with *T. turgidum* ssp. *durum* as the female parent producing hexaploid lines (2n = 42) with the genetic constitution AABBRR (see wheat Table 1.5). For more information on ploidy, see Chapter 3. The hexaploid lines have been the more successful. While their agronomic performance has been good, triticale varieties have not yet fulfilled breeders' aspirations for a food grain.

1.2.9.2 Cultivation

Triticale can be effectively cultivated on weaker soils that support rye as well as on the most fertile soils suited to its other parent. It requires less fertilizer than wheat and can outyield rye and barley grown under similar conditions.

Seedbed requirements are for a depth of 25–300 cm, similar to that for wheat, as are other agricultural operations during the growth and harvesting of the crop, although disease prevention measures may be less stringent (see following).

1.2.9.3 Diseases and pests

Triticale usually has a very low incidence of disease problems compared to other cereals. It has shown good resistance to rusts and to powdery mildew but shares many other minor diseases in common with other cereal crops. Reported levels of ergot have been low, and no major pests have been recorded.

1.2.9.4 Uses

Triticale is used mainly as a feed ingredient but has also been used in speciality breads, breakfast cereals and snacks. Good leavened bread cannot be made from triticale flour alone as the gluten produced on hydration does not hold gas well.

1.2.9.5 Global production

Because triticale was introduced only in the 1960s, it is not surprising that quantities produced in the early years of its availability were very small. However, increase in area harvested since the 1980s has been very significant, rising from less than half a million hectares to more than 3.5 million ha in 2013. With the accompanying rise in yield over the same period, production has risen proportionately faster than any other cereal.

In common with rye, while production is not restricted to Europe, that continental region has been responsible for around 90% of production (Fig. 1.54). The highest yields have consistently been recorded in Europe (Fig. 1.55).

Comparing the trends of triticale and rye suggests that the former may be increasingly favoured by growers of the latter (Fig. 1.56).

FIGURE 1.54 Proportional contributions to world triticale production by continental regions (mean of 2010–14 values). *Based on data from FAOSTAT.*

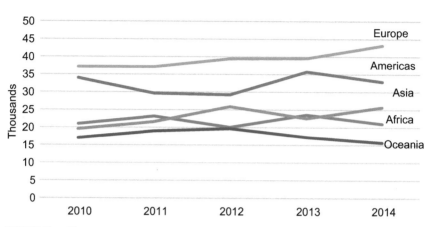

FIGURE 1.55   Variation in yield of triticale in continental regions (hg/ha). *Based on data from FAOSTAT.*

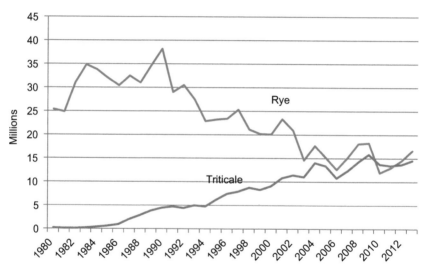

FIGURE 1.56   Comparison of rye and triticale production trends (tonnes). *Based on data from FAOSTAT.*

1.3 PSEUDOCEREALS

The term pseudocereals describes an ill-defined group of species that are not closely related to each other and even less closely related to cereals. But, like cereals, they have starchy, dry seeds. As explained earlier (p. 5), cereals are all monocotyledonous species, classified by taxonomists, on the basis of the characteristics that they share, in the grass family, Poaceae. Plants usually included in the nonsystematic grouping of pseudocereals are dicotyledonous plants belonging to different families; thus buckwheat (*Fagopyron esculentum*) is placed in the Polygonaceae family, quinoa (*Chenopodium quinoa*) is placed in the Chenopodiaceae and grain amaranth is placed in the Amaranthaceae. The justification for grouping the three species as pseudocereals lies in the fact that they produce fruits that share some characteristics with cereal grains and those characteristics render them suitable as substitutes for cereals in some contexts. The main shared characteristics relate to the composition of their fruits as they contain, as storage materials, starch and protein, in approximately the same relative proportions as those found in cereals. Like cereals, pseudocereals can be cultivated and their fruits harvested, dried and ground into meal or flour to form a major food ingredient.

Although current production of pseudocereals is small compared with true cereals, they are considered to be an underutilized resource worthy of promotion as a contributor to the drive to feed the world's expanding population. In comparison with true cereals, pseudocereals have the advantage of not succumbing to the diseases of cereals and being less demanding of growing conditions. Although international interest in pseudocereals is a relatively recent phenomenon, some of the species were cultivated for food thousands of years ago.

1.3.1 Buckwheat

1.3.1.1 Origin and types

Common buckwheat (*Fagopyron esculentum*) has been cultivated in China for about 3000 years. It is a temperate plant but will tolerate tropical conditions at high altitude. Although this is not the only cultivated *Fagopyron* species, it is the most important.

The plant grows to 60–80 cm tall and has a growing season of 70–90 days. Harvesting is difficult because ripening is not uniform and crops are cut arbitrarily when 75% of the grains have matured. Losses are consequentially high. Fruits are three-sided achenes, each weighing 25–30 g, with a hull that must be removed. Beneath the hull lies the seed coats and, within that, the endosperm and embryo.

1.3.1.2 Uses

Complete shelled grain is used as a nutritive supplement for stews, while flour is used for noodle making in Japan. Elsewhere ground products are used in preparation of porridge or may be mixed with wheat or rye for making bread with higher digestion value (Popovic et al., 2014).

1.3.1.3 Global production

More than half the world production of buckwheat has been in Europe with around 35% being grown in Asia (Fig. 1.57). Yields of the crop grown in the Americas have been consistently greater than in other continental regions (Fig. 1.58).

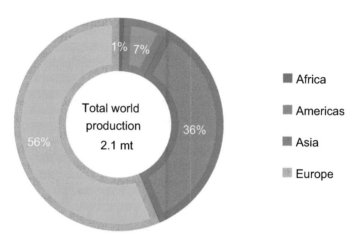

FIGURE 1.57   Proportional contribution by continental regions to production of buckwheat (mean of 2010–14). *Based on data from FAOSTAT.*

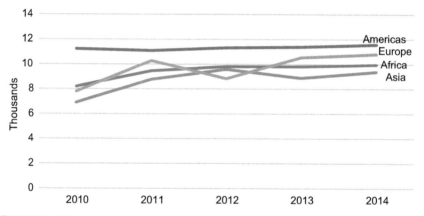

FIGURE 1.58   Variation in buckwheat yields in continental regions (hg/ha). *Based on data from FAOSTAT.*

1.3.2 Quinoa

1.3.2.1 Origin and types

Chenopodium quinoa is thought to have been domesticated in the Altiplano region of the Andes at least 5000 years ago. It was grown by the Incas and Indian cultures in South America.

The crops that are currently grown thrive in temperatures ranging from −8 to 38°C at sea level or 4000 m above, and they are not impacted by low moisture. They are tolerant of poor soils. Plants grow to a height of 0.4–3 m in a 150–220-day growth cycle (Black et al., 2006).

Fruits are disc-like and 1–3 mm in diameter. Quinoa pericarp is rich in saponins, which are considered to be antinutritional and hence need to be removed or reduced before human consumption of the products refined from the fruits. This is achieved by washing or toasting the fruits and then removing the outer layers by grinding.

1.3.2.2 Uses

In addition to traditional uses as flour for making porridge, bread and similar foods, quinoa has been promoted as a 'health food'. The year 2013 was designated the 'International Year of Quinoa' by the United Nations to raise awareness, outside the areas where it has been traditionally consumed, of its nutritional, economic, environmental and cultural values.

Saponins are not waste products as they have uses in pharmaceutical products and detergents.

1.3.2.3 Global production

Quinoa can be found natively in all countries of the Andean region, from Colombia to Argentina, to the south of Chile with almost all production in the hands of small farmers and associations. Most of the world's production comes from Bolivia, Peru and Ecuador. In 2009, production in the Andean region amounted to approximately 70,000 tonnes.

Quinoa cultivation is spreading and now occurs in more than 70 countries, including France, England, Sweden, Denmark, Holland and Italy. It is also being developed successfully in Kenya, India and the United States (FAOSTAT).

Although production quantities remain small in comparison with most cereals, there has been a significant increase since the 1990s (Fig. 1.59). Both increased area harvested (Fig. 1.60) and improved yields (Fig. 1.61) have contributed.

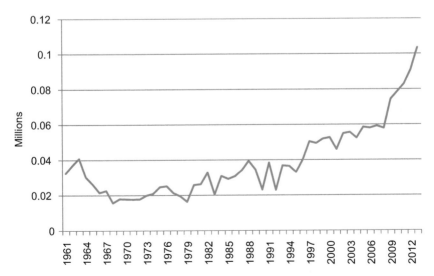

FIGURE 1.59 Variation in world production of quinoa (essentially Bolivia and Peru) (tonnes). *Based on data from FAOSTAT.*

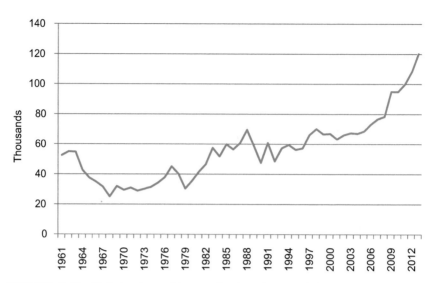

FIGURE 1.60 Variation in world area of quinoa harvested (ha). *Based on data from FAOSTAT.*

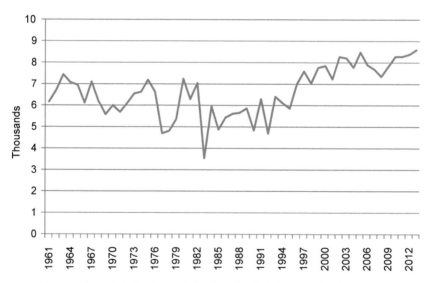

FIGURE 1.61 Variation in yield of quinoa (hg/ha). *Based on data from FAOSTAT.*

1.3.3 Grain amaranth

1.3.3.1 Origin and types

Grain amaranth has been grown in Latin America for thousands of years and is thought to have been domesticated at least as early as 4000 BC. In pre-Columbian times in the New World it was almost as important as maize and beans, and it is thought to have provided 80% of the caloric consumption of the Aztecs. Around 20,000 tonnes were sent annually as a tribute to the Aztec ruler (Raven et al., 1992).

Seventy species of grain amaranth are recognized and some are difficult to distinguish. Many grow as weeds. Two species, *Amaranthus hypochondriacus*, and *Amaranthus cruentus*, are native to Mexico and Guatemala, and *Amaranthus caudatus* is native to the Andes of Peru. The photosynthetic pathway of amaranth is C4 and the chromosome complement is 2n == 32 (*A. caudatus* and *A. hypochondriacus*) or 2n = 34 (*A. cruentus*) (Grubben and Partohardjono, 1996).

A. hypochondriacus plants grow up to 3 m high and those of the other two species reach heights of 2 m. Seed heads can weigh up to a kilogram and may contain half a million fruits. Fruits are only 0.9–1.7 mm in diameter and can vary from cream to gold or pink to black.

The leaves of several species of amaranth are valued as vegetables in many parts of the world, including South America, Asia and Africa, and where multicropping is practiced young leaves are harvested several times before plants flower and fruit. The fruits are then harvested as a grain crop.

1.3.3.2 Uses

Interest in the possible development of grain amaranth as an important food crop was revived in the 1970s. Most grain amaranth is consumed in Latin America today as a popped cereal mixed with molasses, but it can be milled into a flour for inclusion in bread and other baked products. It reportedly contains twice the level of calcium in milk, five times the level of iron in wheat, higher potassium, phosphorus and vitamins A, E, C and folic acid than cereal grains.

It has no functional gluten protein and consequently amaranth flour cannot be used on its own to produce leavened bread. However, it is said to impart a nutty flavour when included in wheat bread, the nutritional value of which is thereby enhanced. Consumption is not restricted to Latin America; it is also consumed in other parts of the world including China and Kenya.

1.3.3.3 Global production

Countries that grow grain amaranth in significant quantities include Mexico, Russia, China, India, Nepal, Argentina, Peru and Kenya. It grows very rapidly, especially under conditions of high temperatures, bright light and dry soil thereby tolerating dry conditions. Quantitative information on world production grain amaranth is sparse and FAOSTAT publishes no records of production.

References

Akesson, N.B., Yates, W.E., 1976. The Use of Aircraft in Agriculture. www.agaviation.org/rice.

Anon., September 2015. Northumberland grower breaks world wheat yield record. Farmers Weekly 21.

Ashley, W., 1928. The Bread of Our Forefathers: An Enquiry in Economic History. Clarendon Press, Oxford, UK.

Bechtel, D.B., Abecassis, J., Shewry, P.R., Evers, A.D., 2009. Development, structure and mechanical properties of the wheat grain. In: Khan, K., Shewry, P.R. (Eds.), Wheat Chemistry and Technology, fourth ed. American Association of Cereal Chemists International Inc., St. Paul, MN, USA, pp. 51–96.

Black, M., Bewley, J.D., Harmer, P. (Eds.), 2006. The Encyclopedia of Seeds, Science, Technology and Uses. CABI, Cambridge, MA, USA.

Bressani, R., 1991. Protein quality of high-lysine maize for humans. Cereal Foods World 36, 806–811.

Brink, M., Belay (Eds.), 2006. Plant Resources of Tropical Africa: 1. Cereals and Pulses. PROTA Foundation. Blackhuys Publishers, Wageningen, Netherlands.

Chamberlain, N., Collins, T.H., Elton, G.A.H., 1962. The Chorleywood bread process. Baker's Digest 36 (5), 52.

CIMMYT, 1987. Institutional Multimedia Publications Repository. http://repository.cimmyt.org/ (CIMMYT Wheat Common cereal viruses).

Dubcovsky, J., Loukoianov, A., Fu, D., Valarik, M., Sanchez, A., Yan, L., 2006. Effect of photoperiod on the regulation of wheat vernalization genes VRN1 and VRN2. Plant Molecular Biology 60 (4), 469–480.

FAO grassland species profile used for table of millets. http://www.fao.org/ag/agp/agpc/doc/gbase/default.htm.

FAOSTAT, 2016. http://faostat.fao.org/site/339/default.aspx.

Grubben, G.J.H., Partohardjono, S. (Eds.), 1996. Plant Resources of South-East Asia No. 10 Cereals. Backhuys Publishers, Leiden, Netherlands.

Hunter, H., 1928. The Barley Crop. Crosby Lockwood, London, UK.

Iltis, H.H., Doebley, J.F., 1980. Taxonomy of *Zea* (Gramineae). II. Subspecific categories in the *Zea mays* complex and a generic synopsis. American Journal of Botany 67, 994–1004.

Jat, M.L., Gathala, M.K., Ladha, J.K., Saharawat, Y.S., Jat, A.S., Kumar, V., Sharma, S.K., Kumar, V., Gupta, R., 2009. Evaluation of precision land leveling and double zero-till systems in the rice-wheat rotation: water use, productivity, profitability and soil physical properties. Soil and Tillage Research 105, 112–121.

Jones, M.P., Dingkuhn, M., Aluko, G.K., Semon, M., 1997. Interspecific *Oryza sativa* L. x *O. glaberrima* Steud. Progenies in upland rice improvement. Euphytica 94, 237–246.

Kaddus, M.A., Haque, A., Douglass, M.P., Clarke, B., 2002. Parboiling of rice Part 1: effect of hot soaking time on quality of milled rice. International Journal of Food Science and Technology 37, 527–537.

Kent-Jones, D.W., Amos, A.J., 1947. Modern Cereal Chemistry, fourth ed. Northern Publishing Co. Ltd., Liverpool, UK.

Khan, Z.R., Charles, A.O., Midega, A.O., Amudavi, D.M., Hassanali, A., Pickett, J.A., 2008. On-farm evaluation of the 'push–pull' technology for the control of stemborers and striga weed on maize in western Kenya. Field Crop Research 106, 224–233.

Kifle, Z., Belay, G., Bekele, A., 2013. Population size, group composition and behavioural ecology of geladas (*Theropithecus gelada*) and human-gelada conflict in Wonchit Valley, Ethiopia. Pakistan Journal of Biological Sciences 16, 1248–1259.

Marshall, E.J., Moonen, A.C., 2002. Field margins in northern Europe: their functions and interactions with agriculture. Agriculture, Ecosystems and Environment 89, 5–21.

Moura Duarte, J., Patto Pacheco, C.A., Teixeira Guimarães, C., Evaristo de Oliveira Guimarães, P., Paiva, E., 2004. Evaluation of high quality protein maize (QPM) hybrids obtained by conversion of normal inbred lines. Crop Breeding and Applied Biotechnology 4, 163–170.

OECD, 2015. OECD Agricultural Codes and Schemes. OECD, Paris, France.

Owens, B., McCann, M.E.E., Park, R., McCracken, K.J., 2007. Defining feed wheat quality for broilers. HCGA Research Report 423 118 pp.

Percival, J., 1921. The Wheat Plant. Duckworth, London (Reprinted 1975).

Popovic, V., Skora, V., Berenji, J., Filipovic, V., Dolijanovic, Z., 2014. Analysis of buckwheat production in the world and Serbia. EkonomikaPoljoprivrede 61 (1), 53–62.

Raven, P.H., Evert, R.F., Eichhorn, S.E., 1992. Biology of Plants, fifth ed. Worth Publishers Inc., New York.

Schruben, L.W., 1979. Principles of wheat protein pricing. In: Wheat Protein Conference, Agriculture Research Manual, ARM9 Science and Education Administration. USDA, Washington, DC, USA.

Tottman, D.R., Makepeace, R.J., Broad, H., 1979. An explanation of the decimal code for the growth stages of cereals, with illustrations. Annals of Applied Biology, 93 (2), 221–234.

Unkovich, M.J., Baldock, J.A., Forbes, M., 2010. Variability in harvest index of grain crops and potential significance for carbon accounting. Advances in Agronomy 105 (1), 173–219.

Further Reading

Ainebyona, R., Mugisha, J., Kwikiriza, N., Nakimbugwe, D., Masinde, D., Ombui Nyankanga, R., 2012. Economic evaluation of grain amaranth production in Kamuli district, Uganda. Journal of Agricultural Science and Technology 2, 178–190.

Bonnett, O.T., July 16, 1954. The inflorescences of maize. Science 120 (3107), 77–87.

Campbell, C.G., 1997. Buckwheat, Fagopyrum Esculentum Moench. Promoting the Conservation and Use of Underutilized and Neglected Crops 19. Institute of Plant Genetics and Crop Plant Research and Gatersleben/International Plant Resources Institute, Rome, Italy.

Coffman, F.A., 1977. Oat History, Identification and Classification. Technical Bulletin No. 1516. United States Department of Agriculture.

De Groote, H., Kimeju, S.C., August 2008. Comparing consumer preferences for color and nutritional quality in maize: application of a semi-double-bound logistic model on urban consumers in Kenya. Food Policy 33 (4), 362–370.

Doggett, H., 1988. Sorghum, second ed. Wiley, New York, USA.

Engels, J.M.M., Hawkes, J.G., Worede, M., 1991. Plant Genetics Resources of Ethiopia. Cambridge University Press, Cambridge NY, USA.

FAO, 1986. Small farm equipment for developing countries. In: Proceedings International Conference on Small Farm Equipment for Developing Countries: Past Experience and Future Priorities. US Agency for Industrial Development/International Rice Research Institute.

Fuller, D.Q., Denham, T., Arroyo-Kalin, M., Lucas, L., Stevens, C.J., Qin, L., Allaby, R.G., Purugganan, M.D., 2014. Convergent evolution and parallelism in plant domestication revealed by an expanding archaeological record. Proceedings of the National Academy of Sciences of the United States of America 111, 6147–6152.

Gooding, M., 2009. The wheat crop. In: Khan, K., Shewry, P.R. (Eds.), Wheat Chemistry and Technology, fourth ed. American Association of Cereal Chemists International Inc., St. Paul, MN, USA, pp. 19–50.

Kauffman, C.S., Weber, L.E., 1990. Grain amaranth. In: Janick, J., Simon, J.E. (Eds.), Advances in New Crops. Timber Press, Portland, OR, USA.

Lu, H., Zhang, J., Liu, H., Zhang, K., Wu, N., Li, Y., Zhou, K., Yie, M., Zhang, T., Zhang, H., Yang, X., Shan, L., Xu, D., Li, Q., 2009. Earliest domestication of common millet, (*Panicum milliaceum*) in east Africa extended to 10,000 years ago. Proceedings of the National Academy of Science of the United States of America 106 (18), 7367–7372.

Metheny, K.B., Beaudry, M.C. (Eds.), 2015. Archaeology of Food: An Encyclopedia. AltaMira Press, Lanham, MD, USA.

Murphy, K., Matariguihan, J., 2015. Quinoa: Sustainable Production, Variety Improvement and Nutritive Value in Agroecological Systems. John Wiley & Sons Inc., Hoboken, New Jersey, USA.

Nesbitt, M., 2005. Grains. In: Prance, G., Nesbitt, M. (Eds.), The Cultural History of Plants. Routledge, New York, USA, pp. 45–60.

Nyvall, R.F., 1989. Field Crop Diseases Handbook, second ed. Springer Science and Business Media, New York, USA.

Prasanna, B.M., Vasal, S.K., Kassahun, B., Singh, N.N., 2001. Quality protein maize. Current Science 81 (10), 1308–1317.

Rooney, W.L., Smith, W., 2000. Technology for developing new cultivars. In: Sorghum: Origin, History, Technology and Production. John Wiley & Sons Inc., New York.

Sweeny, M., McCouch, S., 2007. The complex history of the domestication of rice. Annals of Botany 100, 951–957.

U.S. Department of Agriculture – Economic Research Service, 2015. Adoption of Genetically Engineered Crops in the US.

Vaughan, D.A., Morishima, H., Kadowaki, K., 2003. Diversity in the *Oryza* genus. Current Opinion in Plant Molecular Biology 6, 139–146.

Williams, P., 2011. A Practical Introduction to Cereal Chemistry. PDK Projects Inc., Namaimo, BC, Canada.

Zohary, D., Hopf, M., Weiss, E., 2012. Domestication of Plants in the Old World: The Origin and Spread of Domesticated Plants in Southwest Asia, Europe, and the Mediterranean Basin. Oxford University Press, Oxford.

Production, uses and trade of cereals by continental regions

In Chapter 1 worldwide cereals production and the relative contributions by continental regions to the total are considered. In this chapter yield, production and trade are discussed for each cereal as well as the broad uses to which the products are put within the regions. The grain types are listed in order of production quantity and results are given for individual countries, rather than trading blocs such as the European Union, where possible.

Trade

In trading cereals, as with all commodities, trust between vendor and purchaser is paramount, and to ensure this in both domestic and export markets grading schemes have been devised. An internationally agreed series of standards has been produced by the Food and Agriculture Organization (FAO) of the United Nations. These standards are published as the Codex Alimentarius (www.fao.org/fao-who-codexalimentarius/standards/list-of-standards).

The International Organization for Standardization (ISO) also produces specifications.

Some countries set their own standards and define grades and classes so that customers can decide the qualities they require and pay the appropriate price. The limits of moisture content are defined. Standards for quality are also defined by official bodies within some importing countries or trading blocs.

Coarse grains

In the context of trading the term 'coarse grains' is often used. It is described by the Organisation for Economic Co-operation and Development (OECD) as referring to grains other than rice and wheat. In OECD countries these are used mainly for animal feed and brewing.

2.1 MAIZE (CORN)

The standard applying to whole maize is Codex Stan 153-1985. The standard defines its scope and provides a 'Description of the product'. It defines the 'Essential composition and quality factors' and sets limits for different types of contaminant. It also sets standards for 'Hygiene, packaging, labelling' and defines the methods of sampling and analysis.

Codex Stan 154 refers to maize meal and 155 to degermed maize and maize grits.

The ISO standard for maize, ISO/CD 19942, is currently under development.

2.1.1 Americas

The United States is the world's largest corn producer and currently exports between 10% and 20% of its annual production (USDA www.ers. usda.gov/topics/crops/corn/background.aspx).

The production quantities over 5 years for the top producers in the Americas are shown in Fig. 2.1.

Clearly the United States of America is by far the greatest producer, regularly producing six times the quantity produced by the second-highest producer, Brazil.

Most maize production in the United States occurs in the loosely defined 'Corn Belt' located in the MidWestern region. In 2008 the four main producing states, Iowa, Minnesota, Nebraska and Illinois, contributed more than half of total production. Other producing states include Indiana, Michigan, Ohio, Kansas and Missouri.

Brazil, the second largest producer in the Americas, has two seasons for corn production. First-season corn is planted in September and harvested in March. The safrinha crop is planted in January and February and harvested between June and August; it often accounts for over 65% of national production. Mato Grosso is the largest producer, accounting for 34% of the safrinha crop. Other key areas are the states of Paraná, Mato Grosso do Sul and Goiás (USDA, 2016).

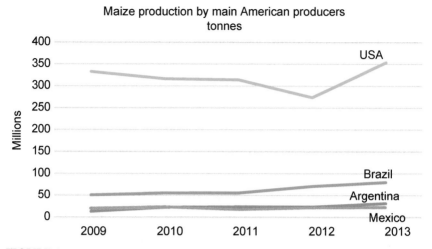

FIGURE 2.1 Variation in maize production by top four producers in Americas (tonnes). *Based on data from FAOSTAT.*

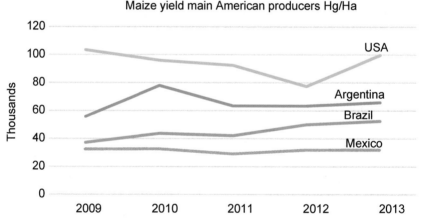

FIGURE 2.2 Variation in maize yields in main producing American countries. *Based on data from FAOSTAT.*

Argentina's maize belt is located in the vast lowland Pampas area (Krishna, 2013).

Yield

There have been considerable differences in yields among all the main producers in the region (Fig. 2.2), with the United States being consistently the highest at around 10 tonnes/ha.

Uses

The principal use of maize is as feed, but substantial consumption is recorded as human food and by industry. Proportionate uses are shown for major consuming countries in Figs. 2.3–2.6.

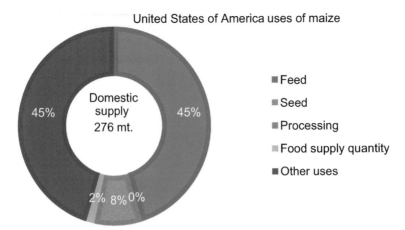

FIGURE 2.3 Reported proportional domestic uses of maize in the United States of America (mean of 2008–11). *Based on data from FAOSTAT.*

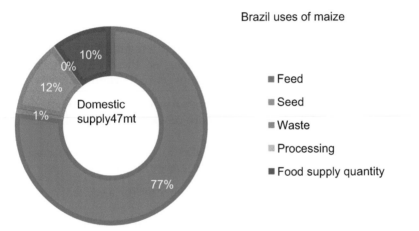

FIGURE 2.4 Reported proportional domestic uses of maize in Brazil (mean of 2008–11). *Based on data from FAOSTAT.*

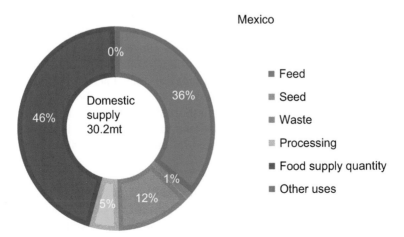

FIGURE 2.5 Reported proportional domestic uses of maize in Mexico (mean of 2008–11). *Based on data from FAOSTAT.*

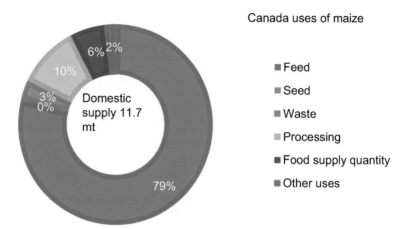

FIGURE 2.6   Reported proportional domestic uses of maize in Canada (mean of 2008–11). *Based on data from FAOSTAT.*

Maize is processed for human consumption and other industrial uses by either wet or dry milling, depending on the desired end products (http://www.ers.usda.gov/topics/crops/corn/background.aspx).

- Wet millers process corn into high-fructose corn syrup (HFCS), glucose and dextrose, starch, corn oil, beverage alcohol, industrial alcohol and fuel ethanol.
- Dry millers process corn into flakes for cereal, corn flour, corn grits, corn meal and brewers' grits for beer production.

Uses in Brazil and Canada are prominently for feed, while in Mexico feed usage is second to food supply.

Between 2008 and 2011 an average of 276 mt of maize was used each year in the United States. Traditionally it has been regarded as a feed crop, but increasingly it is used as a feedstock for ethanol plants producing bio-fuels. This use constitutes the major proportion of 'other uses' (Fig. 2.3).

As ethanol production increases, the supply of ethanol coproducts also increases. Both the dry-milling and wet-milling methods of producing ethanol generate a variety of economically valuable coproducts, the most prominent of which is 'distillers' dried grains with solubles' (DDGS), which can be used as a feed ingredient for livestock. Each 56-pound bushel of corn used in dry-mill ethanol production generates about 17.4 pounds of DDGS. In the United States, cattle (both dairy and beef) have been the primary users of DDGS as livestock feed, but increasingly larger quantities are making their way into the feed rations of pigs (hogs) and poultry.

Exports

It is not surprising that, as by far the greatest producer (Fig. 2.7), the United States of America is the largest exporter of maize. However, during the period considered a marked decline in exports occurred. In 2012 Brazil and Argentina each exported up to half the quantity recorded for the United States (Fig. 2.8).

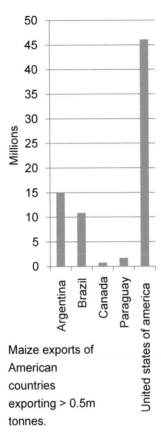

FIGURE 2.7 Maize exports by main exporting nations of the Americas (tonnes). *Based on data from FAOSTAT.*

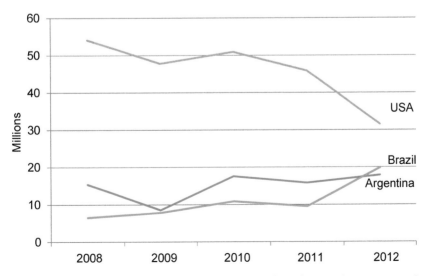

FIGURE 2.8 Variation in exports of maize and products by top three exporters in Americas (tonnes). *Based on data from FAOSTAT.*

Grading in the United States is based on criteria defined by the United States Department of Agriculture (USDA). Full details can be found at https://www.gipsa.usda.gov/fgis/standards/810corn.pdf. The system defines three classes, yellow, white and mixed, and five principal grades based on test weight, proportion of damaged kernels and broken corn and foreign materials. Within the damaged kernels proportion, a limit is defined for the proportion of heat-damaged kernels. In addition, a sample grade is recognized for samples that do not conform to the criteria of the numbered grades or suffer from other shortcomings. Special grades are recognized for flint corn, flint and dent corn and waxy corn. Further explanation of the system has been provided by Evans et al. (https://www.extension.purdue.edu/extmedia/ay/ay-225.pdf).

Argentina's grading system evaluates only three criteria: damaged and broken kernels and foreign material. Three grades are thus defined (https://www.princeton.edu/~ota/disk1/1989/8917/891703.PDF).

In Brazil maize quality is classified as type 1, 2 or 3 according to the presence of impurities and broken, hollow or mouldy grains (Tardin, 1991).

Imports

Mexico is by far the greatest importer in the region, consistently requiring about one-third of its domestic supply from other countries (Fig. 2.9). Import levels for both of the highest importers have been relatively consistent (Fig. 2.10).

Despite being major exporters of maize, both the United States of America and Brazil were also importers, although the imported quantities were much lower than the exports (Fig. 2.11). It is reported that much of the imported maize to the United States has been to fulfil the demand for nonGM (genetically modified) grain.

2.1.2 Asia

Production and yield

Mainland China is the largest producer of maize in Asia Fig. 2.12, producing around two-thirds of the United States' total and recording a progressive increase between 2009 and 2013 (Fig. 2.13), partly as a result of a 20% increase in area harvested.

Maize is grown in every province of China, but the 'Maize Belt' that traverses the country from northeast to southwest is predominant for production. The three regions of northwest China, north China and the Yellow River valley account for 70% of harvested area and nearly 75% of production (Meng et al., 2006).

Other significant producers are India and Indonesia (Fig. 2.14) although yields in these countries do not match those of China (Fig. 2.15).

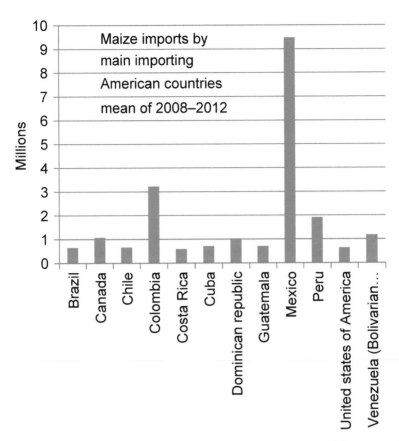

FIGURE 2.9 Maize and products imports by main importing American countries (tonnes), mean of 2008–12. *Based on data from FAOSTAT.*

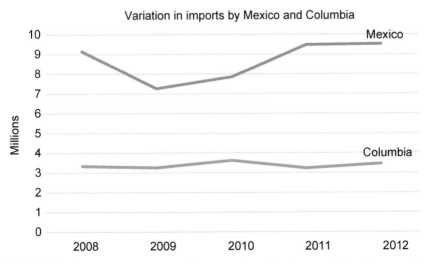

FIGURE 2.10 Variation in imports of maize and products by two highest importers in Americas (tonnes). *Based on data from FAOSTAT.*

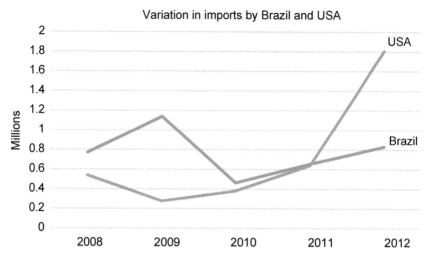

FIGURE 2.11 Variation in imports of maize and products by Brazil and the United States (tonnes). *Based on data from FAOSTAT.*

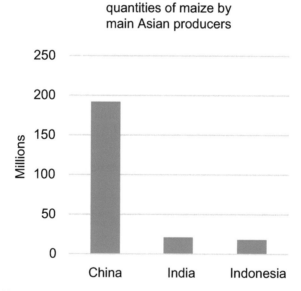

FIGURE 2.12 Relative production quantities of maize by Asian countries (tonnes). *Based on data from FAOSTAT.*

More than half the maize of India is produced in the four states of Madhya Pradesh, Andhra Pradesh, Karnataka and Rajasthan. Madhya Pradesh is the largest producer in India, contributing over 14% of total maize. Andhra Pradesh and Karnataka have emerged as important producers; in both the

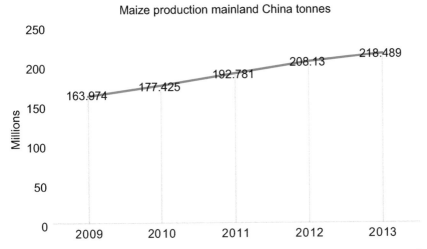

FIGURE 2.13 Relative production quantities of maize by main Asian producers (tonnes). *Based on data from FAOSTAT.*

FIGURE 2.14 Variation in maize production by India and Indonesia (tonnes). *Based on data from FAOSTAT.*

states, maize is irrigated properly. The arid lands of Rajasthan are especially suited to maize. Rajasthan has the largest area under maize cultivation and gives the lowest yields among all the major maize-producing states of India (www.yourarticlelibrary.com/cultivation/maize-cultivation-in-india).

In Indonesian maize is grown in all provinces, but East Java, Central Java and Lampung are the leading producers, while South Sulawesi, North Sumatra, West Java and Gorontalo are the second-tier producers (pangan.litbang.pertanian.go.id/files/IMC-PDF/01).

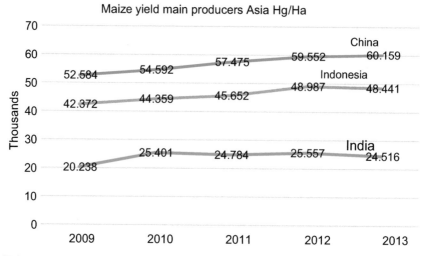

FIGURE 2.15   Variation in yields of maize in main producing Asian countries (hg/ha). *Based on data from FAOSTAT.*

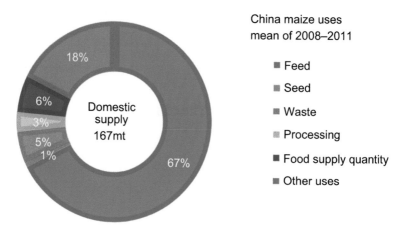

FIGURE 2.16   Reported proportional domestic uses of maize by mainland China, mean of 2008–11. *Based on data from FAOSTAT.*

During the 1990s most maize was grown in Java and contributed about 61% to national maize production. Although maize continues to be most widely grown in Java, the maize area has tended to decline slightly over time. On the other hand, the maize area outside Java grew at a rate of 1.97% per year during the period of 1970–2000 (Swastika et al., 2004).

Domestic uses

While in China feed accounts for the majority of the domestic supply, a significant amount found other uses (Fig. 2.16). Other uses (mainly

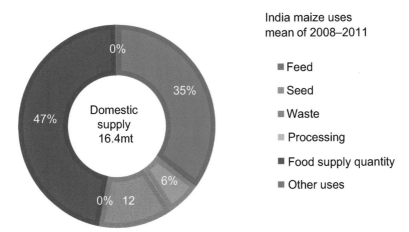

India maize uses
mean of 2008–2011

■ Feed

■ Seed

■ Waste

■ Processing

■ Food supply quantity

■ Other uses

FIGURE 2.17   Reported proportional domestic uses of maize by India, mean of 2008–11. *Based on data from FAOSTAT.*

starch production) were also significant in Indonesia (Fig. 2.18), but here the greatest consumption was as food. An even greater proportion was used as food in India (Fig. 2.17), where no 'other uses' were reported. In Japan feed and processing accounted for more than 90% of supply (Fig. 2.19).

Exports

Compared with the countries of the Americas region, export quantities by Asian countries are low. India exports most (Fig. 2.20), but quantities have been quite variable (Fig. 2.21). Of the remaining Asian countries, only Thailand exported as much as half a million tonnes annually.

In India quality assurance is provided by the legally enforceable AGMARK system whereby certification is given to agricultural products conforming to a series of food safety and quality criteria set out in the Prevention of Food Adulteration Rules, 1955. In addition, AGMARK standards differentiate quality by having four grades (I–IV) for maize. The grades are differentiated on the basis of moisture, foreign matter, other edible grains, admixture of different varieties, damaged grains, immature and shrivelled grains and weevilled grains (http://agmarknet.nic.in/Maize_manual.htm).

Thailand recognizes two grades based on proportions of other-colour kernels, broken and immature kernels, partly broken kernels, heavily damaged kernels, weevilled kernels and foreign matter (Alavi, 2012).

Imports

The main importers of maize in Asia are shown in Fig. 2.22.

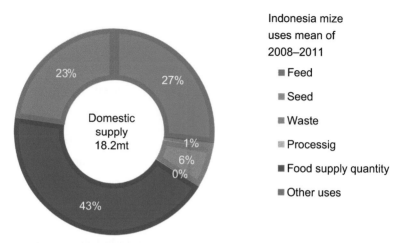

FIGURE 2.18   Reported proportional domestic uses of maize in Indonesia, mean of 2008–11. *Based on data from FAOSTAT*

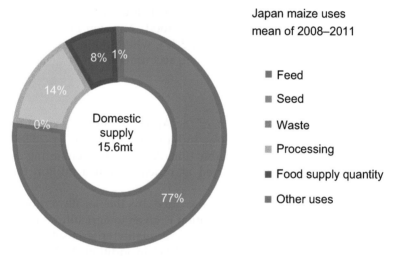

FIGURE 2.19   Reported proportional domestic uses of maize by Japan mean of 2008–11. *Based on data from FAOSTAT.*

2.1.3 Europe

Production and yield

The production values for the five main maize-producing countries in Europe between 2009 and 2013 are shown in Fig. 2.23. Ukraine stands out as trebling its production during that period, while the others showed no obvious trends. Much of this increase resulted from increasing the area harvested (Fig. 2.24).

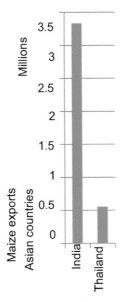

FIGURE 2.20 Maize and products exports by the top two exporting countries (tonnes), mean of 2008–12. *Based on data from FAOSTAT.*

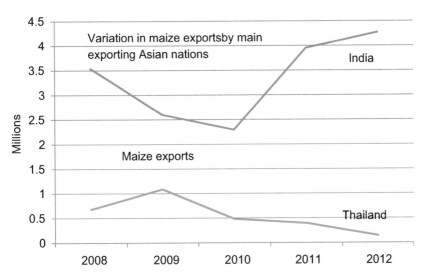

FIGURE 2.21 Variation in maize and products exports by top two Asian exporting countries (tonnes), *Based on data from FAOSTAT.*

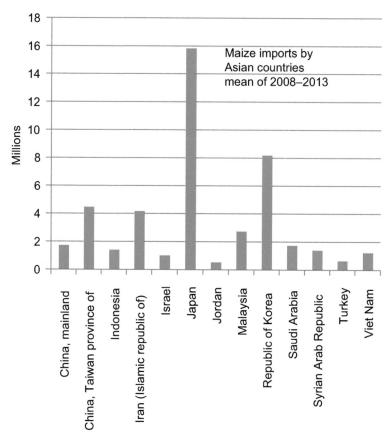

FIGURE 2.22   Maize and products imports by main maize-importing countries in Asian region (mean of 2008–13). *Based on data from FAOSTAT.*

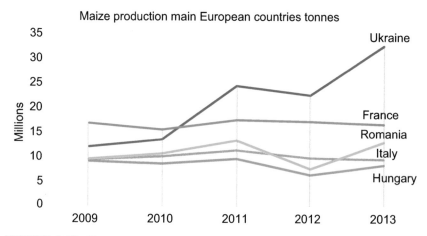

FIGURE 2.23   Variation in maize production by main producing European countries (tonnes). *Based on data from FAOSTAT.*

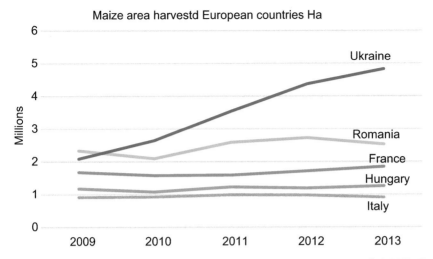

FIGURE 2.24 Variation in maize area harvested in main producing countries (ha), 2008–13. *Based on data from FAOSTAT.*

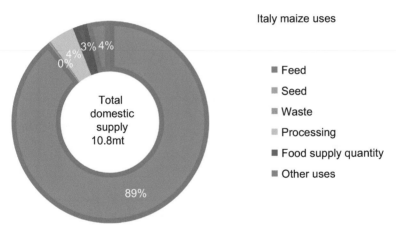

FIGURE 2.25 Reported proportional domestic uses of maize by Italy (mean of 2008–11). *Based on data from FAOSTAT.*

Ukraine is divided for administrative purposes into 24 'oblasts'. The principal regions for maize production are the Eastern and Southern oblasts (www.ukraine-arabia.ae/economy/agriculture).

Domestic uses

Feed was the predominant use in Europe. However, there were variations in the amount used for food and processing, and waste was higher in Ukraine than in France and Italy (Figs. 2.25–2.27).

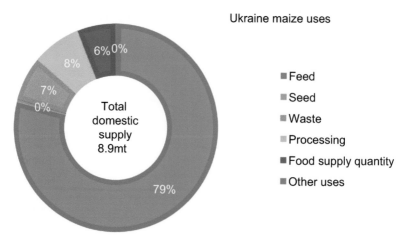

FIGURE 2.26   Reported proportional domestic uses of maize by Ukraine (mean of 2008–11). *Based on data from FAOSTAT*

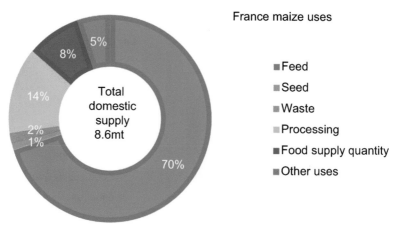

FIGURE 2.27   Reported proportional domestic uses of maize by France (mean of 2008–11). *Based on data from FAOSTAT.*

Exports

Ukraine, France and Hungary are the main exporters, with approximately half their production being exported in 2013 (Fig. 2.28). Ukraine's exports have increased in line with production and its share of the maize export market has increased in recent years, to make it the world's third-largest exporter after the United States and Brazil (Fig. 2.29). Information on Ukraine's grading system is unavailable.

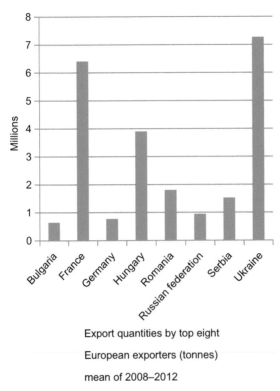

Export quantities by top eight

European exporters (tonnes)

mean of 2008–2012

FIGURE 2.28 Maize and products export quantities by top eight European exporting countries (tonnes), mean of 2008–12. *Based on data from FAOSTAT.*

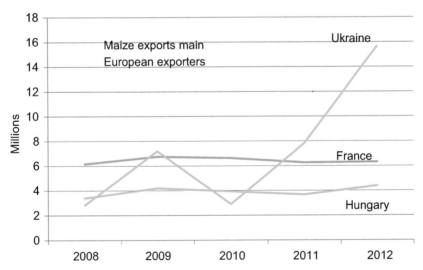

FIGURE 2.29 Variation in maize and products exports by top three exporting countries (tonnes). *Based on data from FAOSTAT.*

Imports

Spain, the highest importer, imported marginally more than it produced during the period 2008–12. The Netherlands, Italy and Germany were the next largest importers, but several other countries imported in excess of 0.5 million tonnes (Figs. 2.30 and 2.31).

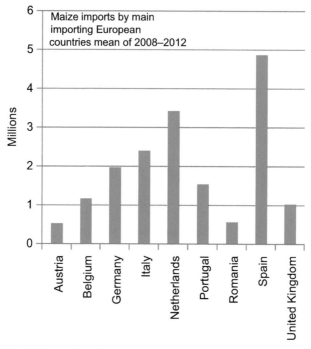

FIGURE 2.30   Maize and products import quantities by top nine European importing countries (tonnes). *Based on data from FAOSTAT.*

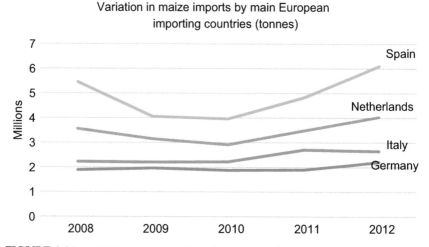

FIGURE 2.31   Variation in maize and products imports by top four European importers (tonnes). *Based on data from FAOSTAT.*

2.1.4 Africa

Production and yield

Of the African countries, South Africa produces the greatest quantity of maize grain, around half of which is white (Fig. 2.32). However, yields in that country are approximately half those achieved in Egypt (Fig. 2.33), where yields compare favourably with those found in Europe. Yields in other African countries have been even lower than South Africa, but increases have been consistent and substantial in Ethiopia.

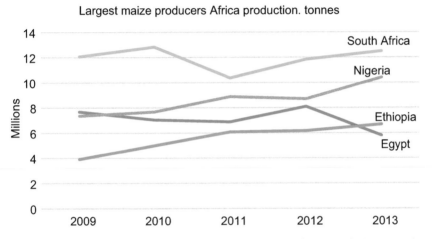

FIGURE 2.32 Variation in maize production by top four African producing countries (tonnes). Based on data from FAOSTAT.

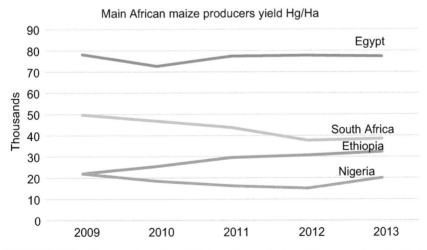

FIGURE 2.33 Variation in maize yield in main producing African countries (hg/ha). Based on data from FAOSTAT.

Maize is produced in most parts of South Africa, but mainly in North West province, the Free State, the Mpumalanga Highveld and the KwaZulu-Natal Midlands.

In Nigeria maize is grown throughout the country, but the main producing areas lie in the central states such as Niger, Kaduna, Taraba, Plateau and Adamawa.

In Ethiopia maize is grown primarily in Amhara, Oromia and the Southern Nations, Nationalities and Peoples region.

The agricultural land in Egypt is confined to the Nile valley and delta, with a few oases and some arable land in Sinai. The total cultivated area represents only 3% of the total land area. Irrigation is the norm except for some rain-fed areas on the Mediterranean coast.

White maize production is of paramount importance in Africa. In this region, which produces about one-third of the global white maize crop, it represents about 90% of the total regional maize output. Producers include Kenya, Malawi, Tanzania, Zambia, Zimbabwe (countries in which white maize represents between 66% and 90% of total cereals production), Egypt, Ethiopia and Nigeria, where white maize constitutes from 15% to 35% of total cereals production.

Domestic uses

In all major maize-using countries in Africa food constitutes one of the main uses, and in Kenya this accounts for over 90%. The other major use is feed, but there is considerable variation among countries in the proportional divide between the two major uses. In South Africa most of the feed grains are yellow and most of the food grains are of the white variety (Figs. 2.34–2.37).

Exports

South Africa is the main exporting country, but between 2008 and 2012 export quantities did not exceed 2.5 million tonnes. No other African country achieved annual maize exports of 0.5 million tonnes.

South African grading is done according to the factors defined in the South African grading regulation: defective kernels above and below a 6.35 mm sieve, total defective kernels, foreign matter, other colour, total deviation and pinked kernels (http://agbizgrain.co.za/uploads/documents/prospectus-or-maize-industry.pdf).

Imports

Egypt, Algeria and Morocco are the main importing countries (Fig. 2.38). Despite variations, particularly in Egypt's record, the order of the top three importers remained consistent between 2008 and 2012 (Fig. 2.39).

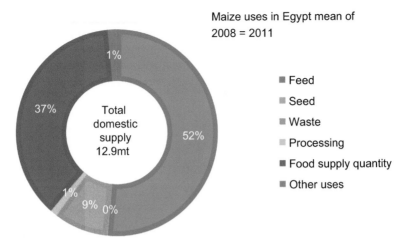

FIGURE 2.34 Domestic uses of maize by Egypt (mean of 2008–11). *Based on data from FAOSTAT.*

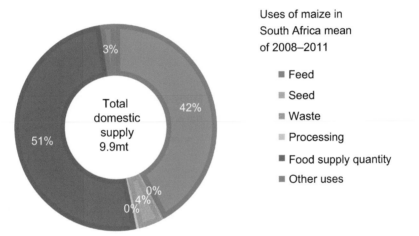

FIGURE 2.35 Domestic uses of maize by South Africa (mean of 2008–11). *Based on data from FAOSTAT.*

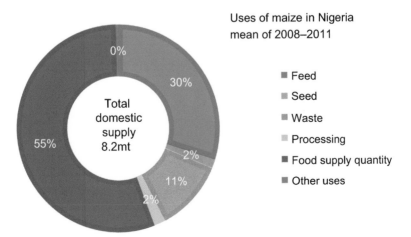

FIGURE 2.36 Domestic uses of maize by Nigeria (mean of 2008–11). *Based on data from FAOSTAT.*

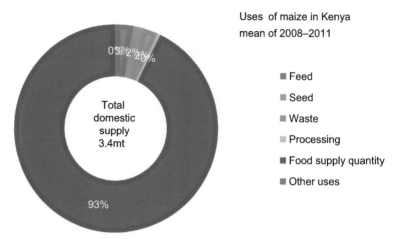

FIGURE 2.37   Domestic uses of maize by Kenya (mean of 2008–11). *Based on data from FAOSTAT.*

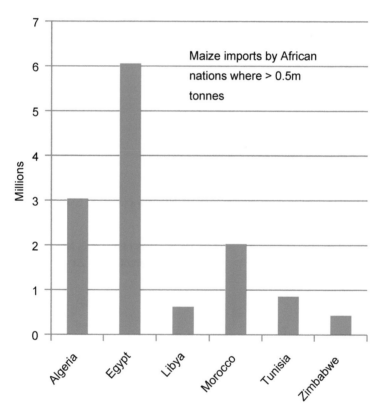

FIGURE 2.38   Maize and products imports by main importing African countries (tonnes). *Based on data from FAOSTAT.*

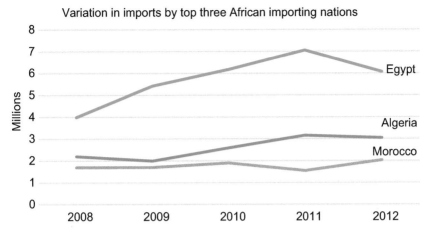

FIGURE 2.39 Variation in maize and products import quantities by top two African importing countries (tonnes). *Based on data from FAOSTAT.*

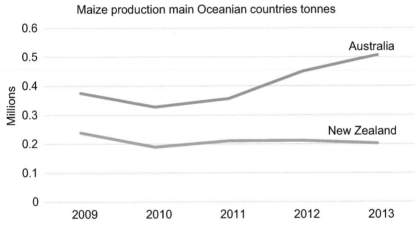

FIGURE 2.40 Variation in maize production in Oceanian countries (tonnes). *Based on data from FAOSTAT.*

2.1.5 Oceania

Production and yield

Maize is not a major cereal in Oceanian countries. Even Australia, the main producer, grows only around 0.5 million tonnes annually, and New Zealand's production is less than this (Fig. 2.40). Yields in New Zealand, however, are high by world standards (Fig. 2.41).

The main growing areas in Australia are in Queensland (Atherton tableland, Burnett and Darling downs) and New South Wales (Murrumbidgee, Murray and Lachlan river valleys).

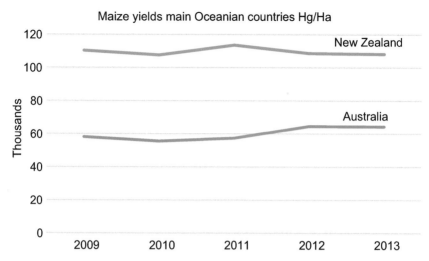

FIGURE 2.41   Variation in maize yields in Oceanian countries (hg/ha). *Based on data from FAOSTAT.*

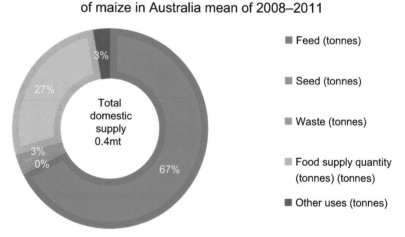

FIGURE 2.42   Reported proportional domestic uses of maize in Australia (mean of 2008–11 values). *Based on data from FAOSTAT.*

Grain Trade Australia reviews and defines standards and grades as each harvest is assessed, according to a large number of detailed criteria. Two major categories, prime and feed, are defined. Within the feed category Feed No. 1 and Feed No. 2 maize exist, but others may be added where appropriate (maize standards at www.graintrade.org.au/commodity_standards).

Domestic uses

Uses in Australia are shown in Fig. 2.42.

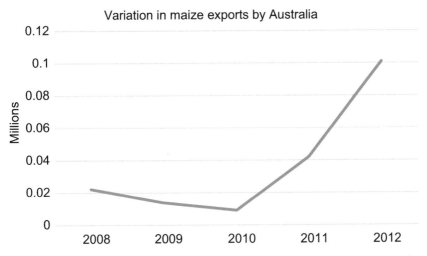

FIGURE 2.43 Variation in maize and products exports by Australia (tonnes). *Based on data from FAOSTAT.*

Exports

Australia exported variable quantities between 2008 and 2011, but never more than 0.1 million tonnes (Fig. 2.43).

Imports

In 2012 New Caledonia and French Polynesia were the only countries that imported over 3000 tonnes. Australia imported 2000 tonnes and New Zealand and Papua New Guinea each imported less than 500 tonnes.

2.2 RICE

2.2.1 Asia

Production and yield

Rice production (and consumption) is highly localized. In some Asian countries, e.g., Cambodia, up to 90% of agricultural land is dedicated to rice production. Asian production increased steadily during the years 2003–13 by more than 25% (Fig. 2.44).

In spite of producing much less rice in total than Asia, other continental regions have also enjoyed substantial increases in production (Fig. 2.45). Thus African production had increased by 50% through consistent growth between 2003 and 2013. The increase in Oceania was less consistent but nevertheless substantial, rising from less than half a million tonnes in 2003 to over a million in 2013.

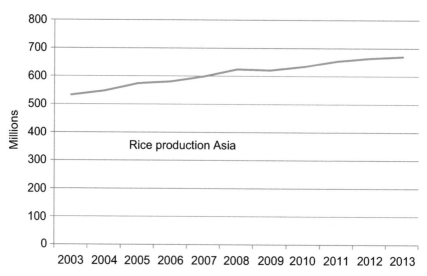

FIGURE 2.44   Variation in paddy-rice production in Asia (tonnes). *Based on data from FAOSTAT.*

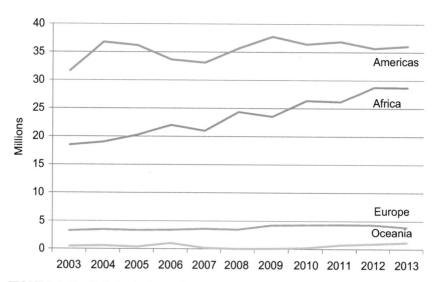

FIGURE 2.45   Variation in rice production in continental regions other than Asia (tonnes). *Based on data from FAOSTAT.*

Within Asia, mainland China is the largest producer, with India being the second largest. Together they account for more than half of the world's total. Increases in production in these two countries have also contributed significantly to the total increase in Asia (Fig. 2.46).

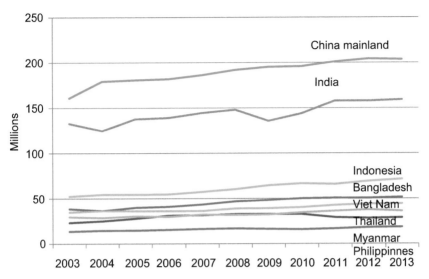

FIGURE 2.46 Variation in paddy-rice production in main producing Asian countries (tonnes). *Based on data from FAOSTAT.*

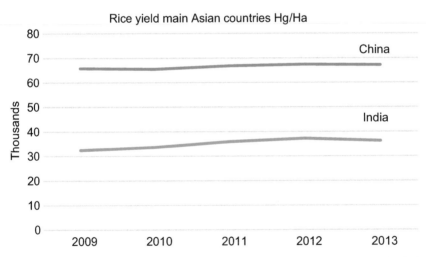

FIGURE 2.47 Variation in rice yield in the two main producing countries in Asia (hg/ha). *Based on data from FAOSTAT.*

There are considerable differences in yield within and among producing countries, as illustrated by the values for China and India (Fig. 2.47). However, in both cases any trend, even over such a short period as 5 years, has been upwards.

In China the vast majority of rice is irrigated, with only around 10% being rain-fed. Much is produced in double-cropping systems, either with two rice crops or in a rotation with wheat or another crop species. Because potassium (K) fertilizers are in short supply, continuous use of the nitrogen (N) and phosphorus (P) fertilizers needed for such cropping has led to an imbalance in some areas.

Although most production occurs in Hunan province and the southeast of the country, particularly in the Yangtze valley where water is abundant, it has been increasing in the northern provinces also. Jilin, Liaoning, Henan Shandong and even the most northerly province of Heilongjiang now produce rice. The varieties adopted in the north tend to be of the semiglutinous japonica type, which have a higher value. In the double-cropping areas of the south the first crop is sown from February to April, followed immediately by the second crop which is harvested in October to November. Where single cropping is practised, March to June are the sowing months for an October–November harvest.

In India the states that produce most rice lie in strips on the east coast and close to the northern borders. The wet monsoon provides ideal conditions of high rainfall and high temperature in the eastern coastal region, and two to three crops per year can be produced. Crops are described in terms of their harvest period, hence an autumn crop is sown in May to June and harvested in September and October, the winter crop is sown in June and harvested in November to December and the summer crop is sown in November to December and harvested in March and April. In the northwest winters are much colder and only one crop is possible. In order of production volume the states are West Bengal, Uttar Pradesh, Andhra Pradesh and Telangana, Punjab, Bihar, Odisha (Orissa) Chhattisgarh, Assam, Tamil Nadu and Haryana (www.mapsofindia.com/top-ten/india-crops/rice.html). A narrower strip lies on the west coast and multiple crops are also possible here. River deltas and extensive networks of canals provide the essential water to both east and west coastal regions.

In Indonesia rice is cultivated in both lowlands and uplands throughout the archipelago, with the upland crop typically being rain-fed and receiving only low amounts of fertilizer. Irrigated lowland rice is both well watered and heavily fertilized. Approximately 84% of the total rice area in Indonesia in 2012 was irrigated, while the remaining 16% relied on rainfall. Production is heavily concentrated on the islands of Java and Sumatra; nearly 60% of total production emanates from Java alone. Rice is grown by approximately 77% of all farmers in the country (25.9 million) under predominantly subsistence conditions. The average farm size is less than 1 ha, with the majority of the farmers cultivating landholdings of 0.1–0.5 ha. Rice is grown year round, with some farmers able to produce three crops a year, but it is common to grow two rice crops a year (http://ricepedia.org/indonesia).

In Bangladesh the major rice ecosystems are upland irrigated (mainly dry season), rain-fed lowland (mostly monsoon-season transplanted), medium-deep stagnant water (50–100 cm), deep water (>100 cm), tidal saline and tidal nonsaline. Bangladesh receives about 400 mm of rain during the premonsoon months of March to May, which enables farmers to grow a short-duration drought-resistant crop (http://ricepedia.org/bangladesh).

Most of Vietnam's rice is produced in the Mekong River delta. The other rice-growing regions are the Red River delta, northeast and the north–central coast. The Mekong delta has three major cropping seasons: spring or early season; autumn or midseason; and winter, with a long-duration wet-season crop. The largest rice area is cropped during the autumn season, followed by a spring crop; only a small area is cropped in winter.

In Thailand rain-fed lowland is the predominant growing regime, followed by irrigated, deep water and upland. Thailand is divided into four regions: central, north, northeast and south. Each region has different rice-growing environments. Almost half of the rice land of Thailand is located in the northeastern region, where the average size of rice farms is smaller than in other regions. Irrigation potential in the region is limited due to undulating topography. The northeastern region produces both long-grain and glutinous rice.

The central region is an intensively cultivated alluvial area. During the rainy season, rice covers the major part of the region. The average farm size is large, and a large proportion of the rice land has access to irrigation facilities, allowing many farmers to grow two rice crops during the year. Almost 75% of the dry-season rice grown under irrigated conditions is located in this region. Farm operations are almost entirely mechanized. This region produces mostly long-grain rice. Upland rice is grown in the lower altitudes of high hills and in upland areas of the northern region. Lowland rice is grown mainly in lower valleys and on some terraced fields where water is available. The southern region, touching the west and east coasts of the peninsula, grows only 6% of the country's rice (http://ricepedia.org/thailand).

In Myanmar late-sown rain-fed lowland is sown during the monsoon period; winter (Mayin) rice can be transplanted only after the monsoon when floodwater recedes. Rain-fed lowland (the largest of the ecosystems) and deep-water rice are confined to the delta region and coastal strip of Rakhine state. Because of rainfall and hydrologic patterns, irrigation is critical in Myanmar's central dry zone, whereas in the delta there is more concern about drainage and flood protection. The country's upland area is mostly in Mandalay, Sagaing and Shan states (http://ricepedia.org/myanmar).

In the Philippines almost 70% of the total rice area is irrigated and the remaining 30% is rain-fed and upland. Much of the country's irrigated rice is grown on the central plain of Luzon, the country's ricebowl. Rain-fed rice is found in the Cagayan valley in northern Luzon, in Iloilo province and on the coastal plains of Visayas and Ilocos in northern Luzon. Upland rice is grown in both permanent and shifting cultivation systems scattered throughout the archipelago on rolling to steep lands (http://ricepedia.org/philippines).

Domestic uses

Rice is essentially a food grain, and in Asia around 79% is consumed by humans. In China the figure is over 80% (Fig. 2.48). In Bangladesh the proportion consumed in this way is 88% and in India it is 92% (calculated from FAOSTAT values for 2008–11).

In Indonesia 'other uses' account for 15%, thus reducing food use to 72% (Fig. 2.49).

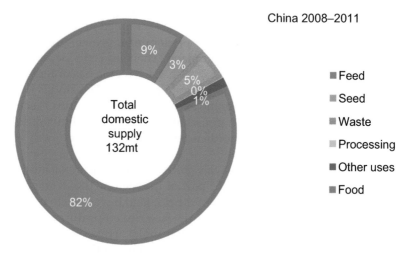

FIGURE 2.48   Reported proportional domestic uses of rice in China (mean of 2008–11). *Based on data from FAOSTAT.*

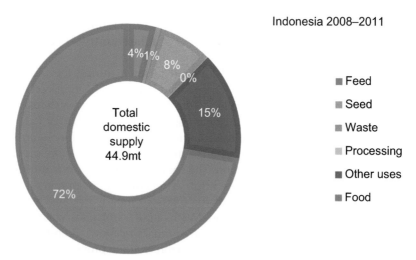

FIGURE 2.49   Reported proportional domestic uses of rice in Indonesia (mean of 2008–11). *Based on data from FAOSTAT.*

Exports

Only a small fraction of world rice production (~8%) enters international trade, as most production is used to meet local demand. Main individual exporters between 2008 and 2011 were Thailand, Vietnam, India and Pakistan (Fig. 2.50), but the relative proportions enjoyed by these countries are highly variable. The huge increase in exports by India is the most notable change (Fig 2.51).

Several standards and grades for rice in Asian countries are discussed by Afsar et al. (2001).

The Thai Agricultural Standard TAS 4000 2003 defines three classes of long grain rice, based on length and one class of short-grained rice. Four types and several grades are recognized: white (13 grades), cargo (i.e., dehusked) (6 grades), white glutinous (3 grades) and parboiled (9 grades). Quality is assessed as either extra well milled, well milled, reasonably well milled or ordinarily milled.

Thai jasmine white rice, also called fragrant rice or 'hom mali' rice, belongs to the *indica* (long-grain) category and can be divided into three

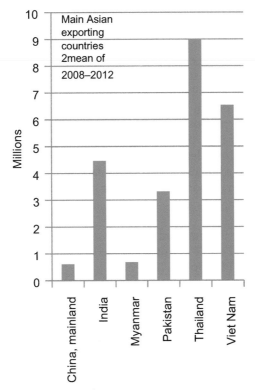

FIGURE 2.50 Rice and products exports by main exporting Asian countries. *Based on data from FAOSTAT.*

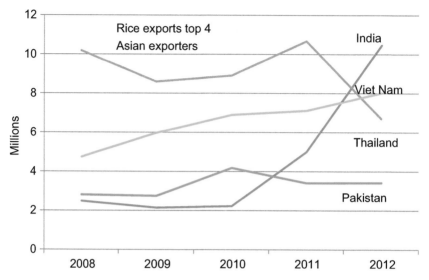

FIGURE 2.51   Variation in rice exports by main exporting Asian countries (tonnes). *Based on data from FAOSTAT.*

main grades according to quality: prime, superb and premium (www. pechsiam.com/allabout_ricetype.htm).

A Vietnamese rice standard for export was established by the Standardization Meteorology and Quality Control Centre, but little information exists as to detail.

Indian grading proposed by the government in 1967 was based on aspect ratios of grains, thus five grades were proposed: long slender, medium slender, short slender, long bold and short bold.

Basmati rice has a higher price than regular *indica* rice, and as adulteration has been detected, methods of DNA testing have been devised.

Imports

Rice imports by the main importing Asian countries are shown in Fig. 2.52. Quantities for some countries are extremely variable, as shown for Indonesia and the Philippines (Fig. 2.53).

2.2.2 Americas

Production and yield

Within the American continental region, the two main producers are Brazil and the United States of America, both consistently growing between 8 and 13 million tonnes per year. Peru and Colombia each produce 2–3 million tonnes and Argentina and Uruguay produce 1.3–1.5 million tonnes (Fig. 2.54).

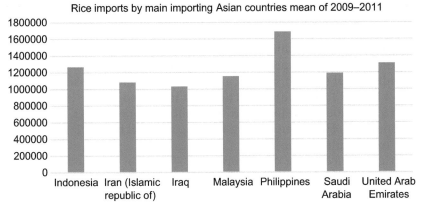

FIGURE 2.52 Rice imports by main importing Asian countries (mean of 2009–11), tonnes. *Based on data from FAOSTAT.*

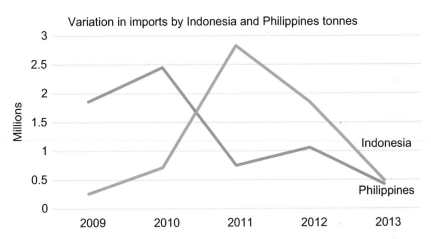

FIGURE 2.53 Variation in rice and products imports by Indonesia and the Philippines (tonnes). *Based on data from FAOSTAT.*

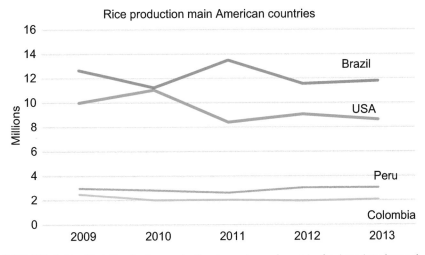

FIGURE 2.54 Variation in rice production by main producers in the Americas (tonnes). *Based on data from FAOSTAT.*

In Brazil rice is cultivated in almost all 558 microregions of the country, but 81.8% of production was concentrated in the states of Rio Grande do Sul (61.1%), Santa Catarina (9.2%), Mato Grosso (6.1%) and Maranhão (5.2%). Rice production has increased in the last 20 years, from 9.5 million tonne in 1991 to 13.5 million tonne in 2011, an increase of about 30%. However, the rice-farming area fell from 4.1 million ha in 1991 to 2.7 million ha in 2011.

Irrigated rice is the most important system of production. In 2010 irrigated crops accounted for 78.0% of production, upland rice produced 21.4% of the total and rain-fed lowland rice produced 0.6% of the total. Upland rice area decreased, associated with its low productivity, in comparison with lowland rice, attributed to dry spells during the crop season, low soil fertility and high risks due to rainfall dependence.

In the United States rice is grown in the southern states, and in 2015 around half the total crop was produced in Arkansas. California contributed around 20% and Louisiana 15%. Missouri, Mississippi and Texas contribute between 4.5% and 6.5% each. Most of the production is of long grain *indica* rice, but California produces medium- and short-grain *japonica* varieties. Typically, in Arkansas seed is planted in March/April and harvesting occurs in August/September. Growing methods throughout the country are highly mechanized, with seeding by seed drill and harvesting by combine harvester. Laser levelling and aerial seeding and fertilizer applications are also widely practised.

In Peru irrigated rice accounts for two-thirds of the rice area and most of the supply, mainly in the forest rim in the regions of Cajamarca and Amazonas, in the higher forest of San Martín and in the coastal regions of Tumbes, Piura, Lambayeque, La Libertad and Arequipa. Upland rice is grown in the high and low forest, and lowland rice in the low forest in Ucayali and Loreto.

Irrigated rice is sown by hand using seedlings or pregerminated seeds; in the lowland system, transplanting is used, while in uplands rice is sown by broadcasting seedlings or laying them in rows on dry soil. Broadcasting seedlings and mechanized transplanting are preferred (http://ricepedia. org/peru).

In Colombia rice is cultivated in five zones: central, eastern, low Cauca, north coast and the Santanderes. Tolima, Huila, Meta and Casanare departments are responsible for 72% of national rice production, and the two Santanderes have the lowest production. The two main rice environments are irrigated and rain-fed (mostly lowland), using both mechanized and manual/traditional methods; about 70% of the rice-growing area is mechanized (http://ricepedia.org/colombia).

Yields of 8–9 tonnes/ha achieved in the United States are among the highest for any country (Fig. 2.55).

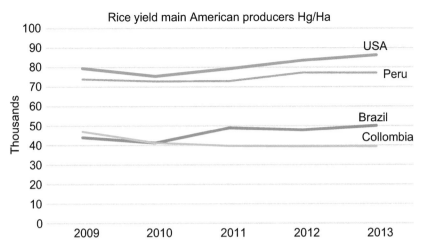

FIGURE 2.55 Variation in yields by main rice producers in the Americas (tonnes). *Based on data from FAOSTAT.*

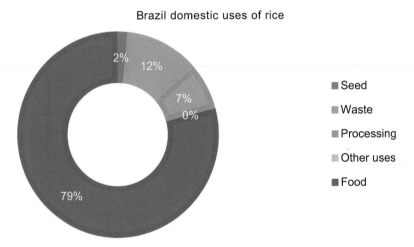

FIGURE 2.56 Reported proportional domestic uses of rice in Brazil (mean of 2008–11). *Based on data from FAOSTAT.*

Domestic uses

As elsewhere in the world, food use predominates in the American region. Both major producers reported relatively higher quantities of wastage, and Peru reported relatively high 'other uses' (Figs. 2.56–2.59).

Exports

The United States is the largest American exporter and its exports have at times reached. one-third of its production. Few other American

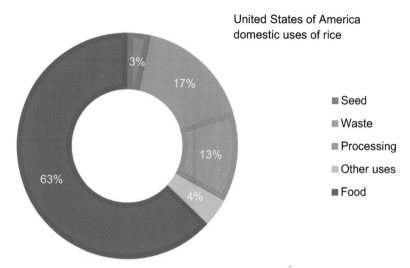

FIGURE 2.57 Reported proportional uses of rice in the United States (mean of 2008–11). *Based on data from FAOSTAT.*

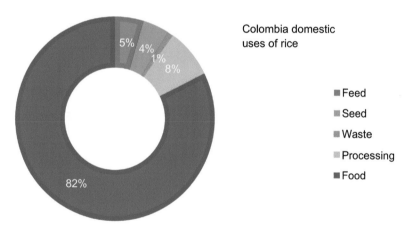

FIGURE 2.58 Reported proportional domestic uses of rice in Colombia (mean of 2008–11). *Based on data from FAOSTAT.*

countries exported more than 0.5 million tonnes per year between 2008 and 2012 (Fig. 2.60).

Standards cover four classes: long grain, medium grain, short grain and mixed. Within each class, trading occurs in various levels of processing, thus standards apply for rough rice (defined as containing 50% or more paddy kernels of rice), brown rice (defined as containing more than 50% of brown rice and which is intended for processing to milled rice) and milled rice (defined as whole or broken kernels of rice from which the hulls and

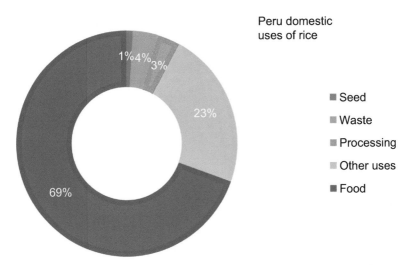

FIGURE 2.59 Reported proportional domestic uses of rice in Peru (mean of 2008–11). *Based on data from FAOSTAT.*

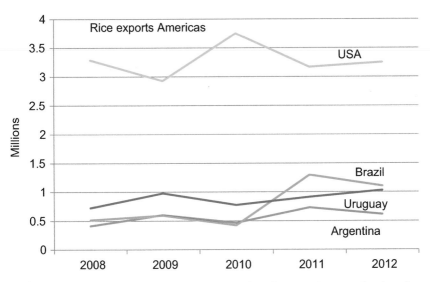

FIGURE 2.60 Variation in exports of rice and products by main rice-exporting American countries (tonnes). *Based on data from FAOSTAT.*

at least the outer bran layers have been removed and which contain not more than 10% of seeds, paddy kernels or foreign material, either singly or combined).

Grading is applied to samples at all processing levels, such that for rough rice and milled rice six grades plus a sample grade are defined

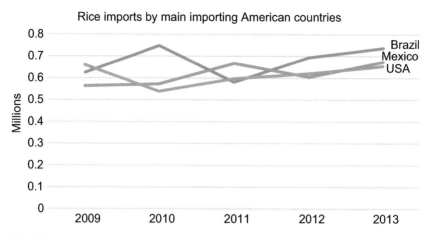

FIGURE 2.61 Variation in imports of rice and products by main importing American countries (tonnes). *Based on data from FAOSTAT.*

and for brown rice there are five grades plus a sample grade. Grades are defined on the basis of presence or absence of contaminants including red rice, broken kernels, opaque kernels, other types and, in the case of brown and milled rice, quality of processing. Additional classes exist for second head, screenings, milled rice and brewers' milled rice. The full details of rice (*Oryza sativa* L.) grading can be found at https://www.gipsa.usda.gov/fgis/standards/ricestandards.pdf.

Imports

No American country imported more than 1 million tonnes, but both Brazil and the United States were among the highest importers and exporters, while Mexico imported three times its own production (Fig. 2.61).

2.2.3 Africa

Production and yield

Africa is a relatively small producer of paddy rice, although seven countries produced between 1 million and 6 million tonnes each year from 2009 to 2013 (Fig. 2.62).

In Egypt rice production takes place only in the lower valley of the Nile River. Due to the intrusion of seawater, about 25%–30% of the land in the lower Nile valley is affected by different degrees of salinity. In these areas, rice production helps to leach the salt from upper soil layers and thus reclaim the land for agricultural activities.

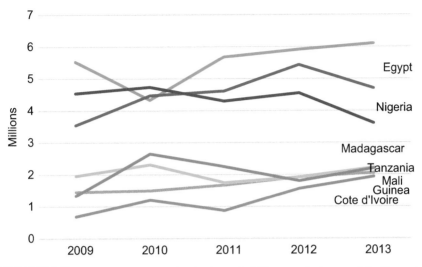

FIGURE 2.62 Variation in rice production by main African producing countries (tonnes). *Based on data from FAOSTAT.*

Most of the planted rice varieties are of the *japonica* type. The Egyptian rice yield is one of the highest in the world. Because of limited water resources the government of Egypt has tried to limit rice growing, but cultivation has continued to expand due to its high profitability and Egypt is today a major rice exporter. Rice is sown from the beginning of May until the end of June, and the harvest season lasts from the middle of August to the middle of October.

In Nigeria rice is grown in paddies or on upland fields; there is limited mangrove cultivation. The fields cannot be ploughed until after the first rain, generally in May or June. Most farmers produce one rice crop each year, but some have made irrigation channels which allow them to reap two or even three harvests in a year.

In Madagascar there are four principal types of rice growing: irrigated rice, rain-fed lowland rice, upland rain-fed rice (called tanety) and rice as a first crop after slash and burn (called tavy). In terms of cultivated area, irrigated rice is the most important, covering 82% of all land under rice in 2008. About 60% of irrigated rice is transplanted. Rice is grown in six zones of Madagascar: the north, northwest and central–western regions; the central part of the Malagasy highlands; the east; and the central–eastern part, including Lake Alaotra, with its swampy areas, plains and valleys suited for rice (http://ricepedia.org/madagascar).

Yields of paddy rice in Egypt are consistently exceptionally high by world standards and around twice as high as other major producers (Fig. 2.63).

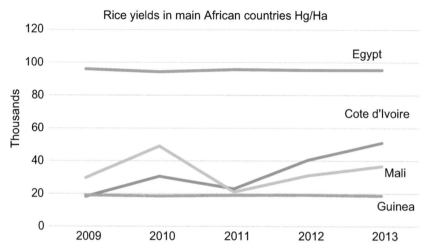

FIGURE 2.63 Variation in rice yields by main producing countries (tonnes). *Based on data from FAOSTAT.*

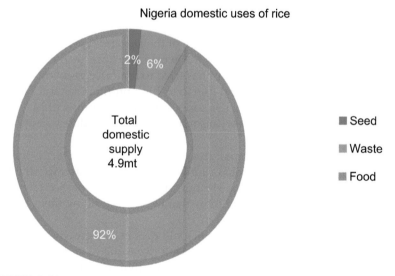

FIGURE 2.64 Reported proportional domestic uses of rice in Nigeria (mean of 2008–11). *Based on data from FAOSTAT.*

Domestic uses

Nigeria is the second-largest producer but the greatest user of rice. All its available stock is reported to be consumed as food (Fig. 2.64). In Egypt small quantities find other uses and a small amount is used as feed (Fig. 2.65). Madagascar uses more for feed, reducing its food use to 70% (Fig. 2.66).

No other African country has an annual consumption greater than 1 million tonnes.

Egypt domestic uses of rice

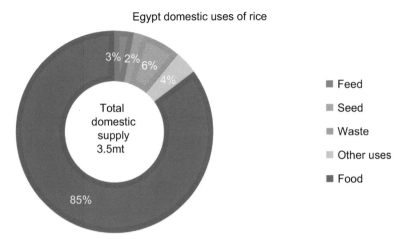

FIGURE 2.65   Reported proportional domestic uses of rice in Egypt (mean of 2008–11). *Based on data from FAOSTAT.*

Madagascar domestic uses of rice

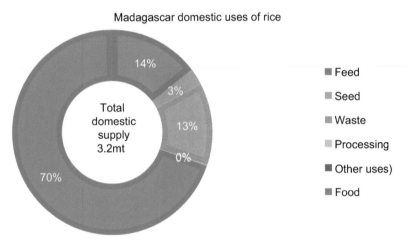

FIGURE 2.66   Reported proportional domestic uses of rice in Madagascar (mean of 2008–11). *Based on data from FAOSTAT.*

Exports

Throughout Africa exports of rice are relatively low, with no country exporting as much as 1 million tonnes. The example of two of the larger exporters shows that quantities traded can be very variable (Fig. 2.67).

The Egyptian specification for rice defines six grades distinguished by maximum limits (from 1 to 6) for broken grains (3%–40%), red grains (1.5%–4.0%), damaged and yellow grains (0.25%–2.50%), immature grains (2.0%–12.0%), foreign matter (0.05%–0.70%) and paddy (0.0%–0.20%) (www.ditp.go.th/contents_attach/77122/77122.pdf).

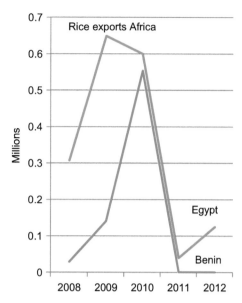

FIGURE 2.67   Variation in rice and products exports by African countries. *Based on data from FAOSTAT.*

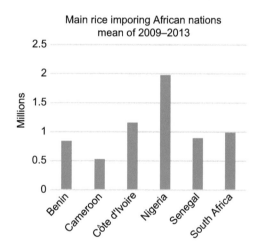

FIGURE 2.68   Rice and products imports by main rice importing African countries (tonnes), mean of 2009–13. *Based on data from FAOSTAT.*

Imports

Nigeria was responsible for the largest mean import quantity, but five other countries imported in excess of 0.5 million tonnes (Fig. 2.68). Considerable year-on-year variation occurred (Fig. 2.69).

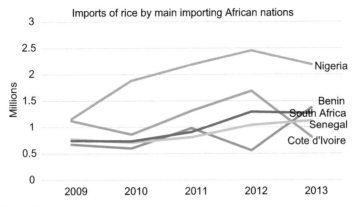

FIGURE 2.69 Variation in rice and products imports by main importing African countries. *Based on data from FAOSTAT.*

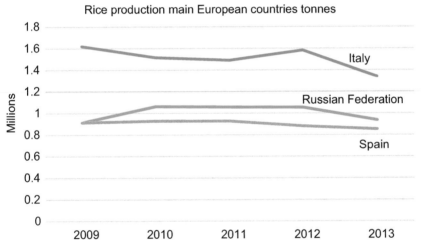

FIGURE 2.70 Variation in rice production by main European producing countries (tonnes). *Based on data from FAOSTAT.*

2.2.4 Europe

Production and yield

The mean total rice production in Europe between 2009 and 2013 was a little over 4 million tonnes per annum. The three largest producers accounted for around 80% of this (Fig. 2.70). Production quantities were relatively consistent throughout the period, as were the yields reported. Yields fell within the range of those achieved in the main producing American countries (Fig. 2.71).

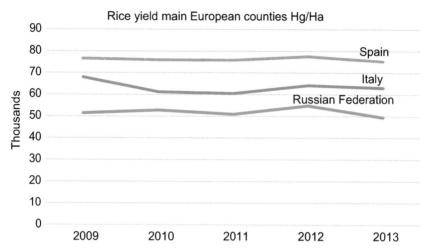

FIGURE 2.71   Variation in rice yields in main producing European countries (hg/ha). *Based on data from FAOSTAT.*

In Italy rice is grown mainly in the northern Po valley, using wet paddyfield system. The sowing season is March to April and harvesting takes place in September/October. Quality rice production is dominated by *japonica* varieties.

In the Russian Federation rice is cultivated mostly in the areas near the Black Sea and Ukraine, with the largest production occurring in Krasnodar (70%), Rostov (18%) and Actpakhan (12%). Rice is planted only under irrigated conditions, in medium to large and highly mechanized farms, in June and July for an August to September harvest.

In the Kuban River delta within Krasnodar territory in the mid-1960s, wide-scale research showed the possibility and usefulness of rice cultivation in this zone: 30 years' experience showed that after 5–7 years of growing rice, soils lose excess salinity and, as a result, can be used for cultivating cereals, fodder crops and vegetables.

In Spain rice is cultivated in nine autonomías, the main administrative divisions, but only in six of them does rice reach the 1500 ha level. These rice areas are geographically separated with different local market preferences. Andalucía, in the south of Spain, has the biggest rice area cultivated in the province of Sevilla when there is enough water supply. However, in a year with severe water shortage the rice area can drop. Main varieties are of the *indica* type. Extremadura, in the southwest of Spain, also suffers from water shortage. Leading varieties are medium grain and long grain. Cataluña in the northeast, in the delta of the Ebro River, has plenty of water and the leading varieties are medium grain. In the Comunidad Valenciana, in the east, water supply can be problematic

Uses of rice in Europe mean of 2008–2011

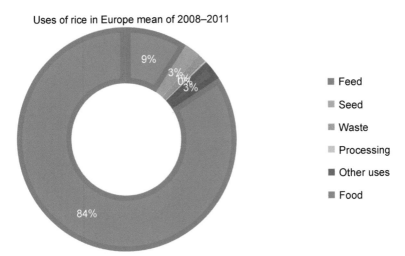

Legend:
- Feed
- Seed
- Waste
- Processing
- Other uses
- Food

9%
3%
0%
3%
84%

FIGURE 2.72 Reported proportional domestic uses of rice in Europe (mean of 2008–11). *Based on data from FAOSTAT.*

in dry years. Leading varieties are medium grain. Aragón and Navarra are colder regions needing varieties shorter in cycle than those cultivated in the other regions.

Domestic uses

Reported uses other than food were small (Fig. 2.72).

Exports

With such small levels of production, it is not surprising that export quantities throughout the region are low. As the largest exporter by far, Italy's export performance varied between 0.7 million and 0.8 million tonnes of milled rice between 2008 and 2012 (Fig. 2.73).

Imports

The United Kingdom imported the most rice between 2009 and 2011, but quantities did not rise above 1 million tonnes (milled equivalent) (Fig. 2.74).

2.2.5 Oceania

Production and yield

In the period 2009–13 the output of the larger regional producers remained stable, but production in Oceania increased substantially (Fig. 2.75). This was partly the result of impressive increases in yield during the period (Fig. 2.76), which led to Oceania reporting the highest yields (see Fig. 1.36).

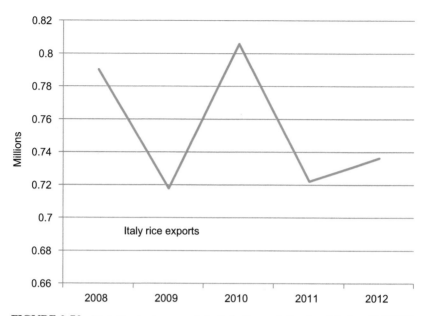

FIGURE 2.73 Variation in rice exports by Italy (tonnes). *Based on data from FAOSTAT.*

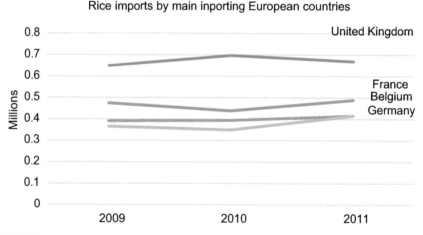

FIGURE 2.74 Variation in rice imports by European countries. *Based on data from FAOSTAT.*

However, Oceania also increased the area devoted to rice growing by almost tenfold. It should be borne in mind that in the context of rice Oceania is almost synonymous with Australia, as that country accounts for 98% of rice production.

In Australia the rice growing areas lie in the Murrumbidgee valley of New South Wales and the Murray valley of New South Wales and Victoria.

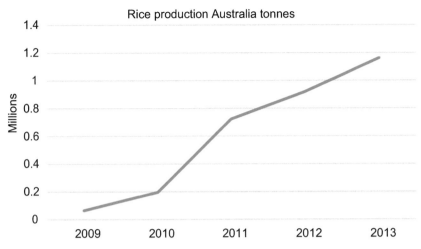

FIGURE 2.75 Variation in rice production in Australia (tonnes). *Based on data from FAOSTAT.*

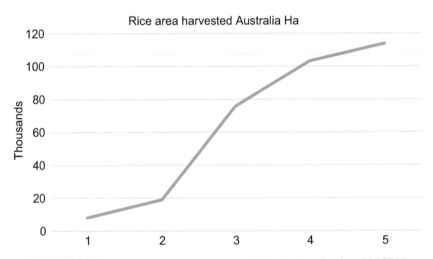

FIGURE 2.76 Variation in rice area harvested (ha). *Based on data from FAOSTAT.*

Rice is a summer cereal crop in the region, with planting in October. Various methods of sowing are used, with water applied after sowing, or crops can be aerial sown using pregerminated seed into bays already containing water. Crops are grown in 5–25 cm of water, depending on the plant's growth stage, which provides moisture for the plant and protects it from fluctuations in temperature. Using water to insulate the plant from cold overnight temperatures during January is particularly important for yield.

Domestic uses

Oceania conformed to the global pattern of the vast majority of rice being consumed as food (Fig. 2.77).

Exports

Australian milled-rice exports increased substantially in line with its increased production. In 2012 almost half of total production was exported (Fig. 2.78).

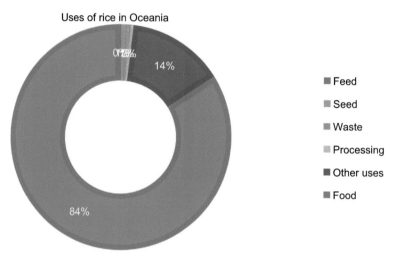

FIGURE 2.77   Reported proportional domestic uses of rice in Oceania (mean of 2009–13). *Based on data from FAOSTAT.*

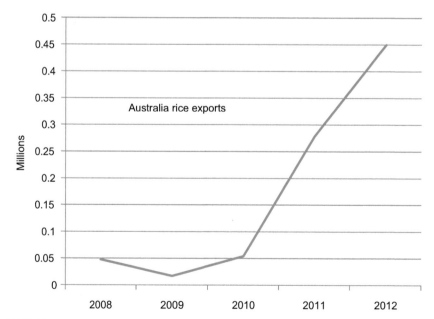

FIGURE 2.78   Rice and products exports by Australia (tonnes). *Based on data from FAOSTAT.*

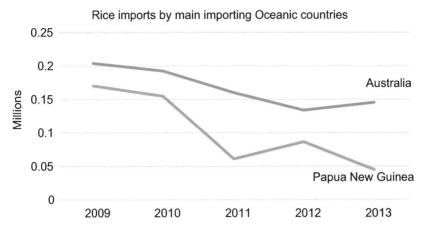

FIGURE 2.79 Variation in rice and products imports by main importing Oceanian countries (tonnes, milled equivalent). *Based on data from FAOSTAT.*

Imports

Imports of rice to Oceania were small, and in the main importing countries the quantities declined between 2009 and 2013 (Fig. 2.79).

2.3 WHEAT

2.3.1 Asia

Production and yield

Asia grows nearly half of the world's wheat (see Fig. 1.38). Over a 10-year period from 2003 there was a marked increase in production quantity (see Fig. 2.139), and even in the last 5 years to 2013 there was a perceptible rise in production. Such an increase has not been detectable in other continental regions (Fig. 2.80).

The highest yields of wheat have been reported in Europe and the lowest in Oceania, where yields may be as low as half the European average (Fig. 2.81).

Mainland China and India recorded by far the highest levels of production in the region, and together contributed two-thirds of total Asian production (Fig. 2.82). The overall increase in Asian production is due to increases in these two countries.

In general, Chinese wheat has acceptable protein content but weak gluten quality, and thus is not suitable for mechanized food production (He et al., 2012).

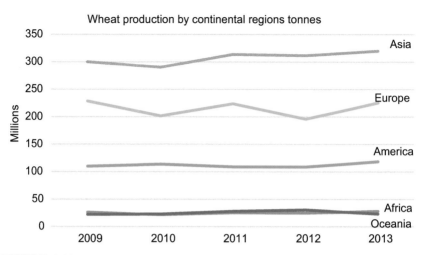

FIGURE 2.80 Variation in wheat production by continental regions (tonnes). *Based on data from FAOSTAT.*

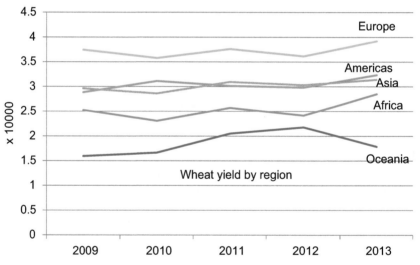

FIGURE 2.81 Variation in wheat yields in continental regions (hg/ha). *Based on data from FAOSTAT.*

In present-day China the main producing provinces are, in order of production quantity, Henan, Shandong, Hebei, Anhui and Jiangsu. In 2014 these provinces each grew between 3.3 and 1.2 million tonnes. Other provinces grew less than 1 million tonnes (www.statista.com/statistics/242630/wheat-production-in-china-by-province). More than 95% of wheat is sown in the autumn. A double-cropping system is used in the

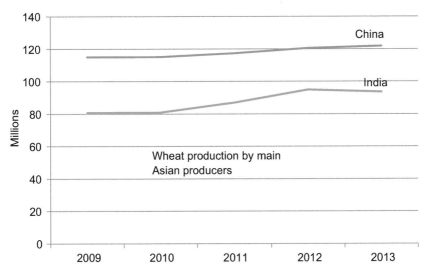

FIGURE 2.82 Variation in wheat production by main Asian producers (tonnes). *Based on data from FAOSTAT.*

Yellow River and Huai River valleys in which wheat is rotated with maize. In the Yangtze valley it is rotated with rice.

In India most wheat production occurs in the north and northwest of the country. Utah Pradesh in the north grew around 30 million tonnes in 2012–13, almost twice the production of the next-largest producer, Punjab (16 million tonnes). Madhya Pradesh in the centre of the country grew 13 million tonnes and Haryana in the northwest grew 11 million tonnes. Other significant producing states – Rajasthan and Gujarat in the west, Bihar (northeast) and Maharashtra (southwest) – grew between 9.3 and 1.2 million tonnes each. West Bengal, the most easterly producer of note, grew 900,000 tonnes. Most of the wheat grown in India is of the soft to medium-hard, medium-protein bread type, but just over 1 million tonnes of durum are grown in Madhya Pradesh.

Several other Asian countries grow substantial quantities of wheat. Most have been quite consistent, but the performance of Kazakhstan has been particularly variable (Fig. 2.83).

In Pakistan the province of Punjab ranks top in the production of wheat. The upper Indus plain accounts for 70% of the total wheat of the country. Canal-fed fields produce two-thirds and the rest comes from rain-fed fields.

The province of Sindh ranks second in wheat production. The Kabul, Kurram and Gomal river valleys provide a substantial proportion of the wheat acreage. Some wheat is also grown in the plateau areas and the valleys of the hilly regions of Baluchistan. The Northwest Frontier province of Khyber Pakhtunkhwa contributes little towards wheat production.

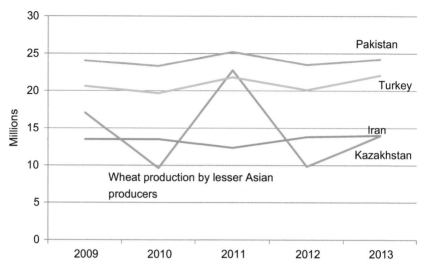

FIGURE 2.83   Variation in wheat production by lesser Asian producers (tonnes). *Based on data from FAOSTAT.*

Turkey has a wide variety of geomorphology, topography and climate, but cereals feature strongly in the economy of most regions (www.fao.org/ag/agp/agpc/doc/counprof/Turkey/Turkey.htm). Gummadov et al. (2015) identified seven regions with distinctive wheat-growing preferences. The largest is Central Anatolia, where irrigation is provided to 20% of the winter wheat. Winter wheat is also grown in Eastern Anatolia, where 35% is irrigated. In Marmara region more than half the wheats grown are winter varieties and the remainder are autumn-sown spring varieties. Autumn-planted spring wheats are grown in the Aegean and Mediterranean regions, with 30%–40% being irrigated. Southeast Anatolia is characterized by facultative and spring wheats, mostly varieties of durum (35% irrigated) and the Black Sea region grows autumn-sown spring wheats.

Kazakhstan produces high-quality spring wheats, mainly (70%) in the Northern region. The Southern region produces 10% and the Eastern and Central regions produce around 7% each. Large enterprises are responsible for producing around 90% of the country's wheat (www.fao.org/ag/agp/agpc/doc/field/wheat/asia/kazakhstan.htm).

In Iran irrigated wheat covers one-third of the total wheat area, thus the bulk of the wheat crop depends on the performance of seasonal precipitation. Most of the rain-fed wheat crop is concentrated in the west and northwestern regions of the country. Wheat is sown from September to early November and harvested in June and July.

Domestic uses

Throughout Asia wheat is used largely as food, with a small amount devoted to feed. The distribution in China is very similar to that of the

Wheat uses china mean of 2008–2011

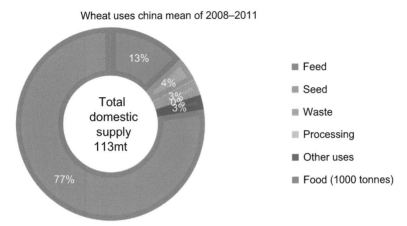

FIGURE 2.84 Reported proportional domestic uses of wheat in China (mean of 2008–11). *Based on data from FAOSTAT.*

Wheat uses india 2008–2011

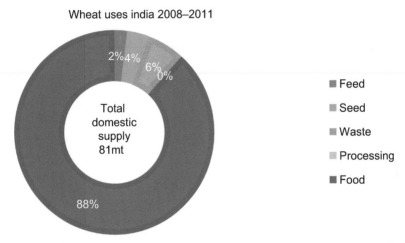

FIGURE 2.85 Reported proportional domestic uses of wheat in India (mean of 2008–11). *Based on data from FAOSTAT.*

whole region, while in India there is a considerably reduced amount used as feed (Figs. 2.84 and 2.85).

Exports

The main exporting country in Asia is Kazakhstan, whose exports have reached more than half its total production (Fig. 2.86). In line with its production total there have been large variations in export quantities. Export performance by India and Pakistan appears to be less related to production quantity (Fig. 2.87).

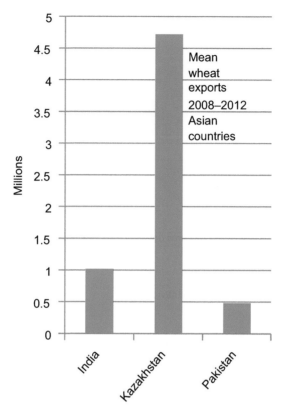

FIGURE 2.86   Wheat and products exports by main exporting countries (mean of 2008–12). *Based on data from FAOSTAT.*

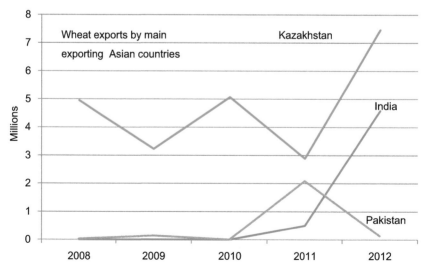

FIGURE 2.87   Variation in wheat and products exports by main exporting Asian countries (tonnes). *Based on data from FAOSTAT.*

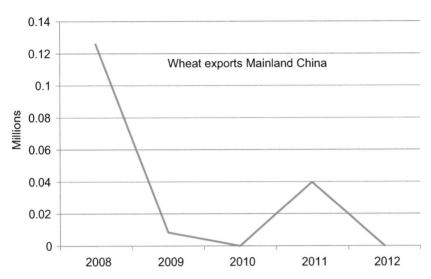

FIGURE 2.88 Variation in wheat and products exports by mainland China (tonnes). *Based on data from FAOSTAT.*

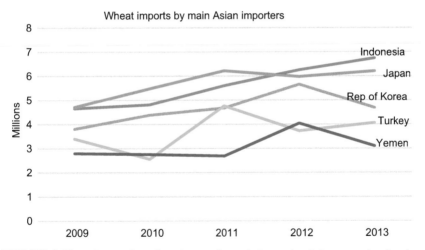

FIGURE 2.89 Wheat and products imports by main importing Asian countries. *Based on data from FAOSTAT.*

The largest producer, China, exports only a very small and variable proportion of its production (Fig. 2.88).

Imports

Several Asian countries imported in excess of 1 million tonnes. Only the largest are shown in Fig. 2.89. Short-term upward trends are apparent in some cases (Fig. 2.89).

2.3.2 Europe

Production and yield

Europe produced more than 225 million tonnes of wheat in 2013. The Russian Federation is the largest producer and, like its much smaller western neighbour Ukraine, is subject to considerable variation in output. Although many European countries produce more than 1 million tonnes annually, the five largest producers contribute around two-thirds of the total (Fig. 2.90).

Wheat production in the Russian Federation extends eastwards from Georgia in the west to China in the east. It reaches northwards along the Russian border with Ukraine, Belarus, Latvia and Estonia in the west and even further in the central region. Wheat is even grown close to the north Finnish border in the northwest. The Baltic exclave of Kaliningrad oblast also produces wheat (http://wheatatlas.org/country/map/RUS/0). Spring wheats contribute 70% of production. Winter wheat is cultivated in the North Caucasus and spring wheat in the Don basin, the middle Volga region and southwestern Siberia. Durum is included among the winter wheats. Harvest begins in the south in early July and progresses northwards until completion in mid to late August.

In France wheat is grown widely, but the main growing regions are Centre (16% of the total) and Picardie (10%). Very little is grown in Limousin and the southeast (www.spectrumcommodities.com/education/commodity/statistics/wheat.html).

Winter wheat is the predominant type produced in France. It is planted from the beginning of October to the end of November. The harvest season begins around the start of July and is usually concluded by the end of August.

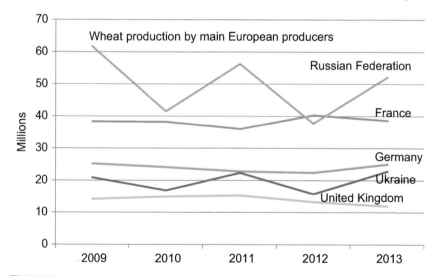

FIGURE 2.90 Wheat production by main European producers (tonnes). *Based on data from FAOSTAT.*

Durum wheat is cultivated widely, but most areas devoted to durum are located around Orleans, Toulouse, Montpellier, Poitiers, Marseilles and Nantes.

In Ukraine major production areas of winter wheat exist throughout the entire country except the northern region, where production is less. Most of the harvest is winter wheat, with spring types accounting for only around one-quarter of production.

In the United Kingdom wheat is grown mainly in the east and the east Midlands, followed by Yorkshire and Humberside, the southeast and southwest. Less is produced in Scotland, the northeast, the northwest, Wales and Northern Ireland. Winter wheat is the highest-yielding cereal in Britain and covers the most acreage. Sown in late September to November (but it can be sown until February), it comes to harvest in August (September in Scotland). Spring wheat is sown from February to April and harvested in September. Spring varieties are lower yielding than winter varieties but usually have higher quality for breadmaking. Traditionally the United Kingdom has been known for its biscuit or 'soft' wheats. More recently, plant breeders have developed United Kingdom cereal varieties suitable for breadmaking, allowing United Kingdom millers to use more home-grown wheat in place of imports.

Wheat yields vary considerably among European countries and those of western countries tend to be higher than their eastern counterparts. Belgium produced only around two million tonnes per annum between 2009 and 2013 but yields in that country were nearly 20% above those of the next highest and fourfold those of the largest producer, the Russian Federation (Fig. 2.91).

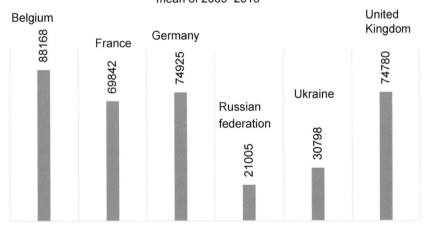

FIGURE 2.91 Wheat yields in Belgium and main wheat-producing countries (hg/ha). *Based on data from FAOSTAT.*

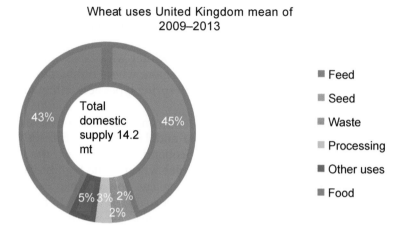

FIGURE 2.92   Reported proportional domestic uses of wheat in the United Kingdom (mean of 2009–13). *Based on data from FAOSTAT.*

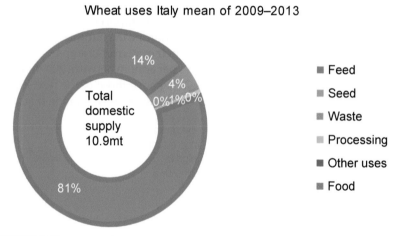

FIGURE 2.93   Reported proportional domestic uses of wheat in Italy (mean of 2009–13). *Based on data from FAOSTAT.*

Domestic uses

Uses for the whole of Europe are approximately the same as uses in the United Kingdom (Fig. 2.92). In the Russian Federation (not shown) the main difference lies in a larger proportion set aside for seed (15%), reflecting the lower yields experienced there. There is a correspondingly lower allocation to feed.

Italy has an unusually large allocation to food, and the Netherlands has an unusually high allocation to feed (Figs 2.93 and 2.94).

Uses of wheat Netherlands

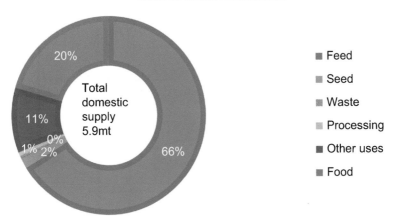

FIGURE 2.94   Reported proportional domestic uses of wheat in the Netherlands (mean of 2009–13). *Based on data from FAOSTAT.*

Wheat exports by main European countries mean of 2008–2012

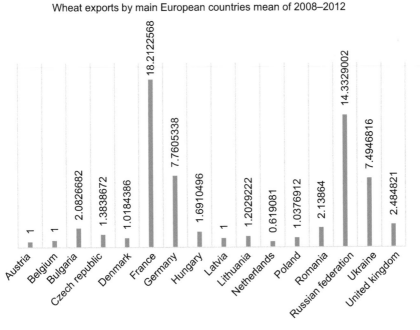

FIGURE 2.95   Wheat and products exports by main exporting European countries (mean of 2008–12). *Based on data from FAOSTAT.*

Exports

A large number of countries have wheat exports exceeding 1 million tonnes (Fig. 2.95), but the largest exporters by far are the largest producers: the Russian Federation, which regularly exports between a

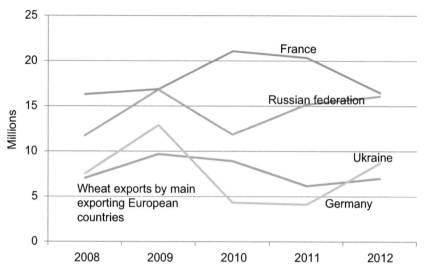

FIGURE 2.96   Variation in wheat and products exports by main exporting European countries (mean of 2008–13). *Based on data from FAOSTAT.*

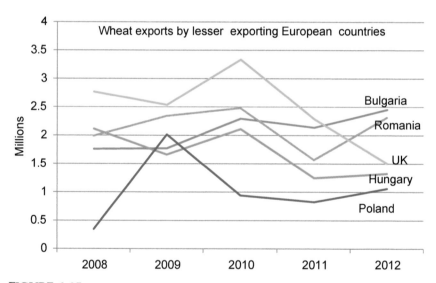

FIGURE 2.97   Variation in wheat and products exports by lesser exporting European countries (tonnes). *Based on data from FAOSTAT.*

quarter and a third of its harvest, and France, which exports almost half of its production.

Export quantities have been quite variable among the larger and smaller exporters (Figs. 2.96 and 2.97).

Wheat imports by main importing European countries mean of 2009–2013 tonnes

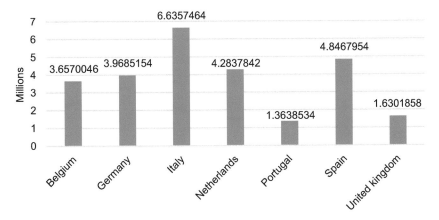

FIGURE 2.98 Wheat and products imports by main importing European countries (tonnes), mean of 2009–13. *Based on data from FAOSTAT.*

Imports

Wheat imports into Europe are relatively high even among countries with high wheat production themselves. Belgium, a relatively small producer, was a net importer of around 3 million tonnes towards a total domestic supply of less than 4 million tonnes (Fig. 2.98).

2.3.3 Americas

Production and yield

The United States was the largest producer, with quantities double that of the next largest, Canada. Argentina was the largest producer in South America, growing about half the annual production of Canada (Fig. 2.99).

In the United States wheat is grown in all states, but different types are associated with specific regions. Thus of the spring wheats, hard red spring (HRS) is grown in the northern states of North and South Dakota and Montana; in North and South Dakota durum is included, as it is in Wisconsin and eastern Montana. Hard red winter (HRW), the most-grown wheat, is planted in Kansas, Colorado, Oklahoma and Texas. Soft white winter (SWW) comes mainly from Washington state, Oregon and southern Idaho. Hard white winter (HWW) is grown mainly in Kansas and Colorado.

Winter wheats are sown from late August to end October (early September to mid-November for SWW) and harvested from August

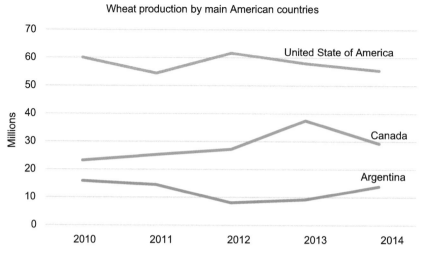

FIGURE 2.99   Variation in wheat production by main producing American countries (tonnes). *Based on data from FAOSTAT.*

to end October (July to early September for SWW). Spring wheats are sown during April and May and harvested from mid-July to mid-September (http://www1.agric.gov.ab.ca/$department/deptdocs.nsf/all/sis5219).

In Canada the majority of wheat is grown in the prairie provinces of western Canada: Saskatchewan, with 46% of total production, Alberta with 30% and Manitoba with 14% (based on the 5-year average of 2005–09). Prairie wheats are mainly dark northern spring (DNS) varieties, but around a quarter are durum varieties. In eastern Canada production is led by Ontario with 9%; Quebec grows 1% and the Atlantic region produces less than 1% of total wheat production. Wheats of these eastern provinces are winter type. Spring wheats in Canada are sown in May and harvested between August and mid-October; winter varieties are sown in October and harvested from late June to early October.

In Argentina the majority of wheat is produced in the southern regions of the country, centred on the province of Buenos Aires which produces 63% of the nation's wheat. The next-largest wheat-producing province in Argentina is Santa Fe, accounting for 17% of the nation's total. Argentinean wheats are all spring types, normally planted from early May to the end of July, with harvest season beginning around mid-November and running until mid-January.

Yields in the three main producing countries of the Americas were all similar, but yields in Chile were considerably higher, being comparable with some of the highest-yielding major producing European countries (Fig. 2.100).

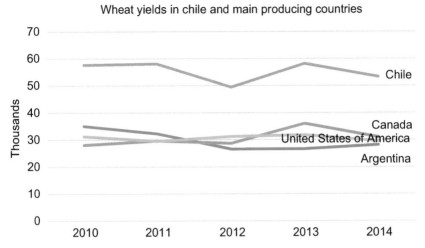

FIGURE 2.100 Variation in wheat yields by Chile and main producing American nations (tonnes). *Based on data from FAOSTAT.*

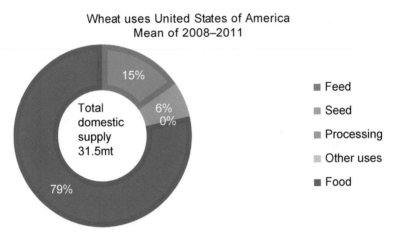

FIGURE 2.101 Reported proportional domestic uses of wheat in the United States (mean of 2008–11). *Based on data from FAOSTAT.*

Domestic uses

The United States had the largest domestic supply of wheat in the Americas, the majority of it being consumed as food (Fig. 2.101). Argentina consumed a similar amount as food, but whereas the United States uses 15% as feed, Argentina, with a domestic supply of 5.3 million tonnes, suffered 11% waste with only 2% being used for feed.

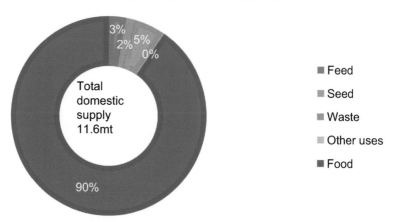

FIGURE 2.102   Reported proportional domestic uses of wheat in Brazil (mean of 2008–11). *Based on data from FAOSTAT.*

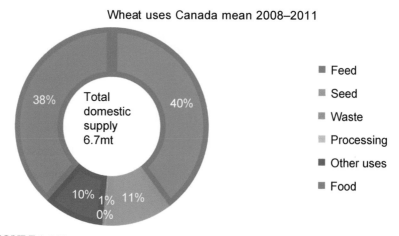

FIGURE 2.103   Reported proportional domestic uses of wheat in Canada (mean of 2008–11). *Based on data from FAOSTAT.*

The country with the second-highest domestic supply was Brazil, where an even higher proportion was consumed as food (Fig. 2.102). Canada, by contrast, used equal proportions for food and feed and a relatively large amount (20%) for processing and other uses (Fig. 2.103).

Exports

The order of the three main exporters of wheat and products in the Americas follows the same order as seen for production quantities, namely the United States followed by Canada and Argentina (Fig. 2.104)

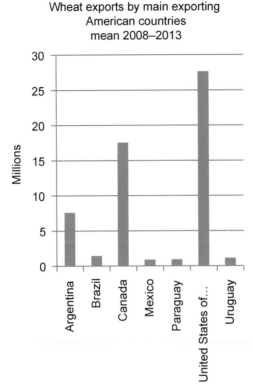

Wheat exports by main exporting American countries mean 2008–2013

FIGURE 2.104   Wheat and products exports by main exporting American countries (tonnes, mean of 2008–13). *Based on data from FAOSTAT.*

Although other American countries engage in wheat exporting, some of them achieving over 1 million tonnes, the quantities are a fraction of those relating to these three countries.

The 'United States Standards for Wheat' is published by the Grain Inspectors, Packers and Stockyards Administration Service (https:// www.gipsa.usda.gov). The system recognizes eight classes: durum wheat, HRS wheat, HRW wheat, soft red winter (SRW) wheat, hard winter (HW) wheat, soft white (SW) wheat, unclassed wheat and mixed wheat.

Durum wheat subclasses of hard amber durum and amber durum wheats are distinguished on the basis the proportion of hard and vitreous kernels of amber colour present.

HRS subclasses, dark northern spring, northern spring and red spring wheats are distinguished on the basis of the proportion of dark hard and vitreous kernels present.

SW subclasses, soft white, white club and western white are distinguished on the basis of the proportion of club (*T. aestivum* ssp. *compactum*) wheat present among other white wheats.

Grades 1–6 are defined by minimum test weight (pounds per bushel: 58–50 for HRS and white club and 60–51 for other classes) and the presence of the following (the ranges of maximum permitted levels are shown in parenthesis):

- Damaged kernels: heated (0.2%–3.0%) total (2.0%–15.0%)
- Foreign material (0.4%–5.0%)
- Shrunken and broken kernels (3.0%–20.0%)
- Wheat of other classes (1.0%–10.0%)
- Wheat of contrasting classes (3.0%–10.0%)
- Stones (0.1%)

Limits apply to all grades (counts per kg) for animal filth (1), castor beans (1), glass (0), stones (3), unknown foreign substances (4), a total of 4. Insect-damaged grains are limited to 31 per 100 g.

An inferior sample grade also exists, as do special grades for ergoty wheat, garlicky wheat, light smutty wheat, smutty wheat and treated wheat.

Canadian grades are available at https://www.grainscanada.gc.ca/oggg-gocg/ggg-gcg-eng.htm. Twenty classes are defined and several grades are recognized for each, as detailed below.

Western classes: Canada western red spring (CWRS) (three grades), Canada hard white spring (CHWS) (three grades), Canada western amber durum (CWAD) (five grades), Canada western red winter (CWRW) (three grades), Canada western soft white spring (CWSWS) (three grades), Canada western red winter (CWRW) (three grades), Canada western extra strong (CWES) (two grades), Canada prairie soft white (CPSW) (two grades) Canada prairie soft red (CPSR) (two grades), Canada western special purpose (CWSP) (two grades). Northern class: Canada northern hard red (CNHR) (three grades). Eastern classes: Canada eastern red (CER) (three grades), Canada eastern red spring (CERS) (three grades), Canada eastern hard red winter (CEHRW) (three grades), Canada eastern soft red winter (CESRW) (three grades), Canada amber durum (CAD) (four grades), Canada eastern hard white winter (CEHWW) (three grades), Canada eastern white winter (CEWW) (three grades), Canada eastern soft white spring (CESWS) (three grades), Canada eastern hard white spring (CEHWS) (three grades), and Canada eastern feed (CEF) (one grade) and Canada eastern H.

Considerable variation occurred in quantities exported by the major exporters (Fig. 2.105), but variation was relatively greater among the lesser exporting countries, with Brazil's performance being the most variable (Fig. 2.106).

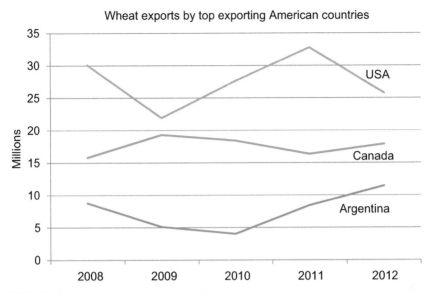

FIGURE 2.105 Variation in wheat and products exports by main exporting American countries (tonnes). *Based on data from FAOSTAT.*

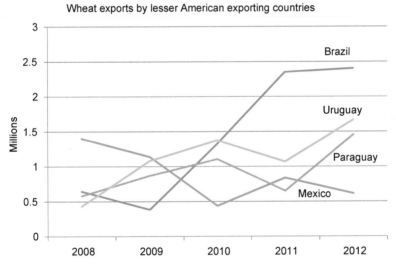

FIGURE 2.106 Variation in wheat and products exports by lesser exporting American countries (tonnes). *Based on data from FAOSTAT.*

Imports

Of all American countries Brazil recorded the highest level of imports of wheat and products during the period 2009–11, with its imports far exceeding its exports. The reverse was true of the United States, the second-highest importer, during the period (Fig. 2.107).

2.3.4 Africa

Production and yield

As is the case with rice, Egypt was the largest producer of wheat in the African region (Fig. 2.108). During the period 2010–14 an upward trend was evident, as it was for the second- and third-highest producers, Morocco and Ethiopia (Fig. 2.109).

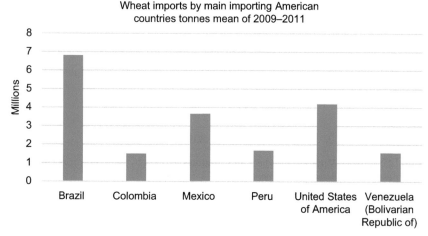

FIGURE 2.107   Wheat and products imports by main importing American countries (mean of 2009–11 values). *Based on data from FAOSTAT.*

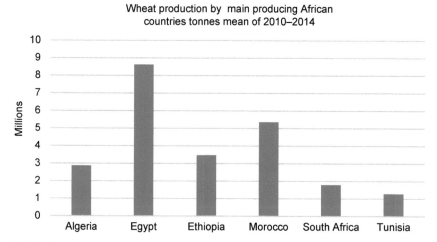

FIGURE 2.108   Wheat production by main producing African countries (tonnes) (mean of 2010–14). *Based on data from FAOSTAT.*

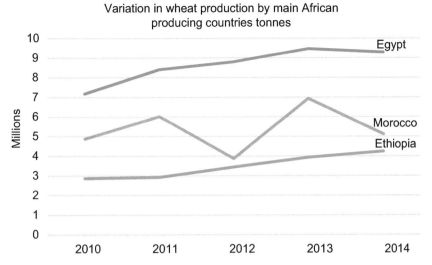

FIGURE 2.109 Variation in wheat production by main African producers (tonnes). *Based on data from FAOSTAT.*

In Egypt wheat is produced in a very limited area along the Nile River and its delta plane. Throughout the Middle East, wheat planting begins in early September and usually runs to the end of November. Harvest begins around the start of April and continues until the end of August. All bread wheat grown in Egypt is white.

In Morocco most wheat is spring sown, a small proportion is facultative and a larger proportion durum. The largest growing region lies in the northwest of the country, but production is widespread.

In Ethiopia wheat is primarily grown in the Amhara, Oromia, Tigray and Southern Nations, Nationalities and Peoples regions. These regions account for more than 90% of national wheat production; 65% of the wheat produced is of the bread wheat type while 35% is of the durum wheat type (http://ethioagp.org/wheat-3/). Bread wheat varieties have proved, in recent years, to be more susceptible to stripe and stem rusts than durum varieties, and efforts are now being made to increase the proportion of durum (http://www.icarda.org/blog/durum-wheat-key-ethiopian-food-security). It is estimated that over 4.5 million farmers are annually involved in wheat production.

Wheat yields in Egypt were considerably higher than elsewhere in the region, and were comparable to some of the yields recorded for the main European countries (Fig. 2.110).

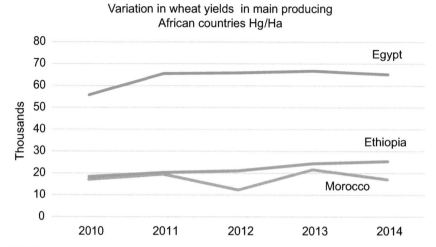

FIGURE 2.110   Variation in wheat yields in main producing African countries (hg/ha). *Based on data from FAOSTAT.*

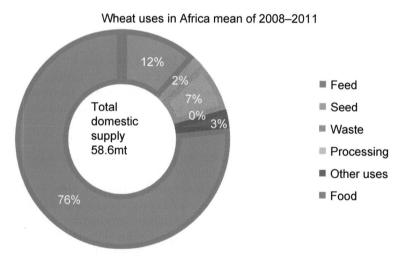

FIGURE 2.111   Reported proportional domestic uses of wheat in Africa (mean of 2008–11 values). *Based on data from FAOSTAT.*

Domestic uses

Wheat uses in Africa were very similar to those in Asia (Fig. 2.111), with three-quarters of the domestic supply being consumed as food and half the remainder being used for feed.

In Egypt a relatively high proportion for the region is devoted to feed (Fig. 2.112).

Domestic wheat uses in Egypt mean of 2008–2011

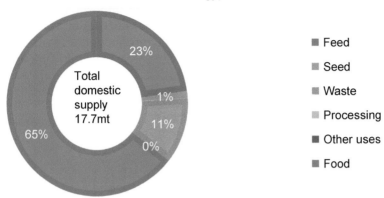

FIGURE 2.112 Reported proportional domestic uses of wheat and products in Egypt (mean of 2008–11). *Based on data from FAOSTAT.*

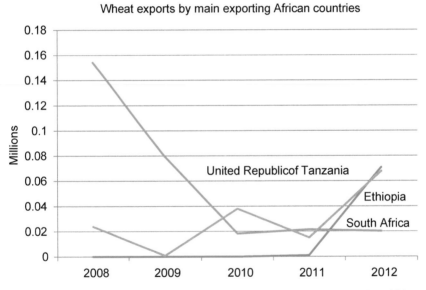

FIGURE 2.113 Variation in wheat and products exports by main exporting African countries (tonnes). *Based on data from FAOSTAT.*

Exports

Very little wheat and its products were exported by African countries, and the small amounts involved were inconsistent (Fig. 2.113).

Imports

Africa imported more than twice as much wheat and wheat products as it produced during the period 2008–12. During the period an upward trend occurred (Fig. 2.114). The main producer of wheat in the region was also the main importer, with imports approximating to home production. The same relationship was true of Morocco, but imports to other countries exceeded their production (Figs. 2.115 and 2.116).

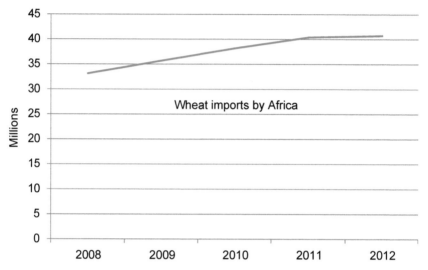

FIGURE 2.114   Variation in total African imports of wheat and products (tonnes). *Based on data from FAOSTAT.*

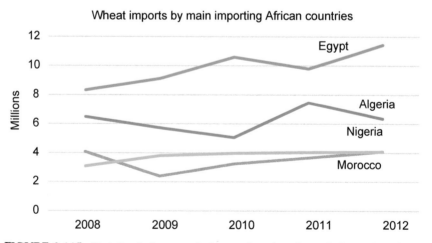

FIGURE 2.115   Variation in imports of wheat and products by main importing African countries (tonnes). *Based on data from FAOSTAT.*

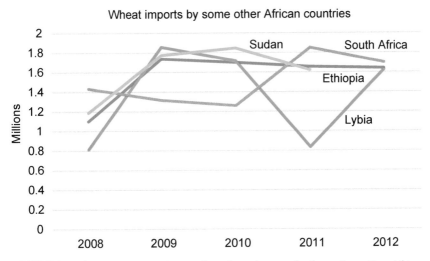

FIGURE 2.116 Variation in wheat and products imports by lesser importing African countries (tonnes). *Based on data from FAOSTAT.*

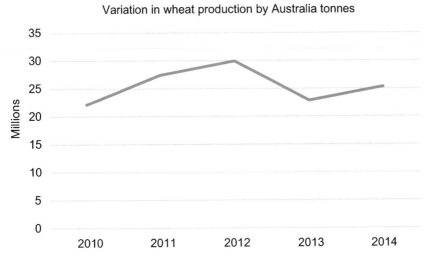

FIGURE 2.117 Variation in wheat production by Australia (tonnes). *Based on data from FAOSTAT.*

2.3.5 Oceania

Production and yield

Australia is by far the largest producer of wheat in Oceania (Fig. 2.117). The second largest, New Zealand, has not produced more than 0.5 million tonnes per annum during the period for which records are available.

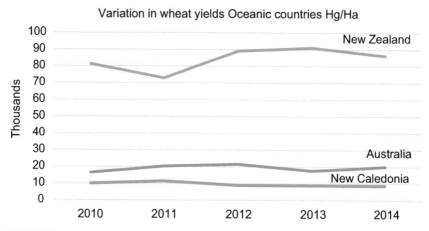

FIGURE 2.118   Variation in wheat yields in main producing Oceanian countries (hg/ha). *Based on data from FAOSTAT.*

However, yields recorded in New Zealand have consistently exceeded those elsewhere in the region by a considerable margin (Fig. 2.118).

In Australia wheat is grown in the southwest of Western Australia, the south and east of New South Wales, the south of South Australia, the southwest of Victoria and the southeast of Queensland. A small amount is produced in Tasmania.

Wheat is strictly graded according to quality, type and end use. Some grades such as Australian hard (AH, 15%–20% of production), Australian premium white (APW, 30%–40% of production) and Australian soft white (ASW, 20%–25% of production) are grown in all wheat-growing areas. Others have a more localized production area, thus Australian prime hard (APH, 5%–10% of production) is grown in northern New South Wales and Queensland, and Australian premium white noodle (APWN, 5%–10% of production) and Australian soft (ASFT, <5% of production) are grown only in Western Australia. Australian noodle wheat (ANW, <5% of production) is grown mainly in Western Australia. Australian prime durum (APD, <5% of production) is grown on both sides of the Queensland/New South Wales border and in Victoria. Some irrigation of wheat is practised where and when possible.

All the *T. aestivum* wheats grown in Australia are white. In spite of lacking a vernalization requirement, most are grown as winter-sown crops and only half of the Queensland wheat and a small proportion of the New South Wales crop are spring sown. Planting is from April to June and harvest takes place between October and January, with a peak in November.

In New Zealand most wheat is produced in the South Island with only 2% coming from the North Island. The Canterbury region in the eastern central area produces nearly 90% and the remainder is grown further south. Winter and spring and hard and soft varieties are grown.

Domestic uses

Uses of wheat and products in Australia were biased towards feed, with a relatively small amount allocated to food and a significant proportion of 'other uses' (Fig. 2.119). The relatively large amount used as seed reflects the low yields noted above. In New Zealand the proportion used as food was higher, with significant use in processing (Fig. 2.120).

Exports

Compared to production quantities the amount of wheat exported by Australia is exceptionally large (Fig. 2.121). In 2012 total production was 29.9 mt, of which 23.5 mt was exported.

Grain Trade Australia publishes wheat trading standards (www.graintrade.org.au/commodity_standards). A large number of detailed criteria are grouped under the main headings of variety, protein, moisture

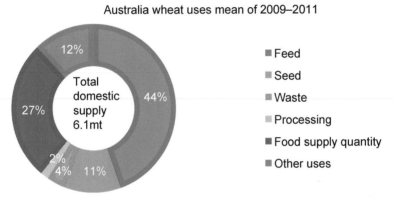

FIGURE 2.119 Reported proportional domestic uses of wheat and products in Australia (mean of 2009–11). *Based on data from FAOSTAT.*

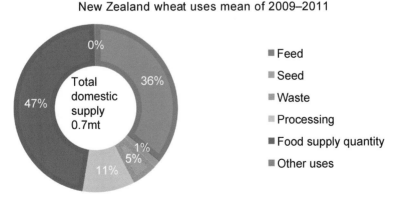

FIGURE 2.120 Reported proportional domestic uses of wheat and products in New Zealand (mean of 2009–11). *Based on data from FAOSTAT.*

content, screenings and unmillable material above screen, falling number, defective grains, foreign seeds and other contaminants. The classes are described at the beginning of this section.

Imports

By world standards imports of wheat and products are very small in Oceania. The three largest importers are shown in Fig. 2.122.

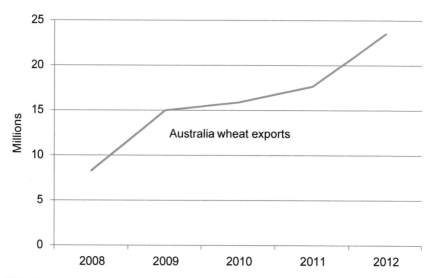

FIGURE 2.121 Variation in wheat and products exports by Australia (tonnes). *Based on data from FAOSTAT.*

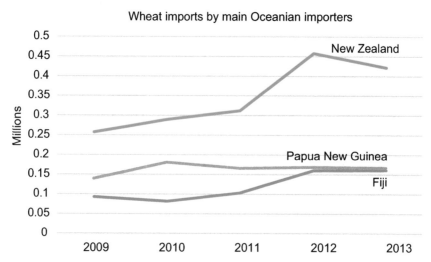

FIGURE 2.122 Variation in wheat and products imports by main importing Oceanian countries (tonnes). *Based on data from FAOSTAT.*

2.4 BARLEY

2.4.1 Europe

It can be seen from Fig. 2.123 that barley is largely a feed grain, with very little (5%) being consumed as food, although much of the relatively large proportion allocated to processing is destined for human consumption because barley is the principal world grain involved in malting for the production of alcoholic drinks. It is also clear from Figs. 1.40 and 1.41 (Chapter 1) that barley is a cereal mostly grown in Europe.

Barley production has been consistent in recent years in the main producing regions, with a tendency to increase in its principal region of production, Europe (Fig. 2.124).

World uses of barley mean of 2009–2011

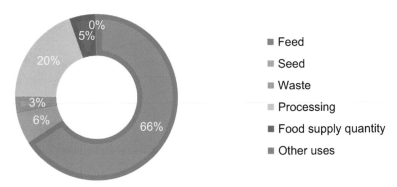

FIGURE 2.123   Reported proportional world uses of barley and products (2009–11). *Based on data from FAOSTAT.*

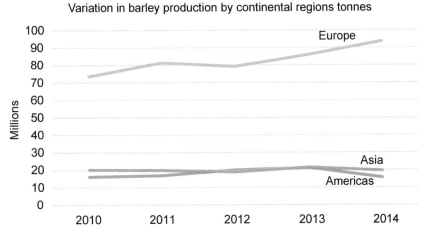

FIGURE 2.124   Variation in barley production by main producing continental regions (tonnes). *Based on data from FAOSTAT.*

Production and yield

Within Europe large variations in production quantities have occurred, particularly in the largest contributor, the Russian Federation (Fig. 2.125). Much of the variation appears to have arisen as a result of changes in area harvested, as the reported changes in yield are inadequate to account for the variation (Fig. 2.126).

Within the continental region the northwestern countries are characterized by the highest yields.

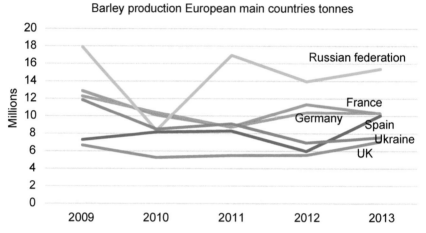

FIGURE 2.125 Variation in barley production by main producing European countries (tonnes). *Based on data from FAOSTAT.*

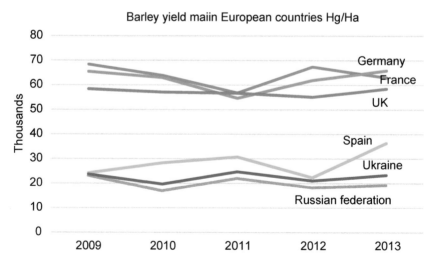

FIGURE 2.126 Variation in yields of barley in main producing European countries (hg/ha). *Based on data from FAOSTAT.*

Domestic uses

Uses of barley in the main producing European countries (Fig. 2.127) do not vary much from those of the entire world, although in France processing accounted for less than elsewhere (Fig. 2.128). By contrast the Czech Republic devoted 30% of its domestic supply of 1.3 million tonnes to processing. The large proportion retained for seed in the Russian Federation accords with the low yields there.

Exports

The Russian Federation, the largest producer of barley, was not the largest exporter between 2008 and 2011 (Fig. 2.129). However, the record of barley exports was very variable during that period (Fig. 2.130).

Russian federation

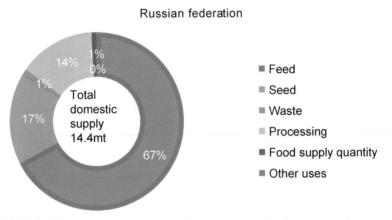

FIGURE 2.127 Reported proportional domestic uses of barley and products in the Russian Federation (mean of 2009–11). *Based on data from FAOSTAT.*

France

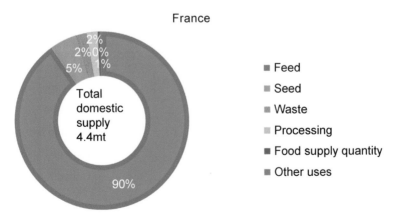

FIGURE 2.128 Reported proportional domestic uses of barley and products in France (mean of 2009–11). *Based on data from FAOSTAT.*

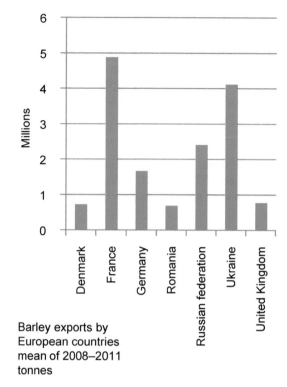

Barley exports by
European countries
mean of 2008–2011
tonnes

FIGURE 2.129 Barley and products exports by main exporting European countries (tonnes) (mean of 2008–11). *Based on data from FAOSTAT.*

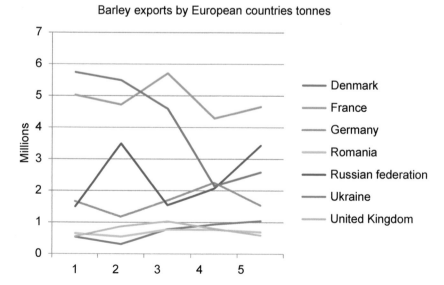

FIGURE 2.130 Variation in barley and products exports by main exporting European countries (tonnes). *Based on data from FAOSTAT.*

Imports

No country averaged more than 2 million tonnes of barley and products during the period 2008–12 (Fig. 2.131), and apart from Germany, the third-largest importer, considerable variation has occurred (Fig. 2.132).

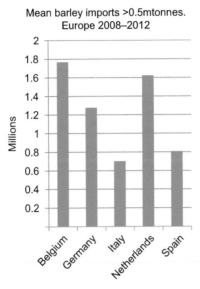

FIGURE 2.131 Barley and products imports by main importing European countries (tonnes). *Based on data from FAOSTAT.*

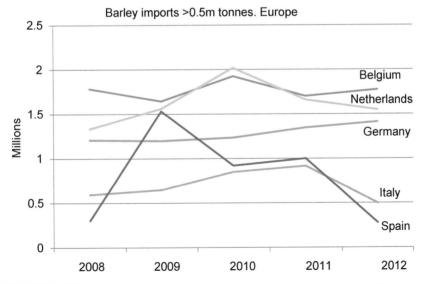

FIGURE 2.132 Variation in barley and products imports by main importing European countries (tonnes). *Based on data from FAOSTAT.*

2.4.2 Asia

Production and yield

Turkey is the main producer of barley in Asia (Fig. 2.133), with production similar to that in Spain, Ukraine and the United Kingdom in Europe, and it has matched these countries for consistency. It has also recorded similar yields, but China's yields have exceeded these (Fig. 2.134).

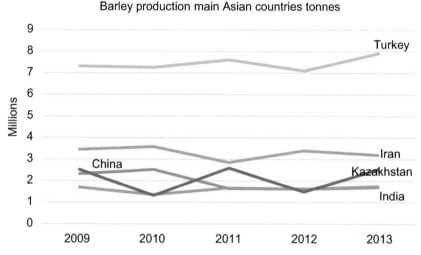

FIGURE 2.133 Variation in barley production by main producing Asian countries (tonnes). *Based on data from FAOSTAT.*

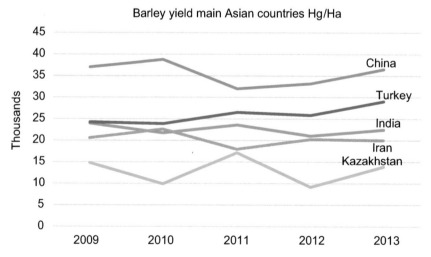

FIGURE 2.134 Variation in barley yields in main producing Asian countries (tonnes). *Based on data from FAOSTAT.*

Domestic uses

Uses for the whole of Asia are similar to those of the entire world (see Fig. 2.123), but the Asian country with the second-highest domestic supply uses almost all of this for feed (Fig. 2.135).

Exports

Exports by Asian countries are small and variable (Fig. 2.136).

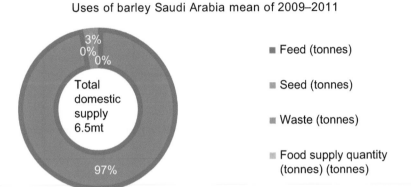

FIGURE 2.135   Reported proportional domestic uses of barley in Saudi Arabia (mean of 2009–11). *Based on data from FAOSTAT.*

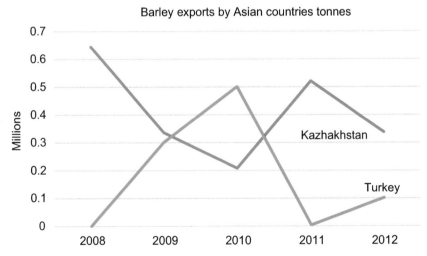

FIGURE 2.136   Variation in barley exports by main exporting Asian countries (tonnes). *Based on data from FAOSTAT.*

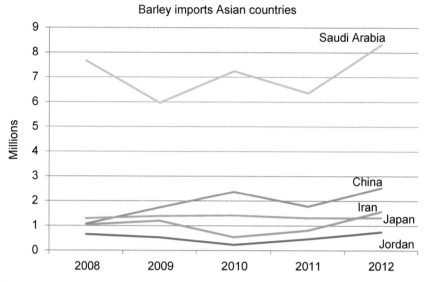

FIGURE 2.137   Variation in barley and products imports by main importing Asian countries (tonnes). *Based on data from FAOSTAT.*

Imports

By contrast with exports, imports are relatively high, particularly in the case of Saudi Arabia, where home production has been only around 20,000 tonnes (Fig. 2.137).

2.4.3 Americas

Production and yield

Canada is the leading producer of barley in the American region, producing a total similar to that of Turkey in Asia. Argentina and the United States (Fig. 2.138) each produced approximately half the total of Canada, and yields in these three main American producers were similar and equal to around two-thirds of the best in Europe (Fig. 2.139).

Domestic uses

The proportion of the domestic supply of barley in the Americas used for processing is the highest for any continental region. Unusually, it exceeds feed usage (Fig. 2.140). The extreme processing usage applies to the United States (Fig. 2.141), but Argentina's processing proportion is also high (Fig. 2.142). Canada's proportional uses are more like the rest of the world, with a high proportion being devoted to feed (Fig. 2.143).

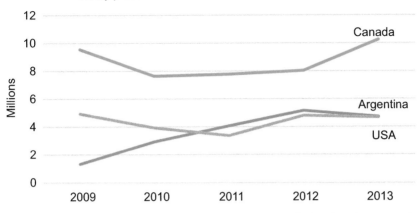

FIGURE 2.138 Variation in barley production by main producing American countries (tonnes). *Based on data from FAOSTAT.*

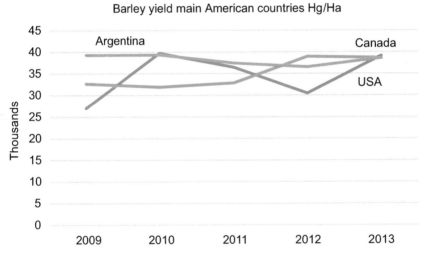

FIGURE 2.139 Variation in barley yield by main producing American countries (tonnes). *Based on data from FAOSTAT.*

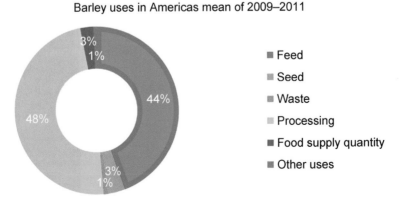

FIGURE 2.140 Reported proportional domestic uses of barley in Americas (mean of 2009–11). *Based on data from FAOSTAT.*

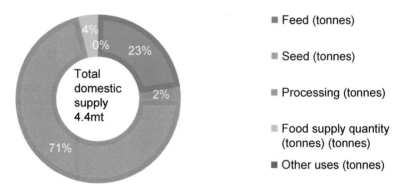

FIGURE 2.141   Reported proportional domestic uses of barley and products by the United States (mean of 2009–11). *Based on data from FAOSTAT.*

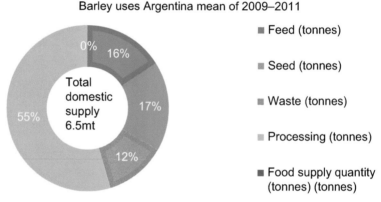

FIGURE 2.142   Reported proportional domestic uses of barley and products by Argentina (mean of 2009–11). *Based on data from FAOSTAT.*

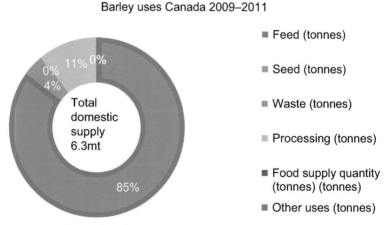

FIGURE 2.143   Reported proportional domestic uses of barley in Canada (mean of 2009–11). *Based on data from FAOSTAT.*

Exports

The three major producers are responsible for most of the export trade in the region but the consistency of performance is very different, with the United States exporting small quantities consistently and Canada and Argentina exporting greater quantities but with high levels of variation (Fig. 2.144).

Details of the Canadian grading system are published by the Canadian Grains Commission as Chapter 6 Barley, Official Grain Grading Guide (www.grainscanada.gc.ca/oggg-gocg/06/oggg-gocg-6e-eng.htm). Three classes are defined – malting, food and general purpose – and four types: covered, hull-less, two-row and six-row. Primary grade requirements are set for Canada Western (CW) and Canada Eastern (CE) malting grades, including limits (several are nil) for various types of damage (adherent hulls, fireburnt, frost, fusarium, heated, rotted and severely mildewed) and foreign materials: ergot, excreta, sclerotinia, stones, inseperable seeds and large oil-bearing seeds.

Details of United States grades are accessible at www.gipsa.usda.gov/fgis/standards/810barley97.pdf.

In summary, the term barley does not include hull-less barley and black barley. Two classes with three subclasses in each are recognized: malting barley (subclasses six-rowed malting barley, six-rowed blue malting barley and two-rowed malting barley) and barley (subclasses six-rowed barley, two-rowed barley and barley). For the six-rowed malting barleys there are four classes based on minima for test weight, suitable malting types and sound barley, and maxima for damaged kernels, foreign materials, other grains, skinned and broken kernels and thin barley. For the two-rowed malting barleys there

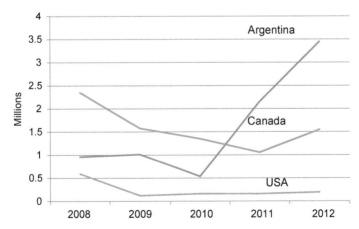

FIGURE 2.144 Variation in barley and products exports by main exporting American countries (tonnes). *Based on data from FAOSTAT.*

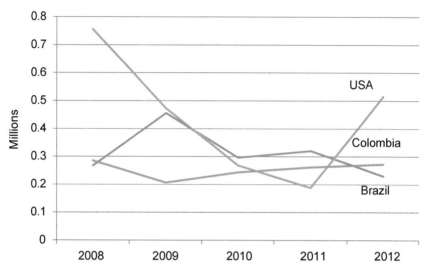

FIGURE 2.145   Variation in imports of barley and products by main importing American countries (tonnes). *Based on data from FAOSTAT.*

are also four classes based on similar minima, but a maximum for wild oats replaces maxima for other grains and damaged kernels.

The barley class has five grades plus a sample grade based on minima for test weight and sound barley and maxima for damaged kernels, heat-damaged kernels, foreign material, broken kernels and thin barley.

Imports

Import levels of barley by American countries have been low (Fig. 2.145).

2.4.4   Oceania

Production and yield

Africa and Oceania produced similar quantities of barley between 2003 and 2013, but the distribution of production was very different. In Africa several countries produce small amounts, while in Oceania the vast majority is produced by a single country, Australia (Fig. 2.146) The only other Oceanian producer, New Zealand, grew less than 0.5 million tonnes on average, while Australia's average was between 7 and 8 million. Nevertheless, yields in New Zealand were three times as high as those recorded for Australia (Fig. 2.147).

Domestic uses

Australian uses of its domestic supply have been dominated by feed, with under 10% used for processing (Fig. 2.148).

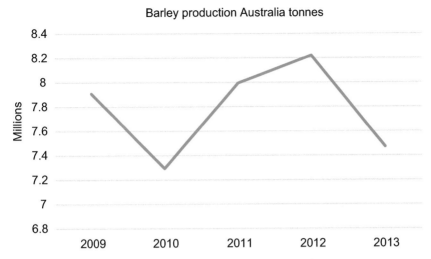

FIGURE 2.146 Variation in barley production by Australia (tonnes). *Based on data from FAOSTAT.*

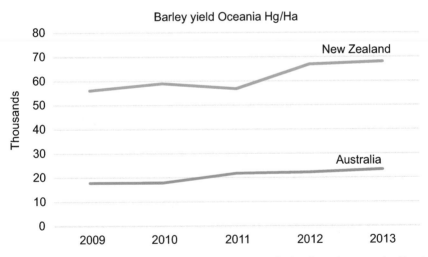

FIGURE 2.147 Variation in yields of barley in main producing Oceanian countries. *Based on data from FAOSTAT.*

Exports

It follows from the distribution of production in Oceania that only Australia can have been a substantial exporting country, and its exports equated to more than half of its production between 2009 and 2013 (Fig. 2.149).

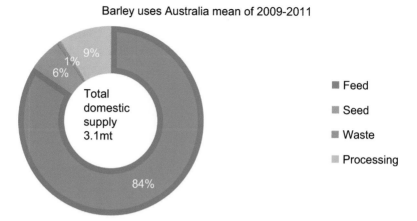

FIGURE 2.148   Reported proportional domestic uses of barley and products in Australia (mean of 2009–11). *Based on data from FAOSTAT.*

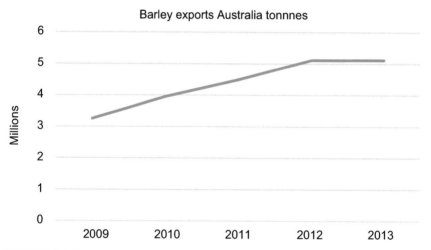

FIGURE 2.149   Variation in barley and products exports by Australia (tonnes). *Based on data from FAOSTAT.*

Imports

New Zealand was the largest barley-importing country, but levels have been low and variable (Fig. 2.150).

2.4.5 Africa

Production and yield

Three African countries have dominated barley production in the region, with Morocco growing the most (Fig. 2.151). However, yields in

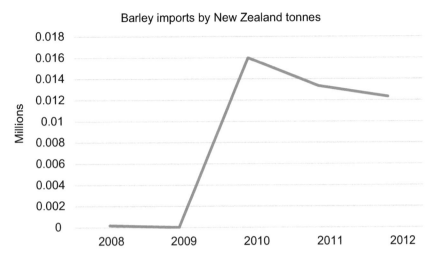

FIGURE 2.150 Variation in barley and products imports by New Zealand (tonnes). *Based on data from FAOSTAT.*

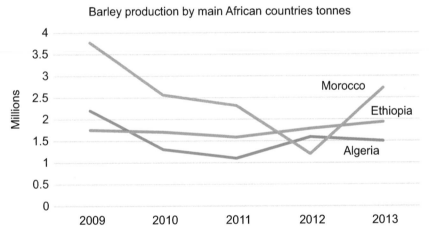

FIGURE 2.151 Variation in barley production by main producing African countries (tonnes). *Based on data from FAOSTAT.*

that country have been more variable than in the other two producing countries, Ethiopia and Algeria (Fig. 2.152).

Domestic uses

Uses reported in Africa are substantially different from those in other continental regions, as in Africa food uses are greater than feed or processing uses (Fig. 2.153). Algeria reported the lowest proportion of food usage (Fig. 2.154) and Ethiopia the highest (Fig. 2.155). In Nigeria and South Africa nearly all available barley was used for processing.

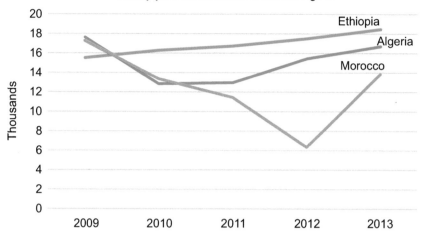

FIGURE 2.152 Variation in barley yields in main producing African countries. *Based on data from FAOSTAT.*

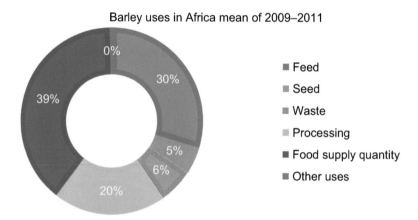

FIGURE 2.153 Reported proportional domestic uses of barley and products in Africa (mean of 2009–11). *Based on data from FAOSTAT.*

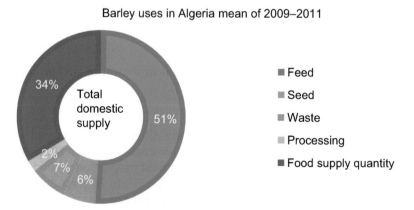

FIGURE 2.154 Reported proportional domestic uses of barley and products in Algeria (mean of 2009–11). *Based on data from FAOSTAT.*

Exports

The main exporting countries in Africa are not among the major producers, and quantities traded are very small (Fig. 2.156).

Imports

Total imports for Africa in 2012 were around 1.6 million tonnes, but no individual country imported as much as 0.5 million tonnes on average in the 5 years up to that year (Fig. 2.157) and importing records were variable (Fig. 2.158).

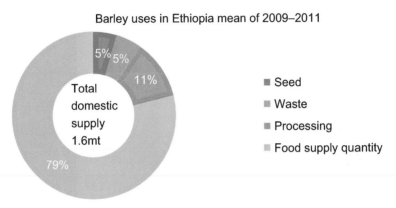

FIGURE 2.155 Reported proportional domestic uses of barley and products in Ethiopia (mean of 2009–11). *Based on data from FAOSTAT.*

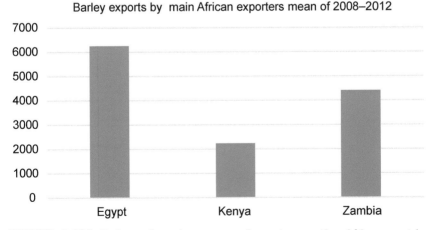

FIGURE 2.156 Barley and products exports by main exporting African countries (tonnes), mean of 2008–12. *Based on data from FAOSTAT.*

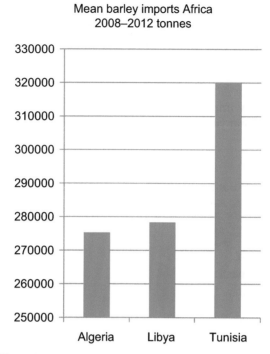

FIGURE 2.157   Barley and products imports by main importing African countries (tonnes), mean of 2008–12. *Based on data from FAOSTAT.*

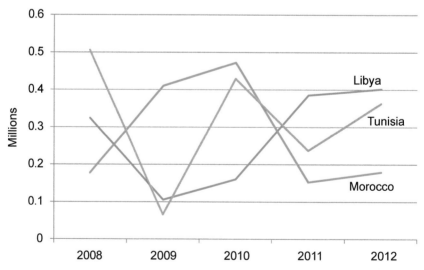

FIGURE 2.158   Variation in barley and products imports by main importing African countries (tonnes). *Based on data from FAOSTAT.*

2.5 SORGHUM

In 2014 global production of sorghum reached around 68 million tonnes. Of this Africa produced around 29 million tonnes, marginally more that the Americas, whose output had been similar to that of Africa for several years up to that date (see Fig. 1.43, Chapter 1).

2.5.1 Africa

Production and yield

Production quantities showed considerable year-on-year variation, and this was particularly marked in Sudan (Fig. 2.159). This resulted from variations in both yield and area harvested (Figs. 2.160 and 2.161).

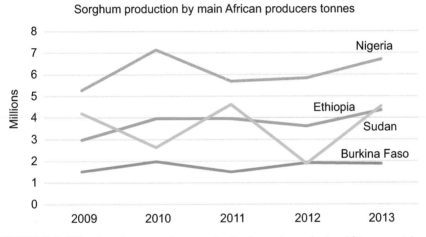

FIGURE 2.159 Variation in sorghum production by main producing African countries (tonnes). *Based on data from FAOSTAT.*

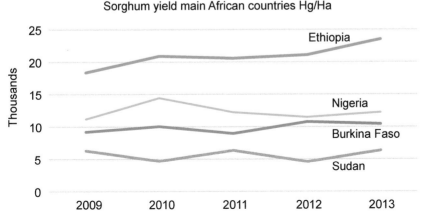

FIGURE 2.160 Variation in yields of sorghum in main producing African countries (hg/ha). *Based on data from FAOSTAT.*

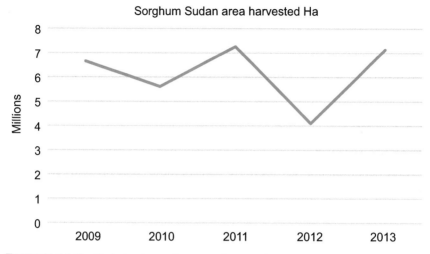

FIGURE 2.161 Variation in sorghum area harvested in Sudan (ha). *Based on data from FAOSTAT.*

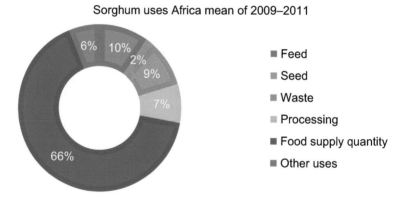

FIGURE 2.162 Reported proportional domestic uses of sorghum and products in Africa (mean of 2009–11). *Based on data from FAOSTAT.*

Domestic uses

Throughout Africa the main use of sorghum has been as food (Fig. 2.162).

Exports

Export quantities have been low and variable (Fig. 2.163).

Imports

While many African countries have imported small amounts, Ethiopia was the largest importer between 2009 and 2012. Considerable variation is evident (Fig. 2.164).

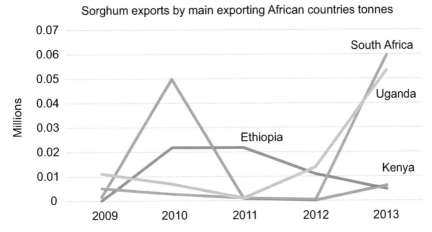

FIGURE 2.163 Variation in exports of sorghum and products by main exporting African countries (tonnes). *Based on data from FAOSTAT.*

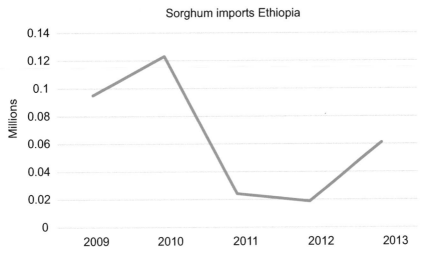

FIGURE 2.164 Variation in sorghum and products imports by Ethiopia (tonnes). *Based on data from FAOSTAT.*

2.5.2 Americas

Production and yield

Although the United States has been responsible for the highest levels of production within the Americas, levels have varied year on year. Production in the main South American countries has been more consistent (Fig. 2.165). Yields have also shown variations (Fig. 2.166).

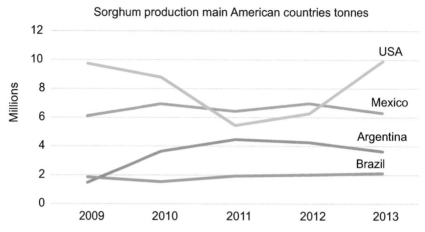

FIGURE 2.165 Variation in sorghum production main producing American countries (tonnes). *Based on data from FAOSTAT.*

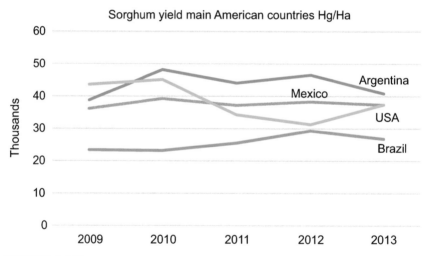

FIGURE 2.166 Variation in yields of sorghum in main producing American countries (hg/ha). *Based on data from FAOSTAT.*

In the United States sorghum is grown mainly in the Great Plains area, chiefly in Texas, where it is the most important cereal, and Kansas.

Domestic uses

In contrast to African priorities, the main use of sorghum in South and Central America is as feed. All the main producers used more than 95% of domestic supply for this purpose (Fig. 2.167). As well as being by far the greatest user, Mexico also devoted the highest proportion to feed. Uses of

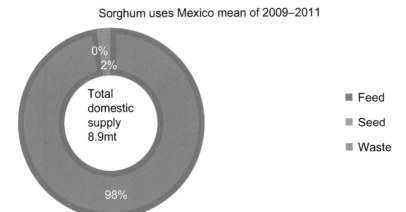

FIGURE 2.167 Reported proportional domestic uses in Mexico (mean of 2009–11). *Based on data from FAOSTAT.*

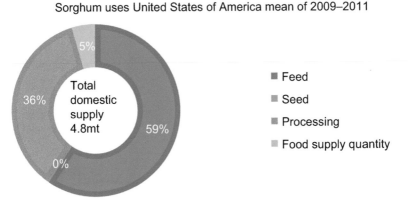

FIGURE 2.168 Reported proportional domestic uses of sorghum in United States (mean of 2009–11). *Based on data from FAOSTAT.*

sorghum in the United States differ from southern American uses in the amount devoted to processing (Fig. 2.168).

Exports

Sorghum exports have not reached high levels and the main exporters have shown considerable variation (Fig. 2.169).

Imports

In spite of being the second-largest grower in the Americas, Mexico has been the major importer of sorghum in the region (Fig. 2.170).

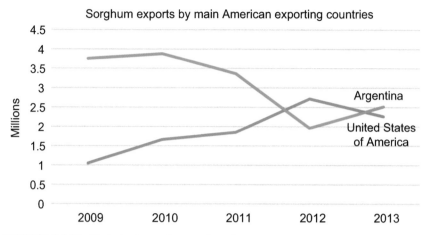

FIGURE 2.169   Variation in sorghum and products exports by main exporting American countries (tonnes). *Based on data from FAOSTAT.*

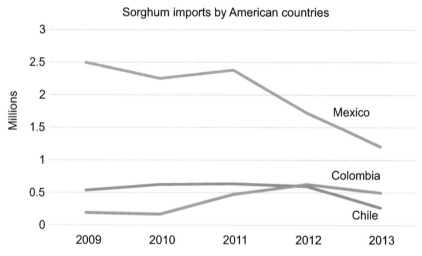

FIGURE 2.170   Variation in sorghum and products imports by main importing American countries (tonnes). *Based on data from FAOSTAT.*

2.5.3 Asia

Production and yield

Of the 9.5 million tonnes produced throughout Asia in 2013, India produced 5.3 million tonnes and mainland China 2.9 million tonnes. In the 5-year period up to that date India's production fell and that of China increased (Fig. 2.171). Yields in both of the major producing countries

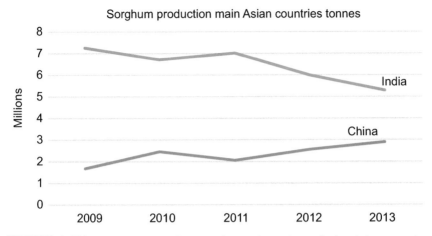

FIGURE 2.171 Variation in sorghum production by main producing Asian countries (tonnes). *Based on data from FAOSTAT.*

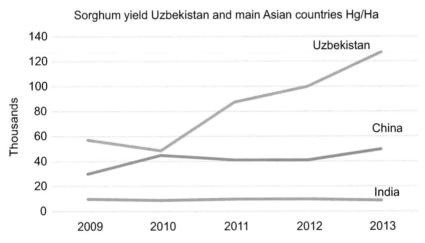

FIGURE 2.172 Variation in sorghum yields in Uzbekistan and main producing Asian countries. *Based on data from FAOSTAT.*

remained fairly consistent, but that of a minor producer, Uzbekistan, is reported to have doubled (Fig. 2.172).

Domestic uses

In India sorghum has been used mainly as food, while in mainland China feed and food uses were approximately equal. In Japan only use as feed is reported (Figs. 2.173–2.175).

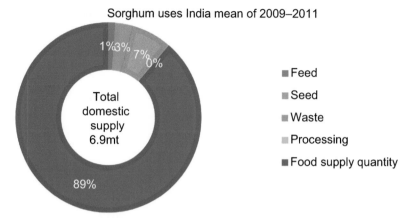

FIGURE 2.173   Reported proportional domestic uses of sorghum and products in India (mean of 2009–11). *Based on data from FAOSTAT.*

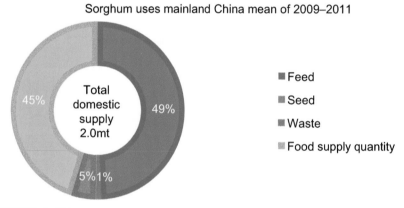

FIGURE 2.174   Reported proportional domestic uses of sorghum and products in mainland China (mean of 2009–11). *Based on data from FAOSTAT.*

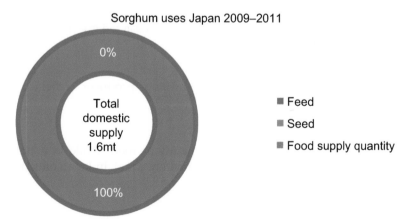

FIGURE 2.175   Reported proportional domestic uses of sorghum and products in Japan (mean of 2009–11). *Based on data from FAOSTAT.*

Exports

Exports of sorghum by Asian countries are low. The largest exporters are the largest producers, and while China's performance has been fairly consistent, that of India has varied widely (Fig. 2.176).

Imports

Japan is the largest Asian importer of sorghum. Sorghum is not grown in Japan, so its entire domestic supply is imported (Fig. 2.177).

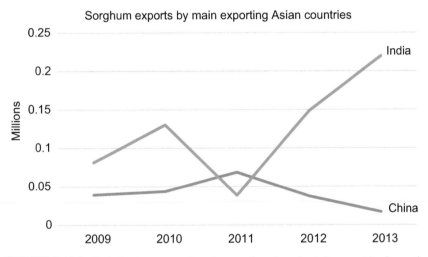

FIGURE 2.176 Variation in exports of sorghum and products by Asian countries (tonnes). *Based on data from FAOSTAT.*

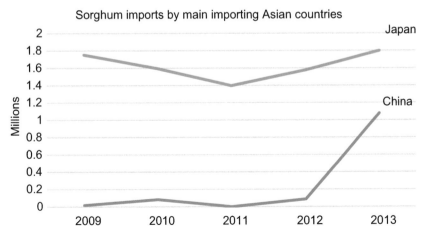

FIGURE 2.177 Variation in sorghum and products imports by main importing Asian countries (tonnes). *Based on data from FAOSTAT.*

2.5.4 Europe

Production and yield

Sorghum is a very minor cereal crop in Europe, with no country producing as much as 0.5 million tonnes. The marked increase in production by Ukraine (Fig. 2.178) resulted from a substantial (approximately sixfold) increase in area harvested.

Domestic uses

Feed use accounts for 99% of sorghum and products in Europe (Fig. 2.179).

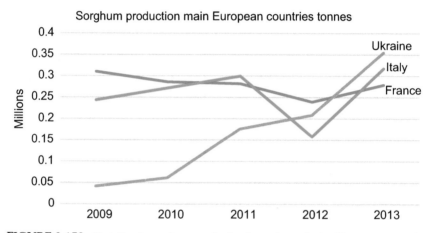

FIGURE 2.178 Variation in sorghum production by main producing European countries (tonnes). *Based on data from FAOSTAT.*

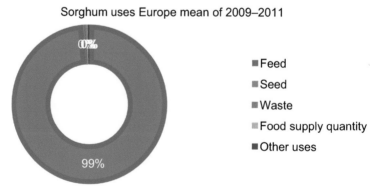

FIGURE 2.179 Reported proportional domestic uses of sorghum and products in Europe (mean of 2009–11). *Based on data from FAOSTAT.*

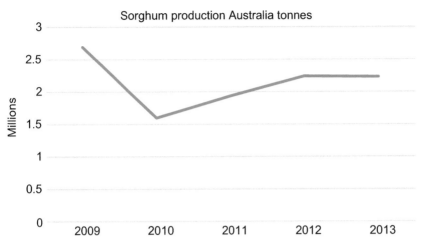

FIGURE 2.180 Variation in sorghum production by Australia (tonnes). *Based on data from FAOSTAT.*

Exports and imports

No substantial exports or imports were recorded, although several countries exported small quantities. The same applies to import quantities.

2.5.5 Oceania

Production and yield

As is true of several other cereals, the only significant producer of sorghum in the region is Australia. Production has been variable (Fig. 2.180).

Domestic uses

Uses of sorghum and products are similar to those recorded for Europe.

2.6 MILLET

2.6.1 Asia

Production and yield

More than half of the world's millet production occurred in Asia, where, as with several other cereals, India and mainland China dominate production (Fig. 2.181). Though substantially lower than India in terms of production, China has recorded higher yields (Fig. 2.182).

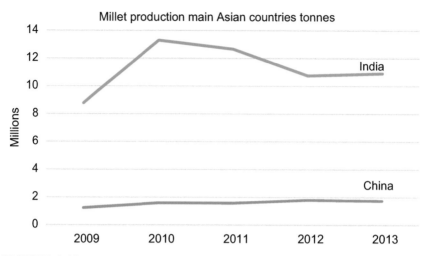

FIGURE 2.181   Variation in millet production by main producing Asian countries (tonnes). *Based on data from FAOSTAT.*

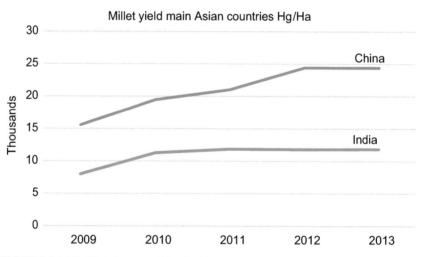

FIGURE 2.182   Variation in yields of millet in main producing Asian countries. *Based on data from FAOSTAT.*

Domestic uses

Millet and products are important as food in India, where large quantities are available and food is the principal use (Fig. 2.183). In China half the available supply is used as feed (Fig. 2.184).

Millet uses in India mean of 2009–2013

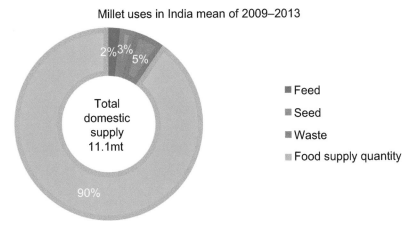

FIGURE 2.183 Reported proportional domestic uses of millet and products in India (mean of 2009–13). *Based on data from FAOSTAT.*

Millet uses in China mean of 2009–2013

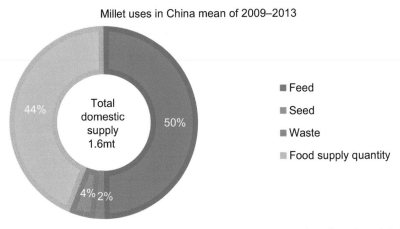

FIGURE 2.184 Reported proportional domestic uses of millet and products in mainland China (mean of 2009–13). *Based on data from FAOSTAT.*

Exports

Exports account for little of the supply of millet in Asia (Fig. 2.185).

Imports

Imports are low in quantity throughout Asia. The main countries importing millet and products and the quantities involved are shown in Table 2.1.

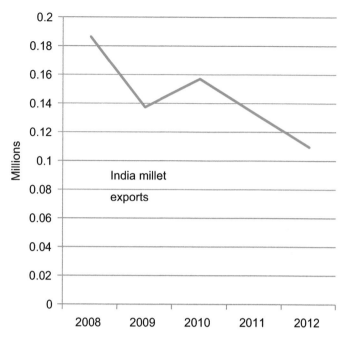

FIGURE 2.185   Variation in millet exports by India (tonnes). *Based on data from FAOSTAT.*

TABLE 2.1   Millet imports by Asian countries (mean of 2009–13)

Importing country	Yemen	Republic of Korea	Nepal
Import quantity (tonnes)	16,666	15,885	14,634

Data from FAOSTAT.

2.6.2 Africa

Production and yield

Nigeria is the main African producer of millet (Fig. 2.186). Yield in most producing countries has varied year on year, accounting for most of the variation in production (Fig. 2.187).

Domestic uses

Uses in the main millet consuming African countries are similar to those for the region (Fig. 2.188).

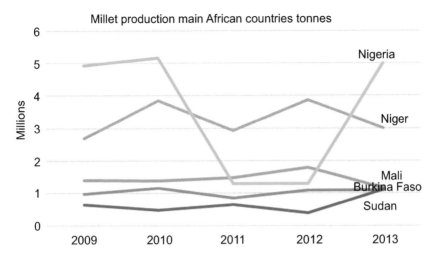

FIGURE 2.186 Variation in millet production by main producing African countries (tonnes). *Based on data from FAOSTAT.*

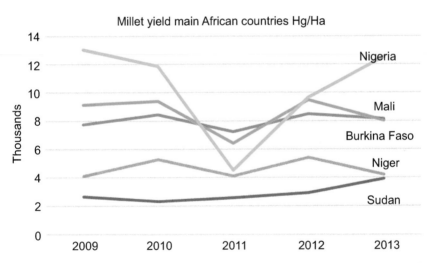

FIGURE 2.187 Variation in yields of millet in main producing African countries (tonnes). *Based on data from FAOSTAT.*

Exports and imports

No African country exported annual average millet quantities above 2000 tonnes between 2009 and 2013.

The largest importers between 2009 and 2013 were Kenya and South Africa, with average import quantities between 5000 and 1 million tonnes.

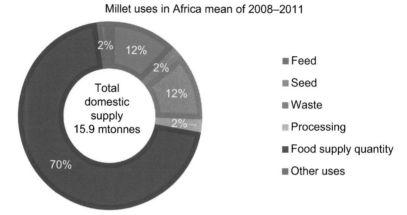

FIGURE 2.188 Reported proportional domestic uses of millet and products in Africa (mean of 2008–11). *Based on data from FAOSTAT.*

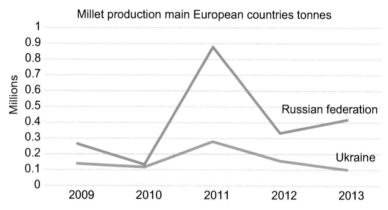

FIGURE 2.189 Variation in millet production by main producing European countries (tonnes). *Based on data from FAOSTAT.*

2.6.3 Europe

Production and yield

Millet production in Europe has contributed only 3% of world production and averaged under 1 million tonnes between 2009 and 2013.

The main producers are the Russian Federation and Ukraine (Fig. 2.189). Yields are comparably variable in the two countries (Fig. 2.190).

Domestic uses

The majority of millet in Europe has been used as feed. The proportions for the whole region are similar for the Russian Federation (Fig. 2.191). In Ukraine equal amounts were used as food and feed (Fig. 2.192).

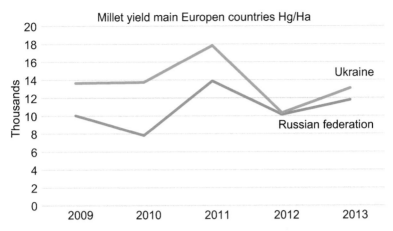

FIGURE 2.190 Variation in yields of millet in main producing European countries (hg/ha). *Based on data from FAOSTAT.*

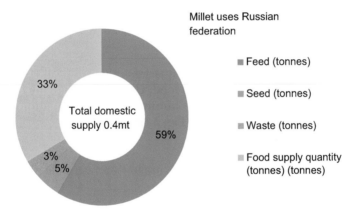

FIGURE 2.191 Reported proportional domestic uses of millet in Europe (mean of 2009–13). *Based on data from FAOSTAT.*

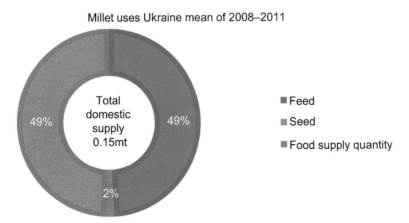

FIGURE 2.192 Reported proportional domestic uses of millet in Europe (mean of 2008–11). *Based on data from FAOSTAT.*

Exports and imports

No trades above 50,000 tonnes took place between 2009 and 2013.

2.7 OATS

2.7.1 Europe

Production and yield

Europe is responsible for producing more than 60% of the world's oat supply, and together with the Americas accounts for nearly 90%. Production was relatively stable between 2010 and 2014.

Within Europe, the Russian Federation contributed around one-third of total production, with few other countries producing in excess of 1 million tonnes (Fig. 2.193).

The highest yields have been achieved consistently in the United Kingdom, where they have been four to five times those of the main producer (Fig. 2.194).

Domestic uses

In Europe as a whole about three-quarters of the domestic supply has been used as feed, with only 10% as food. Uses in the Russian Federation were somewhat similar (Fig. 2.195) but in the United Kingdom a larger proportion was consumed as food (Fig. 2.196).

Exports

As with other minor cereals, exports are low and variable (Fig. 2.197).

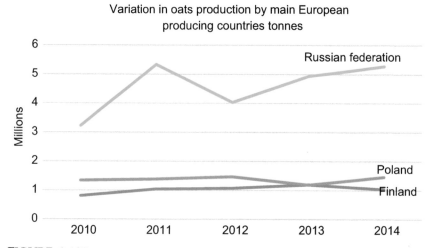

FIGURE 2.193 Variation in oats production by main producing European countries (tonnes). *Based on data from FAOSTAT.*

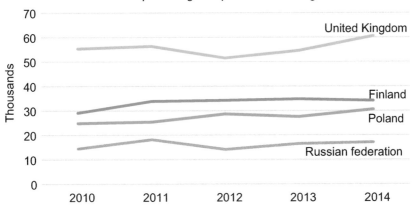

FIGURE 2.194 Variation in yields of oats by United Kingdom and main producing European countries (hg/ha). *Based on data from FAOSTAT.*

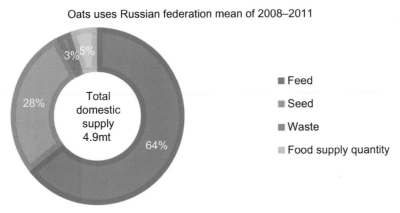

FIGURE 2.195 Reported proportional domestic uses of oats in the Russian Federation (mean of 2008–11). *Based on data from FAOSTAT.*

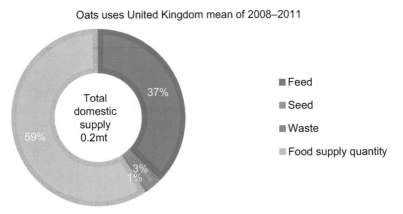

FIGURE 2.196 Reported proportional domestic uses of oats in the United Kingdom (mean of 2008–11). *Based on data from FAOSTAT.*

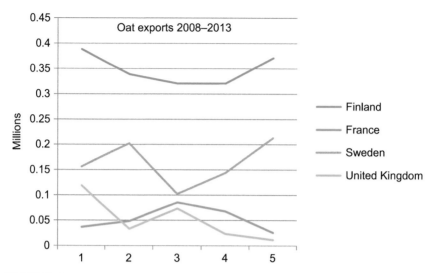

FIGURE 2.197   Variation in oats and products exports by main European exporting countries (tonnes), 2008–13. *Based on data from FAOSTAT.*

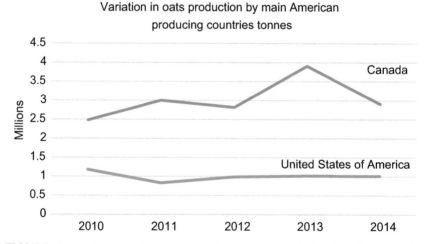

FIGURE 2.198   Variation in oats production by main producing American countries (tonnes). *Based on data from FAOSTAT.*

Imports

Total imports for Europe are around 100,000 tonnes.

2.7.2 Americas

Production and yield

Canada is the largest producer in the Americas (Fig. 2.198) but the highest yields are achieved in Chile, where they approach the highest in Europe (Fig. 2.199).

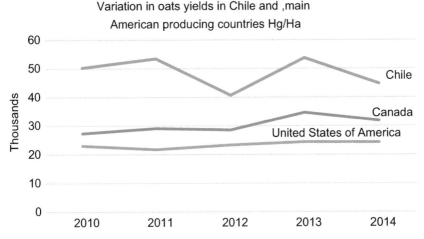

FIGURE 2.199 Variation in oats yields in Chile and main producing American countries (hg/ha). *Based on data from FAOSTAT.*

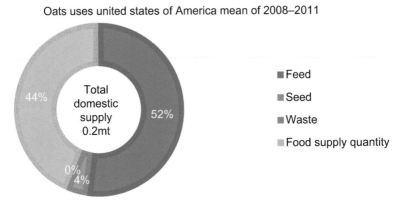

FIGURE 2.200 Reported proportional domestic uses of oats in the United States (mean of 2008–11). *Based on data from FAOSTAT.*

Domestic uses

Feed constitutes the almost exclusive use in Canada (domestic supply 1.2 mt), while in Mexico (domestic supply 7000 mt) around one-quarter of supply was consumed as food. In the United States consumption as food used half the supply (Fig. 2.200).

Exports and imports

Canadian Grain Commission standards (www.grainscanada.gc.ca/oggg-gocg/07/oggg-gocg-7e-eng.htm) refer to Canada Western and Canada Eastern oats. Four grades (CW1–4 and CE 1–4) are defined according to minimum test weight and soundness and maximum proportions of hulled and hull-less, damage (fireburnt, frost, fusarium and heated/rotted), other cereal grains and wild oats (four types plus total), large seeds, sclerotinia, stones, ergot and excreta.

The United States distinguishes four grades (US 1–4) of oats based on minimum test weight and sound oats and maximum proportions of heat-damaged kernels, foreign material and wild oats (www.gipsa.usda.gov/fgis/standards/810oats.pdf).

Canadian exports and United States imports have followed similar trends (Figs. 2.201 and 2.202).

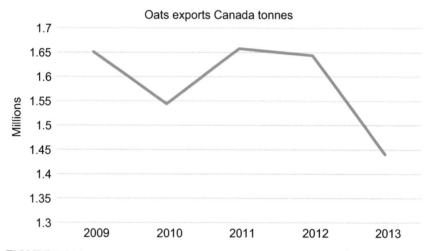

FIGURE 2.201   Variation in oats and products exports by Canada (tonnes). *Based on data from FAOSTAT.*

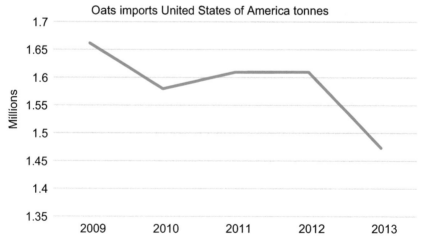

FIGURE 2.202   Variation in oats and products imports by United States (tonnes). *Based on data from FAOSTAT.*

2.7.3 Oceania

Australia is the largest producer in the region, but oats represent only 3% of total cereals production (mean of 2010–14).

2.8 RYE

The majority of rye production (around 90%) occurred in Europe and much of the remainder was in Asia (Fig. 2.203).

Yields were also highest in these continental regions (Fig. 2.204).

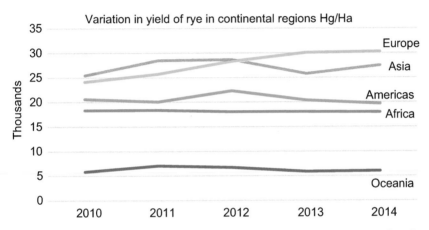

FIGURE 2.203 Variation in yields of rye in continental regions (hg/ha). *Based on data from FAOSTAT.*

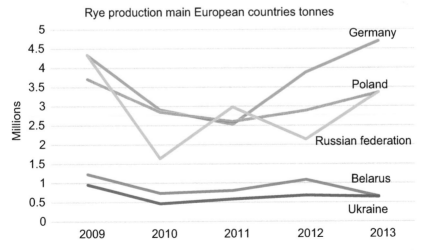

FIGURE 2.204 Variation in rye production by main producing European countries (tonnes). *Based on data from FAOSTAT.*

2.8.1 Europe

Production and yield

Production by the main European producers has been erratic (Fig. 2.203). Yields reported among European countries were very different, with Germany's far exceeding those of other producers, but annual variation in yield contributed significantly to the erratic production (Fig. 2.205).

Domestic uses

In Europe as a whole, main uses of rye were shared between food and feed (Fig. 2.206). Both world uses and uses by the largest user, the

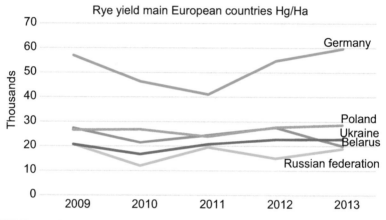

FIGURE 2.205   Variation in rye yields in main producing European countries (hg/ha). *Based on data from FAOSTAT.*

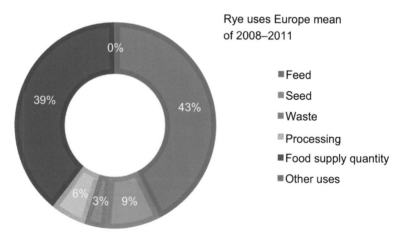

FIGURE 2.206   Reported proportional domestic uses of rye in Europe (mean of 2008–11). *Based on data from FAOSTAT.*

Russian Federation (total domestic supply 3.4 million tonnes), are similar to those shown for Europe. Germany used a greater proportion as feed and Ukraine used a greater proportion for food, but Ukraine is not a major user (Figs. 2.207 and 2.208).

Exports and imports

Export and import quantities between 2009 and 2013 were small and most trade involved European countries (Fig. 2.209). Within Europe the largest net exporter was Poland and the largest net importer was the

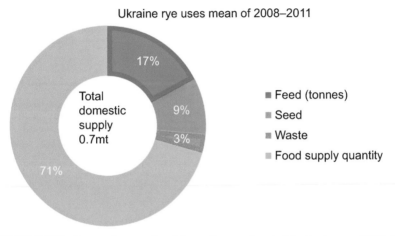

FIGURE 2.207 Reported proportional domestic uses of rye in Ukraine (mean of 2008–11). *Based on data from FAOSTAT.*

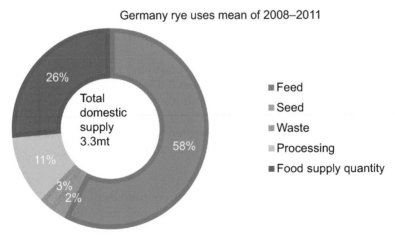

FIGURE 2.208 Reported proportional domestic uses of rye in Germany (mean of 2008–11). *Based on data from FAOSTAT.*

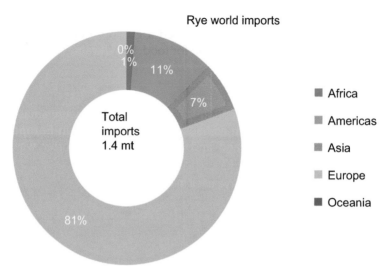

Rye world imports

Total imports 1.4 mt

0%
1%
11%
7%
81%

■ Africa
■ Americas
■ Asia
■ Europe
■ Oceania

FIGURE 2.209 Proportional contributions to world rye imports by continental regions (mean of 2009–13). *Based on data from FAOSTAT.*

Netherlands. Germany's exports were almost balanced by imports. No country's trade exceeds 0.5 million tonnes.

2.9 TRITICALE

Production and yield

Europe is the continental region from which the largest share (around 90%) of world production comes. Asia contributes the largest share of the remainder (see Fig. 1.54). An upward trend describes the change in production between 2010 and 2014 (Fig. 2.210).

Four European countries together produced more than 70% of the world's triticale stock between 2009 and 2013 (Fig. 2.211).

In Oceania and the Americas quantities grown declined progressively between 2010 and 2014 (Fig. 2.212).

Yields are highest and most consistent in Europe, but differences in yields are not as marked as they are with many cereals (Fig. 2.213).

Domestic uses

Accredited reports of uses of triticale throughout the world are currently unavailable. However, it is generally understood that although there are many small-scale specialist triticale products for human consumption, its main application is as feed.

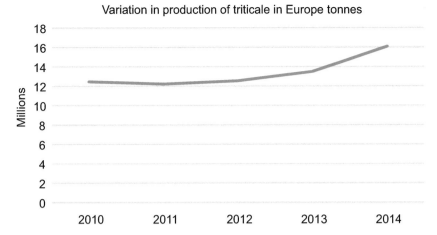

FIGURE 2.210 Variation in triticale production in Europe (tonnes). *Based on data from FAOSTAT.*

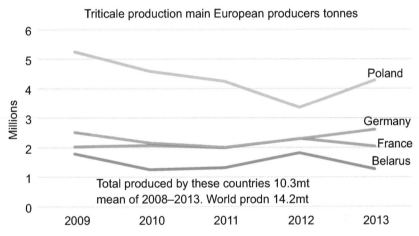

FIGURE 2.211 Variation in triticale production by main European countries (tonnes). *Based on data from FAOSTAT.*

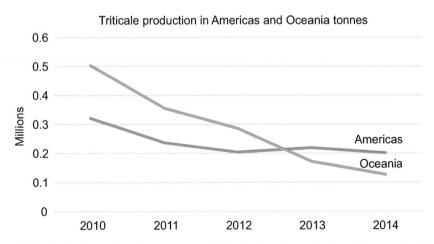

FIGURE 2.212 Variation in triticale production in Americas and Oceania (tonnes). *Based on data from FAOSTAT.*

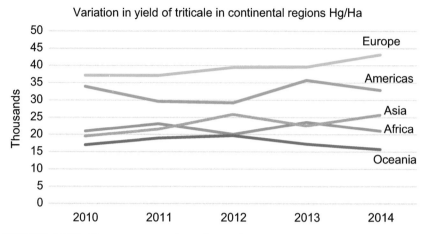

FIGURE 2.213   Variation in triticale production by main producing European countries (tonnes). *Based on data from FAOSTAT.*

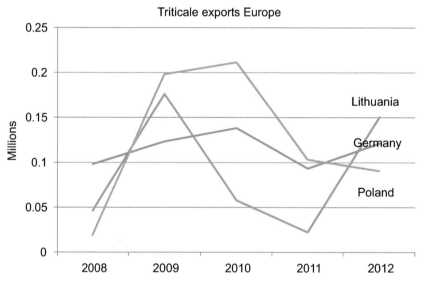

FIGURE 2.214   Variation in triticale and products exports by main exporting European countries (tonnes). *Based on data from FAOSTAT.*

Exports and imports

In 2012 world trade in triticale was around 0.5 million tonnes and most of this was in Europe. Export records by individual countries showed high levels of variation (Fig. 2.214).

2.10 BUCKWHEAT

Production and yield

Europe is the continental region from which the largest share of world production comes. Asia contributes another substantial share and the Americas produce a small amount (see Fig. 1.57). Production performance of these continental regions between 2010 and 2014 are shown in Fig. 2.215. The main producing countries have been the Russian Federation, China, Ukraine and France, but none produced as much as 1 million tonnes between 2009 and 2013 (Fig. 2.217).

Yields are highest and most consistent in Americas, but differences in yields are not as marked as they are with many cereals (Fig. 2.216).

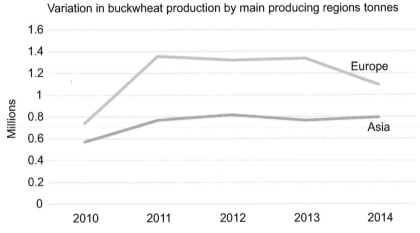

FIGURE 2.215 Variation in buckwheat production by main producing regions (tonnes). *Based on data from FAOSTAT.*

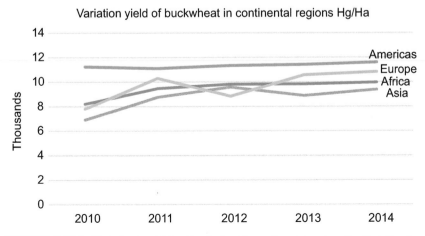

FIGURE 2.216 Variation in yields of buckwheat in continental regions (hg/ha). *Based on data from FAOSTAT.*

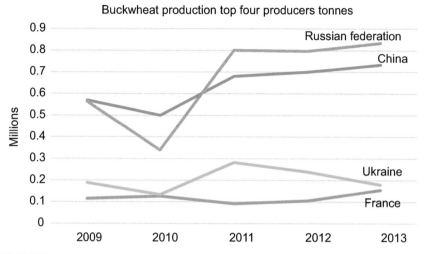

FIGURE 2.217   Variation in buckwheat production by main producing countries (tonnes). *Based on data from FAOSTAT.*

2.11 QUINOA

Production and yield

Quinoa is grown in many parts of the world on a very small scale, but production is both increasing and expanding. However, most is grown in the Andes of South America, where it originated. Thus 99% of the world crop is grown in Bolivia and Peru (Fig. 2.218). Reported world production equates to that reported for the South American countries. Quantities have been very small compared with cereals but an upward trend is apparent (Fig. 2.219).

Domestic uses

Quinoa has been promoted as human food, for which its nutritious and health-promoting properties have been emphasized. The fact that it is produced mainly by organic methods also appeals to the affluent consumers to whom marketing is directed.

Exports

Although quantities and value are small compared with major cereals, export trade has increased rapidly in recent years. In 2002 1000 tonnes was traded outside the production areas, whereas in 2012 the equivalent figure was 37,000 tonnes; 75% of exports were from Bolivia, and Peru furnished a further 23%.

Proportional contribution to world
production of quinoa. Mean of 2009–2013

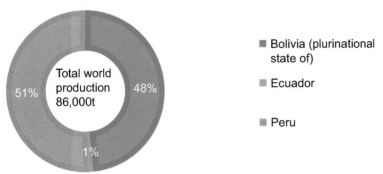

FIGURE 2.218 Reported contribution to world production by Bolivia, Peru and Ecuador. *Based on data from FAOSTAT.*

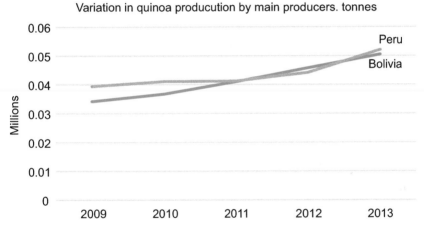

FIGURE 2.219 Variation in quinoa production by main producing countries (tonnes). *Based on data from FAOSTAT.*

Imports

The United States imports 56%, France 10%, the Netherlands 8% and Germany 6% of traded stocks, and smaller markets include other European countries as well as Canada and Australia (Furche et al., 2013).

References

Afsar, A.K., Baki, M.N., Rahman, M., Rouf, M.A., 2001. Grades, Standards and Inspection Procedures of Rice in Bangladesh. FMRSP Working Paper No. 28 Bangladesh. Food Management & Research Support Project Ministry of Food, Government of the People's Republic of Bangladesh International Food Policy Research Institute.

Alavi, H.R., 2012. Trusting Trade and the Private Sector for Food Security in Southeast Asia. The World Bank, Washington, DC, United States of America (Published in the Royal Gazette vol. 114, Section 31 D. 17 April B.E. 2540).

Furche, C., Salcedo, S., Krivonos, K., Rabczuk, P., Jara, B., Fernandez, D., Correao, F., 2013. International quinoa trade. In: Bazile, D., Boterop, D., Nieto, C. (Eds.), State of the Art Report on Quinoa Around the World in 2013. Officina Regional de la FAO para Americo Latina y el Caribe, pp. 316–329.

Gummadov, N., Keser, M., Akin, B., Cakmak, M., Mert, Z., Taner, S., Oztark, I., Topal, A., Yazar, S., Morgannov, A., 2015. Genetic gains in wheat in Turkey: winter wheat for irrigated conditions. The Crop Journal 3, 507–516.

He, Z.H., Xia, X.C., Ziang, Y., Chan, X.M., 2012. Wheat quality improvement in China: progress and prospects. In: Zonghu, H., Wang, D. (Eds.), Proceedings of the 11th International Gluten Workshop, Beijing, China, August 12–15, 2012. International Maize and Wheat Improvement Center (CIMMYT), Mexico DF, Mexico, pp. 85–90.

Krishna, K.R., 2013. Maize Agroecosystem, Nutrient Dynamics and Productivity. Apple Academic Press, Toronto, ON, USA.

Meng, E.C.H., Hu, R., Shi, X., Zhang, S., 2006. Maize in China: Production Systems, Constraints, and Research Priorities. CIMMYT, Mexico, DF, Mexico.

Swastika, D.K.S., Kasim, F., Suhariyanto, K., Sudana, W., Hendayana, R., Gerpacio, R.V., Pingali, P.L., 2004. Maize in Indonesia: Production Systems, Constraints, and Research Priorities. CIMMYT, Mexico, DF, Mexico.

Tardin, A.C., 1991. Programa de controle de qualidade para rações produzidas na granja. In: Anais do Simpósio Técnico de Produção de Ovos; Campinas, São Paulo, Brasil, pp. 50–72.

USDA, 2016. Foreign Agricultural Service Circular Series WAP 2-16 February World Agricultural Production.

Further reading

Hettiarachchy, N., Ju, Z.Y., Siebenmorgan, T., Rice, S.R.N., 2000. Production, processsing and utilization. In: Kulp, K., Ponte, J.G. (Eds.), Handbook of Cereal Science and Technology. Marcel Dekker Inc., New York, USA, pp. 203–222.

FAO, 2009. Agribusiness Handbook Barley, Malt, Beer. FAO, Rome. http://www.scielo.br/scielo.php?script=sci_arttext&pid=S1516-635X2014000300002.

Longtau, S.R., 2003. Multi-agency Partnership in West African Agriculture: A Review and Description of Rice Production Systems in Nigeria. EDO/ODI, Nigeria, Jos, Plateau State.

Popović, V., Sikora, V., Berenji, J., Filipović, V., Dolijanović, Ž., Ikanović, J., Dončić, D., 2014. Analysis of buckwheat production in the world and Serbia. Economics of Agriculture 1, 53–62.

USDA, 2015. Crop Production 2014 Summary. National Agricultural Statistics Service.

CHAPTER

3

Botanical aspects

Kent's Technology of Cereals, Fifth Edition
http://dx.doi.org/10.1016/B978-0-08-100529-3.00003-7

3.1 TAXONOMY

The names commonly applied to individual species are sometimes confusing as many different names may be used for the same plant. Worse still, the same name may be applied to more than one plant or its fruit. Theoretically the application of systematic nomenclature should remove all such confusion and, indeed, in practice it undoubtedly clarifies communications. However, experts are not agreed on all details and several systems of scientific nomenclature exist.

3.1.1 Classification

The primary objective of plant classification is the grouping of plants and populations into recognizable units with reasonably well-defined boundaries and stable names. Modern taxonomists strive to establish a phylogenetic arrangement of the taxa, based on known or presumed genetic relationships.

Living organisms are classified in a hierarchical system in which descending groupings indicate progressively closer relationships. The lowest taxonomic level to which all cereals belong is the family. A family may be divided into subfamilies, each of which is further divided into tribes. Within a tribe there may be several genera with several species within a genus. The species is the highest level at which routine natural breeding among members would be expected.

Within a species there may be several cultivars, which, if accepted by appropriate authority may be recognized as commercial varieties. At the species level binomial designation applies, the first part of the name being that of the genus, for example, 'Hordeum'. Addition of the 'specific epithet' (trivial name) 'vulgare' completes the species name: *Hordeum vulgare*. (In the case of barley, the two-rowed and four-rowed types are distinguished at the 'convar' level – an informal term that lies between species and cultivar in the taxonomic hierarchy.) It is customary to print specific names in italics or to underline them. Designation of taxonomic status is somewhat arbitrary. Those competent taxonomists responsible for establishment of species are credited by their name being suffixed to the species name, either in full or in shortened form. The most frequent name suffixed is Linnaeus or L., crediting the Swedish biologist who devised the system.

Several classifications of the Poaceae (previously Gramineae) family exist. In Fig. 3.1 a suggested family tree is shown. The diagram also indicates the photosynthetic pathway adopted by members of some of the groupings. The C3 is typical of temperate plants, and the C4 is appropriate to tropical plants.

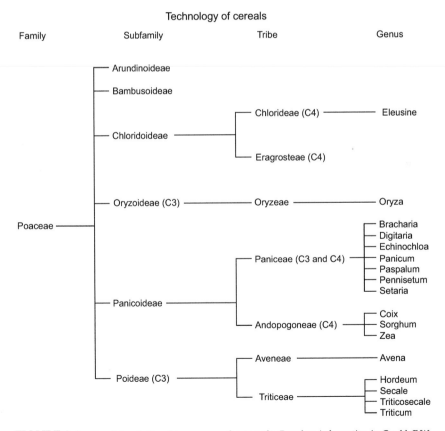

Technology of cereals

FIGURE 3.1   Possible relationships among the cereals. *Based on information in Gould, F.W., Shaw, R.B., 1983. Grass Systematics. Texas A and M University Press.*

3.2 PLANT REPRODUCTIVE MORPHOLOGY

3.2.1 The inflorescence

The generalized structures of the vegetative and reproductive parts of grasses, including their fruits, are described in Chapter 1, and, while all cereals can be related to those descriptions, significant differences among species exist. Those species that conform most closely to the generalized vegetative form depicted are rice, wheat, barley rye, triticale and oats, while the resemblance of some of the features of maize, sorghum and the millets is less readily detected.

Full botanical descriptions of most known grass species are given in Clayton et al. (2012)

The structural elements of the ear have been given names to avoid confusion, however, there is not complete consistency among authorities. What follows is widely accepted.

The stalk that bears the inflorescence is called the peduncle. Beyond the peduncle, the main axis of the inflorescence, is the rachis, from which branches may spring. Florets, containing the sexual elements of the inflorescence, occur within small clusters known as spikelets, and the axis of a spikelet is the rachilla. Florets are attached, usually alternately to the rachilla, with the first formed closest to the base. Spikelets may be attached to the rachis or its branches by a stalk known as a pedicel, but pedicels are not always present and when absent spikelets are described as sessile. Unbranched inflorescences of grasses are described as racemes or spikes, depending on whether spikelets are stalked or sessile. The elements are perhaps most readily identifiable in the lax paniculate cereals, rice, some types of sorghum and millet, and oats, in which the rachis bears groups of lateral axillary branches arising from nodes; these may bear secondary branches and tertiary branches may also be present.

3.2.1.1 Panicles

The structure of the oat panicle is shown in Fig. 3.2.

Within the spikelet of oats, two to five florets are borne alternately (Fig. 3.3); the two closest to the base are similar in size but florets become progressively smaller toward the tip. Flowering (otherwise described as 'anthesis' – when anthers emerge from between lemma and palea) occurs mainly between 2 and 5 p.m. beginning in the upper spikelets and taking 5–7 days for completion (Poehlman, 1987).

Rice inflorescences are also panicles but rice spikelets contain only one floret. Glumes are mostly insignificant small scales. Rice florets are unlike those of other cereals in having six stamens (Fig. 3.4).

Varietal differences in the number of spikelets borne on an inflorescence are dependent on variation in the number of secondary branches present, and considerable interest has been shown recently in the control of inflorescence

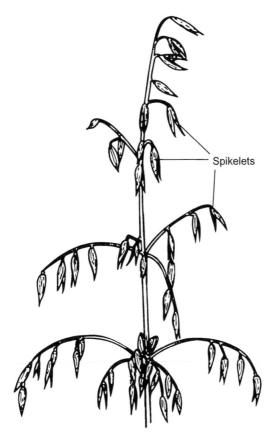

Spikelets

FIGURE 3.2 The oat panicle. *Reproduced from Poehlman, J.M., 1987. Breeding Field Crops, third ed. AVI Publishing Company Ltd., Westport, Connecticut, USA, by courtesy of AVI Publishers, New York. Permission sought from Springer via website 29.08.16.*

architecture (e.g., Gao et al., 2015) as a yield determinant. Anthesis occurs between 10 a.m. and 2 p.m. (Grubben and Partohardjono, 1996).

In sorghums, the inflorescences are panicles but the structural details vary according to cultivar group. The peduncle may be erect (Fig. 3.5 (1, 2, 4 and 5)) or recurved (Fig. 3.5 (3)). The rachis may be long or short, and primary, secondary and sometimes tertiary branches may be present, bearing racemes of spikelets. Inflorescence shape is determined by the length and closeness of branches, and may be densely packed, conical or ovoid, or spreading and lax (Grubben and Partohardjono, 1996).

Spikelets occur in pairs, one is sessile and the other borne on a short pedicel. The sessile spikelet contains two florets, one perfect and fertile and the other sterile. The pedicelled spikelet is either sterile or develops male organs only (Fig. 3.6).

Millets are an extremely diverse group; there are many different species belonging to several different tribes (see Fig. 3.1) so generalizations are hazardous. Only the most-grown species are considered here.

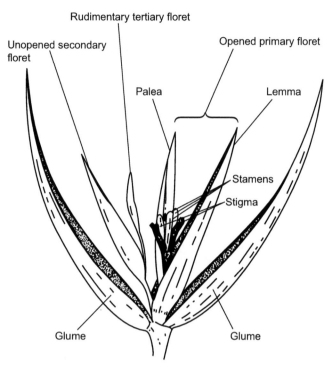

Rudimentary tertiary floret

Unopened secondary floret

Opened primary floret

Palea

Lemma

Stamens

Stigma

Glume

Glume

FIGURE 3.3   Spikelet of oat. *Reproduced from Poehlman, J.M., 1987. Breeding Field Crops, third ed. AVI Publishing Company Ltd., Westport, Connecticut, USA, by courtesy of AVI Publishers, New York.*

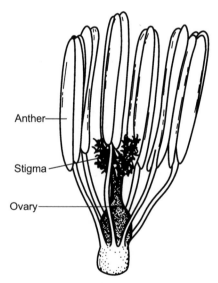

Anther

Stigma

Ovary

FIGURE 3.4   Reproductive organs of a rice floret. Note the six stamens present. *Reproduced from Poehlman, J.M., 1987. Breeding Field Crops, third ed. AVI Publishing Company Ltd., Westport, Connecticut, USA, by courtesy of AVI Publishers, New York.*

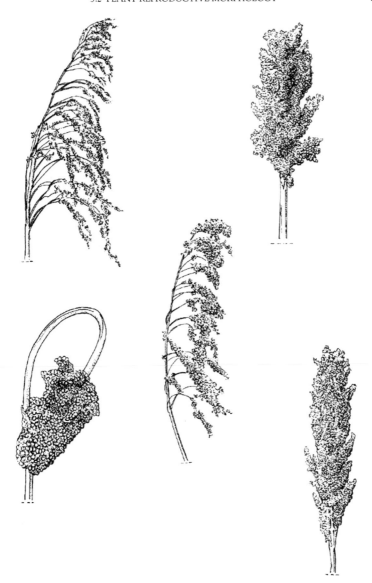

FIGURE 3.5   Inflorescences of various types of *Sorghum bicolor* (L.) Moench. *Reproduced from Grubben, G.J.H., Partohardjono, S. (Eds.), 1996. Plant Resources of South-East Asia No. 10 Cereals. Backhuys Publishers, Leiden, with permission.*

Pearl millet (*Pennisetum glaugum* (L.) R.Br.) has a stiff, contracted, spike-like panicle; in most types, it is either cylindrical or ellipsoid with a length between a few centimetres and over a metre long with an average of 1600 densely packed spikelets per panicle. Spikelets occur in clusters of 1–4 pairs, each surrounded by 25–90 bristles (Fig. 3.7). The lower floret

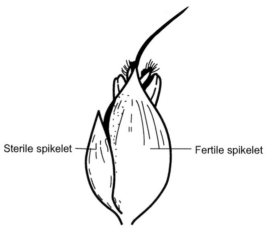

Sterile spikelet ———— Fertile spikelet

FIGURE 3.6 A pair of spikelets of sorghum. *Reproduced from Poehlman, J.M., 1987. Breeding Field Crops, third ed. AVI Publishing Company Ltd., Westport, Connecticut, USA, by courtesy of AVI Publishers, New York, Now Springer, see Fig. 3.2.*

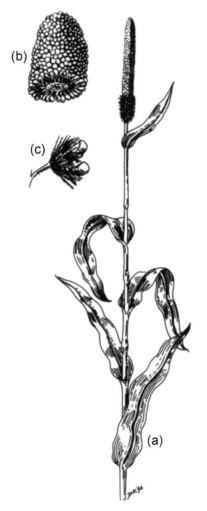

(b)

(c)

(a)

FIGURE 3.7 Pearl millet: (a) spike (b) part of spike enlarged, (c) spikelet.

of each pair is male only, or sterile, while the upper one bears both male and female structures. Lemmas and paleas do not clasp the caryopsis and lodicules are absent. Flowering begins at the top of the panicle with the hermaphrodite florets, which are protogynous, the stigmas becoming evident about 3 days earlier than the anthers of the same floret. When anthesis has reached the lowest bisexual florets, anthesis of the male-only florets begins. Flowering occurs mainly between 8 a.m. and 2 p.m., with a peak at 10 a.m. A useful presentation is available at www.authorstream. com/Presentation/chhabra61-532441-flower-structure-of-pearl-millet.

Like that of pearl millet, the inflorescence of foxtail millet (*Setaria italica* (L.) P. Beauvois) is a spike-like panicle, but it may be erect or pendulous. In length it varies from 25 to 32 cm. Several orders of branching may be present (Doust and Kellogg, 2002) but they are much reduced in length. Spikelets are subtended by 1–3 bristles, which extend twice as far as the spikelet itself and have high contents of silicon (Hodson et al., 1982). Each spikelet bears two florets, the lower of which is sterile and the upper is hermaphrodite. The caryopsis is tightly enclosed by palea and lemma.

Flowering occurs late at night and early in the morning and lasts for 10–15 days. The earliest florets to flower are on the lower-order branches at the apex, and flowering proceeds down the panicle and down the branching order.

Proso millet (*Panicum miliaceum* L.) inflorescences are slender panicles of up to 45 cm length. They may be compact or open and erect or drooping. Spikelets contain pairs of florets, the upper of which is hermaphrodite and the lower infertile.

The male inflorescence of maize is a spike-like panicle; it is described following with the female inflorescence, which is a spike.

3.2.1.2 Spikes (racemes)

Spikes are characteristic of wheat, barley, rye and triticale, and similarities among these cereals are immediately obvious. The maize female inflorescence is also a spike but its structure is less readily appreciated.

Consideration of the variation among inflorescences of different wheats provides an insight into the changes that have been brought about by selection and breeding. It also provides an opportunity to consider types of wheat in which interest is increasing, even though their yields and field characteristics make them unattractive as large-scale commercial crops.

Reference has been made in Chapter 1 to the free-threshing character of modern commercial wheat types, a condition otherwise described as naked. This describes the freedom of the grain from adherent glumes, palea and lemma. In contrast, some wheat types, the importance of which declined many years ago, do have adherent palea and lemma, and are hence described as hulled. Although this is the main anatomical distinction, there are other major differences that affect processing – these are the tendency for the entire spikelets of hulled types to remain entire, owing to their tougher floral components; and the tendency of the rachis to fragment, so that, on being subjected to threshing, each spikelet retains its attachment to the relevant intermodal segment of rachis. In contrast, grains of free-threshing wheats are readily separated from other floral parts and the rachis, which remains intact (Fig. 3.8).

Before conventional processing, hulled wheats have to undergo a dehulling stage, similar to that applied to oats. Milling of hulled wheats, especially spelt, to produce a white flour has enjoyed a revival in interest as a consequence of some consumers seeking greater 'naturalness' in foods

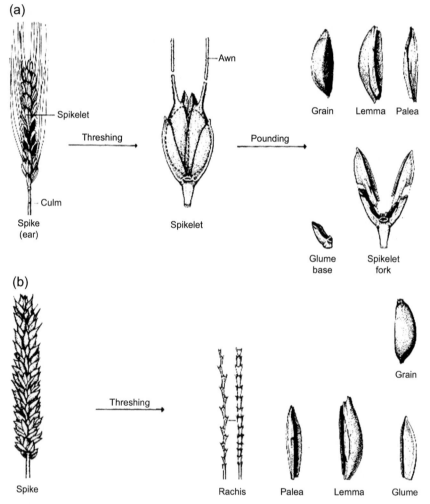

FIGURE 3.8   The anatomy of hulled (a) and free-threshing (b) wheat. *Reproduced from Nesbitt, M., Samuel, D., 1996. From staple crop to extinction? The archaeology and history of the hulled wheats. In: Padulosi, S., Hammer, K., Heller, J. (Eds.), Hulled Wheats. Proceedings of the First International Workshop on Hulled Wheats 21–22 July, 1995. International Genetic Resources Institute, Rome, with permission.*

and a possible taste and nutritional advantage; other uses as human food involve whole or broken, pearled grains or wholemeal. As with other cereals, hulled wheats can be malted for alcoholic beverage production. The hulled-wheat species of interest in this context are spelt, emmer and einkorn. In Italy the term 'farro' includes grains and products of all of them. As in several cereals whose inflorescences are spikes, the lemmas of hulled wheats are extended into long thin barbed projections known as awns. While green they contain chloroplasts and contribute to photosynthesis in the ear. In hulled wheats the unit of dispersal is the entire spikelet and in the wild the awns play a role in determining the manner in which dispersal occurs. They also assist the penetration of the spikelet into the soil (Elbaum et al., 2007). Even when cultivated, the entire spikelet is sown as 'seed' although awns are mostly absent through breakage. Further information on hulled wheats is available in Padulosi et al. (1996).

In the wheats widely grown commercially today each spikelet contains up to six florets (Fig. 3.9); but it is unusual for all six florets in a spikelet to be fertile and those at the extremes of the inflorescence may bear only one or even no fertile florets. Common or bread wheats may be awned or 'awnless' (i.e., with reduced awns or 'awnlets'). The merits of awns in relation to yield

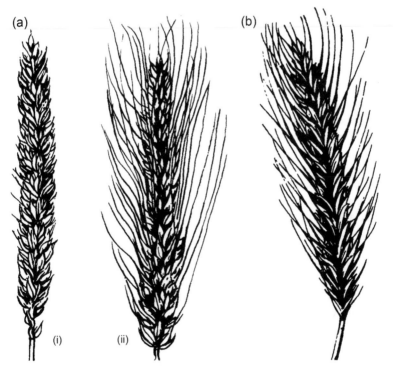

FIGURE 3.9   Spikes of (a) wheat and (b) rye. Wheat may be awned (bearded) (i) or awnless (ii).

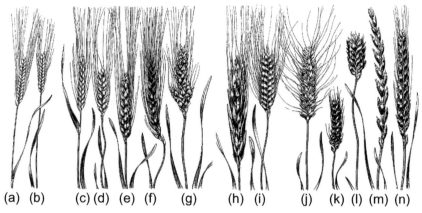

(a) (b)   (c)(d) (e) (f)   (g)   (h) (i)   (j)   (k) (l) (m) (n)

FIGURE 3.10   Variations in the appearance of heads of wheat species, one of many morphological characteristics used for their taxonomic classification. The wheat species are (including their genome assignments and common names) (a) *Triticum boeoticum* (2*x*: wild einkorn); (b) *T. monococcum* (2*x*: einkorn); (c) *T. dicoccoides* (4*x*: wild emmer); (d) *T. dicoccum* (4*x*: emmer); (e) *T. durum* (4*x*: macaroni wheat); (f) *T. carthlicum* (4*x*: Persian wheat); (g) *T. turgidum* (4*x*: river wheat); (h) *T. polonicum* (4*x*: Polish wheat); (i) *T. timopheevii* (4*x*: Timopheev's wheat); (j) *T. aestivum* (6*x*: bread wheat); (k) *T. sphaerococcum* (6*x*: shot wheat, Indian dwarf wheat); (l) *T. compactum* (6*x*: club wheat); (m) *T. spelta* (6*x*: spelt wheat); and (n) *T. macha* (6*x*: macha wheat). The diploid A-genome species, *T. urartu*, is not shown here. 2*x*=diploid; 4*x*=tetraploid; 6*x*=hexaploid. *Adapted from Mangelsdorf, P.C., 1953. Wheat. Scientific American 189, 50–59 (New York).*

have been researched widely and recent research on near-isogenic lines led to a conclusion that spikelets of awned varieties tend to have fewer fertile florets and consequently produce fewer grains but with greater mean size and fewer unacceptably small grains (Rebetzke et al., 2016).

Variants of wheat spikes are illustrated in Fig. 3.10 and excellent photographs of many wheat types can be found in Percival (1921) (republished and available as a free download).

The distribution of grain size within a spike was studied by Bremner and Rawson (1978); all grains in the centre of the spike are larger than their equivalents at the extremes and, while those in the two basal florets of each spikelet are the heaviest, the second floret from the base is the heavier of these. The variation in size occurs as a function of the ability of each grain to compete for nutrients but also of the period of development, the earliest to flower being those in the basal florets of the central spikelets (Fig. 3.11).

In the case of barley, the vast majority of the commercial crop is hulled. As in wheats, the rachis is flattened and adopts a zigzag form, glumes are very small but the lemma and palea fully surround the grain and remain closely adherent to it even after threshing. The lemma tapers to a long awn that does break off during threshing. When grains are malted the husk's components remain intact and can form a useful filtration medium. References have been made to two-rowed and six-rowed barleys in earlier chapters. Although they differ, one from the other, in

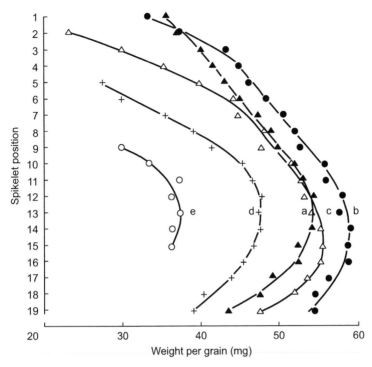

FIGURE 3.11 Typical profiles of wheat grain weights within a spike. The grain in the basal floret is designated *a*, and represented by ▲. Other grains are lettered progressively toward the tip of the spikelet, and represented by ●, △, + and O.O

appearance, their fundamental architecture is very similar. Both have inflorescences that are true spikes with three rows of sessile spikelets alternating on opposite sides of a rachis, and in both cases spikelets contain a single floret; however, while in the six-rowed type all three spikelets bear fertile florets, and consequently at maturity bear grains, in the two-rowed type only the floret in the central spikelet is fertile and the others are insignificant (Figs 3.12 and 3.13).

Many variants of the barley spike are illustrated by Briggs (1978).

In maize the male panicles, occurring at the top of the culm, bear spikelets in pairs, one being sessile and the other pedicellate. Both types contain two florets, each with three anthers. The entire male inflorescence is known as the tassel and each panicle may be up to 40 cm long (Fig. 3.14).

On the female inflorescences the spikelets are aligned in longitudinal double rows along the length of a spongy axis. Each carries two florets, only one of which is fertile; the upper functions while the lower one aborts Fig. 3.15. Each fertile floret contains a single ovary; its style is not of the feathery type typical of most cereals but a long thread-like structure (silks) covered in fine hairs that entrap wind-borne pollen. A single ear may contain 8 to 20 rows and up to 800 fertile florets each with a silk, up to 45 cm long.

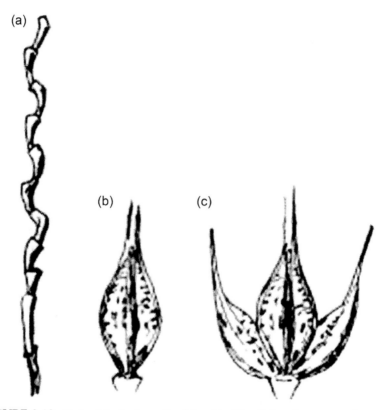

FIGURE 3.12 Parts of the barley spike: (a) rachis, (b) spikelet of a two-rowed type and (c) spikelet of a six-rowed type.

The ear or cob is wrapped by 8–13 modified leaf sheaths forming husks or shucks, and the silks emerge together from the distal open end of the protective husk (Fig. 3.16).

Perhaps the protection afforded by the husk obviates the need for enclosure of the reproductive structures by bracts, lemma and palea. These are insignificant in maize and as a result grains are not separated one from another on the cob. In some cases their mutual pressure imposes an angular form on them.

The concentration of the sexual organs on separate spikes encourages cross-pollination, which is the norm for maize. Its outbreeding habit fits it admirably for F1 hybrid production, whereby yields have been increased dramatically through the heterosis or hybrid vigour that results. While most cereals are dependent to some degree on cultivation for their survival, maize has the ultimate dependence on man since no mechanism for dispersal of its seeds remains.

Finger millet (*Eleusine coracana* (L.) Gaertner) is the exception among the most-grown millets in that its inflorescence is not a panicle but a cluster of dense sessile, finger-like spikes (racemes) (www.kew/data/grasses-db/

(b)

(a)

FIGURE 3.13   Spikes of barley, showing: (a) the two-rowed and (b) the six-rowed forms.

www/gen00227.htm). The 'fingers' are only about 1 cm diameter and between 3.5 and 15 cm long. They may be branched or not, and straight or incurved. Spikelets containing 6–12 florets are arranged alternately in two parallel rows, each attached by a thin, zigzag rachilla. Florets contain structures of both sexes but terminal florets may be sterile or male only. Flowering proceeds from the top to the bottom of the spikes but within the spikelet the basal florets flower earliest and flowering proceeds upwards. Anthesis occurs in the early hours of the morning.

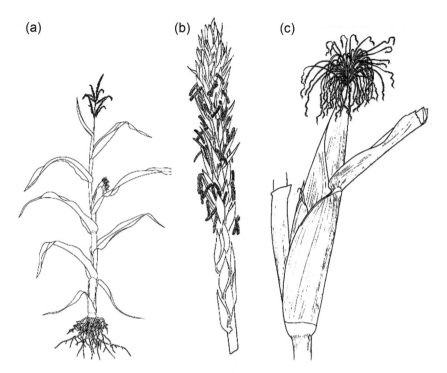

(a)   (b)   (c)

FIGURE 3.14   (a) Entire plant showing terminal tassel of male panicles, and axial female inflorescence, (b) detail of a panicle and (c) female inflorescence showing thread-like styles (silks) emerging
from the husk. *From Mackean, D.G., www.biology-resources.com. Reproduced by permission.*

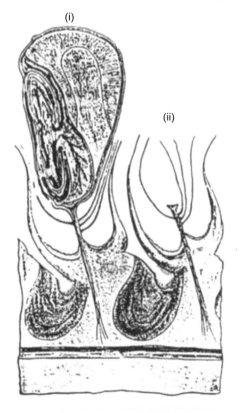

(i)

(ii)

FIGURE 3.15 Radial section of a
maize cob, showing (i) a perfect (fertile)
and (ii) a rudimentary (empty) floret.
*Based on Winton, A.L., Winton, K.B., 1932.
The Structure and Composition of Foods,
vol. I. Cereals, Starch, Oil Seeds, Nuts, Oils,
Forage Plants. Chapman and Hall, London.*

FIGURE 3.16   Cob of maize, showing the protective husk.

3.2.2 Pseudocereals

All three species of pseudocereals covered here, namely buckwheat, quinoa and grain amaranth, are dicotyledons but they are classified in different families. While the C3 photosynthetic pathway is found in buckwheat, it is the C4 pathway that is utilized by quinoa and grain amaranth. Within each species considerable variation exists, so the descriptions given are general in nature. Full botanical information can be found on the searchable database at http://www.prota4u.org.

Buckwheat (*Fagopyrum esculentum* Moench) (Fig. 3.17) is a member of the Polygoneaceae. Plants have a chromosome complement of 2n = 16 or 32. Plants grow to 120 cm tall. They are branched and the degree of branching increases with the space available to each plant. All aerial parts of the plant contain the flavonoid rutin, which has beneficial medicinal properties. Inflorescences are compound and borne terminally and in leaf axils. They consist of clusters of flowers combined in corymbous pseudospikes (a corymb exhibits the same branching pattern as a panicle but its lower branches are longer so that all branches terminate with flowers at the same level giving a flat top). Individual flowers have a short pedicel and 5 tepals (elements of the perianth that cannot be distinguished as petals or sepals). The 8 anthers and 3 styles present exhibit variation in their relative lengths. Pollination is by insects and cross-fertilization is essential as self-incompatibility exists.

Fruits are 3-sided achenes approximately 3 × 6 mm in size and grey to black in colour. Yields are around 2 tonnes/ha and the harvest index is about 0.44.

Quinoa (*Chenopodium quinoa*) is classified in the Chenopodiaceae. It has a chromosome number of 2n = 36. Plants are erect and grow to 0.7–3 m tall, with leaves on long petioles alternating up the stem. Responses to day length are variable. Flowering occurs between 50 and 70 days after emergence and the period between anthesis to maturity varies, from 90

FIGURE 3.17 *Fagopyrum esculentum* Moench. 1, Flowering branch; 2, flower; 3, unwinged fruit; 4, winged fruit; 5, top view winged fruit. *Reproduced from Grubben, G.J.H., Partohardjono, S. (Eds.), 1996. Plant Resources of South-East Asia No. 10 Cereals. Backhuys Publishers, Leiden, with permission.*

to 110 days. Inflorescences occur terminally and in leaf axils; they are panicles with clusters of flowers occurring on the secondary axes. No petals are present but five tepals are, and so are the same number of anthers, stigmas have two branches. Self-fertilization is the norm but 10%–15% crossing occurs, with insects as pollinators. Fruits are about 2 mm diameter and their colour varies with white, orange, pink, yellow, red, brown and black examples occurring. Grain yields are higher in temperate regions where they might reach 3 tonnes/ha but in South America they can be as low as 0.8 tonne/ha.

Grain amaranth (*Amaranthus caudatus*) (Fig. 3.18) is classified in the Amaranthaceae, a family closely related to the Chenopodiaceae, which includes quinoa. It has a chromosome complement of 2n = 32. Plants, which are characterized by much branching, grow to a height of up to 2.5 m. Leaves are arranged spirally. Flowers are unisexual with 3–5 tepals and equivalent numbers of stamens or three stigmas. Pollination is effected

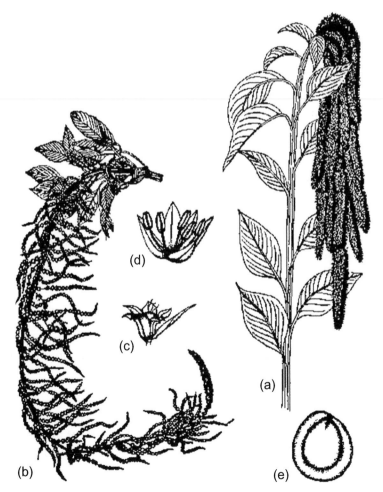

FIGURE 3.18 *Amaranthus caudatus.* (a) Plant, (b) inflorescence, (c) female flower, (d) male flower, (e) fruit. *From Stafford, W.E. (Ed.), 1917. Proceedings of the 19th Congress of Americanists 1917.*

by insects and self-pollinating is the norm but nearly 30% outcrossing has been recorded. Inflorescences arise terminally and in leaf axils. They are large, long pendulous and complex and with many flowers maturing over a long period; pollination takes place for a long time, leading to uneven ripening. Full height is reached 4–6 weeks after emergence and harvesting can take place 70–130 days after emergence.

3.3 CARYOPSIS INITIATION AND GROWTH

The beginning of each new generation of cereals occurs when fertilization is achieved as a result of pollen produced in the anther contacting the stigma on the carpel of another, or even the same, floret. Once on the stigma, pollen grains have a mechanism whereby a pollen tube is produced. The tube progresses toward the micropyle and, having effected access by this route, it allows nuclei from the pollen grain to pass into the ovule and fuse with nuclei present there. The primary fusion is of the sperm nucleus with the egg nucleus. The product is a cell, the successive divisions of which and its daughters produce the embryo. Separate nucleic fusions produce the first endosperm nucleus. Three, not two, nuclei are involved, one from the pollen and two polar nuclei from the ovule. With such different origins, there is some justification for regarding the embryo and endosperm as different organisms (Olsen, 2004).

All endosperm cells are ultimately derived from this first endosperm cell and each inherits chromosomes from three nuclei rather than the more usual two. Endosperm cells thus have one and a half times as many chromosomes as do cells elsewhere in the plant (in fact many endosperm cells undergo endoreduplication during grain development and hence their ploidy is increased even further). The fundamental details of endosperm development in grasses are given by Olsen (2004) and Sabelli and Larkins (2009)and aspects of development of individual cereals are described in relevant texts (Kiesselbach, 1980; Bushuk, 2001; Watson, 1987; Hulse et al., 1980; Percival, 1921; Bechtel et al., 2009; Palmer, 1989; Hoshikawa, 1967).

3.3.1 Grain anatomy

A description of a generalized cereal grain has been provided in Chapter 1. In this section attention is given to more detail of the grain components and differences among grains of individual species.

3.3.1.1 Embryo

The embryonic axis and the scutellum constitute the embryo. Together they contribute between less than 2% to more than 10% of the total grain weight (Table 3.1). The embryonic axis is the plant of the next generation.

TABLE 3.1 Proportions of parts of cereal grains (%)

Cereal	Hull	Pericarp + testa	Aleurone	Starchy endosperm	Embryo Embryonic axis	Scutellum
Wheat:						
Thatcher	–	8.2	6.7	81.5	1.6	2.0
Vilmorin 27	–	8.0	7.0	82.5	1.0	1.5
Argentinian	–	9.5	6.4	81.4	1.3	1.4
Egyptian		7.4	6.7	84.1	1.3	1.5
Barley:						
Whole grain	13	2.9	4.8	76.2	1.7	1.3
Caryopsis	–	3.3	5.5	87.6	1.9	1.5
Oat:						
Whole grain	25	9.0		63.0	1.2	1.6
Caryopsis (groat)	–	12.0		84.0	1.6	2.1
Rye	–	10		86.5	1.8	1.7
Rice						
Whole grain	20	4.8		73.0	2.2	
Caryopsis	–					
Indian	–	7.0		90.7		
Egyptian	–	5.0		91.7		
					0.9	1.4
					3.3	
Sorghum	–	7.9		82.3	9.8	
Maize:	–					
Flint		6.5	2.2	79.6	1.1	10.6
Sweet	–	5.1	3.3	76.4	2.0	13.2
Dent	–	6				
				82	12	
Proso millet	16	3	6	70	5	

Sources as in Kent, N.L., 1983. Technology of Cereals, third ed. Pergamon Press Ltd, Oxford.

It includes primordial roots and shoots with leaf initials. It is connected to and couched in the shield-like scutellum, which lies between it and the endosperm. There is some confusion about the terminology of the embryo as the term 'germ' is also used by cereal chemists to describe part or all of the embryo. If the botanical description is adopted as above and 'germ' reserved for the embryo-rich fraction produced during milling, then there can be no confusion.

The scutellum behaves as a secretory and absorptive organ, serving the requirements of the embryonic axis when germination occurs. It consists mainly of parenchymatous cells, each containing nucleus, dense cytoplasm and oil bodies or spherosomes. The layer of cells adjacent to the starchy endosperm consists of an epithelium of elongated columnar cells arranged as a palisade. Cells are joined only near their bases.

Exchange of water and solutes between scutellum and starchy endosperm is extremely rapid. Secretion of hormones and enzymes and absorption of solubilized nutrients occurs across this boundary during germination. The embryonic axis is well supplied with conducting tissues of a simple type and some conducting tissues are also present in the scutellum (Swift and O'Brien, 1970).

The ability of scutellar epithelial cells to generate somatic embryos has led to the importance of the tissue in creation of genetically modified cereal plants (Christou, 1994).

Although the fusions of nuclei, occurring during sexual fertilization and leading to the formation of embryo and endosperm, respectively, take place approximately at the same time, the development of the embryo tissue, by cell division, is relatively delayed. When the embryo does enlarge, it compresses the adjacent starchy endosperm tissue giving rise to a few layers of crushed, empty cells, the contents of which have either been resorbed or failed to develop. The crushed cells are described variously as the cementing layer, depleted layer or fibrous layer.

3.3.1.2 Endosperm

The endosperm is the largest tissue of the grain. It has an origin that is independent of that of the embryo (see Section 3.3). In the mature caryopsis the endosperm comprises two major components that are clearly distinguished and two minor components. The majority, a central mass described as starchy endosperm, consists of thin-walled cells packed with nutrients that can be mobilized to support growth of the embryonic axis during germination and seedling growth. Nutrients are stored in insoluble form, the major component being the carbohydrate, starch. Next in order of abundance is protein. In all cereals there is an inverse gradient involving these two components, the protein percentage per unit mass of endosperm tissue, though not necessarily per cell (Evers, 1970) increasing toward the periphery (Fig. 3.19).

P

A

SA

I

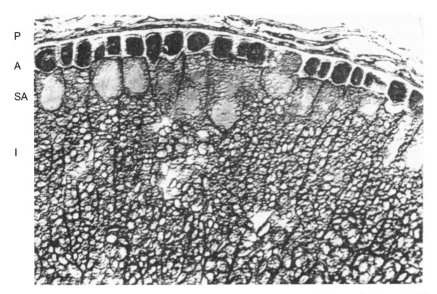

FIGURE 3.19 Part of a transverse section of a grain of Hard Red Winter wheat, 14.4% protein content, showing concentration of protein in subaleurone endosperm. Protein concentration diminishes toward the central parts of the grain in all cereals. *A*, aleurone layer; *I*, inner endosperm; *P*, pericarp; *SA*, subaleurone endosperm.

Cell size also diminishes toward the outside and this is accompanied by increasing cell wall thickness. The walls of the starchy endosperm of wheat are composed mainly of arabinoxylans, while in barley and oats (1–3) and (1–4) β-D glucans predominate. Cellulose contributes little to cereal endosperm walls except in the case of rice. The presence of lignin in cell walls of developing maize starchy endosperm has been reported, although higher levels were found in transfer cells of the same grains (Rocha et al., 2014).

Surrounding the starchy endosperm is the other major endosperm tissue, the aleurone, consisting of one to three layers of thick-walled cuboid cells with dense contents and prominent nuclei. The number of layers present is characteristic of the cereal species, wheat, rye, oats, maize and sorghum having one, barley having three, and rice having a variable number, between one and three, according to location. Unlike the tissue they surround, aleurone cells contain no starch but they have a high-protein content and they are rich in lipid. They are extremely important in both grain development, during which they divide to produce starchy endosperm cells, and during germination, when in most species they are a site of synthesis of hydrolytic enzymes, responsible for solubilizing the reserves. Unlike starchy endosperm cells, which undergo programmed cell death, aleurone cells remain alive and continue to respire for indefinite periods after harvest.

In addition to the major endosperm cell types there are two additional tissues consisting of transfer cells and cells adjacent to the embryo. Transfer cells are functional during endosperm development, when they promote passage of nutrients from conducting tissue, present in the maternal elements of the grain, to the growing starchy endosperm. This function is facilitated by their characteristically great surface area. Transfer cells lie close to the conducting tissue whereby nutrients are transported from the parent plant. In grains where a ventral crease is present (wheat, rye, barley and oats) the conducting tissues are located in the crease region and hence the transfer cells lie at the endosperm periphery in that region. They have been referred to as modified aleurone or thick-walled cells in some reports. In maize, sorghum and millets they lie near the conducting tissue at the point of attachment to the mother plant. In rice, they lie close to the main vascular tissue on the dorsal side of the grain.

Cochrane (1994) drew attention to the tissue described as 'germ aleurone' between the scutellum and the starchy endosperm of barley, and others have described cells close to the embryo as having a compressed appearance with dense cytoplasm. They have been most studied in maize where they have been shown to exhibit complex membrane systems and to be rich in small vacuoles. They become metabolically active at the onset of germination when they supply the embryo with sugars. They may also have a defensive role against pathogens (Sabelli and Larkins, 2009).

3.3.1.3 Seed coats

Surrounding the endosperm and embryo lie the remains of the nucellus, the body within the ovule in which the cavity known as the embryo sac develops. Following fertilization, the embryo and endosperm expand at the expense of the nucellus, which is broken down except for a few remnants of the tissue and a single layer of squashed empty cells of the nucellar epidermis. Epidermal cells in many higher plants secrete a cuticle, and a cuticle is present on the outer surface of the nucellar epidermis of many cereals.

The outermost tissue of the seed is the testa, which develops from the integuments. The testa may consist of one or two cellular layers; in some varieties of sorghum a testa may be absent altogether. Where two layers are present the long axes of their elongated cells lie at approximately 90 degrees to each other. Frequently the testa accumulates corky substances that reduce permeability and pigments that confer colour on the grain. A cuticle, thicker than that of the nucellar epidermis, is typical, and this also plays a role in regulating water and gaseous exchange.

Both testa and nucellus once formed part of the ovule of the mother plant. They are thus of an earlier generation than the endosperm and embryo, which they surround and to which they closely adhere.

3.3.2 Pericarp

The pericarp (or fruit coat) comprises several complete and incomplete layers. In all cereal grains the pericarp is dry at maturity, consisting of mainly empty cells. During development it serves to protect and support the growing endosperm and embryo. The innermost layer of the pericarp is the inner epidermis. In many cereals this is an incomplete layer. Its cells are elongate and thus termed 'tube cells'. They are sometimes squashed flat in mature grains. Outside this layer lies the 'cross cell layer'. Unique to grasses, this layer takes its name from the fact that in wheat, rye and barley, the long axes of its elongate cells lie at right angles to the grain's long axis. Cells are arranged side by side, in rows. In the immature green grain cross cells and tube cells contain chloroplasts. The 'mesocarp', outside the cross cells, is not found as a true layer in mature grains but during development its cells were the sites of small starch granule accumulation. In some types of sorghum the cells and the starch granules within persist, but in most mature cereal grains the cells are empty and squashed or broken down.

The outermost layer of the pericarp and indeed of the caryopsis is the outer epidermis or 'epicarp'; it is one cell thick and adherent to the 'hypodermis', which may be virtually absent as in oats or some millets, one or two layers thick as in wheat, rye, sorghums and pearl millet or several layers thick as in maize. The outer epidermis has a cuticle that controls water relations in growing grains but generally becomes leaky on drying (Radley, 1976). Hairs or trichomes are present at the nonembryo end of wheat, rye, barley, triticale and oats. They are collectively known as the 'brush' and have a high-silicon content. Trichomes have a spiral sculptured surface that differs among cereals, allowing their origin to be determined when they are found detached from the grain (Bennet and Parry, 1981).

In some cereal grains, such as oats, barley, rice, hulled wheat species and some millets, the lemma and palea are not removed by threshing. They may be adherent to the grain surface as in barley or firmly held on to the other, as in rice and oats. The tissues described earlier are present in cereal grains even if they are of the naked or free-threshing type, although some may be less robust in the hulled species. For example, the pericarp of hulled wheats is thinner than that found in naked types (Nesbitt and Samuel, 1996).

Where hulls are not removed by threshing they form part of the grain as traded, and their additional contribution to grain mass has to be borne in mind when comparing the relative proportions of nutrients in different species (Table 3.1).

3.3.3 Grain characteristics of individual cereals and pseudocereals

In spite of structural similarities, there are wide variations among cereal grains in size and shape. Comparisons in size and form are shown diagrammatically in Chapter 1, Fig. 1.8, and size and weight differences are given in Table 3.2.

TABLE 3.2   Dimensions and weight per 1000 grains of the cereals

Cereal	Dimensions		Weight per 1000 grains	
	Length (mm)	Width (mm)	Average (g)	Range (g)
Millets				
Teff	1.0–1.5[a]	0.5–1.0[a]		0.2–0.5[a]
Proso	Up to 3[a]	Up to 2[a]	6	4.7–7.2[a]
Pearl	2.5–6.5[a]	1.0–2.5[a]	7	2.5–16[a]
Finger	<2 diam[a]			2–3[a]
Rye	4.5–10	1.5–3.5	21	8–50
Sorghum	3–5	2–5	28	
Rice[b]	5–10	1.5–5	27	
Oats[b]	6–13	1.0–4.5	32	
Triticale			36	28–45
Barley[b]	8–14	1.0–4.5	35	32–36
Wheat	5–8	2.5–4.5	37	CWRS: 27
				English:48
Maize	8–17	5–15	324	150–600

[a] *Protobase.*
[b] *Includes husk.*
Sources as in Kent, N.L., 1983. Technology of Cereals, third ed. Pergamon Press Ltd, Oxford.

The grains of different cereal species can be distinguished in several ways, but the presence or absence of a ventral crease has some significance and individual grain types are grouped in the next section according to that distinction.

3.3.4 Caryopses without a crease

3.3.4.1 Maize (dent corn)

Although there are many types of maize and their morphology and anatomy vary (Watson, 1987), it is possible only to describe one type here; dent corn is the most abundantly grown and this explains its selection. The maize grain is the largest of cereal grains. Its structural features are shown in Fig. 3.20. The basal part (embryo end) is narrow, the apex broad. The embryonic axis and scutellum are relatively large.

The pericarp is thicker and more robust than that of the smaller grains (Fig. 3.21). (Although an old publication, The Structure and Composition

Maize

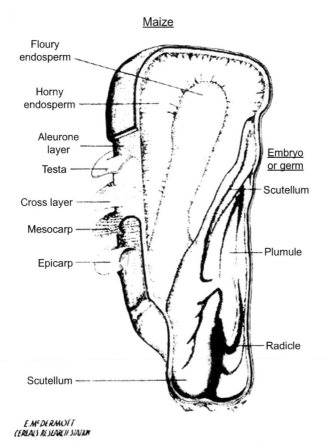

Floury endosperm

Horny endosperm

Aleurone layer

Testa

Cross layer

Mesocarp

Epicarp

Embryo or germ

Scutellum

Plumule

Radicle

Scutellum

E. McDERMOTT
CEREALS RESEARCH STATION

FIGURE 3.20   Diagrammatic longitudinally cut face of a maize grain.

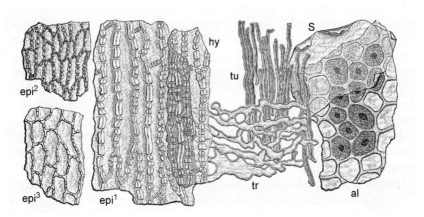

FIGURE 3.21   Maize. Bran coats in surface view. *epi³*, epicarp at base; *epi²*, epicarp at apex. Layers on dorsal side: *epi¹*, epicarp; *hy*, hypoderm; *tr*, cross cells; *tu*, tube cells; *S*, speroderm testa cuticle; *al*, aleurone cells. ×160. (A.L.W.)

of Foods, by Winton and Winton (1932), from which this illustration is taken, provides the best collection of detailed descriptions of many cereal tissues.) It is known as the hull (it should not be confused with the hull formed from lemma and palea in some cereal species), and the part of the hull overlying the embryo is known as the tip-cap. An epidermis is present as the outermost layer; no hairs are present. Beneath the epidermis lie up to 12 layers of hypodermis, which appear increasingly compressed toward the inside. Both tube cells and cross cells are present, cross cells occurring in at least two layers. Cell outlines are extremely irregular and there are many spaces among the anastomosing cells. No cellular testa layer is present but a cuticular skin persists to maturity. The same applies to the nucellus.

In spite of the great size of the endosperm of maize, individual aleurone cells are small, comparable to those of oats and rice. One layer of them is present. In the grains of coloured varieties, it is anthocyanin pigmentation present in the aleurone cells that provides the colouration. In the starchy endosperm many small starch granules (average $10\,\mu m$) occur. Protein (zein) also occurs in tiny granular form. Horny endosperm occurs as a deep cap surrounding a central core of floury endosperm. The designation 'dent' results from the indentation in the distal end of the grain that contracts on drying. The dent is not found in other types of maize such as flint maize, popcorn and sweetcorn. The most significant differences among maize types lie in the endosperm character and shape. The covering layers are similar but other types may have fewer hypodermis layers than dent corn.

3.3.4.2 Rice

The lemma and palea of rice are removed from the grain only with difficulty, as they are locked together by a 'rib and groove' mechanism (Fig. 3.22). Once they are removed the outer epidermis of the pericarp is revealed as the outer layer of the caryopsis. The rice grain is laterally compressed and the surface is longitudinally indented where broader ribbed regions of the lemma and palea restricted expansion during development. The proportion of husk in the rice grain averages about 20%. Varieties of rice are classified according to grain weight, length and shape, which is described as round, medium or long, and defined by their aspect ratio.

In shape, grains of the *indica* type are short, broad and thick, with a round cross-section; grains of *japonica* rice are long, narrow and slightly flattened in shape.

The structural relationships within the grain are shown in Fig. 3.23.

Distinctively, in all except one of the tissue layers (the tube cells) surrounding the endosperm, the cells are elongated transversely (in other

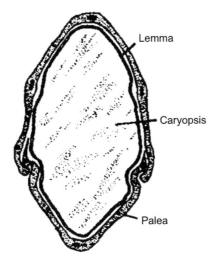

FIGURE 3.22 Transverse section through a rice grain showing the 'locking' mechanism of the lemma and palea.

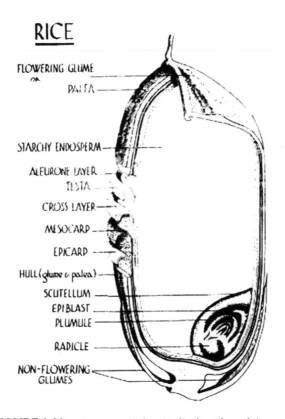

FIGURE 3.23 Diagrammatic longitudinal cut face of rice grain.

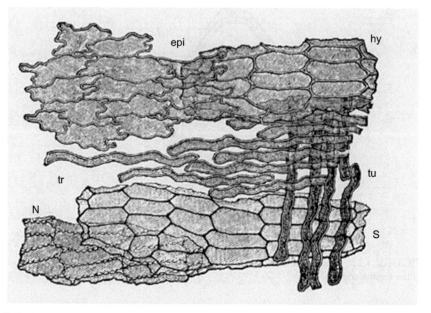

FIGURE 3.24   Rice. Bran elements in surface view. *epi*, epicarp; *hy*, hypoderm; *tr*, cross cells; *tu*, tube cells; *S*, spermoderm (testa); *N*, nucellar epidermis. ×160. (A.L.W.). *From Winton, A.L., Winton, K.B., 1932. The Structure and Composition of Foods, vol. I. Cereals, Starch, Oil Seeds, Nuts, Oils, Forage Plants. Chapman and Hall, London.*

grains only the cross cells are elongated in this direction) (Fig. 3.24). Cells of the epidermis have wavy walls making them quite distinct from those of the hypodermis, with their flat profiled walls. Cross cells have many intercellular spaces and the cells are worm-like; they are similar to, but lying at right angles to, the tube cells. The testa has one cellular layer, with an external cuticle. Cells of the nucellar epidermis are similar to those of the testa, but the walls have a beaded appearance in section. Aleurone cells range from 130×50 to $560 \times 105\,\mu m$ but the number of layers varies around the grain from one to three.

Compound starch granules are present in the starchy endosperm cells. In one report compound granules were $7\text{--}25\,\mu m$ in diameter and comprised 20–60 polygonal granuli, each between 4 and $6\,\mu m$ in size (Hayakawa et al., 1980). A broader survey reported granuli between 1.8 and $8\,\mu m$ (Wani et al., 2012). Protein bodies of three types, distributed around the compound granules, were distinguished on the basis of their size and staining properties. Each type was specific to a location within the endosperm (central or peripheral). Unusual for cereal grains, the embryo of rice is not firmly attached to the endosperm.

3.3.4.3 *Sorghum*

While the structural details of sorghum grain conform largely to the generalized architecture shown in Chapter 1 (Fig. 1.8), considerable variation exists among different varieties in the composition of the tissues surrounding the endosperm.

Sorghum grains are near-spherical with a relatively large embryo that varies in the degree to which it is embedded into the endosperm and hence the degree to which it protrudes. Close to the lower (radicle) end of the embryo is the 'hilum' or area of earlier attachment to the parent plant. At the distal end are the remains of the style. A single outer cuticularized epidermis of the pericarp (epicarp) surrounds the grain, but the stylar area is reported to be vulnerable to penetration by water and microorganisms. Dark pigmentation ('black layer') in the stylar area at maturity in some types provides an indication of ripeness.

Within the epidermis are two or three hypodermis layers that surround the mesocarp, a tissue that shows considerable variation. In thin-pericarp varieties mesocarp cells are empty, flattened and disrupted as described for other cereals, while in thick-pericarp varieties they are more isodiametric and filled with small starch granules. At 6 μm diameter or less they are smaller than those in the endosperm. As part of the milling process bran is removed by abrasion (decortication) and the presence of mesocarp starch makes the bran more friable, thus increasing the possibility of flour being contaminated by bran fragments. Most of the mesocarp starch passes into the bran fraction and does not contribute to the flour. Mesocarp cells are not pigmented but they can contribute to the way colours in other tissues are perceived as light scattering at the starch interfaces contributes a degree of opaqueness.

Cross cells do not form a complete layer; they are elongated vermiform cells aligned in parallel but separated by large spaces. Tube cells on the other hand are numerous and closer packed. Seed coats are subject to considerable variation among types of sorghum.

In early publications such as Winton and Winton (1932) reference is made to a nucellar epidermis in the mature grains of some varieties. In more recent publications references to a nucellus are few, although Earp et al. (2004) showed a cuticle lying between the testa and the aleurone layer, presumably originating on the outer face of the nucellar epidermis.

Aleurone cells are similar in size and appearance to those of maize as are the inner endosperm cells and the starch granules that they contain. Aleurone may or may not contain pigmentation. Transfer aleurone cells are present during development in the region of the placental sac (close to the point of attachment) (Zheng and Wang, 2010) and they remain distinctive at maturity. Peripheral endosperm cells are less clearly distinguished than in maize, from the aleurone cells to which they are adjacent.

The starchy endosperm comprises a floury central region surrounded by a corneous peripheral region, although the proportions vary among soft, intermediate and hard varieties. Hard varieties resist breakage during decortication and hence permit more distinct separation of bran and endosperm fractions. Harder varieties are also more resistant to attack by insects and pathogens. Endosperm may be colourless or yellow due to the presence of carotenoids. Waxy and high-amylose starch types, which resemble variations found in other cereals, also exist.

Grain colour is an important factor affecting acceptability; it is genetically complex with at least 10 pairs of genes having been shown to exert influence. The epicarp can be red, lemon yellow or white according to which R and Y genes are expressed, but pericarp colour is also modified by the influence of I (intensifier) genes (Rooney, 2000).

In the testa pigmentation is under the control of B genes but modification occurs in some cases as a result of the influence of the S (spreader) gene. Spreader genes are so named because in certain genetic backgrounds they control the 'spread' of pigmentation from testa to pericarp. Microscopic studies have shown that tannins in Type II sorghums are deposited in vesicles while those of Type III varieties are dispersed along cell walls of the testa, and, in some cases, also of the pericarp (Earp et al., 2004). Endosperm also exhibits colour variation and may be white, yellow or cream according to the level of carotenoids present.

Sorghum varieties are classified on the basis of the presence or absence of pigmented testa, and those with a pigmented testa are further divided on the basis of the extractability of the tannins.

Type I sorghums have no pigmented testa and very low levels or even no tannins.

Both Type II and Type III sorghums have a pigmented testa and contain tannins.

Tannins of Type II sorghums are extractable with acidified methanol (1% HCl methanol) while those of Type III sorghums are extractable with methanol (Dykes and Rooney, 2006). There are reports of the testa being incomplete in some Type II varieties.

Types with significant levels of tannins are known as 'bird proof' or 'bird resistant'. They are attacked less by birds than are nontannin types because of the unpalatability of the condensed tannins (proanthocyanidins) that they contain. The presence of tannins also offers protection against bacterial and fungal pathogens. While these effects of tannins are recognized as beneficial, there are also disadvantages in that when fed to monogastric animals they can reduce protein availability, and they also inhibit enzyme activity in the malting process. Interest has been shown in sorghum as a potential source of tannins for health-promoting pharmaceutical products. Information on the tannins present in sorghum and other cereals can be found in Salunkhe et al. (1990).

3.3.4.4 Millets

Pearl millet

The grain of pearl millet is about one-third the size of sorghum grain although size and shape show considerable variation (see McDonaugh and Rooney, 1989). Grain colours range from near white to brown and purple, but yellow and grey are the most common. There are 15,000 lines of pearl millet in the World Germplasm collection. In general the embryo represents a higher proportion of the grain than in most cereal grains (17.4%; Abdelrhaman et al., 1984). The cuticularized pericarp is somewhat similar to that of sorghum. The epidermis/hypodermis combination is thicker than that of sorghum and cells may contain con-centrically layered pigmented tissue (McDonaugh and Rooney, 1989). Variations in the depth of collapsed cells of the mesocarp determine whether varieties are classified as thick- or thin-pericarp type, but starch is absent from the mesocarp in the mature grain of both types. Cross cells and tube cells are spongy and the seed coat is thin and lightly pigmented. It tends to be incomplete in grey varieties where its pres-ence corresponds with the areas of greyness as perceived in the whole grain. A single layer of aleurone cells surrounds a starchy endosperm with horny peripheral and floury central regions. As in other cereals the aleurone contains lipid, protein and phytin. Pigmentation occurs in the pericarp, testa and aleurone. That in the aleurone can render the milled products unacceptable. Pigmentation in the aleurone and endo-sperm is not always perceived in the whole grain as in thick-pericarp types it can be masked. The aleurone cells have conspicuous knobbly thickening. Similar cells in foxtail millet have been noted and assigned a transfer function, like that of modified aleurone cells in other cereals. Tannins are not found in pearl millets.

Finger millet

Finger millet grains are 1–2 mm diameter, with colour varying from light brown to dark brown. Most recent reports about their detailed structure are taken from older references (mainly Winton and Winton, 1932). The pericarp is thin and papery, but, unusually for cereals, the testa layer remains cellular in mature grain and forms a hard coat. Its outer layer comprises isodiametric cells with wavy walls and warty protuberances on the outer cuticularized surface. The inner layer is less robust but more deeply coloured. No mention is made of any nucellar tissue remaining to maturity. Aleurone cells are small and in a single layer. Endosperm is generally friable. Tannins are present in finger mil-let, but levels of phenolics and tannins are considerably lower in light-coloured varieties. The location of tannins in finger millet is the testa (Siwela et al., 2007).

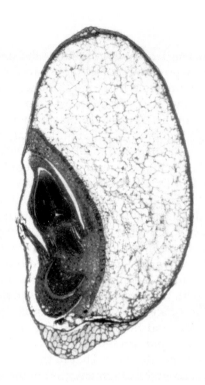

FIGURE 3.25 Longitudinal section of teff grain. *Image provided by Dr. Mary Parker, IFR Norwich, UK.*

Foxtail millet

In contrast to finger millet, the testa of foxtail millet is insignificant or rudimentary. A nucellar epidermis may be present, but in some types only its cuticle remains. Only the inner and outer epidermis of the pericarp remain.

Teff

Teff caryopses are ovate, they are very small, being 1–1.2 mm long and weighing around 0.7 mg. The embryo occupies more than half the length on one side (Fig. 3.25). Grain colours are white, red or brown, the pigmentation in coloured types lying in the testa where polyphenols or tannins are located. Some reports (Umeta and Parker, 1996) suggest that starch granules occur in the mesocarp of the mature grain while others refer to only the epidermis remaining (Winton and Winton, 1932). Endosperm comprises a single aleurone layer and an inner floury core of starch-filled starchy endosperm cells, surrounded by a corneous peripheral region.

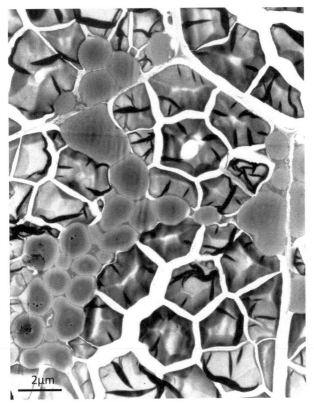

FIGURE 3.26 Transmission electron micrograph of teff endosperm showing compound starch granules and protein bodies. *Image provided by Dr Mary Parker. IFR Norwich, UK.*

Endosperm starch granules are compounds with individual granuli being angular and 2–6 μm in size (Fig. 3.26). Spherical protein bodies are present but no gluten is formed on hydration.

3.3.5 Caryopses with a crease

The crease is an elongated, central, reentrant region parallel to the grain's long axis on its ventral side – that is the opposite side to the embryo. It is a significant feature, not just in a botanical context but also in relation to processing. In the milling of grain, one of the objectives is to separate starchy endosperm from other tissues, and it is arguable that the gradual reduction process of roller milling evolved because of the presence of a crease. Had a crease not been present in wheat, the initial stages of flour milling might well have been a pearling stage as in the milling of rice and other grains without a crease (a pearling stage is now adopted by some processors, but the crease remains a difficult area for separation to be achieved). Indeed, over a very long period, interest has been shown in the possible breeding of 'creaseless' cereals, and since it has become possible to understand developmental mechanisms, attention has turned to the genetic control of grain shape (e.g., Hands et al., 2012). Although a crease is characteristic of several cereal species, and indeed grasses, there are considerable variations in the morphological features, such as depth of crease, exhibited.

3.3.5.1 Wheat

It is in the possession of a crease that wheat differs most from the generalized cereal grain shown in Chapter 1 (Fig. 1.8). The crease can be seen on the right side of the diagrammatic longitudinal cut face in Fig. 3.27 and on the lower side of a transverse section in Fig. 3.28).

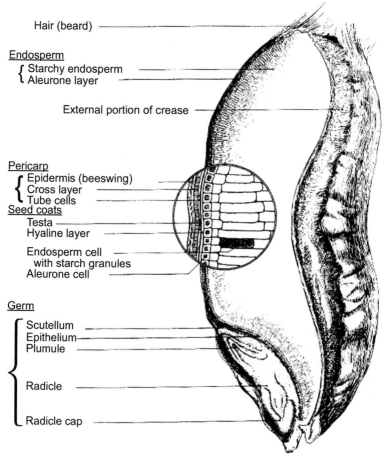

FIGURE 3.27   Diagrammatic longitudinal cut face of a wheat grain.

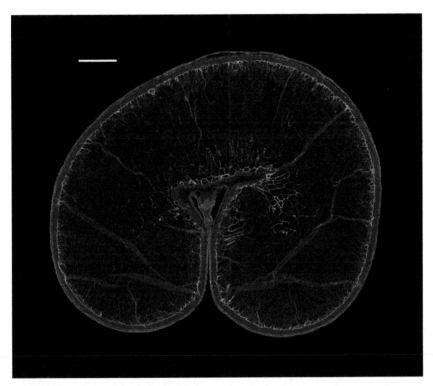

FIGURE 3.28 Fluorescence image of wheat grain transverse section. Stained and observed as Fig. 3.33. Fluorescence of cell wall is much less than in barley cell walls due to differences in β-glucans content. Scale bar represent 500 μm. *Image provided by Dr. Ulla Holopainen-Mantila, VTT Technical Research Centre. Reproduced with permission.*

The cellular nature of the outer caryopsis tissues is shown in Fig. 3.29.

The pericarp comprises a cuticularized epidermis with trichomes forming a 'brush' at the distal end of the grain; a hypodermis is also present. Some remnants of the mesocarp remain as 'intermediate cells'. The disintegration of most of the mesocarp leaves a space between cross cells and outer epidermis, which is thus only loosely attached and easily becomes removed as 'beeswing', except in the crease, where the mesocarp persists. The tube cells are separated by many large gaps except where they cover the tips of the grain. A cuticle is present on the outer surface of the testa, which comprises two layers, derived from the inner and outer integuments. The crossing of the elongated cells of the respective layers at right angles gives the impression of a net-like structure. Pigment of the testa layer in red wheats, reported to be phlobophene or proanthocyanidin, is responsible for the grain colour (Himi et al., 2005). A nucellar epidermis, sometimes described as the hyaline layer, is present and has a cuticle on its outer surface. More details of the structure of seed coats can be found in Evers and Reed (1988).

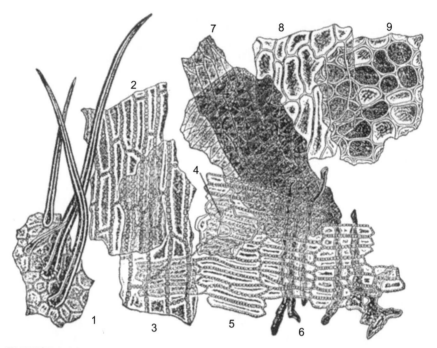

FIGURE 3.29   Bran tissues of wheat in surface view. 1–6 are pericarp components. 1, Fragment of epidermis with brush hairs; 2, epidermis on body of grain; 3, hypodermis; 4, intermediate cells; 5, cross cells, 6, tube cells; 7, outer and inner testa layers; 8, nucellar epidermis. 9, aleurone cells.

Whereas elsewhere in the grain the nucellar epidermis is virtually all that remains of that tissue, at the inner margin of the crease, between the testa and the endosperm, lies a column of nucellar tissue described as the nucellar projection. It carries the remains of vascular tissue by which nutrients were transported into the developing grain, and the final passage into the developing endosperm was *via* the nucellar cells themselves. In the mature grain they are rarely recognizable as transfer cells but their identity as such in developing grain has now made their function clear. The modified aleurone cells adjacent to them have also been identified as transfer cells by their characteristic appearance during development. To the outside of the nucellar projection, between the insertions of the testa, lies a bundle of corky cells forming the pigment strand. Its name derives from the dark colour that it exhibits in red wheats. It is not highly pigmented in white wheats.

Between the nucellar projection and the endosperm lies a void known as the endosperm cavity. Although obvious and relatively large in developing grains, the cavity is less pronounced in dry mature grains. The change occurs as a result of contraction, as it dries, of a hydrophilic gel within the cavity, bringing the endosperm/nucellar interfaces closer together and

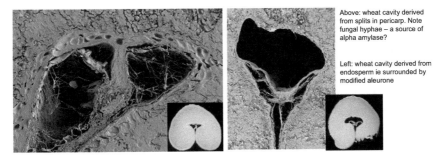

Above: wheat cavity derived from splits in pericarp. Note fungal hyphae – a source of alpha amylase?

Left: wheat cavity derived from endosperm ie surrounded by modified aleurone

FIGURE 3.30 Details of the special relationships between nucellar projection and endosperm of (left image) dry grain section, where adhesion remains, and (right) hydrated grain section, where separation has occurred, creating an endosperm cavity. Inset are images of the respective entire sections.

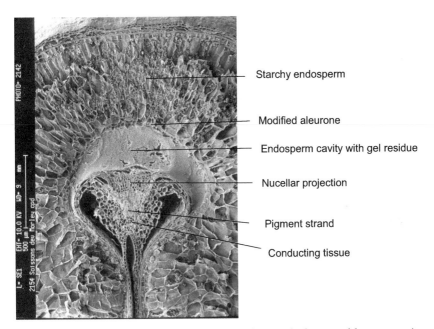

Starchy endosperm

Modified aleurone

Endosperm cavity with gel residue

Nucellar projection

Pigment strand

Conducting tissue

FIGURE 3.31 Transverse section of a wheat grain showing the features of the crease region. *Scanning electron micrograph provided by Mary Parker Institute of Food Research, Norwich, UK.*

cementing them one to the other. Thus the cavity is eclipsed, but it expands again if and when hydration occurs (Mary Parker, personal communication). The effect of hydration can be seen in adjacent thin transverse sections, the first of which was photographed dry and the second after hydration (Fig. 3.30). In the scanning electron microscope image of a grain that had been hydrated (Fig. 3.31) the presence of the gel residue can be seen.

Wheat aleurone tissue is one cell thick, and apart from where they are modified in the crease and adjacent to the scutellum, the cells are approximately cubic with 50 μm sides.

Among cereals wheat has a relatively high-protein content in the starchy endosperm (say 8%–16%). Two readily distinguishable populations of starch granules are present, two-thirds of the starch mass being contributed by A-type, large lenticular granules between 8 and 30 μm and one-third by near-spherical B-type granules of less than 8 μm diameter (see Chapter 4). A third population of granules even smaller than B type has been reported but their appearance is similar to the B type (Bechtel et al., 1990).

Endosperm texture is determined by the degree of adhesion between storage protein and starch granule surfaces, and softness results from the presence of a surface protein complex named friabilin. Two of the main components of friabilin are the isoform polypeptides puroindoline-a and puroindoline-b. The presence of both isoforms appears to be necessary for the expression of softness (Bechtel et al., 2009). Genetic control of puroindoline expression is now well understood in wheat, and its relevance to other cereals has also been demonstrated.

In durum wheat, amber endosperm is a desirable trait and carotenoid (particularly lutein) content has been shown to correlate well with yellowness. However, it is recognized that other unknown components also contribute significantly (Chung et al., 2009).

3.3.5.2 Barley

Barley grains of most varieties have a hull, formed of adherent lemma and palea, which is removable only with difficulty. It amounts to about 13% (average) of the grain weight, the proportion ranging from 7% to 25% according to type, variety, grain size and latitude where the barley is grown. Winter barleys have more hull than spring types; six-row (12.5%) more than two-row (10.4%). The proportion of hull increases as the latitude of cultivation approaches the equator, e.g., 7%–8% in Sweden, 8%–9% in France, 13%–14% in Tunisia. Large and heavy grains have proportionately less hull than small, lightweight grains.

Grains are generally larger and more pointed than wheat though they are less broad. They have a ventral crease that is shallower (Figs 3.32 and 3.33) than those of wheat and rye, and its presence is obscured in hulled types by the adherent palea. Cross cells are in a double layer and their walls do not appear beaded. Only one cellular testa layer is present. Two to four (mostly three) aleurone layers are present, cells being smaller than those in wheat, about 30 μm in each direction. Blue colour may be present due to anthocyanidin pigmentation. Barley grains contain high amounts of phenolic compounds (0.2%–0.4%). Polyphenols,

BARLEY

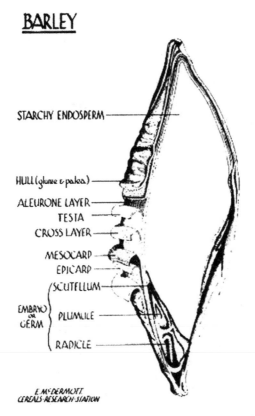

STARCHY ENDOSPERM

HULL (glume & palea)
ALEURONE LAYER
TESTA
CROSS LAYER
MESOCARP
EPICARP
SCUTELLUM
EMBRYO or GERM
PLUMULE
RADICLE

E. McDERMOTT
CEREALS RESEARCH STATION

FIGURE 3.32   Diagrammatic longitudinal section of a barley grain.

phenolic acids, proanthocyanidins and catechins reside in the hull, testa and aleurone.

In the starchy endosperm two populations of starch granules are present in most types, though in some mutants, exploited for their chemically different starch, only one population may be present.

Barley starchy endosperm cell walls have a high content of mixed-linkage β-glucans and, when stained with Calcofluor, fluoresce under incident ultraviolet irradiation, making the pattern of cells easy to discern (Fig. 3.33).

3.3.5.3 Rye

Rye grains are slenderer and more pointed than wheat grains but they also have a crease (Fig. 3.35) and indeed share many of the features described for wheat. A diagrammatic longitudinal section of a rye grain is shown in Fig. 3.34.

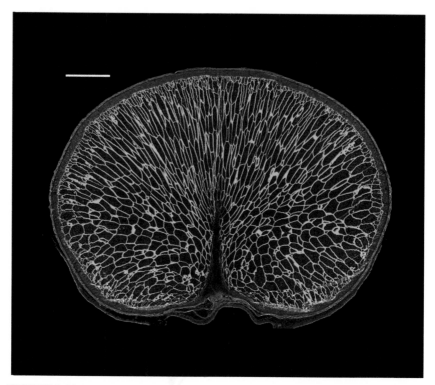

FIGURE 3.33   Fluorescence image of barley grain transverse section. Scale bar represents 500 µm. Stained with Calcofluor White, β-glucans in cell walls fluoresce brightly (excitation 400–410 nm; emission >455 nm). *Image provided by Dr. Ulla Holopainen-Mantila, VTT Technical Research Centre. Reproduced with permission.*

The beadlike appearance of cell walls of the pericarp is less distinct than in wheat. Rye grains may exhibit a blue-green cast due to pigment present in the aleurone cell contents. Cell walls of the starchy endosperm, like those in barley, contain mixed-linkage β-glucans.

Two populations of starch granules are present as in wheat; the larger granules, seen under the microscope, often display an internal crack. Protein bodies are present in rye endosperm as in wheat but, unlike wheat, rye proteins do not form coherent gluten on hydration.

3.3.5.4 Oats

Although naked varieties exist, the oat grain characteristically has a lemma and palea that are not removed during threshing. As they do not adhere to the groat (the name describing the actual caryopsis of the oat) within, they can be removed mechanically. Oats are traded with the husk in place. In this condition they have an extremely elongated appearance, and even with the husk removed groats are long and narrow (Fig. 3.36).

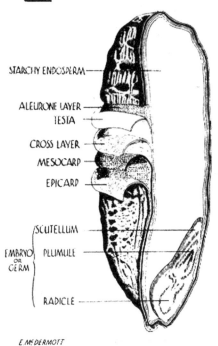

STARCHY ENDOSPERM

ALEURONE LAYER
TESTA
CROSS LAYER
MESOCARP
EPICARP

SCUTELLUM
EMBRYO
OR
GERM
PLUMULE

RADICLE

E. McDERMOTT

FIGURE 3.34 Diagrammatic longitudinal section of rye grain.

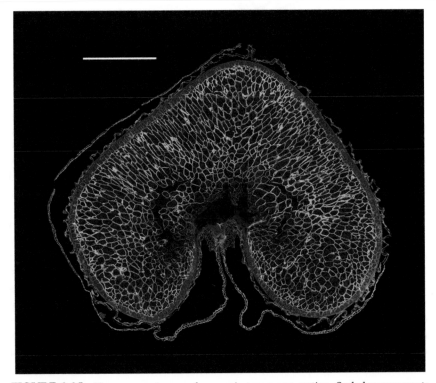

FIGURE 3.35 Fluorescence image of rye grain transverse section. Scale bar represents 500 µm. Stained and observed as for Fig. 3.33. *Image provided by Dr. Ulla Holopainen-Mantila, VTT Technical Research Centre. Reproduced with permission.*

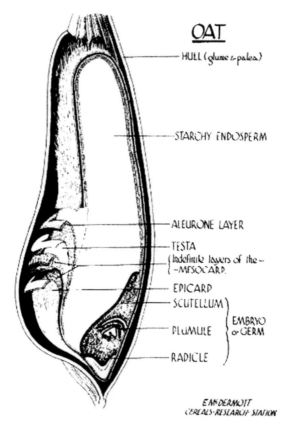

FIGURE 3.36   Diagrammatic longitudinal section of an oat grain.

Within a spikelet the two grains present differ in form, the upper being convex on the contact face while the lower grain has a concave face (Fig. 3.37). A transverse section of a primary grain is shown in Fig. 3.38. The grain in the primary floret is larger and heavier than that in the secondary floret and consequently two subpopulations are present in the harvested bulk. A further subpopulation of even smaller tertiary grains may also be present (Marshall et al., 2013).

The groat's contribution to the entire grain mass varies from 65% to 81% in cleaned British-grown oats (average 75%). Differences are due to both variety and environment. The groat contribution tends to be higher in winter-sown than in spring-sown types, in Scottish-grown than in English-grown samples, and in small, third grains than in the large, first (main) grains of varieties with three grains per spikelet (Hutchinson et al., 1952). Because of the starchy endosperm cell wall composition oats are a source of β-glucans, which generally have a higher solubility than those of barley. Oil content, with high levels of linoleic acid, is also high and oat protein contains more lysine than most

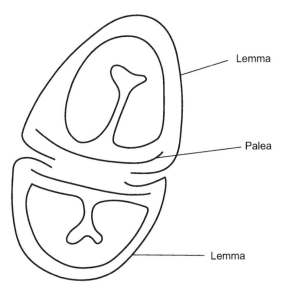

FIGURE 3.37 Diagrammatic transverse section through an oat spikelet showing the relationship between the grains within, accounting for their different shapes.

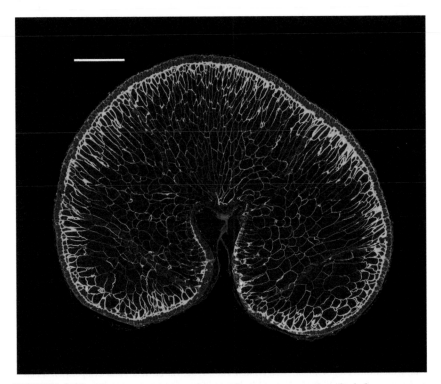

FIGURE 3.38 Fluorescence image of oat grain transverse section. Scale bar represents 500 μm. Stained and observed as for Fig. 3.33. *Image provided by Dr. Ulla Holopainen-Mantila, VTT Technical Research Centre. Reproduced with permission.*

cereal proteins. Antioxidants are present including α-tocotrienol, α-tocopherol and, found almost exclusively in oats, avenanthramides. These have medical and pharmaceutical applications (Marshall et al., 2013).

Naked oats *Avena nuda* L. is a type of oat with an unlignified husk that is readily lost during threshing, thus obviating any need for a special dehulling stage in milling. Naked oats have high-protein and oil contents, and a high-energy value, and they are used in high-value feeds for monogastric animals. In spite of continuing improvements, naked oats provide lower yields than conventional types.

Only two pericarp layers can be distinguished in oats, an epidermis with many trichomes on the outer surface (unlike the Triticeae fruits, these are not restricted to the nonembryo end; they can be irritants when fed to some animals) and a hypodermis, consisting of an irregular, branching collection of worm-like cells with long axes lying in all directions (Fig. 3.39).

The testa comprises a single cellular layer with cuticle. In cross-section the nucellar epidermis can be seen as a thin colourless membrane, its cellular structure cannot be discerned and a cuticle separates it from the testa.

In the endosperm there is a single aleurone layer, the cell walls of which are not thick, as they are in wheat and rye. Conversely, the starchy endosperm cells have thicker walls than in wheat. Starch granules are polyadelphous (compound) consisting of many tiny granuli that fit together to form a spherical structure. Possibly 80 or more granuli up to 10 μm diameter constitute a single compound granule. Individual free granuli are also

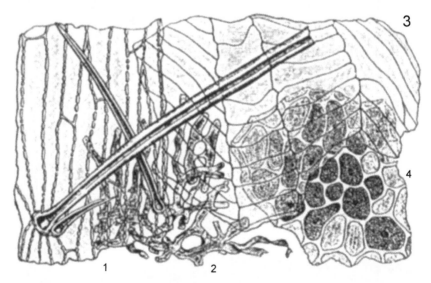

FIGURE 3.39 Bran coats of oat in surface view. 1, Outer epidermis; 2, hypodermis; 3, testa; 4, aleurone. *From Winton, A.L., Winton, K.B., 1932. The Structure and Composition of Foods, vol. I. Cereals, Starch, Oil Seeds, Nuts, Oils, Forage Plants. Chapman and Hall, London.*

present in the spaces among the aggregates. Endosperm cells of the oat have a relatively high-oil content.

3.3.6 Pseudocereals

The fruits of pseudocereals bear little resemblance to true cereal grains save for the facts that they bear a single seed and their storage products are starch and protein. Buckwheat, in common with cereals, stores these in an endosperm, which also has an aleurone layer, but in quinoa and grain amaranth the site of starch storage is the perisperm (nucellus), a tissue of which little or nothing remains in mature cereals grains. The embryo of all three pseudocereals has two cotyledons, unlike the true cereals, in which the scutellum is considered to be a modified single cotyledon.

3.3.6.1 Buckwheat

Fruits are pointed, broad at the base, and triangular to near-round in cross-section (Fig. 3.40). Their size varies from 2mm wide by 4mm long to 4×6mm (6×9mm according to some sources). The hull consists of a

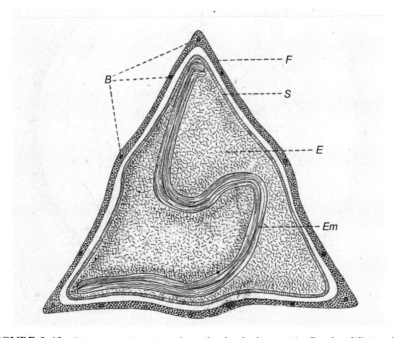

FIGURE 3.40 Diagrammatic section through a buckwheat grain. *Based on Winton, A.L., Winton, K.B., 1932. The Structure and Composition of Foods, vol. I. Cereals, Starch, Oil Seeds, Nuts, Oils, Forage Plants. Chapman and Hall, London.*

thin or thick pericarp, according to variety, within which lies a seed coat derived from the integuments. The degree of adhesion between pericarp and testa varies with variety, as does the colour from light grey to auburn. An endosperm comprising a single aleurone layer around a starchy endosperm, containing starch and protein, surrounds the embryo.

3.3.6.2 *Quinoa*

Fruits are cylindrical, conical or ellipsoid, 1–3 mm in diameter and mostly pale yellowish with occasional magenta grains owing to betacyanin pigmentation (Taylor and Parker, 2002). The pericarp is two-layered, the outer layer consisting of large papillose cells, the inner being discontinuous with cells tangentially stretched. The inner layer is adpressed to a two-layered testa, with the outer layer cells being large and thick walled and the inner having cells whose inner walls have thickenings. Testa colour may be red, brown, grey or black. A curved embryo surrounds a floury perisperm (Fig. 3.41), which contributes 40% of the grain, and comprises

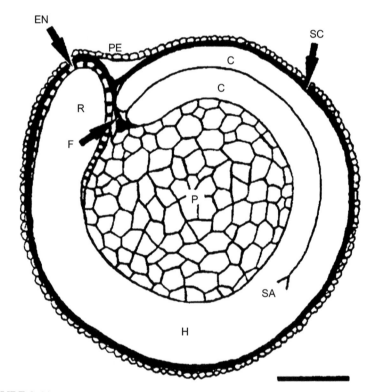

FIGURE 3.41 Median longitudinal section of a quinoa grain. *EN*, endosperm; *F*, funicle; *P*, perisperm; *PE*, pericarp; *SC*, seed coat. The embryo comprises *C*, cotyledon; *H*, hypocotyl; *R*, radicle; *SA*, shoot apex (2). Bar = 500 μm. *From Prego, I., Maldonado, S., Otegui, M., 1998. Seed structure and localization of reserves in Chenopodium quinoa. Annals of Botany 82, 481–488.*

thin-walled cells, rich in starch but with little protein. The endosperm is rich in protein, oil and phytin, but it is small and restricted to the micropyle area. Details of the ultrastructure of quinoa can be found in Varriano-Marston and De Francisco (1984) and Prego et al. (1998).

Both pericarp and testa contain saponins, which are antinutritional and have to be removed or reduced by washing or toasting before consumption.

3.3.6.3 Grain amaranth

The fruit is 0.9–1.9 mm diameter, weighing about 1 mg and coloured cream to gold or pink to black. The seed coat is thin and smooth. The embryo contributes about one-quarter of the grain weight, and it surrounds a central mealy perisperm containing starch and protein.

3.4 GERMINATION

The term 'germination', as it relates to grasses, covers those events occurring between the start of uptake of water by a seed and the emergence of the radicle through the testa and pericarp (and the palea/lemma in hulled types). For completion it requires that a ripe, fertile grain is subjected to damping to an adequate moisture content at an appropriate temperature in the presence of adequate oxygen. Because of its importance to both agriculture and processing, germination and the factors affecting it have been much studied and found to be complex. Details of these, as they apply to all seeds, can be found in, e.g., Black et al. (2006). The important events involved in germination are shown in the flow diagram (Fig. 3.42)

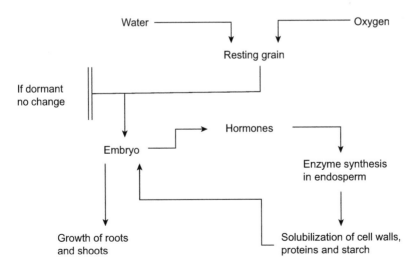

FIGURE 3.42   The main events involved in germination of a seed.

from which it can be seen that the process involves not only the embryo but also the tissues within the endosperm.

In grains destined for use as seed or for malting the requirement is for ready and vigorous germination as soon after harvest as possible, but this must be combined with resistance to premature germination, or sprouting, prior to harvest. Resistance to preharvest sprouting is, in fact, desirable in all cereals, irrespective of their intended use because the growth of the embryonic axis is accompanied by the production of hydrolytic enzymes that render stored nutrients in the endosperm soluble, thus reducing the amount of starch and protein harvested. Additionally, the presence of high-germination-enzyme levels in cereal flours intended for baked goods production gives rise to excessive hydrolysis during processing. Bread-making flours are particularly sensitive to such high-enzyme levels as processing conditions are well suited to enzyme-catalyzed hydrolysis.

Aleurone and scutellum are sites of synthesis for the major hydrolytic enzymes that contribute to digesting reserves held in the starchy endosperm. Several enzymes are involved, but as starch is the prevalent storage component, it is the starch-degrading enzymes on which most emphasis is placed. Two major types of amylase are involved: alpha- and beta-amylase. Although details may vary among cereal species, the general pattern of alpha-amylase synthesis is for it to be initiated in the scutellum first, followed by a longer period of much greater synthesis in the aleurone. Synthesis occurs as a result of promotion by hormones including gibberellic acid transmitted from the embryonic axis (*via* the scutellum in the case of the aleurone). By contrast, β-amylase is synthesized during caryopsis development and remains inactive and partly in a bound form in the endosperm, where it is associated with the surfaces of starch granules (Khan and Shewry, 2009).

3.5 BREEDING

The process of fertilization described in Chapter 1 is typical of outbreeding species. However, some of the most important species of cereal do not conform to this pattern. Although other mechanisms may also be involved, one certain barrier to cross-pollination is the failure of anthers to emerge from the flowering bracts before shedding their pollen. Pollen thus falls on the stigma within the same floret, leading to self-fertilization.

Barley is about 99.5% self-pollinated as may be inferred from the strong adhesion of its flowering bracts. Rice, oats and wheat are 97%–98%, and sorghum about 94% self fertilized. Finger millet and foxtail millet also have a high degree of self-pollination. Rye, on the other hand, is almost entirely cross-pollinated and indeed may be self-sterile; maize engages in about 95% crossing and pearl millet about 80% crossing. Unlike cereals, pseudocereals are pollinated by insects; buckwheat is an obligate cross-pollinating crop on account of sporophyte self-incompatibility and cross-pollination is assisted by the existence of different floral morphologies, some plants producing pin-eyed and some producing thrum-eyed flowers. Quinoa is mainly self-pollinated but 10%–15% crossing has been reported (Risi and Galwey, 1989). Grain amaranth is predominantly self-pollinated, but variable levels of crossing are said to occur.

3.5.1 Breeding objectives

To the grower the most important characters of a cultivar may be related to the responses of the whole plant to environmental pressures, while to the processor – the miller, the maltster or the starch manufacturer – it is the grain characteristics that are most important. Poehlman (1987) lists objectives for a number of cereals, acknowledging that the emphasis may vary from one part of the world to another.

The primary objectives for all species are yield potential, yield stability (including optimum maturity date, resistance to lodging and shattering, tolerance to drought and soil stress, and resistance to disease and pests) and grain quality. The factors involved in yield stability are to some extent linked, for example, resistance to lodging (the tendency of standing crops to bend at or near ground level, thus collapsing) might involve the height of the plant, the size and weight of the ear in relation to culm strength and resistance to stem rot pathogens and to insects such as the corn borer. Reduction in height of crop plants also has the potential to increase yield by improving two factors: harvest index (see Chapter 1) and ability to stand in conditions of high soil fertility. They were, to some extent, the benefits of what has been styled the 'green revolution' of the 1960s, and their delivery ranks among the greatest breeding achievements of the 20th century.

In some cereals additional objectives include the production of hull-less varieties of oats and barley and the adaptation to mechanized harvesting of sorghum and millets. In barley, low yield remains a problem, and varieties with smooth awns are also desirable. Some winter-sown cereals such as wheat, oats and barley provide herbage for stock grazing on vegetative parts of immature plants in the autumn, and good fodder properties may thus be desirable. Fodder aspects also feature in breeding programmes for millets and sorghum.

Agronomic characters of importance to breeding programmes include the seasonal habit. Winter-sown crops of a given cereal produce greater yields than those sown in spring. Spring-sown types need to grow rapidly in order to complete the necessary phases of growth in a relatively short period. They must not require a period of cold treatment in order to germinate or to produce reproductive structures in the first flowering season (some cereals are naturally perennials although they are treated as annuals in agriculture). A more ambitious objective may be the breeding of crops with special qualities previously not associated with the species concerned. Examples of the successful achievement of this objective lie in the breeding of types with high-amylose or high-amylopectin starch, the production of golden rice to combat vitamin A deficiency, and maize with higher proportions of essential amino acids.

In the case of pseudocereals, the global distribution of which remains limited, breeders are seeking to breed varieties that are capable of adaptation to new environments. Among the top priorities for other improvements are increased yield, resistance to pests and disease and resistance to shattering. Also, in some cases existing varieties are too tall for convenient machine harvesting so shorter plants are desirable.

3.5.2 Variation

Because new genetic material is not naturally introduced into self-pollinating plants, they are homozygous. That is, the chromosomes derived from pollen bear genes that not only control the same characters (traits) as those derived from the ovum but also influence them in the same way. Continued self-pollination in successive generations would not therefore produce variation. Variation can thus be introduced by crossing with a compatible plant with different genes or by mutation – a change to genes induced by chemical, radiation or other influence.

3.5.3 Hybridization

Crossing to produce heterozygosity and selection of a pure line from an advanced line following the cross is known as hybridization. It should not be confused with the breeding of hybrids in which the F1 generation is grown commercially (see Section 3.5.4). The F1 generation is the first filial generation or that produced immediately from the cross Following a cross, heterozygosity is reduced by one half with each successive self-fertilization. Complete homozygosity is theoretically unattainable by conventional methods but a practical state of uniformity is normally reached after five to eight generations of selfing.

A more rapid method of achieving homozygosity is by creating double haploids. Haploid embryos can produce plants that spontaneously diploidize, but induced dipoidization usually involves the use of colchicine. Colchicine disrupts the mitotic process at the point where daughter chromosomes migrate to opposite poles of the spindle. Instead of migrating, the daughter chromosomes remain together in the same cell, which can then divide, in culture, without mitotic disruption, to produce a new plant with two sets of identical chromosomes. While both male and female gametes can form the basis of double haploid production the source material is usually pollen as it is abundant and easily isolated.

Strategies for hybridization are necessarily different in self-pollinated types, in which true breeding lines can be easily established, and

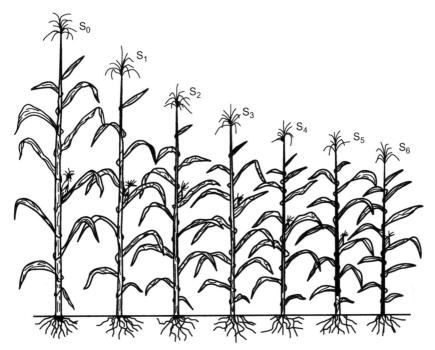

FIGURE 3.43   Reduction in vigour in maize with successive generations of inbreeding. S_0 represents the original inbred plant and S_{1-6} successive inbred generations. *Reproduced from Poehlman, J.M., 1987. Breeding Field Crops, third ed. AVI Publishing Company Ltd., Westport, Connecticut, USA, by courtesy of AVI Publishers, New York.*

out-crossing types where this is difficult and may even lead to progressive deterioration of stocks (Fig. 3.43).

3.5.4 Heterosis

Whereas continued inbreeding of outbreeding species leads to loss of vigour, F1 hybrids produced from inbred lines exhibit heterosis or hybrid vigour. This is the increase in size or vigour of a hybrid over the average of its parents, which is referred to as the mid-parent value. In practice, hybrid vigour is the increase in size and vigour over the better parent as this is the only achievement of interest to growers or processors. The mechanism by which heterosis enhances yield is not fully understood, but a comprehensive analysis of its effects has been provided by Blum (2013).

When growing hybrid maize, the farmer generally obtains fresh F1 hybrid seed each year from growers who specialize in its production. The specialized grower chooses an isolated field on which he grows two inbred lines of maize in alternate strips, one row of male parent plants to about four rows of female parent plants. At the appropriate time, the female parent plants are detasselled and then subsequently fertilized by pollen from the adjacent male parent plants. Seed is later collected from the female parent only. An alternative method of producing hybrid maize is by use of male sterility, which may be genetic or cytoplasmic, the latter being induced by the extranuclear (mitochondrial) genome.

Before 1970, over 90% of maize seed in the United States produced in Texas utilized male-sterile cytoplasm; but after 1970, when the fungus *Helminthosporium maydis* race T caused much damage, cytoplasmic male sterility was not used for a few years.

In sorghum, both male and female reproductive parts occur in the same spikelets, and commercial production of hybrid sorghum seed awaited development of cytoplasmic male-sterile types. Hybrid sorghum was first grown commercially in 1956, and by the late 1950s, 58% and 22% increases in yield over that of the better parent were reported. A further increase of 40% was later recorded but only slightly lower increases also occurred in the inbred parents over the same period (Rooney and Smith, 2000).

Rice hybridization began in China in 1964 and following the discovery of male sterility in a wild rice, and transferring it to cultivated types, commercial hybrid rices became a reality in 1974. By 1976 hybrid production was massive. In recent years hybrid rice accounts for nearly 60% of rice area in China. Yields are regularly 20% higher than those of conventional rice varieties and a record of 15 tonnes/ha have been recorded. Hybrid rice has now been grown in 40 countries (Yuan, 2014). Some reports of serious disease susceptibility in adverse weather conditions appeared in the Chinese press in 2016 (China Daily, 2016).

Early cytoplasmic-genetic hybrids of pearl millet produced in India in the 1960s became widely accepted (around 50% of area) but fell victim to downy mildew (*Sclerospora graminicola*) disease. Replacement by new types became driven by the quest for disease resistance rather than improved yield. However, in the early 2000s, 60% of the pearl millet area was sown to more than 70 varieties, bred to suit the wide variety of growing conditions.

Male sterility, both chemically induced and due to cytoplasmic-genetic factors, has been exploited in production of commercial wheat F1 hybrids, but factors including high cost of seed production and limited perceived advantage have prevented wide adoption. Similarly, commercial barley hybrids have had little impact on world barley production (Ullrich, 2010). Little activity has been reported for F1 oat production.

3.5.5 Evolution of breeding strategies

Breeding techniques used are sophisticated; they have progressed from simple selection to complex crossing programmes and use of biotechnology and induced mutation. The progressive development of plant-breeding strategies has been reviewed by Bresseghello and Coelho (2013), and their analysis is summarized in Table 3.3.

TABLE 3.3   Plant breeding methods

Breeding strategy	Essential features and benefits	Limitations
PLANT BREEDING BASED ON OBSERVED VARIATION		
Plant domestication: The origin of crops	Rescue and protection of naturally selected mutants by farmers.	Failure of the selections to retain some useful genes and traits, e.g., disease and pest resistance.
Intuitive farmer selection: The origin of landraces	Development of varieties adapted to local conditions as a result of continuous isolation.	Suited to conditions and preferences of local area but some undesirable characteristics may be retained, e.g., tall habit.
Pure line selection and mass selection: The origin of cultivars	Incorporation of knowledge of inheritance laws. Parents selected from landraces for desired traits.	Pure lines less able to resist stress than diverse populations.
PLANT BREEDING BASED ON CONTROLLED MATING		
Pedigree breeding: Playing with parents	Crossing parents and generating segregating populations conducted through generations of self-pollination and selection. The majority of released varieties have resulted from this strategy.	Yield only evaluated at the end of the process when sufficient seed available.
Idiotype breeding: Playing with traits	A refinement of pedigree breeding. Improving complex traits by changing simpler ones associated with them.	Unfavourable genetic correlations can offset advantages brought by the traits that make up the idiotype.
Population breeding: playing with genetic variance	Also known as 'recurrent selection breeding'. Increasing the frequency of favourable alleles controlling traits of interest.	Must maintain large effective population to prevent genetic drift overriding other factors. Some issues with self-pollinating types.
Hybrid breeding: playing with heterosis	Gaining advantage from heterosis (superiority of hybrids over inbred individuals).	Less easily applied to self-pollinated types. Need to create and maintain populations sufficiently distinct for strong heterosis.
PLANT BREEDING BASED ON MONITORED RECOMBINATION		
Mapping genes of interest: finding needles in the haystack	Using quantitative trait loci (QTL) maps to determine the number of loci controlling a trait. Markers used in selection.	Access to expanding gene banks and gene libraries required.

Continued

TABLE 3.3   Plant breeding methods —cont'd

Breeding strategy	Essential features and benefits	Limitations
PLANT BREEDING BASED ON OBSERVED VARIATION		
Marker-assisted selection: building tailored genotypes	As above but several genes monitored simultaneously.	Dependent on previous QTL etc. research.
Genomic selection: speeding up genetic progress	Using all information available on markers in a plant to predict its breeding value, without having to evaluate phenotypes.	Depends on results from a 'training population'.

Based on Breseghello, F., Coelho, A.S.G., 2013. Traditional and modern plant breeding methods with examples in rice (Oryza sativa L.). Agricultural and Food Chemistry 6, 8277–8286.

3.5.6 Ploidy

Genes capable of coding for all plant characters are accommodated on a relatively small number of chromosomes. In wheat the minimum number is seven. Somatic cells are diploid, that is they contain two sets of chromosomes, they thus have two genes for each character. In the simplest case, one gene of each pair is dominant and it is this that is expressed, however, other degrees of cooperative expression also occur.

Organisms with an exact multiple of the haploid (gametic) number of chromosomes for the species are termed 'euploid'. When an organism lacks an individual chromosome or has an additional one its condition is described as 'aneuploid'. The vast majority of animals and plants are diploid and fall into the euploid category. Within the category also some plants have two or three or even four pairs of chromosomes per somatic cell, and these are termed polyploid. Polyploid plants may arise as a result of duplication of the inherent chromosomes to form autopolyploids (or autoploids), or by combining genomes from two or more species, leading to allopolyploids (or alloploids). Polyploids of both types can occur in nature or they can be induced by breeders. Phenotypically, they are larger than either parent and generally also exhibit greater vigour. An example of a natural allopolyploid is bread wheat, *Triticum aestivum*. It is hexaploid, with six of each type of chromosome in each somatic cell. Details of its genetic constitution and that of other wheats are given in Chapter 1. An example of a 'man-made' allopolyploid cereal is triticale, and this is also discussed in Chapter 1.

3.5.7 Genetic enhancement, genetic modification and genetically modified organisms

Whether crosses are achieved naturally or by design, crossing results in an enormous range of progeny and, while only genes, and hence characters (traits), originating in at least one of its parents are expressed in each of the resulting organisms, the suites of companion genes present in each organism are bound to vary. Consequently, selection has to be made in order that only those displaying acceptable combinations of characters are multiplied for establishment of new varieties. In nature selection may occur purely by ability to survive, but for the plant breeder, additional requirements are involved, and these are defined by the objectives conceived when the parents were chosen.

As a means of illustrating the benefits of genetic engineering, it is helpful to consider a scenario in which a breeder is in the happy position of having a parent available that perfectly matches his objectives except for one or more characters. The breeder's target is thus to introduce those missing traits into the 'perfect' background. Using conventional breeding methods, crosses would be made with a second parent in which the target traits are expressed, but it may be flawed in other ways. Because of the genetic background of the second parent, many of the progeny would display less-desirable characters, even though the target traits may be present. Hence, an extensive programme would be needed to select a 'perfect' organism. It is in such a case that genetic engineering offers a solution.

Genetic modification is essentially a technique that allows one or more genes to be isolated from one organism and introduced into the genetic complement of another. Where more than one gene is introduced the term 'gene stacking' is applied. It allows several components required for the generation of the missing character to be introduced, e.g., multiple enzymes involved in the production of a valuable metabolite or nutrient such as β-carotene in golden rice.

As with any novel complex process, a specific vocabulary has developed; genetic enhancement (GE) and genetic modification (GM) are used as alternative terms to mean much the same thing. The term 'biotechnology' or 'biotech' is also frequently applied in the same context as GM, although it has a broader definition, including the adoption of biological processes into industrial applications. The term 'transformation' describes the process of introducing isolated or cloned genetic material into the genome. It can clearly bring benefit to the illustrative example just discussed, but there are additional benefits because isolation of a gene removes obstacles to compatibility that prevent genes from one species becoming part of the genetic complement of another. In fact, because of the universality of DNA as the inheritance mechanism, compatibility can be achieved across many taxonomic boundaries. Such manipulation is described as 'transgenic', while transformations with genes from closely related species are described as 'cisgenic'. The term transgenic is also used by some as a more general name for any artificially transferred gene.

To describe as genetically modified only new varieties created without mating or natural recombination is to imply that no genetic modification occurred before the introduction of transformation by breeders. This of course is untrue because modification to genomes occurs naturally and it has been the basis of breeding since it was first practiced. Among the unique features of what is understood by GM are the means by which new genes are introduced into cells of the target organism. These include the use of microbial vectors or microprojectile bombardment (biolistics), two techniques that have been applied successfully to cereals. From the former category one of the most successful vectors used is *Agrobacterium tumefaciens*. It is an unusual pathogen in that, following infection, it transfers part of its own DNA (called T-DNA) into the host genome (an example of transformation occurring in nature). Scientists are able to create strains of *A. tumefaciens* in which the pathogen's T-DNA is replaced with DNA, isolated from another organism, that codes for traits that they wish to introduce into the host (Institute of Medicine and National Research Council for the National Academies, 2004). Biolistics may appear to be a less-sophisticated technique, but it is effective and widely adopted for stable transformation of monocotyledonous species. Microscopic pellets of gold or tungsten are coated with DNA capable of being advantageously incorporated into the host genome and fired from a 'gene gun' into appropriate host tissue. It is perhaps significant that in GM it is usually the chromosomes of somatic cells that are changed so that variation is achieved without the need for the normal fertilization process being involved.

Many stages are involved in the production of transformed crops, including several that have been in use in conventional breeding programmes such as progeny testing and backcrossing to ensure that new desired traits are consistently expressed in the stable desired background.

Reference has been made in Chapter 1 to the success of transgenic maize in the United States and some other countries, but in many countries there is considerable resistance among the public and governments to GM crops. While the benefits of GM maize are enjoyed mainly by the growers, a transgenic rice was developed to benefit consumers. The series of 'golden rice' varieties was developed to counter vitamin A deficiency, which is widespread in parts of the developing world. Genes coding for synthesis of betacarotene, which is a precursor of vitamin A, were isolated from daffodil and the bacterium *Erwinia uredovara* and inserted into the rice genome (Black et al., 2006). In spite of the potential benefits to large numbers of people, distribution has been delayed by individuals and agencies opposed to the method of production. There are some naturally occurring varieties of pearl millet with yellow endosperm, due to its betacarotene content, which have the same potential for improving nutrition.

Although wheat proved to be the most difficult of the major cereals to transform, it has now been demonstrated that viable GM wheats can be created. Scientists involved have identified many potential agronomic, technological and nutritional improvements that might be achieved

(Jones et al., 2009), but incentives to invest in them are limited in view of the low level of public acceptance.

3.5.7.1 *Genome editing*

Genome editing is a more recently introduced technique, whereby any changes to a selected chromosome can be made with precision at a predetermined point. A requirement for the technique is that the chromosome has been mapped so that a nuclease can be constructed to recognize the sequence at a point where a separation is deemed desirable. The introduction of the nuclease then brings about a break in the double-stranded DNA at the chosen point. Deletions can thus be made precisely at the breakpoints or insertions of new genetic material can be made equally precisely. Induced mutations at chosen points are also possible. Further information on the technique can be found in, e.g., Huang et al. (2016).

References

Abdelrhaman, A., Hoseney, R.C., Varriano-Marston, E., 1984. The proportions and chemical compositions of hand-dissected anatomical parts of pearl millet. Journal of Cereal Science 2, 127–133.

Bechtel, D.B., Zayas, I., Kalekau, L., Pomeranz, Y., 1990. Size distribution of wheat starch granules during endosperm development. Cereal Chemistry 67, 59–63.

Bechtel, D.B., Abecassis, J., Shewry, P.R., Evers, A.D., 2009. Development, structure, and mechanical properties of the wheat grain. In: Khan, K., Shewry, P.R. (Eds.), Wheat Chemistry and Technology, fourth ed. American Association of Cereal Chemists Inc., St. Paul, MN, USA, pp. 51–96.

Bennet, D.M., Parry, D.W., 1981. Electron probe microanalysis studies of silicon in the epicarp hairs of the caryopses of *Hordeum sativum* Jess., *Avena sativa* L., *Secale cereale* L. and *Triticum aestivum* L. Annals of Botany 6, 645–654.

Black, M., Bewley, J.D., Harmer, P. (Eds.), 2006. The Encyclopedia of Seeds, Science, Technology and Uses. CABI, Cambridge, MA, USA.

Blum, A., 2013. Heterosis, stress and the environment: a possible road map towards a general improvement of crop yield. Journal of Experimental Botany 64 (16), 4829–4837.

Bremner, P.M., Rawson, H.M., 1978. The weights of individual grains in the wheat ear in relation to their growth potential, the supply of assimilate and interaction between grains. Australian Journal of Plant Physiology 5, 61–72.

Breseghello, F., Coelho, A.S.G., 2013. Traditional and modern plant breeding methods with examples in rice (*Oryza sativa* L.). Agricultural and Food Chemistry 6, 8277–8286.

Briggs, D.E., 1978. Barley. Chapman and Hall, London, UK.

Bushuk (Ed.), 2001. Rye: Production, Chemistry and Technology, second ed. American Association of Cereal Chemistry Inc., St. Paul, MN, USA.

China Daily, May 17, 2016.

Christou, P., 1994. Rice Biotechnology and Genetic Engineering. Technomic Publishing Company Inc., Lancaster, PA, USA.

Chung, O.K., Ohm, J.-B., Ram, M.S., Howitt, C.A., 2009. Wheat lipids. In: Khan, Shewry, P.R. (Eds.), Wheat Chemistry and Technology, fourth ed. American Association of Cereal Chemists Inc., St. Paul, MN, USA.

Clayton, W.D., Vorontsova, M.S., Harlman, K.T., Williamson. 2012. Grassbase – The Online World Grass Flora. Royal Botanic Gardens, Kew.

Cochrane, P., 1994. Observations on the germ aleurone of barley. Phenol oxidase and peroxidase activity. Annals of Botany 73 (2), 121–128.

Doust, A., Kellogg, A., 2002. Inflorescence diversification in the panicoid "bristle grass" clade (Panicoideae, Poaceae): evidence from molecular phylogenies and developmental morphology. American Journal of Botany 89, 1203–1222.

Dykes, L., Rooney, L.W., 2006. Sorghum and millet phenols and antioxidants. Journal of Cereal Science 44, 236–251.

Earp, C.F., McDonough, C.M., Awika, J., Rooney, L.W., 2004. Testa development in the caryopsis of *Sorghum bicolor* (L) Moench. Journal of Cereal Science 39, 303–311.

Elbaum, R., Zaltman, L., Burgert, I., Fratzi, P., 2007. The role of wheat awns in the seed dispersal unit. Science 316, 884–886.

Evers, A.D., 1970. Development of the endosperm of wheat. Annals of Botany 34, 547–555.

Evers, A.D., Reed, M., 1988. Some novel observations by scanning electron microscopy on the seed coat and nucellus of the mature wheat grain. Cereal Chemistry 65, 81–85.

Gao, F., Wang, K., Liu, Y., Chen, Y., Chen, P., Shi, Z., Luo, J., Jiang, D., Fan, F., Zhu, Y., Li, S., 2015. Blocking miR396 increases rice yield by shaping inflorescence architecture. Nature Plants. 2:15196. www.nature.com/nplants2015196.

Gould, F.W., Shaw, R.B., 1983. Grass Systematics. Texas A and M University Press, USA.

Grubben, G.J.H., Partohardjono, S. (Eds.), 1996. Plant Resources of South-East Asia No. 10 Cereals. Backhuys Publishers, Leiden, Netherlands.

Hands, P., Kourmpetli, S., Sharples, D., Harris, R.G., Drea, S., 2012. Analysis of grain characters in temperate grasses reveals distinctive patterns of endosperm organization associated with grain shape. Journal of Experimental Botany 63, 6253–6266.

Hayakawa, T., Seo, S.W., Igave, I., 1980. Electron microscopic observations of rice grain. Part 1. Morphology of rice starch. Journal of the Japanese Society of Starch Science 27, 173–179.

Himi, E., Nisar, A., Noda, K., 2005. Colour Genes (R and Rc) for grain and coleoptile upregulate flavinoid biosynthesis genes in wheat. Genome 48 (4), 747–754.

Hodson, M.J., Sangster, A.G., Wymm Parry, D., 1982. Silicon deposition in the inflorescence bristles of *Setaria italica* (L) Beauv. Annals of Botany 50, 843–850.

Hoshikawa, K., 1967. Studies on the development of rice 2. Process of endosperm tissue formation with special reference to the enlargement of cells and 3. Observations the cell division. Proceedings of the Crop Science Society of Japan 36, 203–216.

Huang, S., Weigel, D., Beachy, R.N., Li, J., 2016. A proposed regulatory framework for genome-edited crops. Nature Genetics 48, 109–111.

Hulse, J.H., Laing, E.M., Pearson, O.E., 1980. Sorghum and the millets: their composition and nutritive value. Academic Press Inc., London, UK.

Hutchinson, J.B., Kent, N.L., Martin, H.F., 1952. The kernel content of oats. Variation in percentage kernel content and 1,000 kernel weight within the variety. Journal of the National Institute of Agricultural Botany 6, p.149.

Institute of Medicine and National Research Council of the National Academies, 2004. Methods and mechanisms for genetic manipulation of plants, animals and microorganisms. In: Safety of Genetically Engineered Foods. The National Academies Press, Washington, DC, USA.

Jones, H.D., Sparks, C.A., Shewry, P.R., 2009. Transgenic manipulation of wheat quality. In: Khan, K., Shewry, P.R. (Eds.), Wheat Chemistry and Technology, fourth ed. American Association of Cereal Chemists International Inc., St. Paul, Minnesota, USA, pp. 437–452.

Kent, N.L., 1983. Technology of Cereals, third ed. Pergamon Press Ltd, Oxford.

Khan, K., Shewry, P.R., 2009. Wheat Chemistry and Technology, fourth ed. American Association of Cereal Chemists Inc., St. Paul, MN, USA.

Kiesselbach, T.A., 1980. The Structure and Reproduction of Corn. University of Nebraska Press, Lincoln, NE, USA (Reprint of 1949 edn.).

Mangelsdorf, P.C., 1953. Wheat. Scientific American 189, 50–59 (New York).

Marshall, A., Cowan, S., Edwards, S., Griffiths, I., Howarth, C., Langdon, T., White, E., 2013. Crops that feed the world 9. Oats- a cereal crop for human and livestock feed with industrial applications. Food Security 5 (1). http://dx.doi.org/10.1007/s12571-012-0232-x. Springer (online).

McDonaugh, C.M., Rooney, L.W., 1989. Structural characteristics of Pennisetum americanun (pearl millet) using scanning electron microscopy and fluorescence microscopy. Food Structure 8, 137–149.

Nesbitt, M., Samuel, D., 1996. From staple crop to extinction? The archaeology and history of the hulled wheats. In: Padulosi, S., Hammer, K., Heller, J. (Eds.), Hulled Wheats. Proceedings of the First International Workshop on Hulled Wheats 21–22 July, 1995. International Genetic Resources Institute, Rome, Italy.

Olsen, O.-A., 2004. Nuclear endosperm development in cereals and Arabidopsis thaliana. The Plant Cell 16 (Suppl. 1), S214–S227.

Padulosi, S., Hammer, K., Heller, J.r (Eds.), 1996. Hulled Wheats. Proceedings of the First International Workshop on Hulled Wheats 21–22 July, 1995. International Genetic Resources Institute, Rome, Italy.

Palmer, G.H., 1989. Cereals in malting and brewing. In: Palmer, G.H. (Ed.), Cereal Science and Technology. Aberdeen University Press, Aberdeen, Scotland, pp. 61–242 (Chapter 3).

Percival, J., 1921. The Wheat Plant. Duckwoth and Co., London, UK (Reprinted 1975).

Poehlman, J.M., 1987. Breeding Field Crops, third ed. AVI Publishing Company Ltd, Westport, Connecticut, USA.

Prego, I., Maldonado, S., Otegui, M., 1998. Seed structure and localization of reserves in Chenopodium quinoa. Annals of Botany 82, 481–488.

Radley, M., 1976. The development of wheat grains in relation to endogenous growth substances. Journal of Experimental Botany 27, 1009–1021.

Rebetzke, G.J., Bonnett, D.G., Reynolds, M.P., 2016. Awns reduce grain number to increase grain size and harvested yield in irrigated and rainfed spring wheat. Journal of Experimental Botany 67 (9), 2573–2586.

Risi, J., Galwey, N.W., 1989. Chenopodium grains of the Andes: a crop for the temperate latitudes. In: Wickens, G.E., Haq, N., Day, P. (Eds.), New Crops for Food and Industry. Chapman and Hall, New York, USA.

Rocha, S., Monjardino, P., Mendonca, D., Machardo, A., Fernandes, R., Sampaio, P., Salema, R., 2014. Lignification of developing maize (Zea mais L.) endosperm transfer cells and starchy endosperm cells. Frontiers in Plant Science. 5 (102), 71–80. www.frontiersin.org.

Rooney, L.W., 2000. Genetics and cytogenetics. In: Smith, C.W., Frederiksen, R.A. (Eds.), Sorghum, Origin, History, Technology and Production. John Wiley & Sons Inc., New York, USA.

Rooney, W.L., Smith, C.W., 2000. Techniques for developing new varieties. In: Smith, C.W., Frederiksen, R.A. (Eds.), Sorghum, Origin, History, Technology and Production. John Wiley & Sons Inc., New York, USA.

Sabelli, P.A., Larkins, B.A., 2009. The development of endosperm in grasses. Plant Physiology 149, 14–26.

Salunkhe, D.K., Chavan, J.K., Kadan, S.S., 1990. Dietary Tannins: Consequences and Remedies. CRC Press Inc., Boca Ratan, FL, USA.

Siwela, M., Taylor, J.R.N., deMilliano, A.J., Duodo, K.G., 2007. Occurrence and location of tannins in finger millet grain and antioxidant activity of different grain types. Cereal Chemistry 84, 169–174.

Stafford, W.E. (Ed.), 1917. Proceedings of the 19th Congress of Americanists 1917.

Swift, J.G., O'Brien, T.P., 1970. Vascularization of the scutellum of wheat. Australian Journal of Botany 18, 45–53.

Taylor, J.R.N., Parker, M.L., 2002. Quinoa. In: Belton, P.S., Taylor, J.R.N. (Eds.), Pseudocereals and Less Common Cereals. Springer Verlag, Berlin, Germany.

Ullrich, S.E., 2010. Barley, Production, Improvement and Uses. Wiley Blackwell, Oxford, USA.

Umeta, M., Parker, M.L., 1996. Microscopic studies of the major macro-components of seeds, dough and *injera* from tef (*Eragrostis tef*) SINET. Ethiopian Journal of Science 19, 141–148.

Varriano-Marston, E., De Francisco, A., 1984. Ultrastructure of quinoa fruit (*Chenopodium quinoa* Willd). Food Microstructure 3, 165–175.

Wani, A.A., Singh, P., Shah, M.A., Schweiggert-Weisz, V., Gul, K., Wani, I.A., 2012. Rice starch diversity: effects on structural, morphological, thermal and physicochemical properties – a review. Comprehensive reviews in food science and food safety. Institute of Food Technologists 11, 417–436.

Watson, S.A., 1987. Structure and composition. In: Watson, S.A., Ramstat, P.E. (Eds.), Corn: Chemistry and Technology. American Association of Cereal Chemists Inc., St. Paul, MN.

Winton, A.L., Winton, K.B., 1932. The Structure and Composition of Foods. Cereals, Starch, Oil Seeds, Nuts, Oils, Forage Plants, vol. I. Chapman and Hall, London, UK.

Yuan, L.-P., 2014. Development of hybrid rice to ensure food security. Science Direct 21 (1), 1–2.

Zheng, Y., Wang, Z., 2010. Structural character of sorghum endosperm transfer cells and their relationship with embryo and endosperm. International Journal of Plant Biology 1 (2) Zheng (open access).

Further Reading

Becraft, P.W., Yi, G., 2010. Regulation of aleurone development in cereal grains. Journal of Experimental Botany 62, 1669–1675.

Bonnett, O.T., 1966. Inflorescences of Maize, Wheat, Rye Barley and Oats: Their Initiation and Development. Bulletin 721. University of Illinois College of Agriculture, USA.

Das, S., 2016. Amaranthus: A Promising Crop of Future. Springer Science-Business Media, Singapore.

Evers, T., Miller, S., 2002. Cereal grain structure and development: some implications for quality. Journal of Cereal Science 36, 261–284.

Gomez-Pando, L., 2015. Quinoa breeding. In: Murphy, K., Matariguihan, J. (Eds.), Quinoa: Sustainable Production, Variety Improvement and Nutritive Value in Agroecological Systems. John Wiley & Sons Inc., Hoboken, New Jersey, USA (Chapter 6).

Cheng, S., Sun, Y., Halgreen, L., 2009. The relationships of sorghum kernel pericarp and testa characteristics with tannin content. Asian Journal of Crop Science 1 (1), 1–5.

Lu, B.S. (Ed.), 1991. Rice Volume II Utilization, second ed. Van Norstrand Rheinhold, New York.

Rooney, L.W., Murty, D.S., 1982. International Symposium on Sorghum Grain Quality. ICRISAT, Patancharu, India.

Sato-Nagasawa, N., Nagasawa, N., Malcomber, S., Sakai, H., Jackson, D., 2006. A trehalose metabolic enzyme controls inflorescence architecture in maize. Nature 441, 227–230.

Scott, M.P., 2009. Transgenic Maize: Methods and Protocols. New York, Springer/Humana Press, USA.

Vaughan, J.G., Geissler, 1997. The New Oxford Book of Food Plants. Oxford University Press, Oxford, UK.

C H A P T E R

4

Chemical components and nutrition

4.1 INTRODUCTION

An understanding of the chemical composition of cereal grains is critical to their effective utilization. Nutrient components of the seeds are important to the germination and growth of new plants. Storage and processing can alter composition, which may not be desirable, although with some processing operations chemical changes are required. Ultimately the chemistry of cereal grains and the products which are made from them affects the nutrition and health of the humans and animals consuming them.

PART 1. CHEMICAL COMPOSITION

4.2 CARBOHYDRATES

The terminology surrounding carbohydrates frequently serves to confuse rather than to clarify. Archaic and modern conventions are often intermixed and definitions of some terms are inconsistent in their use. Even the term carbohydrate itself is not entirely valid. It originated in the belief that naturally occurring compounds of this class could be represented formally as hydrates of carbon, i.e., $C_x(H_2O)_y$. This definition is too rigid, however, as important deoxy sugars like rhamnose, the uronic acids and compounds such as ascorbic acid would be excluded and acetic acid and phloroglucinol would qualify for inclusion. Nevertheless, the term carbohydrate remains to describe those polyhydroxy compounds which reduce Fehlings solution either before or after hydrolysis with mineral acids (Percival, 1962).

It is customary to classify carbohydrates according to their degree of polymerization; thus monosaccharides (1 unit), oligosaccharides (2–20 units) and polysaccharides (>20 units).

4.2.1 Monosaccharides

Monosaccharides are the simplest carbohydrates; most of them are sugars. Monosaccharides may have between three and eight carbon atoms, but only those with five carbons (pentoses) and six carbons (hexoses) are common. Both pentoses and hexoses exist in a number of isomeric forms – they may be polyhydroxyaldehydes or polyhydroxyketones. Structurally they occur in ring form, which may be six-membered (pyranose form) or five-membered (furanose form).

In mature cereal grains the monomers are of little importance in their own right, but as components of polymers they are of extreme importance in their contribution to both the structural and storage components of the grain and the behaviour of grains and their products during processing. In this context the most important monosaccharide, because of its abundance, is the six-carbon polyhydroxyaldehyde D-glucose. It is the monomeric unit of starch, cellulose and β-D-glucans.

The most important pentoses are the polyhydroxyaldehydes D-xylose and L-arabinose, because of their contribution to cell wall polymers. The structures of these compounds and of some other monosaccharides found in cereals are shown in Fig. 4.1.

The most abundant derivatives of monosaccharides are those in which the reducing group forms a glycosidic link with the hydroxyl group of another organic compound (as in Fig. 4.2), frequently another molecule of the same species or another monosaccharide. Sugar molecules may be joined to form short or long chains by series of glycosidic links, thus producing oligosaccharides or polysaccharides.

4.2.2 Oligosaccharides

The smallest oligosaccharide, the disaccharide, comprises two sugar molecules joined by a glycosidic link. Although this may appear to be a simple association, it is capable of considerable variation according to the configuration of the glycosidic link and the position of the hydroxyl group involved in the bonding. Three important variants among disaccharides involving only α-D-glucopyranose are shown in Fig. 4.3.

In these compounds the reducing group of only one of the monosaccharide molecules is involved in the glycosidic link and the reducing group of the other remains functional.

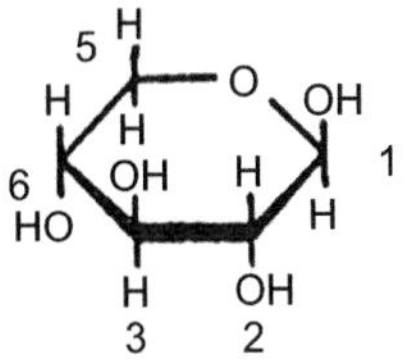

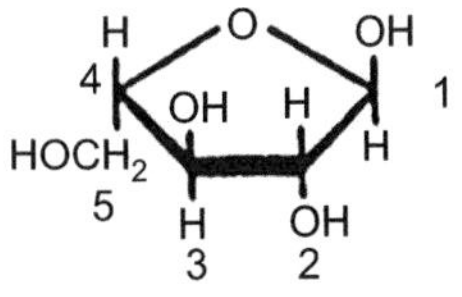

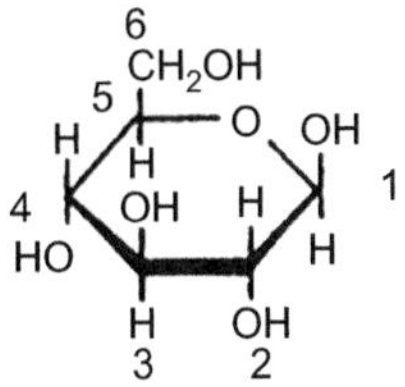

FIGURE 4.1 Structural representations of (1) xylose (β-D-xylopyranose), (2) arabinose (α-L-arabinofursnose), (3) glucose (β-D-glucopyranose), (4) fructose (β-D-fructofuranose), (5) D-galacturonic acid, (6) ribose (β-D-ribofuranose), (7) deoxyribose (β-D-deoxyribofuranose) and (8) mannose (α-D-manno-pyranose).

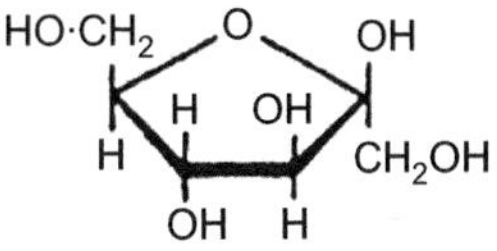

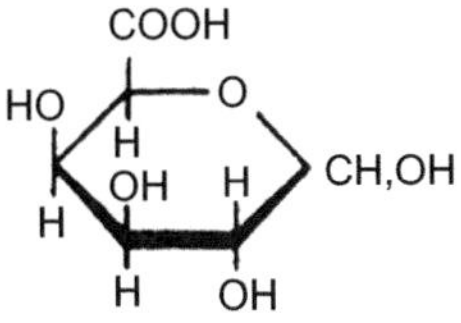

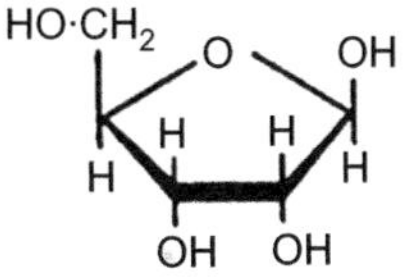

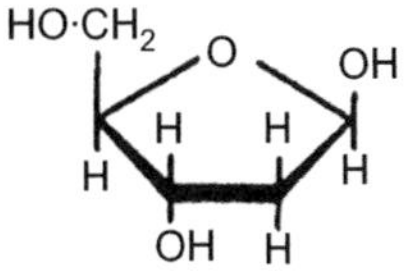

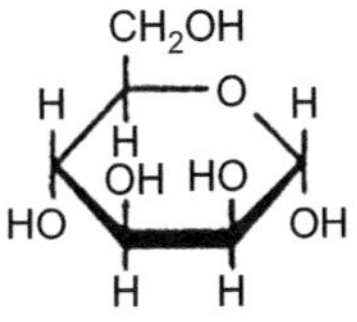

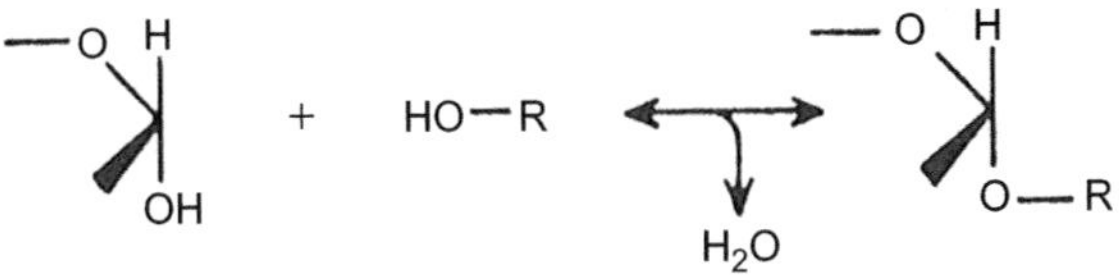

FIGURE 4.2 Formation of the glycosidic link.

In sucrose, another important disaccharide found in plants, fructose and glucose residues are joined through the reducing groups of both; hence their reducing properties are lost. Sucrose is readily hydrolysed under mildly acid conditions, or enzymically, to yield its component monomers, which of course again behave as reducing sugars. Sucrose is the main carbon compound involved in translocating photosynthate to the grain. It features prominently during development rather than in the mature grain because it is converted during maturation, to structural and longer-term storage carbohydrates such as starch. In sweetcorn the sucrose content is higher by a factor of two to four throughout grain development than in other types of maize at similar stages, as the rate of conversion is slower (Boyer and Shannon, 1983).

Literature values for sugars in cereals vary with methods of analysis and varieties examined, and in consequence tables which bring together results of different authors can be misleading. Henry (1985) analysed two varieties of each of six cereal species. All results were obtained by the same methods and are thus comparable. Values for free glucose and total glucose (including that in sucrose and trisaccharides) are given in Table 4.1.

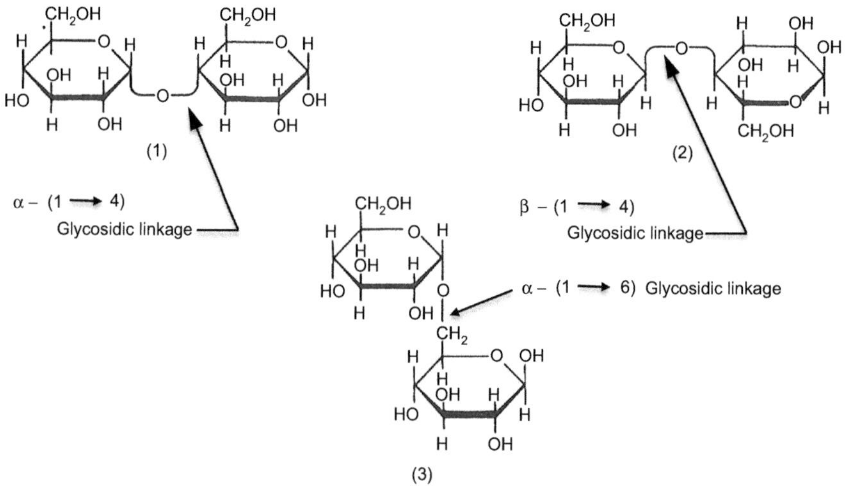

FIGURE 4.3 Structural conformation of (1) maltose (α-D-glucopyronosyl-(1→4)-α-D-glucopyranose), (2) cellobiose (β-D-gluco-pyranosyl-(1→4)-α-D-glucopyranose) and (3) isomaltose (α-D-glucopyranosyl-(1→6)-β-D-glucopyranose).

TABLE 4.1　Total soluble glucose and fructose in two varieties of each of six cereals

	Barley	Oat	Rice	Rye	Triticale	Wheat
Glucose	0.17	0.12	0.14	0.21	0.25	0.11
	0.09	0.13	0.19	0.29	0.21	0.11
Fructose	2.31	1.01	0.84	5.79	3.22	1.73
	1.98	1.00	0.75	5.11	3.05	2.46

Data from Henry, R.J., 1985. A comparison of the non-starch carbohydrates in cereal grains. Journal of the Science of Food and Agriculture 36, 1243–1253.

TABLE 4.2 Proportions of free sugars in the anatomical fractions of maize grain

Grain part	% of dry matter
Endosperm	0.5–0.8
Embryo	10.0–12.5
Pericarp	0.2–0.4
Tip cap	1.6
Whole grain	1.61–2.22

Data from Watson, S.A., 1987. Structure and composition. In: Watson, S.A., Ramstad, P.E. (Eds.), Corn: Chemistry and Technology. Amer. Assoc. of Cereal Chemists Inc., St. Paul, MN, USA, pp. 53–82.

TABLE 4.3 Proportions of soluble sugars in mill fractions of rice

Mill fraction	% of dry matter
Rough	0.5–1.2
Brown	0.7–1.3
Milled	0.22–0.45
Hull	0.6
Bran	5.5–6.9
Embryo	8.0–12.0

Data from Juliano, B.O., Bechtel, D.B., 1985. The rice grain and its gross composition (Chapter 2). In: Juliano, B.O. (Ed.), Rice Chemistry and Technology. Amer. Assoc. of Cereal Chemists Inc., St. Paul, MN, USA, pp. 17–57.

Free sugars are not distributed uniformly throughout the grain. The distribution in maize grain is shown in Table 4.2. The embryo also has the highest concentration of free sugars in other cereals. This is reflected in the distribution among mill fractions, as illustrated with respect to rice in Table 4.3.

4.2.3 Polysaccharides

Oligomers and polymers in which glucose residues are linked by glycosidic bonds are known as glucans. The starch polymers, amylose and amylopectin, are glucans in which the α-(1→4)-link features, as in maltose (Fig. 4.2). Additionally, in amylopectin the α-(1→6)-link, as in isomaltose (Fig. 4.3), occurs, giving rise to branch points. The same linkages are present in the other main storage carbohydrate found in sweetcorn. The product, known as phytoglycogen, is highly branched with α-(1→4) unit chain lengths of 10–14 glucose residues and outer chains of 6 to 30 units (Marshall and Whelan, 1974). Unlike the true starch polymers, phytoglycogen is largely soluble in water and as a result the soluble saccharides of sweetcorn contribute about 12% of the total grain dry weight. The starch polymers are discussed at greater length later in this chapter.

In cellulose the β-(1→4) form of linkage occurs, as present in cellobiose (Fig. 4.3). β-Links are also involved in the other important cell wall components, (1→3,1→4)-β-D-glucan. These polymers contribute about a quarter of the cell walls of wheat aleurone but are particularly important in oat and barley grains in the starchy endosperm, of which they may contribute as much as 70% (Fincher and Stone, 1986). With water they form viscous gums and contribute significantly to dietary fibre. Both the ratio of (1→3) to (1→4) links and the number of similar bonds in an uninterrupted sequence differ between the species. Extraction and analysis of the mixed linkage compounds are particularly difficult in the presence of such large excesses of α-glucan (Wood, 1986).

4.2.4 Pentosans

While glucans are polymers of a single sugar species, the common pentosans (polymers of pentose sugars) comprise two or more different species, each in a different isomeric form. Thus arabinoxylans, found in endosperm cell walls of wheat and other cereals, have a xylanopyranosyl backbone to which are attached single arabinofuranosyl residues (Fig. 4.4).

4.2.5 Starch

Starch is the most abundant carbohydrate in all cereal grains, constituting about 64% of the dry matter of the entire wheat grain (about 70% of the endosperm), about 73% of the dry matter of the dent maize grain and 62% of the proso millet grain. It occurs as discrete granules of up to 30 μm diameter and characteristic of the species in shape.

Starch granules are solid, optically clear bodies that appear white when seen as a bulk powder because of light scattering and the starch–air interface. They have a refractive index of about 1.5. Specific gravity depends upon moisture content but is about 1.5. The mysteries of granule structure, development and behaviour have exercised the minds of scientists

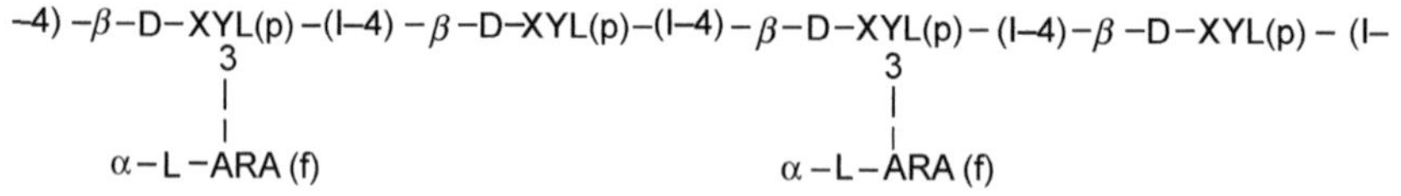

$$-4) -\beta-D-XYL(p)-(I-4) -\beta-D-XYL(p)-(I-4)-\beta-D-XYL(p)-(I-4)-\beta-D-XYL(p)-(I-$$
$$\underset{\alpha-L-ARA\,(f)}{\overset{3}{\underset{|}{\big|}}}\underset{\alpha-L-ARA\,(f)}{\overset{3}{\underset{|}{\big|}}}$$

FIGURE 4.4 Structure of arabinoxylan of wheat aleurone and starchy endosperm cell walls – p, represents the pyranose or six-membered ring form; f, represents the furanose or five-membered ring form.

for hundreds of years and continue to do so. Granules from different species differ in their properties, and there is even variation in form among granules from the same storage organ. Shape is determined in part by the way that new starch is added to existing granules, in part by physico-chemical conditions existing during the period of growth and in part by composition.

4.2.5.1 Composition

The main way in which composition varies is in the relative proportions of the two macromolecular species of which granules consist (Fig. 4.5).

Amylose comprises linear chains of (1–4) linked α-D-glucopyranosyl residues. Amylopectin has, in addition, (1–6) tri-O-substituted residues acting as branch points. Amylose has between 1000 and 4400 residues, giving it a molecular weight between 1.6×10^5 and 7.1×10^5. In solution amylose molecules adopt a helical form and may associate with organic acids, alcohols or, more importantly, lipids to form complexes in which a saturated fatty acid chain occupies the core of the helix. Binding of poly-iodide ions in the core in the same way is responsible for the characteristic blue colouration of starch by iodine.

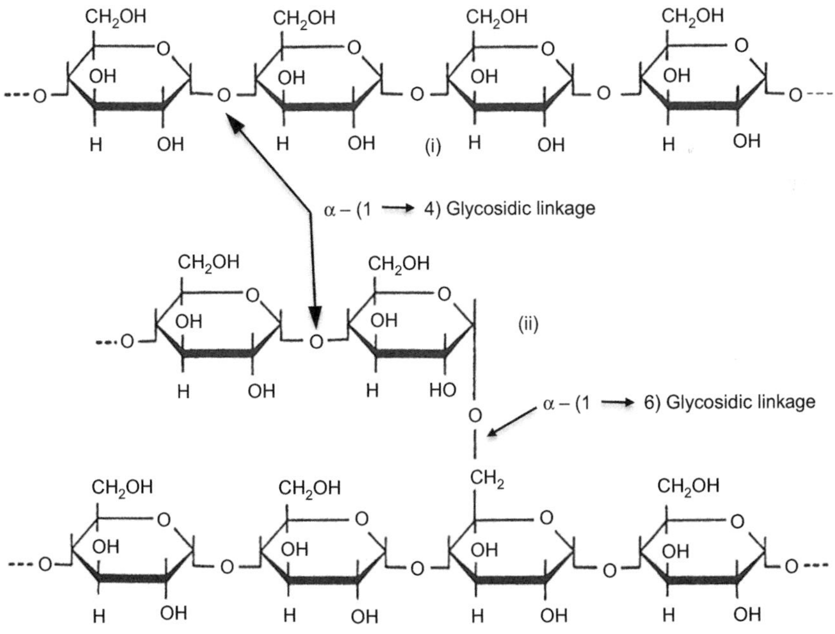

FIGURE 4.5 Structural representation of parts of amylose (i) macromolecule and amylopectin (ii) macromolecule, showing a branch point.

The average length of amylopectin branches is 17–26 residues. As their reducing groups are involved in bonding, only one is exposed. The molecule is generally considered to consist of three types of chain (Fig. 4.6):

- A chains: side chains linked only via their reducing ends to the rest of the molecule;
- B chains: those to which A chains are attached;
- C chains: chains which carry the only reducing group of the molecule.

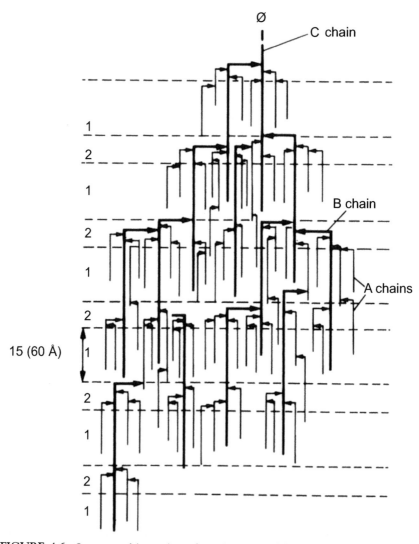

FIGURE 4.6 Structure of (potato) amylopectin proposed by Robin et al. (1974). Bands marked (1) are considered to be crystalline while alternating (2) bands are amorphous. *Reproduced by courtesy of American Association of Cereal Chemists Inc.*

The amylose content of most cereal starches lies between 20% and 35%, but mutants have been used in breeding programmes to produce cultivars with abnormally high or low amylose contents. It is in diploid species such as maize and barley that such breeding has been most successful, as polyploid species are more conservative, with single mutations having less chance of expression. High-amylopectin types are generally described as waxy, as the appearance of the endosperms of the first mutants discovered suggested a waxy composition. Waxy maize cultivars have up to 98% amylopectin (100% according to some references). High-amylose maize starches consist of up to 80% amylose.

4.2.5.2 *Granular form*

Although some variation exists within species, there are many characteristic features by which an experienced microscopist can identify the source of cereal starch, either from observation of an aqueous suspension at room temperature or with the additional help of observed changes when the suspension is heated, leading to gelatinization at a temperature characteristic of the species and type (Snyder, 1984). The characteristic blue staining reaction with iodine/potassium iodide solution does not occur with waxy granules, which contain virtually no amylose and stain brownish red to yellow. It is characteristic for amylose percentage to increase during endosperm development, and consequently staining reactions change during growth.

Granules of cereals from the Triticeae tribe are of two distinct types. The larger ones are biconvex, while the smaller ones are nearly spherical (Fig. 4.7). Granules from most other cereals are similar in shape to the smaller population of Triticeae granules, but rice and oats have some compound granules in which many granuli fit together to produce large ovoid wholes. Shapes of high-amylose granules are varied and may be related to their individual composition. The later developers tend to be filamentous, some resembling strings of beads. Characteristics of starch granules from cereals are shown in Table 4.4.

Within the endosperm of a species small differences in granule shape may arise as a result of packing conditions. These can be seen in grains as mealy and vitreous (or horny) regions. In mealy regions, packing is loose and granules adopt what appears to be their natural form. In horny regions close packing causes granules to become multifaceted as a result of mutual pressure. Small indentations can also arise from other endosperm constituents such as protein bodies (Fig. 4.8).

Pitting on the surface can be caused by enzymatic hydrolysis, and it is possible to find such granules in some cereal grains in which germination has begun or insect damage has occurred (Fig. 4.8). There is no evidence that these two physical modifications to granule form in themselves change the chemical properties of the granules.

Fig. 4.9 shows scanning electron micrographs of starch granules from various sources, including potato, rice, wheat, mung bean, maize, waxy corn, tapioca, shoti and leaf starch.

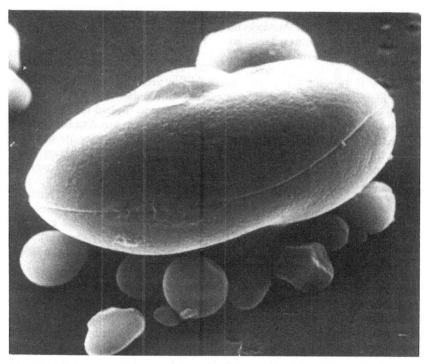

FIGURE 4.7 Scanning electron micrograph of one large starch granule and numerous small starch granules of wheat. The large granule shows the equatorial groove. *From Evers, A.D., Stärke, 1971 23, 157. Copyright by Leica UK. Reproduced by permission of the editor of* Die Stärke.

TABLE 4.4   Characteristics of starch granules of cereals

Cereal	Shape and diameter (µm)	Features
Barley	Large, lenticular: 10–30	As wheat
	Small, spherical: 1–5	
Maize	Angular	In flinty endosperm
	Both types 2–30, average 10	
	Spherical	In floury endosperm
Millet, pearl	Spherical/angular: 4–12, average 7	As maize
Oats	Compound, ovoid: up to 60	Comprising up to 80 granuli
	Simple, angular: 2–10	
Rice	Compound granules	Comprising up to 150 angular granuli: 2–10 µm
Rye	Large lenticular: 10–40	As wheat, often displaying radial cracks
	Small, spherical: 2–10	Visible hilum
Sorghum	Spherical/angular: 6–20, average 15	As maize
Triticale	Large, lenticular: 1–30	As wheat
Wheat	Large, lenticular: 15–30	Characteristic equatorial groove
	Small, spherical: 1–10	Angular where closely packed

Based on Kent, N.L., 1983 Technology of Cereals, third ed. Pergamon Press Ltd., Oxford, UK.

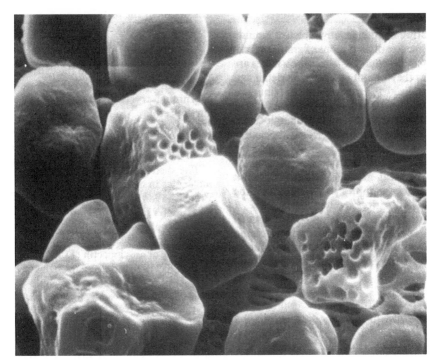

FIGURE 4.8 Scanning electron micrograph of maize starch granules of spherical and angular types. Some angular granules show indentations due to pressure from protein bodies.

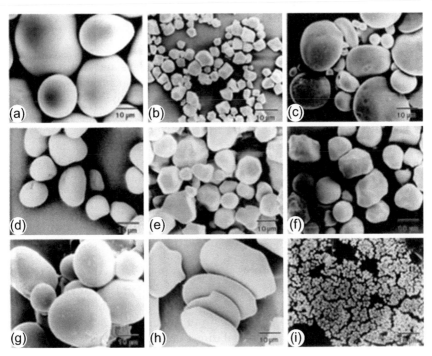

FIGURE 4.9 Scanning electron micrographs of several starch granules from various sources: (a) potato; (b) rice; (c) wheat; (d) mung bean; (e) corn; (f) waxy corn; (g) tapioca; (h) shoti; (i) leaf starch. *From Robyt, J.F., 2008. Starch: structure, properties, chemistry, and enzymology. In: Fraser-Reid, B., Tatsuta, K., Thiem, J. (Eds.), Glycoscience. Springer-Verlag, Berlin, Heidelberg, p. 1443. http://dx.doi.org/10.1007/978-3-540-30429-6_35, with permission.*

As granules are transparent some manifestations of internal structure can be detected, even if their significance cannot be fully appreciated. One such internal feature is the hilum exhibited by granules of some species. It is a small air space, considered to represent the point of initiation around which growth occurs (Hall and Sayre, 1969). This assumes that granules grow by deposition of new starch material on the outer surface of existing granules, and indeed this has been demonstrated by detection of radioactively labelled precursors incorporated into growing granules (Badenhuizen, 1969). Such a system of growth allows for the change in shape that occurs in starches of the Triticeae by preferential deposition on some parts of the surface. As a result they change from tiny spheres to larger lentil-shaped granules (Evers, 1971).

Some structures not evident in untreated granules can be revealed or exaggerated by treatment with weak acid or amylolytic enzymes. In cereal starches a lamellate structure results from removal of more susceptible layers and persistence of more resistant layers. Layers may be spaced progressively more closely towards the outside. The number of rings appears to coincide with the number of days for which a granule grows (Buttrose, 1962). Lamellae cannot be revealed in granules from plants grown under conditions of continuous illumination (Evers, 1979). Fig. 4.10 illustrates the internal structure of starch granules, which ultimately consists of matrices of amylose and amylopectin chains which build out from a central hilum.

4.2.5.3 Size distributions

The literature contains many tables of size ranges and distributions of granules from different botanical sources. While such tables are useful guides, they do not all accord in detail and some fail to indicate the nature of the distribution. For example, the bimodal distribution of the Triticeae is not always indicated, although this is an important characteristic by which the source of a starch may be recognized. In wheats the proportional relationship between large biconvex and small spherical granules is fairly constant (approximately 70% large granules w/w), and this is the same for rye and triticale.

In barley there is a wider variation, in part due to the existence of more mutant types (Goering et al., 1973). Among 29 cultivars, small granules accounted for between 6% and 30% of the total starch mass.

4.2.5.4 Granule organization

Under crossed polarizers starch granules exhibit birefringence in the form of a Maltese cross (Fig. 4.11). This indicates a high degree of order within the structure. The positive sign of the birefringence suggests that molecules are organized in a radial direction (French, 1984). Amylomaize

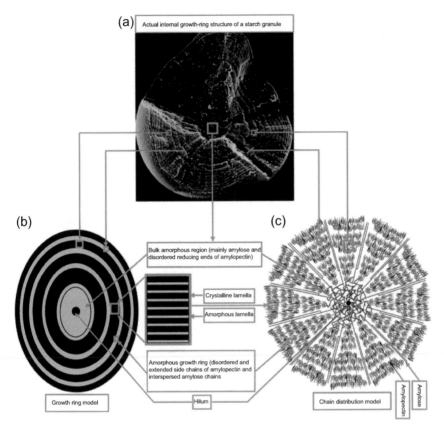

FIGURE 4.10 Internal structure of starch granules consists of matrices amylose and amylopectin chains which build out from a central hilum. *From Wang, S., Copeland, L., 2013. Molecular disassembly of starch granules during gelatinization and its effect on starch digestibility: a review. Food & Function 4, 1564–1580. http://dx.doi.org/10.1039/C3FO60258C, with permission.*

starch (high-amylose maize) exhibits only weak birefringence of an unusual type (Gallant and Bouchet, 1986).

Starch granules exhibit X-ray patterns, indicating a degree of crystallinity. Cereal starches give an A pattern, tuber, stem and amylomaize starches give a B pattern and bean and root starches a C pattern. The C pattern is considered to be a mixture of A and B. It is accepted that the crystallinity is due to the amylopectin, as it is shown by waxy granules. Furthermore, amylose can be leached from normal granules without affecting the X-ray pattern. The A and B patterns are thought to indicate crystals formed by double helices in amylopectin. The double helices occur in the outer chains of amylopectin molecules, where they form regions or clusters. The crystalline parts of starch granules

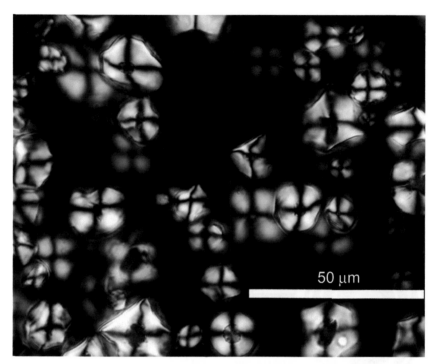

FIGURE 4.11   Birefringence of maize starch appears as a Maltese cross. *Based upon Mcginn, J., 2014. Microscopic advancements aid food QC labs. Lab Equipment. Available online:* http://www. laboratoryequipment.com/article/2014/09/microscopic-advancements-aid-food-qc-labs.

are responsible for many of the physical characteristics of the granules' structure and behaviour. Nevertheless, they involve less than half the total starch present. Some 70% is amorphous; this comprises all the amylose but must also include much of the amylopectin. The evidence of biochemical studies and electron microscopy has pointed to the existence of structures with a periodicity of 5–10 nm, reflecting the alternating crystalline and amorphous zones of amylopectin.

4.2.5.5 *Granule surface and minor components*

The distribution of amylose and amylopectin molecules in a starch granule was estimated by French (1984): for one spherical granule 15 μm in diameter, with a mass of 2.65×10^{-9} g, there would be about 2.5×10^{9} molecules of amylose (DP = 1000, 25% of total starch) and 7.4×10^{7} molecules of amylopectin (DP = 100,000 u, 75% of starch). If the molecular chains are perpendicular to the surface of the granule there would be about 14×10^{8} molecular chains terminating at the surface. Of these, 3.5×10^{8} would be amylose molecules and the rest would be amylopectin chains.

Surface characteristics of granules are also affected by the minor components of starches. Bowler et al. (1985) reviewed developments in work on these, although they point out that this is an underresearched area. Nonstarch materials present in commercial starch granules can arise from two sources: they may be inherent components of the granules in their natural condition, or may arise as deposits of material solubilized or dispersed during the process by which the starch is separated.

The main nonstarch components of starch granules are protein and lipid. Amounts vary with starch type: in maize 0.35% of protein ($N \times 6.25$) is present on average, and slightly more is present in wheat starch (0.4%). The most significant proteins in terms of their recognized effects on starch behaviour are amylolytic enzymes bound to the surface. Even traces of α-amylase can have drastic effects on pasting properties through hydrolysing starch polymers at temperatures up to the enzymes' inactivation temperatures.

SDS PAGE (sodium dodecyl sulphate, polyacrylamide gel electrophoresis) have shown that surface proteins of wheat starch have molecular masses under 50K, while integral proteins were over 50K. Altogether 10 polypeptides have been separated between 5K and 149K. The major 59K polypeptide is probably the enzyme responsible for amylose synthesis. It has been shown to be concentrated in concentric shells within granules. Two other polypeptides of 77K and 86K are likely to be involved in amylopectin synthesis. Perhaps the most interesting of the surface proteins is that in the 15K band. This has been found in greater concentration on starches from cereals with soft endosperm than on those from cereals with hard endosperm. The protein has been called 'friabilin', because of its association with a friable endosperm (Greenwell and Schofield, 1989).

Phosphorus is another important minor constituent of cereal starches. It occurs as a component of lysophospholipids. They consist of 70% lysophosphatidyl choline, 20% lysophosphatidyl ethanolamine and 10% lysophosphatidyl glycerol. The proportion of lysophospholipids to free fatty acids varies with species: in wheat, rye, triticale and barley over 90% occurs as lysophospholipids, in rice and oats 70% and in millets and sorghum 55%. In maize 60% occurs as free fatty acids (Morrison, 1985).

Removal of lipids from cereal starches reduces the temperatures of gelatinization-related changes and increases peak viscosity of pastes – in other words, they become more like the lipid-free potato starch.

4.2.5.6 Technological importance of starch

Much of the considerable importance of starch in foods depends upon its nutritional properties, as it is a major source of energy for humans and domestic herbivorous and omnivorous animals. In the human diet it is usually consumed in a cooked form wherein it confers attractive textural qualities to recipe formulations. These can vary from those of gravies and sauces, custards and pie fillings to pasta, breads, cakes and biscuits (cookies). Much of the variation in texture depends upon the degree of gelatinization, which in turn depends upon the temperature, and the amount of water available during cooking. Digestibility in the intestines of single-stomached animals is also increased by gelatinization.

During industrial processing of food products, and particularly industrial alcohol and other chemicals, starch can be challenging to process when in the form of a dispersion. Starch dispersions behave as a shear thickening fluid (i.e., to transport the fluid at faster flow rates, greater forces and thus larger pumps are required; also known as dilatancy). And at some point pumping actually becomes impossible. To remedy this problem, this behaviour can be modified into shear thinning (i.e., lower viscosities at higher flow rates) by inducing gelatinization, via increasing temperature above 80°C or altering pH, which will greatly reduce pumping requirements (Rao et al., 1997).

4.2.5.7 Gelatinization

This is a phenomenon manifested as several changes in properties, including granule swelling and progressive loss of organized structure (detected as loss of birefringence and crystallinity), increased permeability to water and dissolved substances (including dyes), increased leaching of starch components, increased viscosity of the aqueous suspension and increased susceptibility to enzymatic digestion (Fig. 4.12).

At room temperature starch granules are not totally impermeable to water – in fact water uptake can be detected microscopically by a small increase in diameter. The swelling is reversible and the wetting and drying can be cycled repeatedly without permanent change. If the temperature of a suspension of starch in excess water is raised progressively, a condition is reached around 60°C at which irreversible swelling begins, and continues with increasing temperature. The change is endothermic and can be quantified by thermal analysis techniques.

Typical heats of gelatinization in cals per gram of dry starch are wheat 4.7, maize 4.3, waxy maize 4.7 and high-amylose maize 7.6 (Maurice et al., 1983). Swelling involves increased uptake of water and can thus lead to increased viscosity by reducing the mobile phase surrounding the granules; accompanying leaching of starch polymers into this phase can further increase viscosity. The swelling behaviour of starch heated in water is often followed using a continuous automatic viscometer, such as the Brabender amylograph (Shuey and Tipples, 1980). Upon heating a slurry of 7%–10% starch w/w in water at a constant rate of 1–5°C/min, starch eventually gelatinizes and begins to thicken the mixture. The temperature

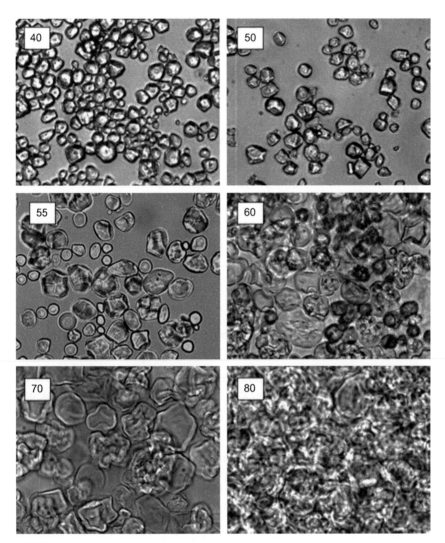

FIGURE 4.12 Images of maize starch granules at different temperatures (°C) show the swelling and rupture of the granules as denaturing progresses. *From Ratnayake, W.S., Jackson, D.S., 2006. Gelatinization and solubility of corn starch during heating in excess water: new insights. Journal of Agricultural and Food Chemistry 54 (10), 3712–3716, with permission.*

at which a rise in consistency is shown is called the pasting temperature. The curve then generally rises to a peak, called the peak viscosity. When the temperature reaches 95°C, that temperature is maintained for 10–30 min and stirring is continued to determine the shear stability of the starch. Finally the paste is cooled to 30°C and the increase in consistency is called set-back (Fig. 4.13).

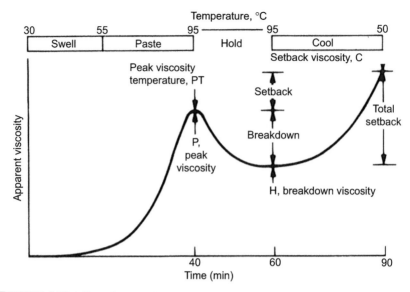

FIGURE 4.13   Chart showing characteristics recorded by the Brabender amylograph. This curve is characteristic of wheat starch at concentrations of 8%–12% (dry basis). *Reproduced with permission from Dengate, H.N., 1984. Swelling, pasting and gelling of wheat starch. Advances in Cereal Science and Technology 7, 49–82.*

Gelatinization typically occurs over a unique temperature range, depending on the starch: 61–72°C for maize, 53–64°C for wheat and 65–73°C for rice (Fennema, 1985).

4.2.5.8 Retrogradation

Suspensions of gelatinized granules containing more than 3% starch form a viscous or semisolid starch paste which on cooling sets to a gel. Three-dimensional gel networks are formed from the amylose containing starches by a mechanism known as 'entanglement'. The relatively long amylose molecules that escape from the swollen granules into the continuous phase become entangled at a concentration of 1%–1.5% in water. On cooling the entangled molecules lose translational motion, and the water is trapped in the network. Crystallites eventually begin to form at junction zones in the swollen discontinuous phase, causing the gel slowly to increase in rigidity (Osman, 1967). When starch gels are held for prolonged periods, retrogradation sets in. As applied to starch this means a return from a solvated, dispersed, amorphous state to an insoluble, aggregated or crystalline condition. Retrogradation is due largely to crystallization of amylose, which is much more rapid than that of amylopectin. It is responsible for hardening of cooked rice, shrinkage and syneresis of starch gels and possibly firming of bread. Although regarded as crystalline, retrograded gels are susceptible to amylolysis, but a fraction has been found that is resistant to enzyme attack. Known as resistant starch, it behaves as dietary fibre and is most abundant in autoclaved amylomaize starch suspensions (Berry, 1988).

4.2.5.9 *Starch damage*

Granule damage of a particular type alters the properties of starch in some ways similar to gelatinization. Defining the exact type of damage is difficult, and this accounts for the continued use of the general term. The essential characteristics associated with damaged starch are somewhat similar to gelatinized granules, but there are differences also. Thus mechanical damage results in:

- Increased capacity to absorb water, from 0.5-fold starch dry mass when intact to 3–4-fold when damaged (gelatinized granules absorb as much as 20-fold);
- Increased susceptibility to amylolysis;
- Loss of organized structure manifested as loss of X-ray pattern, birefringence, differential scanning calorimetry gelatinization endotherm;
- Reduced paste viscosity;
- Increased solubility, leading to leaching of mainly amylopectin (in gelatinized granules amylose is preferentially leached (Craig and Stark, 1984)).

At a molecular level the organization of granules appears to be accompanied by fragmentation of amylopectin molecules during damage, whereas gelatinization achieves loss of organization without either polymer being reduced in size.

Controlling the starch damage level during milling of wheat flour is important, as it affects the amount of water needed to make a dough of the required consistency (Evers and Stevens, 1985).

4.2.6 Cell walls

The older literature describes the components of cereal grain cell walls as pentosans and hemicelluloses. Pentosans are defined earlier in this chapter, but hemicelluloses are more difficult to define and indeed the term is even now used loosely. Hemicelluloses were originally assumed to be low molecular weight (and therefore more soluble) precursors of cellulose. Coultate (1989) writes that the name as applied now covers the xylans, the mannans and glucomannans, and the galactans and arabinoxylans, but others use the name also to include β-glucans (Hoseney, 1986).

Cell walls are important in several contexts.

- As a structural framework with which the grain is organized.
- As a physical boundary to access by enzymes produced outside the cell.
- As a source of energy in ruminants and of dietary fibre in single-stomached animals, including humans.
- They or their derivatives affect processing of raw and cooked cereal products.

Cell walls of different cereals have some common components, but composition is not consistent among species. Cellulose is one component present in all cell walls, and is the material of the simplest and youngest structures. In most cases additional carbohydrates of varying complexity are deposited as a matrix, and some protein also becomes included. Lignin is a common component of secondary thickening in the pericarp of all cereal grains. It is found in the pales, but this is relevant to processing only in those grains of which they remain part after threshing (i.e., oats, barley and rice). The walls of nucellus and seed-coat are generally unlignified; they may contain some corky material. The pigment strand, which is continuous with the seed-coat in grains where a crease is present, is lignified and later becomes adcrusted with material of unknown composition. Similar unidentified material adcrusts testa cell walls on their inner surfaces (Zee and O'Brien, 1970). The more precise composition of cell walls has been reviewed by Fincher and Stone (1986). Walls of cereal endosperm (aleurone and starchy endosperm) consist predominantly of arabinoxylans and $(1\rightarrow3, 1\rightarrow4)$-$\beta$-glucans, with smaller amounts of cellulose, heteromannans, protein and esterified phenolic acids. They are unlignified and contain little, if any, pectin and xyloglucan, or hydroxyproline-rich glycoprotein, all of which are common components of other primary cell walls.

Walls with high β-D-glucans content, such as endosperm cell walls in oats and barley, can be identified under a fluorescence microscope because of a specific precipitation reaction with calcofluor white MR (Fig. 4.14). The blue fluorescence is intense, and excitation by a wide range of wavelengths is possible (Fulcher and Wong, 1980).

Rice is exceptional in containing significant proportions of pectin and xyloglucan, plus small amounts of hydroxyproline-rich protein. The cellulose content of rice cell walls is also unusually high (25%–30%), and mannose-containing sugars in some may contribute as much as 15%.

A significant noncarbohydrate molecule intimately associated with arabinoxylans in wheat and other cereal cell walls is ferulic acid, a phenol carbonic acid very abundant in plant products. It is esterified to the primary alcoholic group of the arabinose side chain (Amado and Neukom, 1985). The formation of diferulic acid cross-links is at least partly responsible for the gelation of aqueous flour extracts or solutions

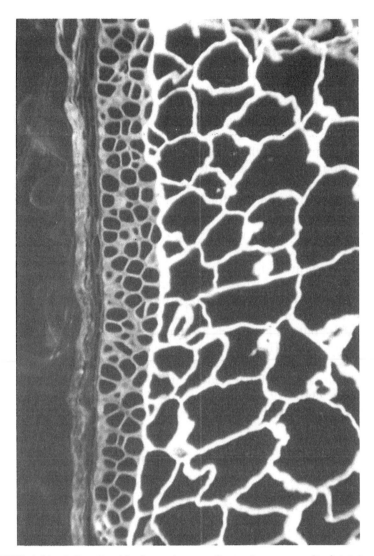

FIGURE 4.14   Cell walls of barley endosperm fluorescing as a result of staining with calcofluor white MR. The bright fluorescence of the starchy endosperm cell walls contrasts with that of the walls in the triple aleurone layer. *Courtesy: Dr. S. Shea Miller, Centre for Food and Animal Research, Agriculture Canada, Ottawa.*

of cereal pentosans in the presence of oxidizing agents. Ferulic acid exhibits intensive blue fluorescence when irradiated with light of 365 nm. The reaction is particularly marked in aleurone cell walls, and thus can be used for identifying these in ground cereal products under a fluorescence microscope.

Fractionation into water-soluble and insoluble pentosans by various protocols is common analytical practice, as it has been found to distinguish different functional properties. The water-soluble pentosans (mainly arabinoxylans) of wheat have a very high water-absorbing capacity. They are linear molecules, while those of the insoluble fraction are highly branched. The backbone of arabinoxylans consists of D-glucan units linked by $\beta(1-4)$ glycosidic bonds. Single α-L-arabinofuranose residues are attached randomly to the xylan and cause the water solubility of the arabinoxylans. As much as 23% of water in a bread dough may be associated with pentosans (Bushuk, 1966). It has been suggested (Hoseney, 1984) that pentosans reduce the rate of CO_2 diffusion through the dough, behaving in this way similarly to gluten.

4.2.6.1 Fibre

Extraction of individual cell wall components is complex and unsuitable for routine analysis. Nevertheless an estimate of cell wall content is often required, particularly in relation to nutritional attributes of a product. Analytical procedures have been devised to determine undigestible material as 'fibre', but not all experts are agreed as to which chemical entities should be included. The term is frequently qualified to reflect the method of analysis employed, because different methods produce different values (it should also be noted that some methods are themselves inconsistent). The following types of fibre may be encountered; the definitions are based on a glossary by Southgate et al. (1986).

Crude fibre

The residue left after boiling the defatted food in dilute alkali and then in dilute acid.

The method recovers 50%–80% of cellulose, 10%–50% of lignin and 20% of hemicellulose. Results are inconsistent.

Acid detergent fibre

The cellulose plus lignin in a sample; it is measured as the residue after extracting the food with a hot dilute H_2SO_4 solution of the detergent cetyl trimethylammonium bromide.

Neutral detergent fibre

The residue left after extraction with a hot neutral solution of SDS, also known as sodium lauryl sulphate. It is designed to divide the dry matter of feeds very nearly into those which are nutritionally available by the normal digestive process and those which depend on microbial fermentation for their availability.

Dietary fibre

All the polymers of plants that cannot be digested by the endogenous secretions of the human digestive tract.

The last definition differs from those that precede it, in that it is not based on the method by which it is determined but represents the actual value that the analytical methods seek to achieve.

Other terms are in use, such as 'non-starch polysaccharides', 'unavailable carbohydrates' and 'plantix' (seldom used now), which depart from the indication that only fibrous material (i.e., occurring as fibres) is included. Instead they suggest a matrix of plant materials. Over the years there has been a lack of agreement on whether a functional or compositional definition is more appropriate – a review of historical definitions is provided by Jones (2014).

In recent years, however, an international agreement on what specifically constitutes dietary fibre has finally been established by the United Nations Food and Agriculture Organization (FAO) (Codex Alimentarius, 2010). From the Codex Alimentarius (2010):

Dietary fibre denotes carbohydrate polymers with 10 or more monomeric units, which are not hydrolysed by the endogenous enzymes in the small intestine of humans and belong to the following categories:

- Edible carbohydrate polymers naturally occurring in the food consumed,
- Carbohydrate polymers, which have been obtained from food raw material by physical, enzymatic or chemical means and which have been shown to have a physiological benefit to health, as demonstrated by generally accepted scientific evidence to competent authorities,
- Synthetic carbohydrate polymers that have been shown to have a physiological benefit to health, as demonstrated by generally accepted scientific evidence to competent authorities.
- Includes also lignin and other compounds if quantified by AOAC 991.43.
- Decision on whether to include carbohydrates with a degree of polymerization from DP 3 to 9 should be left to national authorities.

Various methods that distinguish several classes of indigestible material are therefore useful. Southgate et al. (1986) distinguish among cellulose, noncellulosic polysaccharides and lignin. Asp et al. (1983) distinguish soluble and insoluble fibre. Insoluble components include galacto- and gluco-mannans, cellulose and lignin, and the soluble class includes galacturonans (pectins), (1-3,1-4)-β-glucans and arabinoxylans.

McCleary et al. (2013) reviewed 15 laboratory methods that can be used to determine components of dietary fibre. Only one method, however, the AACC International Approved Method 32-45.01 (AOAC method 2009.01), can be used to determine all the dietary fibre components used in the Codex definition.

4.3 PROTEINS

Although an enormous range of proteins exist in nature, they are all composed of the same relatively simple units: amino acids. The diversity of proteins comes about because the amino acids are arranged in different sequences and those sequences are of different lengths. There are only 20 amino acids commonly found in proteins. Cereal proteins are important in human and animal nutrition, and provide the unique gas-retaining qualities in wheat-flour doughs and bread, but in all organisms proteins are present which function as enzymes. Within the growing plant the genetic code is interpreted through the synthesis and activation of enzymes, providing the means by which characteristics of individual species are expressed. When seen in the context of this function it is perhaps easier to appreciate the subtlety of the differences in behaviour that can be achieved among what, at first sight, appear to be molecules of relatively simple construction. The subtle functional differences are possible because of the diversity of the properties of the amino acids and the relationships in which they are capable of engaging with other amino acids or even with lipids, carbohydrates and other molecules. Additional variation comes about as a result of the environment in which a protein finds itself. A change in pH, temperature or ionic strength can lead to a single protein species behaving in different ways.

4.3.1 Structure

All amino acids have in common the presence of an α-amino group (NH_2) and a carboxyl group (-COOH). It is through the condensation of these groups that neighbouring amino acids are joined by a peptide bond, as in Fig. 4.15.

A sequence of a large number of units linked by peptide bonds is called a polypeptide.

The differences among amino acids lie in the side chains attached to the carbon atom lying between their carboxyl and amino groups. Side chains may be classified according to their capacity for interacting with other

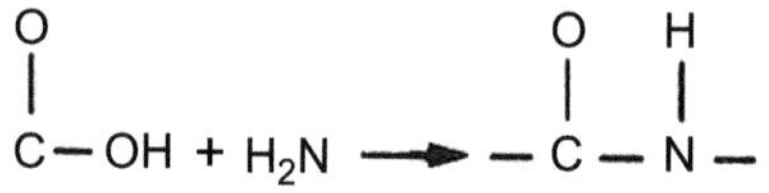

$$
\begin{matrix}
O & & & O & H \\
\| & & & \| & | \\
C-OH + H_2N & \longrightarrow & -C-N-
\end{matrix}
$$

FIGURE 4.15   The peptide bond.

amino acids by different mechanisms. The types of interaction and the amino acids capable of engaging in them are listed in Table 4.5.

The order in which amino acids occur in a polypeptide defines its 'primary structure' (Fig. 4.16). Because of the range of interactions that can occur among the side chains, different sequences are capable of different interactions, giving rise to a 'secondary structure' which consists of α-helices and β-sheets. The unit's 'tertiary structure' defines the three-dimensional conformation adopted as a result of side-chain interactions. The secondary and tertiary structures of a protein change in response to the environment, but the primary structure remains unaltered unless its length is reduced by hydrolysis.

All the interactions listed in Table 4.5 can contribute to tertiary structure, but the most stable types are the covalent disulphide bonds formed by oxidation of sulphydryl groups on interacting cysteine/cystine residues.

TABLE 4.5 Grouping of amino acid residues according to their capacity for interacting within and between protein chains

Type of interaction	Amino acid
1. Covalent – disulphide bonding Dissociated by oxidizing and reducing agents, e.g., performic acid 2-mercaptoethanol	Cysteine/cystine
2. Neutral – hydrogen bonding Dissociated by strong H-bonding agents, e.g., urea, dimethyl formamide	Asparagine Glutamine Threonine Serine Cysteine
3. Neutral – hydrophobic interaction Dissociated by ionic and nonionic detergents, e.g., sodium salts of long-chain fatty acids	Tyrosine Tryptophan Phenylalanine Proline Methionine Leucine Isoleucine Valine Alanine[a] Glycine[a]
4. Electrostatic – acid hydrophilic – basic hydrophilic Dissociated by acid, alkali or salt solutions	Aspartic acid Lysine Arginine Histidine Glutamic acid

[a] Amino acids with short, aliphatic side-chains show very little hydrophobicity. Both glycine and alanine are readily soluble in water.

From Simmonds, D.H.., 1989. Wheat and Wheat Quality in Australia. CSIRO, Australia, with permission.

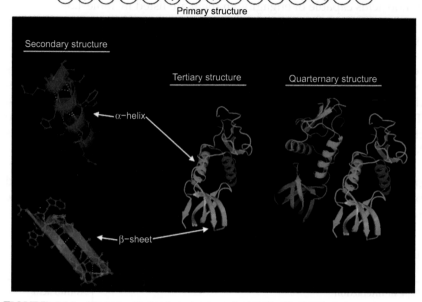

FIGURE 4.16 Diagram of protein structures showing examples of primary, secondary, tertiary and quaternary structures. *Based upon* https://en.wikipedia.org/wiki/Protein_structure, *2016*.

Such bonds also occur between cysteine/cystine residues on different polypeptides, giving rise to a stable structure involving more than one polypeptide. Interpeptide links can thus produce in a protein a fourth or 'quaternary' level of structure.

Disulphide bonds are stronger than noncovalent bonds, but are nevertheless capable of entering into interchange reactions with substances containing free sulphydryl groups. Such reactions are of great importance in dough formation.

4.3.2 Cereal proteins

The complexity of cereal proteins is enormous, and the determination of the structure of gluten – the protein complex responsible for the dough-forming capacity of wheat flour – has been described as one of the most formidable problems ever faced by the protein chemist (Wrigley and Bietz, 1988). To simplify their studies cereal chemists have sought to separate the proteins into fractions that have similarities in behaviour, composition and structure. As protein studies have proceeded and knowledge has accumulated, the validity of earlier criteria of classification has been, and continues to be, challenged.

One of the most significant means of classifying plant proteins is that which Osborne (1907) made on the basis of solubility. Water-soluble proteins were described as 'albumins', saline soluble as 'globulins', aqueous alcohol soluble as 'prolamins' and those that remained insoluble as 'glutelins'.

There are differences in amino acid composition between proteins in the Osborne classes but there is also heterogeneity within each class, and this may be as significant as differences between classes. Newer analytical methods have shown that the solubility classes overlap considerably. The distinction formalized by Osborne that remains unquestionably valid is that between albumins and globulins on the one hand and prolamins and glutelins on the other. In composition there is a marked difference due mainly to the extremely high content of proline and glutamine in the less-soluble fractions (the name 'prolamine' reflects this characteristic). An extremely low lysine content is also characteristic of insoluble cereal proteins.

Both content (Fig. 4.17) and composition of protein are affected by the contributions of different anatomical parts of the grain; and, in the endosperm, by the contributions of the different protein fractions (albumins, globulins, prolamins and glutelins).

The insoluble fractions are particularly deficient in lysine, as illustrated in the comparison of the solubility fractions of wheat endosperm (Table 4.6).

Among samples of the same cereal there are considerable variations in protein content. Because there is a consistent relationship between protein content and the proportions of the fractions present, protein composition in whole grains also varies. Nevertheless, the differences among samples of the same cereal are generally less than differences among cereal species,

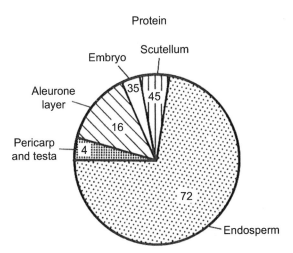

FIGURE 4.17 Distribution of total protein in wheat grain. The figures show the percentage of the total protein found in the various anatomical parts. *Based on microdissections by Hinton, J.J.C. From the Research Association of British Flour Millers, 1923–1960.*

TABLE 4.6   Amino acid composition of wheat proteins: glutenin, gliadin, albumin, globulin (g amino acid/16 g nitrogen)

Amino acid	Glutenin[a]	Gliadin[a]	Albumin[b]	Globulin[c]
Isoleucine	3.9	4.5	4.1	1.4
Leucine	6.9	7.2	10.7	9.2
Lysine	2.3	0.7	11.0	12.2
Methionine	1.7	1.5	0	0.4
Phenylalanine	4.8	5.6	5.0	3.2
Threonine	3.3	2.3	2.9	4.5
Tryptophan	2.1	0.7	nd	nd
Valine	4.5	4.4	8.1	2.2
Cystine	2.5	3.1	6.7	12.6
Tyrosine	3.6	2.6	3.4	2.3
Alanine	3.1	2.3	5.6	4.3
Arginine	4.2	2.7	7.5	14.5
Aspartic acid	3.9	3.0	7.9	6.3
Glutamic acid	34.1	40.0	17.7	5.9
Glycine	4.5	1.8.	3.1	5.6
Histidine	2.4	2.3	4.3	2.2
Proline	11.0	14.7	8.4	3.3
Serine	5.9	5.1	4.7	9.1

nd, not determined.

[a] *From Ewart, J.A.D., 1967. Amino acid analysis of glutenins and gliadins. Journal of the Science of Food and Agriculture 18, 111, recalculated. Original data are given as moles of anhydro amino acids per 10^5 of recovered anhydro amino acids.*

[b] *From Waldschmidt-Leitz, E., Hochstrasse, K., 1961. Grain proteins. VII. Albumin from barley and wheat. Zeitschrift fuer Physiologische Chemie 324, 423.*

[c] *From Fisher, N., Redman, D.G., Elton, G.A.H., 1968. Fractionation and characterization of purothionin. Cereal Chemistry 45, 48.*

and proteins characteristic of individual species can be described. It must, however, be remembered that values cited in comparisons are only averages at best, representing points within a range typical of the species.

Values for the proportions of classified amino acids in whole grains of cereals are given in Table 4.7.

4.3.2.1 Soluble proteins

Soluble proteins are found in starchy endosperm, aleurone and embryo tissues of cereals. They account for approximately 20% of the total protein

TABLE 4.7 Amino acid content of cereal grains (g amino acid/16 g nitrogen)

Amino acid	Wheat	Triticale	Rye	Barley	Oats	Rice
Essential						
Isoleucine	3.8	4.1	3.6	3.8	4.2	3.9
Leucine	6.7	6.7	6.0	6.9	7.2	8.0
Lysine	2.3	3.0	2.9	3.5	3.7	3.7
Methionine	1.7	1.9	1.2	1.6	1.8	2.4
Phenylalanine	4.8	4.8	4.5	5.1	4.9	5.2
Threonine	2.8	3.1	3.3	3.5	3.3	4.1
Tryptophan	1.5	1.6	1.2	1.4	1.6	1.4
Valine	4.4	5.0	4.9	5.4	5.6	5.7
Nonessential						
Cysteine + cystine	2.6	2.8	2.3	2.5	3.3	1.1
Tyrosine	2.7	2.3	1.9	2.5	3.0	3.3
Alanine	3.3	3.6	3.7	4.1	4.6	6.0
Arginine	4.0	4.9	4.2	4.4	6.6	7.7
Aspartic acid	4.7	5.9	6.5	6.1	7.8	10.4
Glutamic acid	33.1	30.9	27.5	24.5	21.0	20.4
Glycine	3.7	3.9	3.6	4.2	4.8	5.0
Histidine	2.2	2.5	2.1	2.1	2.2	2.3
Proline	11.1	10.7	10.4	10.9	4.7	4.8
Serine	5.0	4.6	4.3	4.2	4.8	5.2
Protein[a]	16.3	12.1	17.8	14.5	17.9	11.1

			Millets		
Amino acid	Maize	Sorghum	Pearl	Foxtail	Proso
Essential					
Isoleucine	4.0	3.8	4.3	6.1	4.1
Leucine	12.5	13.6	13.1	10.5	12.2
Lysine	3.0	2.0	1.7	0.7	1.5
Methionine	1.8	1.5	2.4	2.4	2.2
Phenylalanine	5.1	4.9	5.6	4.2	5.5

Continued

TABLE 4.7   Amino acid content of cereal grains (g amino acid/16 g nitrogen)—cont'd

| Amino acid | Maize | Sorghum | Millets | | |
			Pearl	Foxtail	Proso
Threonine	3.6	3.1	3.1	2.7	3.0
Tryptophan	0.8	1.0	1.4	2.0	0.8
Valine	5.2	5.0	5.4	4.5	5.4
Nonessential					
Cysteine + cystine	2.5	1.1	1.8	1.4	1.0
Tyrosine	4.4	1.5	3.7	1.6	4.0
Alanine	7.7	9.5	11.3		
Arginine	4.7	2.6	3.3	2.3	3.2
Aspartic acid	6.4	6.3	6.4		
Glutamic acid	18.8	21.7	22.2		
Glycine	3.9	3.1	2.3		
Histidine	2.8	2.1	2.3	1.2	2.1
Proline	8.8	7.9	6.9		
Serine	4.9	4.3	6.9		
Protein[a]	10.6	10.5	13.5	12.4	12.5

[a] $N \times 5.7$, d.b. Original data for maize and sorghum given as $N \times 6.25$, viz. 11.6% and 11.5% respectively. Data for wheat, barley, oats, rye, triticale and pearl millet from Tkachuk, R., Irvine, G.N., 1969. Amino acid composition of cereals and oilseed meals. Cereal Chemistry 46, 206; data for rice (except tryptophan) from Juliano, B.O., Baptista, G.M., Lugay, J.C., Reyes, A.C., 1964. Rice quality studies on physicochemical properties of rice. Journal of Agricultural and Food Chemistry 12, 131; data for maize (except tryptophan from Busson, F., Fauconnau, G., Pion, N., Montreuil, J., 1966. Acides aminés. Annales de la Nutrition et de l'Alimentation 20, 199; data for sorghum from Deyoe, C.W., Shellenberger, J.A., 1965. Amino acids and proteins in sorghum grain. Journal of Agricultural and Food Chemistry 13, 446; data for foxtail and proso millets from Casey, P., Lorenz, K., Feburary 1977. Millet – functional and nutritional properties. Baker's Digest 45; tryptophan data for rice and maize calculated from Hughes, B.P., 1967. Amino acids. In: McCance, R.A., Widdowson, E.M. (Eds.), The Composition of Foods, Med. Res. Coun. Spec. Rep. Ser. No. 297, 2nd Impression. H.M.S.O., London, UK. All data are for whole grains, except oats and rice (hulled grains) and rye (dark rye flour), ash 1.1% db.

of the grain. Albumins are usually more prevalent than globulins. The amino acid composition of soluble proteins is similar to that of proteins found in most unspecialized plant cells suggesting that they include those that constitute the cytoplasm found in most cells. They are a complex mixture including metabolic enzymes, hydrolytic enzymes, enzyme inhibitors and phytohaemaglutenins (proteins that clot red blood cells).

Globulins may also arise as storage proteins occurring in protein bodies, particularly in oat and rice endosperm. In other cereals storage proteins arising in protein bodies are exclusively of the insoluble types (Payne and Rhodes, 1982).

The number of individual proteins in the soluble categories is large. By two-dimensional electrophoresis 160 components have been separated in aqueous extracts from wheat endosperm, and a different pattern of 140 components has been separated from the 0.5 M NaCl extracts (Lei and Reeck, 1986).

4.3.2.2 Enzymes

Enzymes may be considered in the context of the stage of a grain's life cycle. Thus most enzyme activity during maturation is concerned with synthesis, particularly the synthesis of storage products. Some hydrolytic enzymes involved in breakdown of starch and protein stored in the pericarp are found before maturity and may persist (Fretzdorf and Weipert, 1990).

In the mature grain enzyme levels are relatively low if the grain is sound and dry. If damaged, as in milling, lipids become exposed to lipase. This is particularly relevant to oats and germ and bran fractions of other grains.

On adequate damping, germination begins and enzymes concerned with solubilization are produced. Cell walls are hydrolysed, permitting greater access to storage products by enzymes that catalyse hydrolysis of starch and protein.

Technologically the highest-profile enzyme is α-amylase, as large quantities are essential to successful malting and brewing and small quantities are necessary in breadmaking. Excessive α-amylase in milling wheats is disastrous, leading to dextrin production during baking and making the crumb sticky. Polyphenol oxidases can lead to production of dark specks in stored flour products. Other classes of enzymes of technological importance, found or used in cereals and cereal processing are β-amylases, proteases, β-glucanases, lipases, lipoxygenase and phytase.

4.3.2.3 Amylases

Both α- and β-amylases are α-(1-4)-D-glucanases; by definition catalysing the hydrolysis of the same bonds within starch molecules. Their action is synergistic, because β-amylase gains greater access to the substrate through the activity of α-amylase. As this last observation implies, their modes of action are quite different: α-amylase is endoacting while β-amylase is exoacting (Fig. 4.18).

Exoenzymes catalyse removal of successive low molecular weight products from the nonreducing chain end, and the product removed through β-amylase activity is maltose due to the hydrolysis of alternate α-(1-4) glycosidic bonds. β-Amylase is inactive on granular starch but is capable of rapid action when the substrate is in solution. As the exoaction produces a large number of small sugars with reducing power, the reducing capacity of the solution increases rapidly. When the substrate is amylose the iodine staining reaction is reduced only slowly, as the chain lengths, on which iodine binding depends, are reduced slowly.

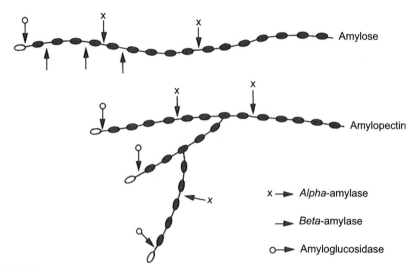

FIGURE 4.18 Diagrammatic representation of hydrolytic cleavage catalysed by *alpha*-amylase, *beta*-amylase and amyloglucosidase, respectively. *From Simmonds, D.H., 1989, Wheat and Wheat Quality in Australia. CSIRO, Australia. Reproduced by courtesy of CSIRO.*

By contrast, endoaction of α-amylolysis through random fragmentation reduces iodine staining relatively quickly in relation to the increase in reducing power. A further consequence of the rapid reduction in molecular size resulting from α-amylolysis is a marked reduction in viscosity of a starch suspension. This is exploited in laboratory tests for the enzyme. Assaying for β-amylase is more difficult because the rate of maltose production is influenced by the presence of α-amylase, and the enzymes are almost always present together. Even in well-washed starch they are absorbed on the granule surface (Bowler et al., 1985).

Grain quality is more influenced by the α-enzyme, as its amount is more variable according to the condition of the grain. β-Amylase is present in resting grain and increases only a few times on germination through release of a bound form.

α-Amylase is actually synthesized during germination and activity increases progressively, as germination proceeds, by several hundredfold. In different cereals the site of synthesis of α-amylase varies; in wheat, rye and barley it occurs first in the scutellum and later in the aleurone, while in maize only the scutellum is involved. Several isoenzymes of the α-amylase type exist in most cereals, and fall into two groups depending upon their isoelectric points. The Triticeae cereals contain two groups, while sorghum, millet, maize, oats and rice have only one (Kruger and Reed, 1988).

Even the combined action of α- and β-amylases cannot completely digest solubilized starch. Neither of them can catalyse hydrolysis of α-(1-6) bonds, and hence branch points remain intact. Also, those α-(1-4) bonds close to branch points resist hydrolysis, hence only about 85% of starch is converted to sugars. To increase yield of sugars in commercial processes (such as industrial alcohol or chemical fermentations), debranching enzymes may be used. Amyloglucosidase (also known as glucoamylase) from fungal sources (primarily *Aspergillus niger*) is a popular expedient; it catalyses hydrolysis of both α-(1-4) and α-(1-6) bonds, leaving glucose as the ultimate product. Many alcohol production processes (especially fuel ethanol) use this enzyme, and saké production is dependent upon it (see the discussion on industrial alcohol production elsewhere in this book).

4.3.2.4 β-Glucanases

These enzymes assume greatest importance in processing of barley, in which β-glucans contribute 70% of cell walls. There are two endo-β-glucanases in barley malt, both synthesized during germination. Each catalyses hydrolysis of β-(1-4) linkages adjacent to β-(1-3) links, ultimately producing a mixture of oligosaccharides containing three or four glucosyl units (Woodward and Fincher, 1982). The two isoenzymes are synthesized in different sites, one in the scutellum and the other in the aleurone. Before being susceptible to these enzymes it is thought that another enzyme, β-glucan solubilase, renders the substrate soluble (Bamforth and Quain, 1989).

4.3.2.5 Proteolytic enzymes

Although proteolytic enzymes may be important technologically in baking, their significance is usually masked by the greater effects of α-amylase. In brewing their role is better understood. Both endopeptidases and exoenzymes (the carboxypeptidases which catalyse cleavage of single amino acids from the carboxyl terminus) are present.

Proteolysis increases access by amylases to starch granules, as well as producing nitrogenous nutrients for the growing embryo in nature and for yeast during fermentation for beer production.

4.3.2.6 Lipid-modifying enzymes

Enzymes of two types are important in catalysing breakdown of lipids: lipase and lipoxidase. Both are capable of causing rancidity in cereals, thus both hydrolytic and oxidative rancidity are recognized. Lipoxidase can only catalyse degradation of free fatty acids and monoglycerides, and therefore follows lipolysis. Lipolysis proceeds slowly in the dry state; enzymatic oxidation occurs rapidly on wetting.

The problem of rancidity is potentially greatest in oats, which have a high oil content (4%–11%, average 7%). Maize also has a relatively high oil content because of its large embryo (about 4.4%) and brown rice contains about 3%, but other cereals contain only 1.5%–2.0%. Problems caused by hydrolysis catalysed by lipase are prevented in the case of processed oats by 'stabilization', a steaming process which inactivates the enzyme. In other milled cereals potential storage problems can be avoided in starchy endosperm if it is separated from other grain parts where enzyme and substrate are concentrated. This is common practice in the cases of sorghum and maize grits, in which the embryo presents the greatest hazard, and in wheat and rice, in which the aleurone layer also has a high lipid content. In wheat, lipase activities in the embryo and aleurone layer are 10–20-fold that of the endosperm (Kruger and Reed, 1988). The storage life of bran, germ and wholemeal flour is considerably less than that of white flour for this reason.

As well as true lipases, esterases are present in cereals, and in most studies the two classes have not been distinguished. Like other hydrolases, they are synthesized during the early stages of germination, although oats are exceptional in having high lipase activity in resting grain.

Lipases catalyse hydrolysis of triglycerides to produce diglycerides and free fatty acids; of diglycerides to give monoglycerides and free fatty acids; and of monoglycerides to give glycerol and free fatty acids. The unsaturated fatty acids can be converted to hydroperoxides, which in turn are changed to hydroxy acids by lipoxygenase, lipoperoxidase and other enzymes, as well as by nonenzymatic processes (Youngs, 1986).

Lipoxygenase is an effective bleaching agent; a coupled oxidation destroys the yellow pigments in wheat endosperm. Cosmetically, this is beneficial in bread doughs but undesirable in pasta products, in which a yellow colour is valued (Hoseney, 1986).

4.3.2.7 Phytase

Phytase catalyses hydrolysis of phytic acid (inositol hexaphosphoric acid) to inositol and free orthophosphate. Much of the phosphorus in cereal grains is stored in the form of phytic acid – which can lead to absorption problems in humans and animals, as they do not naturally produce phytase (discussed later). In wheat its activity increased sixfold on germination, and more activity was found in hard wheats than in soft (Kruger and Reed, 1988). In oats the activity is much lower than in wheat, rye and triticale (Lockhart and Hurt, 1986).

In rice the phytin level dropped from 2.67 to 1.48 mg/g of dry mass after 1 day of germination, then to 0.44 mg/g after 5 days. Phytase activity levelled off after 7 days (Juliano, 1985a).

4.3.2.8 Phenol oxidases

In the mature wheat grain several polyphenol oxidases are present in the starchy endosperm, and are more concentrated in the bran. On germination an increase occurs, including new isoenzyme synthesis, mainly in the coleoptile and roots. Monophenolase also increases. Durum wheat has less activity than other wheat types (Kruger and Reed, 1988).

4.3.2.9 Catalase and peroxidase

Catalase and peroxidase are haemoproteins. Peroxidase is involved in the degradation of aromatic amines and phenols by hydrogen peroxidase. Its activity is greater in wheat than in other cereals. Catalase catalyses degradation of hydrogen peroxide to water and oxygen. Its physiological function is not understood but it increases during germination (Kruger and Reed, 1988).

4.3.2.10 Insoluble proteins

The state of knowledge of many insoluble cereal proteins has now advanced even to complete sequencing of their amino acids. This is true of prolamins of maize, which are known as zeins, since they come from *Zea*. Barley prolamins are hordeins, rye prolamins are secalins and oat prolamins are avelins. A different basis for nomenclature is applied to the naming of wheat prolamins, which are called gliadins.

The cereal prolamins were reviewed by Shewry and Tatham (1990). On the basis of sequencing, four major groups of zeins have been defined. The groups differ in their amino acid content as well as the sequence in which they occur. As prolamins they are by definition rich in proline and glutamine, and low in lysine and tryptophan. The groups are designated α-, β-, γ- and δ-. The β- and δ-groups are relatively rich in methionine, and the δ-group is also rich in cysteine and histidine.

4.3.2.11 α-Zeins

The predominant group is the α-, contributing 75%–80% of the total insoluble fraction of zein. By electrophoresis the apparent molecular weights of the two major α-zeins are 19,800 and 22,000. They can be separated by isoelectric focussing into a series of monomers and oligomers, though some of the latter can be extracted only after reduction of the S—S bonds by which the monomers are held together. It is frequently found in peptide sequences that the domain in the centre is quite different from those at the C- and N-terminal parts. In α-zeins the C-terminal domain consists of 10 amino acids in a unique sequence, similarly the N terminus has a unique sequence of 36–37 residues in which one or two cysteine residues are present. The central domain comprises repetitive blocks of 20 residues that are rich in leucine and alanine. The tertiary structure of α-zeins, divined from circular dichroism, suggests a high content of α-helix and low β-sheet content.

4.3.2.12 β-Zeins

β-Zeins contribute 10%–15% of total prolamin. They are rich in methionine and cysteine and can be extracted only in the presence of a reducing agent, indicating mutual association through disulphide bonds. No sequences are repeated and all differ from those in α-zeins. The tertiary structure consists mainly of β-sheet, and there is an aperiodic structure (β turns and random coils).

4.3.2.13 γ- and δ-Zeins

γ-Zeins account for 5%–10% of the total prolamin. Like β-zeins, they require the presence of a reducing agent for extraction. Eight hexapeptide sequences in the central domain are flanked by unique N- and C-terminal regions. The repeat sequences are very hydrophilic, rendering the proteins very soluble when reduced.

δ-Zeins also require reduction before extraction: no sequences are repeated in the central region, but between 17 and 29 methionine residues occur here.

4.3.2.14 Other tropical cereals

Although only the prolamins of maize among the tropical cereals have been studied extensively, available evidence indicates that sorghum, pearl millet and Job's tears contain essentially similar groups.

4.3.2.15 Temperate cereals

Under the classical nomenclature the necessity to reduce disulphide bonds would define β-, γ- and δ-zeins as glutelins. Indeed, to regard all insoluble cereal proteins as prolamines is not universally accepted among protein chemists. Such a classification can be extended to temperate cereals, but at the present time the traditional Osborne classification is more widespread, especially in wheat proteins where the functional aspects are particularly important. From two-dimensional electrophoresis studies it has been established that up to 20 different polypeptides are found in glutenins. An even greater number – up to 50 – may be found in gliadins. One argument advanced for distinguishing between wheat prolamins and glutenins is the different physical properties of the two classes when hydrated: gliadins behave as a viscous liquid and glutenins as a cohesive solid. Although both influence gluten behaviour, it is the larger polymeric glutenins that wield the greater influence.

One of the most attractive theories concerning the relationship between glutenin structure and function is the linear gluten hypothesis (Fig. 4.19). It envisages a series of polymeric subunits joined head to tail by interchain disulphide bonds. The essential features of the subunits are terminal α-helices and central regions of many β-turns (β-turns also occur in the body tissue protein elastin; they are capable of much extension under tension and can return to their former folded condition on subsequent relaxation of the tension). In general the sulphur-containing cysteine residues occur in the α-helical regions, so the disulphide bonds form between these regions in adjacent polypeptides. β-Turns thus remain unencumbered by interchain bonds that might otherwise restrict their extension. Molecular weights of glutenins are upward of 10^5.

The unusually high content of the amino acids asparagine and glutamine found in gluten proteins may be significant in providing stability of

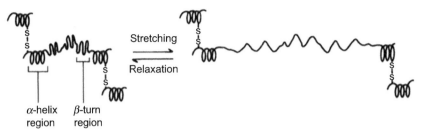

α-helix region β-turn region

FIGURE 4.19 Schematic representation of a polypeptide subunit of glutenin within a linear concatenation. The subunits are joined head to tail via S–S bonds to form polymers with molecular weights of up to several million. The subunits are considered to have a conformation that may be stretched when tension is applied to the polymers, but when the tension is released the native conformation is regained through elastic recoil. The N- and C-terminal ends of some high molecular weight subunits, where interchain S–S bonds are located, are now thought to be α-helix-rich domains, whereas the central domains are thought to be rich in repetitive β-turn structures. The presence of repetitive β-turn structures may result in a β-spiral structure, which may confer elasticity. *From Schofield, J.D., 1986. Flour proteins: structure and functionality in baked products. In: Blanshard, J.M.V., Frazier, P.J., Galliard, T. (Eds.), Chemistry and Physics of Baking. Roy. Soc. of Chem., London, UK. Reproduced by courtesy of the Royal Society of Chemistry, London, UK.*

gluten, through their tendency to become involved in hydrogen bonding. Hydrophobic and electrostatic reactions associated with other amino acid side chains also contribute.

The relative importance of glutenins and gliadins varies in wheats from different parts of the world. In Australian and Italian wheats gliadin variations have the strongest association with bread quality. In European wheats high-molecular-weight glutenin subunits with apparent molecular weight of 90–150K are paramount in determining quality. Each wheat possesses a complement of three to five types and a variety of individual subunits (allelic forms) may represent each type, giving rise to variation in baking properties. Gliadins are thought to behave as plasticizers, and the proportional relationship between them and glutenins is an important factor. Too low a gliadin content leads to inhibition of bubble expansion, while the reverse results in excessive expansion and collapse.

Gliadin complements are characteristic of individual cultivars, and these, revealed through PAGE, are exploited in establishing the varietal identity of wheat cultivars and detecting adulteration of *Triticum durum* products with *Triticum aestivum* additions (Fig. 4.20).

While this technique may also be useful in other species, it has not been developed to the same degree as in wheat. An even more sensitive method of identifying protein components is high-performance liquid chromatography, which is faster and capable of greater resolution than PAGE and has become widely used in recent years.

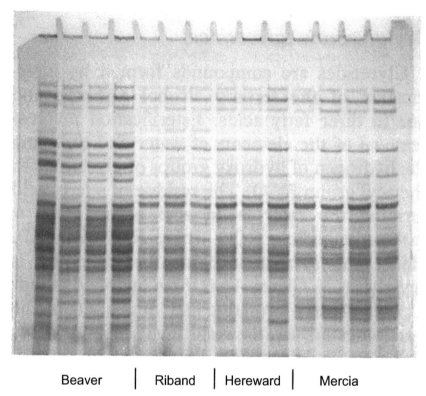

| Beaver | | Riband | | Hereward | | Mercia |

FIGURE 4.20   PAGE electrophoretogram showing distinctive gliadin patterns of four United Kingdom wheat varieties. *Courtesy of FMBRA, Chorleywood, UK.*

4.3.2.16 Mutations and technologies

To produce cereals with a better balance of proteins, from a nutritional point of view, breeders have exploited mutants with high lysine and high arginine contents. It is the storage proteins that are deficient in these amino acids, so the mutants selected frequently achieve improved balance through a deficiency in storage proteins (Hoseney and Variano-Marston, 1980). In the 'opaque' varieties of maize, high lysine content is associated with 'opaque' (O_2) and 'floury' (fl_2) genes being doubly recessive, and the consequent inhibition of zein synthesis (Watson, 1987). Thus the 'high-lysine' varieties of maize, barley, sorghum and pearl millets have lower yields than their conventional counterparts.

In recent years, in response to needs of cereal-processing companies, seed companies have used breeding programmes as well as genetic engineering technologies to produce a wide variety of traits and properties that are targeted at specific applications. For example, not only can cereal grains be grown which have higher lysine contents, but also higher starch,

TABLE 4.8 Average yield and composition of maize in Iowa, USA, from 2000-2015.

Year	Yield (bu/ac)	Protein (%)	Oil (%)	Starch (%)
2000	150.4–187.4	6.6–8.6	2.8–4.0	58.9–61.6
2005	201.9–236.6	6.5–7.9	3.1–3.7	60.6–62.3
2010	169.7–209.5	6.1–7.1	2.9–3.8	61.1–63.0
2015	145.2–236.0	5.8–7.4	2.9–3.9	60.6–62.5

higher extractable starch and higher fermentable starch; the type of starch (amylose vs amylopectin) can also be specified, as can higher oil content. Additionally, modern cereal crops continually have improved yields, weed resistance, insect resistance, fungal resistance, stalk strength and height and other traits that lead to improved agronomic performance on the farm. Not only can the hybrids/technology traits affect the seed composition, but also maturity at time of harvest, growing conditions, growing environment, geographical location and soil fertility.

Changes in composition over time are sometimes difficult to decipher, however, as considerable variability exists across geography and time, due in large measure to the factors mentioned above. Here is an example relating to maize from one county in Iowa, United States (www.iastate.edu) (Table 4.8).

4.4 LIPIDS

Lipids have been defined as substances which:

- Are insoluble in water,
- Are soluble in organic solvents such as chloroform, ether or benzene,
- Contain long-chain hydrocarbon groups in their molecules, and
- Are present in or derived from living organisms (Kates, 1972).

This covers a wide range of compounds, including long-chain hydrocarbons, alcohols, aldehydes and fatty acids, and derivatives such as glycerides, wax esters, phospholipids, glycolipids and sulpholipids. Also included are substances which are usually considered as belonging to other classes of compounds, for example the 'fat-soluble' vitamins A, D, E and K and their derivatives, as well as carotenoids and sterols and their fatty acid esters (Kates, 1972).

The terms lipid, fat and oil are often used loosely, but, applied strictly, 'lipids' include all three while only triglycerides (triacylglycerols) are described as fats and oils. Fats are solid at room temperature, while oils are liquid. Although many fats and oils originate in living organisms (where they function as a means of storing energy), this is not a feature of their definition as it is for lipids.

4.4.1 Nomenclature

With many series of compounds several conventions by which they are named coexist. The earlier 'trivial' names may have been chosen to reflect the original source or another arbitrary connection, and provide no indication of the structure of the molecules. As knowledge increases and more compounds of the series are identified, the need for a systematic system of names and the means of achieving it increase. Such is the case with lipids, and a systematic convention for their nomenclature was recommended by the Union of Pure and Applied Chemists (IUPAC) (Sober, 1968).

4.4.2 Fatty acids

Fatty acids present in cereal lipids mainly consist of a long hydrocarbon chain covalently linked to a carboxylic acid group (Fig. 4.21).

A fatty acid in which all bonds are single is said to be saturated. In the absence of two adjacent hydrogens, a double bond is formed and the resultant fatty acid is described as unsaturated. Where more than one double bond is present the term polyunsaturated is applied. The systematic description of the compound depends on where double bonds are substituted. If the remaining hydrogens are on the same side of the chain, the conformation is called 'cis-'; if on different sides a 'trans-' double bond exists (Fig. 4.22).

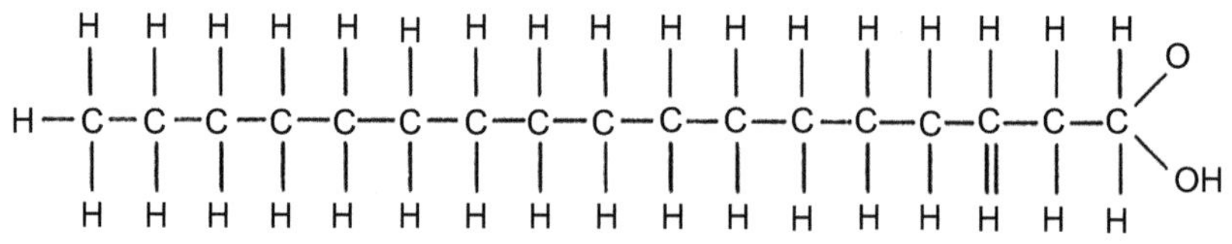

FIGURE 4.21   Generalized structure of a fatty acid.

Cis–   Trans–

FIGURE 4.22   Cis- and trans-configurations.

As well as a systematic nomenclature, a shorthand way of describing the fatty acid may be used. Thus cis-9-octadecenoic acid has a shorthand description of C18:1.9 cis, indicating 18 carbons (octadec-) and a double bond (-en-) in the cis form between the 9th and 10th positions, counting from the functional group carbon.

4.4.2.1 Acylglycerols (glycerides)

Glycerides are compounds formed by esterification of the tertiary alcohol glycerol and one to three fatty acids. Esterification involves removal of the elements of water and replacing the hydrogen of hydroxy groups of glycerol with the acyl group RCO. The residue of a fatty acid forming the ester is an acyl group (acyl=carboxylic radicle RCO where R is aliphatic). Hence the systematic name for glycerides is acylglycerols. Glycerol has three hydroxyl groups capable of ester formation and, depending on the number esterified, the resulting compounds may be monoacylglycerols, diacylglycerols or triacylglycerols. Plants usually store lipids as triacylglycerols, and cereal grains conform to this plant characteristic. The highest triacylglycerol levels occur in aleurone and scutellar tissue, but there are appreciable quantities in cereal embryonic axes and the endosperm of oats (Morrison, 1983). They are the main lipid stored in all cereal endosperm, and in wheat the endosperm contributes about 12% of the total in the grain. Monoacylglycerols and diacylglycerols occur only in small quantities as intermediates in the biosynthesis of triacylglycerols or products of their breakdown.

4.4.2.2 Phosphoglycerides (phospholipids)

The principal phosphoglycerides in cereal grains are phosphatidyl-choline, phosphatidyl-ethanolamine, phosphatidyl-inositol, N-acylphosphatidyl-ethanolamine and its monoacyl (lyso-) derivative. The monoacylphospholipids lysophosphatidyl-choline, lysophosphatidyl-ethanolamine and lysophosphatidyl-glycerol are the major internal starch lipids. Monoacylphospholipids are also formed from diacylphospholipids by enzymatic hydrolysis (Morrison, 1983).

The structures of diacylphosphoglycerides are shown in Fig. 4.23, in which **R, R'** and **R″** are acyl groups.

4.4.2.3 Glycosylglycerides (glycolipids)

Monoglycosyldiglyceride and diglycosyldiglyceride are the major glycolipids, with some monoglycosylmonoglycerides and diglycosylmonoglycerides. Triglycosyldiglyceride and tetraglycosylglycerides have also been reported. In wheat and most other cereals the sugar is mainly galactose, sometimes with small amounts of glucose or none. Other minor glycolipids include sterylglycosides (Morrison, 1983).

The structure of gycosyldiglycerides is shown in Fig. 4.24, where **R** and **R'** are acyl groups and **S** is a sugar (monosaccharide or tetrasaccharide).

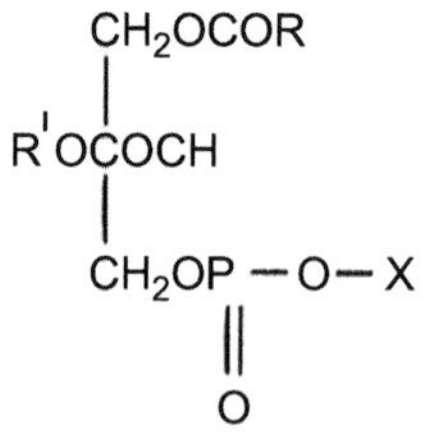

$$CH_2OCOR$$
$$R'OCOCH$$
$$CH_2OP-O-X$$
$$\parallel$$
$$O$$

In phosphatidyl-choline X = $CH_2CH_2N(CH_3)_3$

In phosphatidyl-ethanolamine   X = $CH_2CH_2NH_2$

In phosphatidyl-inositol   X =

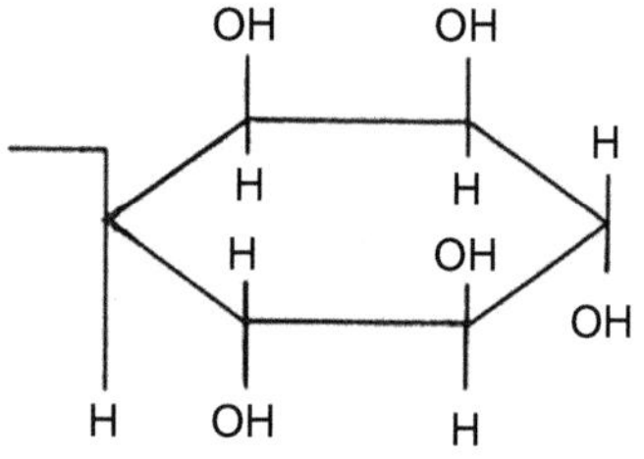

In N-acylphosphatidyl-ethanolamine X = CH_2CH_2NHCOR''

In lyso-phospholipids R or R' = H

FIGURE 4.23   General formulae of diacylphosphoglycerides.

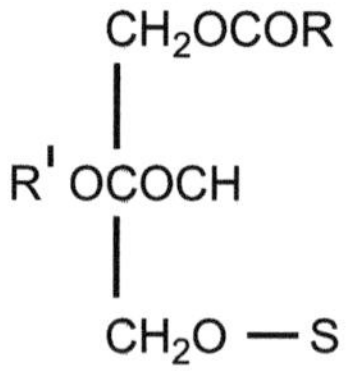

$$CH_2OCOR$$
$$R'OCOCH$$
$$CH_2O-S$$

FIGURE 4.24   General formula of glycosyl-diglycerides in glycosyl-monoglycerides **R** or **R′=H**.

4.4.2.4 *Distributions*

All cereal grain lipids are rich in unsaturated fatty acids (see Table 4.9). Palmitic (16:0) is a major saturated fatty acid and linoleic (18:2) is a major unsaturated fatty acid in most cereals, exceptions being brown rice and oats, which are rich in oleic acid (18:1). Millets are richer in stearic acid (18:0) than other cereals. No plant oils contain cholesterol.

TABLE 4.9   The fatty acid composition of cereal lipids

Material	At saturated			At unsaturated		
	Myristic $C_{14.0}$ (%)	Palmitic $C_{16.0}$ (%)	Stearic $C_{18.0}$ (%)	Oleic $C_{18.1}$ (%)	Linoleic $C_{18.2}$ (%)	Linolenic $C_{18.3}$ (%)
Barley						
Two-row	1.0	11.5	3.1	28.0	52.3	4.1
Six-row	3.3	7.7	12.6	19.9	33.1	23.1
Maize	–	14.0	2.0	33.4	49.8	1.5
Millet						
Pearl	–	17.8	4.7	23.9	50.1	3.0
Foxtail	0.6	11.0	14.7	21.8	38.2	6.4
Proso	–	11.5	–	25.8	50.6	7.8
Oats	0.5	15.5	2.0	43.5	35.5	2.0
Rice	–	17.6		47.6	34.0	0.8
Rye	–	21.0	–	18.0	61.0	–
Sorghum	0.4	13.2	2.0	30.5	49.7	2.0
Triticale	0.7	18.7	0.9	11.5	61.2	6.2
Wheat						
Grain	0.1	24.5	1.0	11.5	56.3	3.7
Germ	–	18.5	0.4	17.3	57.0	5.2
Endosperm	–	18.0	1.2	19.4	56.2	3.1

The values in this table provide only an indication of the relationships among fatty acids. They are not definitive and wide ranges have been reported for most values (Chung, 1991).
Based upon Kent, N.L., 1983. Technology of Cereals. third ed. Pergamon Press Ltd., Oxford, UK.

Two advantages of rice-bran oil are the low content of linolenic acid and its high content of tocopherols, both important from the point of view of oxidative stability. Its high content of linoleic acid make it a good source of essential fatty acid.

Oat groats contain 7% oil, pearl millet 5.4%, maize 4.4%, sorghum 3.4%, brown rice 2.3%, barley 2.1% and wheat 1.9%.

The hard, high-melting fraction of rice-bran wax has lustre-producing qualities similar to those of carnauba wax. It has been approved by the United States Food and Drug Administration (FDA) as a constituent of food articles up to 50 mg/kg and for use as a plasticizer for chewing gum at 2% (Juliano, 1985c).

Maize-germ oil is rich in essential fatty acids (about half of its fatty acid content is linoleic). It is often used as a salad oil and for cooking.

4.5 MINERALS

About 95% of the minerals in the actual fruits of cereals (i.e., the grain without adherent pales in the case of husked types) consist of phosphates and sulphates of potassium, magnesium and calcium. The potassium phosphate is probably present in wheat mainly in the form of KH_2PO_4 and K_2HPO_4. Some of the phosphorus is present as phytic acid. Important minor elements are iron, manganese and zinc, present at a level of 1–5 mg/100 g, and copper, at about 0.5 mg/100 g. Besides these, a large number of other elements are present in trace quantities. Representative data from the literature are collected in Table 4.10. The content of mineral matter in the husk of barley, oats and rice is higher than that in the caryopses, and the ash is particularly rich in silica (Table 4.11).

4.6 VITAMINS

Vitamins comprise a diverse group of organic compounds. They are necessary for growth and metabolism in the human body, but humans are incapable of making them in sufficient quantity to meet the body's needs, hence the diet must supply them to maintain good health. Most vitamins are known today by their chemical descriptions rather than the earlier identification as vitamin A, B, C, etc. A table of equivalence relates the two methods of nomenclature (Table 4.12).

Vitamins are sometimes classified according to solubility: A, D, E and K are fat soluble, and B and C are water soluble. Fat-soluble vitamins are more stable in cooking and processing.

It is clear from Table 4.12 that it is the B vitamins, more specifically thiamin, riboflavin, pyridoxine, nicotinic acid and pantothenic acid, and vitamin E that are most important in cereal grains. The average contents of B vitamins are shown in Table 4.13.

The table also includes values for inositol and para amino benzoic acid. Although essential for some microorganisms, these substances are no longer considered essential for humans. Their status as vitamins is thus dubious. Choline and inositol are by far the most abundant, but cereals are not an important source as many foods contain them and deficiencies are rare (Bingham, 1987).

4.6.1 Distribution of vitamins in cereals

Variation in content from one cereal to another is remarkably small except for niacin (nicotinic acid), the concentration of which is relatively much higher in barley, wheat, sorghum and rice than in oats, rye, maize and the millets.

TABLE 4.10 Mineral composition of cereal grains (mg/100 g d.b.)

			Oats						Rice		
Element	Wheat	Barley	Whole grain	Groat	Rye	Triticale	Paddy	Brown	White
Main									
Ca	48	52	94	58	49 36	37	15	22	12
Cl	61	137	82	73	524	–	15	–	19
K	441	534	450	376	138 10	485	216	257	100
Mg	152	145	138	118	428	147	118	187	31
Na	4	49	28	24	165	9	30	8	6
P	387	356	385	414	6	487	260	315	116
S	176	240	178	200		–	–	–	88
Si	10	420	639	28		–	2047	70	10
Minor									
Cu	0.6	0.7	0.5	0.4	0.7	0.8	0.4	0.4	0.2
Fe	4.6	4.6	6.2	4.3	4.4	6.5	2.8	1.9	0.9
Mn	4.0	2.0	4.9	4.0	2.5	4.2	2.2	2.4	1.2
Zn	3.3	3.1	3.0	5.1	2.0	3.3	1.8	1.8	1.0

Continued

TABLE 4.10 Mineral composition of cereal grains (mg/100 g d.b.)—cont'd

| Element | Wheat | Barley | Oats | | Rye | Triticale | Rice | | |
			Whole grain	Groat			Paddy	Brown	White
Trace									
Ag	0.05	0.005	—	—	—	—	0.02	—	—
Al	0.4	0.67	0.6	0.6	0.56	—	0.9	—	—
As	0.01	0.01	0.03	—	0.01	—	0.007	—	—
B	0.4	0.2	0.16	0.08	0.3	—	0.14	—	—
Ba	0.7	0.5	0.4	0.008	—	—	1.2	—	—
Br	0.4	0.55	0.3	—	0.19	3.3	0.1	—	—
Cd	0.01	0.009	0.02	—	0.001	—	—	—	0.005
Co	0.005	0.004	0.006	0.02	0.01	—	0.007	0.007	0.0006
Cr	0.01	0.01	0.01	—	—	—	0.06	—	0.003
F	0.11	0.15	0.04	0.04	0.1	—	0.07	—	0.04
Hg	0.005	0.003	—	—	—	—	0.001	0.002	0.002
I	0.008	0.007	0.007	0.06	0.004	—	—	—	—
Li	0.05	—	0.05	—	0.017	—	0.5	—	0.02
Mo	0.04	0.04	0.04	—	0.03	—	0.07	—	—
Ni	0.03	0.02	0.15	—	0.18	—	0.08	0.1	—
Pb	0.08	0.07	0.08	—	0.02	—	0.003	—	—

Continued

Rb	–	–	–	–	–	0.3	0.4	–	–
Sb	–	–	–	–	–	–	0.05	–	0.03
Sc	–	–	–	–	–	–	0.005	0.04	0.03
Se	0.05[a]	0.21	0.2	0.01	0.23	–	0.01	–	–
Sn	0.3	0.065	0.21	–	0.19	–	0.03	–	–
Sr	0.1	0.2	0.21	–	–	–0.5	0.02	–	–
Ti	0.15	0.1	0.2	–	0.08	–	1.4	–	–
V	0.007	0.005	0.1	–	–	–	0.01	–	–
Zr	–	–	–	–	–	–	0.007	–	–
Ash (%)	1.9	3.1	2.9	2.1	2.2	2.1	7.2	1.8	0.6

Element	Maize	Sorghum	Pearl	Millets Foxtail	Proso	Kodo	Finger
Main							
Ca	20	30	36	29	13	37	352
Cl	55	52	32	42	21	13	51
K	342	277	454	273	177	165	400
Mg	143	148	149	131	101	128	180
Na	40	11	11	5	7	5	16
P	294	305	379	320	221	245	323

TABLE 4.10 Mineral composition of cereal grains (mg/100 g d.b.)—cont'd

Element	Maize	Sorghum	Pearl	Foxtail	Proso	Kodo	Finger
					Millets		
S	145	116	168	192	178	156	184
Si	–	200	–	–	–	–	–
Minor							
Cu	0.4	1.0	0.5	0.7	0.5	1.0	0.6
Fe	3.1	7.0	11.0	9.0	9.0	3.0	4.5
Mn	0.6	2.6	1.5	2.0	2.0	–	1.9
Zn	2.0	3.0	2.5	2.0	2.0	–	1.5
Trace							
Ag	–	<0.005	<0.005	–	–	–	0.4
Al	0.057	1.8	1.7	–	–	–	–
As	0.03	–	0.01	–	–	–	<0.05
B	0.3	0.13	0.19	–	–	–	2.2
Ba	3.0	0.08	0.04	–	–	–	–
Br	0.26	0.14	0.38	–	–	–	–
Cd	0.012	–	–	–	–	–	<0.01
Co	0.008	<0.05	0.05	–	–	–	0.02

Cr	0.004	0.05	0.03	—	—	—	—
F	0.04	—	—	—	—	—	—
I	0.2	0.07	0.0016	—	—	—	0.2
Li	0.005	0.2	0.01	—	—	—	0.2
Mo	0.03	0.3	0.07	—	—	—	0.02
Ni	0.04	0.11	0.11	—	—	—	0.6
Pb	0.01	0.12	0.02	—	—	—	0.2
Rb	0.3	—	0.34	—	—	—	—
Sc	0.01	0.07	—	—	—	—	0.006
Sn	0.01	0.18	0.004	—	—	—	3.3
Sr	0.02	0.1	0.03	—	—	—	0.03
Ti	0.17	0.05	0.02	—	—	—	0.04
V	0.01	0.05	<0.01	—	—	—	—
Zr	0.02	—	—	—	—	—	—
Ash (%)	1.7	1.7	2.4	3.7	2.2	3.0	2.2

NB. A dash in the table indicates that no reliable information has been found.

[a] Level found in wheat growing in normal soils. Much higher values, e.g., up to 6 mg/100 g, can be found in wheat growing in seleniferous soils.

Based upon Kent, N.L., 1983. *Technology of Cereals.* third ed. Pergamon Press Ltd., Oxford, UK.

TABLE 4.11   Ash and silica in the husk of cereal grains

Material	Yield of ash (%)	Silica in ash (%)
Barley husk	6.0	65.8
Oat husk	5.2	68.0
Rice husk	22.6	95.8

Based upon Kent, N.L., 1983. Technology of Cereals. third ed. Pergamon Press Ltd., Oxford, UK.

TABLE 4.12   Vitamins and their occurrence in cereals

Vitamin	Chemical name	Concentration in cereals
A	Retinol and carotene	
B$_1$	**Thiamin**	Embryo (scutellum)
B$_2$	**Riboflavin**	Most parts
B$_6$	**Pyridoxin**	Aleurone
B$_{12}$	–	
	Nicotinic acid (niacin)	Aleurone (not maize)
	Folic acid	
	Biotin	
	Pantothenic acid	Aleurone, endosperm
	Choline	
	Carnitine	
C	Ascorbic acid	
D	Cholecalciferol and ergocalciferol	
E	**Tocopherol and tocotrienol**	Embryo
K	Phylloquinone	

Those in bold type occur in cereals in significant quantities (in relation to daily requirements).

Details of the distribution in grains were worked out by Hinton and associates (Heathcoate and Hinton, 1952; Hinton, 1944, 1947, 1948, 1953; Hinton and Shaw, 1953), who assayed the dissected morphological parts of wheat, maize and rice. Their results for wheat are shown in Table 4.14. The distribution of these vitamins in the wheat grain is also shown diagrammatically in Fig. 4.25.

Typical vitamin and mineral compositions for various types of rice are provided in Table 4.15. The vitamin contents of milled (unconverted) and parboiled (converted) rice are shown in Table 4.16. The folic acid content

TABLE 4.13 Vitamin B content of the cereal grains (μg/g)

Cereal	Thiamin	Riboflavin	Niacin	Pantothenic acid	Biotin
Wheat					
Hard	4.3	1.3	54	10.0	0.10
Soft	3.4	1.1	45		
Barley	4.4	1.5	72	5.7	0.13
Oats (whole)	5.8	1.3	11	10.0	0.17
Rye	4.4	2.0	12	7.2	0.05
Triticale	9.2	3.1	16	7.5	0.06
Rice (brown)	3.3	0.7	46	9.0	0.10
Maize	4.0	1.1	19	5.3	0.10
Sorghum	3.5	1.4	41	11.0	0.19
Millet					
Pearl	3.6	1.7	26	11.4	–
Foxtail	5.9	0.8	7	–	–
Proso	2.0	1.8	23	–	–
Finger	3.6	0.8	13	–	–

Cereal	Pyridoxin	Folic acid	Choline	Inositol	p-Amino benzoic acid
Wheat					
Hard	4.5	0.5	1100	2800	2.4
Soft					
Barley	4.4	0.4	1000	2500	0.5
Oats (whole)	2.1	0.5	940	–	–
Rye	3.2	0.6	450	–	–
Triticale	4.7	0.7	–	–	–
Rice (brown)	4.0	0.5	900	–	–
Maize	5.3	0.4	445	–	–
Sorghum	4.8	0.2	600	–	–
Millet					
Pearl	–	–	–	–	–
Foxtail	–	–	–	–	–
Proso	–	–	–	–	–
Finger	–	–	–	–	–

A dash indicates that reliable data have not been found.
The above data are similar to those given by Holland et al. (1991), who also provide comprehensive tables of vitamin contents of cereal products.
Sources of data are as quoted in Kent, N.L., 1983. Technology of Cereals. third ed. Pergamon Press Ltd., Oxford, UK.

TABLE 4.14 Distribution of B vitamins in wheat grain[a]

Part of grain	Thiamin[a]		Riboflavin[b]		Niacin[b]		Pyridoxin[b]		Pantothenic acid[b]	
	Concentration (μg/g)	%	Concentration (μg/g)	%	Concentration (μg/g)	%	Concentration (μg/g)	%	Concentration μg/g	%
Pericarp, testa, hyaline	0.6	1	1.0	5	25.7	4	6.0	12	7.8	9
Aleurone layer	16.5	32	10	37	741	82	36	61	45.1	41
Endosperm	0.13	3	0.7	32	8.5	12	0.3	6	3.9	43
Embryonic axis	8.4	2	13.8	12	38.5	1	21.1	9	17.1	3
Scutellum	156	62	12.7	14	38.2	1	23.2	12	14.1	4
Whole grain	3.75		1.8		59.3		4.3		7.8	

Concentration in μg/g and % of total in parts.
[a] Wheat variety Vilmorin 27.
[b] Wheat variety Thatcher.
Based upon Clegg, K.M., 1958. The microbiological determination of pantothenic acid in wheaten flour. Journal of the Science of Food and Agriculture 9, 366; Clegg, K.M., Hinton, J.J.C., 1958. The microbiological determination of vitamin B6 in wheat flour, and in fractions of the wheat grain. Journal of the Science of Food and Agriculture 9, 717; Heathcote, J.G., Hinton, J.J.C., Shaw, B., 1952. The distribution of nicotinic acid in wheat and maize. Proceedings of the Royal Society B139, 276; Hinton, J.J.C., 1947. The distribution of vitamin B1 and nitrogen in the wheat grain. Proceedings of the Royal Society B134, 418; Hinton, J.J.C., 1953. The distribution of protein in the maize kernel in comparison with that in wheat. Cereal Chemistry 36, 19.

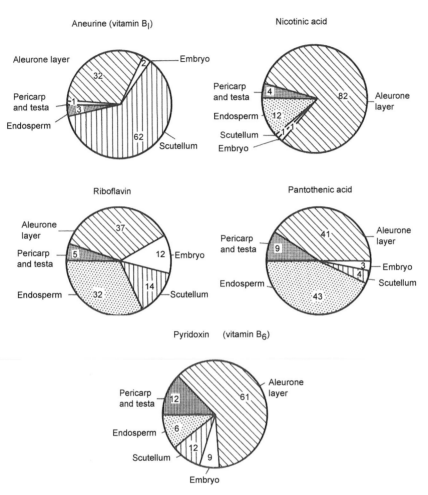

FIGURE 4.25 Distribution of B vitamins in wheat grain. The figures show the percentages of the total vitamins in the grain found in the various anatomical parts. *Based on microdissections by Hinton, J.J.C. From the Research Association of British Flour Millers, 1923–1960.*

of rice is increased from 0.04 to 0.08 µg/g by conversion, while that of the polishings is reduced from 0.26–0.40 µg/g to 0.12 µg/g (De Caro et al., 1949).

The proportions of total thiamin and niacin are shown for rice and maize in Table 4.17.

The distributions of thiamin in rice and wheat are quite similar: it is concentrated in the scutellum, though not to the same degree as in rye and maize. The embryonic axis of rice, which has a relatively high concentration of thiamin, contains over 10% of the total in the grain, a larger proportion than that found in the other cereals (see Table 4.18).

TABLE 4.15 Nutrient composition of rice (mg/100 g)

Product	Ca	P	Fe	Na	K	Thiamine	Riboflavin	Niacin	Tocopherol
Brown rice	32	221	1.6	9	214	0.34	0.05	4.7	29
White rice									
Unenriched	24	94	0.8	5	92	0.07	0.03	1.6	—
Enriched	24	94	2.9	5	92	0.44	—	3.5	—
Parboiled rice									
Enriched	60	200	2.9	9	150	0.44	—	3.5	

Based upon USDA, 1963. Agricultural Handbook, vol. 8. U.S. Dept. of Agriculture, Agricultural Research Service, Washington, DC, USA.

TABLE 4.16 Vitamin contents of rice as affected by processing (μg/g)

Material	Thiamine	Riboflavin	Niacin
Paddy rice	3.5–4.0	0.4–0.5	52.3–55.0
Converted, parboiled rice	1.9–3.1	0.3–0.4	31.2–47.8
Milled, polished, nonparboiled rice	0.4–0.8	0.15–0.3	14.0–25.0

Based upon Kik, M.C., 1943. Thiamin in products of commercial rice milling. Cereal Chemistry 20, 103 and Kik, M.C., Van Landingham, F.B., 1943. Riboflavin in products of commercial rice milling and thiamin, riboflavin and niacin content of rice. Cereal Chemistry 20, 563–569.

TABLE 4.17 Distribution of thiamin in rice and niacin in rice and maize

Part of grain	Percentage of total thiamin in rice	Percentage of total niacin		
		In rice (var: Indian)	In maize	
			Flint	Sweet
Pericarp, testa, hyaline	34	5.0	2	3
Aleurone layer	–	80.5	63	59
Endosperm	8	12.3	20	26
Embryonic axis	11	0.6	2	2
Scutellum	47	1.6	13	10

Based upon Heathcote, J.G., Hinton, J.J.C., Shaw, B., 1952. The distribution of nicotinic acid in wheat and maize. Proceedings of the Royal Society B139, 276 and Hinton, J.J.C., Shaw, B., 1953. The distribution of nicotinic acid in the rice grain. British Journal of Nutrition 8, 65.

TABLE 4.18 Thiamin in embryonic axis and scutellum of cereal grains

Cereal	Weight of tissue (g/100 g grain)		Thiamin concentration (μg/g)		Proportion of total thiamin in grain (%)	
	Embryonic axis	Scutellum	Embryonic axis	Scutellum	Embryonic axis	Scutellum
Wheat (average)	1.20	1.54	12.0	177	3.0	59
Barley (dehusked)	1.85	1.53	15.0	105	8.0	49
Oats (groats)	1.60	2.13	14.4	66	4.5	28
Rye	1.80	1.73	6.9	114	5.0	82
Rice (brown)	1.00	1.25	69.0	189	11.0	47
Maize	1.15	7.25	26.1	42	8.0	85

Based upon Hinton, J.J.C., 1944. The chemistry of wheat with particular reference to the scutellum. The Biochemical Journal 38, 214 and Hinton, J.J.C., 1948. The distribution of vitamin B1 in the rice grain. British Journal of Nutrition 2, 237.

Cereals, except maize, contain tryptophan, which can be converted to niacin in the liver in the presence of sufficient thiamin, riboflavin and pyridoxin. The distributions of niacin in wheat, rice and maize are similar, and it is concentrated in the aleurone layer. About 80% of the niacin in the bran of cereals occurs as niacytin, a complex of polysaccharide and polypeptide moieties biologically unavailable unless treated with alkali (Carter and Carpenter, 1982).

Riboflavin and pantothenic acid are more uniformly distributed. Pyridoxin is concentrated in aleurone and other nonstarchy endosperm parts of the grain.

The uneven distribution of the B vitamins throughout the grain is responsible for considerable differences in vitamin content between the whole grains and the milled or processed products.

Vitamin E is essential for maintaining the orderly structure of cell membranes; it also behaves as an antioxidant, particularly of polyunsaturated fatty acids. These are abundant in nervous tissue, including the brain. Deficiency symptoms, which are rare as stores of the vitamin in the body are large, include failure of nervous functions. Other functions of vitamin E are claimed by some, but these are not well substantiated by experiment (Bingham, 1987).

In the United Kingdom cereals contribute about 30% of vitamin E requirement. Wheat contains α-, β-, γ- and δ-tocopherols, with the total tocopherol content being 2.0–3.4 mg/100 g; α-, β- and γ-tocotrienols are also present. The biological potency of β-, γ- and δ-tocopherols is 30%, 7.5% and 40%, respectively, of that of the α-tocopherol. The total tocopherol contents of germ, bran and 80% extraction flour of wheat are about 30, 6 and 1.6 mg/100 g, respectively (Moran, 1959). α-Tocopherol predominates in germ and γ-tocopherol in bran and flour, giving α-equivalents of 65%, 20% and 35% for the total tocopherols of germ, bran and 80% extraction flour, respectively.

Quoted figures for the total tocopherol content of other cereal grains are (in mg/100 g) barley 0.75–0.9, oats 0.6–1.3, rye 1.8, rice 0.2–0.6, maize 4.4–5.1 and millet 1.75 (the latter two mostly as γ-tocopherol) (Science Editor, 1970; Slover, 1971).

The oil of cereal grains is rich in tocopherols: quoted values (in mg/g) are wheat-germ oil 2.6, barley oil 2.4, oat oil 0.6, rye oil 2.5 and maize oil 0.8–0.9 (Green et al., 1955; Slover, 1971). Wheat tocopherols have particularly high vitamin E activity (Morrison, 1978).

PART 2. NUTRITIONAL ASPECTS

Nutrition in most adults is concerned with the supply and metabolism of those components of the diet needed to maintain normal functioning of the body (water and oxygen are also necessary, but these are not generally regarded as nutrients). In the young and in pregnant and lactating mothers additional nutritional requirements are imposed by the need to support growth or milk production. In recent years it has become widely recognized that diet has a great impact on human health and disease prevention.

Nutrients – the substances that provide energy and raw materials for the synthesis and maintenance of living matter in the diet of humans and other animals – comprise protein, carbohydrate and fat, all of which can provide energy, plus minerals and vitamins. Those nutrients that cannot be made in sufficient quantities by conversion of other nutrients in the body are called 'essential'. They include some vitamins, minerals, essential amino acids and essential fatty acids. An insufficiency of an essential nutrient causes a specific deficiency disease. Deficiency diseases are now declining globally where food is plentiful, but remain a challenge in developing nations, where natural disasters and conflict frequently lead to malnutrition and even starvation. Aid provided for the relief of such disasters always includes a high proportion of cereals, demonstrating their high nutritional value.

Although cereals make an important contribution to the diet they cannot alone support life because they are lacking in vitamins A (except for yellow maize), B_{12} and C. Whole cereals also contain phytic acid, which may interfere with the absorption of iron, calcium and some trace elements. And cereal proteins are deficient in certain essential amino acids, notably lysine. Cereals are rarely consumed alone, however, and nutrients in foods consumed together may mutually compensate for each other's deficiencies.

While it is indisputable that individuals and populations should consume the right amounts of nutrients to avoid symptoms of deficiency and excess, defining those 'right amounts' is not easy, not least because the requirements vary from one individual to another and from country to country. The British government for many years issued standards in the form of *Recommended Intakes for Nutrients* (DHSS, 1969) and recommended daily amounts (RDAs) of food energy and nutrients (DHSS, 1979). In revising the recommendations for the dietary requirements of the nation, the Committee on Medical Aspects of Food Policy noted that the standards were frequently used in a way that was never intended, in that they were used to assess the adequacy of the diets of individuals. To ensure that deficiencies were avoided, the RDAs represented at least the minimum requirements of individuals with the greatest need. In terms of the population as a whole, therefore, they were overestimates, and individuals ingesting less than the RDA for any nutrient may be far from deficient. Instead of revised RDAs, *Dietary Reference Values (DRV) for Food Energy and Nutrients for the United Kingdom* was issued (DH, 1991). The latest recommendations were published in 2015 (https://www.nutrition.org. uk/attachments/article/234/Nutrition%20Requirements_Revised%20 Nov%202015.pdf). They apply to energy, proteins, fats, sugars, starches, nonstarch polysaccharides, 9 vitamins and 11 minerals, and comprise the following categories.

- *Estimated average requirement (EAR)* – an estimate of the average requirement or need for food energy or a nutrient (50% of the population will require more; 50% will require less).
- *Reference nutrient intake (RNI)* – enough of a nutrient for almost every individual (defined as 97.5% of the population), even someone who has high needs for the nutrient.
- *Lower reference nutrient intake (LRNI)* – the amount of a nutrient that is enough for only a small number (2.5% of the population) of people with low needs.
- *Safe intake* – a term normally used to indicate the intake of a nutrient for which there is not enough information to estimate requirements. A safe intake is one which is judged to be adequate for almost everyone's needs but not so large as to cause undesirable health effects.

In recognition of the fact that people of different sexes, ages (e.g., infants, children, adults) and conditions (e.g., pregnant and lactating mothers) have different requirements, DRVs appropriate to the different groups were defined.

Deficiency diseases, such as rickets, stunting, deformities and anaemia, are now rare in developed countries and, in considering the relationship between food and disease, emphasis has shifted to other diseases that are thought to be diet related: these include cancer of the colon (associated with animal protein intake, particularly meat), breast cancer (associated with fat intake) and stroke and heart disease, associated with consumption of salt and animal (saturated) fat (Bingham, 1987). In recent years diabetes and obesity have come to the forefront of diet-related diseases.

4.7 RECOMMENDATIONS

Recommendations appropriate to the United Kingdom situation are as follows.

Cereals, particularly whole grain, and potatoes should be eaten in generous amounts at each main meal to satisfy appetite.

Three or more portions of fresh vegetables or fruit, preferably green or yellow, should be eaten per day and two or more portions of low-fat foods containing high protein.

Low-fat dairy foods should be chosen in preference to high-fat ones, and all sugar, refined starches and foods made from them, such as biscuits, cakes, sweets, etc., should be used sparingly.

Dietary fibre intake should be increased, while saturated fat and sugar should be reduced.

More than 80 g/day for men and 50 g/day for women of alcohol is considered excessive and should be avoided. So also should table salt and foods cooked or preserved in excessive salt.

In the United States the USDA/USDHHS publishes guidelines every few years to promote a healthy diet (1985). These were updated in 2015 (https://www.choosemyplate.gov/dietary-guidelines) to include the following recommendations.

- Make half your grains whole grains.
- Make half your plate fruits and vegetables.
- Focus on whole fruits.
- Vary your vegtables.
- Move to low-fat and fat-free milk and yogurt.
- Vary your protein routine.
- Drink and eat less sodium, saturated fat and added sugars.

The reader is referred to the official policy document for more information (USDA/USDHHS, 2015).

Both British and American recommendations acknowledge the importance of cereals, particularly as a source of energy and nonstarch polysaccharides. In addition, cereals provide many other valuable nutrients, including proteins, vitamins and minerals, as several of the tables and figures in this chapter demonstrate.

4.8 CEREALS IN THE DIET

For the majority of the world's human population, cereal-based foods constitute the most important daily source of energy and other nutrients. In the poorest parts of the world starchy foods, including cereals, may supply 70% of total energy. In the wealthiest nations the proportion obtained from cereals has declined fairly rapidly: in the United States during the last few decades the proportion of total energy provided by cereals dropped from 40% to approximately 20%. But total yearly per capita consumption of cereal-based products rose from 155 lb/person/year in the 1950s to nearly 200 lb/person/year in 2000 (http://www.usda.gov/factbook/chapter2.pdf, 2016). The proportions of some important nutrients derived from cereals and products in Britain are shown in Table 4.19.

4.8.1 Attributes of cereals as foods

4.8.1.1 Starch

Cereals are a particularly rich source of starch, as it constitutes by far the most abundant storage product in the endosperm. Starch is an important source of energy and is found only in plants (although the related compound glycogen occurs in animal tissues). In the past starch has been

TABLE 4.19 Contributions (%) made by cereal products to the nutritional value of household food in Britain (1990)[a]

		Bread		Cakes, pastries, biscuits	Breakfast cereals
	Cereals	White	Brown and wholemeal		
Energy	31.5	7.2	3.4	8.2	3.4
Fat	12.8	1.0	0.8	7.6	0.5
Fatty acids					
Saturated	12.1	0.6	0.3	9.0	0.3
Polyunsaturated	13.7	2.2	1.4	5.0	1.4
Sugars	18.7	1.8	0.7	10.0	2.8
Starch	73.0	20.4	9.0	9.9	8.5
Fibre	45.5	7.4	11.6	5.0	9.9
Calcium	24.6	7.5	2.8	4.0	1.0
Iron	49.0	8.7	7.7	6.7	15.4
Sodium	39.4	12.6	6.5	5.0	5.4
Vitamin C	1.5	–	–	–	1.0
Vitamin A	1.1	–	–	0.5	–
Vitamin D	12.3	–	–	1.3	9.6

A dash in the table indicates nil.

[a] *Values calculated from appendix B, Table 14,* Household Food and Expenditure 1990 *MAFF, HMSO, London 1991.*
Reproduced with the permission of the Controller of Her Majesty's Stationery Office.

undervalued by nutritionists, who have emphasized its association with obesity and recommended reduction in, for example, bread consumption by those wishing to control their weight. However, starch is preferable as an energy source to fat, and a further advantage of starch consumed as part of a cereal-based food is that it is accompanied by vitamins, minerals, fibre and protein. In the best-balanced diet starch would probably contribute rather more than the 20% it provides in the average United Kingdom diet today (nearly 40% comes from fat, and 13% from sugar). Note that the value of 31.5% for energy contributed by cereals refers only to foods consumed in the home.

Most starch is consumed in cooked products, in the majority of which the starch granules are gelatinized, making them readily digestible by amylase enzymes present in the gut. For this to occur, however, abundant water is required, as starch can absorb more than 20 times its own mass during gelatinization. In some baked products, such as shortbread, much fat and little water are present; consequently few of the granules are gelatinized. Other factors, such as osmotic conditions, affect gelatinization, and these are much affected by the amount of sugar in a recipe: in high-sugar conditions water activity is low and gelatinization takes place at an elevated temperature.

Energy is released from starch by digestion of starch polymers to produce glucose, which is absorbed into the bloodstream. Glucose yields 16 kJ or 4 kcal/g (Joules are now the preferred unit in which to express energy, 4.184 J = 1 cal).

Starches from 'amylo' mutant types of cereals (mainly maize), which have a higher than usual amylose content, are less readily digested. After cooking at high temperatures the indigestibility may be enhanced, giving rise to a small proportion of resistant starch. Even in other cooked cereal products some resistant starch can arise; it behaves like fibre, passing unchanged through the gut. The method of cooking is important in determining the amount of resistant starch formed. For example, in corn flakes produced by extrusion cooking the proportion of resistant starch is less than in conventionally produced flakes.

Resistant starch has been receiving considerable attention from the food industry due to its positive health benefits, including a lower glycemic response, which is important for persons with diabetes, but also due to improved functionality vis-à-vis fibre in high-fibre, low-fat food products, which can result in greater consumer acceptability and better palatability. A comprehensive review is provided by Sajilata et al. (2006).

Energy is a vital requirement of every healthy individual, but energy that is not expended in physical or physiological activity is stored either as adipose tissue or glycogen. These provide a necessary store from which energy may be released when required. The superiority of starch as a dietary energy source does not derive from a particularly high calorific value; in fact the value of fat is higher than that of starch, at 37 kJ (9 kcal) per gram, as is alcohol at 29 kJ (7 kcal) per gram.

4.8.1.2 Protein content and quality

Cereals, including bread, contribute approximately 25% of the protein in the average adult diet in the United Kingdom. Three thin slices of bread contribute as much protein as an egg. Between 2007 and 2014 consumers in the United Kingdom purchased less bread (all types, average of 10% less) but more cereals (+2.3% more), probably in response to food price changes (DEFRA, 2015).

In nutritional terms there are two factors of prime importance in relation to protein: the total protein content, and the contribution that essential amino acids make to the total.

There are eight essential amino acids (out of a total of 20 or so): methionine, tryptophan, threonine, valine, isoleucine, leucine, phenylalanine and lysine. Two other amino acids are sometimes classified as essential but can be made in the body: tyrosine from phenylalanine and cysteine/cystine from methionine. Their presence in foods reduces the requirements of the relevant 'parent essential amino acids'. In foods derived from plants in general, the sulphur-containing amino acids methionine and cysteine are most likely to be limiting, but this is not true of cereal grains. In cereals, lysine is the first to be limiting: rice, oat and rye are relatively rich among wholegrain cereals but are deficient in relation to the FAO/WHO (1973) reference amino acid pattern, in which the lysine content is 5.5 g/16 g of N. Maize protein is also limiting in tryptophan, based on the reference value of 1.0 g/16 g of N, which the other cereals just reach. A comparison between wheat protein and protein from other food sources is shown in Fig. 4.26.

In the past much was made of the superiority of animal-derived proteins, containing as they do the correct proportions of essential amino acids. However, protein types are rarely eaten alone and tend to complement each other; for example, bread may be eaten with cheese, a good source of lysine. Even in vegetarian diets, many legumes and nuts supply essential amino acids. A good combination is rice and peas.

4.8.1.3 Fibre

The laxative properties of fibre or 'roughage', as it was previously (and picturesquely) described, have been well known for many hundreds of years (Hippocrates around 400 BC advocated it!), but within the last 20 or so attention has been focused on them by the assertion that the high fibre of African diets prevents many chronic noninfective diseases common in the West, where refined carbohydrates are more commonly consumed. Some of the more extreme claims for the beneficial effects of fibre, emanating from the surge of activity consequent upon these assertions, have now been seriously challenged, as populations of developing countries eating lower-fibre diets but not showing high incidence of the relevant diseases have been discovered. Nevertheless, dietary fibre has been shown to have palliative effects on diseases, particularly those of the gut, and diabetes mellitus (especially type 2, because dietary fibre has been shown to reduce postprandial glucose elevations in the blood (Tabatabai and Li, 2000)).

4.8.1.4 Cholesterol

Cholesterol has been much publicized as an indicator of potential health problems over the years, particularly for heart disease, strokes and blocked arteries, but it is in fact not all bad. Some cholesterol is necessary in the body as a precursor to hormones and bile acids. Cholesterol is transported in the blood in three principal forms: free cholesterol, or bound to lipoprotein as either high-density lipoprotein (HDL) or low-density

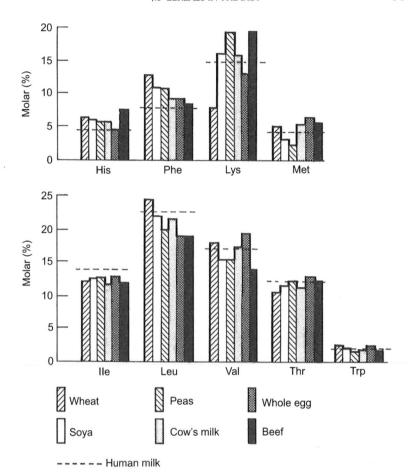

FIGURE 4.26 The essential amino acids in food proteins. The relative proportions of each essential amino acid are shown expressed as the percentage of the total essential amino acids. *From Coultate, T.P., 1989. Food, The Chemistry of its Components, second ed. Roy. Soc. of Chemistry, Letchworth, London, UK by courtesy of the Royal Society of Chemistry.*

lipoprotein (LDL). HDL confers some protection against heart disease, so reduction below a threshold actually increases risk. About 80% of blood cholesterol is associated with LDL, however, and it is this form which is believed to deposit in the arteries. It is also this form that increases as a result of consumption of saturated fats.

Some studies have found considerable reductions in blood cholesterol levels in response to increased cereal fibre in the diet. Several mechanisms have been proposed to account for the hypocholesterolaemic effect of soluble fibres: viscous soluble fibre may exert an effect by physically entrapping cholesterol or bile acids in the digestive tract, thereby preventing their absorption and resulting in their increased excretion. Alternatively, β-glucans may be fermented by colonic bacteria to short-chain fatty acids.

Several of these compounds have been suggested as inhibitors of cholesterol synthesis. In relation to the first possibility, it has been found that different bile acids are bound more effectively by different types of fibre; pentosans of different types of rice even vary in their effectiveness and their 'preferred' bile acids (Normand et al., 1981). In an extensive metaanalysis of 67 published studies, Brown et al. (1999) determined that consumption of soluble fibre did indeed lower overall blood cholesterol levels, and LDL levels in particular, but the effects were small; HDL levels were largely unchanged.

It has been found that eating bran can be an effective cure for constipation and diverticular disease. It is more doubtful whether dietary fibre is as effective in preventing problems other than constipation. An important factor, though not the only one, contributing to the beneficial effects of bran is its ability to hold water, thus increasing stool weight and colonic motility. The relative water-holding capacities of cereal brans and other sources of fibre, given by Ory (1991), are shown in Table 4.20.

TABLE 4.20   Water-holding capacity of cereal brans and other fibre-containing foods

Fibre source	Water-holding capacity (g/100 g d.b.)
Sugar-beet pulp	1449
Apple pomace	235–509
Apple, whole fruit	17–46
All bran	436
Wheat bran	109–290
Rice bran	131
Oat bran	66
Maize bran	34
Cauliflower	28
Lettuce	36
Carrot	33–67
Orange, whole fruit	20–56
Orange pulp	176
Onion, whole	14
Banana, whole fruit	56
Potato, minus skin	22

(values are reported as 'corrected' for fresh weight basis). Reprinted with permission from Ory, R.L., 1991. Grandma Called it Roughage. Amer. Chem. Soc., Washington, DC. Copyright 1991, American Chemical Society.

4.8.1.5 Fats

Apart from essential fatty acids, the liver is able to make all the fat that the body requires from carbohydrates and protein, provided these are eaten in sufficient quantities. Depending on age and gender, 20–50 g of essential fatty acids are needed every day by the human body (USDA/USDHHS, 2015), but in the United Kingdom and the United States the average consumption is currently higher than recommended levels.

Nearly all the fat in the diet is composed of triacylglycerols (triglycerides). Saturated fatty acids, found mainly in animal fats and hardened fats, include lauric, myristic (14:0) and palmitic (16:0) acids, all three of which have been implicated in raising the levels of cholesterol in blood. Palmitic acid (16:0) is the most commonly occurring fatty acid, comprising 35% of animal fats and palm oil and 17% of other plant and fish oils. The most commonly occurring monounsaturated fatty acid is oleic acid (18:1), which contributes 30%–65% of most fats and oils.

4.8.1.6 Minerals

At least 15 minerals are required by humans. Of these, deficiencies are unlikely to occur in phosphorus, sodium, chlorine or potassium, even though daily requirements are relatively high. Anaemia, due to iron deficiency, is one of the most common nutritional disorders, particularly in premenopausal women. Iron from exhausted red blood cells is reused in new cells, so that almost the only requirement is to replace blood that has been lost. Whole-grain cereals contain sufficient iron to supply a large proportion of an adult's daily requirement, but there is some doubt as to whether it can be absorbed from cereal and legume sources because of the presence of phytic acid. About 900 mg of calcium are present in the average United Kingdom diet, and of this 25% is supplied by cereals. Growing children and pregnant and lactating mothers have a higher requirement, of about 1200 mg/day. The aged have an enhanced requirement for calcium, as it may be depleted by insufficient vitamin D. Adequate calcium consumed during growth affords some protection against osteoporosis in later years. A further protective function served by adequate calcium in the diet concerns the radioactive isotope strontium 90 (Sr^{90}), produced as part of the fall-out of nuclear explosions, which can arise from weapons or accidents in nuclear power stations. Sr^{90} can replace calcium in bones, causing irritation and disease, but in the presence of high calcium levels this is less likely. Flours other than wholemeals, malt flours and self-raising flours (which are deemed to contain sufficient calcium) are required to be supplemented with chalk (calcium carbonate) in the United Kingdom, but it is doubtful if this is necessary. If the exception of wholemeal might seem illogical (since wholemeal, of all types of flour, contains the largest amount of phytic acid and would seem to require the largest addition of chalk), it must be remembered that consumers of this particular product are concerned to an exceptional extent with the concept of absence of all additions.

Bran and wheat germ are good sources of magnesium, but, as with other minerals, absorption can be impaired by the phytate also present.

In addition to the above, the body requires much smaller 'trace' amounts of iodine, copper, zinc, manganese, molybdenum, selenium and chromium, and even smaller quantities of silicon, tin, nickel, arsenic and fluorine may be needed.

Whole-grain cereals can contribute to the supply of zinc, although its absorption might be impaired by phytic acid. The selenium content of grain depends upon the selenium status of the soil on which the crop was grown. In North America many selenium-rich soils support cereals, and wheat imported from that continent has relatively large amounts present, enabling up to half the daily requirement to be met from cereals. In recent years several studies have investigated the relationships between soil selenium content and bioavailability for uptake by various wheat varieties (Lee et al., 2011). They found a complex interaction among environmental variables: beyond high soil selenium content, they found a strong nonlinear relationship between wheat selenium content and protein content. Soils in the United Kingdom have less selenium and hence the grains grown on them are lower in the element. Symptoms of selenium deficiency have been reported in countries with notably deficient soils, including New Zealand and some areas of China.

4.8.2 Effects of processing

Some animals, including poultry, may be fed grains of various types in a totally unmodified form. Other stock may consume grains that have received only minimal modification, such as crushing, and some elements of the human diet, like muesli, may also contain these. Nevertheless, most cereals are consumed only after various degrees of processing, and this affects the nutritive value of the products. Changes in nutritional properties of cereals result from several types of processing, including refinement, cooking and supplementation.

4.8.2.1 Refinement

Refinement includes processes involved in milling that separate anatomical parts of the grain to produce palatable foodstuffs. The most palatable (lowest fibre) and most stable (lowest fat) parts of the grains are not necessarily the most nutritious, and if only these are consumed, much of the potential benefit can be lost. For example, 80% of the vitamin B_1 is removed when rice is milled and polished. This results from the fact that many of the nutrients reside in the embryo and outer parts of the grains

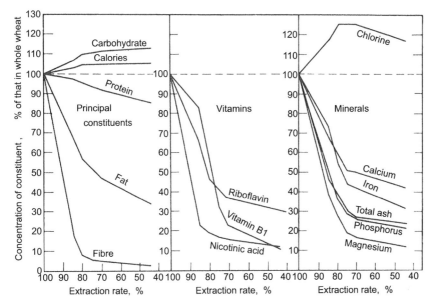

FIGURE 4.27 Nutrient composition of flours of various extraction rates in relation to that of whole wheat. *From Kent, N.L., 1983. Technology of Cereals, third ed. Pergamon Press Ltd., Oxford, UK.*

(mainly in the aleurone tissue). The degree of change depends upon the degree of separation that occurs. The effects of varying extraction rates in wheat milling are illustrated diagrammatically in Fig. 4.27, and the effects of varying extraction rates on wheat and rye milling products are shown in Table 4.21. Compositions of fractions obtained by milling other cereals are shown in Table 4.22.

In the fermentations, filtrations and distillations involved in the production of beers and spirits, some of the nutrients in the original grains are removed – mainly starches and sugars. Some nutrients are derived from other ingredients, notably vitamin B_{12} from yeast. The nutrients in several types of beer are shown in Table 4.23.

4.8.2.2 Cooking/heat treatment

Changes resulting from cooking are complex, and in only a few cases, such as boiled rice and pasta, are cereal products cooked substantially on their own. More frequently they are included in a recipe or formulation, so that differences between the raw cereal ingredient and the final product reflect not only changes due to cooking but also to dilution and interactions with other ingredients.

In the traditional method of tortilla production, cooking and refinement are combined as the heating of whole maize grains in alkaline water loosens the pericarp and embryo, allowing the endosperm to be concentrated;

TABLE 4.21 Composition of flour and milling by-products at various extraction rates

Material	Yield (%)	Protein (%)	Oil (%)	Ash (%)	Crude fibre (%)	Thiamin (µg/g)[a]	Niacin (µg/g)[a]
Wheat flour[b]							
85% extraction	85.0	12.5	1.5	0.92	0.33	3.42	
80% extraction	80.5	12.0	1.4	0.72	0.20	2.67	19
70% extraction	70.0	11.4	1.2	0.44	0.10	0.70	10
Fine wheatfeed							
85% extraction	10.0	12.6	4.7	5.1	10.6	6.0	
80% extraction	12.5	14.3	4.7	4.7	8.4	10.4	191
75% extraction	20.0	15.4	4.7	3.5	5.2	14.0	113
Bran							
85% extraction	5	11.1	3.7	6.1	13.5	4.6	
80% extraction	7	12.4	3.9	5.9	11.1	5.0	302
70% extraction	10	13.0	3.5	5.1	8.9	6.0	232
Rye flour[c]							
60% extraction	60	5.7	1.0	0.5	0.2		
75% extraction	75	6.9	1.3	0.7	0.5		

85% extraction	85	7.5	1.6	1.0	0.8
100% extraction	100	8.2	2.0	1.7	1.6
Brans[d]					
Fine		14.0	3.2	4.2	5.0
Coarse		16.6	5.2	3.8	9.4
Germ[d]		35.5	10.3	4.8	3.4

[a] Naturally occurring.
[b] Jones, C.R. 1958. The essentials of the flour-milling process. The Proceedings of the Nutrition Society 17, 7.
[c] Neumann, M.P., Kalning, H., Schleimer, A., Weinmann, W., 1913. Die chemische Zusammensetzung des Roggens und seiner Mahlprodukte. Z Ges Getreidew 5, 41.
[d] Mccance, R.A., Widdowson, E.M., Moran, T., Pringle, W.J.S., Macrea, T.F., 1945. The chemical composition of wheat and rye and of flours derived therefrom. The Biochemical Journal 39, 213.

TABLE 4.22   Composition of milling products of various cereals (%, d.b.)

Cereal and fraction	Protein	Fat	Ash	Crude fibre	Carbohydrate
BARLEY					
Pearl barley[a]	9.5	1.1	1.3	0.9	85.9
Barley flour[a]	11.3	1.9	1.3	0.8	85.4
Barley husk[b]	1.6	0.3	6.2	37.9	53.9
Barley bran[c]	16.6	4.0	5.6	9.6	64.3
Barley dust[c]	13.6	2.5	3.7	5.3	74.9
OATS[B]					
Oatmeal	12.9	7.5	2.1	1.2	75.0
Rolled oats	13.3	7.6	2.0	0.9	74.7
Oat flour	14.2	7.9	2.0	1.1	73.4
Oat husk	1.4	0.4	4.5	37.8	0.9
Oat dust	10.6	5.0	6.3	22.8	10.6
Meal seeds	8.6	3.8	3.1	18.7	28.8
Oat feed meal	3.5	1.5	3.8	30.6	5.8
RICE					
Brown rice[d]	8.5	2.2	1.4	1.0	86.9
White rice[d]	7.6	0.4	0.6	0.3	91.0
Parboiled rice[d]	8.2	0.3	0.8	0.2	90.1
Hulls[e]	1.1	0.4	27.3	34.1	34.1
Bran[f]	14.4	21.1	8.9	10.0	45.6
Bran, extracted[f]	17.8	0.6	11.1	12.2	57.8
Polishings[g]	12.1	9.9	5.5	2.2	70.3
MAIZE					
Maize grain[h]	11.2	4.80	1.7	1.9	80.4
Dry-milling products					
Grits[i]	10.5	0.90	–	–	–
Meal[h]	10.1	5.70	1.2	1.4	83.7
Flour[h]	8.1	1.50	0.7	1.0	88.7
Germ meal[j]	14.6	14.00	4.0	4.6	62.8
Hominy feed[h]	10.6	0.80	0.3	0.4	87.9

TABLE 4.22 Composition of milling products of various cereals (%, d.b.)—cont'd

Cereal and fraction	Protein	Fat	Ash	Crude fibre	Carbohydrate
Wet-milling products					
Corn flour[k]	0.8	0.07	0.1	–	99.0
High-protein corn gluten meal[l]	68.8	6.20	2.0	1.3	21.7
Corn gluten feed[l]	24.8	4.20	8.0	8.9	54.1
Corn germ meal[l]	25.1	5.10	4.1	10.6	55.1
SORGHUM					
Dry-milling products[m]					
Whole sorghum	9.6	3.4	1.5	2.2	
Pearled sorghum	9.5	3.0	1.2	1.3	
Flour, crude	9.5	2.5	1.0	1.2	
Flour, refined	9.5	1.0	0.8	1.0	
Brewers' grits	9.5	0.7	0.4	0.8	
Bran	8.9	5.5	2.4	8.6	
Germ	15.1	20.0	8.2	2.6	
Hominy feed	11.2	6.5	2.7	3.8	
Wet-milling products[n]					
Germ	11.8	38.8			18.6
Fibre	17.6	2.4			30.6
Tailings	39.2	–			25.3
Gluten	46.7	5.1			42.8
Squeegee	14.0	0.6			81.6
Starch	0.4	–			67.3
Solubles	43.7	–			–
MILLET[N]					STARCH
Wet-milling products					
Germ	10.4	45.6			10.4
Fibre	11.8	6.0			13.5
Tailings	34.1	1.9			34.2
Gluten	37.8	9.0			44.0
Squeegee	17.7	0.8			75.5

Continued

TABLE 4.22   Composition of milling products of various cereals (%, d.b.)—cont'd

Cereal and fraction	Protein	Fat	Ash	Crude fibre	Carbohydrate
Starch	0.7	0.1			57.5
Solubles	46.1	–			–

[a] Chatfield, C., Adams, G., 1940. Food Composition. US Dept Agric Circ 549, Washington, DC, USA.

[b] Original third edition.

[c] Watson, S.J., 1953. The quality of cereals and their industrial uses. The uses of barley other than malting. Chemical Industry, 95.

[d] USDA, 1963. Agricultural Handbook, vol. 8. U.S. Dept. of Agriculture, Agricultural Research Service, Washington, DC, USA.

[e] Houston, D.F., 1972. Rice Chemistry and Technology. Amer. Assoc. of Cereal Chemists, St. Paul, MN, USA.

[f] Australian Technical Millers Association, April 1980. Current use and development of rice by-products. Australasian Baker & Millers' Journal, 27.

[g] Fraps, G.S., 1916. The composition of rice and its by-products. Texas Agricultural Experiment Station 191, 5.

[h] Woods, C.D., 1907, Food Value of Corn and Corn Products. U.S Dept. Agric. Farmers Bull. 298, Washington DC, USA.

[i] Stiver Jr., T.E., 1955. American corn-milling systems for degermed products. Bull Ass Oper Millers, 2168.

[j] Woodman, H.E., 1957. Rations for Livestock. Ministry of Agriculture, Fisheries and Food, Bull. 48. H.M.S.O., London, UK.

[k] Boundy, J.A., 1957. Quoted in Kent-Jones, D.W. and Amos, A.J. (1967).

[l] Reiners, R.A., Hummel, J.B., Pressick, J.C., Morgan, R.E., 1973. Composition of feed products from the wet-milling of corn. Cereal Science Today 18, 372.

[m] Hahn, R.R., 1969. Dry milling of sorghum grain. Cereal Science Today 14, 234.

[n] Freeman, J.E., Bocan, B.J., 1973. Pearl millet: a potential crop for wet milling. Cereal Science Today 18, 69.

nutrients, including vitamins and fibre, are lost from the principal product as a result. In spite of this, niacin availability in tortillas is higher than in uncooked maize. Protein availability is generally reduced due to a number of cross-linking reactions (Rooney and Serna-Saldivar, 1987). Fortification may more than compensate for the deficiencies.

In the case of parboiling of rice the nutritional properties of the refined product are improved, as nutrients such as vitamins and minerals in the parts of the grain that are subsequently removed by milling migrate with water into the endosperm. The loss of vitamins on washing of milled rice is reduced, but so is protein availability (Bhattacharya, 1985).

Cooking losses vary according to the amount of water used, and are greater when excess water is present and least when a double boiler is used (Table 4.24).

As recipes involving cereal products abound, a comprehensive illustration of changes during cooking is not possible here. A selection of examples of raw and cooked products is given in Table 4.25. Typical values for the chemical composition of brown rice, white (polished) rice and parboiled rice are given in Table 4.26.

A more comprehensive analysis of the common breads in the United Kingdom is given in Table 4.27.

TABLE 4.23 Average nutrients in 100 g (1/6 pint) of beer[a]

Nutrient	Draft bitter	Draft mild	Keg bitter	Bottled lager	Bottled pale ale	Bottled stout
Energy (kJ)	132	104	129	120	133	156
Protein (g)	0.3	0.2	0.3	0.2	0.3	0.3
Alcohol (g)	3.1	2.6	3.0	3.2	3.3	2.9
Sugars (g)	2.3	1.6	2.3	1.5	2.0	4.2
Sodium (mg)	12	11	8	4	10	23
Potassium (mg)	38	33	35	34	49	45
Calcium (mg)	11	10	8	4	9	8
Magnesium (mg)	9	8	7	6	10	8
Iron (mg)	0.01	0.02	0.01	0	0.02	0.05
Copper (mg)	0.08	0.05	0.01	0	0.04	0.08
Zinc (mg)	–	–	0.02	–	–	–
Riboflavin (mg)	0.04	0.03	0.03	0.02	0.02	0.04
Niacin (mg)	0.60	0.40	0.45	0.54	0.52	0.43
Pyridoxin (mg)	0.02	0.02	0.02	0.02	0.01	0.01
Vitamin B_{12} (µg)	0.17	0.15	0.15	0.14	0.14	0.11
Folic acid (µg)	8.8	4.5	4.6	4.3	4.1	4.4
Pantothenic acid (mg)	0.1	0.1	0.1	0.1	0.1	0.1
Biotin (µg)	0.5	0.5	0.5	0.5	0.5	0.5

[a] *Zero or trace amounts in all beers of fat, starch, dietary fibre, vitamins A, C and D and thiamin.*
Based upon Bingham, S., 1987. The Everyday Companion to Food and Nutrition. J.M. Dent and Sons Ltd., London.

TABLE 4.24 Vitamin b losses from milled rice cooked by different methods

Nutrient	% Loss on cooking		
	Excess water	Absorbable water	Double boiler
Thiamin	47	19	4
Riboflavin	43	14	7
Niacin	45	22	3

Based upon Juliano, B.O., 1985b. Production and utilization of rice (Chapter 1). In: Juliano, B.O. (Ed.), Rice Chemistry and Technology. Amer. Assoc. of Cereal Chemists Inc., St. Paul, MN, USA, pp. 1–16.

TABLE 4.25 Composition of cereal grains and some products

Product	Water (%)	Protein (%)	Fat (%)	Carbohydrate (%)	Energy value (kJ/100 g)
Wheat (and wholemeal)	14	12.7	2.2	63.9	1318
Flour white breadmaking	14	11.5	1.4	75.3	1451
Plain	14	9.4	1.3	77.7	1450
Brown	14	12.6	1.8	68.5	1377
Bran	8.3	14.1	5.5	26.8	872
Germ	11.7	26.7	9.2	(44.7)	1276
Bread white sliced	40.4	7.6	1.3	46.8	926
Toasted	27.3	9.3	1.6	57.1	1129
Brown	39.5	8.5	2.0	44.3	927
Toasted	24.4	10.4	2.1	56.5	1158
French stick	29.2	9.6	2.7	55.4	1149
Hamburger buns					1192
Crusty	26.4	10.9	2.3	57.6	
Soft	32.7	9.2	4.2	51.6	1137
Croissants	31.1	8.3	20.3	38.3	1505
Cream crackers	4.3	9.5	16.3	68.3	1857
Semisweet biscuits	2.5	6.7	16.6	74.8	1925
Cake (cake mix)	31.5	5.3	3.3	52.4	1951
Sponge cake	15.2	6.4	26.3	52.4	1920
Madeira cake	20.2	5.4	16.9	58.4	1652
Pastry (short)	20.0	5.7	27.9	46.8	1874
Durum Wheat					
Macaroni					
Raw	9.7	12.0	1.8	75.8	1483
Boiled	78.1	3.0	0.5	18.5	365
Spaghetti					
Raw	9.8	12.0	1.8	74.1	1465
Boiled	73.8	3.6	0.7	22.2	442

TABLE 4.25 Composition of cereal grains and some products—cont'd

Product	Water (%)	Protein (%)	Fat (%)	Carbohydrate (%)	Energy value (kJ/100 g)
Wholemeal					
Raw	10.5	13.4	2.5	66.2	1379
Boiled	69.1	4.7	0.9	23.2	485
Rye					
Grain and wholemeal	15	8.2	2.0	75.9	1428
Rye bread	37.4	8.3	1.7	45.8	923
Crispbread	6.4	9.4	2.1	70.6	1367
Rice					
Brown					
Raw	13.9	6.7	2.8	81.3	1518
Boiled	66.0	2.6	1.1	32.1	597
White, easy cook					
Raw	11.4	7.3	3.6	85.8	1630
Boiled	68.0	2.6	1.3	30.9	587
Oats					
Oatmeal	8.2	11.2	9.2	66.0	1567
Porridge (made with water)	87.4	1.5	1.1	9.0	1381

Data from Holland, B., Welch, A.A., Unwin, I.D., Buss, D.H., Paul, A.A., Southgate, D.A.T., 1991. McCance and Widdowson's the Composition of Foods, fifth ed. Roy. Soc. Chem. and MAFF, Cambridge, UK. Reproduced with permission of the Royal Society of Chemistry.

TABLE 4.26 Nutrient composition of rice (g/100 g)

Product	Moisture	Protein	Fat	Crude fibre	Carbohydrate	Ash	Calories
Brown	12	7.5	1.9	0.9	76.5	1.2	360
White	12	6.7	0.4	0.3	80.1	0.5	363
Parboiled	10	7.4	0.3	0.2	81.1	0.7	369

Based upon USDA, 1963. Agricultural Handbook, vol. 8. U.S. Dept. of Agriculture, Agricultural Research Service, Washington, DC, USA.

TABLE 4.27 Nutrient composition of bread in the UK

Nutrient	White average	White sliced	Brown	Germ bread	Wholemeal
Water (g)	37.3	40.4	39.5	40.3	38.3
Protein (g)	8.4	7.6	8.5	9.5	9.2
Fat (g)	1.9	1.3	2.0	2.0	2.5
Fatty acids					
Saturated (g)	0.4	0.3	0.4	0.3	0.5
Monosaturated (g)	0.4	0.2	0.3	0.3	0.5
Polyunsaturated (g)	0.5	0.4	0.6	0.7	0.7
Sugars (g)	2.6	3.0	3.0	1.8	1.8
Carbohydrate (g)	49.3	46.8	44.3	41.5	41.6
Starch (g)	46.7	43.8	41.3	39.7	39.8
Dietary fibre (g)					
Southgate	3.8	3.7	5.9	5.1	7.4
Englyst	1.5	1.5	(3.5)	3.3	(5.8)
Energy (kJ)	1002	926	927	899	914
Thiamin (mg)	0.21	0.2	0.27	0.8	0.34
Riboflavin (mg)	0.06	0.05	0.09	0.09	0.09
Nicotinic acid (mg)	1.7	1.5	2.5	4.2	4.1
Potential nicotinic acid (mg)	1.7	1.6	1.7	1.9	1.8
Pyridoxin (mg)	0.07	0.07	0.13	0.11	0.12
Folic acid (µg)	29	17	40	39	39
Pantothenic acid (mg)	0.3	(0.3)	0.3	(0.3)	0.6
Biotin (µg)	1	(1)	3	(2)	6
Vitamin E (mg)	Tr	Tr	Tr	N	0.2
Sodium (mg)	520	530	540	600	550
Potassium (mg)	110	99	170	200	230
Calcium (mg)	110	100	100	120	54
Magnesium (mg)	24	20	53	56	76
Phosphorus (mg)	91	79	150	190	200
Iron (mg)	1.6	1.4	2.2	3.7	2.7

Tr, trace; figures in parentheses ( ) are estimated; N, the nutrient is present in significant quantities but there is no reliable information available.
Data from Holland, B., Welch, A.A., Unwin, I.D., Buss, D.H., Paul, A.A., Southgate, D.A.T., 1991. McCance and Widdowson's the Composition of Foods, fifth ed. Roy. Soc. Chem. and MAFF, Cambridge, UK.

During extrusion processing, many nutrients will change due to the simultaneous influence of heat, pressure, shear and time. These interactions will vary with each chemical component, and the alterations will be affected by moisture content in the ingredient matrix. Depending upon the conditions experienced, lysine is often the first amino acid to degrade (because it is heat sensitive), proteins may denature, digestibility (for humans or animals) may decline, vitamins may degrade and colour may change (Camire et al., 1990). On the positive side, starch can be gelatinized, trypsin inhibitors can be reduced in soy products and microbes may be reduced or even destroyed. Singh et al. (2007) provides an extensive review of the impacts of extrusion processing on many chemical components in foods.

4.8.2.3 Specific interactions

Some of the more important interactions among ingredients during cooking are described below.

Maillard reactions

Reactions between sugars and amino groups give rise to browning, which is important in the production of commercial caramel and in providing a colour to the crust of bread and other baked products. The amino acids which engage in Maillard reactions most readily are those with a free amino group in their side chain, particularly lysine, followed by arginine, tryptophan and histidine. While the effects on product palatability are generally valued, there is a nutritional price to pay as the cross-linked sugar prevents access by proteolytic enzymes, obstructing digestion of the essential amino acids involved. This will reduce digestibility in both humans and animals consuming these products. Fortunately, the proportion of amino acids involved in crust browning is relatively small (Coultate, 1989).

Enzymatic changes

Probably the most significant enzymatic changes that occur during processing are those involved in conversion of starch into sugars and the subsequent fermentation mediated by microorganisms, converting sugars to alcohol or lactic acid. Another important example is the reduction in phytin content during fermentation and proving of bread doughs, and the accompanying increase in the more readily absorbed inorganic phosphate, due largely to catalysis by phytases produced by yeasts.

4.8.2.4 Supplementation

This differs from the other types of processing in that change in nutritional properties is its primary purpose, rather than a consequence of an improvement in palatability. In general it changes the eating qualities as little as possible.

Because of the staple nature of cereal products, their contributions to the diets of the populations of most of the world's nations are frequently perceived by governments as important in ensuring adequate nutritive standards, not only through their natural composition but also through the addition of nutrients from other sources. Various terms are applied to such additions, including restoration, fortification and enrichment. Restoration implies the replacement of nutrients lost during processing, such as milling, to the level found in the original grain, while fortification and enrichment suggest addition of nutrients not originally present or enhancement of originally present nutrients.

Supplementation policies usually arise in response to disasters which bring about shortages of essential foods through conflict or poverty. Thus World War II was responsible for the formulation of policy in the United Kingdom, and the Great Depression of the 1930s precipitated the institution of the fortification programme in the United States. A supplementation policy may also apply specifically to cereal products exported as part of an aid programme to populations in which a particular deficiency prevails.

In the United States the Food and Drug Administration (FDA) has defined in the U.S. Code of Federal Regulations (CFR Title 21, Chapter 1, Subchapter B, Part 104, Subpart B; http://www.accessdata.fda.gov/scripts/cdrh/cfdocs/cfcfr/cfrsearch.cfm?fr=104.20) the purposes for which addition of nutrients is appropriate.

- Correction of a recognized dietary deficiency.
- Restoration of nutrients lost during processing, handling and storage.
- Balancing the nutrient content in proportion to the caloric content.
- Avoidance of nutritional inferiority of new products replacing traditional foods.
- Compliance with other programmes and regulations.

For any supplementation initiative to succeed it is necessary that the added nutrient is physiologically available; it does not create an imbalance of essential nutrients; it does not adversely affect the acceptability of the product; and there is reasonable assurance that intake will not become excessive.

4.8.2.5 Addition of nutrients

In the United Kingdom the composition of flour is controlled by the Bread and Flour Regulations 1984 (SI 1984, No. 1304), as amended by the Potassium Bromate (Prohibition as a Flour Improver) Regulations 1990 (SI 1990, No. 339), and amended in 1998 (https://www.food.gov.uk). Flour derived from wheat and no other cereal, whether or not mixed with other flour, must contain the nutrients in the amounts specified in the Table 4.28.

TABLE 4.28   U.K. requirements for wheat-based flour

Nutrient	Required quantity (mg/100 g)
Calcium carbonate	Not less than 235 and not more than 390
Iron	Not less than 1.65
Thiamin	Not less than 0.24
Nicotinic acid[a]/nicotinamide[a]	Not less than 1.60

[a] Also known as niacin.

The requirement concerning calcium carbonate does not apply to wholemeal flour, self-raising flour with a calcium content of not less than 0.2% or wheat malt flour; while iron, thiamin and nicotinic acid or nicotinamide should be naturally present in wholemeal and need be added to flour other than wholemeal only where not already present in the specified amounts.

In Canada only enriched flour can be sold; in the United States the FDA does not require enrichment but, as many states require it, virtually all affected products sold across state borders are enriched. The FDA specifies what products advertised or labelled as 'enriched' must contain.

The current requirements for Canada and the United States, compared to whole grains, are given in Table 4.29.

Addition of vital wheat gluten to bread flour may be regarded as nutrient supplementation, but this is not its primary purpose. The rationale behind its addition is purely to improve the breadmaking properties of the flour.

In the United Kingdom and Canada addition of nutrients to flour, where addition is required, is made at the flour mill. In the United States enrichment is permitted either at the mill or at the bakery.

Enrichment of whole grains requires a more subtle approach than that of ground cereal products, as nutrients added to the raw grains can easily be removed during cooking. In Japan several strategies have been adopted for addition of vitamins, minerals and lysine. In the multinutrient method two stages of addition are involved. The first consists of soaking rice for up to 24 h in acetic or hydrochloric acid solution (0.2%, 1% or 5% at 39°C) containing thiamin, niacin, riboflavin, pantothenic acid and pyridoxine; the grains are then steamed and dried, after which they are coated with vitamin E, calcium and iron in separate layers. Finally a protective coating is applied, consisting of an alcoholic (or propan-2-ol) solution of zein, palmitic or stearic acid and abietic acid. In addition to retaining the majority of the added nutrients during washing and cooking, coated rice retains the natural nutrients to an improved degree (Misaki and Yasumatsu, 1985). Losses are approximately 10% for coated rice and up to 70% for milled rice. Coated rice also provides a nutritious and palatable alternative to the unpopular brown or parboiled types in Puerto Rico.

Wheat and maize flours (as well as sugar, cow's milk, chocolate drinks and other food products) provide vehicles for delivery of vitamin A and iron to consumers in developing countries who are vulnerable to deficiencies in these nutrients. In Venezuela both nutrients are added in maize flour and iron is added to wheat flour. Chile, Egypt, El Salvador, Grenada, Guatemala, Jamaica, Kyrgyzstan, St Vincent/Grenadines and Sri Lanka fortify wheat flour for all or certain uses, and other nations are giving consideration to fortification as well (www.unicef.org/pon96/nufortif.htm).

TABLE 4.29 Standards for flour and bread in the United States and Canada, and enrichment standards for cereals in the United States compared with whole cereal complements (mg/100 g)

Product	Thiamin	Riboflavin	Niacin	Iron	Calcium (+)
UNITED STATES					
Enriched flour	0.64	0.4	5.3	4.4	(211)
Enriched bread	0.4	0.24	3.3	2.75	(132)
Enriched farina	0.4–0.6	0.26–0.33	3.5–4.4	2.9	(110)
Maize meal and grits	0.44–0.66	0.26–0.4	3.52–5.3	2.86–5.73	(110–165)
Self-raising maize meal	0.44–0.66	0.26–0.4	3.52–5.3	2.86–5.73	(110–385)
Rice	0.44–0.88	0.26–0.53	3.52–7.05	2.86–5.73	(101–220)
Pasta (macaroni and noodles)	0.88–1.10	0.37–0.24	5.95–7.49	2.86–3.63	(110–137)
CANADA					
Flour and enriched flour	0.44–0.77	0.27–0.48	3.5–6.4	2.9–4.3	(110–140)
Enriched bread and enriched white bread	0.24	0.18	2.20	1.76	(66)
WHOLE GRAIN					
Wheat	0.43	0.13	5.0	4.6	48
Maize	0.40	0.11	1.9	3.1	20
Brown rice	0.33	0.07	4.6	1.9	22
White rice	0.07	0.03	1.6	0.9	12

(+) Brackets indicate an optional addition.

Based upon Kent, N.L., 1983. Technology of Cereals, third ed. Pergamon Press Ltd., Oxford, UK; Ranum, P., 1991. Cereal enrichment (Chapter 22). In: Lorenz, K.J., Kulp, K. (Eds.), Handbook of Cereal Science and Technology. Marcel Dekker, Inc., New York, NY, USA, pp. 833–861, US CFR Title 21 (http://www.ecfr.gov; 2016).

4.8.3 Negative attributes

In spite of the fact that cereals are among the safest and most important foodstuffs, there are some antinutritional aspects of their composition (some of these affect only a small proportion of consumers). Also there are hazards associated with their storage and handling. Awareness of these allows monitoring to be carried out, providing a means of ensuring their safe and proper use.

4.8.3.1 Phytic acid

An important constituent of cereals, legumes and oilseed crops is phytic acid. The systematic name of phytic acid in plant seeds is myoinositol-1,2,3,5/4,6-hexakis (dihydrogen phosphate) (IUPAC-IUB, 1968). Phytic acid in the free form is unstable, decomposing to yield orthophosphoric acid, but the dry salt form is stable The terms phytic acid, phytate and phytin refer respectively to the free acid, the salt and the calcium/magnesium salt, but some confusion arises in the literature, where the terms tend to be used interchangeably. The salt form, phytate, accounts for 85% of the total phosphorus stored in many cereals and legumes. In cereals its distribution varies: in maize the majority lies in the embryo, while in wheat, rye, triticale and rice most of the phytate is found in the aleurone tissue. In pearl millet the distribution is apparently more uniform (Table 4.30).

TABLE 4.30   Distribution of phytate in the morphological components of cereals

Cereal	Fraction	Phytate (%)	% of total in grain
Maize	Endosperm	0.04	3 (4)
	Embryo	6.39	87 (95)
	Hull	0.07	<0.1 (1)
Wheat	Endosperm	0.004	2 (1)
	Embryo	3.91	12 (29)
	Aleurone	4.12	86 (70)
Rice	Endosperm	0.01	1 (3)
	Embryo	3.48	7.6 (26)
	'Pericarp'	3.37	80.0 (71)
Pearl millet	Endosperm	0.32	(48)
	Embryo	2.66	(31)
	Bran	0.99	(11)

The values in parentheses are calculated from the table values, using the proportions given by Kent (1983). *Values from Reddy, N.R., Pierson, M.D., Sathe, S.K., Salunkhe, D.K., 1989. Phytates in Cereals and Legumes. CRC Press Inc., Boca Raton, FL, USA.*

The negative nutritional attributes of phytate derive from the fact that it forms insoluble complexes with minerals, possibly reducing their bioavailability and leading to failure of their absorption in the gut of animals and humans, and thus to mineral deficiencies.

An estimate of daily phytate intake in the United Kingdom is 600–800 mg. Of this, 70% comes from cereals, 20% from fleshy fruits and the remainder from vegetables and nuts (Davies, 1981).

The reduced bioavailability of minerals due to phytate depends on several factors, including the nutritional status of the consumer, concentration of minerals and phytate in the foodstuff, ability of endogenous carriers in the intestinal mucosa to absorb essential minerals bound to phytate and other dietary substances, digestion or hydrolysis of phytate by phytase and/or phosphatase in the intestine, processing operations and digestibility of the foodstuff. The 'other substances' referred to include dietary fibre (nonstarch polysaccharides) and polyphenolic compounds.

Many metal ions form complexes with phytate, including nickel, cobalt, manganese, iron and calcium; the most stable complexes are with zinc and copper. The phosphorus of the phytate molecule itself is also only partly available to nonruminant animals and humans; estimates vary from 50% to 80% and depend upon several factors, including the calcium–phytic acid ratio (Morris, 1986).

The presence of phytate may also influence the functional and nutritional properties of proteins, with the nature of the phytate–protein complexing being dependent on pH. The mechanisms involved are complex and ill-understood, but it is greatest at low pH because under these conditions phytic acid has a strong negative charge and many plant proteins are positively charged. Very little complexing between wheat proteins and phytate has been found, and complexing between rice-bran phytate and rice-bran proteins occurred only below pH 2.0 (Reddy et al., 1989).

Inhibition of the activity of enzymes such as trypsin, pepsin and α-amylase by phytic acid has been reported. In the case of the amylase enzyme it is not clear whether this results from the phytic acid complexing with the enzyme itself, or chelation of the Ca^{++} ions required by the enzyme (Isaksen, 2006).

Industrially, the reduced efficacy of α-amylase can reduce cereal grain fermentation efficiency (see the chapter on fermentation). Phytate can also alter slurry viscosity, increase scaling and fouling of processing equipment and reduce process throughput (Shetty et al., 2008). Additionally, because fermentation can increase nonstarch components by up to three times in the coproduct grains, the reduced P digestibility in animals (primarily monogastrics, such as pigs and poultry, which do not produce phytase endogenously) will be magnified. One option is to use phytase enzymes to improve fermentation and other plant operations, as well as digestibility in livestock (Hruby, 2011).

4.8.3.2 *Tannins*

Tannins are phenolic compounds of the flavonoid group – they are derivatives of flavone. They are considered in the 'negative' attributes section of this chapter because it is alleged that they reduce protein digestibility through phenol–protein complexing. Some flavonoids have been implicated as carcinogens; on the other hand, they have also been credited with the ability to stimulate liver enzymes which offset the effects of other carcinogens. Clearly we know little about the nutritional effects of tannins (Bingham, 1987), but there is no doubt that they and their derivatives can adversely affect flavour and colour, thus reducing palatability.

Although present in all cereals, tannins are particularly associated with sorghums of the 'birdproof' type. All sorghums contain phenolic acids and other flavonoids, but only brown types contain procyanidin derivatives, otherwise known as condensed tannins. Table 4.31 shows their relative abundance in cereal species.

In recent years many conflicting studies have indicated potential mutagenicity or carcinogenicity due to tannins. Some implicate the tannins themselves, while others have found evidence for associated molecules. Some studies (such as for tea) have even found evidence for cancer-fighting properties of tannins and other polyphenols, due to antioxidative properties (Chung et al., 1998).

4.8.3.3 *Harmful effects of alcoholic drinks*

Most beers contain 2%–4% w/w of alcohol, and barley wine contains 6%. Lagers contain less alcohol than ales and stouts, but it is absorbed more rapidly from lagers because of their greater effervescence. Spirits, whether produced from grains, other fruits or starch, contain 33% or more alcohol by volume.

Low to moderate consumption of alcohol increases life expectancy, reduces blood cholesterol concentration and (in France) is associated with low incidence of coronary vascular disease. However, alcohol is a drug that can become addictive and it is rapidly absorbed into the blood, one-fifth of the amount being ingested through the stomach wall and the remainder from the small intestine. Following ingestion of alcohol, blood vessels become dilated and body heat loss is accelerated, overriding the

TABLE 4.31   Percentage of tannins in cereal grains at 14% moisture content[a]

Brown rice	Wheat	Maize	Rye	Millet	Barley	Oat	Sorghum
0.1	0.4	0.4	0.6	0.6	0.7	1.1[b]	1.6-(5)

[a] *Data from Juliano, B.O., 1985a. Production and utilization of rice (Chapter 1). In: Juliano, B.O. (Ed.), Rice Chemistry and Technology. Amer. Assoc. of Cereal Chemists Inc., St. Paul, MN, USA, pp. 1–16.*
[b] *Data from Collins, F.W., 1986. Oat phenolics: structure, occurrence and function (Chapter 9). In: Webster, F.H. (Ed.), Oats Chemistry and Technology. Amer. Assoc. of Cereal Chemists Inc., St. Paul, MN, USA, pp. 227–295.*

body's thermoregulatory mechanisms – sometimes dangerously. From the blood, alcohol diffuses widely. It is a diuretic, promoting the production of urine and giving rise to dehydration – one of the causes of the hangover. It also has an inhibiting effect on the nervous system, including the brain, inducing euphoria and depressing judgement. The amount of alcohol required to induce intoxication depends on the individual, influential factors being body weight, sex and drinking habits, with regular heavy consumers developing a higher threshold than infrequent drinkers. It is generally considered that symptoms of intoxication are apparent when blood contains 100 mg/100 mL alcohol (a 70 kg man drinking 1.6 L (3 pints) of beer containing 50 g ethanol achieves this level). Reactions are impaired below this level, however, and in England and Wales the limit for someone in charge of a motor vehicle is 80 mg/100 mL by blood test, 35 mg/100 mL by breath test or 107 mg/100 mL by urine test; in Scotland the level is 50 mg/100 mL by blood test (www.drinkaware.co.uk). In the United States the legal limit is 80 mg/100 mL by blood test (www.drinkingandyou.com), and each state has specific laws and penalties. In some countries the limit is zero. The intoxicant effect on women is usually greater than on men, as their bodies metabolize alcohol more slowly and their body weight is generally lower than that of men. Moderate drinking is regarded as less than 50 g of alcohol per day for men and 30 g for women.

Long-term effects of alcohol consumption are damage to the heart, liver (cirrhosis is a condition in which some liver cells die) and brain. Cancers of the mouth, throat and upper digestive system in particular are statistically associated with regular heavy consumption of alcohol.

4.8.3.4 Allergies

An allergy is an unusual immune response to a natural or manmade substance that is harmless to most individuals. Hypersensitivity is a nonimmune response to a substance which generates responses whose symptoms can be similar to allergies. Allergies involve abnormally high reactions of the body's natural immune mechanisms to invasion by foreign substances. Substances that stimulate an immune response are known as antigens, and when the response is abnormal the antigens involved are called allergens. The subject of cereals as allergens was reviewed by Baldo and Wrigley (1984).

Allergies arise to substances that enter the body as a result of inhalation or ingestion, or can be a response to skin contact. Cereal pollens and fragments of 1–5 μm present in the dust raised when grains or other plant parts are being handled constitute the inhaled antigens. While everyone is likely to inhale some pollen grains of cereals and many other species, people involved in handling cereals are more likely to inhale grain dust.

One of the best-documented and longest-established allergies associated with cereals is bakers' asthma. As the name implies, bakers' asthma and rhinitis are most prevalent in, but not exclusive to, those who habitually handle flour. It is now established as an IgE-mediated reaction (i.e., it is a true allergy – IgE is an immunoglobulin associated with an allergic response) arising from inhalation of airborne flour and grain dust. As an occupational disease it is declining as a result of improved flour-handling methods creating progressively less dust. Nevertheless, significant proportions of bakery and mill workforces exhibit some sign, though not necessarily asthmatic symptoms, of the allergy.

Many studies have been made on the harmful effects of inhaling grain dusts; for example, comparisons of grain handlers with city dwellers in an American study showed that inhalation of grain dust gave rise to serious problems in the lungs. Chronic bronchitis was significantly higher in grain handlers (48%) than in controls (17%), as was wheezing at work ('occupational asthma') and airway obstruction. Compared with smoking, grain-dust exposure had a greater effect on the prevalence of symptoms but the same or less effect on lung function (Rankin et al., 1979). In other studies skin-prick tests, which are a means of detecting allergic responses, confirmed their occurrence following exposure to grain dust (Baldo and Wrigley, 1984).

While it is clear that inhalation of dusts is harmful and immune responses are demonstrable, the degree of responsibility for the symptoms attributable specifically to allergy cannot easily be established, as similar symptoms have been detected in the absence of allergies.

Coeliac (celiac) disease is an autoimmune pathological condition leading to loss of villi and degeneration of the intestine wall, induced by gliadin and gliadin-like proteins. Other names by which it is known are gluten sensitive enteropathy, sprue, nontropical sprue and idiopathic steatorrhea. Symptoms of the disease are malabsorption and resulting deficiencies of vitamins and minerals, loss of weight and many others. Not everyone with gluten sensitivity has been diagnosed with coeliac disease. Noncoeliac gluten sensitivity is not well understood, and there are no formal tests or biomarkers that can be used to diagnose this condition.

Its cause remains unknown, although there is some evidence that it results from a genetic defect characterized by the absence of an enzyme necessary for gliadin digestion. Alternatively it may involve an allergic response in the digestive system. Many coeliac patients are children, with the symptoms showing when cereals are first introduced in their diet. In a normal condition cells lining the small bowel absorb nutrients, which are duly passed into the blood, but in coeliac patients the cells are irritated and become damaged. Their resulting failure to absorb leads to vomiting, passing of abnormal stools containing the unabsorbed nutrients, and symptoms such as pot belly, anorexia, anaemia, rickets and abnormal growth. Milder forms may pass unnoticed, but in adult life may result in general ill health, weight loss, tiredness and osteomalacia. Newly diagnosed adults were probably born with the disease, but their symptoms presented only in later life.

Treatment currently consists of strict adherence to a gluten-free diet. All untreated wheat flour must be avoided, and rye, triticale, oat and barley products are also excluded. Maize and rice (and probably millet and sorghum) are accepted as being totally gluten-free. Because cereal flours are so versatile, gluten is unexpectedly present in many foods and labelling requirements are inadequate to indicate its presence in all cases. Manufacturers and consumers of gluten-free products can confirm the absence of gluten using antibody test kits. Some successful kits use antibodies raised to wheat omega-gliadins. Although these are not considered to be toxic (it is the alpha- and beta-gliadins that are), they retain their antigenicity after heating and are thus suitable for use on cooked products as well as raw ingredients (Skerrit et al., 1990).

Many food companies now have processing and packaging systems solely dedicated to the production of gluten-free products, and the availability of gluten-free foods in the marketplace has drastically increased in recent years. Research has accelerated for development of more gluten-free foods (Arendt and Dal Bello, 2009; Casper and Atwell, 2014). Although medical understanding of coeliac disease is growing and several potential therapies are under development, to date no vaccines are publicly available (https://celiac.org/celiac-disease/research/future-therpies-celiac-desease/). To aid the consumer, food companies now state potential allergens on the ingredient label – in the case of gluten-containing ingredients, the packaging states that the product 'contains wheat'.

4.8.3.5 Schizophrenia

Evidence of the implication of cereal prolamins as a contributory factor to the incidence of schizophrenia in genetically susceptible patients was reviewed by Lorenz (1990). Much of the evidence in favour of the association is epidemiological, but some clinical trials support it. Coeliac disease is reported to occur considerably more frequently in schizophrenic patients than would be expected by chance alone, and administration of gluten-free diets has in some cases led to remission of both disorders. Not all studies support a relationship between gluten ingestion and schizophrenia, and the difficulties involved in designing definitive experiments in this area make a rapid conclusion to the question unlikely.

There is some evidence that many patients with schizophrenia or autism may suffer due to absorption of exorphins, which are formed in the intestine from incomplete digestion of gluten (Reichelt et al., 1986; Cade et al., 2000).

There is also evidence that there can be benefits to gluten withdrawal in a subset of affected persons. An extensive review of published literature found that coeliac disease and schizophrenia may be related, but correlation versus causation has not yet been completely established (Kalaydjian et al., 2006).

4.8.3.6 Dental caries

Dental caries is an infectious disease that leads to tooth decay through the production of organic acids by bacteria present in the oral cavity. The bacteria are supported by fermentable sugars and starch, and foods containing high proportions of these nutrients have been investigated in view of their perceived potential to encourage development of the disease.

Lorenz (1984) reviewed work on the cariogenic and protective effects of diet patterns, with particular reference to cereals. There is a general consensus that sugary foods, particularly sticky ones, which remain in the mouth for a long time do lead to increased occurrence and severity of caries in susceptible subjects. There are reports that the disease is prevalent among bakers and bakery workers, but bread is not considered to be cariogenic and no difference between white and wholemeal breads has been established in this context. Cereal products with a high sucrose content, such as cakes and sweet biscuits, can increase caries, but the relationship is not simple and many factors are involved. Maize-based breakfast cereals have been shown to be more cariogenic than those made from wheat or oats, but conflicting results have emerged from different trials in which sucrose-coated breakfast cereals were compared with uncoated equivalents.

Cariogenicity of a food, especially ready-to-eat cereals, may be related not just to sucrose content but also the time of consumption, the conditions under which it was eaten and other factors (Glass and Fleisch, 1974). Lorenz (1984) concluded that few unquestionable guidelines can at present be offered as to the avoidance of cariogenic foods or to reduce cariogenicity.

4.8.3.7 β-Glucans

A number of negative nutritional effects have been associated with β-glucans. They may impair absorption of mineral and fat-soluble vitamins, but additional research is required in this area. Perhaps the most extensively studied negative nutritional factor involves reduced growth rates of chickens fed barley-based diets. Barley is of low digestibility and low energy value when fed to chickens, resulting in poor growth and production. Low production values are attributed to β-glucans, as supplementation with β-glucanase improves the performance of barley-based diets (Pettersson et al., 1990). Presumably β-glucans in the endosperm cell wall physically limit the accessibility to starch and protein and increase digesta viscosity.

However, evidence is mounting that there may be some health benefits from ingestion of β-glucans. For example, interactions among oat β-glucans and other food components have been shown to alter starch digestibility and consequently reduce glycemic response – this is useful for people suffering from obesity and diabetes (Tahvonen, 2007; Regand et al., 2011).

4.8.4 Noncereal hazards

Adverse effects on health can arise from the consumption of foods contaminated by toxic substances. In the handling of all food raw materials, some risk of contamination exists and cereals are no more susceptible than other natural products to these. Similarly, all natural products are associated with particular risks from, for example, infection with specific diseases. Hazards of this sort are considered here.

4.8.4.1 Ergotism

The disease gangrenous ergotism, caused by consumption of products containing 'ergot' – the sclerotia of the fungus *Claviceps purpurea* – is discussed earlier in this book.

4.8.4.2 Mycotoxins

Mycotoxins are toxic secondary metabolites produced by fungi and moulds. Technically, the toxins in poisonous mushrooms and ergot are mycotoxins, but poisoning by mushrooms or ergot requires consumption of at least a moderate amount of the fungus tissues containing the toxins, whereas the mycotoxins in, for example, nuts or cereal grains may be present in the absence of any obvious mould. Strictly, substances like penicillin and streptomycin are mycotoxins, but the term is usually reserved for substances toxic to higher animals (including humans). The term 'antibiotics' is generally applied to compounds produced by fungi that are toxic to bacteria.

Over 100 species of fungus are associated with mycotoxins giving rise to symptoms in domestic animals (Brooks and White, 1966). Of these over half are in the genera *Aspergillus*, *Penicillium* and *Fusarium*.

Mycotoxins cause concern in cereals as well as in peanuts, brazil nuts, pistachio nuts and cottonseed – the risks depending to some extent on their place of origin. Maize is an excellent substrate for the growth of *Aspergillus flavus* and *A. parasiticus*, two fungal species that produce aflatoxins. The importance of mycotoxins in the cereal context was reviewed by Mirocha et al. (1980).

Aflatoxins are highly carcinogenic. There are six important aflatoxin isomers: B_1, B_2, G_1, G_2, M_1 and M_2; the initial letter of the first four being that of the colour of the fluorescence which they exhibit under ultraviolet light, i.e., blue and green. The M indicates milk, in which these isomers were first found. All isomers are toxic and carcinogenic, but aflatoxin B_1 is the most abundant and presents the greatest danger to consumers. Several derivatives produced by the microorganisms or in the consuming animal are also toxic.

In the United States it has been found that infection of maize with *Aspergillus* spp. occurs in the field, usually after damage to the grains by insects. Degree of infection varies geographically, growth of the organism being favoured by high temperatures and drought conditions. It is thus most common in the south and southeast United States, and tropical maize-growing countries such as Thailand and Indonesia, and relatively rare in the cooler Corn Belt of North America. Other genera are more associated with infection during storage. Aflatoxins have been detected in small amounts in wheat, barley, oats, rye and sorghum. It must be noted that mould and fungi infection can occur anywhere if the growing conditions are appropriate.

Other mycotoxins include zearalenones, produced mainly by species of *Fusarium*, the major producers being members of the *F. roseum* complex and *F. graminearum*. These often occur in cereals during high humidity and low temperatures. Zearalenone mimics oestrogen and its effects primarily involve the genital system. Most affected are pigs; the effects on humans are not well documented. Although reported in wheat, barley, oats and sorghum, maize is the cereal most affected.

Trichothecenes are a group of over 200 compounds associated with *Fusarium* spp. (especially *F. graminearum* and *F. sporotrichioides*) but also produced by other fungi, such as *Stachybotrys*, *Trichodrema* and *Trichothecium*. These mycotoxins are implicated as causes of alimentary toxic aleukia as well as other immunosuppressive effects. Occurrence of the disease in Russia between 1931 and 1943 was associated with consumption of proso millet that had been left in the field over winter and had become seriously infected with a number of fungal species. Trichothecenes have been isolated from other cereal species, but mainly maize, especially that stored on the cob.

Deoxynivalenol (also known as vomitoxin) is one of the most common trichothecenes. It is produced by *F. graminearum*, often due to damp weather conditions, and has particular impacts on monogastric animals.

Ochratoxin is a mycotoxin produced by *Penicillium viridicatum* and *Aspergillus* sp. It has been found in stored (often overheated) maize and wheat in North America and Europe, as well as in tropical countries. Nine ochratoxins have been identified to date, but ochratoxin A is the most common, while B and C also occur often. Ochratoxin is associated with spontaneous nephropathy in pigs. It is also a carcinogen.

Citreoviridin, citrinin and other mycotoxins are described as 'yellow-rice toxins', as they cause yellowing of rice in storage. They are produced mainly by *Penicillium citreoviridin*, *P. citrinum* and *P. islandicum*, and are associated with inhibition of mitochondrial ATP synthetase and diseases of the heart, liver and kidneys.

Limits on mycotoxin levels have been set by many countries for cereals and other commodities intended for feed or food. Actual exposures that will produce cancer in humans have been established, and therefore limits for mycotoxins are based on both scientific principles and concepts of safety (Stoloff et al., 1991).

The United States FDA has established regulatory levels for various mycotoxins in cereal grains. These are not maximum regulatory limits, but levels at which the FDA may take enforcement action for human and animal safety. Tables 4.32–4.34 provide action levels for aflatoxin (which are precise levels at which the FDA may take action to remove the grain from the market for animal and human safety, as it deems levels higher than these to be unsafe), as well as advisory levels for deoxynivalenol and fumonisin, respectively, which are guidance to the industry as the FDA deems these levels will provide a margin of safety for the protection of humans and animals.

TABLE 4.32   FDA action levels for *Aflatoxin* intended for use in human foods and animal feeds

Intended use	Grain, grain by-product, feed or other products	Aflatoxin level (parts per billion (ppb))
Human consumption	Milk	0.5 ppb (aflatoxin M1)
Human consumption	Foods, peanuts and peanut products, brazil and pistachio nuts	20 ppb
Immature animals	Corn, peanut products and other animal feeds and ingredients, excluding cottonseed meal	20 ppb
Dairy animals, animals not listed above or unknown use	Corn, peanut products, cottonseed and other animal feeds and ingredients	20 ppb
Breeding cattle, breeding swine and mature poultry	Corn and peanut products	100 ppb
Finishing swine 100 pounds or greater in weight	Corn and peanut products	200 ppb
Finishing (i.e., feedlot) beef cattle	Corn and peanut products	300 ppb
Beef, cattle, swine or poultry, regardless of age or breeding status	Cottonseed meal	300 ppb

Based upon National Grain, Feed Association, 2011. FDA Mycotoxin Regulatory Guidance: A Guide for Grain Elevators, Feed Manufacturers, Grain Processors and Exporters. National Grain and Feed Association, Washington, DC.

TABLE 4.33 FDA advisory levels for *Deoxynivalenol* (*Vomitoxin*) for grains and by-products intended for use in animal feeds

Intended use	Grain or grain by-products	Vomitoxin levels in grains or grain by-products and (complete diet in parentheses) (parts per million (ppm))
Human consumption	Finished wheat products	1 ppm
Swine	Grain and grain by-products not to exceed 20% of diet	5 ppm (1 ppm)[b]
Chickens	Grain and grain by-products not to exceed 50% of diet	10 ppm (5 ppm)[b]
Ruminating beef and feedlot cattle older than 4 months	Grain and grain by-products[a]	10 ppm (10 ppm)[b]
Ruminating dairy cattle older than 4 months	Grain and grain by-products not to exceed 50% of diet[a]	10 ppm (5 ppm)[b]
Ruminating beef and feedlot cattle older than 4 months and ruminating dairy cattle older than 4 months	Distillers' grains, brewers' grains, gluten feeds and gluten meals[a]	30 ppm (10 ppm beef/feedlot, 5 ppm dairy)[b]
All other animals	Grain and grain by-products not to exceed 40% of diet	5 ppm (2 ppm)[b]

[a] 88% dry-matter basis.
[b] Complete diet figures shown in parentheses.
Based upon National Grain, Feed Association, 2011. FDA mycotoxin regulatory guidance: a guide for grain elevators, feed manufacturers, grain processors and exporters. National Grain and Feed Association, Washington, DC.

TABLE 4.34 FDA guidance levels for *Fumonisin* for corn and corn products intended for use in human foods and animal feeds

Product	Total fumonisins (FB1, FB2 and FB3) (parts per million (ppm))
HUMANS	
Degermed dry-milled corn products (e.g., flaking grits, corn grits, corn meal, corn flour with fat content of <2.25%, dry-weight basis)	2 ppm
Cleaned corn intended for popcorn	3 ppm
Whole or partially degermed dry-milled corn products (e.g., flaking grits, corn grits, corn meal, corn flour with fat content of ≥2.25% dry-weight basis)	4 ppm
Dry-milled corn bran	4 ppm
Cleaned corn intended for masa production	4 ppm

continued

TABLE 4.34   FDA guidance levels for *Fumonisin* for corn and corn products intended for use in human foods and animal feeds—cont'd

Class of animal	Grain or grain by-products	Total fumonisin (FB1, FB2 and FB3) levels in grain or grain by-products (complete diet in parentheses) (parts per million (ppm))
ANIMALS		
Equids and rabbits	Corn and corn by-products not to exceed 20% of diet[b]	5 ppm (1 ppm)
Swine and catfish	Corn and corn by-products not to exceed 50% of diet[b]	20 ppm (10 ppm)
Breeding ruminants, breeding poultry and breeding mink[a]	Corn and corn by-products not to exceed 50% of diet[b]	30 ppm (15 ppm)
Ruminants ≥3 months old being raised for slaughter and mink being raised for pelt production	Corn and corn by-products not to exceed 50% of diet[b]	60 ppm (30 ppm)
Poultry being raised for slaughter	Corn and corn by-products not to exceed 50% of diet[b]	100 ppm (50 ppm)
All other species or classes of livestock and pet animals	Corn and corn by-products not to exceed 50% of diet[b]	10 ppm (5 ppm)

[a] *Includes lactating dairy cattle and hens laying eggs for human consumption.*
[b] *Dry-weight basis.*
Based upon National Grain, Feed Association, 2011. FDA mycotoxin regulatory guidance: a guide for grain elevators, feed manufacturers, grain processors and exporters. National Grain and Feed Association, Washington, DC.

References

Aldrick, A.J., 1991. The Nature, Sources, Importance and Uses of Cereal Dietary Fibre. HGCA Review No. 22. Home Grown Cereals Authority, London, UK.

Amado, R., Neukom, H., 1985. Minor constituents of wheat flour: the pentosans. In: Hill, R.D., Munck, L. (Eds.), New Approaches to Research on Cereal Carbohydrates. Elsevier, Amsterdam, Netherlands.

Arendt, E.K., Dal Bello, F., 2009. The Science of Gluten-free Foods and Beverages. American Association of Cereal Chemists Inc., St. Paul, MN, USA.

Asp, N.-G., Johansson, C.-G., Hallmer, H., Uiljestrom, M., 1983. Rapid enzymatic assay of insoluble and soluble dietary fibre. Journal of Agricultural and Food Chemistry 31, 476–482.

Australian Technical Millers Association, April 1980. Current use and development of rice by-products. Australasian Baker & Millers' Journal 27.

Badenhuizen, N.P., 1969. The Biogenesis of Starch Granules in Higher Plants. Appleton-Crofts, NY, USA.

Baldo, B.A., Wrigley, C.W., 1984. Allergies to cereals. Advances in Cereal Science and Technology 6, 289–356.

Bamforth, C.W., Quain, D.E., 1989. Enzymes in brewing and distilling. In: Palmer, G.H. (Ed.), Cereal Science and Technology. Aberdeen University Press, Scotland.

Berry, C.S., 1988. Resistant starch – a controversial component of 'dietary fibre'. BNF Nutrition Bulletin 54, 141–152.

Bhattacharya, K.R., 1985. Parboiling of rice (Chapter 8). In: Juliano, B.O. (Ed.), Rice Chemistry and Technology, second ed. Amer. Assoc. of Cereal Chemists Inc., St. Paul, MN, USA, pp. 289–348.

Bingham, S., 1987. The Everyday Companion to Food and Nutrition. J.M. Dent and Sons Ltd., London, UK.

Boundy, J.A., 1957. Quoted in Kent-Jones, D.W. and Amos, A.J. (1967).

Bowler, P., Towersey, P.J., Waight, S.G., Galliard, T., 1985. Minor components of wheat starch and their technological significance. In: Hill, R.D., Munck, L. (Eds.), New Approaches to Research on Cereal Carbohydrates. Elsevier, Amsterdam, Netherlands, pp. 71–79.

Boyer, C.D., Shannon, J.C., 1983. The use of endosperm genes for sweet corn improvement. In: Janick, J. (Ed.), Plant Breeding Reviews. vol. 1. Avi Publishing Co., Westport, CT, USA, pp. 139–161.

Brooks, P.J., White, E.P., 1966. Fungus toxins affecting mammals. Annual Review of Phytopathology 4, 171–194.

Brown, L., Rosner, B., Willett, W.W., Sacks, F.M., 1999. Cholesterol-lowering effects of dietary fiber: a meta-analysis. The American Journal of Clinical Nutrition 69 (1), 30–42.

Bushuk, W., October 1966. Distribution of water in dough and bread. Baker's Digest 40, 38–40.

Busson, F., Fauconnau, G., Pion, N., Montreuil, J., 1966. Acides aminés. Annales de la Nutrition et de l'Alimentation 20, 199.

Buttrose, M.S., 1962. The influence of environment on the shell structure of starch granules. Journal of Cell Biology 14, 159–167.

Cade, R., Privette, M., Fregly, M., Rowland, N., Sun, Z., Zele, V., Herbert Wagemaker, H., Edelstein, C., 2000. Autism and schizophrenia: intestinal disorders. Nutritional Neuroscience 3 (1), 57–72.

Camire, M.E., Camire, A., Krumhar, K., 1990. Chemical and nutritional changes in foods during extrusion. Critical Reviews in Food Science and Nutrition 29, 35–57.

Carter, E.G.A., Carpenter, K.J., 1982. The bioavailability for humans of bound niacin from wheat bran. The American Journal of Clinical Nutrition 36, 855–861.

Casey, P., Lorenz, K., Feburary 1977. Millet – functional and nutritional properties. Baker's Digest 45.

Casper, J.L., Atwell, W.A., 2014. Gluten-free Baked Products. American Association of Cereal Chemists Inc., St. Paul, MN, USA.

Chatfield, C., Adams, G., 1940. Food Composition. US Dept Agric Circ 549, Washington, DC, USA.

Chung, O.K., 1991. Cereal lipids. In: Lorenz, K.J., Kulp, K. (Eds.), Handbook of Cereal Science and Technology. Marcel Decker, Inc., New York, NY, USA, pp. 497–553.

Chung, K.-T., Wong, T.Y., Wei, C.-I., Huang, Y.-W., Lin, Y., 1998. Tannins and human health: a review. Critical Reviews in Food Science and Nutrition 38 (6), 421–464.

Clegg, K.M., 1958. The microbiological determination of pantothenic acid in wheaten flour. Journal of the Science of Food and Agriculture 9, 366.

Clegg, K.M., Hinton, J.J.C., 1958. The microbiological determination of vitamin B_6 in wheat flour, and in fractions of the wheat grain. Journal of the Science of Food and Agriculture 9, 717.

Codex Alimentarius, 2010. Guidelines on Nutrition Labelling CAC/GL 2-1985. Joint FAO/WHO Food Standards Programme, Secretariat of the Codex Alimentarius Commission, FAO, Rome, Italy.

Collins, F.W., 1986. Oat phenolics: structure, occurrence and function (Chapter 9). In: Webster, F.H. (Ed.), Oats Chemistry and Technology. Amer. Assoc. of Cereal Chemists Inc., St. Paul, MN, USA, pp. 227–295.

Coultate, T.P., 1989. Food: The Chemistry of Its Components, second ed. Roy. Soc. of Chemistry, Letchworth, London, UK.

Craig, S.A.S., Stark, J.R., 1984. The effect of physical damage on the molecular structure of wheat starch. Carbohydrate Research 125, 117–125.

Davies, K., 1981. Proximate composition, phytic acid, and total phosphorus of selected breakfast cereals. Cereal Chemistry 58, 347.

De Caro, L., Rindi, G., Casella, C., 1949. Contents in thiamine, folic acid, and biotin in an Italian converted rice. In: Abst. Communs. 1st Int. Cong. Biochem., p. 31.

DEFRA (Department for Environment, Food, Rural Affairs), 2015. Family Food 2014. Department for Environment, Food and Rural Affairs, London, UK. www.gov.uk/defra.

Dengate, H.N., 1984. Swelling, pasting and gelling of wheat starch. Advances in Cereal Science and Technology 7, 49–82.

Deyoe, C.W., Shellenberger, J.A., 1965. Amino acids and proteins in sorghum grain. Journal of Agricultural and Food Chemistry 13, 446.

DH (Department of Health), 1991. Dietary Reference Values for Food Energy and Nutrients for the United Kingdom (Report on Health and Social Subjects; 41). H.M.S.O., London, UK.

DHSS (Department of Health, Social Security), 1969. Recommended Intakes of Nutrients for the United Kingdom (Reports on Public Health and Medical Subjects; 120). H.M.S.O., London, UK.

DHSS (Department of Health, Social Security), 1979. Recommended Daily Amounts of Food Energy and Nutrients for Groups of People in the United Kingdom (Reports on Health and Social Subjects; 15). H.M.S.O., London, UK.

Evers, A.D., 1971. Scanning electron microscopy of wheat starch. 3. Granule development in endosperm. Stärke 23, 157–162.

Evers, A.D., 1979. Cereal starches and proteins. In: Vaughan, J.G. (Ed.), Food Microscopy. Academic Press, London, UK.

Evers, A.D., Stevens, D.J., 1985. Starch damage. Advances in Cereal Science and Technology 7, 321–349.

Ewart, J.A.D., 1967. Amino acid analysis of glutenins and gliadins. Journal of the Science of Food and Agriculture 18, 111.

Fennema, O.R., 1985. Food Chemistry. Marcel Dekker, Inc., New York, NY, USA.

Fincher, G.B., Stone, B.A., 1986. Cell walls their components in cereal grain technology. Advances in Cereal Science and Technology 8, 207–295.

Fisher, N., Redman, D.G., Elton, G.A.H., 1968. Fractionation and characterization of purothionin. Cereal Chemistry 45, 48.

Food and Agriculture Organization/World Health Organization, 1973. WHO Technical Report Series No. 522, FAO Nutrition Meetings Report Series No. 52. FAO/WHO, Geneva.

Fraps, G.S., 1916. The composition of rice and its by-products. Texas Agricultural Experiment Station 191, 5.

Freeman, J.E., Bocan, B.J., 1973. Pearl millet: a potential crop for wet milling. Cereal Science Today 18, 69.

French, D., 1984. Physical and chemical organization of starch granules. In: Whistler, R.L., Bemiller, J.N., Paschall, E.F. (Eds.), Starch Chemistry and Technology. Academic Press Inc., NY, USA.

Fretzdorf, B., Weipert, D., 1990. Enzyme activities in developing triticale compared to developing wheat and rye. In: Ringlund, K., Mosleth, E., Mares, D.J. (Eds.), Fifth International Symposium on Pre-harvest Sprouting in Cereals. Westview Press Inc., Boulder, CO, USA, pp. 156–164.

Fulcher, R.G., Wong, S.I., 1980. Inside cereals-a fluorescence microchemical view. In: Inglett, G.E., Munck, L. (Eds.), Cereals for Food and Beverages. Academic Press, NY, USA.

Gallant, D.J., Bouchet, B., 1986. Ultrastructure of maize starch granules. Food Microstructure 5, 141–155.

Glass, R.L., Fleisch, S., 1974. Diet and dental caries: dental caries incidence and the consumption of ready-to-eat cereals. Journal of the American Dental Association 88 (4), 807–813.

Goering, K.J., Fritts, D.H., Eslick, R.F., 1973. A study of starch granule size distribution in 29 barley varieties. Stärke 25, 297–302.

Green, J., Marcinkiewicz, S., Watt, P.R., 1955. The determination of tocopherols by paper chromatography. Journal of the Science of Food and Agriculture 6, 274.

Greenwell, P., Schofield, J.D., 1989. What makes cereals hard or soft? In: Proc. SAAFOST. 10th Congress and Cereal Sci. Symp., vol. 2. Natal Technikon Printers, Durban, RSA.

Hahn, R.R., 1969. Dry milling of sorghum grain. Cereal Science Today 14, 234.

Hall, D.M., Sayre, J.G., 1969. A scanning electron microscopy study of starch. Root and tuber starches. Textile Research Journal 39, 1044–1052.

Heathcote, J.G., Hinton, J.J.C., Shaw, B., 1952. The distribution of nicotinic acid in wheat and maize. Proceedings of the Royal Society B139, 276.

Henry, R.J., 1985. A comparison of the non-starch carbohydrates in cereal grains. Journal of the Science of Food and Agriculture 36, 1243–1253.

Hinton, J.J.C., 1944. The chemistry of wheat with particular reference to the scutellum. The Biochemical Journal 38, 214.

Hinton, J.J.C., 1947. The distribution of vitamin B_1 and nitrogen in the wheat grain. Proceedings of the Royal Society B134, 418.

Hinton, J.J.C., 1948. The distribution of vitamin B_1 in the rice grain. British Journal of Nutrition 2, 237.

Hinton, J.J.C., 1953. The distribution of protein in the maize kernel in comparison with that in wheat. Cereal Chemistry 36, 19.

Hinton, J.J.C., Shaw, B., 1953. The distribution of nicotinic acid in the rice grain. British Journal of Nutrition 8, 65.

Holland, B., Welch, A.A., Unwin, I.D., Buss, D.H., Paul, A.A., Southgate, D.A.T., 1991. McCance and Widdowson's the Composition of Foods, fifth ed. Roy. Soc. Chem. and MAFF, Cambridge, UK.

Hoseney, R.C., January 1984. Functional properties of pentosans in baked foods. Food Technology 38, 114–117.

Hoseney, R.C., 1986. Principles of Cereal Science and Technology. Amer. Assoc. of Cereal Chemists Inc., St. Paul, MN, USA.

Hoseney, R.C., Variano-Marston, E., 1980. Pearl millet: its chemistry and utilization. In: Inglett, G.E., Munck, L. (Eds.), Cereals for Food and Beverages. Academic Press Inc., NY, USA.

Houston, D.F., 1972. Rice Chemistry and Technology. Amer. Assoc. of Cereal Chemists Inc., St. Paul, MN, USA.

Hruby, M., 2011. Improved and new enzymes for fuel ethanol production and their effect on DDGS. In: Liu, K., Rosentrater, K.A. (Eds.), Distillers Grains: Production, Properties, and Utilization. CRC Press, Taylor and Francis, Boca Raton, FL, USA.

Hughes, B.P., 1967. Amino acids. In: McCance, R.A., Widdowson, E.M. (Eds.), The Composition of Foods. Med. Res. Coun. Spec. Rep. Ser. No. 297, 2nd Impression, H.M.S.O., London, UK.

Isaksen, M.F., 2006. Phytic acid effect on endogenous alpha-amylase activity and starch degradation. International Poultry Scientific Forum Abstracts T111, 35–36.

IUPAC-IUB, 1968. The nomenclature of cyclitols. European Journal of Biochemistry 5, 1.

Jones, C.R., 1958. The essentials of the flour-milling process. The Proceedings of the Nutrition Society 17, 7.

Jones, J.M., 2014. CODEX-aligned dietary fiber definitions help to bridge the 'fiber gap'. 13, 34.

Juliano, B.O., 1985c. Rice bran (Chapter 18). In: Juliano, B.O. (Ed.), Rice Chemistry and Technology. Amer. Assoc. of Cereal Chemists Inc., St. Paul, MN, USA, pp. 647–687.

Juliano, B.O., Bechtel, D.B., 1985. The rice grain and its gross composition (Chapter 2). In: Juliano, B.O. (Ed.), Rice Chemistry and Technology. Amer. Assoc. of Cereal Chemists Inc., St. Paul, MN, USA, pp. 17–57.

Juliano, B.O., 1985a. Biochemical properties of rice. In: Juliano, B.O. (Ed.), Rice Chemistry and Technology. Amer. Assoc. of Cereal Chemists Inc., St. Paul, MN, USA, pp. 175–206.

Juliano, B.O., 1985b. Production and utilization of rice (Chapter 1). In: Juliano, B.O. (Ed.), Rice Chemistry and Technology. Amer. Assoc. of Cereal Chemists Inc., St. Paul, MN, USA, pp. 1–16.

Juliano, B.O., Baptista, G.M., Lugay, J.C., Reyes, A.C., 1964. Rice quality studies on physico-chemical properties of rice. Journal of Agricultural and Food Chemistry 12, 131.

Kalaydjian, A.E., Eaton, W., Cascella, N., Fasano, A., 2006. The gluten connection: the association between schizophrenia and celiac disease. Acta Psychiatrica et Neurologica Scandinavica 113 (2), 82–90.

Kates, M., 1972. Techniques of Lipidology: Isolation, Analysis and Identification of Lipids. Elsevier, NY, USA.

Kent, N.L., 1983. Technology of Cereals, third ed. Pergamon Press Ltd., Oxford, UK.

Kik, M.C., 1943. Thiamin in products of commercial rice milling. Cereal Chemistry 20, 103.

Kik, M.C., Van Landingham, F.B., 1943. Riboflavin in products of commercial rice milling and thiamin, riboflavin and niacin content of rice. Cereal Chemistry 20, 563–569.

Kruger, J.E., Reed, G., 1988. Enzymes and color. In: Pomeranz, Y. (Ed.), Wheat Chemistry and Technology. Amer. Assoc. of Cereal Chemists Inc., St. Paul, MN, USA, pp. 441–500.

Lee, S., Woodard, H., Doolittle, J., 2011. Selenium uptake response among selected wheat (*Triticum aestivum*) varieties and relationship with soil selenium fractions. Soil Science and Plant Nutrition 57 (6), 823–832.

Lei, M.-G., Reeck, C.G.R., 1986. Two dimensional electrophoretic analysis of wheat kernel proteins. Cereal Chemistry 63, 111–116.

Lockhart, H.B., Hurt, H.D., 1986. Nutrition of oats. In: Webster, F.H. (Ed.), Oats Chemistry and Technology. Amer. Assoc. of Cereal Chemists Inc., St. Paul, MN, USA, pp. 297–308.

Lorenz, K., 1984. Cereal and dental caries. Advances in Cereal Science and Technology 6, 83–137.

Lorenz, K., 1990. Cereals and schizophrenia. Advances in Cereal Science and Technology 10, 435–469.

Marshall, W.R., Whelan, W.J., 1974. Multiple branching in glycogen and amylopectin. Archives of Biochemistry and Biophysics 161, 234–238.

Maurice, T.J., Slade, L., Sirett, R.R., Page, C.M., 1983. Polysaccharide-water interactions - thermal behaviour of rice starch. In: 3rd Int. Symp. on Properties of Water in Relation to Food Quality and Stability, Baune, France.

Mccance, R.A., Widdowson, E.M., Moran, T., Pringle, W.J.S., Macrea, T.F., 1945. The chemical composition of wheat and rye and of flours derived therefrom. The Biochemical Journal 39, 213.

Mccleary, B.V., Sloan, N., Draga, A., Lazewska, I., 2013. Measurement of total dietary fiber using AOAC method 2009.01 (AACC international approved method 32-45.01): evaluation and updates. Cereal Chemistry 90 (4), 396–414.

Mcginn, J., 2014. Microscopic advancements aid food QC labs. Lab Equipment. Available online: http://www.laboratoryequipment.com/article/2014/09/microscopic-advancements-aid-food-qc-labs.

Mirocha, C.J., Pathre, S.V., Christensen, C.M., 1980. Mycotoxins. Advances in Cereal Science and Technology 3, 159–225.

Misaki, M., Yasumatsu, K., 1985. Rice enrichment and fortification (Chapter 10). In: Juliano, B.O. (Ed.), Rice Chemistry and Technology. Amer. Assoc. of Cereal Chemists Inc., St. Paul, MN, USA, pp. 389–401.

Moran, T., 1959. Nutritional significance of recent work on wheat, flour and bread. Nutrition Abstracts and Reviews 29, 1.

Morris, E.R., 1986. Phytate and dietary mineral bioavailability. In: Graf, E. (Ed.), Phytic Acid: Chemistry and Applications. Pilatus Press, Minneapolis, MN, USA, pp. 57–76.

Morrison, W.R., 1978. Cereal lipids. Advances in Cereal Science and Technology 2, 221–348.

Morrison, W.R., 1983. Acyl lipids in cereals. In: Barnes, P.J. (Ed.), Lipids in Cereal Chemistry. Academic Press Inc., London, UK, pp. 11–32.

Morrison, W.R., 1985. Lipids in cereal starches. In: Hill, R.D., Munck, L. (Eds.), New Approaches to Research on Cereal Carbohydrates. Elsevier Science Publ. B.V., Amsterdam, Netherlands.

National Grain, Feed Association, 2011. FDA Mycotoxin Regulatory Guidance: a Guide for Grain Elevators, Feed Manufacturers, Grain Processors and Exporters. National Grain and Feed Association, Washington, DC, USA.

Neumann, M.P., Kalning, H., Schleimer, A., Weinmann, W., 1913. Die chemische Zusammensetzung des Roggens und seiner Mahlprodukte. Z Ges Getreidew 5, 41.

Normand, F.L., Ory, R.L., Mod, R.R., 1981. Interactions of several bile acids with hemicelluoses from several varieties of rice. Journal of Food Science 46, 1159.

Ory, R.L., 1991. Grandma Called it Roughage. Amer. Chem. Soc., Washington, DC.

Osborne, T.B., 1907. The Proteins of the Wheat Kernel. Publ. 84. Carnegie Inst., Washington, DC, USA.

Osman, E.M., 1967. Starch in the food industry. In: Whistler, R.L., Paschal, E.F. (Eds.). Whistler, R.L., Paschal, E.F. (Eds.), Starch: Chemistry and Technology, vol. 2. Academic Press Inc., NY, USA.

Payne, P.I., Rhodes, A.P., 1982. Cereal storage proteins: structure and role in agriculture and food technology. Encyclopedia of Plant Physiology, New Series 14A, 346–369.

Percival, E.G.V., 1962. In: Percival, E.G.V. (Ed.), Structural Carbohydrate Chemistry, second ed. Garnet Miller Ltd., London, UK.

Pettersson, D., Graham, A.H., Aman, P., 1990. Enzyme supplementation of broiler chicken diets based on cereals with endosperm cell walls rich in arabinoxylans and mixed linkage ß-glucans. Animal Production 51, 201–207.

Rankin, J., Dopico, G.A., Reddan, W.G., Tsiatis, A., 1979. Respiratory disease in grain handlers. In: Miller, B.S., Pomeranz, Y. (Eds.), Proc. Int. Symposium on Grain Dust. Kansas State Univ. Manhattan, KS, USA, p. 91.

Ranum, P., 1991. Cereal enrichment (Chapter 22). In: Lorenz, K.J., Kulp, K. (Eds.), Handbook of Cereal Science and Technology. Marcel Dekker, Inc., New York, NY, USA, pp. 833–861.

Rao, M.A., Okechukwu, P.E., Da Silva, P.M.S., Oliveira, J.C., 1997. Rheological behaviour of heated starch dispersions in excess water: role of starch granule. Carbohydrate Polymers 33 (4), 273–283.

Ratnayake, W.S., Jackson, D.S., 2006. Gelatinization and solubility of corn starch during heating in excess water: new insights. Journal of Agricultural and Food Chemistry 54 (10), 3712–3716.

Reddy, N.R., Pierson, M.D., Sathe, S.K., Salunkhe, D.K., 1989. Phytates in Cereals and Legumes. CRC Press Inc., Boca Raton, FL, USA.

Regand, A., Chowdhury, Z., Tosha, S.M., Wolever, T.M.S., Wood, P., 2011. The molecular weight, solubility and viscosity of oat beta-glucan affect human glycemic response by modifying starch digestibility. Food Chemistry 129 (2), 297–304.

Reichelt, K.L., Saelid, G., Lindback, T., Boler, J.B., 1986. Childhood autism: a complex disorder. Biological Psychiatry 21 (13), 1279–1290.

Reiners, R.A., Hummel, J.B., Pressick, J.C., Morgan, R.E., 1973. Composition of feed products from the wet-milling of corn. Cereal Science Today 18, 372.

Robin, J.P., Mercier, C., Charbonniere, R., Guilbot, A., 1974. Lintnerized starches. Gel filtration and enzymatic studies of insoluble residues after prolonged acid treatment of potato starch. Cereal Chemistry 51, 389–406.

Rooney, L.R., Serna-Saldivar, S.O., 1987. Food uses of whole corn and dry-milled fractions (Chapter 13). In: Watson, S.A., Ramsted, P.E. (Eds.), Corn Chemistry and Technology. Amer. Assoc. of Cereal Chemists, St. Paul, MN, USA, pp. 399–429.

Sajilata, M.G., Singhal, R.S., Kulkarni, P.R., 2006. Resistant starch – a review. Comprehensive Reviews in Food Science and Food Safety 5, 1–17.

Schofield, J.D., 1986. Flour proteins: structure and functionality in baked products. In: Blanshard, J.M.V., Frazier, P.J., Galliard, T. (Eds.), Chemistry and Physics of Baking. Roy. Soc. of Chem., London, UK.

Science Editor, 1970. Tocopherols and cereals. Milling 152 (4), 24.

Shetty, J., Paulson, B., Pepsin, M., Chotani, G., Dean, B., Hruby, M., 2008. Phytase in fuel ethanol production offers economical and environmental benefits. International Sugar Journal 110 (1311), 2–12.

Shewry, P.R., Tatham, A.S., 1990. The prolamin storage proteins of cereal seeds: structure and evolution. The Biochemical Journal 267, 1–12.

Shuey, W.C., Tipples, K.H., 1980. The Amylograph Handbook. Amer. Assoc. of Cereal Chemists Inc., St. Paul, MN, USA.

Simmonds, D.H., 1989. Wheat and Wheat Quality in Australia. CSIRO, Australia.

Singh, S., Gamlath, S., Wakeline, L., 2007. Nutritional aspects of food extrusion: a review. International Journal of Food Science and Technology 42 (8), 916–929.

Skerritt, J.H., Devery, J.M., Hill, A.S., 1990. Gluten intolerance: chemistry, celiac-toxicity, and detection of prolamins in foods. Cereal Foods World 35 638–639, 641–644.

Slover, H.T., 1971. Tocopherols in foods and fats. Lipids 6, 291.

Snyder, E.M., 1984. Industrial microscopy of starches. In: Whistler, R.L., Bemiller, J.N., Paschall, E.F. (Eds.), Starch Chemistry and Technology. Academic Press Inc., NY, USA.

Sober (Ed.), 1968. Handbook of Biochemistry. CRC Press Inc., Boca Raton, FL, USA.

Southgate, D.A.T., Spiller, G.A., White, M., Mcpherson-Kay, R., 1986. Glossary of dietary fiber components. In: Spiller, G.E. (Ed.), Handbook of Dietary Fibre in Human Nutrition. CRC Press Inc., Boca Raton, FL, USA.

Stiver Jr., T.E., 1955. American corn-milling systems for degermed products. Bull Ass Oper Millers 2168.

Stoloff, L., Van Egmond, H.P., Park, D.L., 1991. Rationales for the establishment of limits and regulations for mycotoxins. Food Additives and Contaminants 8, 213–222.

Tabatabai, A., Li, A., 2000. Dietary fiber and type 2 diabetes. Clinical Excellence for Nurse Practitioners 4 (5), 272–276.

Tahvonen, R., 2007. The effect of bold italic beta-glucan on the glycemic and insulin index. European Journal of Clinical Nutrition 61, 779–785.

Tkachuk, R., Irvine, G.N., 1969. Amino acid composition of cereals and oilseed meals. Cereal Chemistry 46, 206.

U.S. Code of Federal Regulations, Title 21, Part 104B, Fortification Policy.

USDA, 1963. Agricultural Handbook, vol. 8. U.S. Dept. of Agriculture, Agricultural Research Service, Washington, DC, USA.

USDA/USDHHS, 1985. Nutrition and Your Health: Dietary Guidelines for Americans. U.S. Dept. of Agriculture and U.S. Dept. of Health and Human Services, Washington, DC, USA.

USDA/USDHHS, December 2015. 2015–2020 Dietary Guidelines for Americans, eighth ed. U.S. Department of Health and Human Services and U.S. Department of Agriculture, Washington, DC, USA. Available at: http://health.gov/dietaryguidelines/2015/guidelines/.

Waldschmidt-Leitz, E., Hochstrasse, K., 1961. Grain proteins. VII. Albumin from barley and wheat. Zeitschrift fuer Physiologische Chemie 324, 423.

Watson, S.A., 1987. Structure and composition. In: Watson, S.A., Ramstad, P.E. (Eds.), Corn: Chemistry and Technology. Amer. Assoc. of Cereal Chemists Inc., St. Paul, MN, USA, pp. 53–82.

Watson, S.J., 1953. The quality of cereals and their industrial uses. The uses of barley other than malting. Chemical Industry 95.

Wood, P.J., 1986. Oat ß-glucan: structure, location, and properties. In: Webster, F.H. (Ed.), Oats Chemistry and Technology. Amer. Assoc. of Cereal Chemists Inc., St. Paul, MN, USA, pp. 121–152.

Woodman, H.E., 1957. Rations for Livestock. Ministry of Agriculture, Fisheries and Food, Bull. 48. H.M.S.O., London, UK.

Woods, C.D., 1907. Food Value of Corn and Corn Products. U.S Dept. Agric. Farmers Bull. 298, Washington, DC, USA.

Woodward, J.R., Fincher, G.B., 1982. Substrate specificities and kinetic properties of two (1-3,1-4)-ß-D-glucan hydrolases from germinating barley (*Hordeum vulgare*). Carbohydrate Research 106, 111–122.

Wrigley, C.W., Bietz, J.A., 1988. Proteins and amino acids. In: Pomeranz, Y. (Ed.), Wheat Chemistry and Technology, third ed. Amer. Assoc. of Cereal Chemists Inc., St. Paul, MN, USA, pp. 159–275.

Youngs, V.L., 1986. Oat lipids and lipid related enzymes. In: Webster, F.H. (Ed.), Oats Chemistry and Technology. Amer. Assoc. of Cereal Chemists Inc., St. Paul, MN, USA, pp. 205–226.

Zee, S.-Y., O'brien, T.P., 1970. Studies on the ontogeny of the pigment strand in the caryopsis of wheat. Australian Journal of Biological Sciences 23, 1153–1171.

Further Reading

Alais, C., Linden, G., 1991. Cereals – bread (Chapter 10). In: Food Biochemistry. Ellis Horwood, USA, pp. 119–129.

Alexander, R.J., Zobel, H.F., 1988. Developments in Carbohydrate Chemistry. Amer. Assoc. of Cereal Chemists Inc., St. Paul, MN, USA.

Anon., 2003. Diet, Nutrition and the Prevention of Chronic Disease. World Health Organisation, Geneva, Switzerland.

Aman, P., Newman, C.W., 1986. Chemical composition of some different types of barley grown in Montana, USA. Journal of Cereal Science 4, 133–141.

Becker, R., Hanners, G.D., 1991. Carbohydrate composition of cereal grains. In: Lorenz, K., Kulp, K. (Eds.), Handbook of Cereal Science and Technology. Marcel Decker, Inc., NY, USA.

Bengtsson, S., Aman, P., Graham, H., Newman, C.W., Newman, R.K., 1990. Chemical studies on mixed linkage ß-glucan in hull-less barley cultivars giving different hypocholesterol-emic responses in chickens. Journal of the Science of Food and Agriculture 52, 435–445.

Bock, M.A., 1991. Minor constituents of cereals. In: Lorenz, K.J., Kulp, K. (Eds.), Handbook of Cereal Science and Technology. Marcel Decker, Inc., NY, USA, pp. 555–594.

Banks, W., Greenwood, C.T., 1975. Starch and its Components. Edinburgh Univ. Press, Edinburgh, Scotland.

Bender, A.E., 1990. Dictionary of Nutrition and Food Technology, sixth ed. Butterworth and Co. (Publ.), London, UK.

Davidson, M.H., Dugan, L.D., Burns, J.H., Bova, J., Sorty, K., Drennan, K.B., 1991. The hypo-cholesterolemic effects of ß-glucan in oatmeal and oat bran. A dose controlled study. JAMA 265, 1833–1839.

Dreher, M.L., 1987. Handbook of Dietary Fibre. Marcel Dekker, Inc., NY, USA.

De Ruiter, D., 1978. Composite flours. Advances in Cereal Science and Technology 2, 349–379.

Fowden, L., Miflin, B.J. (Eds.), 1983. Seed Storage Proteins. Roy. Soc., London, UK.

Frolich, W., 1984. Bioavailability of Minerals from Unrefined Cereal Products. In Vitro and in Vivo Studies. University of Lund, Sweden.

Gunstone, F.D., 1992. A Lipid Glossary. The Oily Press Ltd., Ayr, Scotland.

Galliard, T., 1983. Enzymic degradation of cereal lipids. In: Barnes, P.J. (Ed.), Lipids in Cereal Technology. Academic Press Inc., London, UK, pp. 111–147.

Hoffmann, R.A., Kamerling, J.P., Vliegenthart, J.F.G., 1992. Structural features of a water-soluble arabinoxylan from the endosperm of wheat. Carbohydrate Research 226 (2), 303–311.

Home-Grown Cereals Authority, 1991. Grain storage practices in England and Wales. H-GCA Weekly Bull 25 (Suppl.), 38.

Izydorczyk, M., Biladaris, C.G., Bushuk, W., 1991. Physical properties of water-soluble pentosans from different wheat varieties. Cereal Chemistry 68, 145–150.

Johnson, L.A., 1991. Corn: production, processing, and utilization. In: Lorenz, K., Kulp, K. (Eds.), Handbook of Cereal Science and Technology. Marcel Decker, Inc., NY, USA.

James, W.P.T., Ferro-Luzzi, A., Isaksson, B., Szostak, W.B., 1987. Healthy Nutrition. World Health Organisation, Copenhagen.

Kent-Jones, D.W., Amos, A.J., 1967. Modern Cereal Chemistry, sixth ed. Food Trade Press, London, UK.

Klopfenstein, C.F., 1988. The role of cereal beta-glucans in nutrition and health. Cereal Foods World 33, 865–869.

Kruger, J.E., Lineback, D., Stauffer, C.E., 1987. Enzymes and Their Role in Cereal Technology. Amer. Assoc. of Cereal Chemists Inc., St. Paul, MN, USA.

Kumar, P.J., Walker-Smith, J.A., 1978. Coeliac Disease: 100 Years. St. Bartholomews Hospital, London, UK.

Marousis, S.N., Karathanos, V.T., Saracacos, G.D., 1991. Effect of physical structure of starch materials on water diffusivity. Journal of Food Processing and Preservation 15, 161–166.

Mclaughlin, T., 1978. A Diet of Tripe. David and Charles, Newton Abbot, Devon, UK.

Ministry of Agriculture, Fisheries, Food, 1991a. Annual Report of the Working Party on Pesticide Residues: 1989-1990. H.M.S.O., London, UK.

Ministry of Agriculture, Fisheries, Food, 1991b. Household Food Consumption and Expenditure 1990. Annual Report of the National Food Survey Committee. H.M.S.O., London, UK.

Mitchell, J.R., Lockward, D.A. (Eds.), 1986. Functional Properties of Food Macromolecules. Elsevier, London, UK.

Newman, R.K., Newman, C.W., Graham, H., 1989. The hypocholesterolemic function of barley ß-glucan. Cereal Foods World 34, 883–886.

O'Brien, L.O., O'Dea, K., 1988. The Role of Cereals in the Human Diet. Cereal Chem. Division, Royal Australian Chem. Inst., Parkville, Victoria, Australia.

Rayner, L., 1990. Dictionary of Foods and Food Processes. Food Sci. Publ. Ltd., Kenley, Surrey, UK.

Rittenburg, J.H., 1990. Development and Application of Immunoassay for Food Analysis. Elsevier, Barking.

Rooney, L.W., Serna-Saldivar, S.O., 1991. Sorghum. In: Lorenz, K., Kulp, K. (Eds.), Handbook of Cereal Science and Technology. Marcel Decker, Inc., NY, USA.

Sanders, T., Bazalgette, T., 1991. The Food Revolution. Transworld Publishers Ltd., London, UK.

Slattery, J., 1987. The Healthier Food Guide. Chalcombe Publications, Marlow, Bucks, UK.

Southgate, D.A.T., 1986. The Southgate method of dietary fiber analysis. In: Spiller, G.E. (Ed.), Handbook of Dietary Fibre in Human Nutrition. CRC Press Inc., Boca Raton, FL, USA.

Southgate, D.A.T., Waldron, K., Johnson, I.T., Fenwick, G.R., 1990. Dietary Fibre. Roy Soc. Chemistry, Cambridge, UK.

Truswell, A.S., 1992. ABC of Nutrition, second ed. British Medical Journal, London, UK.

Storage, handling and preprocessing

Compared with many other fruits, cereal grains are extremely amenable to storage, principally because their moisture contents at harvest are relatively low (or can be lowered easily) and their composition is such that biodeterioration is slow – if the grains are protected. Harvesting is seasonal, but the need for fresh cereal products (e.g., for food, feed, industrial products, etc.) is continuous throughout the year. The least requirement for storage, therefore, is for the period between harvests. Under appropriate conditions this can easily be met, and storage for many years without serious loss of quality is possible. Even in biblical times long periods of storage were apparently achieved. Indeed, archaeological findings over the years have shown that many ancient societies around the world stored grain for long periods of time.

5.1 STORAGE FACILITIES

Key requirements for safe long-term storage include protection against dampness, caused by weather, leakage, or other sources, microorganisms, destructively high temperatures, insects, rodents, birds, objectionable odours and contaminants, unauthorized disturbance and theft.

Clearly the simplest type of storage (i.e., unprotected piles on the ground), is suitable for short periods only, because the grain is susceptible to all types of deleterious effects mentioned previously. Other simple stores, however, have several advantageous features. For example, underground storage can provide protection from temperature fluctuations; the most successful simple ones are found in hot, dry regions. Containers or vessels are often filled, to leave little air space, and then sealed. This is the concept of hermetic storage, under which insects and moulds rapidly use up oxygen, giving rise to high CO_2 content in the intergrain atmosphere, which thus prevents further mould and insect growth.

In more humid regions ventilation is desirable, as the crop may have to be stored before reaching a safe moisture level. Ventilation may not provide a fast enough moisture reduction before deterioration by moulds occurs, thus drying operations may need to be employed. Clearly the requirements of ventilation and exclusion of insects are not immediately compatible, so careful design is essential.

Storage of maize (corn) on the cobs used to be practised widely (as it provides significant space for air movement within the store), but it is no longer common in developed countries; mechanization and industrialization have changed this approach to storage in much of the world. It is currently practised by small-scale growers, often in developing countries (Fig. 5.1). In the not-too-distant past, though, it was used commonly, even in highly commercial practice, much as small cereal grains were stored unthrashed in ricks.

In the commercial context, storage facilities are needed for three purposes:

1. Holding stocks on the farm prior to sale,
2. Centralization before distribution or processing during the year following harvest,
3. Storage of annual surpluses over a longer period.

Farm storage may consist of any available space that will keep out the elements. The facilities for protection against mould and pests are very variable. Stores range from small wooden enclosures in the barn, to round steel bins holding 20–100 tonnes (or more), to concrete silos of larger capacities, or even steel flat storage buildings (Fig. 5.2). On-farm storage facilities allow farmers to choose the time to sell, to receive the best prices, or to hold the grain that has already been sold under contract until the specified time of delivery.

The degree of centralization depends upon the marketing regime within the country of production. In many countries, *country* elevators (Figs 5.3 and 5.4) and *terminal* elevators (Figs 5.5 and 5.6) with storage capacities up to 1,000,000 tonnes (or more) exist. The country elevators provide local staging en route to terminal elevators, which then distribute grain to processing facilities, livestock feeding operations, export terminals or other

FIGURE 5.1   Common maize storage on subsistence farms in Malawi, and in fact much of Africa, c.2015.

FIGURE 5.2   On-farm storage in many developed countries can consist of a combination of steel bins and steel flat storage buildings, c.2000 in Iowa, USA.

end users. Both country and terminal elevators typically include equipment for cleaning, drying and conditioning of grain. The term 'elevator' is applied to the entire facility, although the origin of this term refers literally to the mechanism (i.e., bucket elevator) by which the grain is raised to a level from which it can be transferred and deposited into the bin or silos. Elevators must be serviced by appropriate transport infrastructure,

FIGURE 5.3   In many countries, small country elevators are the entry point into grain markets, c.2010 in Iowa, USA.

including road, rail, water or all three. Many terminal elevators are capable of loading grain into vessels at a rate of 3000 tonnes/h (or even greater).

Although very similar in construction and operation to grain elevators, grain storage at large processing plants is meant only as a temporary store until the grain can be used to manufacture other products (Figs 5.7 and 5.8).

It is sometimes necessary to provide storage for grain beyond the normal capacity of farm bins, elevator silos or at processing facilities. In such conditions a relatively inexpensive expedient is flat storage. This is essentially a cover for a pile (or piles) of dry grain, each pile conforming to its natural angle of repose (Table 5.1). Most often flat storage entails a steel-shelled building supported by either wood or steel posts, frames and trusses. But very temporary (emergency) stores may make use of inflatable covers for outdoor piles, and sometimes use temporary bunker walls to contain the piles (Fig. 5.9). This often occurs during the harvest season, and the piles are removed as quickly as the grain can be loaded onto trains or barges.

Flat storage buildings are easy to fill via mechanical conveyor but, as they have flat concrete floors, removal of grain can be difficult, usually requiring the use of tractor shovels to push the grain into an unloading conveyor. In contrast, bins and silos may have a floor formed like a conical hopper (either steel or concrete) whose walls make an angle greater than

FIGURE 5.4 Country grain elevator c.2016 in Kenya.

the angle of repose for the grain that will be stored, and the grain thus flows via gravity into an unloading conveyor. Sometimes bins and silos will also have flat concrete floors, but with unloading conveyors located under the floor. In this case gravity is used to move most of the grain into the conveyor, and sweep augers will then be used to clean out any residual grains and place them into the conveyor.

FIGURE 5.5   River terminal elevator in the Sancerre region of France, c.2016.

Piles created by grains falling freely from a central spout are fairly uniform, as whole grains tend to roll from the apex down the sloping surfaces. The angle at which the pile of grain forms vis-à-vis the horizontal
plane is called the 'angle of repose'. It forms due to friction amongst the
individual grains themselves. Angle of repose depends primarily upon
the grain itself, as well as the moisture content of that grain. In the case of

FIGURE 5.6 Terminal elevator near Omsk, Russia, c.2002.

FIGURE 5.7 Grain storage facilities at a corn-based fuel ethanol manufacturing plant in South Dakota, USA, c.2008.

FIGURE 5.8   Grain storage facilities at a flour mill in California, USA, c.2000.

TABLE 5.1   Common angles of repose for various cereal grains and oilseeds[a]

Grain	Angle	
Barley	28	
Buckwheat	25	
Corn	21–23	
Flaxseed	21	Angle of Repose
Oats	21–28	
Rice	19	
Rye	23	
Safflower	20–29	
Soybean	21–28	
Wheat	25–28	

[a] *Based on values found in CEMA Standard 550, Classification and Definitions of Bulk Materials (2015),
Conveyor Equipment Manufacturers Association, Naples, FL, USA.*

wheat, the angle is approximately 25 degrees, for corn (maize), the angle of repose is a bit lower.

Small impurities and broken grains roll less readily and thus become trapped in the central core of the pile. Such a core is described as the spoutline. As the interstice spaces amongst kernels can amount to 30% of the

FIGURE 5.9 Emergency storage piles at a grain elevator in Iowa, USA, c.2016. Outdoor piles are often used for temporary storage during the harvest season, until the grain can be loaded onto trains and shipped to terminal elevators, processing plants or export markets.

occupied space, fines in the spoutline can approach that level. Because air circulation and hence heat loss is prevented, the spoutline can often be associated with early deterioration through overheating and moisture condensation. The diameter of the spoutline is proportional to the width of the bin.

In contrast to tall silo and tower storage facilities, flat stores require much lower strength in the sidewalls. This is due in to the much lower wall height in flat storage, but it is also due to lower wall pressures. In a silo much of the pressure of the column of grain is borne not by the floor but by the sidewalls. This occurs because each grain rests on several grains below it and that some of the weight is distributed laterally via grain-on-grain friction until it reaches the walls, which ultimately supports the majority of the grain column. In all stores some settling occurs, and this varies according to the cereal type and moisture content. Wheat is relatively dense and settling may be only 6% of volume, but oats may pack as much as 28%. Settling is a continuous process arising in part from the collapse of hulls, brush hairs, embryo tips, etc., as well as air voids between the grain kernels. Additionally, settling depends on the geometry of the storage building (i.e., flat vs silo).

5.1.1 Material handling

In modern grain storage and processing facilities, mechanical conveyors are key to moving grain, flour, feed and other products. The most common conveyors are shown in Fig. 5.10. For the most part, bucket

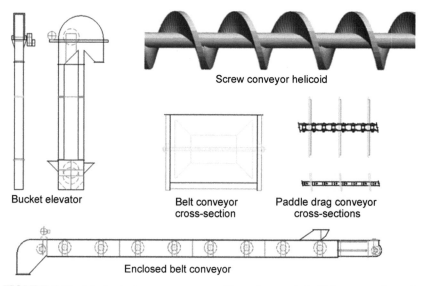

Screw conveyor helicoid

Bucket elevator

Belt conveyor cross-section

Paddle drag conveyor cross-sections

Enclosed belt conveyor

FIGURE 5.10 Mechanical conveyors are used in modern grain facilities to transfer grain; bucket elevators for vertical transfer; and belt, drag and screw conveyors for horizontal transfers.

elevators are used for vertical transfer of grain. Horizontal transfer can be accomplished via paddle drag, belt and screw conveyors. Paddle and belt conveyors can also be used to raise grain at small elevation changes (i.e., conveyors installed at shallow angles), but screw conveyors can actually be used for steep (and even vertical) angles – it can be more expensive to use a screw conveyor instead of a bucket elevator – depending upon the distance lifted and the specific situation. Gravity is also used, either via confined flow (i.e., by chute or spout) or unconfined flow (i.e., open drop, such as filling a silo, bin or container). Specific information about engineering and design of these systems can be found at the Conveyor Equipment Manufacturers Association webpage (http://www.cemanet.org).

5.1.2 Dust explosions

Dust is released whenever grain or particulate matter (i.e., milled grain, feed, flour, etc.) is moved: the atmosphere in grain silos and mills therefore tends to be dusty, leading to conditions under which dust explosions may occur. A suspension of dust in air, within certain limiting concentrations (generally $50–150 \, \text{g/m}^3$), may explode if a source of ignition above a certain temperature is present. In fact, dust explosions require three key components: fuel (i.e., airborne dust), oxygen and an ignition (e.g., flame or heat) source.

A series of dust explosions in US grain elevators in 1978 caused the deaths of at least 50 persons, and led to the initiation of extensive programmes of research into the causes and avoidance of dust explosions. Some of this information can be found at: http://pods.dasnr.okstate.edu/docushare/dsweb/Get/Document-2604/CR-1737web.pdf. An online video of a typical dust explosion can be found at: http://articles.extension.org/pages/63142/grain-dust-explosions. It has been determined that in order to prevent dust explosions, facilities need to:

1. suppress airborne dust
2. avoid sources of ignition

Devices that minimize dust formation include light damping with water or oil (about 1% by wt.), dust-free intake nozzles, dust suppressors, dust collection and removal systems (e.g., bag filters or cyclone filters), or the 'Simporter' system – a method of mechanical grain handling in which the grain is conveyed 'sandwich fashion' between two belts held together by air pressure. By the use of inclined belts operating on this principle, grain can even be conveyed vertically (http://www.vigan.com/en/machines-how-it-work-simporter.php).

Sources of ignition most frequently responsible for dust explosions in grain elevators, feed mills, flour mills and other processing plants are welding and other sources, including flames, hot surfaces, overheated bearings, spontaneous heating, electrical appliances, friction, sparks, static electricity and magnets.

Unfortunately, even though much research has been conducted over the years, and dust control systems and bearing sensors are now used in almost all facilities, fires and explosions continue to occur each year (see, for example, www.feedandgrain.com). The most effective way to prevent dust explosions has proven to be facility cleaning. If the dust is not present, the probability of a dust explosion is minimized.

5.2 DETERIORATION DURING STORAGE

In spite of the diversity of cereal grain types and the ambient conditions throughout the cereal producing and consuming world, the hazards of storage are fundamentally similar, although the relative difficulties involved in their avoidance varies with location (due in large part to weather patterns and ambient conditions throughout the year). Successful storage methods range from primitive to highly sophisticated.

Some of the main hazards besetting cereal grain storage can include:

- Excessive moisture content
- Excessive temperature
- Microbial infestation
- Insect and arachnid infestation
- Rodent predation
- Bird predation
- Biochemical deterioration
- Mechanical damage through handling

The complexity of storage problems results from the matrix of interactions amongst the various hazards. They are considered separately in the text, but their combined and synergistic effects, some of which are shown in Fig. 5.11, should be borne in mind throughout the discussion.

5.2.1 Effects of moisture content and storage temperature

Moisture content is often expressed as a percentage of the wet weight of the grain (this is termed the 'wet basis moisture content'). Safe moisture contents for long-term storage vary according to the type of cereal (as well as storage temperature and humidity), but it is widely assumed that they are roughly equivalent to the equilibrium moisture content of the respective grains at 75% RH and 25°C (Table 5.2). In fact, equilibrium moisture contents of cereal grains (and other biological materials) follow product-specific nonlinear relationships between equilibrium moisture content and equilibrium relative

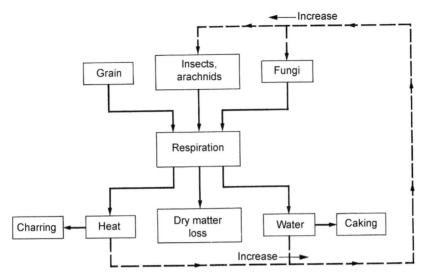

FIGURE 5.11 Schematic diagram of the cumulative effects of grain storage at moisture contents above safe levels. The effects on stored grain itself are shown in bold type.

TABLE 5.2 Equilibrium moisture contents of grains at 75% relative humidity and 25°C[a]

Cereal type	Equilibrium moisture %
Barley	14.3 (25–28°C)
Maize	14.3
Oats	13.4
Rice	14.0
Rye	14.9 (25–28°C)
Sorghum	15.3 (25–28°C)
Wheat	
Durum	14.1 (25–28°C)
Red	14.7 (25–28°C)
White	15.0 (25–28°C)

[a] *Based on values in Bushuk, W., Lee, J.W., 1978. Biochemical and functional changes in cereals: storage and germination. In: Hultin, H.O., Milner, M. (Eds.), Postharvest Biology and Biotechnology. Food and Nutrition Press Inc., Westport, CT, USA, pp. 1–33.*

humidity; these relationships (known as isotherms) are affected by temperature. Generally, equilibrium moisture content decreases as air humidity decreases. And, these nonlinear isotherms shift downward as air temperature increases. Extensive equilibrium isotherm data and regression equations for many cereal grains and oilseeds are provided in ASABE Standards (2009).

In temperate regions, however, the moisture contents at which grain is stored are closer to those described as 'wet' than 'dry'. The significance of moisture contents cannot be considered alone, though, as the deleterious effects of excessive dampness are critically affected by ambient temperature and the composition of the surrounding atmosphere. The increase in relative humidity of the interseed atmosphere with temperature is slight. It amounts to about 0.6%–0.7% moisture increase for each 10°C drop in temperature.

The relationship between moisture content and temperature as they affect storage of wheat is shown in Fig. 5.12.

The relationship depicted takes account only of the maintenance of grain quality as assessed by grain viability vis-à-vis germination. The relationship is also important because of its effects on infesting organisms as Fig. 5.13 shows.

The values used in Figs 5.12 and 5.13 refer to sound, clean grain samples. Broken grains are almost always present to some extent as a result of damage during harvesting or transferring/conveying to storage. In broken grains, endogenous enzymes and their substrates, kept separate in

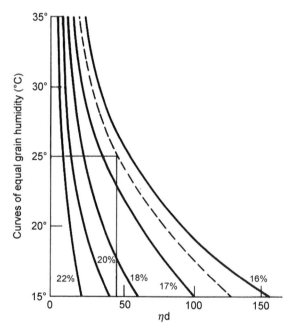

FIGURE 5.12   Potential storage time (number of days, ηd) of wheat grain as a function of temperature (°C) and moisture content (%), the germination rate maintained being 70%. From Guilbot, 1963. Producteur Agricole Francais, Suppl. mai ITCF, Paris.

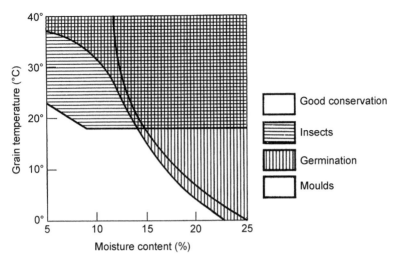

FIGURE 5.13   Risks to which stored cereal grain is exposed as a function of grain temperature (in °C) and moisture content (in %). *From Burges, H.D., Burrel, M.J., 1964, Journal of the Science and Food Agriculture 1, 32–50.*

the whole grain, can achieve contact and lead to necrotic deterioration. Further, the most nutritious elements of the grain, endosperm and embryo, are exposed to moisture, microorganisms,, and animal pests, whereas in the whole grain they are protected by fruit coats, seed coats and possibly husks. Impurities can also reduce storage time in that weeds present in the crop ripen and dry at a different rate from the crop itself. Hence, green plant material – with a relatively high moisture content – can carry excessive moisture into store even when mixed with dry grain.

5.2.2 Changes in grains during storage

5.2.2.1 Respiration

In a natural atmosphere, gaseous exchange will occur in a stored cereal crop. This is due to respiration, and it involves a depletion in atmospheric oxygen and an increase in carbon dioxide with the liberation of water and energy (as heat). Respiration rates measured in a store normally include a major contribution from microorganisms that are invariably present at harvest; nevertheless even ripe dry grain, suitable for storing, contains living tissues in which respiration takes place, albeit at a very slow rate. The aleurone and embryo are the tissues involved, and, like other organisms present, their rate of respiration increases with moisture content and temperature. Respiration is a means of releasing energy from stored nutrients (mainly carbohydrates), and a consequence of long storage is a loss of weight. Under conditions unfavourable to respiration (i.e., modified atmospheres), this may, however, be reduced. Under all storage circumstances it is likely to be of little consequence in relation to other storage hazards. Respiration can be reduced by artificially depleting the oxygen in the atmosphere.

5.2.2.2 Germination

Germination of grain is an essential and natural phase in the development of a new generation of plant. It involves the initiation of growth of the embryo into a plantlet. Roots develop from the radicle, and leaves and stem develop from the plumule. Hydrolytic enzymes are released into the starchy endosperm, and these catalyze the breakdown of stored nutrients into a soluble form available to the developing plantlet.

The conditions required for germination are also conducive to other, more serious hazards such as excessive mould growth. They will rarely occur throughout well-managed storage, but could develop in pockets due to moisture migration or condensation. Deterioration results from loss of weight due to enzyme activity and a loss of quality resulting from excessive enzyme activity downstream in processed products. These problems would apply even if the germination process were terminated through turning the grain. Having germinated in store, the grain will then be useless for seed purposes, as the process cannot be restarted.

5.2.3 Microbial infestation

Fungal spores and mycelia, bacteria and yeasts are present on the surfaces of all cereal crops. During storage they respire and, given adequate moisture, temperature and oxygen, they can rapidly grow and reproduce, causing serious deterioration in grains.

A distinction may be drawn between those that attack developing and mature grain in the field and those that arise during storage. Field fungi thrive in a relative humidity (RH) of 90%–100% while storage fungi require about 70%–90% RH. Several investigations have shown that an RH of 75% is required for germination of fungal spores (Pomeranz, 1974).

Storage fungi are predominately of the genera *Aspergillus*, of which there are five or six groups, and *Penicillium*, the species of which are several types. Some of the more common storage fungi and the minimum relative humidities in which they can thrive are listed in Table 5.3.

As with other spoilage agents dependent upon a minimum moisture content, fungi may be a problem even when the overall moisture content in the store is below the safe level. This can result from air movements leading to moisture migration; warm air moving to a cooler area will give up moisture to grains, thus remaining in equilibrium with them. Unless

TABLE 5.3   Approximate minimum equilibrium relative humidity (RH) for growth of common storage fungi

Mould	RH (%) limit
Aspergillus halophiticus	68[a]
Aspergillus restictus group	70[a]
Aspergillus glaucus group	73[a]
Aspergillus chevalieri	71[b]
Aspergillus repens	71[b]
Aspergillus candidus group	80[a]
Aspergillus candidus	75[b]
Aspergillus ochraseus group	80[a]
Aspergillus flavus group	85[a]
Aspergillus flavus	78[b]
Aspergillus nidulans	78[b]
Aspergillus fumigatus	82[b]
Penicillium spp.	80–90[a]
Penicillium cyclopium	82[b]
Penicillium martensii	79[b]
Penicillium islandicum	83[b]

[a] *Based upon data from Christensen, C.M., Kaufmann, H.H., 1974. Microflora. In: Christensen, C.M. (Ed.), Storage of Cereal Grains and Their Products. Amer. Assoc of Cereal Chemists Inc., St. Paul, MN, USA.*
[b] *Based upon data from Ayerst, G., 1969. The effects of moisture and temperature on growth and spore germination in some fungi. Journal of Stored Products Research 5, 127–141.*
Adapted from Bothast, R.J., 1978. Fungal deterioration and related phenomena in cereals, legumes and oilseeds. In: Hultin, H.O., Milner, M. (Eds.), Postharvest Biology and Biotechnology, Food and Nutrition Press Inc., Westport, CT, USA, pp. 210–243 (Chapter 8).

temperature gradients are extreme the exchanges occur in the vapour phase; nevertheless, variations in moisture content up to 10% within a store are possible.

If mould growth continues in the presence of oxygen, fungal respiration increases, producing more heat and water. If the moisture content is allowed to rise to 30%, a succession of progressively heat-tolerant microorganisms arises. Above 40°C mesophilic organisms give way to thermophiles. Thermophilic fungi die at 60°C and the process is kept going by spore-forming bacteria and thermophilic yeasts up to 70°C.

In recent years much attention has been given to the toxic metabolites of fungi (especially those of *Aspergillus flavus* and *Fusarium moniliforme*), which include aflatoxin and zearalenone, amongst others.

5.2.4 Insects and arachnids

Insects that infest stored grains often belong to the beetle or moth orders; they include those capable of attacking whole grain (primary pests) and those that feed on grain already attacked by other pests (secondary pests). All arachnid pests belong to the order Acarina (mites) and include primary and secondary pests. Most of the common insects and mites are cosmopolitan species found throughout the world where grain is harvested and stored (Storey, 1987). Insects and mites can be easily distinguished as arachnids have eight legs; insects, in their most conspicuous form, have six legs. Reference to the most conspicuous form is necessary as some insects (including those that infest grain) develop through a series of metamorphic forms. There are four stages: the egg, the larva, the pupa and the adult or imago. Although some female insects lay eggs without mating having occurred, this is less usual than true sexual reproduction, and this and dispersion are the two principal functions of the adult. Large numbers of eggs are produced and these are very small. Those of primary pests may be deposited by the female imago inside grains, in holes bored for the purpose prior to the egg laying. Under suitable conditions eggs hatch and from each a larva emerges. The larva is the form most damaging to the stored crop as it feeds voraciously. It grows rapidly, passing through a series of moults during which its soft cuticle is shed, thus facilitating further growth. Finally, pupation occurs; the pupa, chrysalis or cocoon does not eat and appears inactive. However, changes continue and the final metamorphosis leads to the emergence of the adult form. The life cycle of mites is simpler as eggs hatch into nymphs, which resemble the adult form, although there are only six legs present at this stage. By a series of four moults the adult form is achieved. The time taken for development of both insects and mites is influenced by temperature; the greater the temperature the more rapid the development up to the maximum tolerated by the species.

Many common primary pests – those attacking whole grains – are given in Table 5.4.

Three of the most damaging of these pests in the United Kingdom are shown in Fig. 5.14.

Several important secondary insect pests – those feeding only on damaged or previously attacked grains – are given in Table 5.5.

The saw-toothed grain beetle is shown in Fig. 5.15.

TABLE 5.4   Alphabetical list of some primary insect and arachnid pests[a]

Systematic name	Common name	Family
Acarus siro L.	Grain (or flour) mite	Acaridae
Cryptolestes ferrugineus Stephens	Rust red (rusty) grain beetle	Cucujidae
Cryptolestes pusillus Schonherr	Flat grain beetle	Cucujidae
Cryptolestes turacus Grouv	Flour mill beetle	Cucujidae
Latheticus oryzae Waterhouse	Longheaded flour beetle	Tenebrionidae
Oryzaephilus surinamensis L.	Saw toothed grain beetle	Cucujidae
Oryzaephilus mercator Fauv.	Merchant grain beetle	Cucujidae
Prostephanus truncatus (Horn)	Larger grain borer	Bostrichidae
Rhyzopertha dominica F.	Lesser grain borer	Bostrichidae
Sitophilus granarius L.	Granary weevil	Curculionidae
Sitophilus oryzae L.	Rice weevil	Curculionidae
Sitophilus zeamais Motschulsky	Maize weevil	Curculionidae
Sitotroga cerealella Olivier	Angoumois grain moth	Gelechiidae
Tenebroides mauritanicus L.	Cadelle	Trogossitidae
Tribolium castaneum Herbst.	Red flour beetle	Tenebrionidae
Tribolium confusum Duval	Confused flour beetle	Tenebrionidae
Tribolium destructor Uyttenboogaart	Large flour beetle	Tenebrionidae
Trogoderma granarium Everto	Khapra beetle	Dermestidae
Trogoderma granarium Everts	Khapra beetle	Dermestidae

[a] *For more extensive information, see the Canadian Grain Commission's website: https://www.grainscanada.gc.ca/ storage-entrepose/ipi-iir-eng.htm.*

The insects listed in Tables 5.4 and 5.5 are considered major pests. They are particularly well adapted to life in the grain bin and are responsible for most of the insect damage to stored grain and cereal products. Minor pests occur mainly in stores in which grain has started to deteriorate due to other causes, while incidental pests include those that arrive by chance and may not even be able to feed on grains. For further information on minor and incidental insect pests specialist works such as Christensen (1974) and the Canadian Grain Commission's website (https://www.grainscanada.gc.ca/storage-entrepose/ipi-iir-eng.htm) should be consulted.

Among the major primary pests several species develop inside grains. Weevils (i.e., grain, rice and maize) lay eggs inside kernels, while lesser

FIGURE 5.14 *Acarus siro*, the flour mite (top); *Sitophilus granarius*, the grain weevil; and *Cryptolestes ferrugineus*, the rust-red grain beetle (bottom). *Courtesy of Central Science Laboratory 1993. Crown copyright.*

TABLE 5.5   Alphabetical list of some secondary insect pests[a]

Systematic Name	Common name	Family
Ahasverus advena (Waltl)	Foreign grain beetle	Silvanidae
Alphitobius diaperinus (Panzer)	Lesser mealworm	Tenebrionidae
Alphitobius laevigatus (Fabricius)	Black fungus beetle	Tenebrionidae
Atomaria species	Silken fungus beetles Atomaria	Cryptophagidae
Attagenus unicolor (Braham)	Black carpet beetle	Dermestidae
Cadra cautella (Walker)	Almond moth	Pyralidae
Carpophilus, Glischrochilus species	Sap beetle	Nitidulidae
Cartodere constricta (Gyllenhall)	Plaster beetle	Latridiidae
Cryptophagus species	Silken fungus beetles Cryptophagus	Cryptophagidae
Dermestes lardarius L.	Larder beetle	Dermestidae
Dermestids	Dermestid beetles	Dermestidae
Endrosis sarcitrella (L.)	White shouldered-house moth	Oecophoridae
Ephestia elutella Hubner	Tobacco moth	Phycitidae
Ephestia kuehniella Zeller	Mediterranean flour moth	Pyralidae
Gnatocerus cornutus (F.)	Broadhorned flour beetle	Tenebrionidae
Hofmannophila pseudospretella (Stainton)	Brown house moth	Oecophoridae
Lasioderma serricorne (Fabricius)	Cigarette beetle	Anobiidae
Latridius minutus (L.)	Squarenosed fungus beetle	Latridiidae
Lepisma saccharina (L.)	Silverfish (firebrat)	Lepismatidae
Mycetophagus quadriguttatus (Müller)	Spotted hairy fungus beetle	Mycetophagidae
Nemapogon granella (Linnaeus)	European grain moth	Tineidae
Palorus ratzeburgii (Wissmann)	Smalleyed flour beetle	Tenebrionidae
Plodia interpunctella (Hübner)	Indianmeal moth	Pyralidae
Psocids	Psocids	Lachesiilidae, Psyllipsocidae
Ptininae species (~450 species)	Spider beetles	Anobiidae
Ptinus fur (L.)	Whitemarked spider beetle	Anobiidae
Ptinus villiger (Reitter)	Hairy spider beetle	Anobiidae
Pyralis farinalis (L.)	Meal moth	Pyralidae

Continued

TABLE 5.5   Alphabetical list of some secondary insect pests[a]—cont'd

Systematic Name	Common name	Family
Stegobium paniceum (Linnaeus)	Drugstore beetle	Anobiidae
Tenebrio molitor L.	Yellow mealworm	Tenebrionidae
Tenebrio obscurus F.	Dark mealworm	Tenebrionidae
Thermobia domestica (Packard)	Silverfish (firebrat)	Lepismatidae
Tinea species (~11 species)	Clothes moths	Tineidae
Tribolium audax (Halstead)	American black flour beetle	Tenebrionidae
Tribolium madens (Charpentier)	Black flour beetle	Tenebrionidae
Trogoderma glabrum (Herbst)	Glabrous cabinet beetle	Dermestidae
Trogoderma inclusum LeConte	Mottled dermestid beetle	Dermestidae
Trogoderma ornatum (Say)	Ornate carpet beetle	Dermestidae
Trogoderma variabile Ballion	Warehouse beetle	Dermestidae
Typhaea stercorea (Linnaeus)	Hairy fungus beetle	Mycetophagidae

[a] *For more extensive information, see the Canadian Grain Commission's website: https://www.grainscanada.gc.ca/storage-entrepose/ipi-iir-eng.htm.*

FIGURE 5.15 *Oryzaephilus surinamensis.* Saw-toothed grain beetle. *Central Science Laboratory 1993. Crown copyright.*

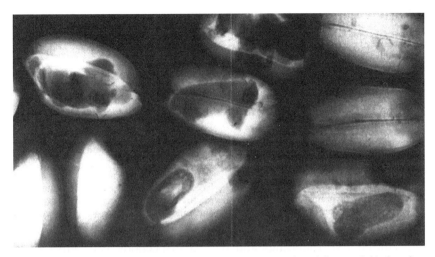

FIGURE 5.16   X-ray photograph of wheat grains, two uninfested (bottom left), the others showing cavities caused by insect infestation. An insect is visible within one of the cavities. *Part of a picture in 1954. Journal of Photographic Science 2, 113 and reproduced by courtesy of Prof. G.A.G. Mitchell and the Editor of Journal of Photographic Science.*

grain borers and Angoumois grain moths deposit eggs on the kernel surfaces, but their newly hatched larvae promptly tunnel into grains. The presence of the insect and the damage it causes may not be evident from outside even though only a hollow bran coat may remain. Detection by means of soft X-rays is possible (Fig. 5.16).

5.2.4.1 Damage caused by insects and mites

Serious grain losses due to consumption of grain by insects and mites occur after prolonged storage under suitably warm conditions. These losses are most serious in hot climates. Other problems caused by insects include creation of hotspots around insect populations where metabolic activity leads to local heating and moisture concentrations. Moisture movements and condensation in cooler areas can result in caking, which further encourages fungal infestation (Fig. 5.11).

Introduction of insects and mites from wheat stores to flour mills can cause serious deterioration in the products. Mite excreta taints flour with a minty smell, and hairs from the animals' bodies can cause skin and lung disorders in workers handling infected flour. Silk from the larvae of the Mediterranean flour moth can web together, causing agglomeration of grains and blockages in handling and processing equipment. In tropical countries termites can weaken the structure of a store, leading to its collapse.

5.2.5 Vertebrate pests

The principal vertebrate pests in cereal stores are rodents and birds. In many countries the three main rodent species involved are:

Rattus norvegicus – the Norway, common or brown rat;
Rattus rattus – the roof, ships or black rat;
Mus musculus – the house mouse.

Apart from consuming grains, particularly the embryo of maize, rodents cause spoilage through their excretions, which contain microorganisms pathogenic to humans. These include hantavirus, salmonellosis, murine typhus, rat-bite fever and Weil's disease. Rodents also damage structural elements of storage facilities, containers, water pipes and electric cables.

In well-managed stores access by rodents is prevented and good-housekeeping practices, such as removal of grain spillages, maintenance of uncluttered surroundings and regular inspections, can help prevent problems. The same is true of birds. These are serious pests only when access is easy, as for example in tropical countries where grain may be left to dry in the sun. Damage to drains and blockage of pipes by nests can give rise to secondary storage problems through promoting local dampness in some stores.

5.2.6 Control of pests and spoilage in grains during storage

Deterioration during storage is less likely if care is taken to ensure that the grain is in a suitable condition for storage. Criteria include a suitably low moisture content, a low mould count, and freedom from insects (all discussed earlier). Wheat containing live insects can be 'sterilized' by passage through an entoleter (Fig. 5.17), run at least 1450 rev/min (BP 965267 recommends speeds of 3500 rev/min for conditioned wheat, and 1700 rev/min for dry wheat). Hollow grains and insects are broken up and can be removed by subsequent aspiration.

The storage building/vessel itself should provide protection from weather (particularly wet) and intrusion by insects and rodents. High temperatures are undesirable and variation should be reduced to a minimum as this can lead to local accumulation of moisture. All spoilage agents depend upon respiration, so a depletion of oxygen can inhibit their proliferation and activity. To achieve this, however, it is necessary to provide a seal around the grain and a minimal headspace. In a sealed storage container, known as hermetic storage, oxygen depletion can be achieved by natural or artificial means. Natural depletion results from respiration, which in most organisms consumes oxygen and produces carbon dioxide. Artificial atmospheric control comes about by flushing of interstitial spaces and headspaces with a gas other than oxygen, usually nitrogen or CO_2, as these are relatively inexpensive. Complete removal of oxygen,

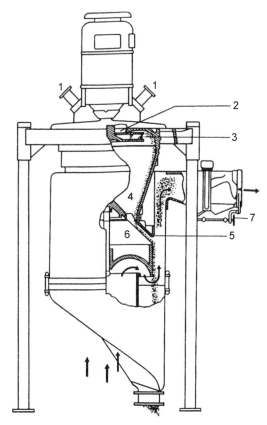

FIGURE 5.17 Diagrammatic section through an entoleter aspirator. 1, Grain inlet; 2, 3, impeller; 4, scouring cone; 5, grain discharge over 6, cone; 7, valve controlling airflow. *Arrows indicate path taken by air. Reproduced from Milling (10 October, 1969), by courtesy of the Editor.*

though, is not possible. Experiments carried out in artificial conditions in the United Kingdom showed that baking properties of wheats were maintained for 18 years in low-oxygen conditions. At ambient temperatures germinative energy was seriously reduced, and although this reduction was prevented by storage at 5°C, this was the only advantage of low temperature in addition to oxygen depletion recorded (Pixton, 1980).

Hermetically sealed conditions, while gaining popularity in parts of the developing world as a solution to maintaining grain quality during storage, are unusual in developed countries. Globally, prevention of spoilage in many cases depends upon carefully maintaining the stored grain condition, typically using fans to provide airflow, and prophylactic treatments with chemicals. Fortunately, nearly all threats to grain quality cause temperature rises, and monitoring of temperature, through incorporation of thermocouples and other temperature-sensing instrumentation, can reveal a great deal about grain condition during storage.

Forced ventilation can reduce temperatures and moisture, but it may be necessary to remove the cause (i.e., insects or moulds) by use of chemical treatments. Such treatments are relevant primarily when the problem is caused by insects. Because of the possible persistence of pesticides on cereals, their use in stores is increasingly becoming regarded as a last resort. In many countries strict codes of practice apply to the types of chemicals allowed and their appropriate use. In the United Kingdom the legislation is contained in the Food and Environment Protection Act 1985, the Control of Pesticides Regulations 1986 and HSE Guidance Notes (E440/85) on Occupational Exposure Limits. In the United States, pesticides are regulated by the Environmental Protection Agency (EPA).

Pesticides used to control insects, during storage of cereal grains are of two types. Those that are designed as a respiratory poison, and are hence applied as gas or volatile liquid, are described as *fumigants*. Those designed to kill by contact or ingestion are described as *insecticides*. They may be applied in liquid or solid form.

Of the gaseous fumigants, methyl bromide and phosphine (PH_3; produced from either aluminium phosphide or magnesium phosphide) are the main products used. Examples of 'liquid' fumigants are mixtures of 1,2-dichloroethane and tetrachloromethane: although the most effective fumigant is methyl bromide, this gas does not penetrate bulk grain well and the use of a carrier gas such as tetrachloromethane is an alternative to the fan-assisted circulatory systems required if methyl bromide is used alone. Methyl bromide use in the United Kingdom, United States, and other countries was banned after 2005 because of its significant impact on depleting the ozone layer. Thus, the primary gaseous fumigant now in use throughout the world is phosphine.

The period of treatment required depends upon the susceptibility of the species of insects present to the fumigant. For example a 3-day exposure to phosphine may eliminate the saw-toothed grain beetle but 6 days at low temperature may be needed to kill the grain weevil.

'Liquid' fumigants penetrate bulks well. The proportions need to be adjusted to suit the depth of the grain stored. Up to 3 m deep a 3:1 mixture of 1,2-dichloroethane:tetrachloromethane is suitable, but for penetration to a depth of 50 m equal proportions are needed. Fumigation requires the stores to be sealed to prevent escape of the toxic fumes.

5.2.6.1 *Pesticide residues*

Some of the various types of pesticides (herbicides, fungicides, insecticides and rodenticides) used in the field or in storage may persist in grains being processed or indeed into foods as consumed. In the United Kingdom the maximum residue limits (MRLs) permitted are mostly defined in EC Directive 86/232 and amendment 88/298, which came into force on 29 July 1988. Additional MRLs came into force in December 1988 referring to pesticides that have been refused approval in the United Kingdom but are

TABLE 5.6 Maximum residue limits for pesticides (mg/kg) in cereals excluding rice (United Kingdom)

Aldrin & Dieldrin	0.01	Captafol	0.05
Carbaryl	0.5	Carbendazim	0.5
Carbon disulphide	0.1	Carbon tetrachloride	0.1
Chlordane	0.02	**Chlorpyrifos methyl**	10
DDT (Total)	0.05	Diazinon	0.05
Dichlorvos	2	Endosulfan	0.1[a]
Endrin	0.01	Ethylene dibromide	0.05
Entrimfos	10	**Fenitrothion**	10
Hexachlorobenzene	0.01	α+β-Hexachlorocyclohexane	0.02
γ-Hexachloro cyclohexane	0.1	Heptachlor	0.01
Hydrogen cyanide	15	Hydrogen phosphide	0.1
Inorganic bromide	50	**Malathion**	8
Mercury compounds	0.02	**Methacrifos**	10
Methyl bromide	0.1	Phosphamidon	0.05
Pirimiphos-methyl	10	Pyrethrins	3
Trichlorphon	0.1		

Bold type indicates limits are set by Good Agricultural Practice. Others are set by detection limit.
[a] 0.2 for maize.
Adapted from Osborne, B.G., Fishwick, F.B., Scudamore, K.A., Rowlands, D.G., 1988. The Occurrence and Detection of Pesticide Residues in UK Grain (HGCA Research Review No. 12). Home-Grown Cereals Authority, London, UK.

used elsewhere; or those that have been consistently found in UK monitoring, where the limits provide a check that good agricultural practice is being observed (Table 5.6).

Monitoring of residues of fumigants applied at levels specified by manufacturers in an experiment revealed that only traces remained in cooked products. They were associated mostly with the bran fractions.

Insecticide residues of laboratory-milled wheat grains, treated at manufacturers' recommended rates, were four times as concentrated in bran and fine wheatfeed as in white flour. In commercial samples the germ contained five times as much as in the white flour. The milling process removed only about 10% of the residue on the whole grain, however, only 50%–70% of that in white flour survived bread baking.

Overall, the results suggested that insecticides applied at recommended rates are unlikely to lead to residues above the MRLs of Codex Alimentarius (Osborne et al., 1988).

In the United Kingdom fewer than one-quarter of stored grain is treated with pesticides (H-GCA, 1992). As part of a surveillance programme, surveys have been carried out since 1987 by the Flour Milling and Baking Research Association, in association with the National Association of British and Irish Millers, on samples representing those purchased by all flour millers. Residues have been low throughout, but a decline in residue level was also noted. Of the 1340 samples tested between April 1990 and March 1992, none approached the MRL for organophosphorus or other classes of pesticide.

Treating bin walls and floors with insecticides has proven a useful preventative measure. Insecticides that are commonly used include malathion, pyrethrin and cyfluthrin. More information can be found at the Canadian Grain Commission (https://www.grainscanada.gc.ca/storage-entrepose/ipi-iir-eng.htm) and Agriculture and Horticulture Development Board (https://cereals.ahdb.org.uk/grainstorage) websites.

5.3 STORAGE AND PREPROCESSING TREATMENTS

Before storage and processing of cereals it is necessary to carry out certain treatments on them. In cases where grains undergo a period of storage before use, some or all the treatments may take place before entering storage, and perhaps after storage as well.

Potential treatments can include drying, cleaning, grading, conditioning, and blending. Of these, drying may be carried out either on the farm or at the grain elevator, while the other treatments likely take place at the elevator, mill or processing plant as appropriate to the marketing system in use.

5.3.1 Drying

Purchasing contracts typically stipulate an acceptable range of moisture contents corresponding to that required for safe storage. These may be considerably lower than the moisture content at harvest so some means of drying must be used.

At its simplest, and in suitable (often tropical) climates, drying can consist of spreading the grain to dry in the sun assisted by frequent turning to expose all grains. Such treatment is unlikely to cause damage to grain beyond that which any handling imposes (as well as insects, rodents and birds). More sophisticated drying methods employing artificially heated air or surfaces have inherent dangers that must be avoided if grain quality is to be maintained, even though moisture reduction can increase the allowable storage life of the grain. In this context quality may be defined in several ways, but the most fundamental is retention of the ability to germinate. This test is particularly relevant to malting barley, as the malting process requires grain to be viable. In other cereals, viability provides an indication that other quality factors have not diminished during drying. In milling wheat, the most damaging change is denaturation of gluten and the consequent deterioration in baking quality. Aside from a germination test other, more rapid, tests include those using 2,3,5-triphenyl tetrazolium chloride. A 0.2% aqueous solution applied to longitudinally bisected grains for 2h in the dark produces, in viable grains, a red colouration around the embryo. The colour results from the activity of the enzyme dehydrogenase, present in respiring grains but inactivated by uncontrolled drying. Another type of test depends on the reduced solubility of heat-damaged proteins in sodium chloride solution (Every, 1987).

The temperature at which damage occurs depends upon the initial moisture content of the grain as well as the type of grain. Recommendations for safe drying conditions are often based on two types of experiment: those in which grains are held at constant moisture during heating in closed containers, and those at which temperature is maintained at a constant value. Neither truly represents the situation in all dryers, and few if any experiments have been carried out on a commercial scale to test these specific conditions. Nevertheless the recommendations are a useful guide, and the sensible practice is to use conditions, as far as practicable, on the safe side of the critical values. The relationships between initial moisture contents and safe drying temperatures for wheat are shown in Fig. 5.18.

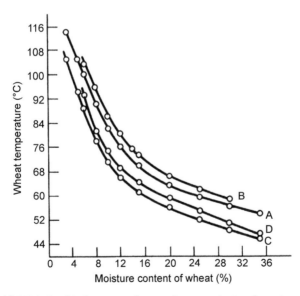

FIGURE 5.18 Relationship between wheat moisture content and maximum safe drying temperature. Curves A and B correspond to no germination after 60 min and 24 min heat treatment, respectively; curves C and D correspond to start of damage after 60 min and 24 h heat treatment, respectively. *From Hutchinson, J.B., 1944. Journal of the Indian Chemical Society, 63, 104 and reproduced by courtesy of the Society of Chemical Industry.*

Similar conditions are suitable for seed grain and malting barley (Nellist, 1978). For grains being fed to livestock, temperatures of 82–104°C may be used as limited denaturation of proteins is not disadvantageous and increased digestibility of starch may arise as a positive advantage.

While drying rice, an additional hazard arises as it is highly susceptible to fissuring during drying and cooling, giving rise to more broken grains. Since intact grains are valued for milled rice, considerable care has to be taken not to remove moisture too rapidly. Although it is customary to harvest rice while the grain is green with high-moisture content, it may undergo a period of drying in the field. This is especially true in less-industrialized growing areas and demands adequate temperatures, but care must be taken to avoid too rapid loss of moisture. Techniques for shading in windrowed crops include overlapping the seeds of one row with the stems and leaves of the next. In more-industrialized production areas, drying is accomplished in specially designed dryers. Grains are dried as paddy or rough rice, that is, with the husk still surrounding them. Control of moisture loss is achieved through the use of a multipass system. Installations may be of the mixing or nonmixing type. The principle of both involves downward flow between two screens. On one side of the screen warm air is generated and passes through the curtain of moving grains; grain nearer to the warm air source tends to become warmer than the more remote grains, and the introduction of a mixing mechanism minimizes the effects of this gradient. The diagrams in Fig. 5.19 show the differences.

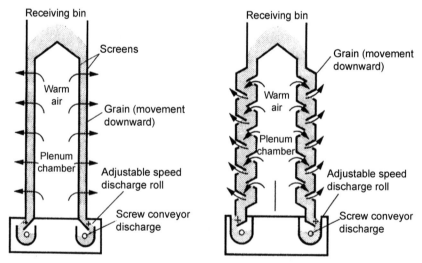

FIGURE 5.19 Nonmixing (left) and mixing (right) type columnar grain dryers (Wasserman and Calderwood, 1972).

TABLE 5.7   Effect of drying temperature on stress cracks in maize kernels (%)[a]

Drying/cooling type	Drying temperature (°C)	% Stress cracks
Low temperature		1–3
High temperature batch/fast cooling	45	37–74
High temperature batch/fast cooling	60	37–64
High temperature batch/fast cooling	80	28–73
High temperature/slow cooling dryeration	60	8–13
High temperature/slow cooling dryeration	80	6–15
High temperature/slow cooling dryeration	100	5–8

[a] *Based upon Brown, R.B., Fulford, G.N., Daynad, T.B., Meiering, A.G., Otten, L., 1979. Effect of drying method on grain corn quality. Cereal Chemistry 56 (6), 529–532.*

All grains, in fact, will fissure if drying/cooling conditions are not optimal (see Table 5.7). Stress cracks will often develop when rapid cooling follows rapid drying. This is especially true when grain is dried with more than five points of moisture removal. Therefore, recommendations include <5 pts./h for corn and sorghum, <4 pts./h for wheat and <2 pts./h for rice.

Many types of hot air dryers exist, they are frequently classified as batch or continuous, but they may also be distinguished on the basis of the direction of airflow in relation of that of the grain (Nellist, 1978). Thus in the *cross-flow* type, air flows across the path of the grains. The layer of grain adjacent to the air inlet is soon dried and its temperature rapidly approaches that of the inlet air. The grain on the exhaust side remains at or near the wet bulb temperature of the drying air for most of the drying period, and its final temperature will normally be much less than the inlet air temperatures. Because of the large moisture gradient, mixing after drying is essential. Cross-flow dryers are the most abundant on both farms as well as in elevators. Historically these types of dryers were constructed in a rectangular form, but now a cylindrical construction has become popular.

In *concurrent flow* dryers, both grain and airflow in the same direction. Thus hot air meets cool grain, but the heat and moisture transfer that occurs ensures that grain temperature does not rise to the inlet air temperature and that the air temperature falls rapidly. For the final phase of drying, the air and grain are almost at the same temperature. Advantages of this design are that high drying-air temperatures can be used to give high initial drying rates without overheating the grain, and the period during which the grain and air are in temperature equilibrium is thought to help temper the grains and reduce potential stress cracking.

In *counter-flow* dryers air travels in the opposite direction to the grain, and the dried grain temperature approaches that of the inlet air. The air tends to exhaust near saturation, and drying can therefore be quite efficient. The temperature of the hot air at inlet and the final grain temperature are almost the same (Nellist, 1978).

5.3.2 Separations

Cleaning may occur before storage, shortly before processing, or both. Cleaning before storage has the advantage that it helps to minimize deterioration in store, which can be aggravated by the presence of broken grains, dust, green plant parts and other impurities. Receival at harvest time however is often accompanied by pressure to rapidly deal with grains (e.g., drying and conveying into storage), hence time for cleaning may not be available. Even grains cleaned before storage may require cleaning again before processing to remove undesirable elements resulting from the storage itself. Impurities, together with damaged, shrunken and broken grains, are collectively known as 'screenings' in the United Kingdom, 'Bezatz' throughout Europe or 'dockage' in the United States. Use of one term in the following includes all.

5.3.2.1 *Impurities*

Frequently encountered components of screenings may be classified according to their composition:

- Vegetable matter
  - Weed seeds, grains of other cereals, plant residues such as straw, chaff and sticks.
  - Fungal impurities – bunt balls, ergot.
- Animal matter – Rodent excreta and hairs, insects and insect frass, mites.
- Mineral matter – Mud, dust, stones, sand, metal objects, nails, nuts, etc.
- Other – String and twine. Miscellaneous rubbish.

Purity of cereal samples can be improved by cleaning and by separating. Cleaning involves the removal of material adhering to the surface of grains (and will be discussed in the next section), while separation involves the removal of freely assorting material. Considerable ingenuity has been demonstrated in the design of devices for eliminating impurities using both methods, and as a result a wide range of machines has been developed by a number of manufacturers. Wide though the range is, the principles involved are few. Thus cleaning depends upon abrasion, attrition or impact, while separation depends upon discrimination by size, shape, specific gravity, composition and texture. No single machine can perform all the necessary operations, and it is customary for parcels of cereal to pass through a series of operations based on the aforementioned principles.

5.3.2.2 Metals

Because of the potential dangers that can be caused to machinery by hard objects, these are usually among the first to be removed. Thus metals are removed by devices capable of detecting their composition, installed in spouting or conveyors through which the grains are directed early in the process. Ferrous metals are attracted by strong magnets. Their removal from the magnetic surface by grain flow cannot be avoided, so magnets must be designed to revolve, allowing removed metal objects to fall into a reservoir out of the grain stream. Otherwise, personnel must manually remove the tramp metal frequently. Nonferrous metals are detected by the interruption that they cause to an electric field when passing through a metal detector. The interruption rapidly activates a mechanism that temporarily diverts the stream containing the metal object.

5.3.2.3 Screening and scalping

Stalks, leaves, stones, sand and string also constitute potentially damaging impurities and hence these are also removed before storage or processing. A single *separator* can be capable of removing many of these and other less hazardous materials, such as some seeds of other cereals. Size and shape are the distinguishing criteria, as this process is essentially one of sieving through a screen coarser than the grain (to remove string, straw and larger objects such as stones and large grains of different species – a process known as 'scalping'), and over a screen finer than the required grains, through which broken grains, small seeds and sand can pass (a process known as 'screening'). Specific gravity separations may also be made on selected material leaving the separator – the grains leave the machine as a curtain and this is subjected to an upward stream of forced air that lifts light material such as chaff, while the denser particles fall under the influence of gravity only. The aspirated air can flow to an expansion chamber where solid particles are deposited; air can then be recycled.

5.3.2.4 Destoners

For more rigorous removal of stones specialised *dry stoning machines* may be used. Destoners have an inclined vibrating deck through which a uniform current of air flows upward. Feed is directed onto the lower end of the deck, which spreads and mixes in response to the vibration. On reaching the region of the bed where the fluidizing effect of air is experienced, the grains are lifted beyond the range of influence of the vibration. They fall under gravity and are discharged at the lower end of the deck. Stones are too dense to be lifted by the air stream and thus continue to the top of the deck where they are discharged independently from the grain.

5.3.2.5 *Unwanted species*

Removal of grains of other cereals and weed seeds surviving the above treatments can be achieved by *disc separators* (Fig. 5.20). This operation is designed to discriminate on the basis of seed length not width. A series of discs are mounted on a single horizontal axle in a trough partially filled with grain. The axle is driven, thus causing the discs to rotate through the bed of grain. Each disc has, on both surfaces, a series of indentations arranged concentrically. The indentations are pockets into which grains of the required length (or smaller) can fit. They are thus lifted out of the grain mass. As the pockets pass the high point of the rotation they fall out of the pockets into channels adjacent to the disc faces, by which they are conducted into a discharge in common with channels from the other discs. Material failing to be picked up in pockets is driven by screw conveyor from the feed end to the discharge end of the machine. A gate at the discharge end can be adjusted to control the residence time spent in the trough. The more impurities in the grain stream, the shorter the time needed for sorting.

By using several machines in series, each with different-sized pockets in its discs, a number of separations may be made in sequence, with the required cereal behaving as the selected material when eliminating larger impurities, or as the rejected material when eliminating small impurities. In some machines several separations can be made within a single unit.

The same principle of separation by pockets on a rotating device is used in *trieur cylinders*. In these machines, the interior surface of the cylinder has the pockets in it. The capacity of disc machines is greater than that of cylinders of the same diameter but the selectivity of cylinders is said to be

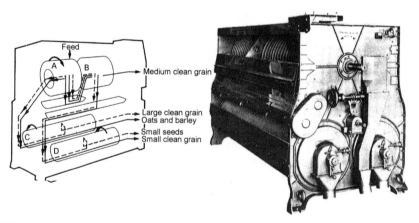

FIGURE 5.20 Carter-Day disc separator for length separation of impurities from wheat, rye and other cereal grains. Exterior view of machines, right; diagram of operation, left. *By courtesy of the Carter-Day Company, Minneapolis, MN, USA.*

better. The loading of discs or cylinders can be reduced if concentrators are used upstream of them.

Concentrators, also known as gravity selectors, combinators or more commonly as gravity tables (Fig. 5.21), serve to effect a preliminary segregation of stocks so that the individual streams can be subjected to treatments appropriate to the impurities they are most likely to contain. Thus a light fraction could be treated to remove small seeds and grains of unwanted small cereals by use of discs or cylinders. A denser stream would not need to pass through these machines but may be routed to the more appropriate destoner instead.

The concentrator operates on a principle similar to the destoner. Grains are stratified according to their specific gravity by the oscillating action of

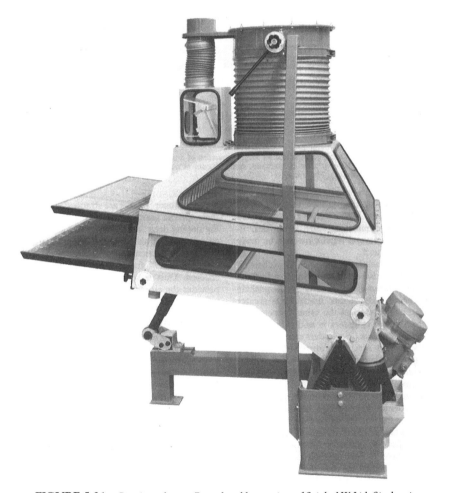

FIGURE 5.21   Gravity selector. *Reproduced by courtesy of Satake UK Ltd. Stockport.*

the sloping deck; lighter seeds are conveyed on a current of air while the heavier material is directed primarily by friction on the oscillating deck. The routes of the various fractions can thus be separately ordered. In fact, depending upon the seed densities, multiple separations can be made along the periphery of the deck. Dust is removed by the airflow that provides the air cushion.

5.3.2.6 Aspiration

The rate at which a particle falls in still air is the resultant of the speed imparted to it by the force of gravity, balanced by the resistance to free fall offered by the air (i.e., drag force). The rate of fall, or 'terminal velocity', depends on the weight of the particle and its surface area:volume ratio. Compact spherical or cubical particles thus have a higher terminal velocity than diffuse or flake-like particles. Instead of allowing the particles to fall in still air, it is more usual to employ an ascending air current into which the grain is introduced. The velocity of the air current can be regulated so that particles of high terminal velocity fall, while those of low terminal velocity are lifted. Utilizing this principle, particles of chaff, straw, small seeds, etc., which have terminal velocities lower than that of grain, can be separated from the grain in an aspirator. In a *duo-separator* the stock is aspirated twice, permitting a more refined separation to be made. In this type of machine, the lifted particles are separated from the air current by a type of cyclone, and the cleaned air is recycled to the intake side of the machine. Such closed-circuit aspirators save energy and minimize effluent problems.

The separations described thus far have focused on the separation of the required species from other cereal species, weed seeds and other materials quite unlike the grains themselves. There are occasions when it is required to make a separation of two or more types of grain of the required species from the same lot of grain. This can mean separation of shrivelled grains or those that have been hollowed out as a result of insect infestation, or it can mean separating various types of the same grain. They are removed mainly by aspiration or other methods depending on specific gravity. For example, in oats the following types can present in a single lot:

1. Double (bosom) oats
2. Pin oats
3. Light oats
4. Other types of oats

In double oats the hull of the primary groat envelops a normal groat, plus a second complete grain. Both groats are usually of poor size. Pin oats are thin, short and contain few or no groats; light oats consist of a husk that encloses only a rudimentary grain. Other types can be twins, discoloured, green and hull-less (Deane and Commers, 1986).

While indented discs and/or cylinders separate on the basis of length, it is sometimes required, particularly with oats, to eliminate narrow grains

FIGURE 5.22   Paddy separator. *Reproduced by courtesy of Satake UK Ltd. Stockport.*

by a separation on the basis of width. This involves the use of screens with elongated but narrow slots, through which only reject grains can pass. The screens are commonly in the form of a *perforated rotating cylinder* through which the bulk flows horizontally. Undersized grains drop through into a hopper. Small round holes recessed into the cylinder walls can be used as an alternative to slots. The recess upends undersized grains directing them through the hole.

Where size and shape are similar and only small differences in specific gravity may exist, sophisticated gravity separators are required. An example of this type of machine is the *Paddy separator* (Fig. 5.22). In this a table mounted on rubber tyres is made to pivot around its horizontal axis. Compartments on the table are formed by a series of zigzags arranged at right angles to the direction of motion. Grains accumulate and stratify within the channels, and the motion of the table causes those light grains that rise to the top to travel toward the upper side of the channel flanks, while the heavier stocks pass down the table. As might be expected with a separation based on small differences, careful adjustment of variables is necessary.

Gravity separations have been found capable of separating satisfactory grains of wheat from sprouting grains. In fact, the sprouting, having led to some mobilization of starch, has rendered the affected grains lighter while not changing their overall size.

5.3.3 Cleaning

Cleaning, in contrast to separating, is performed on *scourers*. As their name suggests, scourers subject grains to an abrasive treatment designed to remove dirt adhering to the surface of grains. Surface layers such as beeswings (outer epidermis of the pericarp of wheat and corn) may also be removed. Scourers propel grain within a chamber by means of rotors to which are fixed beaters or pins. The axis of the rotor may be horizontal or vertical according to type. In the horizontal configuration, beaters rotate within a cylindrical wire mesh. Grains enter at one end and are cast against the mesh by the inclined rotor beaters; cleaning is achieved by the friction of grains against each other or against the mesh. Dust removed passes through the screen and falls into hoppers and is then discharged.

Aspiration typically following scouring removes particles loosened but not removed from the grains. Some vertical machines work on a screen and beater system but others combine a scouring action with aspiration.

Durum wheats are subjected to particularly vigorous treatment with beaters. In addition to cleaning, the beaters also eliminate much of the embryos, considerably facilitating the cleaning of semolina produced and reducing its ash yield.

Scourers have replaced washers in the cleaning programmes of wheat in most countries on account of the following benefits:

1. Problems of pollution control concerning the discharge of effluent,
2. Problems with microbiological control (mainly bacterial),
3. High costs of operating machinery and of water.

Washing may still be practised in some parts of the world, such as in Russia and Eastern Europe, where ultrasonic vibrators were used to assist the cleaning process (Demidov and Kochetova, 1966). Wet cleaning is also used in the United States for removing surface dirt from maize before milling (Alexander, 1987).

5.3.4 Screenroom operation

In practice, grain passes in succession through a series of machines working on the various principles described herein. No single machine can remove all the impurities, but all the machines, working together as a system, remove practically all the impurities, resulting in the loss of very little of the incoming grain.

These systems may be installed at grain storage elevators to clean incoming grain; they may also be installed in flour mills and other processing plants prior to processing. A possible cleaning flow for wheat is shown in Fig. 5.23(a). Fig. 5.23(b) shows an example of a 'cleaning core', which is

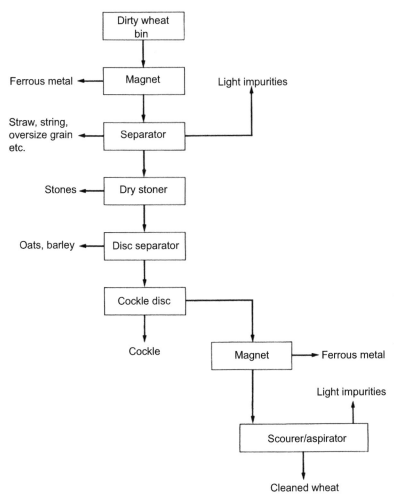

FIGURE 5.23A Schematic flow diagram of a typical flour mill screenroom.

typically housed within a grain storage elevator (note that 'dirty' or unsorted grain is stored in the bins above the core; clean and separated grain is stored in various bins below the core, as is the separated chaff and dust).

The efficiency of operation of screenroom/cleaning core machinery depends on machine design, feed rate and proportion of *cut-off* (i.e., the reject fraction). As the feed rate is reduced, interference between particles decreases, and efficiency of separation can increase. On the other hand, as the proportion of the cut-off increases, the rejection of separable impurities becomes more certain.

In the conventional screenroom arrangement, it is customary to feed the entire feed to each machine in succession, except for the screenings

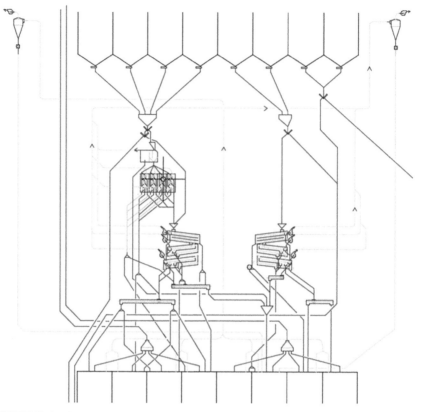

FIGURE 5.23B   Schematic flow diagram of a typical screenroom cleaning core in a storage elevator.

removed stage by stage. As the total quantity of screenings removed frequently amounts to only 1%–1.5% of the feed by wt., every machine in the screenroom flow must have a capacity (i.e., be able to deal with a rate of feed) practically equal to that of the first machine. If, however, other fractions or grains are removed at various stages, then subsequent machines may be sized for smaller capacities.

5.3.4.1 Loop system

The feed rate to many of the machines can be reduced, and the efficiency of separation increased, by an arrangement known as the *loop* or *by-pass* system. In the loop system, the first machine is set to reject a large cut-off (say 10% of the feed) containing all the separable impurities, together with a proportion of clean grain. The remaining 90% or so is accepted as clean, requiring no further treatment. The cut-off is retreated to recover clean wheat. As the cut-off amounts to only a fraction of the total feed, the rate in the retreatment machines can be much reduced, and the efficiency of separation improved. The loop system is frequently applied in the operation of disc separators and trieur cylinders, and is a feature of the Twin-Lift Aspirator, which embodies main aspiration of the whole feed, and re-aspiration of a substantial cut-off to recover clean grain.

The Bühler 'concentrator' operates on the loop system. A cut-off of about 25% by wt. is lifted by aspiration and specific gravity stratification; this contains various impurities but is stone-free, and goes to a gravity table or scourer for separation of the clean wheat from the impurities. The remaining 75% is free of impurities other than stones and needs only to be destoned on a dry stoner or a gravity table.

Individual components of a mixed grist to a flour mill are cleaned separately before blending. Wheat types differ in grain size and shape, and in the types of impurities contained, and so require individual treatments in the screenroom, particularly regarding choice of sieve sizes for the milling separator and of indent sizes for the cylinders and discs.

5.3.5 Damping and conditioning

Damping, as a pretreatment for milling grists, is a long-established practice. It probably originated as part of the cleaning process when washing was as an important treatment. It was adopted as a treatment in its own right when the beneficial effects of washing followed by a period of lying/resting were noted. In current practice the washing of most cereals for cleaning purposes has now been abandoned, but conditioning – or tempering (both terms mean the controlled addition of moisture) – remains an important stage of preprocessing. Conditioning requires much less water than washing, and, as the water will be absorbed by the grains, there is no effluent from this process. Neither is microbial growth usually a problem if lying times are short. If problems do arise they can be minimized by chlorination to levels above that of the water supplied.

The study of the effects of water on the physical properties of cereals, and hence their milling properties, is complex. The cells of the grain kernels have several components that dictate how fast water is absorbed into the kernels, the physical properties of which are altered, each in its own way. Water also affects the adhesion between the components of the cells, the cells themselves, and ultimately the various tissues of the grain. Although applied to almost all cereals, the conditioning process is not singularly appropriate as a preparation for all milling treatments. Even different types of the same cereal undergoing the same treatment may receive different treatments or different degrees of the same treatment. Consequently, each cereal type will be considered separately next.

5.3.5.1 *Water penetration*

Water is absorbed rapidly by capillarity into the empty pericarp cells (especially near the tip cap of the seed). They are capable of taking up 80% of their weight, equivalent to about 8% (w/w) of total grain weight. Once wetted, the bran can behave as a reservoir from which water can pass into the endosperm. The rate of passage is restricted by the presence of impermeable components of the testa and possibly the nucellar epidermis. Since these are thinnest in the nonadherent layers overlying the embryo, this is the area through which most water gains access to the inner components of the grain. It advances toward the distal end of the endosperm, leading to near-uniform distribution within the whole grain. Whereas saturation of the pericarp takes only minutes, equilibration throughout the grain takes hours.

The exact time taken depends on several factors including:

1. The initial moisture content of the grain: damper wheats absorb more rapidly (Moss, 1977) as well as require less damping.
2. The degree of permeability of the testa – particularly in the embryo region. There is little information on this from conditioning studies, but studies on imbibition in relation to germination show this to vary (Wellington and Durham, 1961; King, 1984).
3. The mealiness of the endosperm – water permeates through mealy endosperm more quickly than through vitreous endosperm. The essential feature appears to be the greater proportion of air spaces in mealy tissues. As soft wheats are more mealy than hard, equilibration is achieved more rapidly in softer wheats (Stenvert and Kingswood, 1976). As the continuity of the protein matrix is influenced by total nitrogen content of the endosperm, moisture movement is slower in high-nitrogen wheats than in low.
4. The temperature of the water – a threefold increase in penetration rate was recorded for each 12°C rise by Campbell and Jones (1955) within the range 20–43°C; above 43°C smaller rate increases applied up to 60°C.
5. The uniformity of the water distribution among grains. Where less than 8% water (w/w) is added it is preferentially absorbed into the pericarp of the grains, which it contacts first, leaving little for distribution to more remote grains. Equilibration among grains takes even longer than within a grain.

It is possible to exploit the above to achieve more rapid conditioning. Thus:

1. Multistage damping provides a progressively higher starting temperature for each addition.
2. Abrasion of the grain surface can remove the impermeable layers.

3. Damping causes expansion of the endosperm, but this is reversed as equilibrium is achieved or if drying occurs. Such changes induce stresses, which cause minute stress cracks that increase effective mealiness.
4. Warm (above ambient to 46°C) or hot (above 46°C) conditioning may be practised.
5. Precision placement of water in a uniform fashion and/or vigorous mixing of grains during and immediately following damping may be employed.

Much research has been carried out on the length of time required to achieve equilibrium moisture conditions. With so many factors affecting the rate of penetration, it is not surprising that results are variable. Estimates range from several hours to several days, with more recent work tending to favour the shorter periods. Such revision of ideas has influenced the manner in which conditioning of wheat is carried out (Hook et al., 1984).

5.3.5.2 *Common or bread wheat*

Over the years, the effects of moisture content and distribution have received more attention in wheat than any other cereal. Variation in endosperm textures demands different treatments; in general the amount of water added and the length of lying time increase in proportion to endosperm hardness.

The reasons for conditioning wheat include:

1. To toughen the outer, nonstarchy-endosperm components of the grain so that large pieces survive the milling, and powdering of bran is avoided.
2. To mellow the endosperm to provide the required degree of fragmentation during milling.

The term 'mellow' is often used to describe a desirable state of endosperm but a useful definition of its meaning is difficult to find. The changes induced by water penetration have been better described by Glenn et al. (1991) who found that endosperm strength decreased. 'Strength' as used by these authors has its strict materials-science meaning, i.e., a measure of the stress required to bring about fracture. The toughness (*senso stricto*) of the endosperm (manifested as the energy needed to bring about complete fracture) was found by the same authors to decrease with increasing moisture content, although this was less consistently observed than the increase in strength. The hardness of the endosperm is not the only factor determining the level of water addition. Clearly the grain moisture has an effect on the fractions separated during milling. It influences their surface properties, which are particularly significant in relation to their behaviour on sieves if the covers are made from absorbent fibres such as nylon or (particularly) silk. Moisture can be transferred also to the fabric of the sieves themselves, possibly altering the frictional properties and even changing the tension and the effective aperture size. The ease with which starchy endosperm may be separated from bran is also influenced by moisture content and distribution. Both increase and decrease of difficulty of this separation with increasing moisture content have been imputed (although an increase is the more frequently claimed) but direct evidence is scarce and measurements must be difficult to make. It is generally accepted that, for the milling of high-extraction-rate flour, conditioning to a lower-than-normal moisture level is appropriate. Another factor influencing the degree of damping and postdamping lying is the target moisture content of the final products. Storage life of flour declines with increasing moisture content, and most specifications impose a maximum moisture content – mainly to avoid buying water at flour prices. However, the relative costs of transport and the requirement to absorb adequate water in dough making also exert an influence. In calculating the amount of water required to achieve the target moisture content of stocks, characteristics of the mill and its location have to be considered. These include:

1. The conveying system in use (pneumatic systems incur more moisture loss than elevator systems);
2. Ambient temperature and humidity;
3. Roll temperatures.

Losses of between 1% and 2.5% have been recorded due to evaporation of water from stocks.

5.3.5.3 *Durum wheat*

The purpose of durum milling is to produce not flour but semolina. Hence the purpose of conditioning is solely to toughen the bran. Penetration of water into the endosperm is not required and consequently lying times are short. Less than 4 h is typical and as much as 2% water may be added as a final addition only 20 min before milling.

5.3.5.4 *Rye*

Short lying times are typical of rye conditioning regimes as water penetrates more rapidly into rye grains than into wheat. Lying times depend on the efficiency of the damping system used but even with traditional methods of water addition, such as waterwheel damping, periods of 6–15 h are typical. In the cold winter conditions of parts of North America, warm conditioning may be used to provide uniform milling conditions and to eliminate condensation in break sifters and spouts (Bushuk, 1976). In Western Europe a first temper lasting 5–6 h is typical, while in Eastern Europe a slightly longer first temper lasting 6–10 h may be followed by a second damping an hour or so before first break.

5.3.5.5 Rice

Rice is milled at 14% moisture content with no damping immediately preceding dehulling. For rice subjected to parboiling, however, an elaborate heat/water treatment is involved. Paddy rice has a hull of adherent lemma and palea (Chapter 1) and water penetrates very slowly. Parboiling breaks the tight seal between the two pales, thus removing the main obstacle to entry. This was the original purpose of parboiling, which is an ancient tradition, originating in India and Pakistan. Other changes occur, however, including an improvement in nutritional quality of the milled product. The hot water involved in treatment dissolves vitamins and minerals present in the hull and bran coats, and carries them into the endosperm. Conversely, rice oil migrates outwards during parboiling, reducing oil content of the endosperm and increasing it in the bran. Starch present in the endosperm becomes gelatinized, toughening the grain and reducing the amount of breakage during milling. Aleurone and scutellum adhere more closely so that more of each remains in the milled rice. Some discolouration of grains occurs but susceptibility to insect attack is reduced.

5.3.5.6 Barley

Conditioning for pearling often consists of adjusting the moisture content of the grains to 15%, followed by a rest of 24 h.

5.3.5.7 Maize

The details of the conditioning protocol in preparation for degerming of maize depend upon the process in use: for the Beall degerminator, the moisture content is raised to 20%–22%, while for entoleter or rolling processes an addition of only 2%–3% to stored grain is appropriate. Following damping a rest of 1–2 h is usual but up to 24 h is possible.

5.3.5.8 Conditioning practice

Both phases of conditioning (i.e., addition of water, and lying for a period of equilibration) require specially designed equipment, the design of which has evolved to suit changing circumstances and changing ideas about conditioning.

The amount of water required to be added to achieve the optimum moisture content of a particular type of wheat clearly depends upon the initial moisture content of the wheat. Wheat that travels long distances by sea tends to be traded at lower moisture content than wheat bought locally (water is expensive to transport!) If more than 3%–5% needs to be added, it must be done in more than one stage. The amount capable of being added in one stage depends on the method of addition.

In preparing wheats for milling as a mixed grist, in which components are introduced into the mill at different moisture contents, the individual components are best conditioned separately and blended after lying. Although the time taken for water to penetrate into the grain is reduced as a function of temperature, the practice of conditioning at elevated temperatures has declined as a result of prohibitive costs and the introduction of alternative practices that allow reduced lying times. Details of warm and hot conditioning systems are provided in older textbooks (e.g., Lockwood, 1960; Smith, 1944). They will not be described here.

Principles underlying the design of modern conditioning systems are:

1. Addition of accurately metered quantities of water;
2. Uniform distribution of moisture.

Accurate water addition is achieved by controlling the water addition and flow rate of wheat, possibly using feed-forward and feedback control systems. Uniform distribution is ensured by careful placement of water followed by rapid mixing. Most milling equipment suppliers now produce conditioning equipment with sophisticated control systems. The H_2O-Kay (Satake UK Ltd.) system, consisting of the Kay-Ray controller and a Technovator mixer, employs moisture measurement by microwave attenuation meters. This is the only method by which the moisture of freshly damped wheat can be monitored, allowing corrections to be made to the levels of water addition on a feedback basis. The system has the additional facility for adding steam. A sampling stream is diverted from the main flow, through a cell in which its moisture content is measured. Additionally, flow rate and specific weight are detected by gamma ray absorbance, which is sensitive to mass of wheat in the measurement cell. The required amount of water is computed and added as the wheat stream enters at the lower end of an inclined screw conveyor. Damped wheat is raised along the screw by adjustable paddles whose disposition is set to provide a transit time of 20 s through the screw. Tumbling and mixing for this period is chosen as grain-to-grain transfer of moisture is possible for this time, before being trapped in the pericarp. Measurements similar to those made prior to damping are made on a sampling stream taken from the flow leaving the chute. These measurements can influence the level of water addition. Up to 4.25% of water addition is claimed as the capability of this system.

Even more vigorous mixing is employed in the Bühler Intensive Dampener (Fig. 5.24), as a result of which it is said that additions up to 5% are possible. In this system wheat is mixed and conveyed horizontally, after damping, by blades of a rotor that turns at about 1000 rpm. The dampener is used with a control system in which a microprocessor calculates the required amount of water on the basis of incoming wheat moisture content, flow rate, temperature and target moisture content. Moisture measurement is by capacitance.

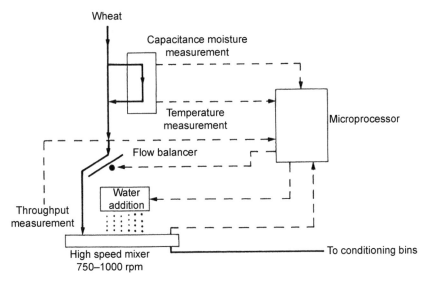

FIGURE 5.24 Schematic diagram showing damping control by the Bühler Intensive Dampener.

5.3.5.9 *Conditioning bins*

In all storage bins used for cereal grains, sound design is important in regard to efficient emptying. It is particularly important for conditioned grain. It is usual for a bin to be filled by deposition from a central spout – grains delivered to the centre tumble toward the edges but the pile remains deepest at a point directly below the delivery spout and a smooth conical profile with an 'angle of repose' characteristic of the particular sample remains at the top. As the bin empties, the profile changes because the grains in the centre fall more quickly than those at the edges. This leads to a reversal of the original surface profile, the centre becoming the lowest point. Such a means of emptying is disorderly; it leads to mixture of grains and a sequence of removal different from that of filling. The sequence of emptying may not be important in general storage considerations, but in conditioned grain, where lying time is controlled, it is important that the grains introduced into the bin first are also the first to leave. The difficulties in achieving this are compounded as damped grain flows less readily than dry and the tendency to nonsequential flow is exaggerated. Also, the shortening of lying times associated with contemporary practice demands greater precision in controlling the lying period, emphasizing the need for sequential flow.

The expedient by which improvements have been made is the deconcentration of outlets. It is now common for bins to have a matrix of

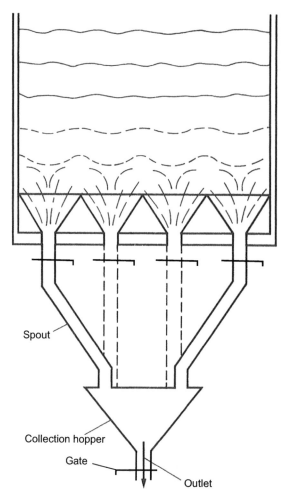

FIGURE 5.25 Manifold discharge from a **First-in-first-out** conditioning bin. *From Screenroom Operations 1. Flour Milling Industry Correspondence Course and by courtesy of Incorporated National Association of British and Irish Millers Ltd.*

funnel-shaped hoppers provided at the base, each leading to spouting that ducts to a common discharge hopper via a manifold (Fig. 5.25).

Existing bins may be converted by imposing an inverted cone-shaped baffle near the top of the discharge hopper. The cone may be 'point up', surrounded by an annular space, directing flow to the periphery of the bin, or 'point down', with an annular space and an orifice in the centre also. In bins of square sections pyramidal baffles replace the conical ones described for cylindrical bins.

References

Alexander, R.J., 1987. Corn dry milling: processes, products, and applications. In: Watson, S.A. (Ed.), Corn: Chemistry and Technology. Amer. Assoc. of Cereal Chemists Inc., St. Paul, MN, USA, pp. 351–376.

ASABE, 2009. D245.6 – Moisture Relationships of Plant-Based Agricultural Products. ASABE Standards – Standards, Engineering Practices, Data. American Society of Agricultural and Biological Engineers, St. Joseph, MI, USA.

Bothast, R.J., 1978. Fungal deterioration and related phenomena in cereals, legumes and oilseeds. In: Hultin, H.O., Milner, M. (Eds.), Postharvest Biology and Biotechnology. Food and Nutrition Press Inc., Westport, CT, USA, pp. 210–243 (Chapter 8).

Brown, R.B., Fulford, G.N., Daynad, T.B., Meiering, A.G., Otten, L., 1979. Effect of drying method on grain corn quality. Cereal Chemistry 56 (6), 529–532.

Bushuk, W., 1976. Rye: Production, Chemistry and Technology. Amer. Assoc. of Cereal Chemists Inc., St. Paul, MN, USA.

Bushuk, W., Lee, J.W., 1978. Biochemical and functional changes in cereals: storage and germination. In: Hultin, H.O., Milner, M. (Eds.), Postharvest Biology and Biotechnology. Food and Nutrition Press Inc., Westport, CT, USA, pp. 1–33.

Campbell, J.D., Jones, C.R., 1955. The effect of temperature on the rate of penetration of moisture within damped wheat grains. Cereal Chemistry 32, 132–139.

Christensen, C.M., 1974. Storage of Cereal Grains and Their Products. Amer. Assoc. of Cereal Chemists Inc., St Paul, MN, USA.

Deane, D., Commers, E., 1986. Oat cleaning and processing. In: Webster, F.H. (Ed.), Oats, Chemistry and Technology. Amer. Assoc of Cereal Chemists Inc., St. Paul, MN, USA.

Demidov, P.G., Kochetova, A.A., 1966. Effect of ultrasonic waves on the technological properties of grain. Izvestiya Vysshikh Uchebnykh Zavedenii, Pishchevaya Tekhnologiya 1, 13.

Every, D., 1987. A simple four-minute protein-solubility test for heat-damage in wheat. Journal of Cereal Science 6, 225–236.

Glenn, G.M., Younce, F.L., Pitts, M.J., 1991. Fundamental physical properties characterizing hardness of wheat endosperm. Journal of Cereal Science 13, 179–194.

Home-Grown Cereals Authority (H-GCA), 1992. Storage Crops. H-GCA, London, UK.

Hook, S.C.W., Bone, G.T., Fearn, T., 1984. The conditioning of wheat. A comparison of UK wheats milled at natural moisture content and after drying and conditioning to the same moisture content. Journal of Science of Food and Agriculture 35, 591–596.

King, R.W., 1984. Water uptake and pre-harvest sprouting damage in wheat: grain characteristics. Australian Journal of Agricultural Research 35, 338–345.

Lockwood, J.F., 1960. Flour Milling, fourth ed. Henry Simon Ltd., Stockport, Cheshire, UK.

Moss, R., 1977. The influence of endosperm structure, protein content and grain moisture on the rate of water penetration into wheat during conditioning. International Journal of Food Science and Technology 12, 275–283.

Nellist, M.E., 1978. Safe Temperatures for Drying Grain Report No. 29. National Institute of Agricultural Engineering (To Home Grown Cereals Authority).

Osborne, B.G., Fishwick, F.B., Scudamore, K.A., Rowlands, D.G., 1988. The Occurrence and Detection of Pesticide Residues in UK Grain. HGCA Research Review No. 12. Home-Grown Cereals Authority, London, UK.

Pixton, S.W., 1980. Changes in quality of wheat during 18 years storage. In: Shejbal, J. (Ed.), Controlled Atmosphere Storage of Grains. Elsevier Scientific Publ. Co., NY, USA, pp. 301–310.

Pomeranz, Y., 1974. Biochemical, functional, and nutritive changes during storage. In: Christensen, C.M. (Ed.), Storage of Cereal Grains and Their Products. Amer. Assoc of Cereal Chemists, St. Paul, MN, USA, pp. 56–114.

Smith, L., 1944. Flour Milling Technology, 3rd. ed. Northern Publishing Co. Ltd., Liverpool, UK.

Stenvert, N.L., Kingswood, K., 1976. An autoradiographic demonstration of the penetration of water into wheat during tempering. Cereal Chemistry 53, 141–149.

Storey, C.L., 1987. Effects and control of insects affecting corn quality. In: Watson, S.A., Ramstad, P.E. (Eds.), Corn: Chemistry and Technology. Amer. Assoc. of Cereal Chemists Inc., St. Paul, MN, USA, pp. 185–200.

Wasserman, T., Calderwood, D.L., 1972. Rough rice drying. In: Houston, D.F. (Ed.), Rice Chemistry and Technology. Amer. Assoc. of Cereal Chemists. Inc., St. Paul, MN, USA, pp. 140–165.

Wellington, P.S., Durham, V.M., 1961. The effect of the covering layers on the uptake of water by the embryo of the wheat grain. Annals of Botany 25, 185–196.

Further Reading

Anon., 1987. Pests of Stored Wheat and Flour. Module 2 in Workbook Series, National Association of British and Irish Millers.

Bakker-Arkema, F.W., Brook, R.C., Lerew, L.E., 1978. Cereal grain drying. Advances in Cereal Science and Technology 2, 1–77.

Bhattacharya, K.R., Ali, S.Z., 1985. Changes in rice during parboiling, and properties of parboiled rice. Advances in Cereal Science and Technology 7, 105–159.

Cornwell, P.B., 1973. Pest Control in Buildings. A Guide to the Meaning of Terms. Rentokill Ltd., E. Grinstead, UK.

FDA Technical Bulletin 4. In: Gorham, J.R. (Ed.), 1991. Ecology and Management of Food-Industry Pests. Assoc. of Official Analytical Chemists, Arlington, VA, USA.

Home-Grown Cereals Authority, 1992. Storage Crops. H-GCA, London, UK.

Poichotte, J.L., 1980. Wheat storage. In: Hafliger, E. (Ed.), Wheat. CIBA Geigy, Ltd., Basle, Switzerland, pp. 82–84.

Robinson, I.M., 1983. Modern Concepts of the Theory and Practice of Conditioning and its Influence on Milling (Gold medal thesis of Nat. Jt. Ind. Council for the Flour Milling Industry).

Sauer, D.B. (Ed.), 1992. Storage of Cereal Grains and Their Products. Amer. Assoc. of Cereal Chemists Inc., St. Paul, MN, USA.

Tkatchuk, R., Dexter, J.E., Tipples, K.H., 1991. Removal of sprouted kernels from hard red spring wheat with a specific gravity table. Cereal Chemistry 68, 390–395.

Dry-milling technology

6.1 INTRODUCTION

Milling schemes are conveniently classified as wet or dry, but this indicates a difference in degree rather than an absolute distinction, as water is used in almost all separations. Damping, 'conditioning' or 'tempering' features even in 'dry' milling; it is considered in detail in Chapter 5 as it is a premilling treatment. This chapter is concerned with so-called dry milling; wet-milling processes are dealt with in Chapter 14. Emphasis is placed on preparation for human consumption, but the term 'milling' also applies to production of animal feeds. A brief description of feed-milling is given in Chapter 13.

Milling, in the context of the production of ingredients for human consumption, involves the transformation of whole grains into forms suitable for conversion into palatable products. The single term covers a wide range of processes. In earlier chapters reference has been made to hulled and naked grains and, while there may be differences among the hulls of different cereal types, it is true of all hulls originating from plant structures surrounding the fruit that they have little or no nutritional value for humans and hence require removal before the more nutritious parts of the grains can be processed. In some cases, such as rice and oat milling, dehulling is part of the milling process while in the milling of hulled wheats, dehulling is regarded as a pretreatment to the milling process. In the case of barley milling the process is almost synonymous with dehulling. Unlike hulls, bran, which comprises those parts of the caryopsis that surround the starchy endosperm and embryo, has value as a food ingredient for humans and livestock, not only on account of its contribution of fibre to the diet but also due to the protein and fat contents of the aleurone. Nevertheless, it is, in many cases, desirable to separate bran from starchy endosperm and milling is the process by which this is achieved. Where endosperm products are to be stored, removal of the embryo as the milling fraction 'germ' is also desirable as oxidation of the high-lipid content of the germ leads to rancidity.

In some cases bran and embryo removal is sufficient, and it is achievable by decortication, the process whereby the outer layers of the grain are removed by abrasion. As the process seeks to remove surface material in a uniform manner, from all areas of the grain, it is simplest and most successful when applied to grains that are near spherical, and it is least successful where grains are long and thin, particularly when they have a crease (see Chapter 3). The texture of the endosperm is also a factor that determines the efficiency of the process as harder endosperms are more resistant to abrasion and fragmentation.

In other cases it is desirable to reduce the starchy endosperm to small particles, and this is achieved through grinding. Grinding, combined with efficient means of separation of bran particles, achieves both purification and size reduction of endosperm.

Dehulling, or decortication, is a process that is also applied to some pseudocereals, and although the anatomy of their fruits is different from cereals, leading to different parts being removed or retained, the principles remain the same.

6.2 MILLING PROCESSES

The changes that occur as the result of milling operations may be summarized as follows:

- The shapes and sizes of grains are changed,
- Train components of different composition, separated in operation 1, are concentrated by fractionation,
- The temperature and/or water content of the stocks are changed.

6.2.1 Treatments that change shape and size

6.2.1.1 Abrasion

Surface abrasion, as applied to naked or dehulled grains, is a relatively gentle process that removes all or part of the fruit coats (pericarp) and possibly the embryo. Grains are brought into contact with an abrasive surface, which may be of natural stone, carborundum, sculptured or perforated metal, or other material. Where perforated screens are used, these also behave as sieves, selectively permitting passage of smaller particles.

Greater severity, or retreatment of already abraded stocks, also removes part of the endosperm. Severe abrasion includes heavy or protracted grinding between surfaces such as those of a pestle and mortar and of grinding stones; it features therefore in many of the simple and historical methods used for preparing ground meals. Such grinding may reduce grains to a range of particle sizes including that of flour.

Impact milling may be regarded as an exaggerated abrasion process. Stocks are fed into a chamber where they are projected at high speed against a surrounding wall, which may contain screens. Other obstacles, such as pins, may also be present. Impacts against the obstacles and other particles cause size reduction, which, in some types of mills, continue until all are small enough to pass the screens. In other types stocks remain in the chamber only for sufficient time for removal of surface structure.

6.2.1.2 Roller milling

Wholegrains or partially milled stocks are passed between rotating rollers for several reasons, including:

- Dehulling. This can be achieved with rubber- or plastic-surfaced rolls.
- Flaking. This is a simple process in which whole or decorticated grains are passed between a pair of rotating, heavy (large-diameter) rolls. Flaking rolls are smooth-surfaced and it is usual for both rolls to turn at the same speed. The purpose is to increase the surface area of one or more of the components of the feedstock, either to facilitate subsequent fractionation (e.g., germ from endosperm) or to impart desired product characteristics, as in porridge oats.
- Grinding. This is a process in which stocks are reduced to particles of smaller size by pressure and shear.

6.2.2 Fractionating

6.2.2.1 Size

Size is an important criterion by which particles are separated. Attempts were made, at least as far back as classical Roman times, to make a white flour for bread making. At the time of the Norman Conquest (AD 1066) in England, stone-ground flour was being sieved into fractions – a fine flour called *smedma* and a coarse flour called *gryth* (Storck and Teague, 1952). Further development led to the production of three flour fractions from stone-ground meal; thus a white flour, bran and a stock of an intermediate particle size known as middlings were produced.

Sieving features at some stages in most modern milling systems that involve grinding, to separate ground stocks as final products, or intermediate stocks that receive further appropriate treatment.

6.2.2.2 Shape

In processes such as oat milling and rice milling, grains are graded on a shape (and size) basis, before treatment, as machine clearances are grain-dimension dependent; and during the milling process, small grains that escape treatment need to be refed. Shape-sensitive fractionating machines include disc separators and trieur cylinders as shown in Fig. 5.20.

6.2.2.3 Specific gravity

Particles differing in density may be separated on a fluid medium such as air or water. In the case of water, the separation usually depends on one or more components being denser than water and others being less dense. Aqueous separations are not appropriate to dry milling stocks. When air is used, the force of an air current supports particles of lesser density to a greater degree than the denser ones, allowing them to be carried upwards and later deposited when the force of the current is reduced. The lighter particles frequently also have a flat shape, which enhances their buoyancy. Winnowing is an example of specific gravity separation. At its simplest a mixture of grains and hulls is thrown into the air. The grains, being denser, fall to the ground under gravity while the hulls are carried away by natural breezes or artificially created air movements. Aspirator machines use a similar, though more controlled, mechanism to remove pearlings from decorticated grains. Aspiration is also one of the principles involved in separating bran from endosperm particles in purifiers.

6.2.2.4 Multiple factors

The paddy separator is an example of a machine in which several grain characteristics are exploited in effecting their separation. Specific gravity, surface roughness and shape all combine to direct grains into appropriate streams on a tilted vibrating table with a cunningly sculptured surface.

6.2.3 Changes in temperature and/or moisture content

Milling processes do not themselves involve intentional heating although the friction involved inevitably leads to machines and stocks becoming hotter. In some cases, as in oat processing, a heating phase precedes the milling, and in maize dry milling heated water and drying phases are included. Stocks produced from grains or intermediates milled at very high moisture need to be dried to permit proper processing or safe storage. Drying is performed by heating, and stocks that have been heated may subsequently require cooling. Exposure of stocks to large volumes of air in pneumatic conveyers can cause reduction or increase in temperature and moisture content depending on ambient conditions. Water can be added with or without substantial change in temperature, it may be added specifically to achieve a calculated combination of physical conditions, as in stabilization of oats, or to change the mechanical properties of the grain components, as in wheat and rice milling.

6.3 PROCESSES IN WHICH THE MAIN PROCESS IS DECORTICATION

6.3.1 Rice milling

Rice milling, as most widely adopted, employs only processes that remove material from the grain surface.

As with processing of most cereals throughout the world, the scale of industrial plants is increasing, but in the case of rice the rate of change is slow and industrial processing accounts for only about half the world crop, the remainder being stored and processed at a domestic or village level. Whatever the level, the objective of the milling is to remove hull, bran and embryo as completely as possible without breaking the remaining cores of endosperm. There are two stages, the first being dehulling and the second 'whitening' or milling. The moisture at which untreated rice is processed is relatively low, around (Serna-Saldivar, 2010) 13%–14% (IRRI Rice Knowledge Bank, 2000), but parboiled rice (see Chapter 5) undergoes heat treatment at 32%–38% m.c. before drying back to 14%. The method of drying is very important as it influences the degree of breakage occurring during milling. Drying in the sun, or with heated air, can lead to cracking and much breakage during milling (Bhattacharya, 1985).

Parboiling (see Chapters 4 and 10) is applied to about one-fifth of the world's rough rice (although there are no official figures to support this estimate). Its effects on milling performance result from changes brought about in both endosperm and hull. In endosperm, starch gelatinization occurs, leading to swelling of the grain, which disrupts the 'locking' of the lemma and palea (cf. Chapter 3). Disadvantages of parboiling lie in the additional costs involved, the increased difficulty in removing bran during the whitening process and a measure of discoloration of the endosperm.

6.3.1.1 Village processing

At a domestic level, hand pounding followed by winnowing is still practiced in some parts of the world, but for machine processing the Engelberg huller is the best known and most widely used small-scale rice mill. It combines the two stages, dehulling and milling, in a single machine. The work is done by a ribbed cylinder rotating on a horizontal axis within a cylindrical chamber, the lower half of which is formed of a screen through which fine material may pass. Rough rice is introduced at one end and passes down the gap between the rotating cylinder and the chamber wall. An adjustable steel blade determines the size of the narrowest gap through which the grain passes and thus determines the level of friction experienced. The rate of flow also serves as a means of controlling the severity of the treatment. The directions of the ribs on the central cylinder vary along its length; the major part of the length has ribs parallel to the cylinder axis, but in the early stages dedicated to dehulling they run in a diagonal curve, helping to feed the stock in as well as to abrade it. The hulls removed in the initial phase help to abrade the grain surface in the later phase. They are ultimately discharged through the slotted screens with fragments of bran, embryo and broken endosperm while the largest endosperm pieces are discharged as over-tails of the screen. The proportion of whole endosperm is as variable as the skills of the operators and the condition of the machine. The fine materials are used for feeding domestic stock and, as such, form a valuable by-product.

Small-scale alternatives to the Engelberg huller include two-stage processors in which dehulling is performed by rubber rollers, typically 150 mm diameter and 150 mm long, and whitening is achieved by friction methods. Both processes may be enclosed within a single housing and are capable of processing quantities of paddy in the order of 0.5–1 tonne/h. The gentler action of the rubber rollers is said to lead to reduced breakage of grains. Mobile versions of several types of small-scale mill serve small communities for whom investment in their own machinery is uneconomic.

6.3.1.2 Industrial processing

On an industrial scale, processing follows the same principles as those employed at the village level, but it is usual for the processes to be carried out in separate machines. Capacities vary, with small mills processing up to 120 t of paddy per 24 h, medium mills up to 360 tonne/day and large mills exceeding that figure, but the scale of machinery does not vary. The larger mills merely have more of the same machines running in parallel. The process is summarized in Fig. 6.1.

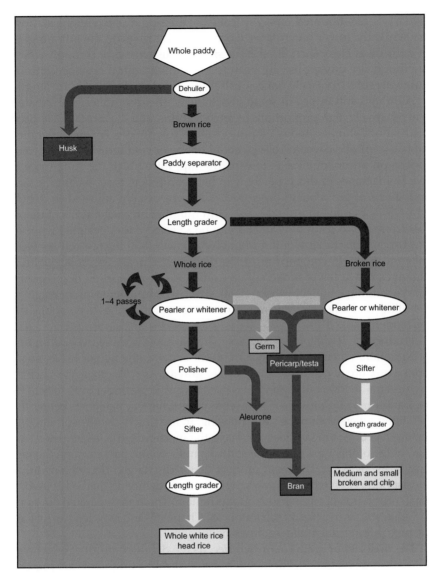

FIGURE 6.1   Schematic diagram of the rice-milling process.

Dehulling

It is important that grain types are not mixed before dehulling so that grain size and shape are as uniform as possible.

Dehulling is possible by use of the disc sheller; it involves abrasion, by discs with surfaces partly covered with coarse emery or carborundum.

Two discs are mounted on a vertical axis; the upper one remains stationary while the lower one rotates. Corresponding rings on the outer region of both discs clear each other by a small adjustable gap, through which the rough rice passes, having been fed in from a central hopper. Optimal hulling results from a clearance of slightly more than half the grain length.

Although still in use in some installations, the disc huller is associated with relatively high proportions of broken grains, and the most widely used method of dehulling today employs rubber roll hullers. A pair of parallel, rubber or plastic covered rolls of equal diameter counterrotate at different speeds on a horizontal axis toward the nip, into which the rough rice is fed. In machines of this type, the gap between rollers is set to suit the size and shape of grain being processed. They are usually adjusted to dehull 90%–95% of grains in a single pass with minimal breakage. Rolls of diameter from 150 to 250 mm diameter can be selected and a peripheral speed of 14 m/s, with the slow roll running 25% slower than the fast roll are typical conditions. In comparison to the disc sheller treatment, roller dehulling gives rise to fewer broken grains, and a lower proportion of the grains escape hulling owing to their small size. A comprehensive coverage of dehulling and other cleaning and processing operations has been published by Tangpinijkul (2010).

Products of dehulling need to be separated, and the usual method for this is aspiration, whereby a forced air current conveys the lighter husks away from a mixture of products falling under gravity. As with other rice machinery, designs abound but aspiration is the basic principle underlying most. The aspiration may be provided by a separate machine or stationed below the dehulling rollers within the same housing. In the closed-circuit aspirator, the air produced by the fan does not leave the machine but is recirculated. An advantage of this type is that the abrasive hulls do not pass through the fan, thus reducing wear. The main product that is discharged from the dehuller consists largely of dehulled grains, but whatever dehulling method is used, a small proportion of entire paddy grains remain. These are separated out from the main stream in a paddy separator (see Chapter 5) and recycled to a further passage through a dehuller set to a smaller gap.

The majority of grains of brown rice, as the dehulled product is known, consists of wholegrains, but inevitably some broken fragments are present and these are selected out by length grader, which might be of the trieur or indented cylinder type. Both fractions from the length separator are further processed to remove bran, but the broken fragments are not included in high-value wholegrain products.

Whitening

The whole brown rice grains are treated to remove bran and embryo. Two principles are involved in design of whitening machines; the most common is the abrasive cone. The cone rotates on a vertical axis, its surface is covered with abrasive material, and it is surrounded by a perforated screen. Rubber brakes are installed at intervals around the cone to provide a smaller clearance than that between the rotor and the screen. The size of the gap influences the retention time on the processing zone. Raising the cone increases the gap between it and the screen, and the brakes can be adjusted independently to determine the severity of the pearling. The abrasive cone machine is used in single or multipass systems, sometimes with the alternative friction-type machine. Friction machines always occur in multipass systems in which abrasive cones also feature. A friction whitener comprises a rotating cylinder within a chamber of hexagonal section, and walls of screening with slotted perforations. The horizontal cylinder has air passages through its centre that pass to the surface as a series of holes in parallel alignment. Rotation of the cylinder causes friction within the grain mass surrounding it, and air from the cylinder cools the stock and supports the discharge of bran. Bran and germ pass out through the perforated screens. It is claimed that the vertical whitener (Fig. 6.2) produces smoother rice with fewer grains breaking during processing.

All whitening processes involve friction and hence heat is produced, but grain temperatures should not be allowed to exceed 43–44°C. Those machines that include strong aspiration with cooling air reduce or eliminate the chance of overheating.

In the 'humidified friction whitener' a small amount of atomized water is injected through the cylinder airways via the pressurized air stream. The process serves a similar purpose to tempering of maize or wheat; it toughens the bran, facilitating its removal. The ultimate result is the reduction in endosperm breakage and the increased recovery of

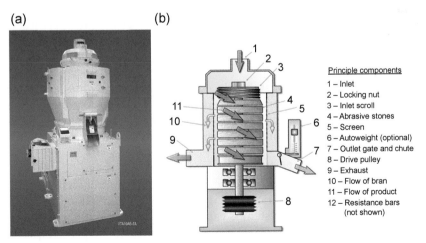

(a)   (b)

Principle components

1 – Inlet
2 – Locking nut
3 – Inlet scroll
4 – Abrasive stones
5 – Screen
6 – Autoweight (optional)
7 – Outlet gate and chute
8 – Drive pulley
9 – Exhaust
10 – Flow of bran
11 – Flow of product
12 – Resistance bars
(not shown)

FIGURE 6.2   (a) The Vertical Rice Whitener and (b) diagram showing the internal components. *By courtesy of Satake Europe Ltd.*

head rice (e.g., Wahid et al., 2006). Humidification may also be combined with cooling before whitening. The cooling hardens the grain, also leading to less breakage.

In the more modern, larger mills a high degree of automation exists in the whitening process, including microcomputer control.

Polishing

The rice emerging from the whitening process has had the outer layers of bran removed but parts of the inner layers remain. These fragments are removed by the process known as polishing or refining ('polishing' is used by the Japanese to describe the milling process – terminology can be very confusing, especially where translation is involved!). Polishing extends the storage life of the product as the aleurone layer is removed, thus reducing the tendency for oxidative rancidity to occur in that high-oil tissue.

In conventional rice mills polishers are like whitening cones, but instead of abrasive coverings, the cone is covered with many leather strips. No rubber brakes are applied. To improve the lustre of white rice, a humidified friction-type whitening machine may be used for the final whitening.

Grading of milled rice products

Rice leaving the milling section of commercial rice mills is a mixture of entire and broken endosperms of various sizes. The number of grades separated depends on the sophistication of the system and of the market. In developing countries only 'brewers' rice' may be separated from the main product. Brewers' rice passes a sieve with 1.4-mm, round perforations (F.A.O. definition). Other markets may justify grading into 'head rice' (grains entire except for the tip at the embryo end), as well as 'large brokens' and 'small brokens'. Separation is by trieurs or disc separators. A full description of products is included in the NFA standard specification for milled rice (FAO, 1980).

Conveying stocks

Rice milling operations take place on a series of machines and hence stocks have to be transported around the mill. Some routes involve raising from a lower to a higher floor, and this can be achieved using bucket elevators, but in modern mills pneumatic conveying is the norm.

Removing nonconforming grains

A relatively recent introduction into rice quality control facilities is the colour sorter. While most rice milling equipment is applied to bulk materials, the colour sorting system is capable of detecting and rejecting individual grains from that bulk. The machine's operation depends upon an optical technique such as a camera to detect the grains of unacceptable colour and a means of removing them such as a jet of compressed air, as the bulk is streamed through a channel. An example of a colour sorter is shown in Fig. 6.3 and the mechanism is shown in Fig. 6.4.

In a well-adjusted modern mill, the approximate proportions of products are shown in Table 6.1.

Rice flour

Rice flour, bakers' cones (a coarser flour) and ground rice are the products of grinding milled rice; the head rice is rarely used as a feedstock because of its high value, but the 'second head' product, comprising endosperm particles of about half grain size, is used.

6.3.2 Sorghum and millet milling

6.3.2.1 *Traditional methods*

Sorghums and millets are considered together as they are both tropical cereals, much of the processing of which remains in the hands of subsistence farmers. The grinding of flour is performed by traditional manual pounding, which involves addition of more water than in most dry-milling methods. A typical scheme for sorghum, described by Rahnavathi (2016), is to add 30%–40% water and to pound for 10–15 min to detach bran, followed by drying and winnowing with the

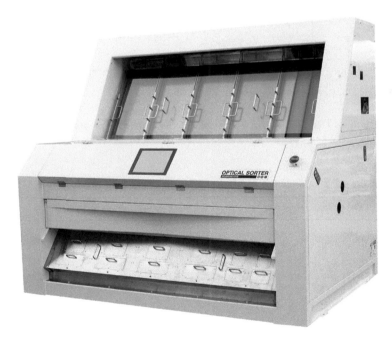

FIGURE 6.3   Colour sorter. *By courtesy of Satake Europe Ltd.*

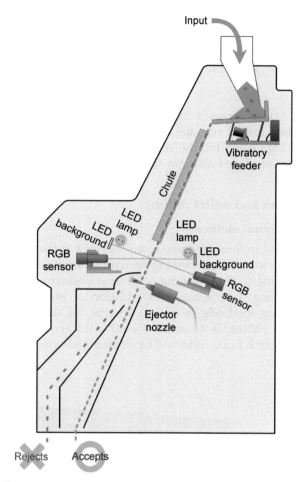

FIGURE 6.4 Diagram of the mechanism of the colour sorter. *By courtesy of Satake Europe Ltd.*

TABLE 6.1 Approximate proportions of products from a large rice mill

Product	Yield as proportion of paddy (%)
Husk	20–30
Bran	7–10
Polishings	3
Head rice	50–60
Brokens	5–10
Small brokens	10–15

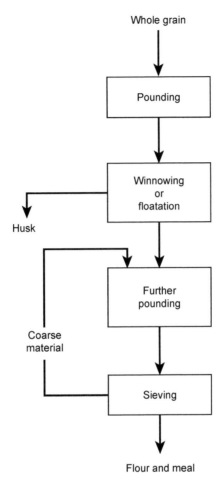

FIGURE 6.5   Schematic diagram of domestic processing of sorghum.

traditional basket winnower. The bran-free fraction is then damped to 20%–30% moisture content and pounded to flour, which is then dried for 18–24 h. The simplicity of the process is illustrated schematically in Fig. 6.5.

The principles illustrated are:

- Use of attrition to break open the grain.
- Separation of endosperm from the surrounding tissues.
- Use of water to aid the separation.
- Sieving to select stocks for appropriate treatment.

Another principle, common to all milling processes, and illustrated by hand milling, is the dependence of the method adopted on the nature of the varieties processed. Variations include the hardness of the endosperm and the thickness of the pericarp. Soft grains break into pieces during decortication and cannot be readily separated from the pericarp. The difficulties are compounded when pericarps are thick (Rooney et al., 1986). Special techniques may be required in cases where one or more components are highly pigmented or flavoured.

Sufficient flour is produced daily for the needs of the family; 3 or 4 h may be required to produce 1.5–1.6 kg of flour from sorghum at an extraction rate of 60%–70% of the initial grain weight. Bran accounts for about 12%, and the same amount is lost. Even longer periods (6 h) may be required to produce a family's daily requirements from millet (Varriano-Marston and Hoseney, 1983). The problem of storage of flour does not arise, and this is just as well as products may have a high moisture content. Further, the continued pounding expresses oil from the embryo and incorporates it into the flour, leading to rancidity on oxidation.

6.3.2.2 Industrial methods

Sorghum

Sorghum has certain nutritional advantages over some other cereals because of antioxidants present and the low glycemic index with which it is associated. It is particularly popular in the gluten-free market. As a consequence of this, as well as progressive industrialization in Africa and Asia, greater interest is being shown in developing and applying mechanized milling methods. In Africa, the relationship between domestic processing and mechanized milling is a delicate balance that is affected by a number of changes occurring on that continent. Processing is justified only if it prolongs the storage period, increases convenience and preserves the nutritional quality of the product (Chinsman, 1984). In Nigeria, where wheat imports have been restricted, wheat mills have been adapted to process sorghum, but this is exceptional for the continent (House et al., 2000).

Because industrial-scale operations are relatively new, these may be regarded as somewhat experimental, and no single method has yet been established as the standard. Whether domestic, small scale of industrial, the objectives include the separation of botanical tissues, one from another, and in some cases reduction in particle size of the endosperm fraction. Flaking is also practiced.

In some cases traditional methods of decortication are combined with 'service' milling. Such combinations can improve on the traditional methods alone by as much as 20% extraction rate.

Wholemeals are produced by use of stone, hammer, pin or roller mills, but almost all processes directed toward white flour production

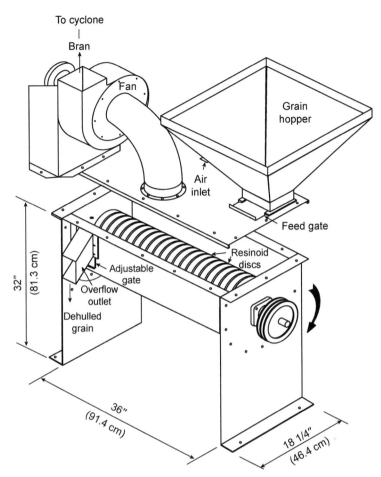

FIGURE 6.6 Schematic diagram of a PRL-type dehuller. *Reproduced with permission Reichert, R.D., 1982. Sorghum dry milling. In: House, L.R., Mughogho, L.K., Peacock, J.M. (Eds.), Sorghum in the Eighties, vol. 2. Patancheru, India. Reproduced by permission of ICRISAT.*

begin with a decortication stage using mills with abrasive discs or carborundum stones. Probably the most widely used abrasive decorticators are based on the PRL dehuller (Prairie Regional Laboratory) (Taylor and Belton, 2002) illustrated in Fig. 6.6. The functional element of the machine is a horizontal barrel in which some 13 evenly spaced carborundum-faced discs rotate at approximately 2000 rpm, progressively abrading off the bran and germ, which are removed and conveyed to a cyclone by means of a fan.

As the amount of pearlings increases, so the composition changes, reflecting the concentration of fibre in the outer layers, and of protein and fibre in the aleurone layer and the peripheral (subaleurone) starchy endosperm (cf. Chapter 3). Oil and protein contents of the pearlings are at a maximum when about 12% of the grain has been abraded.

Sorghum endosperm, used as a brewing adjunct, leaves the mill as coarse grits; they may be produced by impaction following decortication (Rooney and Serna-Saldivar, 1991). Removal of embryos after milling is difficult as they are the same size as some of the grits. They can be separated, however, by virtue of their different densities; floatation on water or use of a gravity table are suitable methods.

Some or all germ may be removed from endosperm cores during debranning, but if it is required to remove the embryo before milling, this may be done using machines designed for degerming maize. Alternatively, special machines have been produced for sorghum itself. One such device consists of a wire brush rotating within a perforated cylinder. Reduction to flour particle size may be achieved by hammer milling, pin milling or roller milling.

Machines developed from the PRL dehuller and other dehulling machines, as well as layouts for mills incorporating them, were reviewed by Bassey and Schmidt (1989), Rice milling equipment such as the vertical whitener (Fig. 6.2) has been used experimentally and may be applied in larger mills (Kebakili, 2010).

A recent development, reported to be gaining acceptance in Africa (Taylor and Belton, 2002), is the compact roller mill, with a capacity of 500 kg/h. The system is devised for maize processing, but it is said to produce good extraction rates combined with good separation of grain components when milling sorghum. The principle of the system is a series of three superposed pairs of rollers disposed in a sequence similar to that adopted in a conventional full-scale roller mill (see later in this chapter), with treatment by coarsely fluted break rolls preceding further treatment on finely fluted rolls with final crushing on smooth rolls. Stocks are sieved before selected stocks are fed to the smooth rolls, after which a further sieving occurs. An example of such a mill is shown in Fig. 6.7.

Millets

Industrial processing of millets is even less common and less developed than that of sorghum. While there are reports of experimental attempts to adapt technology appropriate to other cereals, the concept of industrial-scale millet milling is not well established. Industrial production of flour inevitably imposes a need for distribution facilities and for storage. The inclusion of embryo parts, or even oil expressed from the embryo in flour, reduces storage life, as in sorghum. Any successful process should therefore include a degerming stage. The limited amount of small-scale industrial processing that is carried out consists of abrasive decortication followed by reduction of endosperm with hammer mills or similar

FIGURE 6.7 Compact roller mill designed for small-scale maize milling. *Reproduced by courtesy of Roff Industries (Pty) Ltd.*

devices dependent on attrition. Grains of some millets (e.g., teff) are so small that even decortication is extremely challenging, and it is unlikely that industrial processing will ever feature in preparing them for human consumption.

6.3.3 Barley milling

The main use of barley for human consumption lies in the production of malt, which is used in the production of alcoholic beverages and as a functional or flavouring ingredient in food. Malting is covered in Chapter 12 but reference should be made here to a premalting decortication process that was introduced in the 1970s as a means of reducing malting time. It involved gentle decortication of the grains before steeping in order to permit gibberellic acid, incorporated in the steep liquid, to penetrate into the endosperm, thus supplementing the natural gibberellins produced by the embryo and aleurone (Brown, 1974). The system was widely used in the 1970s and 1980s (Black et al., 2006).

Unmalted barley is also consumed as a food, and because of the beneficial nutritional effects of β-glucans present in barley, widespread attempts have been made to increase the incorporation of barley products into diets (Kiryluk et al., 2000). Nevertheless, its significance remains comparatively low and very variable throughout the world. In Morocco, a traditional consumer of barley-based foods, in 2011, the per capita consumption of barley and products was 38.8 kg/year, while in the United States the comparable figure is under 1 kg (as a comparison, per capita consumption of wheat and products in United States in the same year was 79.5 kg and of rice and products in China was 79.2 kg). Other countries with relatively high barley consumption as food are Latvia (20.2 kg/year) and Ethiopia (14.4 kg/year). Dishes prepared from milled grains tend to be associated closely with the local areas of consumption (Amri et al., 2005), and this variation is accompanied by variation in dry-processing methods. Decortication, as described previously in relation to malting, is a process that has long been used in the preparation of hulled barleys for food use. The degree of pearling varies and different products are defined by the amount of peripheral material removed. It is a necessary treatment not only in the preparation of finished products but also as a precursor to other treatments as the adherent lemma and palea are undesirable in most edible products. An exception is the roasting of whole barley, complete with hulls and its subsequent milling to produce a semolina used to flavour beers or to produce an infusion known as barley tea.

Two-row and six-row barleys may be used, but where it is possible to select barleys for decortication those types to which hulls make the least contribution are desirable. Endosperm hardness is a variable characteristic and it exerts a strong influence on pearling performance. Size reduction proceeds more rapidly in varieties with softer endosperm but more of their endosperm is ground to smaller size particles, reducing the yield of entire pearled cores – the premium product. To achieve uniform pearling, grains of similar size and hardness are desirable as feed to the process. The presence of damaged grains lowers the quality of milling barley. Such grains frequently reveal areas of exposed endosperm where fungal attack may occur, leading to discoloration. Such grains would contribute dark particles to the finished product. Thin grains also lower the milling quality because of the increased proportion of hull. As with the processing of all cereals, cleaning and tempering (to 15.5%) are important premilling treatments. In locations where it is permitted, barleys of varieties with blue-pigmented aleurone may be bleached (Ram and Misra, 2010).

Decorticated grains may be milled on roller mills to produce flour or semolina, or meals may be produced by simpler single-pass techniques as described for sorghum. Flaking and cracking of pearled grains are also practiced by passing between smooth-surfaced rolls. Production of grits may involve a toasting stage before rolling or cutting in a similar fashion to oat cutting (see Section Cutting). Holes of 4.2 mm in the rotating drum used in the cutting process are recommended for barley. Grits produced by cutting may be subjected to a further abrasive process to create rounded particles (Rohde, 2004).

The naming of pearled barley products (even in the English language) and their derivatives is somewhat confused and variable with geographic location. The barley from which the fewest pearlings have been removed (say 5%) may be referred to as blocked, pot, Scotch and in some cases pearl barley, however, pearl barley is usually understood to be barley from which about 11% or more of the original grain has been abraded. The process should achieve a yield of cores representing around 70% of feedstock weight with half that quantity being premium grade (Benton Jones, 2003). While the objective of decortication is the removal of husk (blocking) and bran (pearling), it is not an exact process. The barley grain is elongated and pointed and surface abrasion tends to lead to a near spherical product. Hence the proximal and distal parts of the grain are preferentially removed, possibly including some endosperm. As the embryo is at the proximal end of the wholegrain, it is almost inevitably removed in part or in whole, depending on the degree of abrasion applied. The barley grain has a crease (see Chapter 3) and hence bran tissues here are inaccessible to the abrasive surfaces and are not entirely removed. The degree of pearling that a sample has received can be assessed by the size and colour of the cores, heavily abraded cores being small and creamy white and lightly pearled cores being large and brown.

One alternative to pearling is the inclusion of a stage in which blocked grains are toasted before cracking or cutting to produce grits.

6.3.3.1 Processes in barley milling

Pearling technology and equipment are similar in principle to those described for sorghum. Following cleaning and conditioning, they include an abrasion stage followed by aspiration and screening by which pearlings and cores are separated. Best results are achieved by a multistage process of say three blocking passes and a further three pearling passes. Both batch and continuous processes are in use. Details of a modern barley milling plant are provided by Rohde (2004).

For making barley flakes, barley groats or pearl barley are subjected to:

- Damping
- Steam cooking
- Flaking – on large-diameter flaking rolls
- Drying – on hot-air dryer

Hull-less barleys can be processed in similar ways to hulled or covered barleys but their principal advantage for dry processing lies in the production of flour, because the lemma and palea do not have to be removed by abrasion. Bhatty (1993) produced flours of 70% extraction with acceptable properties in a comparison of hulled and hull-less barleys. He also compared the properties of the bran produced with that of oats.

6.3.3.2 Barley flour

Barley flour is milled from pearled, blocked or hull-less barley. Optimum tempering conditions are 13% m.c. for 48 h for pearl barley, 14% m.c. for 48 h for unpearled, hull-less grain. The milling system uses roller mills with fluted and smooth rolls, and plansifters, in much the same way as in wheat flour milling (see Section 6.4.5). Barley flour is also a by-product of the pearling and polishing processes.

Average extraction rate of 82% of barley flour is obtained from pearl barley representing 67% of the grain, i.e., an overall extraction rate of 55% based on the original wholegrain (including hull). By using blocked grain, an overall extraction rate of 59% of the wholegrain could be obtained, but the product would be considerably less pure than that produced from pearl barley. The chemical composition of milled barley products is shown in Table 6.2.

It is possible to mill a mixture of wheat and hull-less barley in ratios between 90:10 and 80:20. This reduces the tendency for by-products of barley milling to overload the system.

6.3.3.3 Other milled barley products

Barley semolina is used for couscous in North Africa, sometimes after roasting, which gives a nutty flavour (Chakraverly et al., 2003). Malted barley flour is made by grinding or milling barley malt (see Chapter 12).

TABLE 6.2   Chemical composition of milled barley products[a]

MProduct	Protein (%)	Fat (%)	Ash (%)	Crude fibre (%)	Carbohydrate (%)	Energy (kJ/100 g)
Pearl barley	9.5	1.1	1.3	0.9	85.9	1676
Barley flour	11.3	1.9	1.3	0.8	85.4	1693
Barley husk	1.6	0.3	6.2	37.9	53.9	560
Barley bran	16.6	4.0	5.6	9.6	64.3	–
Barley dust	13.6	2.5	3.7	5.3	74.9	–

Dry matter basis.
[a] Values calculated from Kent, N.L., 1983. Technology of Cereals, third ed. Pergamon Press, Oxford.

6.3.4 Oat milling

There are two issues regarding processing of oats in the mill: one arises from the structure of the oat grain, the other from its chemical composition.

The oat grain consists of the groat, or caryopsis, and a surrounding hull or husk. Only the groat is required for the milled products; the hull is indigestible and must therefore be separated and removed in a dehulling (or shelling) stage. Unlike barley and rice the oat hull surrounds but does not adhere to the caryopsis so an impact process rather than abrasion is sufficient for dehulling.

The groat has a lipid content that is between two and five times as high as that of wheat, and the groat also contains an active lipase. Lipids are most highly concentrated in the embryo and scutellum, but the endosperm (aleurone and starchy endosperm) contributes most due to its greater relative size (Youngs et al., 1977). Lipase activity is also highest in the embryo but the aleurone contributes most to the total presence due to its larger contribution to the groat weight. In the intact groat of raw oats the lipase and the lipid do not come into contact with each other and, hence, little or no hydrolysis of the lipid occurs. However, when the grain is milled, the enzyme and substrate come together, and hydrolysis ensues. The substrate consists essentially of acylglycerols and hydrolysis leads to the production of glycerol and free fatty acids, mainly oleic (9-octadecanoic), linoleic (9,12-octadecanoic) and palmitic (hexadecanoic) acids. The glycerol is neutral and stable, but oxidation of the free fatty acids progressively gives the product a bitter flavour. Thus, the inactivation of the enzyme, by a process called stabilization, is an essential step in the milling of oats, a process that, if practiced in relation to other cereals, is applied only to concentrated germ or germ-rich products.

If the lipase is not inactivated, then its effects become apparent in oatcakes baked from oatmeal, fat and water. For making oatcakes, the raising agent frequently used is sodium hydrogen carbonate ($NaHCO_3$). If free fatty acids are present in the oatmeal, they react with the raising agent to form the sodium salts of fatty acids, which are soaps. The resulting oatcakes thus have a soapy flavour. Furthermore, the fat added to the oatmeal in oatcake baking may be animal fat or vegetable fat. The use of certain vegetable fats, e.g., palm kernel and coconut oils, in oatcake baking is particularly undesirable if the oatmeal contains an active lipase because the fatty acids in these fats are chiefly lauric (dodecanoic) and myristic (tetradecanoic); the former, when released by lipase, has an unpleasant soapy taste.

6.3.4.1 Processes in oat milling

The high-lipid content of the oat endosperm contributes to a structure that is ill suited to roller milling or other processes by which the caryopsis components can be separated into flour bran and germ. Instead, the objective of oat milling is to isolate and stabilize the groat and to convert it into a form that is easy to cook. The processes involved are shown schematically in Fig. 6.8.

The process starts with oats that have been through cleaning treatments similar to those used for all industrially processed cereals, however an additional treatment, known as 'clipping', may be included (Decker et al., 2014). Stock is passed through a horizontal rotating perforated cylinder. Blades surround the cylinder, cutting off the tips of grains that protrude through the perforations. The excised tips are separated with other contaminants during the subsequent cleaning stages.

Clipping makes hulling easier at a later stage. A similar process occurs later in the milling process when groats are converted to cut oats on a granulator.

Grading

Oats are subjected to grading because of the presence of 'light grains', which consist of a husk enclosing only a rudimentary groat. The length of these grains is not necessarily less than that of normal grains, so they may defy the selective processes of the cleaning treatments. If not removed, thin grains cause many problems in the later stages of milling. The grading apparatus comprises a perforated cylinder slowly rotating on a horizontal axis. The screen may be slotted or a specially designed mesh. Variation of grain size in oats is a characteristic feature as even within a spikelet large differences exist between primary, secondary and, where present, tertiary grains (see Chapter 3) so a sequence of passages through rotating cylinders, each with smaller holes than the one before, may be used to separate the oats into bulks of grains with similar widths and hence weights. These will then be further processed as separate streams entering dehullers adjusted to suit their physical characteristics (Rohde, 2004).

Shelling or dehulling

Moisture content has a strong influence on shelling efficiency, with higher levels reducing efficiency but also increasing breakage of groats. A moisture content of 12%–13% is considered to provide the best compromise. In most cases hulls are removed with an impact dehuller. The whole oats are fed into the centre of a high-speed rotor fitted with either blades or fins. The grains are thrown outwards and strike, point-first, a hardened ring, in some cases of carborundum, but more often of hardened rubber, attached to the casing of the machine. The combination of high velocity and impact against the ring detaches the lemma and palea from the groat. The speed of the rotor is adjustable (typically 1400–2000 rpm) allowing the process to be optimized for a particular feedstock. Weight, type of oat and moisture content will dictate the conditions required. Factors affecting hulling efficiency were reviewed by Gates (2007).

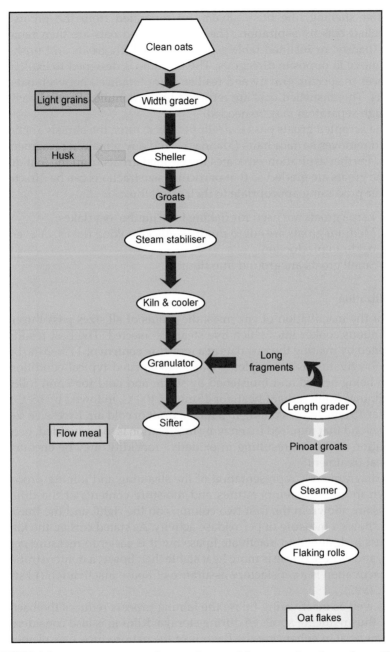

FIGURE 6.8 Schematic diagram showing the essential stages and products of oat milling.

After shelling, the husk slivers are separated from the groats and unshelled oats by aspiration. The few unshelled oats are then removed on a 'paddy' or inclined table separator, on which groats and unshelled oats move in opposite directions. The machine is designed to exploit differences in specific gravity and resiliency (or 'bounce') between oats and groats. The unshelled oats are returned to the dehuller. Several passages through separators may be needed.

The accepted groats pass to a 'clipper' or scourer, the abrasive action of which removes surface hairs ('dannack') and any remaining fragments of husk. Further aspiration separates the cleaned groats from the 'oat dust'.

The groats are graded so that particular size fractions can be directed to further processing appropriate to their sizes, thus:

- Large groats are used for flaking into 'jumbo' oat flakes;
- Medium groats are cut to produce quick-cooking flakes, or even for oat bran production;
- Small groats are ground into flour.

Stabilization

For the inactivation of enzymes, the grains of all sizes pass through a continuous cooker into which live steam is injected. The best results are obtained by treating the raw oats at a moisture content of 14%–20% (below 14% inactivation is incomplete). Gates (2007) cited typical conditions as grain being heated and moistened by steam and held for 9 min followed by kilning, using radiant heat, for 45 min at 100°C, followed by 65°C for a further 15 min before cooling in a forced draft of cold air. Excessive steaming beyond that required for enzyme inactivation is to be avoided, because oxidation of the fat, resulting in oxidative rancidity, may be encouraged by heat treatment.

A diagrammatic representation of the steaming and kilning process is shown in Fig. 6.9, Temperatures and moisture contents at the different stages are shown in the first two columns on the right, and the third column shows a measure of peroxidase activity. As stated earlier, the kilning process is designed to inactivate lipase but it is easier to measure peroxidase, and as peroxidase is more heat stable than lipase, a demonstration of its inactivation is a satisfactory assurance of lipase inactivation (Ekstrand et al., 1992).

As well as inactivating lipase the kilning process reduces the bacterial load, thus reducing spoilage during storage. Kilning is also considered to be beneficial in enhancing the flavour of the oat product. One element of this is the promotion of the Maillard reaction, which occurs between proteins and carbohydrates and gives rise to desirable flavours and browning. It also leads to formation of antioxidant compounds that further stabilize lipids (McClements and Decker, 2008).

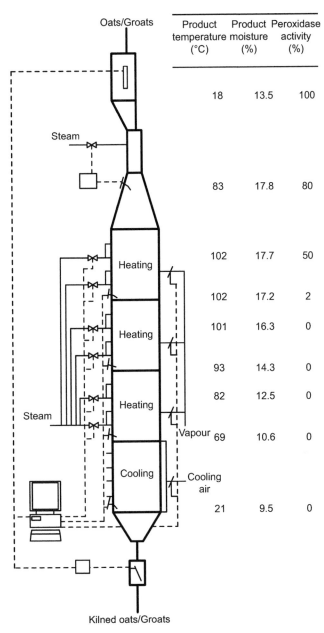

	Product temperature (°C)	Product moisture (%)	Peroxidase activity (%)
	18	13.5	100
	83	17.8	80
	102	17.7	50
	102	17.2	2
	101	16.3	0
	93	14.3	0
	82	12.5	0
	69	10.6	0
	21	9.5	0

FIGURE 6.9 Schematic representation of kilning, showing a typical temperature moisture profile. *From Ganssman, V., Vorwerk, K., 1995. Oat milling, processing and storage. In: Welch, R.W. (Ed.), The Oat Crop: Production and Utilization. Chapman & Hall, pp. 369–408. Reproduced with permission of Springer Science and Business Media.*

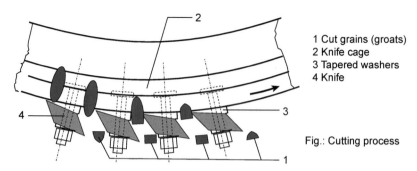

1 Cut grains (groats)
2 Knife cage
3 Tapered washers
4 Knife

Fig.: Cutting process

FIGURE 6.10 The mechanism of cutting in the granulator. *Reproduced by courtesy of Bühler A.G. Uzwil, Switzerland.*

Cutting

Steel-cut groats are the highest volume oat product for human consumption. The machine by which they are produced is the granulator, in which hollow perforated drums rotate on a horizontal axis, typically at 37–40 rpm. On the inner surface, the circular holes (typically 3.7 mm diameter) that pierce the walls are countersunk so that the elongated groats, having been fed into the drums, naturally fall lengthwise into the holes. As the drum rotates they continue to progress through the holes and project beyond the outer profile of the drum. Outside the drum, and parallel to its axis, a series of stationary knives is placed in such a position to execute a series of equidistant cuts on groats passing progressively through the holes, thus producing uniformly sized steel-cut granules (Fig. 6.10).

The holes are kept clear by pins projecting from smaller cylinders mounted parallel to the drums, so that the pins register with the holes at a point on the circumference where the drum is running empty.

The number of groat sections produced depends upon drum speed, but on average groats are cut into three parts. Not all groats are perfectly divided and, consequently, excessively long fragments and whole groats need to be removed, but first the total stock is aspirated to remove husk fragments and floury material. The quantity of fines produced depends on the sharpness of the knives – if sharp, 1.2%–1.3% fines may result but if dull up to 10% is possible (Deane and Commers, 1986). Following granulation, size selection is made with disc separators or trieur cylinders; the good product becomes the liftings and it is now ready for flaking. Oversize portions are returned to the granulator for retreatment.

The cutting process produces a small amount of flour known as 'flow meal'; it is separated by sieving and may be used for making dog biscuits.

Flaking is carried out by passing steam-cooked groats, or parts of groats, between a pair of flaking rolls. In the steamer the temperature is raised to

99–104°C and moisture content rises from 8%–10% to 10%–12%. Steaming performs two functions at this stage: it completes the inactivation of lipase and prepares the groats for flaking; they enter the roll nip while they are still hot, moist and plastic. Flaking rolls are much heavier and larger than standard flour milling rolls; the standard flaking rolls are 500 mm diameter and 800 or 1000 mm long and ideally are cooled with an internal coolant (Rohde, 2004). For rolls run at zero differential, and at 250–450 rpm, roll pressure and roll gap are maintained hydraulically. Flakes are dried, and cooled and passed over a final sifter and metal detector before packaging at a moisture content of about 10%.

The size of the groat or granule that is flaked determines the amount of domestic cooking that the product requires. The smaller the feed to the steam cooker, the greater the proportion of water absorbed and the greater the proportion of starch granules that become gelatinized. The more gelatinization, the less domestic cooking is required. Cooking requirements can also be controlled by varying roll pressures and steam cooking, as thinner flakes are more rapidly cooked. Instant or quick-cooking flakes are about 0.25–0.38 mm thick while traditional rolled oats may be 0.5–0.76 mm thick. The product represents 50%–60% w/w, of the whole oats processed. The composition of oat flakes is much the same as that of the whole groat as no separation is made of the endosperm, pericarp and embryo during processing.

6.3.4.2 Oat bran and oat flours

In recent years, a market for oat bran has developed on account of its high soluble fibre content (see Chapter 4). Because of their high lipid content groats are not suitable for conventional roller milling whereby fluted rolls enable concentration of large bran flakes by scraping off starchy endosperm; however, passing cut groats through roller mills followed by sieving does allow a concentration of branny particles to be achieved (Rohde, 2004). Alternatively, pin milling or hammer milling followed by sieving can be used. The high-fat content of the meal ground from oats has poor flow properties and aggregation occurs readily during grinding and sieving. Enhanced airflow is needed to ensure consistency in flow rate. The coarser fraction is richer in bran but to meet the American Association of Cereal Chemists definition of oat bran it must contain 16% total dietary fibre (dry basis) with one-third of the fibre being soluble and a β-glucan content of at least 5.5% (dry basis) (AACC, 1989).

The finer fraction of the ground meal is called oat flour. The method of flour production has an important influence on the viscosity of flour/water slurries produced in its later applications. Oat flour is used in baby foods and in production of extruded products, which are popular as ready-to-eat breakfast cereals.

6.3.4.3 White groats

This product, used for black puddings and haggis, is made by damping groats and subjecting them to a vigorous scouring action in a barley polisher, pearler or blocker before the moisture has penetrated deeply into the grains. Alternatively, the groats may be scoured without any preliminary damping. A proportion of the pericarp is removed by this process. There is also a certain amount of breakage, with the release of oat flour; the latter becomes pasted over the groats (no attempt being made to remove it), which thus acquire a whitened appearance.

6.3.4.4 Oat hulls

Oat hulls do not have a high monetary value but, when finely ground, can be included as a high-fibre constituent in diets of ruminants and rabbits. They can also be used in bedding for farm animals. A recent application of oat hulls has been as biomass fuel for power plants (Decker et al., 2014).

6.3.5 Buckwheat milling

The buckwheat species that is milled for food use is *Fagopyrum esculentum*. Tartary buckwheat (*Fagopyrum tartaricum*) has a bitter taste but it may be milled on a very small scale, for production of medicinal compounds (Biacs, 2002).

The fruit of buckwheat is known botanically as an achene. It has several similarities with the caryopsis of grasses but there is one major difference that relates to milling. When the term *hull* or *husk* is used in relation to cereals it is understood to mean the palea and lemma, which are not part of the actual fruit. In the case of buckwheat, the term *husk* or *hull* refers to the pericarp, which is part of the fruit.

The thickness and degree of adhesion of hulls of edible buckwheat varies with variety and, since the hull has to be removed as part of the milling process, the milling characteristics of buckwheat are also variable. The dehulled endosperms are known as groats and, as with the processing of other hulled grains, millers are able to choose from several types of machines to perform dehulling, including grinding stones, rollers and centrifugal impact machines. In preparation for passage through any of the dehullers it is advantageous to grade the feedstock according to size and adjust the machine parameters to suit each grade. Dehulling processes may be carried out on grains at their natural moisture content or following exposure to steam. According to Biacs (2002), breakages of groats resulting from 'dry' processing can leave as few as 30% intact, whereas up to 65% may survive when steam treatment is used. Other authors cite a process in which damping without steam is used, after which the dried grains are milled between rubber rollers and 50%–60% whole groat recovery is achieved (Keriene et al., 2016).

These authors found that aflatoxins resulting from the fungal infection *Aspergillus* spp. were reduced in the products of steamed grains. While most of the aflatoxin was associated with the pericarp, its control is important because of the uses to which hulls are put. As well as being used as stock bedding and feed, they are also used for filling pillows and mattresses. Another use of hulls is as a biofuel, which of course is not affected by aflatoxin.

Conditions for steaming have been described as damping to 22% m.c. with water or steam followed by heating (roasting or steaming) to 150–164°C for 10–20 min (Dietrych-Szostak and Oleszek, 1999). During subsequent cooling, hulls split and can be easily removed.

Following dehulling the products are sieved so that hulls and flour and wholegrains requiring further dehulling are separated from the successfully processed groats, which are then milled to produce a straight run flour or a range of flours differing in protein (4.69%–42.87%) and polyphenol (0.46%–7.45%) content (Biacs, 2002).

Flaking, following steaming, converts flours into a suitable form for inclusion in cereal preparations for infant food or breakfast cereals. Flours are also used in pancakes, biscuits (cookies) and extruded products, either at ambient temperature, into noodles or at elevated temperatures into expanded products. These products are dependent on the starch that is present in the flour. The protein is of good nutritional quality but it does not form a gluten on hydration and cannot be used successfully as the only flour in aerated products such as bread. Soba noodles contain a high proportion of buckwheat flour but their texture is improved by inclusion of some wheat flour.

6.3.6 Quinoa milling

World-wide interest in quinoa has increased enormously and it has been promoted, by responsible authorities, as an attractive, healthy and nutritious source of energy and good-quality protein. A comprehensive review of all aspects of quinoa has been published and it is available online (Bazile et al., 2013). Increased international demand has encouraged a trend toward industrial processing in areas of the world where small-scale or domestic processing is the tradition. One example cited for an installation in South America led to an increase of 800% in processing capacity with only an 80% increased demand for electricity and a 36% increase for water.

Water can play an important part in milling of quinoa, but its use is more as a pretreatment made necessary by the presence of saponin in the fruit coat and seed coat (Reichert et al., 1986). There are alternative dry methods for removing saponin and methods that combine wet and dry stages.

In addition to water treatment and abrasion, one traditional method includes a preliminary 30–40 min roasting stage, the effect of which is to make the outermost layers more fragile. While still warm the roasted grain is mixed with abrasive clay and trodden on a rough surface for 30–60 min. Winnowing for 20–40 min is followed by washing for 25–35 min. Drying for 2–4 h is the final process. It is not known what proportion of the original saponin remains, but in many of the traditional methods the absence of foaming in wash water is used as an easily recognized point at which sufficient saponin has been removed.

Several strategies have been employed in developing industrial processing methods most of which involve a combination of washing and pearling. As with all industrial processing, grain samples are cleaned before processing, however, the preliminary cleaning appears not to be sufficient as further removal of foreign matter (e.g., stones and straw) occurs at several stages during the processing. A method developed by the Sustainable Technologies Promotion Centre in Bolivia, which is now said to be applied to 70%–80% of the Bolivian organic 'Quinoa Real' production, involves a horizontal cylinder in which a ribbed rotor revolves at 1200–1600 rpm. Cleaned grain is fed into the gap between the rotor and the cylinder providing intense friction between grains. This mutual attrition is said to lead to uniform removal of episperm. The lower part of the drum is perforated to allow extraction of abraded fragments known as saponin powder, while the pearled grains are retained and they progress through the cylinder until discharged. At this stage of processing 90%–95% of saponins have been removed.

After picking to remove foreign matter the pearled grains are soaked and washed, followed by a second picking, rinses and centrifugation. The grains are dried in warm air, and at this point saponin content is 0.01%. A granulometer sorter (as described above for oats) selects wholegrains, which may then be packed and shipped or rolled to flour or flakes.

It is also possible to obtain a flour that is acceptable to some consumers by removing saponins through prolonged washing followed by drying and milling using, for example, a hammer mill to produce a wholegrain flour (Taylor and Parker, 2002).

6.3.7 Grain amaranth milling

Most reports of dry milling of amaranth grains, and the uses of the resulting flours, are of experimental exercises, and there is no indication that these have been adopted on an industrial scale. The purpose underlying the experimental work is the popularization of grain amaranth as a nutritious food throughout the world, particularly in areas where food of sufficient nutritious value is scarce. Should the campaigns result in widespread success, dry milling may become widespread as several types of milling and uses of flour appear to show promise.

In spite of the small size and asymmetric shape of the amaranth grain, the use of a specific type of barley pearler has produced stocks that were separable into starch-rich and protein-rich fractions representing concentrations of the embryo and perisperm, respectively. Roller milling has also been applied and even scaled up to a 'technical-scale' roller mill. Four fractions were obtained by separation on the basis of particle size. While a wholemeal flour from the same sample had a protein content of 15.25% and a starch content of 67.35%, the fraction with the highest protein content (19% of total products) contained 15.37% protein and the fraction with most starch (38% of total products) contained 70.31% starch (Berghofer and Schoenlechner, 2002). Production of wholemeal flours is possible by hammer milling and other single-passage-type mills.

6.4 DRY MILLING IN WHICH THE MAIN PROCESS IS ROLLER MILLING

6.4.1 Developments in the history of roller milling

In ancient times hulled barleys and wheats had to be dehulled by pounding in a mortar prior to being ground into meal. The action of the pestle stripped off the tough glumes of wheat and the adherent lemma and palea of barley without breaking the grains (Nesbitt and Samuel, 1996).

In the Near East saddle querns were used for grinding grain following dehulling, long before and after the beginnings of farming, and they continue to be used for grinding cereals in sub-Saharan Africa today. A saddle quern is an elongated flat slab ('bed-stone' or 'metate'), often made of stone, stationed on the ground or in an emplacement, or may even be a hollowed area of a much bigger rock. The grinding action is by a smaller hand stone ('mano') being drawn back and forth axially on a bed of grains lying on the slab. A development of the saddle quern is the 'hopper-rubber'. Grinding with the hopper-rubber mill involves a similar action to that used with the quern, but the upper stone is adapted to continuously feedstock onto the grinding surface from a depression or hopper at its centre to hold a small supply of grain, and a hole or slot at the bottom through which the grain falls. Grooves cut in a herringbone or other pattern increase the abrasive friction of the surface. The lever mill is an adaptation of the hopper-rubber in which one end of the hopper stone is attached to a lever arm and the other to a fixed pivot so that, through moving the lever, the operator can impose a backwards and forwards motion in an arc.

A major development came with the introduction of the rotary mill, an example of which is the quern. Querns consisted of two stones – the upper 'capstone' being rotated over the stationary 'netherstone'. Grain was introduced through a hole in the centre of the capstone and it became ground by the abrasive action of the two stones, as it was worked toward the outside, to be collected as a coarse meal. The rotary nature of the process enabled a system that had previously depended entirely on human labour to benefit from the efforts of draught animals. The hourglass mill ('mola asinaria – donkey mill' to the Romans) is an example of such a mill (Fig. 6.11).

Animal power gave way to water power about 2000 years ago, and watermills became common in the medieval period. The dome-shaped stones of the earlier rotary mills gradually evolved into flatter, horizontal millstones. Thereafter, until the development of the roller mill in the mid-19th century, wheat was ground by stone milling. For a comprehensive coverage of mill developmental history, see Lucas (2006).

The stones at the heart of a watermill or a windmill are two discs of hard, abrasive stone, some 1.2 m. in diameter, arranged on a vertical axis. Types of stone used include French burr from La Ferté-sous-Jouarre, Seine-et-Marne, millstone grit from Derbyshire in England, German lava, Baltic flint from Denmark, and an artificial stone containing emery obtained

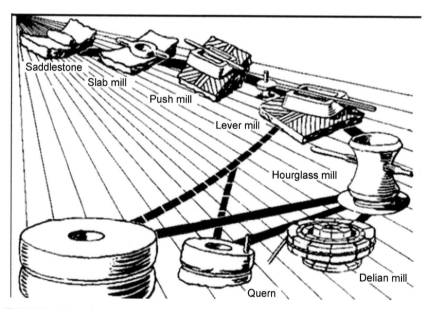

FIGURE 6.11 Types of grinding devices shown in the order of their occurrence (from top left) and possible developmental routes to the stone mill (*heavy lines*). Slab mill and push mill are types of hopper rubber mills. *Reproduced with permission Storck, J., Teague, W.D., 1952. Flour for Man's Bread. Univ. of Minnesota Press, Minneapolis.*

from the island of Paxos, Greece. The opposing surfaces of the two stones, which are separated only by a small gap when operating, are patterned with series of grooves leading from the centre to the periphery. In operation, one stone is stationary while the other rotates. Either the upper stone ('upper runner') or the lower stone ('under runner') may rotate, but usually the former.

Grain fed into the centre ('eye') of the upper stone is fragmented between the two stones, and the ground products issue at the periphery ('skirt').

In western Europe the local soft wheat was ground by a 'low-grinding' process, in which the upper stone was lowered as far as possible toward the lower stone, thereby producing a heavy grind from which a single type of flour was made in a single passage. In 18th-century France and 19th-century Austria, the milling of flours of more than one quality became possible through use of a 'high-grinding' system, in which the upper stone was slightly raised. The resulting gritty material became an intermediate stock from which, by further treatment between stones separated by progressively smaller gaps, flours of diverse quality could be made.

Hard wheat, from the Danube basin, was ideally suited to the gradual reduction achieved through the high-grinding system. Hard wheats were also characteristic of North America and the new process was rapidly adopted in the United States. In the area around Minneapolis and St. Paul, Minnesota, all mills used the new system by 1880. In fact, this area became the premier wheat milling centre of the Western hemisphere (Perren, 2000).

Hungary became the centre of the milling industry through its development and enthusiastic adoption of the roller milling system, and eventually systems evolved in which all stone grinding stages became replaced. Experiments with roller milling began by a Swiss pioneer as early as 1815 and continued in Switzerland, apparently independently of the more widely acknowledged developments in Hungary, culminating with a patent granted in 1876 (Ulmer, 2011).

Roller milling requires less power than stone grinding, but the higher initial power threshold could not always be reliably overcome by water or wind power. Hence the adoption of steam power made a vital contribution to the implementation of the roller system. The adoption of industrial-scale processing that roller milling facilitated was encouraged by the concurrent improvement in transport by water and particularly rail systems. A summary of the early history of flour milling in the United States can be found in Sharrer and Welsh (2003). The years 1878/79 saw the building of the first roller mills in both America and Great Britain, but wider adoption of roller milling in Britain was initially slow. Competition in the form of imported higher quality flours, however, soon led a few British roller milling pioneers to set an example of what could be done with roller milling, which was then followed. The British flour miller was faced with processing softer home-grown wheats as well as imported harder types and consequently evolved special skills and adaptations related to dealing with such mixtures. A full account of the transition from stones to rolls, with special reference to the British industry, is given by Jones (2001).

The gradual reduction system, which started with high grinding with stones and continued with greater sophistication using roller milling, involved many retreatments of stocks. As well as the grain and final products, a range of intermediate stocks had to be automatically conveyed from one treatment to another, and the process could not have developed satisfactorily without the support of other handling techniques. Bucket elevators provided a solution and they are still used in some situations; however, new and remodelled mills are almost always equipped with pneumatic conveyers. Although pneumatic systems were patented in 1909, and used for loading and unloading ships and barges, they were not installed in flour mills until the end of World War II (1945). The development of the purifier, whereby small bran particles are removed from ground stocks, also contributed to the economics of the roller-milling process as valuable stocks that could not be recovered by sieving could be further ground for inclusion in the flour. The early multistage mills consisted of many machines, all driven via belts from pulleys that were themselves driven for a single prime mover. Speeds were controlled by the relative sizes of pulleys on shafts and machines, with ratios being calculated prior to installation. This has now been superseded by systems in which each machine is powered by a dedicated electric motor and controlled by a central electronic system.

The previous description of the development of milling relates mainly to the grinding of wheat, but many of the historical techniques and modern machines are reflected in the processing of maize. Although the pestle and mortar are thought to have been used mainly for dehulling of temperate cereals, indigenous peoples of North and South America used a similar apparatus for grinding of maize grains in ancient times, and it is still used today in parts of Mesoamerica. In eastern North America the hominy block was used, fashioned from a hollowed-out tree stump as a mortar, with the springy limb of another nearby tree serving as a pestle. Prior to pounding, the maize grains were boiled in a solution of ashes and water to remove the tough pericarps. The name 'hominy' was applied to a coarse-ground maize meal mixed with milk or water, and it persists today in names of maize products treated with lye, e.g., 'hominy feed' and 'hominy grits'. Information on the route from the hominy block to stone milling and onwards to roller milling of maize is less readily available than that covering wheat processing.

The sophisticated roller mills of today are suited to industrial societies. However, there are many settings in which processing occurs at a domestic level, and it should not be forgotten that, in some parts of the world, the pestle and mortar remains a vital means of decorticating and grinding of cereal grains, some of which are too small to be processed by industrialized methods, in the quest to produce a major ingredient for a nutritious meal.

It would also be wrong to think that stone grinding has been entirely replaced by roller milling. As well as the working wind- and watermills preserved throughout the world, many roller-milling plants are equipped with facilities for stone grinding, albeit in many cases, using electrically driven stones considerably smaller than the traditional stones, to meet the needs of specialist customers such as consumers of 'organic' foods. Composition of stones varies, and in some modern stone mills the grinding elements are disposed vertically. The entire grinding process may be completed in one pass or a scheme involving rolls for the break stages and discs for reductions, may be followed.

6.4.2 Modern disc mills

Examples of modern milling machines in which stocks are ground between two circular faces are those produced by Engsko United Milling Systems. As with stone mills, one grinding surface rotates and the other is stationary. The grinding elements consist of tiles secured to the periphery of metal discs, the outer of which rotates on a horizontal axis providing a grinding surface between it and a stationary disc. The tiles are made of wolfram (tungsten) carbide and have a pattern of sharp parallel ridges that are disposed almost radially (they are actually parallel to one edge of the tile and the choice of left or right edge influences the milling characteristics). Tiles are arranged in two concentric rows with different ridges. The coarser ridged tiles form the inner ring so that stocks, which pass centrifugally, encounter them before receiving a finer grind by the more finely ridged tiles. The tiles are tapered, with the outside thicker than the inside so stock passes through a narrowing gap between grinding faces (Fig. 6.12).

6.4.3 Modern roller milling

This outline applies mainly to wheat milling but the general principles are relevant to processing of other cereals. The modern roller mill is designed to subject cereal grains and their ground products to a series of grinding treatments that alternate with classifying stages, whereby

(a)

Inlet

Stationary disc

Rotary disc

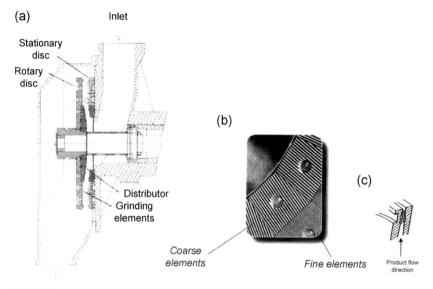

(b)

(c)

Distributor
Grinding elements

Coarse elements

Fine elements

Product flow direction

FIGURE 6.12   Engsko disc mill (a) section through grinding chamber, (b) disposition of tiles at the periphery of a disc, with coarse inner ridging on inner row and finer ridging on outer row, (c) diagram showing taper in grinding area. *Reproduced with permission of Engsko A/S – United Milling Systems.*

stocks are separated according mainly to their sizes. The resulting fractions are conveyed either as intermediate stocks for further rolling treatments or as final products. The initial grinding stages are known as 'breaks' as they break open the grain. After the first break and sieving of products, the larger fractions consist of flattened bran flakes with adhering endosperm, they are conveyed to further break rolls for removal of more endosperm and, in the process, they become smaller flakes. Ultimately, the bran is sufficiently cleared of endosperm and becomes a final product, 'bran'. The smaller particles separated by sieving after each breaking stage, comprise mainly endosperm. They are then conveyed for progressive reduction to flour size, and separation from any remaining bran and germ particles.

6.4.4 Roller mill equipment

Although there are many manufacturers of milling equipment, standards have evolved for several aspects of mill design, thus roll dimensions are usually 1000 or 1250 mm long and 250 mm diameter (some reduction rolls may be 300 mm diameter). Rolls may be cast in vertical moulds comprising a solid cylinder within a larger hollow cylinder. Molten metal is poured into the annular space between the two elements, giving rise to a hollow cylindrical roller. A technically more advanced method is the centrifugal casting method whereby a horizontally positioned permanent mould rotates at high speed while molten metal is poured into it, uniformly along its length. Centrifugal force distributes metal around the inside of the cylinder and it solidifies as it is cooled. Two layers of metal with different characteristics can be applied, with the harder one being introduced first, and hence forming the outside of the cast, thus providing the grinding surface. The inside surface is machined to produce a uniform thickness and to remove any impurities, which concentrate at the inner surface. Engineers have developed special techniques for increasing hardness and durability of rolls, such as subjecting them to rapid chilling after casting. For this reason, they are sometimes referred to as 'chills'. The cast rolls are subjected to further machining to ensure perfect circularity, and the required surface characteristics are applied. Outer surfaces of rolls may be smooth, frosted or corrugated (fluted), and differences are not solely related to the surface profile, as smooth surfaced rolls benefit from a softer surface than that required for fluted rolls. Fluted rolls are appropriate for the breaking stages of the milling process. The profile of part of a fluted roll is shown in Fig. 6.13.

While the general features of fluting are universal there are differences in angles and relative dimensions available, allowing millers to select according to the type of grain being milled. Also flute sizes and hence numbers of flutes per unit length around the circumference vary according to the stage of the milling flow for which the rolls are intended. The rolls with the largest flutes are deployed on first break and flute size diminishes toward later breaks.

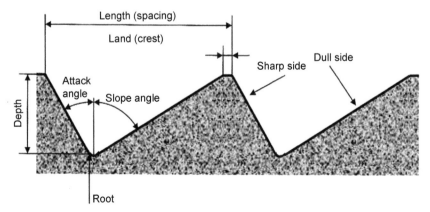

FIGURE 6.13 Corrugation (flute) elements on the roll surface. The profile of each flute resembles an italic V, with one side shorter (sharp side) than the other (dull side).

It is clearly desirable that rolls run parallel one to the other when under load so that all the stock being processed is uniformly subjected to pressure. Rolls are supported by bearings only at the ends and consequently there is a tendency for longer rolls to distort in such a way that they are further apart around the centre. To counteract this tendency, they are finished with a cambered profile in which their diameters at the ends are less than that in the centre with a gradual taper between the extremes. Those rolls that are expected to be subjected to most pressure are limited in length and finished with the greatest taper. Fluted rolls are not subjected to the heaviest pressures and hence they are given a less-pronounced taper (15–20 µm difference between diameters) than smooth rolls (up to 75 µm) (Ulmer, 2011).

Rolls are always operated in pairs and are located within roll stands (Figs. 6.14 and 6.23).

6.4.4.1 Roll stands

Roll stands usually consist of two mills back to back, within the same housing but operating independently and on different feed-stocks. Most roll stands accommodate two pairs of rolls, but in recent years roll stands accommodating four pairs have been incorporated in many mill flows.

The rollers may be disposed diagonally or horizontally. The ground products of each pair are conveyed to a sieving facility for fractionation, and the separated stocks are routed to their next appropriate stage of treatment.

Materials to be ground are introduced, to each side separately, at the top of the stand. A sight glass allows the condition of flow to be monitored visually. Above each pair of grinding rolls is a device that ensures even and consistent feeding of particles into the nip along the length of the

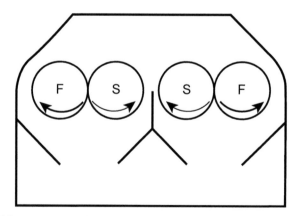

FIGURE 6.14 Disposition of rolls in roller mill stands. *S*, slow; *F*, fast.

rolls. The nature of the device varies according to the type of stock; it may be a single-feed roll for wholegrains or a double-feed roll for finer granular stocks. For feeding voluminous stocks that do not flow easily, such as partly treated bran, a feed roll and a distribution paddle are appropriate.

Pairs of rolls in all modern mill stands are mounted in cassettes that can be easily inserted and withdrawn when the need for replacement arises as a result of wear. Within the pair, rollers rotate toward each other, with the outer roll rotating at a greater speed than the inner. This brings about a situation whereby the slow roll tends to secure grains while the fast roll scrapes them. The difference in roll speed is known as the differential speed or just the differential.

In an 8-roll mill stand the pairs of rolls are disposed one above the other on each side of the stand so that stocks ground by the upper pair fall into the nip of the lower pair. This system, in most cases, means that the entire stock from the first grind is further ground without any selection having taken place, however, at least one manufacturer (Ocrim RMQX double-high roller mill) has incorporated rotary sieves between the two passages so that only the appropriate fraction undergoes further grinding, with remaining stocks being diverted to an appropriate different fate.

Rolls may be smooth or corrugated according to their role and the stage in the process where they function. The corrugations on fluted rolls do not run parallel to the long axis but instead undergo a spiral pattern so that a scissor action results as shown in Fig. 6.15(b).

The angle of the spiral is expressed as a percentage determined as the circumferential displacement of a flute related to the linear distance along the roll axis (Fig. 6.15(a)). Typical values are 4%–14%. For wheat milling spirals toward the lower values are usual.

As the flutes are asymmetric (Fig. 6.13), the rolls can be run with either the steep (sharp) or shallow (dull) profile disposed toward the nip. The relationship between rolls may thus be described as dull-to-dull, sharp-to-sharp, dull-to-sharp or sharp-to-dull. It is conventional to give the fast roll disposition first in such descriptions (Fig. 6.16).

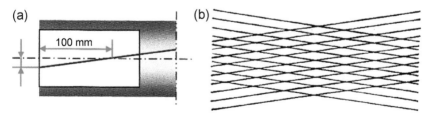

FIGURE 6.15 Fluted rolls. (a) Measurements defining the angle of the spiral. (b) Diagrammatic example of interaction of flutes at the point of grinding.

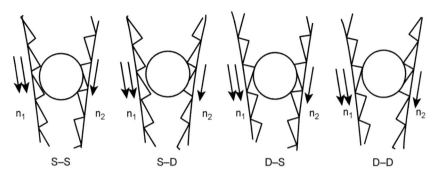

FIGURE 6.16 Four types of roll disposition: sharp to sharp (S–S), sharp to dull (S–D), dull to sharp (D–S), dull to dull (D–D). *Reproduced from Fang, C., Campbell, G.M., 2003. On predicting roller milling performance IV: effect of roll disposition on the particle size distribution from first break milling of wheat. Journal of Cereal Science 37, 21–29, with permission.*

Roller milling is a continuous process and as rolls and stocks are subjected to high pressures these factors lead inevitably to an increase in temperature. High temperatures can be damaging to both equipment and stocks, and for this reason the temperature of those rolls subject to the highest pressures is controlled by the passage of cooling water through the hollow centre.

Another issue arising from continuous running and high pressures is the adhesion of some of the stock being processed to the surface of rolls, leading to blocked flutes in the case of corrugated rolls and irregular increase in effective diameter of smooth rolls. The surface adhesions are removed by scrapers mounted beneath smooth rolls. However, these are not appropriate for corrugated rolls, for which strong-bristled brushes are employed.

6.4.4.2 Decortication

For many years it was considered that decortication would bring no benefits to the roller milling system. However, the inclusion of a preliminary decortication stage, using a modified rice pearling mill, has now been adopted with advantage in many mills. The system is flexible in that the amount of surface material removed can be determined by the severity of the treatment or the number of passes selected. Advantages derive from several factors, some of which are listed here:

- Bacterial, fungal and other contamination associated with the outer layers of grain is removed early in the milling process so that it does not enter flour or semolina, which derive from the inner parts of the grain (Posner, 2009).
- Enzymes associated with germination are concentrated, at an early stage of sprouting, in the outer endosperm layers. Decortication can remove these layers and hence grain samples that would be rejected for conventional milling might be acceptable if decorticated.

- Grains in which no germination enzymes are present can be decorticated in such a way as to remove bran layers but to retain aleurone. The aleurone can then be included in resulting flours, thus improving their nutritional qualities (Atwell et al., 2007).
- A reduced-length roller milling system is adequate for grains that have been decorticated. The coarsely fluted rolls of the conventional first break are not required and they can be substituted with sizing rolls (Posner, 2009).
- Higher extraction rates or flour and semolina with less bran contamination may result when grains are decorticated before milling.

6.4.5 Classifying during the milling process

As stated before, the milling process consists largely of alternating grinding and classifying operations. For the most part, classifying involves sieving. Separation by particle size is effective because, during roller milling, endosperm tends to fragment into small, relatively isodiametric particles while nonendosperm tissues tend to remain or become thinner and flatter and less reduced in size.

All modern plants use plansifters, which consist of a series of flat rectangular-framed sieves held within chests that are driven in a gyratory motion. The chests are suspended on flexible 'canes' in such a way that they can gyrate with stability. Eccentric motion is assured by strategic placement of weights. Many streams of intermediate stocks are sifted and they may come from different roll stands, but sifters are located together and the stocks are conveyed to and from them (Fig. 6.17).

Sieve frames are stacked, like drawers, within the chests, and channels exist so that particles are fed to the top of the sieve appropriate to their sieving requirements. For each sieve two fractions are separated, the particles that pass through (throughs) and those that are too big to pass through (overtails). The stacking of sieves allows for replication of the same sized mesh on a number of frames to accommodate large quantities of stock, as well as for different sized meshes to allow stocks to cascade down the stack for a series of separations. Each final fraction is directed to an appropriate outlet at the base of the chest and thence to the site of further treatment (Fig. 6.18). There are different shapes and sizes of sieves and plansifters, but where square chests are used, up to 10 of them can be accommodated in the same group.

6.4.5.1 Sieve covers

Covers for coarser sieves are constructed from stainless steel fabric with a simple weave. The thickness of the wire used influences

FIGURE 6.17   Plansifters in a modern mill. Each unit comprises eight cabinets, each containing a stack of sieves. *By courtesy of Bühler A.G. Uzwil, Switzerland.*

the 'open' area so thin wires are used to maximize this characteristic. Smaller aperture covers are made from nylon mesh in which the weft and warp are of monofilament yarn. Several gauges and weaving patterns are available, each producing its own open area and roughness characteristics (Ulmer, 2011). Sieve characteristics have been standardized by commercial companies and international and national standards organizations (e.g., Tyler, ASTM, BS, DIN) but relationships are not obvious without reference to comparison tables, several of which can be found online (e.g., http://www.starch.dk/isi/tables/screens.htmhttp://delloyd.50megs.com/moreinfo/mesh.html).

Compared with the traditional fabric for smaller aperture sieves, nylon is less expensive, less subject to change with variations in atmospheric moisture and stronger than silk, allowing thinner yarn to be used and hence increasing the open area of sieves.

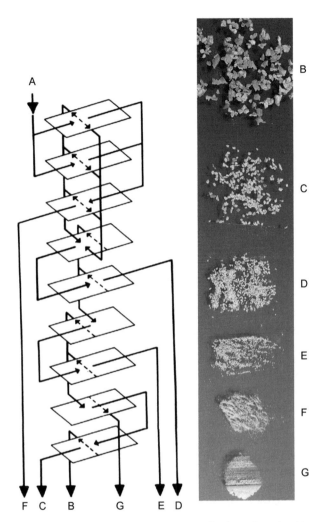

FIGURE 6.18 Diagram showing typical flow of stocks through a plansifter stack where A, inlet and other letters indicate stocks leaving sieve cabinet; B, break coarse; C, break fine; D, semolina coarse; E, semolina fine; F, middlings; G, flour. The relevant stocks are shown at right. *By courtesy of Bühler A.G. Uzwil, Switzerland.*

Sieves are prevented from becoming blinded by shaped, flat, plates of plastic or other material, supported on a grid beneath each cover and able to move freely in a horizontal plane. As a result of the gyratory motion of the sieve frame, the agitation and percussion that they experience are transmitted to the screens above and thus particles are dislodged from the apertures.

(a)   (b)

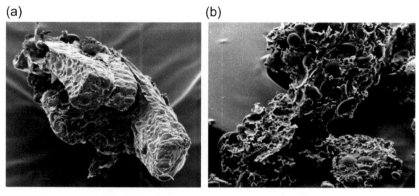

FIGURE 6.19 Scanning electron micrographs of particles of hard (a) and soft (b) wheat endosperm of flour size.

The efficiency of separation by sieving is dependent on several factors, some of which are taken into account in the design of the sieving devices. However, there is one that depends on how the mill is operated, more specifically in the allocation of sifting surface to a stock. When the sieve is optimally loaded, the gyratory motion of the sieve creates movement of stock over the apertures. This, with the friction created, encourages particles of small enough dimensions to pass through, and the weight of the stock above the small particles increases the force that drives them through. Another factor that assists sieving is stratification, the tendency of agitated particles to assort according to their density and size, with the denser, smaller particles thus working to the bottom, and hence the sieving surface, as the stock moves across the sieve space. When the condition of optimal loading is not met, these functions do not occur, or occur only to a limited extent. Therefore, attention to allocation of sieve surface is important at all times.

In regard to sieving, flour particles behave differently according to the texture of the endosperm in the parent cereals. As well as breaking down more readily during grinding, soft endosperm tends to fragment randomly and to release a large number of starch granules from the fenestrated protein matrix. Harder endosperms have a more coherent protein matrix, which is more resistant to breakage. Consequently, hard endosperm tends to fracture along cell boundaries. Examples of hard and wheat endosperm particles are shown in Fig. 6.19. The smoother surface of the hard endosperm particles and the fewer free starch granules contribute to the ease with which hard endosperm particles stratify and pass through sieve apertures.

6.4.5.2 Purifiers

Purifiers provide another important means of classifying milling stocks. Whereas particle separation on sieves is dependant almost entirely on size, separation on purifiers is more complex, depending on both size and terminal velocity.

A purifier consists of a long oscillating sieve, at a slight downward slope from head to tail, which is divided into four or more sections. The cover becomes progressively coarser from the head to tail. Individually controlled air currents rise through the cover, aspirating stocks as they move over the oscillating sieve cover. When particles of similar density but different particle size are shaken together, the heavier ones tend to sink. Grading is effected on the basis of size, density and air resistance. Generally, the more bran and less endosperm there is in a particle of a given size, the lower its density and the higher its air resistance will be. Stratification thus results in the following order from the top:

- light bran and beeswing
- heavy bran
- large composites
- large pure endosperm particles
- small pure endosperm particles

Successful purifying depends upon maintaining a continuous layer of pure endosperm along the entire length of the sieve. Thus, only the finest pure endosperm must be allowed to pass through at the beginning, and progressively larger endosperm particles are allowed through as the stocks progress. The lightest impurities are removed by aspiration while overtails are directed to appropriate further treatment.

The elements of construction of a purifier is shown diagrammatically in Fig. 6.20.

Within a purifier housing there may be two or three screen decks allowing a sequence of purification stages as the denser particles (with the higher value) cascade down. Purifier screens are usually woven from polyester yarn because it produces a rougher surface due to its lesser flexibility then nylon. The upward flow of air lifts the bed of stock as it proceeds across the screen thus reducing wear on it.

Purifiers are widely used in roller milling systems but they are particularly characteristic of durum wheat milling to produce semolina. They are less effective with the products of soft wheats.

Purifier allocation is expressed in mm of purifier width to 100 kg/24 h.

Consideration of factors affecting purifier performance is given by Fowler (2013).

6.4.5.3 Flake disrupters

Flakes of endosperm may form during the milling process. Some stocks are more prone to flaking than others and flaking is also more likely to occur during grinding between smooth rolls at high pressure. Such conditions may be applied intentionally, for example, in order to achieve high

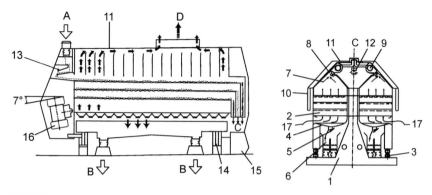

FIGURE 6.20 Schematic diagram of a purifier showing working principles in side and end elevations. 1, Machine frame; 2, sieve box; 3, hollow rubber springs; 4, collection hopper for throughs; 5, deflection gates; 6, collection channels; 7, air flow control; 8, individual air gate; 9, air duct; 10, cover; 11, air hood; 12, machine lighting; 13, feed device; 14, support of collection channel; 15, outlet box; 16, eccentric motors; 17, air aspirator. *Reproduced from Ulmer, K., 2011. Technology and Equipment Grain Milling. Bühler, Uzwil. http://www.buhlergroup.com. By courtesy of Bühler A.G., Uzwil, Switzerland.*

levels of starch damage (see following discussion). Flakes produced in this way often comprise pure or nearly pure endosperm and consequently it is desirable that they should be classified as flour; however, their size prevents them from passing through sieve apertures suited to the pure endosperm fraction and if left untreated the flakes would be classified with the branny material. The problem is solved by introducing flake disrupters into the mill flow where they can be installed in the pneumatic conveying system. Each consists of a pair of rotors mounted on a vertical axis and contained within a housing. The rotors consist of two discs held apart at the periphery by a gallery of pins fixed perpendicular to the faces of the discs. The feed enters at the centre of the plates and is thrust outward by centrifugal force. Flakes are broken by impact against the pins, the housing and each other. A gentler type of flake disrupter is the drum detacher. It comprises a cylindrical metal drum inside which four beater bars, parallel to the long axis of the cylinder, are connected by legs to a central spindle, and when they rotate they are close to the inside wall of the cylinder. Stock enters the cylinder at one end and particles are thrust outwards to impact its inner wall. The profile of the beaters' outer edge is designed to generate maximum impact at the same time as conveying stock to the outlet end of the cylinder. Flake disrupters do not perform any classification function.

6.4.5.4 Bran finishing

In each grinding stage the bran becomes progressively cleaner until, following the last break, it can yield no more endosperm through further grinding. It may become the finished by-product 'bran' at this stage or it may be subjected to treatment in bran finishers. A bran finisher consists essentially of a hollow cylinder, two-thirds of which is perforated. Scalpings are fed into the horizontally disposed cylinder, and finger beaters attached to a central shaft rotate at high speed and propel the bran skins against the perforated cover. Some of the remaining endosperm is rubbed off and passes through the apertures. Clean bran flakes overtail the machine. The throughs contain endosperm, but also small fragments of bran are present and they require further sieving. Because they are 'greasy' and would blind plansifters, they are dressed on special sifters such as vibratory sifters.

6.4.6 Milling of common (bread) wheat

Wheat flour is consumed by farm animals, fish and humans, and for all these purposes grinding the wholegrains into a meal or flour is featured. The description that follows is essentially for producing white flour for human consumption in a range of cooked food types. Different types of white flour are required, and in general it is the type of wheat chosen as feedstock rather than the method of milling that determines the suitability of the flour for the product. In spite of this, the milling system may need to be adapted to provide conditions suited to the type of wheat. This applies particularly to variation in endosperm texture (hardness). An indication is given in Fig. 1.37 of appropriate properties for the more important applications. Many white flour products can also be made from meal in which all grain components are present in their natural proportions (wholemeal), or from a range of mixtures of those components (e.g., brown bread). The character of the products reflects the composition, thus offering a range of flavours and textures from which consumers can choose. The prevalence of white or mixed flour products in a given region often reflects long-established preferences. In the United Kingdom about 75% of bread is made from white flour (NABIM, 2014).

Wheaten white flour has been defined as the product prepared from grain of common wheat (*Triticum aestivum*) or club wheat (*T. aestivum* ssp. *compactum*), by grinding or milling processes in which the bran and embryo are partly removed and the remainder is comminuted to a suitable degree of fineness. Some flour is produced in the milling of durum wheat (*Triticum turgidum* spp. *durum*), but it is not the main product.

The objectives in milling white flour are:

1. To separate the endosperm, which is required for the flour, from the bran and embryo, so that the flour shall be free from bran specks, and of good colour, and so that the palatability of the product shall be improved and its storage life extended.

2. To reduce the maximum amount of endosperm to flour fineness, thereby obtaining the maximum extraction of white flour from the wheat. 'Flour fineness' is an arbitrary particle size, but in practice, most of the material described as white flour would pass through a flour sieve having rectangular apertures of 140 μm length of side.

Table 6.3 shows the relationship between the components of the grain and the end product into which an 'ideal' milling process directs them.

The crease or furrow of the wheat grain is discussed in Chapter 3. It consists of a reentrant region on the ventral side of the grain into which pericarp and testa penetrate, making them inaccessible to processes that remove surface structures by abrasion.

The modern roller-milling process has been designed to solve the problem of the crease by employing a multistage combination of shearing,

TABLE 6.3   Grain components and their destinations in a 'perfect' white flour milling system

Grain component	Mill fraction
Grain (caryopsis)	
(1) Pericarp (fruit coat)	
(a) Outer epidermis (epicarp), Hypodermis, Thin walled cells – remnants over most of grain, but cell walls remain in crease and attachment region; includes vascular tissue in crease	Beeswing
(b) Intermediate cells, Cross cells, Tube cells (inner epidermis)	Bran
(2) Seed	
(a) Seed coat (testa) and pigment strand	
(b) Nucellar epidermis (hyaline layer)	
(c) Endosperm, Aleurone layer, Starchy endosperm	White flour[a]
(d) Embryo, Embryonic axis, Scutellum	Germ

[a] In the case of durum milling, 'semolina' should be substituted for "white flour" in the table.

Kent and Evers (1994).

scraping, crushing and grading of stocks to exploit the differences in mechanical properties between the starchy endosperm, bran and embryo. It is essential, in effecting the required separation, to minimize the production of fine particles of bran, and this basic requirement is responsible for the conditioning process described in Chapter 5, the complex arrangement of modern flour-milling systems, and for the particular design of the specialized machinery used. The main stages in the milling of wheats, whether durum or common wheats, are grinding on roller mills, sieving and purifying.

The modern roller-milling process for making flour is described as a 'gradual reduction process' because the grain and its parts are broken down in a succession of relatively gentle grinding stages. Between grinding stages on roller mills, stocks are sieved and the various fractions conveyed for further grinding if necessary. Although stocks are reground, they are never returned to the machine from which they came or any machine preceding it. Another important principle is that flour or wheatfeed made at any point in the process is separated out as soon as possible (this principle is defied in some 8-roll mill stands as described earlier).

Those fractions that can yield no useful flour are removed from the milling system to contribute to the milling by-products (offals) as bran or 'wheatfeed' ('millfeed' in the United States, 'pollard' in Australia). Flours produced at all stages are also removed from the system to contribute to the finished product.

Essentially the process comprises two stages, a break or breaking stage and a reduction stage. The role of the break system is to separate most of the bran (including germ) in as large flakes as possible; the reduction system reduces endosperm particles to flour fineness, at the same time removing any bran and germ not removed during the break stage. The schematic diagram in Fig. 6.21 shows a simplified flow. The scratch system shown is not universally included but it can increase flour extraction when high levels of starch damage (see following) are not required. It involves the use of finely fluted rolls applied to large chunks of endosperm.

6.4.7 Explanation of terms

6.4.7.1 Stocks and materials

The blend of wheat types entering the milling system is known as the grist. Its composition is usually expressed in percentages of each wheat type present. It is also described by reference to the product for which it is intended, e.g., a bread grist.

Other stocks are described as follows:

- Feed: material fed to, or entering, a machine.
- Grind: the whole of the ground material delivered by a roller mill.

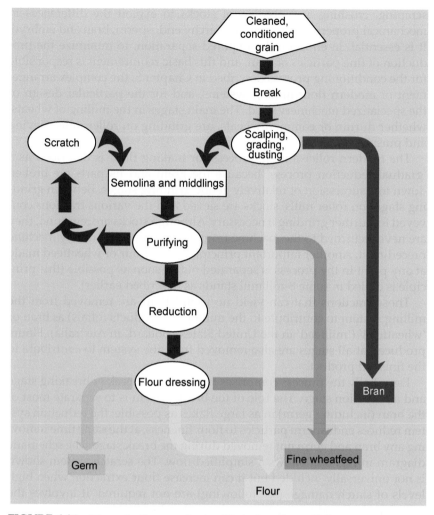

FIGURE 6.21 Schematic diagram of a simplified wheat flour mill flow.

- Break stock: the portion of a break grind overtailing the scalper cover (sieve) and forming the feed to a subsequent grinding stage. The corresponding fraction from the last break grind is the bran.
- Break release: the throughs of the scalper cover, consisting of sizings, semolina (farina), middlings, dunst, and flour.
- Tails, Overtails: particles that pass over a sieve.
- Throughs: particles that pass through a sieve.
- Aspirations: light particles lifted by air currents.
- Sizings, Semolina, Farina: coarse particles of starchy endosperm (pure or contaminated with bran and germ). In the United Kingdom, where durum wheats are not regularly milled, the term 'semolina' is

used to describe a coarse intermediate stock produced from the break system, in the milling of flour from *T. aestivum* wheats. In the United States, where durum, common and club wheats are milled, the term 'semolina' is reserved for the durum product; and the coarse milling intermediate from the other wheats, equivalent to the UK 'semolina' is called 'farina'. Semolina for domestic use (puddings etc.) in the United Kingdom is obtained from durum wheat.

- Middlings, Break middlings: endosperm intermediate between semolina and flour in particle size and purity, derived from the break system.
- Dunst: this term is used to describe two different stocks:
  1. Break dunst: starchy endosperm finer than middlings, but coarser than flour, derived from the break system. This stock is too fine for purification but needs further grinding to reduce it to flour fineness.
  2. Reduction dunst (reduction middlings in the United States): endosperm similar in particle size to break dunst, but with less admixture of bran and germ, derived from semolina by roller milling.
- Flour: starchy endosperm in the form of particles small enough to pass through a flour sieve (say 140 μm aperture). N.B. this is the definition of white flour, the term 'flour' is also extended to include brown and wholemeal flours, whose particle size ranges are less well defined. When used to describe the fine product of other cereals, the name of the parent cereal usually appears as a prefix, e.g., rye flour.

6.4.7.2 *Separation by size*

The milling process includes several sieving stages, each performing a different function; names of particular sieving processes include:

- Scalping: sieving to separate the break stock from the remainder of the break grind.
- Dusting, bolting, dressing: sieving flour from the coarser particles.
- Grading: classifying mixtures of semolina and middlings into fractions of restricted particle size range.
- Purifying: the separation of mixtures of bran and endosperm particles, according to their terminal velocity in air currents. The process is particularly characteristic of durum milling.

6.4.8 Roller-milling operations

6.4.8.1 *Break grinding*

The purpose of break grinding is to separate endosperm from bran and germ. Some flour is produced but the intention is to produce large fragments of endosperm for treatment on purifiers and in the reduction system. The endosperm of soft wheat fragments more readily than that of hard wheats, consequently yielding more flour during break grinding.

The break system consists of four or five break grinds, each followed by a sieving stage (sieving may be omitted between the grinding stages performed on an 8-roll mill stand).

The flutes on the first break rolls are the largest used in the whole process (say 3.5 cm^{-1}). They shear open the grain, ideally along the crease, and unroll the bran coats so that each consists of an irregular, relatively thick layer of endosperm closely adpressed to a thin sheet of bran (see Fig. 6.22). The overtails of the scalping sieve are then conveyed to the second break roller mill in which the rolls are less coarsely fluted, say 5.5 cm^{-1} and the gap between rolls narrower. Again, the largest particles from the sieving stage that follows pass to the third break, to be ground by finer-fluted rolls with a narrower gap, and so on to the fourth and, if the system includes one, the fifth break. The disposition of the fluted rolls used in break milling can be varied and the effects of the options selected have been studied (e.g., Fang and Campbell, 2003), but disposition may also be altered as rolls experience surface wear.

The largest particles sieved from the final break grind are almost ready to become the final product, bran, but because it is still possible to remove small amounts of endosperm, it is passed to a bran finisher where this can be achieved without substantially reducing the size of the bran flakes. The bran finisher performs both a detaching and a separating role; it can also be used earlier in the break sequence, on scalpings destined for break regrinding.

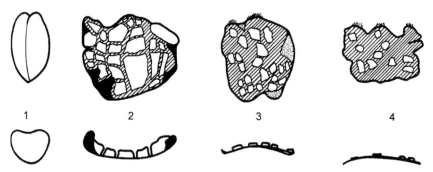

1   2   3   4

FIGURE 6.22 Stages in opening out of the wheat grain and the scraping of endosperm from bran by fluted rolls of the break system. 1, Whole wheat grain; 2, I Bk tails; 3, II Bk tails; and 4, III Bk tails. Upper row: plan view (looking down on the inside of the bran in 2–4), Lower row: side view. In 2–4, endosperm particles adherent to bran are uncoloured; inner surface of bran, free of endosperm, is hatched; outer side of bran curling over is shown in *solid black*; beeswing, from which bran has broken away is shown *dotted*.

The sieving stages are an important part of the flow because particles differing widely in size, in the grind from any one stage of roller milling, also differ in composition – particles of starchy endosperm, which tend to be friable, are generally smaller than particles of bran, which tend to be tough and leathery, due to tempering.

As the branny stocks pass through the break stages they become smaller but also 'cleaner' in the sense that they have less-starchy endosperm attached. Fig. 6.22 shows a representation of the changes.

The scraping action of the break rolls is achieved by a combination of their fluted surface and the speed differential of 2.5:1 at which they run.

Flutes vary in size from mill to mill, but on a 250-mm-diameter roll they might be:

- First break: 3.2–4.1 per $10\,cm^{-1}$
- Second break: 5.1–5.7 per $10\,cm^{-1}$
- Third break: 6.4–7.0 per $10\,cm^{-1}$
- Fourth break: 8.6–9.6 per $10\,cm^{-1}$
- Fifth break: 10.2–10.8 per $10\,cm^{-1}$

Wear on flutes reduces the sharpness of the flutes, leading eventually to the need for refluting.

The break release (throughs of the scalper cover) varies among mills. The criteria determining selected proportions are:

1. the number of break stages,
2. the grist; larger releases on first break are typical for soft wheats, followed by lower releases on succeeding breaks.

Sizing system

In some milling systems, the largest particles of endosperm, known as sizings, coming from the breaks, may be subjected to grinding on finely fluted rolls appropriately termed sizing rolls. Such a system is similar in its objectives to the scratch system. The purpose of sizing is to scrape bran particles from the endosperm chunks so that the two components can be easily separated on purifiers before further reduction of the endosperm particles.

Starch damage

An objective of roller milling that has increased in importance in recent years is the inflicting of high but controlled levels of damage to the starch component of flour. Its increase in importance is a response to technological developments in baking, particularly of bread, and in some countries it accounts for the increased use of hard wheats, as damage occurs more readily in these than in softer wheats. 'Damage' in this context has a very specific meaning (see Chapter 4); it refers to the change resulting to individual starch granules through application of high shear and pressure. One effect of starch damage is to increase water-absorbing capacity. It can be inflicted in all grinding stages but the greater part occurs during reduction.

Starch damage is relevant to the inclusion, or not, of a sizing stage in a mill flow because sizing rolls contribute less to starch damage creation than the alternative option of grinding large endosperm particles on smooth rolls. When this option is selected, the high pressure applied causes some particles to be flattened into flakes, and in order to recover all endosperm as flour, the flakes must be broken into smaller particles. For this purpose, affected stocks are passed through a flake disrupter (see earlier discussion) or a drum detacher before sieving.

6.4.8.2 Reduction system

Apart from the coarsest (bran) fractions separated from the break grinds, the fate of which was already discussed, a series of particle ranges flow from the sieving stages. The intermediate-sized particles are either sizings, farina, semolina, middlings and dunst, or bran snips (some are free bran, some are loaded with endosperm); the smallest are flour. The finer stocks are graded and may be conveyed to be purified in preparation for appropriate further reduction in size, or they may be conveyed directly to reduction mills for grinding. The finest fraction, the flour, is a finished product and is streamed for blending with other machine flours.

The reduction system consists of 6 to 12 grinding stages (Fig. 6.23), interspersed with siftings for removal of the (reduction) flour made by each preceding grind, and a coarse (tailings) fraction from some grinds.

Grinding in the reduction system is carried out on roller mills differing in two important respects from the break roller mills:

1. the roll surfaces are smooth, or more often slightly matt;
2. the speed differential between the rolls is lower, usually 1.25:1, although the fast roll still runs at 500–550 rpm. The grinding effect in reduction mills is one of crushing and shearing (the balance of the two components depending on the smoothness of the roll surface).

In the break system the gap between rolls is an important parameter, and its setting depends on the size of stock being ground. In reduction milling, practices vary according to national and local requirements. Posner (2009) reported studies on optimum grinding of middling stock between 200 and 400 μm involving a roll gap of 0.05 mm, but it is high pressures on rolls that has greater importance in reduction milling especially in production of bakers' flours where high levels of starch damage are required.

While break stages are relatively uniformly assigned in most national descriptions with a sequence of either Roman or Arabic numerals, the

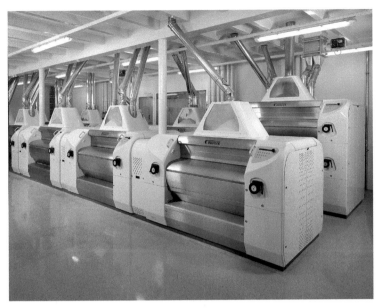

FIGURE 6.23 Part of the roller floor of a modern flour mill. Most roll stands shown in this image are 4-roll but that at rear right is 8-roll. *By courtesy of Bühler A.G. Uzwil, Switzerland.*

identification of reduction passages is less consistent and more opaque (Table 6.4).

The routing of stocks throughout the reduction stages of the milling process and are to be found in specialized milling manuals (e.g., Klabunde, 2004), but a simplified diagram (Fig. 6.24) provides an indication. In summary, the early stages, known as head reductions, grind the largest endosperm particles while the later (tail) reductions are fed from the earlier ones and hence handle increasingly small and increasingly branny particles.

6.4.9 Flours

At each grinding stage in the mill, some particles of flour fineness result, and they are separated from coarser stocks as described previously. Although all these flours are, by definition, similar in particle size, their composition is variable as a result of their originating in different regions of the endosperm tissue. For example, late-break flours, being scraped from the bran after removal of most of the central endosperm, originate in the peripheral (subaleurone) endosperm region. They therefore consist of particles with small starch granules embedded in substantial cores of protein (see Chapter 3). They therefore have a high-protein content. There may also be particles of aleurone present, and both these factors contribute to a

TABLE 6.4 Equivalent designations of roller-milling passages in some countries

American	English	French	German	Italian	Spanish	Swiss
Breaks	**Breaks**	**Broyages**	**Schrotungen**	**Rottura**	**Trituraciones**	**Schrotungen**
1st Brk	I Bk	B1	I Schrot	1 Rott.	T1	B1
2nd Brk	II Bk	B2	II Schr.	2 Rott.	T2	B2
etc.	etc.	etc.	etc.	etc.	etc.	etc.
Sizings	**Reductions**	**Claquage**	**Auflösungen**	**Riduzione**	**Reductiones**	**Auflösungen**
1st Siz f	A	Cl 1 f	1 A a (1Gr.m)	1 Svest f	R 1 f	C1A
1st Siz c	X	Cl 1 g	1 A b (1Gr.g)	1 Svest g	R 1 g	C1B
1st Midds c	B	Cl 2 f	2 A a (1Gr.f)	2 Svest f	R 2 f	C2A
2nd Qu.	B2	Cl 2 g	2 A b (2Gr)	2 Svest g	R 2 g	C2B
1st Tails	F	Cl 3	3 A 1Übg	3 Svest	R 3	C4
2nd Tails	J	Cl 4	2 A 2Übg	4 Svest	R 4	C7
Middlings	**Middlings**	**Convertissage**	**Mahlungen**	**Rimacine**	**Compressions**	**Mahlungen**
1st Midds.f	C	C1	1M (Du)	1 Rim	C 1	C 3
2nd Midds	D	C2	2M	2 Rim	C 2	C 5
3rd Midds	E	C3	3M	3 Rim	C 3	C 6
4th Midds	G	C4	4M	4 Rim	C 4	C 8
5th Midds	H	C5	5M	5 Rim	C 5	C 9
6th Midds	K	C6	6M	6 Rim	C 6	C 10
7th Midds	L	C7	7M	7 Rim	C 7	C 11
1st L.G.	M	C8	8M	8 Rim	C 8	C 12
2nd L.G.	N	C9	9M	9 Rim	C 9	C 13

Reproduced from Klabunde, H., 2004. Mahlverfahren und Diagrammkunde. In: Erling, P. (Ed.), Hanbuch Mehl- und Schälmüllerei. Agrimedia GmbH, Bergen/Dumme, Germany with

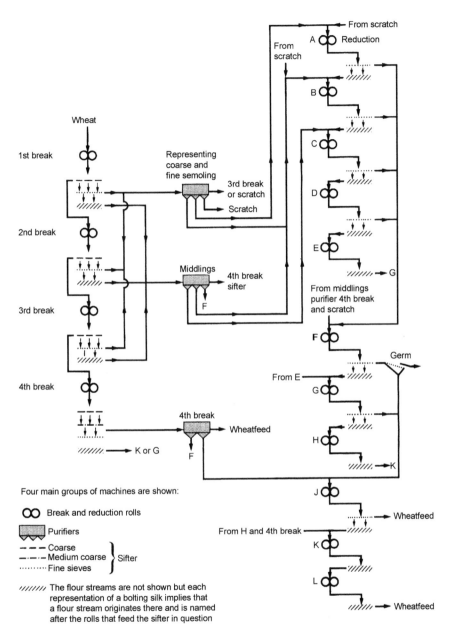

FIGURE 6.24 Diagrammatic representation of flour-milling flow. *Reproduced from Jones, C.R., 1958. The essentials of the flour milling process. Proceedings of the Nutrition Society 17 (1), 7–15. By courtesy of the Cambridge University Press.*

high-protein content. They are also more likely than particles from the central endosperm to contain specks of bran, which, like the aleurone, contributes a high-mineral content. Other flour streams are derived preferentially from the central endosperm and hence will exhibit the characteristics of that region. To create a final flour for packaging or transporting to customers, the miller has a series of options in terms of blending the various flour streams. All the flours can be combined to create the 'straight run' or 'straight grade' flour. Alternatively, flours produced only at selected stages can be combined to produce special characteristics demanded by a customer. Terms such as 'patent' and 'bakers' are used to describe such blends.

A basic specification for bread flours is for an acceptably high-protein content, and where the choice of wheats to be milled is limited, a miller may not be able to meet the specification from the available wheats alone. In such a situation the protein deficit can be corrected by the addition at the mill of an appropriate quantity of 'vital gluten'. Vital gluten is a coproduct of the wheat wet-milling process, in which starch is the main product. The description 'vital' is an indication that the product has not been denatured by, for example, excessive heating and that it thus retains its essential baking properties; however, it is not always included in the description and reference is frequently made just to 'gluten' addition.

As well as adding gluten when needed, the miller is also responsible for adding other additives that are required to meet customers' specifications or, more frequently, relevant national legislation. The latter category includes vitamin concentrates and chalk, both of which can thus be incorporated into the national diet via staple foods that are consumed by most of the population.

6.4.10 Changes to the milling system

Over time the flour-milling procedure has become shorter, incorporating fewer roll stands and passages. This has been made possible as a result of:

1. the introduction of the pneumatic mill. Its cooler stocks allowed more work to be done on reduction rolls without overheating the stock;
2. improvements in the roller mill, allowing rolls to run at higher speeds and pressures;
3. the provision of an efficient cooling system for grinding rolls;
4. improvements in the capacity and dressing efficiency of the plansifter;
5. availability of flake disrupters.

The shorter systems have the advantage that less plant is required for a given capacity. Smaller buildings are less expensive to build and maintain, clean and supervise.

The layout of a modern mill is shown in Fig. 6.25.

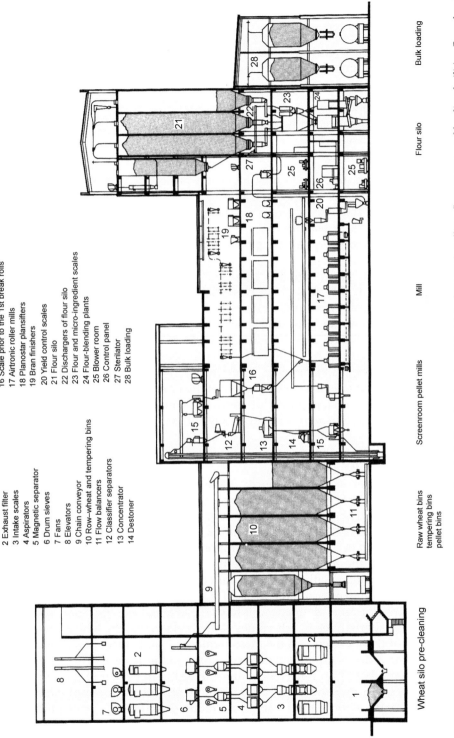

1 Grain intake
2 Exhaust filter
3 Intake scales
4 Aspirators
5 Magnetic separator
6 Drum sieves
7 Fans
8 Elevators
9 Chain conveyor
10 Row–wheat and tempering bins
11 Flow balancers
12 Classifier separators
13 Concentrator
14 Destoner
15 Scourer
16 Scale prior to the 1st break rolls
17 Airtronic roller mills
18 Planostar plansifters
19 Bran finishers
20 Yield control scales
21 Flour silo
22 Dischargers of flour silo
23 Flour and micro-ingredient scales
24 Flour-blending plants
25 Blower room
26 Control panel
27 Sterilator
28 Bulk loading

Wheat silo pre-cleaning   Raw wheat bins   Screenroom pellet mills   Mill   Flour silo   Bulk loading
tempering bins
pellet bins

FIGURE 6.25 An example of a flour mill configuration, including cleaning, tempering (conditioning), milling, flour storage and loading facilities. *Reproduced by courtesy of Bühler A.G. Uzwil, Switzerland.*

6.4.11 Flour extraction

Flour extraction is synonymous with flour yield. It is expressed in different circumstances on the basis of:

- wheat received
- cleaned wheat
- wheat before conditioning
- conditioned wheat, or
- total products

In addition, the values may be stated on an 'as is' basis or a basis whereby all product weights are adjusted to a common moisture content. Commonly used moisture bases are 14%, 12% and dry.

Clearly these possibilities give rise to a range of possible values, therefore, in expressing or evaluating them it is important that the chosen bases are stated.

The wheat grain contains about 82% of white starchy endosperm (see Table 3.1), but it is never possible to separate it exactly from the 18% of bran, aleurone and embryo, and thereby obtain a white flour of 82% extraction rate (products basis). In spite of the mechanical limitations of the milling process, extraction rates well in excess of 75% are now achieved.

It can be argued that expressing extraction rates of white flours or semolinas without reference to the quality of the flour is meaningless. Quality in this context means purity, or freedom from grain components other than starchy endosperm that adversely influence flour appearance and performance.

6.4.11.1 Purity

Methods of expressing purity have been established over a long period. In practice it is the coloured elements of the bran that most concern customers, so it is for the measurement of these that analytical methods have been developed. One way in which bran differs from endosperm is in its higher concentration of minerals, and it is the measurement of these that has become the most widely adopted indicator of contamination with bran. Mineral content is measured by burning a sample and expressing the mass of the ash recovered as a proportion of the dry (or other defined) weight of the original sample. Some countries use alternatives to ash values as their standard; the longest established of these depend on the measurement of some form of reflectance of either a dry sample of flour or a slurry made from it. These methods, being based on reflectance rather than the detection of individual particles, distinguishes them from the more recently introduced method that actually counts the number of coloured particles in a magnified image of flour pressed against a window in a cell. When appropriately illuminated, different elements of the bran can be quantified separately. The particle counting system can be used online in a device that continuously samples a chosen stream or streams. The merits of different methods were discussed by Evers and MacRitchie (2000). Although ash values cannot be measured directly on line, analogue readings based on near infrared reflectance (NIR) spectroscopy can be, and this method is now well established. It has the advantage of being versatile in its capabilities, with calibrations for moisture content, protein content and other parameters of interest having been established. Some instruments combine particle counting by visual and fluorescence imaging and NIR monitoring (e.g., www.branscan.com).

6.4.12 Wholemeal

The terms 'wholemeal' and 'wholegrain' are mostly used to describe flour that includes all the ground products of cleaned grains and nothing else. In some settings the types of wheats that are eligible for inclusion, or more accurately those wheats that are not permitted, are defined. In the United States (where the term 'graham' is also used to denote wholemeal) durum wheats are specifically excluded. Because the presence of germ in wholemeal flour reduces shelf life, its removal may be permitted without rendering the description as wholemeal unacceptable. Canada is an example of a country in which up to 5% of the total products may be removed for that reason.

While the composition of wholemeal may not be defined in law in all countries, it usually equates to 100% extraction. In India and Pakistan a meal known as 'atta' is produced. It is approximately 100% extraction flour, used for making chapattis (see Chapter 13).

Wholemeal may be produced using:

- millstones – either single or multiple passes with intermediate dressing
- roller mills – either a shortened system or by introducing stock diversions in a conventional system
- disc grinders – either metal or stone discs

6.4.13 Brown flours

The presence of bran and germ in wholemeal flour leads to a reduction in gas-retaining properties of products made from it. On the other hand, it is generally accepted that nutritional benefits are derived from consumption of wholemeal products, and their eating qualities are preferred by some consumers. Brown flours represent a compromise that offers baking properties closer to those of white flours and nutritional and organoleptic properties closer to those of wholemeals. Brown flours are comprised of white flour and some of the bran and germ, but the proportions of each are not defined and there are no practical analytical methods that are capable of establishing their exact composition. However, the claimed nutritional benefits of brown over white flours are considered to come mainly from their greater fibre content so it is this factor that is used to evaluate brown flours. Another means of characterizing brown flours is by extraction rates that are usually understood to be between 85% and 98%. Another type of brown bread is made from white or near-white flour dyed with caramel.

6.4.14 Milling by-products

6.4.14.1 Germ recovery

Germ is the milling by-product derived from the embryo (mainly embryonic axis) of the grain. It is not always separated and may be included in wheatfeed. The advantages of separating germ are that:

- it commands a higher price than wheatfeed when of good quality; its presence in reduction stocks can lead to its oils being expressed into flour during grinding, adversely affecting the keeping and baking properties of the flour.

For effective separation of germ, particles from (mainly) first and second break, passing the scalping covers but overtailing covers of approximately 1000 μm (depending on wheat type) are sent straight to the semolina purifiers. Here the germ-enriched stocks are directed to head-reduction rolls. The germ becomes flattened while the endosperm elements fragment, providing a means of concentrating the germ as a coarser fraction. Small pieces of embryo find their way to later reduction where they may be flattened and retrieved or included in wheatfeed.

6.4.15 Mill capacity

Mill capacity is widely expressed in tonnes (of wheat) per 24 h. The accounting unit in the flour-milling industry is 100 kg. Bran and fine wheatfeed are quoted in metric tonnes.

As a rule of thumb, mills processing under 150 tonne/24 h are regarded as small, 150–400 tonne/24 h as medium and over 400 tonne/24 h as large. Some very large mills process 800 or even 1000 tonne/24 h.

In the United States, capacity is expressed as cwt (100 lb = 45.36 kg) of flour per 24 h.

6.4.15.1 *Grinding and sieving surface*

The loading of roller mills can be defined in terms of the length of roll surface devoted to grinding the stocks. It is now conventional to refer to the length applied to 100 kg of wheat ground per 24 h. It is computed by dividing total roll length by the capacity of the mill per 24 h in 100 kgs. An example of a calculation is shown in Table 6.5.

Allocations vary according to the hardness of the wheat being milled and the final product required. These variables also influence grading requirements, particularly as soft wheat products pass through sieve apertures less readily.

Different systems require different lengths and surfaces, but Table 6.6 provides a comparative example.

6.4.16 Automation

Modern flour mills are highly automated, with microprocessor control of many, if not all, systems being standard practice. Entire mills are capable of running completely unattended for several days.

TABLE 6.5 Calculations of break roll allocations for a mill grinding 250 tonnes (i.e., 2500×100 kg) of wheat per 24 h

Passage	No of roll stands	Length of rolls per stand (mm)	Total length (mm)	Loading mm/100 g/24 h
I Bk	2	1250	2500	2500/2500 = 1.0
II Bk	2	1250	2500	2500/2500 = 1.0
III Bk	2	1250	2500	2500/2500 = 1.0
IV Bk coarse	1	1250	1250	1250/2500 = 0.5
IV Bk fine	1	1250	1250	1250/2500 = 0.5
V Bk	1	1000	1000	1000/2500 = 0.4
Total break surface				**4.4**

TABLE 6.6 Specific machine allocation in mills for different classes of wheat

Load level and variable	Hard wheat	Soft wheat
Roll surface mm/100 g/24 h	9–15	10–13
Sifter area m²/100 g/24 h	0.064	0.083–0.088

Data from Posner, E.S., 2009. Wheat flour milling. In: Khan, K., Shewry, P.R. (Eds.), Wheat Chemistry and Technology, fourth ed. American Association of Cereal Chemists Inc., St. Paul.

6.4.17 Quality monitoring during processing

From the previous description it can be seen that flour milling is a complex multistage process in which large quantities of stocks are handled at a rapid rate. The products are expected to meet stringent standards imposed by authorities and customers and, consequently, many tests have traditionally been performed on finished products to ensure compliance. In cases where specifications have not been met, corrections to large bulks have proved expensive and time-consuming, so the ability to monitor for quality parameters during processing is clearly desirable.

Many of the tests on finished products are performed in a laboratory on small samples taken from a bulk. Tests may involve skill and considerable time and samples may be destroyed during the tests, some of which involve use of hazardous chemicals. While these fundamental chemical tests remain as the ultimate standards, major successes have been achieved in developing tests that offer rapid alternatives, which, through calibration, can provide results that accurately reflect the standard test results. The greatest successes in this field are based on NIR spectroscopy. The method depends on the selective absorption of infrared energy at wavelengths characteristic of the different flour components or conditions (Williams (2011) provides a brief introduction to the underlying principles). Measurements of different components can be made simultaneously on dry flour or other stock.

Initial successes led to the rapid adoption of laboratory instruments, and these persist as valuable nondestructive devices in the laboratory and the field, but NIR instruments have now been widely adopted for monitoring qualities of stocks during processing. Measurements made in this way are described as at-, on-, or in-line, and, while the terms are not always applied consistently, they are generally accepted to mean:

- At-line – samples for instrumental analysis are taken manually at one or more points in the process.
- Online – Some stocks are diverted from the main flow into a suitable cell for readings to be taken.
- In-line – Sensors are mounted directly in the process flow.

Instruments available to flour millers are described as both on-line and in-line, and it is claimed that they provide accurate assessments of flour characteristics including moisture content, protein content, ash value, starch damage and water absorption. Some instruments also include a camera and suitable illumination for detecting dark bran specks and aleurone. Values from on- and in-line instruments can be displayed remotely and are capable of being fed into mill control systems.

6.4.18 Energy used in flour production

It has been calculated that the primary energy requirement for milling 1 tonne of flour plus that required to transport the flour in bulk to the bakery is 1.43 GJ, or, if transported in 32 kg bags, 1.88 GJ. The primary energy requirement for the fuel used to transport wheat to the UK flour mill is calculated as 1.46 GJ/tonne for CWRS, or 0.08 GJ/tonne for home-grown wheat. The total primary energy requirement for growing the wheat (30% CWRS, 40% French, 30% home-grown grist), transportation to the UK flour mill, milling the wheat to flour, and transporting this to the bakery is estimated at 6.41 GJ/tonne (Beech and Crafts-Lightly, 1980).

6.4.19 Structure of the milling industry in different parts of the world

European Flour Millers (www.flourmillers.eu) provides up-to-date information on the European industry at 2-year intervals. Information in 2016 identified 3800 milling companies processing 45 mt of wheat and rye to produce more than 600 types of flour. The proportional distribution of flours is indicated in Table 6.7.

Most UK flour-milling companies are members of the National Association of British and Irish Millers (NABIM). The organization publishes information annually on production and uses of flour (www.nabim. org.uk). Flour production is dominated by four companies that produce 65% of total flour milled. There are 10 companies producing significant amounts, and the remainder of the companies are small enterprises mainly serving niche markets. In total there are 30 milling companies and 49 mills. Around 400 different flour types are produced, some of which are included in Table 6.8, which gives comparisons of production and end uses in selected years.

TABLE 6.7 Destinations of flours produced in Europe

Destination	% of total
Bakeries	
Industrial	30
Small	30
Supermarkets	12
Biscuit and Rusk	14
Household	12

Data from European Flour Millers.

TABLE 6.8   Total flour production and proportions of main flour types milled in the United Kingdom in selected years

	1988/89	1998/99	2008/09	2012/13	2013/14	2014/15
Total flour production	3974	4478	4861	5121	4998	5287
FLOUR TYPE %						
Bread making						
White	53.5	53.6	49.2	49.4	50	46.7
Brown	3.5	302	2.4	1.9	1.3	1.1
Wholemeal	6.3	4.3	6.1	6.2	6.4	5.8
Biscuit	14.6	12.7	12.0	10.9	11.5	8.7
Cake	1.9	1.7	1.7	2.2	2.2	2.2
Prepacked household	3.5	2.4	2.5	2.7	2.4	1.9
Self-raising	2.3	1.6	N/A	N/A	N/A	N/A
Food ingredients	N/A	N/A	3.6	3.6	3.6	5.1
Starch manufacture and other	14.5	20.5	22.6	23.2	22.7	28.5

Data from www.nabim.org.uk.

A comprehensive analysis of the flour milling industry in China was presented by Dongson (2012). The numbers of mills of different sizes are shown in Table 6.9, and the distribution of flours into products is shown in Table 6.10.

6.4.20 Heat treatment of wheat

Treatments involving heat are applied to both wheat and flour for a range of purposes, but they are not performed at most mills as the market for treated flours is limited. Steam treatment applied to wheat before milling is performed largely to reduce amylolytic enzyme activity, making the derived flour suitable for production of stable gels in food products such as soups and sauces. When wheat is treated with live steam, alpha-amylase and other enzymes are inactivated. Inactivation of amylase is rapid but not instantaneous; to ensure complete inactivation, the wheat is held at or very near 100°C for 2–4 min in a steamer. To achieve the same objective, flour may be similarly treated, but with indirect heat and with limited amounts of steam, to avoid gelatinization and lumping. Gluten-forming proteins are also denatured making steam-treated

TABLE 6.9 Structure of Chinese milling industry

Number of mills	Tonnes per day
2000–3000	50–200
500–1000	200–400
350–400	400–1000

Data from Dongson, L., 2012. www.flourmillers.eu/uploads/China_Flour_Milling_Industry_BeijingEFM_2012.pdf.

TABLE 6.10 Proportional end uses of flours in China

Product	% of total
Noodles	35
Steamed bread	30
Nang bread	10
Dumplings	8
Biscuits	7
Cake	3
Bread (Western style)	3
Other	4

Data from Dongson, L., 2012. www.flourmillers.eu/uploads/China_Flour_Milling_Industry_BeijingEFM_2012.pdf.

products unsuitable for purposes in which gas retention is important. The thermal and hydrothermal modification of flour has been reviewed by Müller and Schneeweiss (2006).

6.4.21 Fine grinding and air classification

The contents of cells comprising the bulk of storage tissues of many legume cotyledons and cereal endosperms consist essentially of starch granules embedded in a protein matrix (see Chapter 3). In some, such as oats and some legumes, an appreciable amount of oil is also present. The spaces among closely packed spherical or near-spherical starch granules are wedge shaped, and where protein occupies these spaces it is compressed into the same wedge shape. It has thus been called 'wedge protein'. Clearly the size of starch granules determines the sizes of the interstitial wedges. In the case of the Triticeae cereals, the wedges among the larger population of starch granules generally have granules of the smaller population embedded in them.

When wheat endosperm is fragmented by grinding, it is usually reduced to a mixture of particles, differing in size and composition (Greer et al., 1951). These may be classified into three main fractions:

1. Whole endosperm cells (singly or in clumps), segments of endosperm cells and clusters of starch granules and protein (upwards of 35 μm in diameter). This fraction has a protein content similar to that of the parent flour.
2. Large and medium-sized starch granules, some with protein attached (15–35 μm in diameter). This fraction has a protein content one-half to two-thirds that of the parent flour.
3. Small chips (wedges) of protein and detached small starch granules (less than 15 μm in diameter). This fraction has a protein content approximately twice that of the parent flour (Fig. 6.26).

The proportion of medium-sized and small particles (below 35 μm) in flour milled conventionally from soft wheat is about 50% by weight, but in hard wheat flours it is only 10%. The proportion of smaller particles can be increased at the expense of larger ones by further grinding on, for example, a pinned disc grinder, which consists of two steel discs mounted on a vertical axis, each disc being studded on the inward-facing surface with projecting steel pins arranged in concentric rings that intermesh. One disc, the stator, remains stationary while the other, the rotor, rotates at high speed. Feedstock enters the chamber between the discs at the centre and is propelled centrifugally by the air current created. The particles impact against the pins and against each other, as a result of which they are fragmented. The use of a jet mill to similarly reduce particle size has been reported (Létang et al., 2002). The reduction in particle size due to fine grinding further separates the endosperm components, as previously described, allowing increased proportions of starch and protein to be concentrated into different fractions.

Particles below about 80 μm are considered to be in the subsieve range, and for making separations at 15 and 35 μm, the flour as ground, or after fine grinding, is fractionated by air classification. This process involves air elutriation, a process in which particles are subjected to the opposing effects of centrifugal force and air drag. Smaller particles are influenced more by the air drag than by centrifugal force, while the reverse is true of the larger particles. The size at which a separation is made is controlled by varying the amount of air admitted or by adjusting the pitch of baffles that divide or 'cut' the airborne stream of particles.

When practised commercially, air classification is generally carried out in the mill. It is customary to effect separations into a protein-rich fraction of less than 15 μm, a starch-rich fraction of 15–35 μm and a fraction over 35 μm consisting of cells or parts of cells that have resisted breaking into discrete components. Table 6.11 shows typical yields and characteristics of fractions derived from fine-ground and unground flours of hard and soft wheats.

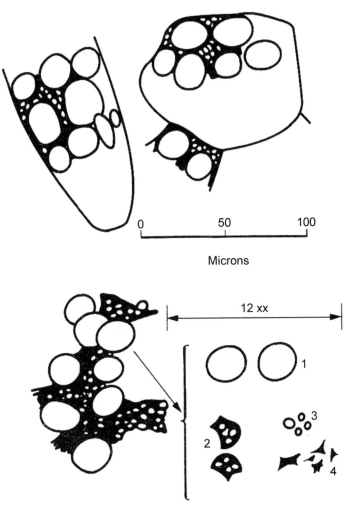

FIGURE 6.26   Above: the two main types of endosperm cell – prismatic (left), polyhedral (right) – showing large and small starch granules (white) embedded in protein matrix (black). Below: exposed endosperm cell contents (left) and products of further breakdown (right): 1, detached large starch granules (about 25 μm diameter); 2, 'clusters' of small starch granules and protein matrix (about 20 μm diameter); 3, detached small starch granules (about 7 μm diameter); 4, fragments of free wedge protein (less than 20 μm diameter). 12xx is the representation to scale of the mesh aperture width of a typical flour bolting cloth. *Redrawn from Jones, C.R., Halton, P.R., Stevens, D.J., 1959. The separation of flour into fractions of different protein contents by means of air classification. Journal of Biochemical and Microbiological Technology and Engineering 1, 77–98 and reproduced by courtesy of Inter-science Publishers.*

TABLE 6.11 Yield and protein content of air-classified fractions of flours with or without pinned-disc grinding[a]

Flour	Parent flour protein content (%)[b]	Fine (0–17 μm)		Medium (17–35 μm)		Coarse (>35 μm)	
		Yield (%)	Protein content[b] (%)	Yield (%)	Protein content[b] (%)	Yield (%)	Protein content[b] (%)
HARD WHEAT							
Unground	13.6	1	17.1	9	9.9	90	13.8
Ground	13.4	12	18.9	41	10.0	47	14.7
SOFT WHEAT							
Unground	7.6	7	14.5	45	5.3	48	8.9
Ground	7.7	20	15.7	71	5.0	9	9.5

[a] *Kent (1965).*
[b] *14% m.c. basis. N × 5.7.*

The term 'protein shift' has been coined to define the degree of protein concentration achieved with a given feed. Protein shift is the amount of protein shifted into the high-protein fraction plus that shifted out of the lower fractions, expressed as a percentage of total protein in the material fractionated.

Applications for which commercial classified fractions might be used are:

- Fine fraction: increasing the protein content of bread flours, particularly those milled from low or medium protein wheats, and in the manufacture of gluten-enriched bread and starch-reduced products.
- Medium fraction: use in sponge cakes and premixed flours.
- Coarse fraction: biscuit manufacture where the uniform particle size and granular nature are advantageous.

Classification can be continued into further fractions by cuts corresponding to larger sizes. As Fig. 6.27 shows, this also produces fractions whose composition varies.

The highest protein fraction above 15 μm is that between 50 and 70 μm, in which are concentrated the cells from the outermost layer of starchy endosperm. This 'subaleurone' layer contains only a few small starch granules embedded in a solid core of protein (Kent, 1966) (see Fig. 3.19).

6.4.22 Milling of spelt wheat

The demand for new foods and flavours, coupled with possible health benefits (Zelinska et al., 2008), in the diets of industrialized nations, has

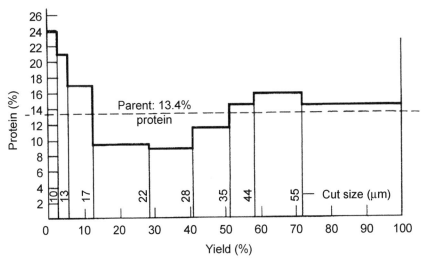

FIGURE 6.27 Results of air classification of flour into nine fractions of varying particle size. The flour, milled from CWRS wheat, was pin-milled one pass, and classified at 10, 13, 17, 22, 28, 35, 44, and 55 μm nominal cut sizes. Protein content of the parent flour was 13.4%.

led to a growing interest in flours milled from ancient, less widely grown wheat types. As a result, a niche market has been established for spelt. It is a hulled wheat and consequently has to undergo dehulling prior to milling. This may be performed at the mill or, more likely, by specialists before delivery to the flour mill. While its chemical, nutritional and functional properties have been investigated, little information is available concerning the milling characteristics, save to confirm their soft milling qualities (Swieca et al., 2014). Cubadda and Marconi (2002) cited reports from Abdel-Aal et al. (1997), who found that flour extraction rates of five varieties were comparable to those of hard red spring wheats. An average of 70% extraction rate was achieved with an average ash value of 0.54 by Marconi et al. (1999), and 65%–70% extraction rate flours were reported by Zeilinski et al. (2008).

6.4.23 Milling of semolina from durum wheat

Durum wheat accounts for less than 10% of the total world wheat area, with almost half of production being in West Asia and North Africa. It is also grown around the Mediterranean, in North America (particularly North Dakota) and Australia. Durum can be milled to flour fineness and used in bread (flatbreads in North Africa) as the only flour, or it can contribute to a composite flour with common-wheat flour. It can also be consumed as a pearled product, 'bulgur'. However, its principal use lies in consumption as couscous or the production of pasta, both of which require endosperm to be ground into particles larger than those in flour. The term 'semolina' has been discussed in relation to the milling of common wheat, (see Section 6.4.7.1) where it may be used mainly to describe an intermediate product. In the case of durum, it is used to describe the final milled product.

Durum is also known as macaroni wheat because of its association with that product and other pastas, and traditionally, the requirement of pasta manufactures has been for a raw material consisting of uniformly coarsely ground endosperm of good colour and high purity. While the requirements for good colour and high purity remain, there has been a trend toward more finely ground products becoming acceptable. The demand for semolina, consisting of particles up to 600 μm is being progressively replaced with one for particles up to 400 μm or even 250–300 μm. For the production of couscous, the requirement remains for large particles of 860–550 μm or 1180–860 μm (Posner, 2009). For pasta, a protein content high enough to provide an elastic gluten, thus ensuring the best eating quality, is required. This is dependent on the grain selected for milling.

Whereas 'colour', as a quality criterion for wheat flour, refers only to the degree of contamination of endosperm with specks of other grain components, in durum products there is an additional requirement for the endosperm itself to be golden yellow. Such a colour gives rise to an amber translucent appearance in the dried pasta that is particularly attractive to manufactures and consumers. The colour is measured by instruments based on the tristimulus principle, which seeks to quantify characteristics of colour as perceived by the human eye.

The presence in durum semolina of nonendosperm parts of the grain is deleterious in pasta production for two reasons: they themselves contribute to a specky appearance and they are associated with enzymes that induce undesirable changes in appearance under conditions of pasta manufacturing. Lipoxygenases in germ oxidize the yellow pigments, and peroxidase and polyphenol oxidase in bran cause browning (Dexter, 2004). Standards and criteria of endosperm purity usually involve the determination of ash value, but other more direct methods may be used; manual counting is routine, but as in the case with bread flour, automated speck assessments are possible (Symons et al., 2009).

Ash values for durum products are generally higher than those in flour. The ash limit defined by the Codex Alimentarius (Standard 178-1991) is 1.3% for semolina and 1.75% for flour (both dry basis), but some standards impose more stringent requirements of, e.g., 0.9% (dry basis). The North Dakota Wheat Commission suggests that ash values of less than 0.5% are indicative of the inclusion of bread wheat flour (http://www.ndwheat.com).

6.4.23.1 *The milling process*

Because of the importance of semolina being speck-free, the cleaning process, particularly the removal of black seeds, has to be very thorough and may involve colour sorting. Because of the extreme hardness of durum wheat, the usual practice is to temper to the high-moisture content of 16%–16.5% before grinding.

Regarding the milling process itself, the main features of durum milling are the extended break and sizing system and the extensive deployment of purifiers. In the milling of flour, advantage can be taken of the different behaviour of bran and endosperm when compressed and sheared between smooth rolls. Endosperm particles are fragmented and bran flakes are flattened and expanded, making the two components easier to separate by sieving. This opportunity is unavailable to the durum miller producing semolina, as all grinding is performed on fluted rolls.

As with flour milling, stocks in the durum mill are conveyed through pneumatic lines, the airflow also serving to cool the mill, particularly the roller mills, which produce heat as a result of grinding.

Grinding

Durum milling is characterized by a long break system, with five or six relatively gentle passages. The gentle grinds of the breaks are designed to release large chunks of endosperm with a minimum of fines. In the early breaks, rolls are disposed dull to dull and the angles of corrugations as well as the spiral angle may differ from those adopted in the milling of common wheat (see Hareland and Shi (1995) for a discussion of optimum conditions).

Later grinding passages are conducted on sizing rolls that reduce large endosperm particles to a uniformly smaller size. The rolls are finely fluted. Smooth reduction rolls are not needed in the durum mill except for the reduction to flour, of particles that are too small to be classed as semolina.

As in the case of common wheat milling, preliminary decortication to remove bran by debranning or peeling is sometimes used in the milling of durum wheat. Such pregrinding treatment removes surface contaminants such as microorganisms and chemicals and allows the number of grinding passages to be reduced (Fowler, 2014). The process described as peeling removes 7%–9% of the grain while the more aggressive debranning removes up to 15%.

Purifying

Durum mills are characterized by the abundance of purifiers (Figs 6.20 and 6.28). They are used at multiple stages during the process to free particles from detached bran. Stocks leaving the roller mills pass to plansifters to be graded and streamed to machines appropriate to their further treatments. Before further grinding sieved stocks are purified and, where possible, stocks that meet the requirements of finished products are removed and accumulated for blending.

The performance of the mill is continuously monitored by strategically sited weighers, indicating any changes in the output of a particular stage of the process and providing a warning of faulty tempering or machine condition.

Specific machine allocation

The greater number of grinding treatments adopted in durum milling is reflected in the roll surface allocation. Posner (2009) cited roll allocations

FIGURE 6.28 Purifiers in a modern durum mill. *By courtesy of Bühler A.G. Uzwil, Switzerland.*

of 11-13 compared with 9-4-15 mm/100 kg/24 h for hard bread wheat. Sifter area allocation was similar for both types but purifier width is considerably higher than those for hard bread wheat (6-8 compared with 3.7).

As Table 6.12 shows, the tendency toward the use of finer semolina in pasta products has led to a reduction in the specific machine allocations.

For the processor, the main advantage to be gained from the use of semolina with a reduced particle size lies in the reduced time needed for water absorption and mixing. Kuenzli (2001) gave a comparison that showed a semolina of 630–125 μm required 15 min while a 350–125 μm equivalent required only 10 min.

Semolina particle size

In Europe several grades of semolina are defined as below (Table 6.13):

Milling extraction rate

The North Dakota Wheat Commission (2016) defines semolina extraction as only that portion of the wheat kernel that is milled into semolina, and it observes that the average for the state is 65% (wheat basis) with ash content between 0.55 and 0.75% (14% m.c.)

This definition suggests that it refers to 'semolina extraction', but the alternative 'total extraction' (semolina plus flour) is also used in some contexts, sometimes without reference to the basis of expression. Hence, comparisons can be misleading.

Dick and Youngs (1988) give commercial averages of 60%–65% for the granular product with about 3.0% flour, for production in the United States. Values of 74%–78% (mean 75.3%) have been quoted (Kent and Evers, 1994). The discrepancy may result from different modes of expression, that is, on the basis of feed weight or product weight.

TABLE 6.12 Specific allocations for mills producing conventional and fine semolinas

	Passage	Specific allocation	
		Conventional semolina mill	Fine semolina mill
Linear roll surface	Breaks	6–8 mm/100 kg/24 h	5–7 mm/100 kg/24 h
	Sizings	6–8 mm/100 kg/24 h	5–7 mm/100 kg/24 h
	Reductions	0.3 mm/100 kg/24 h	1–3 mm/100 kg/24 h
	Total	**13–19 mm/100 kg/24 h**	**12–15 mm/100 kg/24 h**
Sifter surface		0.06–0.07 m²/100 kg/24 h	0.055–0.06 mm/100 kg/24 h
Purifier width		6–8 mm/100 kg/24 h	3–7 mm/100 kg/24 h

Data from Klabunde, H., 2004. Mahlverfahren und Diagrammkunde. In: Erling P. (Ed.), Hanbuch Mehl- und Schämüllerei. Agrimedia GmbH, Bergen/Dumme, Germany.

TABLE 6.13   Particle size ranges and characteristics of European semolina grades

Grade	Particle size range	Comment
Special coarse – (G=gros)	800–600 µm	Usually small quantities produced as cuts for couscous
SSSE (Semoule superieure spezial extra)	600–200 µm	Conventional grade for pasta production and household use
SSF (Semoule superieure fine)	400–125 µm	Fine grade for pasta production
Pasta grade (SSSE reground)	400–125 µm	
Universal grade (straight run)	600–125 µm	
Universal grade (straight run – reground)	400–125 µm	
Fine semolina	300–0 µm	
Durum flour		Ash value=1.6% (dm)
Durum low-grade flour		Ash value=2.0%–3.0% (dm)

Data from Klabunde, H., 2004. Mahlverfahren und Diagrammkunde. In: Erling P. (Ed.), Hanbuch Mehl- und Schämüllerei. Agrimedia GmbH, Bergen/Dumme, Germany.

Klabunde (2004) offers the following as typical extraction values as proportions of wheat to first break: 60% SSSE plus SSF with ash value of 0.88, flour 6% (ash 1.7%) and low-grade flour (ash 3.0% – dry basis).

Rye milling

Compared with wheat, rye grains have a softer endosperm, which breaks down to flour particle size more readily; greater difficulty is experienced in separating rye endosperm from bran.

Possibly as a consequence of the milling and sieving characteristics, rye consumers have developed a taste for products made from a coarser meal with a higher bran content than in the case of wheat flour. The consumption of rye bread has experienced a long-term decline; in Germany, for example, it has dropped from 60% of total bread consumption to 15% in the late 20th century. In the United States, the highest levels of consumption are recorded in areas whose populations have high proportions of descendants of Scandinavian, Jewish and Polish immigrants (Klabunde, 2004).

Rye is milled at a moisture content similar to that of soft wheat: 14.5%–15.5% (hard wheats are conditioned to about 1% higher). Short-conditioning (tempering) periods suffice as water penetrates into the grain more quickly than into wheat; between 6 and 15 h are generally used in North America, where 16–24 h is normal for hard wheats.

The milling methods for rye processing have developed in parallel with those of wheat, but since the end of the 19th century, with the change from stone to roller grinding, there have been significant differences. The reasons for the departure are dependent on both the nature of the grain and the requirements of the consumer. Klabunde (2004) identified the following characteristics that differed from wheat milling:

- Possible use of smooth rolling before first break.
- 300-mm-diameter rolls (cf. 250 mm for wheat).
- Smaller flutes on break rolls.
- Exclusively fluted rolls for reductions.
- Increased linear roll surface.
- Greater plansifter surface.
- Higher power allocation.

Break rolls for rye have shallower and duller grooves to release more break flour. High break releases are desirable in rye milling in contrast to bread wheat milling where the objective is to minimize it. Spirals on rye break rolls are also more exaggerated to increase the cutting action. The use of finely fluted reduction rolls, rather than smooth or frosted rolls, helps to cope with the sticky nature of rye endosperm, which would lead to excessive flaking on smooth rolls due to its high-pentosan content. A rye mill in the United States described by Shaw (1970) has five break passages, a bran duster and nine reduction passages (sizings, tailings and seven middlings reductions). Purifiers do not feature in rye mills. The specific surface in a modern rye mill would be 18–24 mm per 100 kg of rye milled per 24 h.

The yield of reasonably pure flour from rye is 64%–65%. At increasing extraction rates the flour becomes progressively darker and the characteristic rye flavour more pronounced. In the United States a good average yield of rye flour would represent an 83% extraction rate. In Canada 67% has been cited as the average (Lorenz, 2000). Rye meal may be of any extraction rate, but rye wholemeal is 100% extraction.

Rye milling varies considerably according to the extraction rate required and to the geographical region where it is processed. Three major regions have been identified as characteristic: North America, Western Europe and Eastern Europe, including the states of the former Union of Soviet Socialist Republics (Sapirstein and Bushuk, 2016):

- North American practice is to grind higher protein, smaller sized grains into fine flour with no particular quality requirements.
- Western European practice is to grind lower protein, larger sized grains into fine flour with strict quality requirements.
- Eastern European rye mills grind all types of rye into a coarse flour according to local quality specifications.

The quality of rye flour in several European countries is expressed in terms of ash value and in Germany, for example, there are around seven, identified by increasing numbers as ash value rises. In North America fewer grades are recognized, and Canadian and US grades do not correspond exactly. They are shown in Table 6.14.

Rye flour produced in North America may be treated with chlorine (0.19 g–0.31 g/kg) to improve the colour of white flour (Lorenz, 2000).

Wholemeal is produced by two break passages applied to dry grain, but it may also be produced by alternative grinding techniques.

In Europe, particularly Germany, the use of 'combination' or 'swing' mills, grinding both wheat and rye, has proved successful (Bushuk, 1976).

6.4.24 Triticale milling

Milling procedures for triticale resemble those utilized for wheat and rye. Even with the best hexaploid triticales, however, the yield of flour is considerably lower than that of wheat flour. Macri et al. (1986) compared milling performance with Canadian Western Red Spring wheat on a Bühler experimental mill. Triticales produced between 58% and 68% flour of 0.44%–0.56% ash, while the CWRS gave 71.75% at 0.45% ash. Triticales were tempered to 14.5% m.c. Little if any triticale is milled commercially as a separate mill feed.

6.4.25 Dry milling of maize (corn)

Dry milling of maize for food production is a relatively minor industry compared with wet milling. However, in the United States the amount of

TABLE 6.14   North American grades of rye flour

United States of America grade	Extraction rate (% of wholegrain milled)	Ash value (%)	Protein content (%)
White rye	65% (a)	0.6–0.7	7–9
Medium rye	83% (Straight grade) (b)	1.0–1.5	10–13
Dark rye	Clears = (b) – (a)	2.5–3.0[a]	14–16[a]
Rye meal	100%	1.8	10–15
CANADIAN GRADES			
Light rye	75–80	0.7–0.9	–
Medium rye	83–85	1.0–1.2	–
Dark rye	92–95	1.3–1.8	–

[a] *The ash and protein values of dark rye, which is straight grade with white rye flour removed, are higher than those of the wholemeal. This must result from a concentration of aleurone.*

Data from Sapirstein, H.D., Bushuk, W., 2016. Rye grain, its genetics, production and utilization. In: Wrigley, C.W., Corke, H., Seethanramen, K., Faubion, J. (Eds.), Encyclopedia of Food Grains. Academic Press, Oxford.

dry milling has increased in recent years owing to the demand for ground maize as a feedstock to ethanol plants. For this purpose there is little need for separation of the tissues present in a grain, as the wholemeal produced is fermented. Those grain components that are fermentable are converted to soluble compounds, and the nonfermentable components remain and can be incorporated into diets for farm animals. While wet milling is also used to produce a feedstock to ethanol plants, dry milling offers economic benefits and hence 90% of corn bioethanol is processed from dry-milled grain. The other major food use of maize in the United States is the production of starch using a wet-milling process, which is discussed in Chapter 14. Even though a small percentage of grain is processed for food, this amounts to around 5 mt.

Gwirtz and Garcia-Casal (2014) included three types of processing in the category of dry milling. Of these, both nixtamalization and the production of precooked corn flour involve cooking and drying stages after which the dried product is ground. Such grinding is conducted purely to reduce particle size and hence simple one-pass methods such as hammer milling are appropriate.

The following is concerned with the process for which whole cleaned mature maize grains constitute the feedstock. Tempering to a relatively high-moisture content may be necessary for the first stage depending on the method selected for degerming, but this is followed by milling at a lower moisture content.

Maize is consumed by human populations in many forms, some of which are little known outside their countries of origin. In Africa maize is an important staple with national consumption varying from 52 to 328 g/person/day. It is eaten widely as a fine-grit meal 'mealy meal' (spelling varies). Lesotho has the highest consumption. Some Central and South American countries have high consumption also, with Mexico being recorded as consuming 267 g/person/day (Ranum et al., 2014).

In the United States the product with the highest value coming from dry milling is 'grits' but finer products are also demanded.

In the Republic of South Africa (RSA) over 3 mt of maize was used for human consumption and in 2015/16, consumption was recorded as 222 g/person/day. Approximately equal quantities of white and yellow maize are grown in the country, but white maize contributes six- to seven-fold the amount of yellow maize to the human diet. Yellow maize is used for production of grits that are used for polenta, cornflakes and brewing.

Two important characteristics of the maize grain, *viz.* the large embryo and the presence of horny and mealy endosperm in the same grain (see Fig. 3.20), influence the dry-milling processes.

The significance of the large embryo lies not only in its failure to contribute to the grits or flour yield but also in its high-oil content. Inclusion of this oil in the product, either as a component of embryo chunks or through its expression onto the surface of endosperm particles, reduces shelf life through its oxidation and consequent rancidity. The fibrous tip cap located close to the proximal end of the embryo also has to be separated from endosperm fractions. Variation in endosperm texture is important because grits are essentially derived from the horny parts of the endosperm; softer parts of the endosperm readily break down to flour. Grits are mainly used in production of, or consumption as, breakfast cereals in the industrialized world, but they are also used for making fermented beverages.

6.4.25.1 The tempering-degerming system

Describing a process as a tempering-degerming (TD) system defines a principle rather than a universally uniform system. There are many variations in practice, optimized for the required products and influenced by local custom, but one example is shown diagrammatically in Fig. 6.29.

The terms relating to degerming, 'degerminator' and 'degermination', are ill conceived as the process is designed neither to reverse nor to inhibit germination; however, they are universally accepted and understood. The objective of the process is the removal and accumulation of embryos as the milling fraction germ.

Possibly the most important innovation in dry-maize milling was the introduction of the Beall degerminator. It was introduced in 1906, and Bealls, or devices using the same principles, are used in the majority of US dry-milling plants today, as an essential stage in the TD system. The virtue of the Beall lies in its potential to produce ground stocks from which germ can be separated with reasonable efficiency, and a high yield of large-particle-size grits with low-fat and low-fibre content (about 0.5%) suitable for manufacture of corn flakes.

The favoured feedstock to the TD system in the United States is No. 2 yellow dent corn. In Africa yellow maize is also used for grits production. After cleaning and tempering to around 20% moisture content (cited values vary from 18% to 23%), it passes to the Beall. This machine consists of a cast iron cone, rotating at about 750 rev/min on a horizontal axis, within a conical, stationary housing, partly fitted with screens and partly with protrusions on the inner surface. The rotor also has protrusions on its outer surface. The maize is fed into the annular space between rotor and housing at the small end, and it works along to the large end, between the two elements. The protrusions on the rotor and the housing rub off the hull and embryo by abrasive action, also breaking the endosperm into particles of various sizes and degrees of purity. The Beall discharges two types of stock: the tail stocks, which are too large to pass through the screens, consisting mainly of fragmented endosperm, and the through stock, consisting largely of bran and embryo. The Beall can be adjusted for different sized grains and to provide the required proportions of different sized particles (Duensing et al., 2003). A disadvantage of the Beall system

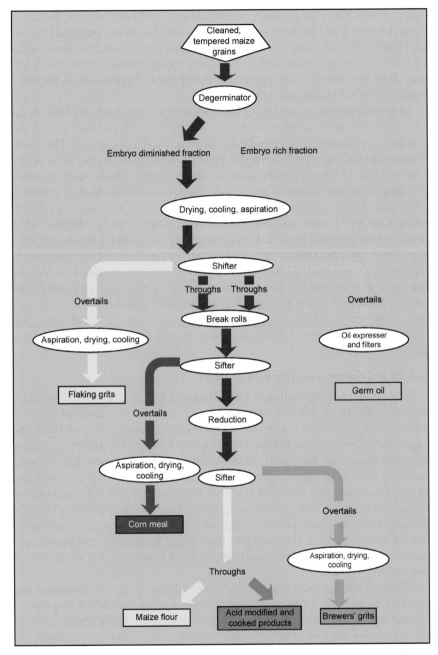

FIGURE 6.29 Schematic diagram of the tempering-degerming process. *Based on Johnson, L.A., 1991. Corn: production, processing, and utilization. In: Lorenz, K.J., Kulp, K. (Eds.), Handbook of Cereal Science and Technology. Marcel Decker Inc., New York.*

is that water has to be added in several stages and may involve the use of heated water. Later in the processing the water has to be removed by heating. Damped stocks have to be processed quickly to ensure optimal water distribution in the grains and to avoid microbial spoilage, and any delays may lead to wastage. The process of damping to high levels is therefore both expensive and inconvenient.

Degerming was carried out even before the introduction of the Beall, and there are also modern alternatives, some of which seek to achieve similar performance to the Beall with reduced damping. The functional elements of one horizontal type of degermer consist of a screen constructed of heavy-gauge wire mesh. A central drive shaft running the length of the mesh cylinder carries rows of beaters that rotate at 1500 rpm. Maize grains that are introduced at one end are impelled to impact on the screen as they pass through the cylinder. Broken fragments of endosperm pass through the screen for separate discharge from the larger retained fragments. An alternative horizontal type of machine also has a screen with a rotor within, but the screen is multiple sided and the rotor is eccentric so that the clearance is continually changing with each revolution Grains are fed into the annulus between rotor and screen and subjected to variable pressure during each revolution as they pass through the chamber. A vertical system based on the same principle is also widely used.

6.4.25.2 *Drying, cooling and grading*

In mill layouts including a Beall, tail stocks are dried to 15%–15.5% or lower moisture content in rotary driers at a temperature of 60–70°C and cooled to 32–38°C. The dried stock is separated into products and coproducts by various methods including gravity separators (as used in grain cleaning) for germ concentration, and aspiration for pericarp removal. The endosperm-rich particles are sifted to produce a number of size fractions. The coarsest fraction, between 3.4 and 5.8 mm, consists of the flaking or hominy grits, originating from the vitreous parts of the endosperm. They may pass through a further aspirator and dryer-cooler, before emerging as a finished product. Finer stocks are combined with coarser fractions of the through stocks from the Beall, for treatment in the milling system.

Through stocks pass through a dryer followed by an aspirator and purifier, to remove bran and germ. Germ may be extracted to produce oil, the remainder being compressed into germ cake. Finer fractions of the through stock consist of endosperm heavily contaminated with bran and germ, the mixture being known as 'standard meal'. The various fractions and appropriate fractions from the tail stocks are combined to produce 'hominy feed'.

6.4.25.3 *Milling*

Degerming is universally involved as a premilling process, but milling systems vary according to the nature of the intended main product. When flour is required as the main product, as in Africa, the system is designed to maximize the production of fine particles. Priorities regarding germ are also different as separated germ for extraction of oil is more important in North America while in Africa the germ is more likely to be included with the bran in animal feed.

Milling for grits production

For the production of grits, as practiced in North America, the system is configured to maximize the proportion of larger endosperm particles and to minimize production of flour.

The feed to the milling system, *viz.* large, medium and fine hominy, germ roll stock, and meal, are mixtures of endosperm, bran and embryo. They are separated by a series of roller milling, sifting and aspiration stages before being dried and emerging as a diverse range of final products. They are fed to the mill, each entering at an appropriate point – the large and medium hominy at the first break, the fine hominy and germ roll stock at the second roll.

The milling is carried out using fluted rolls, a traditional flow containing up to 16 distinct stages. The grindings with fluted rolls (15–23 cuts per cm rotating at a differential of 1.25:1 or 1.5:1) flatten the embryo fragments, allowing them to be removed by sieving. The products are sifted on plansifters and are aspirated. The mill is divided into a break section, a series of germ rolls and a series of reduction and quality rolls. The break system releases the rest of the embryo as intact particles and cracks the larger grits to produce grits of medium size. The whole milling system for maize bears some resemblance to the earlier part of the wheat-milling system (which was described more fully previously) as far as B2 reduction roll (second-quality roll, in the United States) viz. the break, coarse reduction and the scratch systems, but is extended and modified in comparison with this part of the wheat milling to make a more thorough separation of the large quantity of germ present. Modern practice is to use a much shortened system.

The action of the rolls should be less severe at the head end of the mill than at the tail end in order to minimize damage to the germ while simultaneously obtaining maximum yields of oil and oil-free grits.

The finished coarse, medium and fine grits, meal and flour products are dried to 12%–14% moisture content on rotary steam tube dryers. The germ concentrate consists largely of heavily damaged embryos with an oil content of 15%. The fat content of grits and flour is 0.8%–2.7%.

TABLE 6.15 Widely accepted names of dry-milled maize products and their granularity

Source of data	Size range			Yield % by weight
	a	b	c	c
Flaking grits		5600–3360	5800–3400	12
Coarse grits		2000–1410	2000–1400	15
Medium grits		1410 to ~650	1400–1000	20
Fine grits		297–195	1000–650	
Grits	1400–600			23
Coarse meal			650–300	10
Cones		297–177		
Meal	600–300	297 to ~195		
Fine meal	300–212		300–170	10
Flour	<212	<195	<170	5
Germ				14
Hominy feed				11

[a] *Values from Gwirtz, J.A., Garcia-Casal, M.N., 2014. Processing maize flour and corn meal food products. Annals of the New York Academy of Sciences 1312, 66–75. NB. The products listed are derived from degermed maize. Fat content increases from 0.8 to 2.7 with increasing fineness.*
[b] *Based on Serna Saldivar, S.O., 2010. Cereal Grains, Properties, Processing and Nutritional Attributes. CRC Press, Boca Raton.*
[c] *Values from Kent, N.L., Evers, A.D., 1994. Kent's Technology of Cereals, fourth ed. Elsevier Science.*

According to Gwirtz and Garcia-Casal (2014) there is no universal nomenclature to define the different products by their granularity, but they describe a commonly accepted range, which is reproduced in Table 6.15. More detailed but partially incompatible schemes are included for comparison. Approximate expected yields of the defined fractions are included in one case.

Milling for flour production

In the milling system designed for flour production as practiced in the RSA, the feed to the first break comes from a degerming process using the Beall or an equivalent machine. As in the case of grits production, it has to be dried before milling. It consists mainly of large endosperm particles but some bran and germ are also present. Loose bran particles are removed by aspiration before being subjected to grinding. For the production of maize flour, a system of two break or sizing passages is typical; the function of these is to reduce the endosperm to intermediate-sized particles for subsequent grading to facilitate more accurate grinding on reduction rolls. The differences from bread wheat mill conditions reflect the different functions – maize has no crease and consequently no difficult-to-access bran, so the need for opening and scraping is less than in the case of wheat. In other words, the break system is best described as sizing. Very coarse fluting of six cuts per 25 mm with large 'lands (see Fig. 6.13) and shallow spirals (4%) are typical for first break. Large lands are also common on second break rolls although fluting is finer and a more pronounced spiral is chosen. Differentials on break rolls are 1.96:1. Purifiers are rarely used in maize mills but in the South African industry, machines operating on a similar principle, known as concentrators, are widely used for separating germ from endosperm particles.

The reduction system also employs fluted rolls but they are finer than those selected for both breaks and they become finer toward the later passages. A range of 12–24 flutes per 25 mm may be expected between first and fourth reductions. Spirals down to 8% are adopted and higher differentials of 2.6:1 are chosen for the three earliest reduction passages. The final reduction differential reverts to 1.96:1. Plansifters are used for grading stocks between roll passages but purifiers, found in bread flour and particularly durum semolina mills, are replaced by aspirators.

Germ- and bran-rich fractions are subjected to rolling in the germ roll system on the first of which flutes are coarse and shallow, with narrow lands. As with the reduction system, flutes become finer with later germ roll passages. Products from the germ rolls and intervening separations by plansifters, aspirators and concentrators include pure germ, second-grade rice and grits and offals. Some fractions may be returned to the late reduction passages for flour extraction (RSA Grain Milling Federation, 1997).

6.4.25.4 Composition of products

Chemical composition and nutritional value varies among the tissues of the maize grain, hence their separation leads to a concentration of components into different products (see Table 4.21).

Dry-milled maize flour is not to be confused with 'corn flour', the term used in the United Kingdom for maize starch obtained as a product of wet milling.

6.4.25.5 Uses for dry-milled maize products

Flaking grits are used for the manufacture of the ready-to-eat breakfast cereal 'corn' flakes (cf. Chapter 9). Snack foods produced by extrusion cooking are an increasing market for maize products. Other uses for dry-milled maize products are discussed elsewhere: porridge (polenta) and ready-toasted cereals (Chapter 9), for bread and other baked foods (Chapter 8) and for industrial purposes (Chapter 13).

In the RSA, several products suitable for making a porridge known as 'pap' are produced by milling white maize. Their quality differs as follow:

- 'Super' is made from degermed maize and it is mainly starch.
- 'Special' is made from grains that have not been degermed. It is slightly coarser than super and has a higher fat content and hence a shorter shelf life.
- 'Sifted' meal has more bran and germ. It may be nutritionally superior but has a short shelf life and is not popular

The amounts and proportions produced are shown in Tables 6.16 and 6.17.

Other products created from maize in the RSA include:

- Samp – half- and quarter-sized chunks of debranned and degermed grains. Consumed as a vegetable.
- Maize rice – 2–3 mm endosperm chunks consumed as a vegetable as an alternative to rice (RSA Grain Milling Federation, 1997).
- Maize chop – the offals from the milling process. Mainly used as stockfeed.

TABLE 6.16 Food uses of white maize in South Africa (2015/16)

Product	Tonnes produced	% of total
Super	2,239,906	75
Special (mielieli-meal/maize flour)	438,906	14.7
Sifted	30,975	1.0
Unsifted	14,044	0.5
Other	261,132	8.7
Total	2,984,479	

Data from SAGIS, South African Grain Information Service. http://sagis.org.za/.

TABLE 6.17 Food uses of yellow maize South Africa (2015/16)

Product	Tonnes produced	% of total
Maize grits	217,591	63.1
Sifted	30,332	8.8
Maize rice	6634	1.9
Other	90,023	26.1
Total	344,580	

Data from SAGIS, South African Grain Information Service. http://sagis.org.za/.

6.4.25.6 *Impact milling for grits production*

Alternative methods for grits and flour production include the use of impact mills (see Chapter 5). An impact mill comprises a chamber in which an impeller rotates at high speed causing the stock that is introduced to be thrown hard against the surrounding solid chamber wall. Stocks become fragmented but they leave the chamber without any selection having been made. Treatment is performed at natural moisture content of 13%–15%. Separation of germ from endosperm on gravity tables is efficient, but separation of endosperm from bran by aspiration, and of vitreous endosperm from mealy endosperm, is less effective than with the Beall. The following products are obtained from the impact milling (Entoleter) process:

- Maize germs, 1–4.5 mm, with 18%–25% fat.
- Maize grits, 1–4.5 mm, with 1.5% fat, 0.8%–1.2% crude fibre content; about 60% of the original maize.
- Semolina and flour, which may be made by size reduction of the grits. The mealy endosperm, of higher fat content, reduces to flour particle size more readily than does the vitreous endosperm; thus, flour has a higher fat content, about 3%, and semolina a lower fat content, 0.8%–1.3%, than the grits.

6.4.25.7 *Oil extraction from germ*

Extraction of oil from germ rich 'offals' are not part of the dry-milling process, however, the value of the oil contributes significantly to the economics of the entire process. Solvent extraction is generally used in the dry-milling industry, although mechanical pressing, e.g., with a screw press, is sometimes used. The germ from the mill is first dried to about 3% m.c. and then extracted while at a temperature of about 121°C. The oil content of the germ is reduced by extraction from 16% to 25% to about 6% in the germ cake. The extracted oil is purified by filtering through cloth, using a pressure of 552–690 kN/m^2 (80–100 lb/in.^{2}). The oil, which is rich in essential fatty acids (cf. Table 4.8), has a sp. gr. of 0.922–0.925, and finds use as a salad oil. Its high smoke point also makes it suitable for use as a cooking oil.

In the RSA, extraction is performed largely on pellets made at the mill from the mixture of bran and germ that emerges largely from the degerming process. Following extraction, the solvent is removed from the pellets, which are then available as a constituent of balanced feed for animals.

Methods of extraction that avoid the use of nonaqueous solvents have been devised. Some involve the use of enzymes to break down cell wall structure and may bring economic and food safety benefits, however, they currently remain at the experimental stage. These and other methods were compared by Shende and Sidhu (2014).

References

AACC Oat Bran, 1989. http://www.aaccnet.org/initiatives/definitions/Pages/OatBran. aspx.

Abdel-Aal, E.-S.M., Hucl, P., Sosulski, F.W., Bhirvd, P.R., 1997. Kernel, milling and baking properties of spring-type spelt and einkorn wheats. Journal of Cereal Science 26 (3), 363–370.

Amri, A., Ouammou, L., Nassif, F., 2005. Barley based food in southern Morocco. In: Grando, S., Gormez Macphereson, H. (Eds.), Food Barley: Importance, Uses and Local Knowledge. Proceedings of the International Workshop on Food Barley Improvement. ICARDIA, Aleppo, Syria, pp. 22–28.

Atwell, B., von Reding, W., Earling, J., Kanter, K., Snow, K., 2007. Aleurone processing, nutrition, product development and marketing. In: Marquart, L., Jacobs Jr., D.R., McIntosh, G.H., Poutanen, K., Reicks, M. (Eds.), Wholegrains and Health. Blackwell Publishing Professional, Ames, United States.

Bassey, M.W., Schmidt, O.G., 1989. Abrasive-Disk Dehullers in Africa: From Research to Dissémination. IDRC, Ottawa, ON, Canada.

Bazile, D., Bertero, D., Nieto, C., 2013. State of the Art Report on Quinoa Around the World in 2013. FAO, Santiago, Chile. http://www.fao.org/3/contents/ca682370-10f8-40c2-b084-95a8f704f44d/i4042e00.htm.

Beech, G.A., Crafts-Lightly, A.F., 1980. Energy use in flour production. Journal of the Science of Food and Agriculture 31, p830.

Benton Jones, J., 2003. Agronomic Handbook, Management, of Crops, Soils and Their Fertility. CRC Press, Boca Ratan, FL, United States.

Berghofer, E., Schoenlechner, R., 2002. Grain amaranth. In: Belton., P.S., Taylor, J.N. (Eds.), Pseudocereals and Less Common Cereals. Springer-Verlag, Berlin, Germany.

Bhattacharya, K.R., 1985. Parboiling of rice. In: Juliano, B.O. (Ed.), Rice Chemistry and Technology, second ed. American Association of Cereal Chemists Inc., St. Paul, USA.

Bhatty, R.S., 1993. Physico chemical properties of roller-milled barley bran and flour. Cereal Chemistry 7, 397–402.

Biacs, P., 2002. Buckwheat. In: Belton, P.S., Taylor, J.N. (Eds.), Pseudocereals and Less Common Cereals. Springer-Verlag, Berlin Germany.

Black, M., Bewley, J.D., Halmer, P., 2006. The encyclopedia of seeds. CABI, Wallingford.

Brown, C.R., 1974. Improved method for determining the degree of abrasion of barley. Journal of the Institute of Brewing 80, 381–382.

Bushuk, W., 1976. History, world distribution, production and marketing. In: Bushuk, W. (Ed.), Rye, Production, Chemistry and Technology. American Association of Cereal Chemists Inc., St. Paul, USA.

Chakraverly, A., Mujimber, A.S., Rachaven, G.S.R.V., Ramaswamy, H.S., 2003. Handbook of Postharvest Technology. Marcel Dekker Inc., New York, USA.

Chinsman, B., 1984. Choice of technique in sorghum and millet milling in Africa. In: Dendy, D.A.V. (Ed.), The Processing of Sorghum and Millets: Criteria for Quality of Grains and Products for Human FoodICC Symposium, Vienna, .

Cubadda, R., Marconi, E., 2002. Spelt wheat. In: Belton, P.S., Taylor, J.N. (Eds.), Pseudocereals and Less Common Cereals. Springer-Verlag, Berlin, Germany.

Deane, D., Commers, E., 1986. Oat cleaning and processing. In: Webster, F.H. (Ed.), Oats: Chemistry and Technology. American Association of Cereal Chemists Inc., St. Paul, USA, pp. 371–412.

Decker, E.A., Rose, D.J., Stewart, D., 2014. Processing of oats and the impact of processing operations on nutrition and health benefits. British Journal of Nutrition 112, S58–S64.

Dexter, J.E., 2004. Grain, paste products, pasta and Asian noodles. In: Scott-Smith, J., Hui, Y.H. (Eds.), Food Processing, Principles and Applications. Blackwell Publishing Inc., Ames, Iowa, United States.

Dick, J.W., Youngs, V.L., 1988. Evaluation of durum wheat, semolina and pasta in the United States. In: Fabriani, G., Lintas, C. (Eds.), Durum Chemistry, and Technology. American Association of Cereal Chemists Inc., St. Paul, USA.

Dietrych-Szostak, D., Oleszek, W., 1999. Effect of processing on the flavonoid content in buckwheat (*Fagopyrum esculentunm* Moench) grain. Journal of Agricultural and Food Chemistry 47, 4384–4387.

Dongson, L., 2012. www.flourmillers.eu/uploads/China_Flour_Milling_Industry_BeijingEFM_2012.pdf.

Duensing, W.J., Roskens, A.B., Alexander, R.J., 2003. Corn dry milling: processes, products and application. In: White, P.J., Johnson, L.J. (Eds.), Corn Chemistry and Technology, second ed. American Association of Cereals Chemists Inc., St. Paul, MN, USA.

Ekstrand, B., Gangby, I., Åkesson, G., 1992. Lipase activity in oats-distribution, pH dependence and heat inactivation. Cereal Chemistry 69 (4), 379–381.

Evers, A.D., McRitchie, F., 2000. Ash: who needs it. World Grain 18 (6), 56–63.

Fang, C., Campbell, G.M., 2003. On predicting roller milling performance IV: effect of roll disposition on the particle size distribution from first break milling of wheat. Journal of Cereal Science 37, 21–29.

FAO, 1980. NFA Standard Specification for Milled Rice (2nd Revision) TRED SQAD No. 2. http://www.fao.org/docrep/x5048e/x5048e03.htm.

Fowler, M., March 2013. Fine Tuning the Purification Process. World Grain.

Fowler, M., July 2014. The Complexities of Durum Milling. World Grain.

Ganssman, V., Vorwerk, K., 1995. Oat milling, processing and storage. In: Welch, R.W. (Ed.), The Oat Crop: Production and Utilization. Chapman & Hall, UK, pp. 369–408.

Gates, F., 2007. Role of Heat Treatment in the Processing and Quality of Oat Flakes (Thesis). Dept. of Food Technology, University of Helsinki, Finland.

Greer, E.N., Hinton, J.J.C., Jones, C.R., Kent, N.L., 1951. The occurrence of endosperm cells in wheat flour. Cereal Chemistry 28, 58–67.

Gwirtz, J.A., Garcia-Casal, M.N., 2014. Processing maize flour and corn meal food products. Annals of the New York Academy of Sciences 1312, 66–75.

Hareland, G.A., Shi, Y., February 1995. Interaction effects of five milling variables of durum wheat in the first break system. Association of Operative Millers Technical Bulletin 6871–6874.

House, L.R., Gomez, M., Murty, D.S., Sun, Y., Verma, B.N., 2000. Development of some agricultural industries in several African and Asian countries. In: Smith, C.W., Frederiksen, R.A. (Eds.), Sorghum, History, Technology and Production. John Wiley Inc., New York, USA.

IRRI Rice Knowledge Bank (undated). http://knowledgebank.irri.org.

Johnson, L.A., 1991. Corn: production, processing, and utilization. In: Lorenz, K.J., Kulp, K. (Eds.), Handbook of Cereal Science and Technology. Marcel Decker Inc., New York, USA.

Jones, C.R., 1958. The essentials of the flour milling process. Proceedings of the Nutrition Society 17 (1), 7–15.

Jones, C.R., Halton, P.R., Stevns, D.J., 1959. The separation of flour into fractions of different protein contents by means of air classification. Journal of Biochemical and Microbiological Technology and Engineering 1, 77–98.

Jones, G., 2001. The Millers. Carnegie Publishing Ltd., Lancaster, UK.

Kebakile, M.M., 2010. Improvements in Sorghum Milling Technologies. INTSORMIL, USA. http://digitalcommons.unl.edu/cgi/viewcontent.cgi?article=1023&context=intsormilpresent.

Kent, N.L., 1965. Effect of moisture content of wheat and flour on endosperm breakdown and protein displacement. Cereal Chemistry 42, 125.

Kent, N.L., 1966. Subaleurone cells of high protein content. Cereal Chemistry 43, 585–601.

Kent, N.L., 1983. Technology of Cereals, third ed. Pergamon Press, Oxford, UK.

Kent, N.L., Evers, A.D., 1994. Kent's Technology of Cereals, fourth ed. Elsevier Science.

Keriene, I., Manceviciene, A., Bliznikas, S., Cesnuleviene, S., Jablonskyte- Rasce, D., Maiksteniene, S., 2016. The effect of buckwheat processing on the content of mycotoxins and phenolic compounds. CyTA Journal of Food 14, 565–571. http://dx.doi.org/10.1080 /19476337.2016.1176959.

Klabunde, H., 2004. Mahlverfahren und Diagrammkunde. In: Erling, P. (Ed.), Hanbuch Mehl- und Schämüllerei. Agrimedia GmbH, Bergen/Dumme, Germany.

Kiryluk, J., Kawka, H., Gasiorowski, H., Chalcerz, A., Aniola, J., 2000. Milling of barley to obtain β-glucan enriched products. Nahrung 44 (4), S238–S241.

Kuenzli, T., 2001. Particle size requirements of semolina for pasta production. In: Kill, R.C., Turnbull, K. (Eds.), Pasta and Semolina Technology. Blackwell Science Ltd., Oxford, UK.

Létang, C., Samson, M.-F., Lasserre, T.-M., Chaurand, M., Abécassis, J., 2002. Production of starch with very low protein content from soft and hard wheat flours by jet milling and air-classification. Cereal Chemistry 79 (4), 535–543.

Lorenz, K., 2000. Rye. In: Kulp, K. (Ed.), Handbook of Cereal Science and Technology, second ed. Marcel Decker Inc., New York, USA.

Lucas, A., 2006. Wind, Water, Work; Ancient and Medieval Milling Technology. Brill Publishers, Leiden, Netherlands.

McClements, D.J., Decker, F.A., 2008. Lipids. In: Damodarin, S., Parkin, K., Fennema, O.R. (Eds.), Fenema's Food Chemistry. CRC Press, Boca Raton, FL, USA.

Macri, L.J., Ballance, G.M., Larter, E.N., 1986. Factors affecting the breadmaking potential of four secondary hexaploid triticales. Cereal Chemistry 63, 263–267.

Marconi, E., Carcea, M., Graziano, M., Cubadda, R., 1999. Kernel properties and pasta-making property of five European spelt wheat (*Triticum spelta* L.) cultivars. Cereal Chemistry 76 (1), 25–29.

Müller, U., Schneiweiss, V., 2006. Mechanical, thermal and hydrothermal modification of flour. In: Popper, L., Schäfer, W., Freund, W. (Eds.), Future of Flour. Agrimedia GmbH, Bergen, Norway.

Nesbitt, M., Samuel, D., 1996. From staple crop to extinction? In: Padulosi, S., Hammer, K., Hella, J. (Eds.), Hulled Wheats. Proceedings of the First International Workshop on Hulled Wheats 21–22 July 1995. IPGRI, Rome, Italy.

NABIM, 2014. Flour and Bread Consumption. http://www.nabim.org.uk/statistics/ flour-and-bread-consumption.

North Dakota Wheat Commission, 2016. http://www.ndwheat.com/.

Perren, R., 2000. Food processing industries. In: Collins, E.J.T. (Ed.), The Agrarian History of England and Wales. VII 1850–1914. Cambridge University Press, Cambridge, UK.

Posner, E.S., 2009. Wheat flour milling. In: Khan, K., Shewry, P.R. (Eds.), Wheat Chemistry and Technology, fourth ed. American Association of Cereal Chemists Inc., St. Paul, USA.

Rahnavathi, C.V., 2016. Sorghum processing and utilization. In: Ratnvathi, C.V., Patil, J.V., Chavan, U.D. (Eds.), Sorghum Biochemistry – An Industrial Perspective. Academic Press, London, UK.

Ram, S., Misra, B., 2010. Cereals, Processing, and Nutritional Quality. New India Publishing Agency, New Delhi, India.

Ranum, P., Peno-Rosas, J., Garcia-Casal, M.N., 2014. Global maize production, utilization and consumption. Annals of the New York Academy of Sciences 1312, 105–112.

Reichert, R.D., 1982. Sorghum dry milling. In: House, L.R., Mughogho., L.K., Peacock, J.M. (Eds.). House, L.R., Mughogho., L.K., Peacock, J.M. (Eds.), Sorghum in the Eighties, vol. 2. ICRISAT, Patancheru, India.

Reichert, R.D., Tatarynavich, J.T., Tyler, R.T., 1986. Abrasive dehulling of quinoa (*Chenopodium quinoa*): effect on saponin content as determined by an adapted hemolytic assay. Cereal Chemistry 63, 471–475.

Rohde, W., 2004. Schälmüllerei. In: Erling, P. (Ed.), Handbuch Mehl und Schälmüllerei, 2. Auflage. Agrimedia, Bergen, Norway.

Rooney, L.W., Kirleis, A.W., Murty, D.S., 1986. Traditional foods from sorghum: their production, evaluation and nutritional value. Advances in Cereal Science and Technology 8, 317–353.

Rooney, L.W., Serna-Saldivar, S., 1991. Sorghum. In: Lorenz, K.J., Kulp, K. (Eds.), Handbook of Cereal Science and Technology. Marcel Dekker Inc., New York, United States.

RSA Grain Milling Federation, 1997. National Training Scheme Maize Milling Manual, second ed. Grain Milling Federation, Hennopsmeer Republic of South Africa.

SAGIS, 2015/16. South African Grain Information Service. http://sagis.org.za/.

Serna Saldivar, S.O., 2010. Cereal Grains, Properties, Processing and Nutritional Attributes. CRC Press, Boca Raton, FL, USA.

Sapirstein, H.D., Bushuk, W., 2016. Rye grain, its genetics, production and utilization. In: Wrigley, C.W., Corke, H., Seethanramen, K., Faubion, J. (Eds.), Encyclopedia of Food Grains. Academic Press, Oxford, UK.

Sharrer, G.T., Welsh, P.C., 2003. Dictionary of American History. The Gale Group, Framlingham Hils Mchigan, USA. http://www.encyclopedia.com/topic/Flour_mills.aspx.

Shaw, M., 1970. Rye milling in U.S.A. Association of Operative Millers Bulletin 3203–3207.

Shende, D., Sidhu, G.K., 2014. Methods used for the extraction of maize (*Zea mays* L.) germ oil – a review. Indian Journal of Research and Technology 2, 48–54.

Storck, J., Teague, W.D., 1952. Flour for Man's Bread. Univ. of Minnesota Press, Minneapolis.

Swieca, M., Dziki, D., Gawlik-Dziki, U., Rozylo, R., Andruszczak, S., Kraska, P., Kowalczyk, D., Payle, E., Baraniak, B., 2014. Grinding and nutritional properties of six spelt (*Triticum aestivum* ssp. spelta L.) cultivars. Cereal Chemistry 91, 247–254.

Symons, S.J., Venora, G., Schepdael, L., Shahin, M.A., 2009. Measurement of spaghetti speck count, size and color using an automated imaging system. Cereal Chemistry 86 (2), 164–169.

Tangpinijkul, N., 2010. Rice Milling System. http://www.doa.go.th/aeri/files/pht2010/documents_slide/training%20materials/6_rice_milling_system.pdf.

Taylor, J.R.N., Belton, P.S., 2002. Sorghum. In: Belton, P.S., Taylor, J.N. (Eds.), Pseudocereals and Less Common Cereals. Springer-Verlag, Berlin.

Taylor, J.R.N., Parker, M.L., 2002. Quinoa. In: Belton, P.S., Taylor, J.N. (Eds.), Pseudocereals and Less Common Cereals. Springer-Verlag, Berlin.

Ulmer, K., 2011. Technology and Equipment Grain Milling. Bühler, Uzwil. www.buhlergroup.com.

Varriano-Marston, E., Hoseney, R.C., 1983. Barriers to increased utilization of pearl millet in developing countries. Cereal Foods World 28, 392–395.

Wahid, S., Hajar, I., Hashifah, M.A., 2006. Effect of humidified polisher on the physical quality of milled rice during storage. Journal of Tropical Agriculture and Food Science 34, 289–294.

Williams, P., 2011. A Practical Introduction to Cereal Chemistry. PDK Projects Inc., Nanaimo, BC, Canada.

Youngs, V.L., Püskülcü, M., Smith, R.R., 1977. Oat lipids I. composition and distribution of lipid components in two oat cultivars. Cereal Chemistry 54 (4), 803–812.

Zeilinski, H., Ceglinska, A., Michalska, A., 2008. Bioactive compounds in spelt bread. European Food Research and Technology 226, 537.

Further reading

Bushuk, W. (Ed.), 2001. Rye: Production, Chemistry and Technology, second ed. American Association of Cereal Chemists Inc., St. Paul, MA, USA.

Champagne, E.T. (Ed.), 2001. Rice: Chemistry and Technology, third ed. American Association of Cereal Chemists Inc., St. Paul, MA, USA.

Dendy, D.A.V. (Ed.), 1995. Sorghum and Millets: Chemistry and Technology. American Association of Cereal Chemists Inc., St. Paul, MA, USA.

Milling and Grain (Monthly Journal). www.millingandgrain.com.

Shewry, P.R., Ullrich, S.E. (Eds.), 2014. Barley: Chemistry and Technology, second ed. American Association of Cereal Chemists Inc., St. Paul, MA, USA.

Sissons, M., Abecassis, J., Marchylo, B., Carcea, M., 2016. Durum Wheat, Chemistry and Technology. American Association of Cereal Chemists Inc., St. Paul, MA, USA.

Ulmer, K., 2011. Technology and Equipment. Bühler, Uzvil, Switzerland. www.buhlergroup.com.

Webster, F.H., Wood, P.J. (Eds.), 2011. Oats: Chemistry and Technology, second ed. American Association of Cereal Chemists Inc., St. Paul, MA, USA.

World Grain (Monthly Journal). www.World-Grain.com.

Wrigley, C., Corke, H., Seetharamen, K., Faubion, J., 2016. Encyclopedia of Food Grains, second ed. Academic Press, Oxford, UK.

White, P.J., Johnson, L.A. (Eds.), 2003. Corn: Chemistry and Technology, second ed. American Association of Cereal Chemists Inc., St. Paul, MA, USA.

Flour treatments, applications, quality, storage and transport

7.1 INTRODUCTION

In milling cereals by the gradual reduction system (see Chapter 6), flour is produced by every machine in the break, scratch and reduction systems of the normal mill-flow. The stock fed to each grinding stage is distinctive in composition – in terms of proportions of endosperm, embryo and bran it contains, and the region of the grain from which the endosperm is derived – and each machine flour is correspondingly distinctive in respect of baking quality, colour and granularity, contents of fiber and nutrients, and the amount of ash it yields upon incineration.

By far the most abundant flour consumed in the industrialized world is derived from wheat; because of this, and the unique versatility of wheaten flour, the majority of this chapter is devoted to it. Flours from other cereals are, however, given some consideration.

In the United Kingdom and United States today there are no recognized standards for flour grades (for most grains, yes, including wheat, but not for flour): each miller makes products according to customer requirements, and exercises skill in maintaining regularity of quality for any particular product type – several of which are discussed in this chapter.

7.1.1 Flour grades

If the flour streams from all the machines in the break, scratch and reduction systems are blended together in their rational proportions, the resulting flour is known as 'straight-run grade'. Other grades are produced by selecting and blending particular flour streams, frequently on the basis of their ash yield or grade colour (measures of their nonendosperm tissue content). Table 7.1 shows typical proportions of flour streams (expressed as percentages of the wheat) from a well-equipped and well-adjusted mill in the United Kingdom making flours of fairly low ash yield. The ash yield of the individual flour streams is also shown.

Flour streams with the lowest ash yield (e.g., Group 1 in Table 7.1) may be described as 'patent' flour. Those coming from the end of the milling process and having high ash yield are called 'low grade' in the United Kingdom or 'clear' flour in the United States. Clear flour is used industrially in the United States for the manufacture of alcohol, gluten, starch and adhesives.

7.2 TREATMENTS OF WHEAT FLOUR

7.2.1 Bleaching

Flour contains a yellowish pigment, of which about 95% consists of xanthophyll or its esters, with no nutritional significance. Bleaching of the natural pigment of wheat endosperm by oxidation occurs rapidly when flour is

TABLE 7.1   Typical proportions and ash yields of flour streams

Flour streams	Proportions (% of feed to I Bk)	Ash yield (%, d.m.)
GROUP 1: HIGH GRADE		
A	12.0–21.0	0.35–0.38
B	14.0–17.0	0.35–0.38
C	7.0–10	0.38–0.47
Total Group 1	35–40	0.35–0.40
GROUP 2: MIDDLE GRADE		
D	2.5–7.5	0.39–0.70
E	1.7–2.1	0.45–0.89
G	1.3–3.0	0.75–1.47
I Bk	1.5–2.5	0.50–0.72
II Bk	1.5–3.0	0.53–0.69
III Bk	0.0–1.5	0.70–1.00
III Bk bran finisher flour	0.0–2.5	0.70–1.00
X (Scratch)	0.0–0.7	0.70–0.90
I Bk coarse midds	3.0–6.0	0.50–0.82
II Bk coarse midds	1.5–3.5	0.70–0.84
Total Group 2	25–30	0.70–0.80
GROUP 3: LOW GRADE		
B2	1.2–2.5	0.40–0.45
F	0.7–1.2	0.58–1.35
H	0.6–1.2	0.60–1.53
J	0.5–0.7	0.88–2.25
IV Bk	2.0–4.0	1.00–2.00
IV Bk finisher flour	0.0–1.0	1.50–2.00
V Bk	0.0–1.0	1.00–2.50
Total Group 3	8–10	1.80–2.3

exposed to the atmosphere, but more slowly when flour is stored in bulk, and can be accelerated by chemical treatment. The principal agents used, or formerly used, for bleaching flour are nitrogen peroxide, chlorine, chlorine dioxide, nitrogen trichloride, benzoyl peroxide and acetone peroxide. Significant dates in the history of flour bleaching are summarized in Table 7.2.

TABLE 7.2 Significant dates in the history of flour bleaching

1901	Andrews patents flour treatment with NO_2 (chemical process)
1903	Alsop patents flour treatment with NO_2 (electrical treatment)
1909	NO_2 in use
1911	Keswick Convention – unmarked flour to be unbleached
1921	Benzoyl peroxide first used
1921	J.C. Baker patents NCl_3 as flour bleacher
1922	NCl_3 replaces Cl as bleacher for breadmaking flour
1923	Committee appointed to inquire into use of preservatives and colouring matter in food
1924	Committee's activities extended to chemical substances for flour treatment
1927	Committee reported that bleaching and improving agents were in use, and that Cl, NCl_3 and BzO_2 were not among those least open to objection
1949	NCl_3 use discontinued in the United States
1949	ClO_2 first used in the United States
1955	NCl_3 use discontinued in the United Kingdom
1955	ClO_2 first used in the United Kingdom
1961	Acetone peroxide permitted in the United States (not in the United Kingdom)

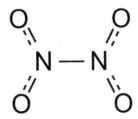

FIGURE 7.1 Chemical structure of nitrogen peroxide (NO_2).

7.2.1.1 Nitrogen peroxide

Nitrogen peroxide (NO_2) (Fig. 7.1), produced by a chemical reaction or the electric arc process, was widely used as a bleaching agent in the early 20th century. Its use has been discontinued (Codex Alimentarius, http://www.fao.org/fao-who-codexalimentarius) except in the United States and Australia, where it is still legally permitted (US FDA 21CFR, https://www.fda.gov).

7.2.1.2 Chlorine

The use of chlorine gas (Cl_2) for treatment of cake flour (except wholemeal) is permitted in the United States up to 2500 mg/kg (https://www.fda.gov).

The chlorine modifies the properties of the starch for high-ratio cake flour. For cake flours the usual level of treatment is 1000–1800 mg/kg. Although listed in the Codex Alimentarius at a level not exceeding 2500 mg/kg, the Bread and Flour Regulations, 1998 forbid the use of chlorine and chlorine dioxide in the United Kingdom (https://www.food.gov.uk/sites/default/files/multimedia/pdfs/breadflourguide.pdf). The use of chlorine is not permitted in most European countries, but it is allowed in flour for all purposes in the United States, Canada, Australia, New Zealand (to 1500 mg/kg) and South Africa (to 2500 mg/kg).

7.2.1.3 Nitrogen trichloride

Nitrogen trichloride gas (NCl_3) (Fig. 7.2), known as 'Agene', was patented as a flour bleach by J.C. Baker in 1921 and replaced chlorine in 1922 as an improving and bleaching agent for breadmaking flour because it was much more effective. Its use was discontinued in the United States in 1949 and in the United Kingdom from the end of 1955, after it had been shown by Mellanby (1946) that flour treated with Agene in large doses might cause canine hysteria (although Agene-treated flour has never been shown to be harmful to human health). NCl_3 reacts with the amino acid methionine, present in wheat protein, to form a toxic derivative, methionine sulphoximine (Bentley et al., 1950).

7.2.1.4 Chlorine dioxide

Chlorine dioxide (ClO_2) (Fig. 7.3), known as 'Dyox', its use has greatly diminished throughout the world, except the United States, where it is still permitted (https://www.fda.gov). It was first used for these purposes in 1949 in the United States and in the United Kingdom in 1955; its use is also permitted in Japan. It is produced by passing chlorine gas through an aqueous solution of sodium chlorite. Dyox gas

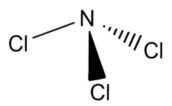

FIGURE 7.2   Chemical structure of nitrogen trichloride (NCl_3).

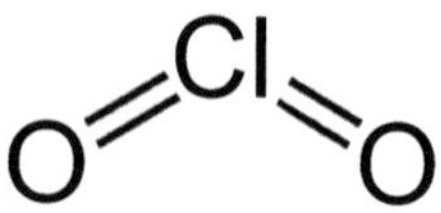

FIGURE 7.3   Chemical structure of chlorine dioxide (ClO_2).

contains a maximum of 4% ClO_2. The chlorine dioxide gas is released by passing air through the solution, and is typically applied to bread-making flour at a rate of 12–24 mg/kg. Chlorine dioxide treatment of flour destroys the tocopherols, unfortunately. The Bread Flour Regulations, 1998 have forbidden the use of chlorine dioxide in the United Kingdom.

7.2.1.5 Benzoyl peroxide

$(C_6H_5CO)_2O_2$ or BzO_2 (Fig. 7.4) is a solid bleaching agent which was first used in 1921. It is supplied as a mixture with inert inorganic fillers such as $CaHPO_4$, $Ca_3(PO_4)_2$, sodium aluminium sulphate or chalk. Novadelox, a proprietary mixture, contains up to 32% of benzoyl peroxide, although 16% is the usual proportion. The dosage rate, normally 45–50 mg/kg, was restricted to 50 mg/kg in the United Kingdom by the Bread and Flour Regulations 1984, but the revised regulations in 1998 have forbidden the use of this bleaching agent. The bleaching action occurs within about 48 h. This bleacher has an advantage over gaseous agents in that only a simple feeder is required, and storage of chemicals presents no hazard; the fact that it has no improving action is advantageous in the bleaching of patent flours. The treated flour contains traces of benzoic acid. BzO_2 is also used in New South Wales, Queensland, the United States, Canada, the Netherlands, New Zealand (up to 40 mg/kg, for pastry flour only) and Japan (up to 300 mg/kg). In the Codex Alimentarius, a maximum level of 60 mg/kg has been stated for use in flours, although in the United States, the FDA has determined that a maximum of one part by weight of benzoyl peroxide must be mixed with not more than six parts of one or any mixture of the following: potassium alum, calcium sulfate, magnesium carbonate, sodium aluminum sulfate, dicalcium phosphate, tricalcium phosphate, starch, or calcium carbonate (https:www.fda.gov).

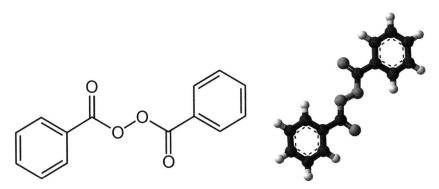

FIGURE 7.4   Chemical structure of benzoyl peroxide ($(C_6H_5CO)_2O_2$).

7.2.1.6 Acetone peroxide

Acetone peroxide (Fig. 7.5) is a dry-powder bleaching and improving agent marketed as 'Keetox' – a blend of acetone peroxides with a diluent such as dicalcium phosphate or starch. The concentration in terms of H_2O_2 equivalent per 100 g of additive plus carrier is 3–10 for maturing and bleaching, and 0.75 for doughmaking. Its use has been permitted in the United States since 1961 (US FDA 21 CFR, https://www.fda.gov), and also in Canada, but it is not as yet permitted in the United Kingdom. It is used either alone or in combination with benzoyl peroxide. The usual dosage rate is 446 mg/kg on a flour mass basis.

7.2.1.7 Flour blending for bleaching treatment

Because the various flour streams differ in their characteristics, the optimum level of bleaching treatment varies correspondingly; the lower grade flours (those nearer the tail end of the break and reduction systems) in general require more treatment than the patent flours. It is therefore customary to group the machine flours according to quality into three or four streams for treatment. A possible grouping is indicated in Table 7.1. Each group would be given an appropriate bleacher treatment: e.g., the lowest 20% of flour might receive treatment at 10 times the rate for the best quality 50%. The final grades are then made up by blending two or more of the groups in desirable proportions.

7.3 FLOURS FOR VARIOUS PURPOSES

Wheat flour is used for making food products of widely varying moisture content (see Table 7.3).

The proportions in which the various ingredients of baked products are present in the recipe, relative to flour (100 parts), are shown in Table 7.4 Biscuit dough is stiff to permit rolling and flattening; bread dough is a plastic mass that can be moulded and shaped; and wafer batter is a liquid suspension that will flow through a pipe.

For comparison with products listed in Table 7.4, a typical wholemeal wheat extruded snack formulation would contain the following amounts of ingredients in relation to 100 g white flour: 7 g soya protein, 14 g wheat

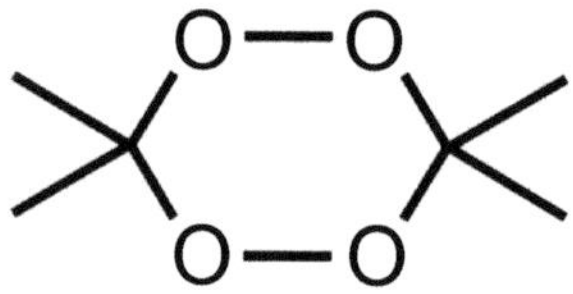

FIGURE 7.5   Chemical structure of acetone peroxide ($C_6H_{12}O_4$).

TABLE 7.3   Flour-based products and their moisture contents[a]

Type of product	Moisture content		Moisture level
	Range (%)	Mean (%)	
Soup	78–80	85	High
Puddings	13–67	45	Medium
Bread	35–40	38	Medium
Cakes	5–30	17	Medium
Pastry		7	Low
Biscuits (cookies, crackers)	1–6	5	Low

[a] *Data extracted from McCance, R.A., Widdowson, E.M., 1967. The Composition of Foods, Med. Res Coun., Spec. Rpt. Ser. 297, 2nd Imp., H.M.S.O., London, UK.*

bran, 1.4 g oil, 0.4 g emulsifier, 23 g water, 7 g sugar, 2 g salt, 2 g dicalcium phosphate and 3.6 g milk powder (Guy, 1993).

The flour content of various flour-containing products, as purchased or consumed, is shown in Table 7.5.

The Codex Alimentarius Commission of the United Nations Food and Agriculture Organization issued standard 152 on flour for human consumption in 1985 (amended in 1995 and 2016). It defines acceptable sources as *Triticum aestivum* L. bread wheat and *T. compactum* club wheat, the required protein and moisture contents (at least 7% and not more than 15.5%, respectively), fat acidity, particle size (98% through a 212 μm sieve) and the protocol for ash determination. Optional ingredients and approved additives are also listed.

In the United Kingdom flour for human consumption should conform to the nutritional requirements set out in the Bread and Flour Regulations 1984 (and updated in 1998; discussed elsewhere in this book).

For each purpose, flour with particular properties is required: these are secured, in the first place, by choice of appropriate wheat grist in terms of strong and weak wheats. Of the total flour milled in the United Kingdom in 1990–1991, 63% was used for bread, 15% for biscuits, 6% for household use, 2% for cakes, 2% for starch manufacture and 12% for 'other products'.

'Other' food products made with wheat flour include pastries, meat pies, sausages, sausage rolls, rusks, pet foods, baby foods, invalid foods, chapattis, buns, scones, teacakes, pizzas, soups, premixes, liquorice, batter (for fish frying), chocolate and sugar confectionery, cereal convenience foods, snack foods, ready-to-eat breakfast cereals, puddings, gravy powders, blancmange and brewing adjuncts. Specific requirements for flours for some of these purposes are described below.

TABLE 7.4 Proportions of constituents in recipes for baked products[a] (relative to flour: 100 parts). Based upon data from Flour Milling and Baking Research Association.

Type of product	Water	Fat	Salt	Whole egg	Raising agent	Milk powder	Sugar
Yeasted products							
Bread, CBP[b]	61	0.7	1.8		(Yeast) 1.8		
Bread, LFP[c]	57	0.7	1.8		1.1		
Cream crackers	32	12.5	1.0		0.1–2.0		
Pastry							
Short	25	50	2.0				
Pie	31	43	2.0				
Steak and kidney pudding	30–36	50	0.7				
Puff	40–50	50–70	0.7				
Choux	125	50		150			
Biscuits							
Hard sweet	20	17	0.7		1.1[a]	(Whey) 2.6	22
Soft	10	32	0.1		0.5[a]	2.0	30
Cake							
Plain	50	40		35	3.5[a]		40
High ratio	70	65	2.0	60	5.0[a]	8	120
Sponge			1.0	170			100
Wafer batter	150	3	0.2		0.3[a]		

[a] Mixtures of sodium and ammonium carbonate or bicarbonate.
[b] Chorleywood bread process.

TABLE 7.5   Flour content of flour-based foods, as purchased or consumed[a]

Food product	Flour content		Parts of product per 100 pt flour
	Range (%)	Mean (%)	
Crispbread		112	90
Biscuit – semisweet	67–82	74	135
Biscuit – ginger nut	43–57	49	205
Bread	53–72	70	145
Short pastry	60–80	65	155
Buns, scones, teacakes	36–57	45	220
Cakes, pastries, chocolate wafer	23–40	33	300
Biscuits (chocolate)	8–46	30	330
Puddings	6–40	25	400

[a] *Based upon data from Flour Milling and Baking Research Association. Flour wt at natural m.c. Product wt at m.c. of final product.*

7.3.1 Bread flour

The predominance of wheat flour for making aerated bread is due to the properties of its protein, which, when the flour is mixed with water, forms an elastic substance called gluten (see Chapter 4). This property is found to a slight extent in rye but not in other cereals.

The ability to produce a loaf of relatively large volume with regular, finely vesiculated crumb structure is possessed by flours milled from wheats described as 'strong'. Protein strength is an inherent characteristic, but the amount of protein present can be influenced by the conditions under which wheats are grown and the cultivar. Protein content is also an important determinant of bread quality, as there is a positive correlation between loaf specific volume (mL/g) and the percentage of protein present.

Typical characteristics of Chorleywood bread process (CBP) flour, bakers' flour (as used in the bulk fermentation process) and rollermilled wholemeal in the United Kingdom are shown in Table 7.6.

7.3.1.1 Maturing and improving agents

The breadmaking quality of freshly milled flour tends to improve during storage for a period of 1–2 months. The improvement occurs more rapidly if the flour is exposed to the action of the air. During such aerated

TABLE 7.6   Typical UK bread flour analysis, 1992. Based upon data from Flour Milling and Baking Research Association.

	CBP	Bakers'	Wholemeal
Moisture	14.6%	14.5%	14.6%
Protein	11.0%	12.1%	14.7%
Grade colour	2.1	2.2	–
Falling number	329	334	330
Alpha-amylase[a]	15FU	22FU	21FU
Starch damage	30FU	34FU	–
Water absorption	60.2%	62.0%	70.2%

[a] *Farrand units, (includes fungal enzyme).FMBRA.*

storage, fat acidity increases at first, owing to lipolytic activity, and later decreases by lipoxidase action; products of the oxidation of fatty acids appear; the proportion of linoleic and linolenic acids in the lipids falls; and disulphide bonds ($-S-S-$) decrease in number.

The change in baking quality, known as maturation or 'ageing', can be accelerated by chemical 'improvers' which modify the physical properties of gluten during fermentation in a way that results in bread of better quality being obtained. Matured flour differs from freshly milled flour in having better handling properties and increased tolerance in the dough to varied conditions of fermentation, and in producing loaves of larger volume and more finely textured crumb.

Improving agents permitted are listed in the UK Bread and Flour Regulations 1984 (SI, 1984, No. 1304), as amended by the Potassium Bromate (Prohibition as a Flour Improver) Regulations and further amended in 1998, 1990 (SI, 1990; No. 399). The Codex Alimentarius lists the following agents which are approved for use in other countries: L-ascorbic acid (300 mg/kg), L-cysteine hydrochloride (90 mg/kg), sulphur dioxide (in flours for biscuit and pastry only) (200 mg/kg), mono-calcium phosphate (2500 mg/kg), lecithin (2000 mg/kg), and the bleaching ingredients discussed above. Besides their improving effect, these substances give a whitened appearance to the loaf because of their beneficial effect on the texture of the crumb. Improving agents do not increase the carbon dioxide production in a fermented dough, but they improve gas retention (because the dough is made more elastic) and this results in increased loaf volume.L

7.3.1.2   *Redox improvers*

The action of improvers is believed to be an oxidation of the cysteine sulphydryl or thiol (–SH) groups present in the wheat gluten. As a result, these thiol groups become unavailable for participation in exchange reactions with disulphide (—S—S—) bonds – a reaction which is considered to release the stresses in dough – and consequently the dough tightens, i.e., the extensibility is reduced. Alternatively, it has been suggested that the oxidation of –SH groups may lead to the formation of new —S—S— bonds, which would have the effect of increasing dough rigidity.

Potassium bromate ($KBrO_3$) (Fig. 7.6) has been used commercially as a bread improver since 1923. The rate of treatment is 10–45 mg/kg on flour weight. The substance acts as an oxidizing agent after the flour has been made into a dough; it increases the elasticity and reduces the extensibility of the gluten. Treatment with bromate has a similar action to that of ageing or maturing the flour, and enables large bakeries to use a constant fermentation period.

Potassium bromate is added to flour after being suitably diluted with an inert filler such as calcium carbonate or calcium sulphate. Proprietary brands of improver contain 6%, 10%, 25% or 90% of potassium bromate. The 6% brand is added at the rate of 0.022%. Higher levels of potassium bromate are used in chemical dough development processes.

In many European countries potassium bromate has never been allowed; it was specifically excluded from the list of permitted additives in the United Kingdom by the Potassium Bromate (Prohibition as a Flour Improver) Regulations 1990, giving rise to considerable initial difficulties in the baking industry. The changes in the use of oxidizing improvers, consequent upon the deletion of potassium bromate, are considered in the next chapter. Its use has also been voluntarily discontinued in Japan and it is now little used in New Zealand. Hazards associated with potassium bromate include the fact that, as a strong oxidizing agent, it can cause fire or explosions. It is also toxic and there is strong evidence for its carcinogenicity. At normal permitted levels of addition, however, it is not considered to persist at a significant level in the baked product.

Potassium bromate remains in use in the United States, although an agreement exists between the government and millers to reduce usage to a minimum. Although permitted in Canada, its use has declined over the years in that country (Ranum, 1992).

Since untreated flour already contains 1–8 mg/kg of bromine (Br), the bread made with untreated flour contains 0.7–5.6 mg/kg of natural Br. Flour

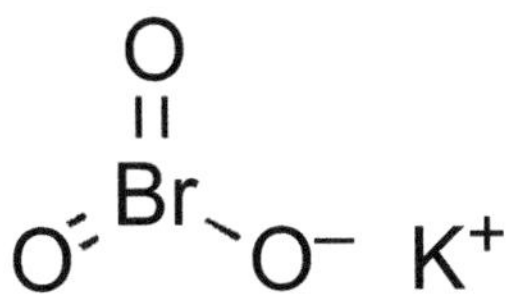

FIGURE 7.6 Chemical structure of potassium bromate ($KBrO_3$).

treatment with 45 mg/kg of bromate leaves a residue of 15 mg/kg of Br in the loaf, increasing the total Br content of the bread to about 18 mg/kg.

Use of potassium bromate is permitted to 75 mg/kg in the United States; to 50 mg/kg in Canada and Sweden; to 40 mg/kg in Russia; and in Eire to about 18 mg/kg. $KBrO_3$ is not allowed in the Netherlands or Australia. The greatest need for bromate occurs in continuous-mix baking, no-time doughs, frozen doughs and overnight sponges, as used in Cuba and other Latin American countries. The typical level of addition in these types of baking approaches the 75 mg/kg maximum (Ranum, 1992).

L-Ascorbic acid (vitamin C) (Fig. 7.7), E300, was first used as a bread improver by Jørgensen in 1935. It can be used for this purpose particularly in mechanical development processes of breadmaking, such as the CBP. The volume increase resulting from use of ascorbic acid is generally less than that obtained with an equivalent weight of potassium bromate, and it is more costly. The improving effect of ascorbic acid is mediated by enzymes present in the flour. The functional form is the oxidized form dehydroascorbic acid (DHA), which is highly effective but cannot be used directly as it is unstable. Ascorbic acid is oxidized to DHA through catalytic action of ascorbic acid oxidase. Injection of oxygen during mixing hastens the oxidation, making ascorbic acid more effective (Chamberlain and Collins, 1977). The oxidation to DHA is improved if the headspace of the mixing machine contains an oxygen-enhanced atmosphere, e.g., a 50/50 mixture of oxygen and air, equivalent to a mixture of 60% oxygen plus 40% nitrogen (Chapter 8). Under these circumstances, ascorbic acid alone is as effective an oxidizing agent as is a combination of ascorbic acid and potassium bromate used when the dough is mixed under partial vacuum. An enzyme 'DHA reductase' is required for oxidation of sulphydryl (–SH) compounds by DHA.

Ascorbic acid strengthens the gluten; gas retention is thus improved and loaf volume augmented. Ascorbic acid does not hasten proving. The maximum permitted levels (1989) are 50 mg/kg in Belgium and Luxembourg, 100 mg/kg in the Netherlands, 200 mg/kg in Canada, Denmark, Italy, Spain,

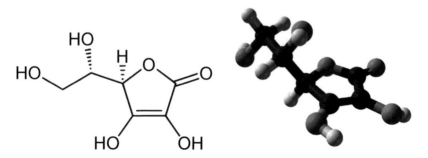

FIGURE 7.7   Chemical structure of L-ascorbic acid ($C_6H_8O_6$).

the United States, and 300mg/kg in France. No maximum level is specified in Australia, Greece, Portugal or Germany since ascorbic acid is reckoned to be quite safe. Assessment by the COT (Committee on Toxicity of Chemicals in Food, Consumer Products and the Environment; https://cot.food.gov.uk), and recently by the European Food Safety Authority (EFSA) has confirmed no adverse effects through use of ascorbic acid (Aguillar et al., 2015). Use of ascorbic acid is also permitted in Japan, New Zealand and Sweden.

Azodicarbonamide (1,1′azobisformamide; NH_2–CONNCONH$_2$; 'ADA') (Fig. 7.8) is a flour maturing agent, marketed as 'Maturox', 'Genitron' or 'ADA 20%' ('AK20'). 'Maturox' contains either 10% or 20% of ADA; 'Genitron' contains 20% or 50% of ADA, dispersed in an excipient, generally calcium sulphate and magnesium carbonate. The particle size of ADA is generally 3–5 µm. It was first used in the United States in 1962 (maximum permitted level 45mg/kg on flour weight). When mixed into doughs it oxidizes the sulphydryl (–SH) groups and exerts an improving action. Oxidation is rapid and almost complete in doughs mixed for 2.5min. Short mixing times are thus appropriate. The residue left in the flour is biurea. Flour treated with ADA is said to produce drier, more cohesive dough than that treated with chlorine dioxide, to show superiority in mixing properties and to tolerate higher water absorption. An average treatment rate is 5mg/kg (on flour weight) in bulk fermentation and low-speed mixing methods of baking, and 20–25mg/kg (on flour weight) in the high-speed mixing CBP. The agent does not bleach, but the bread made from treated flour appears whiter because of its finer cell structure; in fact, it is listed as a bleaching ingredient by the US FDA in 21 CFR. The use of ADA has been permitted, to a maximum level of 45mg/kg, in the United Kingdom since 1972. Its usage is also permitted in Canada, New Zealand and the United States, but not in Australia or in EU countries other than the United Kingdom (1989). In the Bread and Flour Regulations 1998, however, this was no longer allowed for use in the United Kingdom. It is still listed in the Codex Alimentarius at a maximum of 45 mg/kg.

L-Cysteine (Cys) (Fig. 7.9) is a naturally occurring amino acid, is used in the activated dough development process (ADD) (cf. Chapter 8), in which it functions as a reducing agent. The addition of L-cysteine

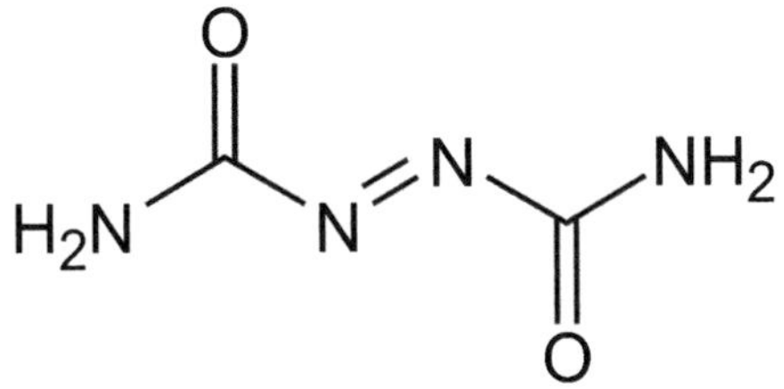

FIGURE 7.8   Chemical structure of azodicarbonamide ($C_2H_4N_4O_2$).

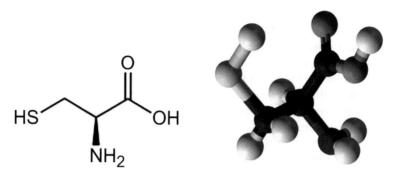

FIGURE 7.9   Chemical structure of L-cysteine (Cys) ($C_3H_7NO_2S$).

(in the form of L-cysteine hydrochloride or L-cysteine hydrochloride monohydrate) to bread doughs for this purpose is permitted in Denmark (up to 25 mg/kg), Germany (up to 30 mg/kg), Belgium (up to 50L mg/kg), Australia, New Zealand and the Netherlands (up to 75 mg/kg), Canada (up to 90 mg/kg) and Sweden (up to 100 mg/kg). L-Cysteine is not mentioned as a permitted additive in France, Greece, Italy, Luxembourg, Portugal, Spain or the United States.

Cysteine accelerates reactions within and between molecules in the dough that lead to an improvement in its viscoelastic and gas-holding properties. These reactions normally take place slowly during bulk fermentation, but with the addition of cysteine the bulk fermentation period can be eliminated. Cysteine, which is a rapid-acting reducing agent, is used in the ADD in conjunction with slow-acting oxidizing agents, such as ascorbic acid and potassium bromate (where permitted) or azodicarbonamide, which complete the 'activation' commenced by the cysteine. The dough-softening action of cysteine reduces the work input required for the production of fully developed dough.

7.3.1.3 Blending for improver treatment

The principles applied to bleaching flours of different grades also apply to improver treatment (Table 7.1).

7.3.1.4 Emulsifiers and stabilizers

'Emulsifiers' and 'stabilizers' are any substances capable of aiding formation of (emulsifiers) or maintaining (stabilizers) the uniform dispersion of two or more immiscible substances. Flours, sold as such, are not allowed to contain emulsifiers but the following are permitted by the Bread and Flour Regulations (1984) to be included in bread:

- E 322 lecithins;
- E460 α-cellulose (Fig. 7.10) (permitted only in bread for which a slimming claim is made);

- E466 carboxymethyl cellulose sodium salt (Fig. 7.11) (permitted only in bread for which a slimming claim is made);
- E471 monoglycerides and diglycerides of fatty acids;
- E472(b) lactic acid esters of monoglycerides and diglycerides of fatty acids;
- E472(c) citric acid esters of monoglycerides and diglycerides of fatty acids;
- E472(e) monoacetyl and diacetyl tartaric acid esters of monoglycerides and diglycerides of fatty acids;
- E481 sodium stearoyl-2-lactylate (Fig. 7.12);
- E482 calcium stearoyl-2-lactylate (Fig. 7.13);
- E483 stearyl tartrate (Fig. 7.14);
- E481 and E482 are subject to a maximum level of 5000 mg/kg.

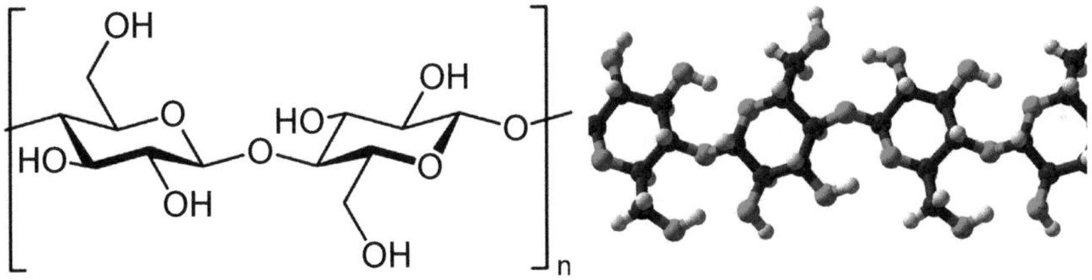

FIGURE 7.10 Chemical structure of cellulose $((C_6H_{10}O_5)_n)$.

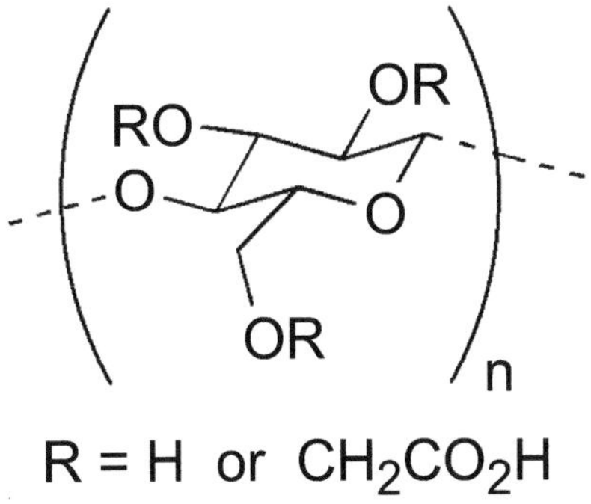

$$R = H \text{ or } CH_2CO_2H$$

FIGURE 7.11 Chemical structure of carboxymethyl cellulose $(C_8H_{15}NaO_8)$.

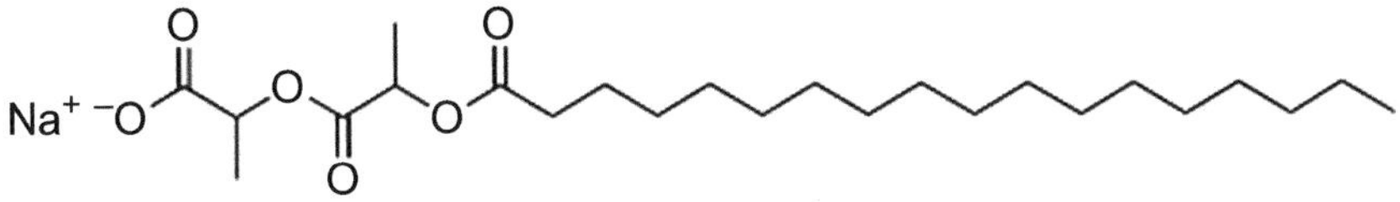

FIGURE 7.12 Chemical structure of sodium stearoyl-2-lactylate $(C_{24}H_{43}NaO_6)$.

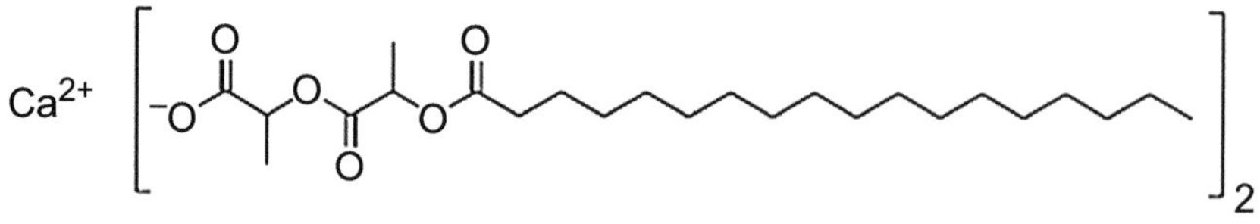

FIGURE 7.13　Chemical structure of calcium stearoyl-2-lactylate ($C_{48}H_{86}CaO_{12}$).

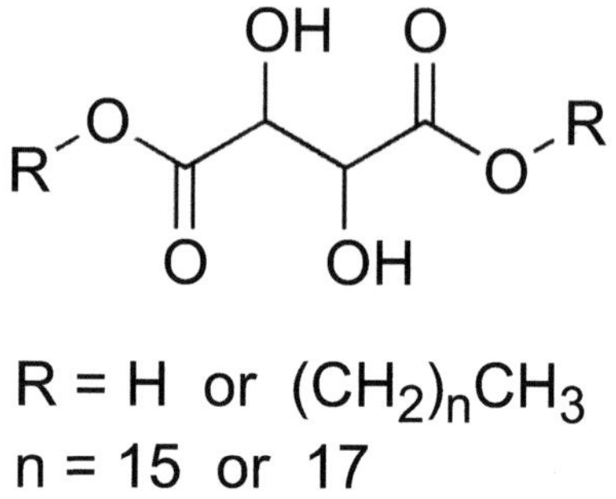

R = H or $(CH_2)_nCH_3$

n = 15 or 17

FIGURE 7.14　Chemical structure of stearyl tartrate ($C_{40}H_{74}O_6$ to $C_{38}H_{78}O_6$).

7.3.2 Biscuit (cookie, cracker) flour

Flours for short and semisweet biscuits are typically produced from grists containing mainly soft wheats, with some hard wheats included to increase the rate of production in the mill. However, hard wheat flours produce thinner biscuits than soft wheats, so it is important to use a narrow range of levels of hard wheat in the flour. The level specified will depend on the manufacturer's preference, since a biscuit plant will have to produce biscuits of particular sizes and weights to suit the packing plant.

There is no developed gluten network in short biscuit doughs, hence neither the level nor the quality of protein is significant in production. However, consistency of quality is critically important in modern high-speed production plants. Flours would normally be specified to have, say, a range of 1% protein within the typical range for flours (8%–10%).

Semisweet biscuits have a developed gluten network which is modified during processing, and for these biscuits low-protein flours (typically 8.5%–9.5%) with weak, extensible glutens are used. At present, sulphur dioxide (SO_2, usually obtained from sodium metabisulphite added at the mixer) is used to increase the extensibility and decrease the elasticity of the doughs. This aids control of the dough sheet and hence biscuit thickness. EU proposals include permission of SO_2 in fine bakery wares, up to 50 mg/kg in the final product. Nevertheless, wheat breeders are seeking to develop varieties which perform well without its use.

For wafers, low-protein flour milled from weak wheat is suitable. Particle size is an important characteristic for this application; ideally about 55% should be below 40 µm, 35% between 40 and 90 µm and not more than 10% of the flour coarser than 90 µm. Too fine a flour produces light, tender, fragile wafers, while too coarse a flour produces incomplete sheets of unsatisfactory wafers.

Gluten development in wafer batters must be avoided, so flours which have a low tendency to produce an aggregated gluten under low shear rates in aqueous flour batters are required. Hence low-protein flours with weak, extensible glutens are normally specified.

Cracker doughs have fully developed gluten networks and protein quality is important in dough processing. Cracker flours with medium protein contents (9.5%–10.5%) made mainly from hard wheats are commonly used. Matzos are water biscuits made from unbleached, untreated flour and water only.

7.3.2.1 Emulsifiers

Emulsifiers are used in biscuits, either as processing aids or as partial replacements for fat. Very low levels (e.g., 0.1% on fat basis) of sodium stearoyl-2-lactylate (E481) in sheeted biscuit doughs produce a smooth, nonsticky surface which aids dough-piece cutting. Lecithin (E322) is commonly used in wafer batters to aid release of the baked wafer from the wafer baking machinery. Emulsifiers which can replace a substantial proportion of fat in biscuits without serious deterioration in product quality are sodium stearoyl-2-lactylate and the diacetyltartaric acid esters of monoglycerides of fatty acids (E472e). Lecithin can be used to replace a low level of fat in biscuits.

7.3.3 Flours for confectionery products

7.3.3.1 Cake flour

Flour in cakes should allow an aerated structure to be retained after the cake has been built up. The stability of the final cake depends largely upon the presence of uniformly swollen starch granules; hence the granules should be undamaged during milling, free from adherent protein and unattacked by amylolytic enzymes. These characteristics are found in flour milled from a soft, low-protein wheat of low alpha-amylase activity.

Typical parameter values for cake flour milled in the United Kingdom for untreated cake flour are 8.5%–9.5% protein and a minimum of particles exceeding 90 µm in size (fine particle size is more important than low protein content for cake quality, giving finer, more even crumb than a coarser flour), while strong cake flour (for fruit cakes) should be 12% protein, 20–25 FU starch damage and 0.18% chlorine treatment.

7.3.3.2 High-ratio flour

In the late 1920s it was discovered in the United States that cake flour which had been bleached with chlorine gas to improve its colour permitted production of cakes from formulae containing levels of sugar and liquid in excess of the flour weight. Such flour for use in high sugar/flour ratio and high liquid/flour ratio formulae is known as 'high-ratio flour'. It should have fine, uniform granularity and low protein content. The chlorination treatment, generally 0.1%–0.15% by weight, besides allowing addition of larger proportions of liquid and sugar, reduces elasticity of the gluten and lowers the pH to 4.6–5.1.

Heat treatment of the grain or the semolina from which the flour is milled has been found to be an effective substitute for the chlorine treatment of high-ratio cake flour (BP Nos 1444173 (FMBRA, Flour Milling and Baking Research Association) and 1499986 (J. Lyons)), and cake flours may be similarly treated.

Typical characteristics of high-ratio flour milled in the United Kingdom would be 7.6%–8.4% protein, 20–25 FU starch damage, granularity such that 70% of the particles were below 32 μm in size and a minimum of particles exceeding 90 μm in size. High-ratio flour is particularly suitable for sponge-type goods.

7.3.3.3 Emulsifiers in cakes

In making cakes, emulsifiers such as glycerol monostearate (GMS) (Fig. 7.15) and monoglycerides and diglycerides of fatty acids (E471) are used in soft fats at levels of up to 10% to produce high-ratio shortenings. Certain emulsifiers such as GMS, polyglycerol esters and lactic acid esters of monoglycerides (E472b) possess remarkable foam-promoting properties, so that when added to sponge batters at a level of about 0.5%–1.0% of batter weight, whisking times can be greatly reduced, all-in mixing methods can be used and liquid egg can be replaced with dried egg.

Foam-promoting emulsifiers such as GMS, polyglycerol esters, propylene–glycol esters (E477) or blends of these, used at about 1% of batter weight, allow a reduction in the fat content of a cake or even substitution of the fat by a smaller quantity of vegetable oil.

Although antistaling effects of emulsifiers in cakes are not as clearly defined as in bread, sucrose esters (E473), sodium stearoyl lactylate (E481) and polyglycerol esters (E475) offer some possibilities as a means of minimizing the effects of staling.

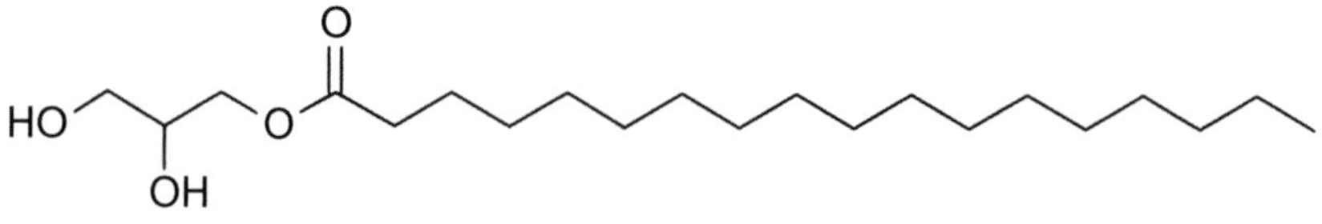

FIGURE 7.15   Chemical structure of glycerol monostearate ($C_{21}H_{42}O_4$).

7.3.3.4 Flour for cake premixes

Some cake premixes contain, in powder form, all the ingredients required for a cake (flour, fat, sugar, baking powder, milk powder, eggs, flavouring and colour), and need only the addition of water before baking. However, some cake premixes, particularly those sold in the United States, omit the eggs and/or the milk, because lighter cakes of larger volume can be made by using fresh eggs instead of dried ingredients.

The type of flour must be suitable for the particular product, with flours of high-ratio type generally being used. The fat must have the correct plasticity and adequate stability to resist oxidation. The addition of certain antioxidants to fat to improve stability is allowed in Britain, the United States and elsewhere. Those allowed in Britain under the Antioxidants in Food Regulations (1978) (SI, 1978, No. 105, as amended) for addition to anhydrous oils and fats and certain dairy products for use as ingredients are propyl (E310), octyl (E311),or dodecyl (E312) gallates (Fig. 7.16) up to 100 mg/kg, and butylated hydroxyanisole (BHA) (E320) (Fig. 7.17) and/or butylated hydroxytoluene (BHT) (E321) (Fig. 7.18) up to 200 mg/kg (calculated on the fat). Those allowed in the United States (with permitted levels based on fat or oil content) are resin guaiac (0.1%), tocopherols (0.03%), lecithin (0.01%), citric acid (0.01%), pyrogallate (0.01%), propylgallate (0.02%) and BHA and/or BHT (0.02%).

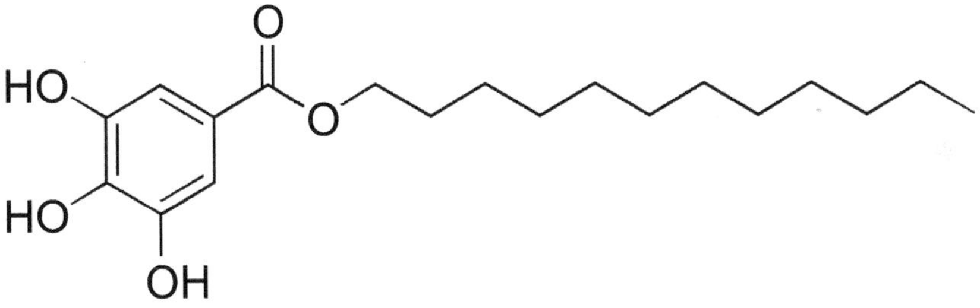

FIGURE 7.16   Chemical structure of dodecyl gallate ($C_{19}H_{30}O_5$).

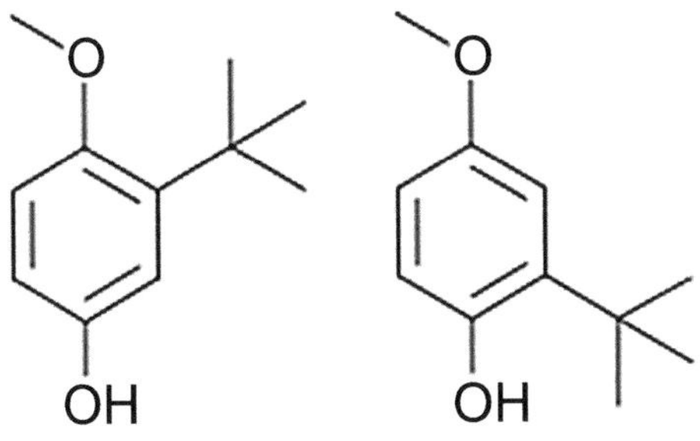

FIGURE 7.17   Chemical structure of butylated hydroxyanisole (BHA) ($C_{11}H_{16}O_2$).

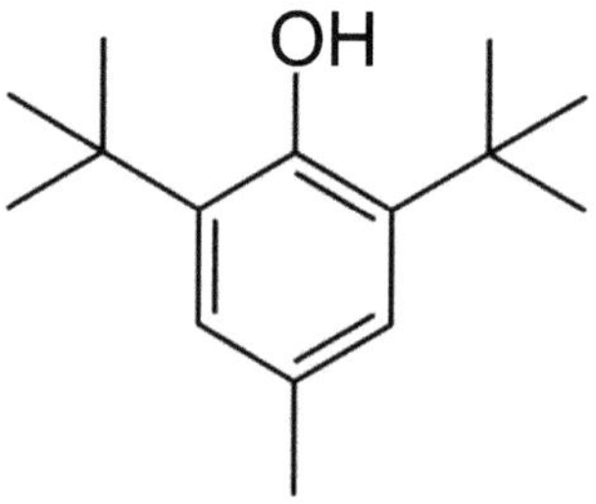

FIGURE 7.18   Chemical structure of butylated hydroxytoluene (BHT) ($C_{15}H_{24}O$).

In preparing the premixes, the dry ingredients are measured out by automatic weighing equipment and conveyed, often pneumatically, to a mixing bin, mixed and then entoleted to ensure freedom from insect infestation. The fat is added, and the mixture packaged. If fruit is included in the formula, it is generally contained in a separate wrapped package (cellophane, plastic, metal can, etc.) enclosed within the carton.

7.3.3.5 Flour for fermented goods

For buns, etc., a breadmaking flour is required. Fermentation time is short; the fat and the sugar in the formula bring about shortening of the gluten.

7.3.3.6 Flour for pastry

A weak or medium-strength flour is needed for the production of sweet and savoury short pastries. Flour strength for puff pastry will vary according to the processing method being used, with rapid processing methods requiring weaker flours than those used with traditional methods of production. In general, flours for puff pastry should have low resistance to deformation (e.g., low Brabender resistance values) but reasonable extensibility.

7.3.4 Flour from steamed wheat

Flour milled from steamed wheat ('stabilized' flour, in which enzymes have been inactivated) is produced for use in manufacture of soups, gravies, crumpets and liquorice and as a thickening agent. For these purposes, the flour should form a thick paste when it is heated with water, and the paste should retain its consistency for some time when heated at 90–95°C. The alpha-amylase activity of normal (nonsteam-treated) flour is usually high enough to degrade swollen starch granules during the cooking process, resulting in loss of water-binding capacity and formation of thin pastes of low apparent viscosity; thus the greater water-absorbing capacity of the flour from steamed wheat makes it a more suitable ingredient for canned foods (both human and pet foods).

As the gluten in the flour from steam-treated wheat has been denatured it is not suitable for breadmaking. So this flour may, for many purposes, be essentially regarded as impure starch, and it is often used to replace starch in certain types of adhesives and as a filler for meat products.

The bacteriological status of flour for soups and other canned foods is important, and must meet the following requirements.

- Not >125 total thermophilic spores per 10 g.
- Not >50 flat sour spores per 10 g.
- Not >5 sulphide spoilage organisms per 10 g.
- Thermophilic anaerobic spores in not >3 tubes out of six.

To ensure food safety, though, all canned foods are heat treated (retorted or using another sterilization procedure) during manufacture.

7.3.5 Quellmehl

Quellmehl, or heat-treated starch, is defined as maize flour or wheat flour in which the starch has undergone hydrothermic (steam) treatment resulting in pregelatinization of the starch, thereby increasing its swelling capacity by at least 50%.

7.3.6 Flour for sausage rusk

A low-protein flour milled from weak wheat, such as UK-grown Riband, or a low-protein air-classified fraction is required for sausage rusk. Desirable characteristics are a low maltose figure (not >1.4 by the Blish and Sandstedt method), low alpha-amylase activity (high Falling number) and high absorbency.

7.3.7 Batter flour

A low-protein flour milled from a grist comprising 90% weak British wheat plus 10% strong wheat is suitable for batter. Alpha-amylase activity should be low. Too high a viscosity in the batter caused by excessive starch damage is to be avoided, and therefore the proportion of hard wheat in the grist should be restricted.

7.3.8 Household flour

Household flour is used for making cookies, puddings, cakes, pastries, etc. With the advent of household bread machines in the 1990s, making bread in the home has become quite popular as well. In the

United Kingdom it is milled from a grist consisting predominantly of weak wheats of low protein, such as British or Western European, with admixture of up to 20% of strong wheat to promote flowability and good mixing. Exclusion of sprouted wheat from the grist is important, as high alpha-amylase activity leads to the production of dextrins and gummy substances during cooking, and to sticky and unattractive baked goods.

7.3.8.1 Self-raising flour

Self-raising flour is a household flour to which raising agents have been added. Choice of sound wheat is important, because evolution of gas during baking is rapid and the dough must be sufficiently distensible, and yet strong enough to retain the gas. The moisture content of the flour should not exceed 13.5% to avoid premature reaction of the aerating chemicals and consequent loss of aerating power.

Distension of the dough is caused by carbon dioxide which is evolved by the reaction between the raising agents (leavening agents in the United States), one alkaline and one acidic, in the presence of water. The usual agents used for domestic self-raising flour include E500, sodium hydrogen carbonate ('bicarbonate') $(NaHCO_3)$ (Fig. 7.19) and acid calcium phosphate (ACP, E341 calcium tetrahydrogen diorthophosphate) $(CaH_4(PO_4)_2)$ (Fig. 7.20). Their use was described by J.C. Walker in BP No. 2973 in 1865. The usual rate of usage is 1.16% bicarbonate plus 1.61% of 80%-grade ACP on a flour-weight basis. A slight excess of the acidic component is desirable, as excess of bicarbonate gives rise to an unpleasant odour and a brownish-yellow discoloration.

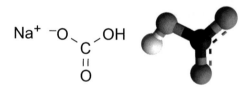

FIGURE 7.19   Chemical structure of sodium hydrogen carbonate $(NaHCO_3)$.

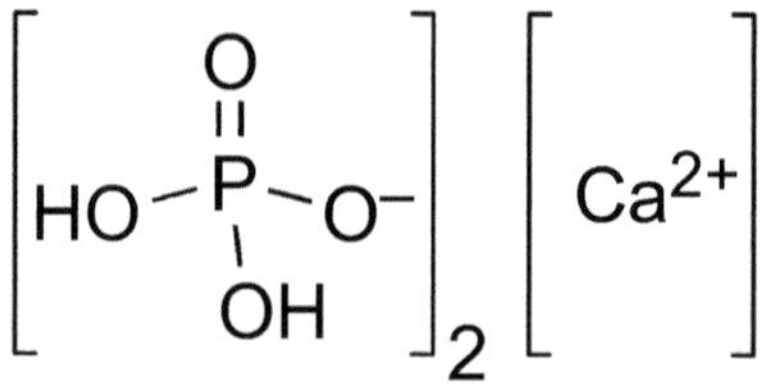

$$\left[HO-\overset{\overset{\displaystyle O}{\|}}{\underset{\underset{\displaystyle OH}{|}}{P}}-O^- \right]_2 \left[Ca^{2+} \right]$$

FIGURE 7.20   Chemical structure of acid calcium phosphate $(CaH_4P_2O_8)$.

Other raising agents can include: E450 disodium dihydrogen diphosphate (sodium acid pyrophosphate (SAPP)), 541 acidic sodium aluminium phosphate (SAP), E575 D-glucono-1,5-lactone and E336 mono potassium-L-(+)-tartrate (cream of tartar).

ACP used at a rate of 1.61% on flour weight adds about 250 mg Ca per 100 g flour; hence self-raising flour is not required to have chalk added to it. Phosphate–starch mixtures are known as cream powders, a commonly used commercial example of which consists of sodium acid pyrophosphate diluted with starch and used in the ratio of 2:1 with sodium hydrogen carbonate at a rate of 4.7% on a flour-weight basis.

7.3.8.2 Instantized or agglomerated flour

This is a form of free-running flour, readily dispersible in water, made by 'clustering' flour particles in an 'instantizer'. Uses include the making of sauces and gravies and thickening and general culinary purposes. In the instantizing process, the flour as normally milled is damped with steam, tumbled in a warm air stream to cause the particles to agglomerate, then dried, sieved, cooled and packed. The US standard for agglomerated flour requires all the flour to pass through a sieve of 840 μm aperture width and not more than 20% to pass through a sieve of 70 μm aperture width. Free-flowing flour is also produced by air classification.

7.3.9 Flour for export

Besides specific requirements according to the purpose for which the flour is to be used, flour for export must have low moisture content to prevent development of mould, taint or infestation during its transportation. As a safeguard, the flour should be entoleted (see Chapter 5). In addition, export flour must conform to any special requirements of the importing country, e.g., regarding the presence or absence of nutrients and improvers, for which the regulations of most other countries differ from those of the United Kingdom.

In January 1988 the US Food and Drug Administration announced guidelines for contamination levels at which flour is 'seizable'. These levels are 75 insect fragments or more per 50 g flour and an average of one rodent hair or more per 50 g flour (see CPG Sec. 578.450 Wheat Flour-Adulteration with Insect Fragments and Rodent Hairs).

7.3.10 Flours from cereals other than wheat

Besides wheat, all other cereals yield flour when subjected to milling processes as outlined in Chapter 6. The uses for these flours, both commercially and domestically, are many and varied, as the following summary indicates.

7.3.10.1 Rye flour

Rye flour of various extraction rates is used extensively in Eastern Europe for making a range of breads – both soft breads and crispbreads – using conventional straight dough or sour dough processes.

Rye flour is also used as a filler for sauces, soups and custard powder and in pancake flour in the United States, and for making gingerbread in France. A mixture of 10% rye flour with 90% wheat flour is used for making biscuits and crackers in the United States. The rye flour is said to improve the quality of the products and is less expensive than wheat flour. Rye flour can be fractionated by air-classification; a flour of 8.5% protein content yielding high- and low-protein fractions of 14.4% and 7.3% protein contents, respectively, is typical. Rye flour is also used in the glue, match and plastics industries.

Rye flour can be used for making gun-puffed and shredded ready-to-eat breakfast cereals.

7.3.10.2 Triticale flour

The use of triticale flour in breadmaking is mentioned in Chapter 8, and its use in making chapattis is referred to in Chapter 11. Other bakery products made with triticale flour include pancakes and waffles.

7.3.10.3 Barley flour

Barley flour is used in the manufacture of flat bread, for infant foods and for food specialities. It is also a component of composite flours used for making yeast-raised bread.

Pregelatinized barley flour, which has high absorbent properties, provides a good binder and thickener. Barley breading is made by combining pregelatinized barley flour with barley crunch.

Malted barley flour is made from barley malt. Malt flour is used as a high diastatic supplement for bread flours which are low in natural diastatic activity, as a flavour supplement in malt loaves and for various other food products.

Malted barley flour can be air-classified to yield protein-rich and protein-poor fractions. The former finds uses in the food industry, while the latter is reported to make a unique beer. The major food uses for malt products and cereal syrups are in bread, biscuits, crackers, crispbread, breakfast cereals, infant and invalid foods, malted food drinks, pickles and sauces, sugar confectionery and vinegar.

7.3.10.4 Oat flour

Oat flour is made by grinding oatmeal on stones and sifting out the fine material. It is also obtained as a by-product of groat cutting. Uses for oat flour include infant foods and ready-to-eat breakfast cereals, e.g., shredded products made by a continuous-extrusion cooking process (Chapter 9) and extruded gun-puffed products (Chapter 9).

A process for the separation of a protein concentrate by air-classification (Chapter 6) of oat flour has been described (Cluskey et al., 1973). An ultra-fine fraction with 85%–88% protein content (N × 6.25, d.b. (dry basis)) was obtained which comprised 2%–5% of the flour by weight. The compound granules of oat starch (Chapter 3) tend to disintegrate upon fine grinding, releasing the individual granules, which measure 2–10 μm. Separation of an almost pure protein fraction would therefore require the use of an extremely fine-cut size – less than 2 μm.

7.3.10.5 Rice flour

Rice flour is used in refrigerated biscuit manufacture to prevent sticking; in baby foods, as a thickener; and in waffle and pancake mixes, as a water absorbent.

The use of rice flour in blends with wheat flour to make bread of acceptable quality is mentioned in Chapter 8, and also its use as a component of composite flours for breadmaking. Rice flour can be used for making pasta products.

7.3.10.6 Maize flour

Maize flour is used to make bread, muffins, doughnuts, pancake mixes, infant foods, biscuits, wafers, breakfast cereals and breadings, and as a filler, binder and carrier in meat products. Dry-milled maize flour is not to be confused with 'cornflour', the term used in the United Kingdom for maize starch obtained as a product of maize wet milling.

The inclusion of maize flour in composite flour used for breadmaking and its use alone to make bread of a sort in Latin America are mentioned in Chapter 8. The use of maize flour, in blends with wheat semolina, to make pasta products is mentioned elsewhere in the book, and for making extrusion-cooked ready-to-eat breakfast cereals.

Industrial uses for maize flour are noted elsewhere, especially in the discussion about alcohol production.

7.3.10.7 Sorghum and millet flours

Sorghum flour is used as a component of composite flour for making bread in those countries in which sorghum is an indigenous crop.

The use of flour or wholemeal from sorghum and some of the millets to make porridge, roti, chapatti, tortillas and other products is described elsewhere.

Sorghum flour finds industrial uses, e.g., as core binder, in resins and adhesives, and in oil-well drilling.

7.3.10.8 Composite flours

The Composite Flour Programme was established by the Food and Agriculture Organization in 1964 to find new ways of using flours other than wheat – particularly maize, millet and sorghum – in bakery and pasta products, with the objective of stimulating local agricultural production and saving foreign exchange in countries heavily dependent on wheat imports (Kent, 1985).

7.4 QUALITY CONTROL AND FLOUR TESTING

Testing protocols and acceptable degrees of reproducibility (agreement between laboratories) and repeatability (agreement between replicate determinations by the same operator using the same equipment) have been established, usually through collaborative testing, by various standardizing organizations. Most countries have national standards organizations (e.g., the British Standards Institute), and international standards are also produced by, for example, the International Association of Cereal Science and Technology (ICC), International Organization for Standardization (ISO), American Association of Cereal Chemists International (AACCI), American Society of Agricultural and Biological Engineers and Association of Official Analytical Chemists International.

Tests may be applicable to whole grains or derived products; in the case of whole grains it may be necessary to grind them to achieve an appropriate particle size distribution for proper instrument operation or sensor reading.

For valid comparisons to be made it is necessary to observe proper sampling procedures (ICC 130, AACC 64) and normalize results of most tests to a constant moisture basis. Either a dry-matter basis or a 14% moisture basis is usually adopted. As testing instrumentation and procedures become more stringent, protocols increasingly demand that test samples contain a consistent dry weight of sample, requiring an adjustment of the actual mass taken to compensate for moisture variation, rather than subsequent correction. Moisture content must thus be determined by an acceptable method, such as determination of weight loss upon drying, when the ground product is heated at 100°C for 5h in vacuo or at 130°C for 1h (flour) or 1.5h (ground wheat) at atmospheric pressure (ICC 101/1), or 135°C for 2h (AACCI). These methods are suitable for moisture contents up to 17%. A summary of recommended AACCI methods is provided in Table 7.7.

Oven methods are known as 'primary' methods, as they determine directly the required parameter (e.g., mass loss, thus moisture content); 'secondary' methods may also be used, which measure a property that

TABLE 7.7 Recommended loss in weight oven-drying methods for determining moisture content of cereal products

AACCI method	Drying temperature (°C)	Drying time (h)	Sample size (g)	Primary applications
44–19	135	2	2	Feed
44–15A	130	1	2–3	Flour, semolina, bread, grains, cereal products, food products
	103	72	15	Corn, soybeans
	103	72	5–7	Flax
44–16	140	0.25	2	Flour, semolina
44–20	103	3	5 or 10	Malt
44–40	98–100	5	2	Flour, semolina, ground grain, feed

varies as a function of the required parameter, and thus require calibration against a standard. Secondary methods are frequently more rapid but somewhat less accurate than primary methods; they include electrical conductivity and near-infrared reflectance spectroscopy (NIRS) methods (ISO 202 covers moisture and protein determinations). NIRS determinations are based on absorption, transmission or reflectance of NIR energy at specific wavelengths by OH bonds in water molecules. The same is true of protein determinations, where the peptide bonds between amino acids define the critical wavelengths. Considerable mathematical processing of signals and measurements at reference wavelengths are necessary to ensure accurate indications of the required parameters. Additionally, accuracy is highly dependent upon proper calibration with samples that have been measured using primary methods, and thus high numbers of samples are critical.

7.4.1 Parameters dependent on the nature of the grains milled

7.4.1.1 Protein content

Protein content of whole grains, meals and flours may be calculated from nitrogen contents determined by the Kjeldahl method (AACCI 46), in which organic matter is digested with hot concentrated acid in the presence of a catalyst. Ammonia, liberated by addition of an excess of alkali to the reaction product, is separated by distillation and then estimated by titration. A convenient apparatus is the Tecator Kjeltec 1030 Auto System, or the FOSS Kjeltec 8100, 8200, or 8400 (https://www.fossanalytics.com/en/products/kjeltec-8400). In recent years combustion instruments (which directly measure N_2 content) have largely replaced

Kjeldahl units. For either type of system, protein content is determined by multiplying by an appropriate conversion factor. For example, in white wheat flour, protein content is estimated by multiplying N_2 content by 5.7, and in many references this factor is used for other wheat products and other cereals as well. However, FAO/WHO (1973) recommended specific factors appropriate to individual foodstuffs. Those relating to cereals are given in Table 7.8.

For routine estimations of protein, moisture and other chemical contents, NIRS methods are now very reliable for whole grains as well as for many of their derivatives. Determinations are often carried out at intake (i.e., grain receiving) and online during processing in many cereals plants throughout the world – grain elevators, flour mills, feed mills, ethanol plants, whiskey plants, etc.

For determining the amount of wheat protein contributing to gluten, the Glutomatic instrument (Falling Number Co.) and method may be used. A dough is prepared from a sample of flour or ground wheat and a solution of sodium chloride. Wet gluten is isolated by washing this dough with a solution of buffered sodium chloride and, after removal of excess water, weighed or dried and weighed according to whether wet or dry gluten content is required (ICC 137, AACCI 56–81B).

7.4.1.2 Sedimentation tests

Sedimentation tests provide a useful indication of the suitability of a flour (it is more usually performed on a ground wheat as opposed to a true flour) for breadmaking. The Zeleny test (Pinkney et al., 1957) (AACC 56–60, ICC 116) has been adopted in a number of countries for protein evaluation. It depends on the superior swelling and flocculating properties, in a dilute lactic acid solution, of the insoluble proteins of wheats with good breadmaking characteristics (Frazier, 1971). In the United Kingdom a better guide to breadmaking properties has consistently been obtained using the SDS (sodium dodecyl sulfate) sedimentation test (Axford et al., 1979) rather than the Zeleny test. The SDS test has been standardized as BS4317, part 19, and has been adopted for evaluation of *T. durum* quality

TABLE 7.8   Factors used for converting Kjeldahl nitrogen (N) to protein values

Wheat fraction	Factor	Cereal	Factor
Wholemeal flour	5.83	Maize	6.25
Flours, except wholemeal	5.70	Rice	5.95
Pasta	5.70	Barley, oats, rye	5.83
Bran	6.31		

as ICC 56–70 and AACC 151. It consists of an initial suspension and shaking of ground material in water, to which sodium dodecyl sulphate is later added. Following a series of carefully timed inversions of the cylinder containing the suspension, it is allowed to stand for 20 min, after which the height of sediment is read. The test is performed under controlled temperature conditions.

7.4.1.3 Enzyme tests

One of the most important enzymes influencing flour quality is alpha-amylase. Its activity may be determined directly, using the method of Farrand (1964) or McCleary and Sheehan (1987), or indirectly, as a result of its solubilizing effect on starch, leading to a reduction in paste viscosity. The most widely adopted method uses the Falling Number apparatus to detect starch liquefaction in a heated aqueous suspension of flour or (more usually) ground grain (ICC Standard Method 107, AACC 56–81B). As enzyme activity can vary dramatically among individual grains, it is essential to grind a large sample of grain (at least 300 g) in preparing a representative meal and to mix it thoroughly before taking the test sample (7 g at 15% moisture content) from it. Wholemeals also require regrinding and thorough mixing, and flours have to be free of lumps. Following the preparation of a suspension in a special tube, this is introduced into the apparatus and the test proceeds automatically. The suspension is heated and stirred at a programmed rate for 60 s, after which a plunger is allowed to fall through it. The Falling Number is the number of seconds from the start of heating to the coming to rest of the plunger.

A similar principle underlies stirring tests with the Rapid Visco-Analyser.

Used on grain or grain products to measure alpha-amylase activity, both instruments probably also respond to hydrolysis of other viscous components, such as proteins and cell-wall components, but the effects of these are usually comparatively small compared to the effects of the enzymes.

7.4.1.4 Heat-damage test for gluten

The effect of overheating on gluten is measured directly by a method introduced by Hay and Every (1990). Described as the glutenin turbidity test, the procedure measures the loss in solubility of the fraction of glutenins normally soluble in acetic acid. Dilute acetic acid extracts are precipitated by addition of alkaline ethanol and the precipitate is quantified by spectrophotometric measurement of turbidity, allowing the degree of damage to be assessed by comparison with standards. Good correlations have been found between loss of turbidity and reductions in baking quality.

7.4.1.5 Pigmentation

The yellow colour of durum semolinas is highly valued. Under ICC 152 the carotenoid pigments are extracted at room temperature with water-saturated butanol for photometric evaluation of optical density of the clear filtrate against a β-carotene standard.

7.4.1.6 End-use tests

Despite much research, no single test has yet been devised which can reliably predict the breadmaking properties of an unknown wheat or flour. Hence for this application, and for many other applications of cereals, the most reliable means of evaluating a sample is to subject it to the intended end use itself, or to a scaled-down version of the same. Several bread baking tests appear under AACC 10 and ICC 131; rye flour is tested by AACC 10–70. Biscuit (cookie) and foam-type cake tests also appear under the 10 heading. Pasta semolinas are also subjected to approved small scale tests (AACC 66-41 and 66-42).

7.4.1.7 Machinability test

In adopting a test for bread wheats eligible for intervention price support, the EU has not standardized a breadmaking test but instead has defined flour of breadmaking quality as flour which produces a dough which does not 'stick' to the blades or the bowl of the mixer in which the dough is mixed, nor to the moulding apparatus.

7.4.1.8 Extraneous matter test ('filth test')

The rodent hair and insect fragment count in flour is determined by digesting the flour with acid and adding the cooled digest to petrol in a separating funnel. The hair and insect fragments are trapped at the petrol/water interface, and can be collected, identified and quantified microscopically.

7.4.2 Tests for characteristics dependent mainly on processing conditions

7.4.2.1 Ash test, BS4317 part 10

The ash test (incineration of the material in a furnace at a specified temperature, under prescribed conditions, for a specific sample mass and the weighing of the resultant ash) is widely used as a measure of milling refinement because pure endosperm yields relatively little ash, whereas bran, aleurone and germ yield much more. The ash test can be carried out very precisely, but, as the endosperms of different wheats vary in mineral content, a given ash value can correspond to different levels of bran content. Also, the test is not suitable for indicating the content of nonstarchy-endosperm components in flours to which chalk has been added.

7.4.2.2 Grade colour

The grade colour test, performed for example with the Kent-Jones and Martin colour grader, can be used to estimate the degree of contamination of white flour with bran particles. In the test, the intensity of light in the 530 nm region reflected from a standardized flour/water paste in a glass cell is compared with that reflected from a paste of a reference flour. The grade colour was said to be unaffected by variation in content of flour pigment (xanthophyll). It has now been demonstrated, however, that this is not so (Barnes, 1978), thus diminishing the value of the test as a means of quantifying bran content.

7.4.2.3 Tristimulus methods

Use of an instrument designed to simulate the visual response of the human eye has found favour in many applications as an alternative to the grade colour system. An instrument (typically either Minolta or Hunter) measures reflectance spectra from a white light source and uses complex mathematical transforms to produce values in three arbitrary spectral ranges. These most commonly include the CIE XYZ colour space (Fig. 7.21) and the Hunter Lab colour space (Fig. 7.21), which cannot be produced by any real lights but do result in highly reproducible results (Hunter and Harold, 1987). Users can derive a series of indices to suit their specific needs (e.g., lightness, yellowness, blueness, greenness, redness, or other colour indices) by selecting from the measured colour space values. Like the grade colour system, colour scores respond to factors other than bran content – often particle size, moisture content, composition, heat damage, etc. – and it thus has somewhat limited value if the user is not careful, especially for highly variable biological materials (Evers, 1993).

7.4.2.4 Damaged starch

The amount of starch that is mechanically damaged influences a flour's ability to absorb water. The level of damaged starch in flour is estimated by methods which measure either the digestibility or the extractability of the starch. Digestibility-based methods measure the amount of hydrolysis effected by added amylase enzymes; extractability-based methods measure the amount of amylose present in an aqueous extract by its reaction with iodine in potassium iodide. The iodine/amylose complex may be assayed colourimetrically (e.g., McDermott, 1980), amperometrically or potentiometrically (e.g., Chopin SD4 method). Only the damaged granules are susceptible to amylase at temperatures below gelatinization temperature, and appreciable leaching of amylose occurs only from damaged granules under the test conditions.

Methods such as those of Farrand (1964), Donelson and Yamazaki (1962) and Barnes (1978) rely on assaying reducing sugars (mainly maltose) produced by the action of alpha-amylase derived from malt flour. Another

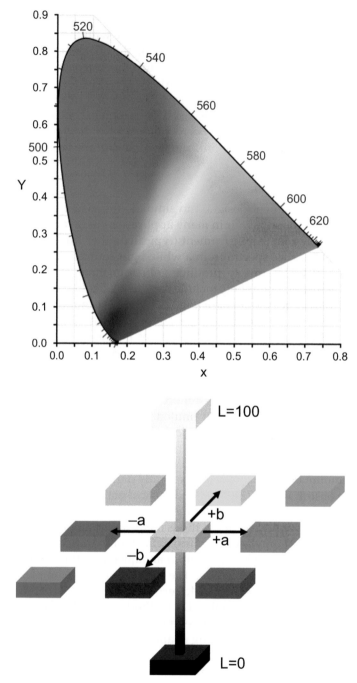

FIGURE 7.21   Tristimulus colour spaces. X-Y-Z colour space (above). L-a-b colour space (below).

method (Gibson et al., 1992) employs amyloglucosidase to hydrolyse oligosaccharides further to glucose, which is then determined by the effect of a derivative on a chromogen.

Damaged starch granules may be recognized microscopically by a red colouration with Congo Red (colour index: 22,120) stain, and have been described, from their microscopic appearance, as 'ghosts' by Jones (1940). Undamaged granules do not stain with Congo Red. A method of damaged-starch determination has been developed, in which damaged and undamaged granules are measured separately taking advantage of their different staining reactions with a fluorescent dye, using image analysis for making the measurements.

7.4.2.5 Particle size analysis

Test sieving by hand or Rotap sieve shakers tends to be somewhat irreproducible for flour, but the Alpine airjet sieve, in which negative pressure below the sieve assists particles through the mesh and clears the mesh with reversed airjets, gives more reproducible analyses. Sedimentation methods depend upon the faster settling rates of larger particles in a nonaqueous solvent; the Andreassen pipette is an example of a simple device using this principle (ICC 127), and another is the Simon sedimentation funnel. Sedimentation methods are not much used today in flour quality control, having been superseded by more reproducible methods. The Coulter counter is a device which rapidly measures the volume of thousands of individual particles as they pass through an orifice. Each particle is measured by the change in electrical resistance that it causes by displacing its own volume of an electrolytic solution (nonaqueous in the case of flour) in which the particles are suspended. Reproducibility is very good with this method (Evers, 1982), which is suitable for flours and starches but not wholemeals. Laser-diffraction-based instruments, and now visual imaging instruments (such as the Horiba Camsizer, which uses multiple high-speed cameras), provide a means of making rapid particle size and shape comparisons among flours (Fig. 7.22 and Table 7.9). There are many instruments of these types available, some capable of operating on dry samples or wet samples. The distributions they indicate may not agree well among different instruments or with those of the other methods described, so calibration and validation are required.

7.4.2.6 Physical tests on doughs and slurries

The physical characteristics of doughs and slurries are important in relation to the uses of flours. Various pseudorheological characteristics of the primary measurement methods follow. All of these instruments are now computer controlled, and have data acquisition and analyses software.

The Brabender Farinograph (D'Appolonia and Kunerth, 1984) measures and records the resistance of a dough to mixing as it is formed from

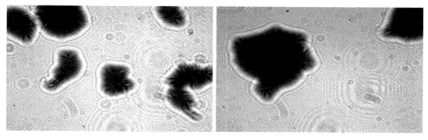

Sample 1   Sample 2

FIGURE 7.22 Examples of visual imaging used to determine particle size and shape properties and their distributions.

TABLE 7.9 Examples of size and shape parameters determined by particle size instruments

Parameter	Unit	Sample	
		1	2
d(0.10)	μm	487.09	641.42
d(0.50)	μm	853.77	1279.21
d(0.90)	μm	1345.23	2054.81
Perimeter	μm	1660.13	1394.13
Area	μm²	190,602.1	189,024.8
Bounding box width (breadth)	μm	466.38	384.91
Bounding box height	μm	439.15	373.29
Bounding box diagonal	μm	646.77	543.40
Bounding box area	μm²	318,714.4	304,568.7
Equivalent diameter (Waddel disk)	μm	393.69	333.10
Equivalent radius	μm	196.85	166.55
Equivalent volume	μm³	98,243,938.4	137,759,440.4
Shape factor (compactness)		0.60	0.59
Sphericity		0.49	0.51
Circumscribing disk	μm	567.56	481.99
Inscribing disk	μm	235.17	198.42
Circularity (roundness)		13.64	13.50
Radius	μm	197.61	166.78
Radius (×2)	μm	395.22	333.56

TABLE 7.9 Examples of size and shape parameters determined by particle size instruments—cont'd

Parameter	Unit	Sample 1	Sample 2
Heywood circularity factor		0.82	0.83
Compactness factor		0.65	0.67
Feret minimum	μm	347.01	294.69
Feret maximum	μm	534.56	455.01
Aspect ratio (elongation)		0.69	0.70
Fibre width (rectangle small side)	μm	175.15	148.68
Fibre length (rectangle big side)	μm	586.38	494.40
Rectangle ratio		2.47	2.20
Fibre curl		0.59	0.56
Orientation	°	88.28	85.86
Convexity (convex perimeter)		0.85	0.80
Convex perimeter	μm	1363.41	1145.68
Solidity (convex area)		0.91	0.91
Convex area	μm²	215,839.66	214,671.76
Convex shape factor		0.86	0.82
Eccentricity (ellipticity)		18,394,911.3	158,538,493.3
Hydraulic radius	μm	75.56	64.15
Hydraulic radius (×4)	μm	302.24	256.60
Maximum intercept	μm	535.20	455.47
Equivalent ellipse major axis (2a)	μm	535.20	455.47
Equivalent ellipse minor axis (2b)	μm	294.55	247.56
Equivalent ellipse area	μm²	190,602.09	189,024.77
Equivalent ellipse volume	μm³	72,959,373.1	102,121,325.5
Equivalent diameter (ellipse volume)	μm³	356.76	301.19
Equivalent ellipse ratio		1.81	1.81
Mean perpendicular intercept	μm	231.3	194.4
Elongation factor		0.74	0.73
Type factor		0.87	0.82

flour and water, developed and then broken down (Fig. 7.22). This resistance is called 'consistency'. The maximum consistency of the dough is adjusted to a fixed value (500 Brabender units) by altering the quantity of water added. This quantity, the water absorption, may be used to determine a complete mixing curve, the various features of which are a guide to the strength of the flour (AACC 54-21, ICC 115).

The Brabender Extensograph (Rasper and Preston, 1991) (ICC 114, AACC 54-10) records the resistance of dough to stretching and the distance the dough stretches before breaking. A flour–salt–water dough is prepared under standard conditions in the Brabender Farinograph and then moulded on the Extensograph into a standard shape. After a fixed period the dough is stretched and a curve drawn, recording the extensibility of the dough and its resistance to stretching (Fig. 7.23). The dough is removed and subjected to a further two stretches. This instrument was developed with bread doughs (especially those with high water content) in mind; the measurements being made to evaluate developed gluten. It will differentiate between strong, medium and weak proteins (as shown in Fig 7.24) and can be used to identify samples that have been partly or extensively heat damaged. Although not necessarily of direct benefit in evaluating the performance characteristics of doughs in which no gluten development occurs, the measurements do aid in assessing the consistency of performance from one batch to another. The Extensograph has replaced the Extensometer in the Brabender instrument range, but the older instrument may still be used for testing biscuit flours.

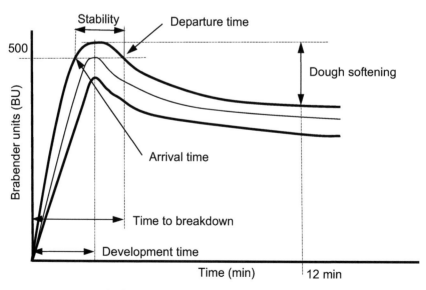

FIGURE 7.23   Typical Farinograph output.

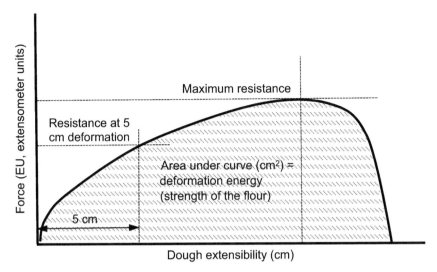

FIGURE 7.24 Typical Extensograph/Extensometer output, which includes force resistance curve (EU) over dough extension, peak force before dough snaps and area under curve (energy), which indicates dough strength. Resistance at a given deformation (5 cm) may also be used, and strength is indicated by the ratio of extensibility (5 cm) to resistance.

The Chopin Alveograph (AACC 54-30) assesses the ability of a dough in which gluten is developed to retain gas (Fig. 7.25). Air pressure inflates a bubble of dough until it bursts; the instrument continuously records the air pressure and the time that elapses before the dough breaks.

The Brabender Amylograph (ICC 126 for wheat and rye flours) continuously measures the resistance to stirring of a 10% suspension of flour in water while the temperature of the suspension is raised at a constant rate of 1.5°C/min from 20 to 95°C and then maintained at 95°C (Fig. 7.26) (Shuey and Tipples, 1980). It is of use in testing flour for soups, etc., for which purpose the viscosity of the product after gelatinization is an important characteristic, and for adjusting the malt addition to flours for breadmaking.

The Rapid Visco-Analyser (RVA), produced by Newport Scientific in Australia, may be regarded as a derivative of the Amylograph. Measurements of viscosity over time (Fig. 7.27) as a function of temperature are made using small samples, containing 3–4 g of starch, in periods which may be as short as 2 min. Use of disposable containers and mixer paddles eliminates the need for careful washing of the parts between tests. As with the Amylograph, the characteristics of starch pastes and the effects of enzymes on them can be recorded by the computer software (Fig. 7.28).

The capacity for measuring the liquefying effects of enzymes on viscous pastes enables the RVA to be used for detecting the products of sprouted grains in cereal meals.

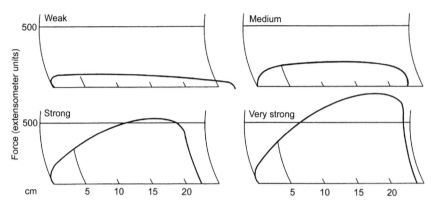

FIGURE 7.25　Extensograph curves of unyeasted doughs showing typical curves for flours of different strengths.

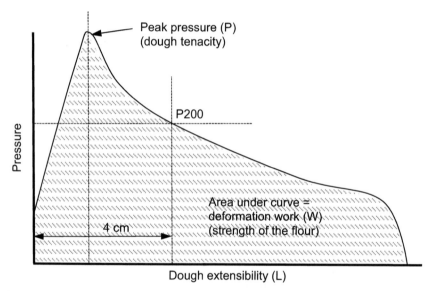

FIGURE 7.26　Typical Alveograph output, which includes pressure curve over dough extension, peak pressure (P) before dough begins to inflate, area under curve (W) and a pressure achieved when 4 cm extension has occurred (P200). Elasticity index (Ie) = P200 × 100/P.

7.4.2.7 *True rheological instruments*

In recent years frustration with instrument-dependent units obtained with some of the above methods, together with the poor reproducibility from one instrument to another of the same type, has led cereal chemists to pursue true rheological measurements. Suitable instruments for use with doughs, slurries and gels derived from flours include the Bohlin VOR (viscometric, oscillation and relaxation), the Carri Med CSL Rheometer and the Rheometrics RDA2.

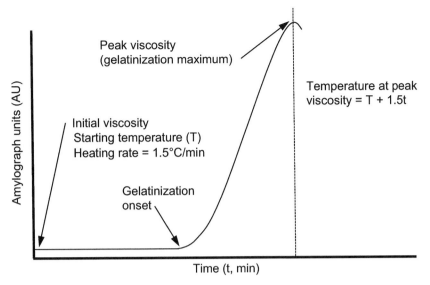

FIGURE 7.27   Typical Amylograph output, which includes viscosity curve over time, as well as temperatures at which gelatinization onset and peak viscosity occur.

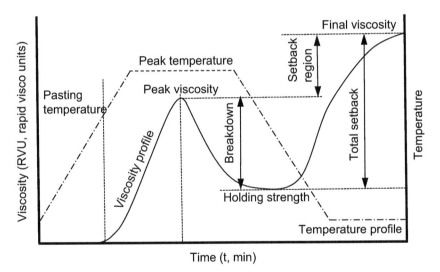

FIGURE 7.28   Typical RVA output, which includes the flour/water viscosity curve as a function of temperature.

Companies such as TA Instruments manufacture a variety of suitable rheometry instruments. In addition to providing excellent reproducibility, these instruments, which are also used for many nonfood and noncereal materials, allow comparisons to be made across a wide range of substances. Although they can be expensive, and may never feature prominently in routine testing, they will undoubtedly enable the development of tests that can be performed on simpler, dedicated instruments (Faridi and Faubion, 1990).

Another advance includes the Mixolab, manufactured by Chopin, which combines some of the previously discussed instruments. Overall, the Mixolab technique combines the functions of the Farinograph, Mixograph, Extensograph and Alveograph (Bloksma and Bushuk, 1988; Chiotelli et al., 2004). The Mixolab analyses a flour's performance throughout the entire breadmaking process, including the mixing, heating and cooling phases. These extra stages thus have the potential to tie research into industry, because the instrument can generate curves to compare differences both between flours and among industrial baking conditions. The Mixolab instrument helps quantify baking performance differences due to starch–protein interactions, enzyme activities, gelatinization, gelling of starch and other potential factors (Saunders et al., 2007).

The Mixolab instrument measures various physical properties of dough, such as stability, strength and pasting. The Mixolab curve is typically divided into five major time periods (Fig. 7.29): development (1), protein reduction (2), starch gelatinization (3), amylase activity (4) and starch gelling (5). Collar et al. (2007) defined α, β and γ on a Mixolab curve as protein breakdown, starch gelatinization and cooking stability rates (i.e., slopes and rates of change), respectively.

Development, stage 1, begins when torque quickly reaches a plateau, and lasts for the duration of the time during heating while the torque is

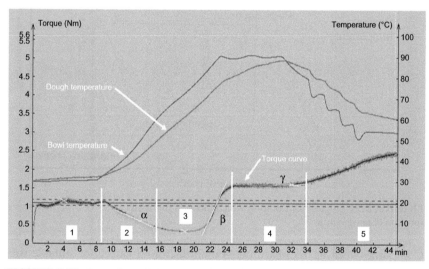

FIGURE 7.29 Typical Mixolab output, which includes temperature curves for the bowl and dough, as well as torque exerted on the rotary blade in the instrument. The main phases of the resulting Mixolab torque curve are denoted by stages 1 through 5; specific slopes of the torque curve, α, β, γ, are also indicated.

kept constant at 1.1 N-m; temperatures are relatively constant during this stage as well, although they will increase slightly (Collar et al., 2007). The water absorption capacity of the flour is determined while at this relatively constant temperature. Properties such as stability and elasticity are also measured. Excessive mixing conditions can cause dough properties to change from smooth and elastic to slack and sticky (Rosell et al., 2007).

In the second stage the temperatures increase rapidly and protein reduction occurs so the torque reading falls, because the dough is subjected to mechanical stress and thermal effects (Collar et al., 2007). Protein weakening occurs when the dough is heated and manipulated. Depending upon flour composition, lower protein qualities are typically signified by greater decreases in consistency. After the initial decrease due to rising temperatures, dough consistency will eventually begin to increase, and the rate of torque fall should decrease during this stage (Kahraman et al., 2008).

This eventual increase in consistency is mostly seen at the beginning of the third stage, due to starch gelatinization. The quality of the starch determines the rate of rise in consistency. Starch granules absorb water and swell, so viscosity increases (Kahraman et al., 2008). Gelatinization occurs when amylose and/or amylopectin molecules are suspended in water and heated. Once heating begins, water is absorbed and the starch granules hydrate. Continued heating weakens hydrogen bonds as molecules swell, resulting in irreversible changes to the starch structure, and eventually some granules will burst. Starch gelatinization results in peak torque concomitant to heating; initial and final pasting are thus determined (Collar et al., 2007).

In the fourth stage the torque curve again plateaus. Endogenous enzymatic activity affects the torque curve, and eventually will impact the torque during the cooling stage (5) when reaching stability (Collar et al., 2007). This activity ultimately determines how large the decrease in consistency will be. Larger decreases are proportional to a greater quantity of enzymatic activity (Kahraman et al., 2008). As discussed previously, the two types of amylases are α-amylase and β-amylase. Alpha-amylase hydrolyses interior α-1,4-glucosidic bonds of starch, glycogen and cyclodextrins; these enzymes are endo-splitting, which acts to increase viscosity. Beta-amylase hydrolyses the α-1,4-glucosidic bonds of starch beginning at the nonreducing end to result in β-maltose; since they are exo-splitting, many bonds need to be hydrolysed before a significant impact on viscosity can be seen.

Finally, the last stage, gel formation, is related to retrogradation, which causes an increase in consistency (and a steady increase in viscosity) within the dough as the temperature decreases during the cooling stage (Kahraman et al., 2008). This cooling allows the starch to retreat, which thus increases product consistency. Retrogradation occurs after maximum viscosity is reached, in which some granules have broken or burst. Upon cooling, some starch granules will partially reassociate to form a gel.

7.5 STORAGE AND TRANSPORT OF FLOUR

It has been recommended that, for long periods of conservation, flour should be stored in a closed atmosphere (Bellenger and Godon, 1972). In these conditions flour acidity increases owing to accumulation of linoleic and linolenic acids, which are slowly oxidized; reduction of disulphide groups (—S—S—) is slow, and there is little increase in sulphydryl groups (–SH); solubility of gluten protein decreases, and as a result changes in baking strength are only minor.

Flour is stored commercially in bags or bulk bins. Bulk bags of flour in the United Kingdom contain 50 kg or 32 kg when packed; typical sizes for consumer sales include 125 g, 250 g, 500 g, 1 kg and 1.5 kg; in the United States, 5 lb (2.3 kg) and 10 lb (4.5 kg) bags are common. These multiwall Kraft paper bags are commonly shrink-wrapped or boxed and stacked, often several tiers high, on pallets. The harshness of treatment to be expected during filling and handling influences the number of plys in the walls of the chosen bags. Using single-spout packers approximately 300–350 bags/h can be filled. Using multispout packers, with up to eight spouts, 600–800 bags/h are possible.

The hazards to flour in storage include those to wheat in storage (mould and bacterial attack, insect and rodent infestation), and also oxidative rancidity and eventual deterioration of baking quality. Thus facility cleanliness and pest control are critical.

Freedom from insect infestation during storage can be ensured only if the flour is free from insect life when put into store and if the store itself is free from infestation. Good housekeeping in the mill and the milling of clean grain should ensure that the milled flour contains no live insects, larvae or eggs, but as a precautionary measure flour is often passed through an entoleter before being bagged or conveyed to bulk bins. The entoleter (BP No. 965267) is a machine consisting of a rotor rapidly rotating within a fixed housing. The flour is fed in centrally and flung with considerable impact against the casing. At normal speeds of operation (2900 rev/min for flour) the machine effectively destroys all forms of insect life and mites, including eggs. The insect fragments, however, are not removed from the flour by the entoleter.

The optimum moisture content for the storage of flour must be interpreted in relation to the length of storage envisaged and the prevailing ambient temperature and r.h. (relative humidity), remembering that flour will gain or lose moisture to the surrounding atmosphere unless packed in hermetically sealed containers. For use within a few weeks, flour can be packed at 14% moisture content, but at moisture contents higher than 13% mustiness due to mould growth may develop, even if the flour does not become visibly mouldy. At moisture contents lower than 12% the risk of fat oxidation and development of rancidity increases. The reactions involved in oxidative rancidity are catalysed by metal ions, such as Cu^{++}.

The expected shelf life of plain (i.e., nonself-raising) white flour packed in paper bags, stored in cool, dry conditions and protected from infestation can be up to 2–3 years. The rate of increase in acidity increases with temperature and with fall in flour grade (i.e., as the ash residue increases). Hence the shelf life of brown and wholemeal flours is shorter than that of white flour.

Stored at 17°C, the shelf life of brown flour of 85% extraction rate and of wholemeal (100% extraction rate) is closely related to the moisture content and temperature. Brown flour, for instance, should keep for 9 months at 14% moisture content, 4–6 months at 14.5% moisture content and 2–3 months at 15.5% moisture content. For wholemeal stored under the most favourable conditions, a shelf life of 3 months may be expected, or, if the product has been entoleted, up to 12 months.

7.5.1 Flour blending

Blending of finished flours is widely practised on the continent of Europe and is also increasingly popular in the United Kingdom. It can ensure greater uniformity in a product and provide flexibility in response to requirements for flours of unusual specification through the blending of separately stored flours of various types. Blending can be performed as a batch process or on a volumetric basis. The simpler volumetric method depends upon flours being discharged from two or more bins into a common conveyor, their discharge rates being controlled to provide the respective proportions required. The more accurate batch method involves the use of weighers to deposit required weights of products from each of the selected bins into a batch mixer. An additional advantage of the batch system is that improvers can be added at the same time as the flours are mixed. The mixers used may be of the ribbon type, whereby the blend is continually tumbled, or the air mixer, in which the blend is agitated by air injected into a holding bin.

7.5.2 Bulk storage and delivery of flour

Storage of flour in bulk bins and delivery in bulk containers (truck, rail) have advantages over storage and delivery in bags. Although construction costs of bulk-storage facilities (bins, air-handling systems, etc.) are high, the running costs are somewhat lower because human labour is much reduced, warehouse space is better utilized and material transfer is much more efficient.

The capacity of bins for storing flour in bulk is often 70–100 tonne or more. Packing pressures inside the bin increases with bin area, not with bin height; a bin area of 5.6 m^2 is often satisfactory. Normally modern bins are constructed of metal, and are mounted on load cells (scales) for bin inventory, although sometimes concrete is still used. Wooden bins are liable to become infested. The choice of construction material is dependent upon the company, but steel is currently most popular as metal bins are cheaper (unless capacity is over 20,000 tonnes), do not crack, are easily installed and relocated and are immediately usable on completion of construction (Anon., 1989). The inner surfaces must be smooth to allow stock to slide down the walls readily. Steel bin walls are usually coated with shellac varnish or other coating, and lower parts may be painted with a low-friction polyurethane paint or coating. Concrete surfaces are ground and coated with several coats of sodium silicate wash to provide a seal. The shape of bins is again a matter of choice. Circular bins are cheaper as lighter-gauge steel may be used, but there is more space wasted between cylindrical bins than between rectangular bins. A problem that can arise when flour is discharged from the bottom of a bin is bridging of stock; this can be avoided by good hopper design (>60° hopper angles) and/or efficient dischargers (live bottoms, air assist, vibratory assist, etc.). Bins are often filled pneumatically (using dilute-phase transport) and emptied using fluidizing dischargers which use $0.8–1.1 \, m^3/min$ of low-pressure air (20–70 kN/ m^2) to fluidize the flour, causing it to behave as a liquid and flow down a reduced gradient to the outlet, or active bin bottoms (rotating bottoms with paddles). Mechanical (worm- or screw-type conveyors) and vibratory dischargers may also be used to assist discharge of flour from bins. When flour and air are present in appropriate proportions there is a risk of dust explosions if a source of ignition is present; thus in all flour storing, conveying and handling situations it is essential to avoid sources of ignition and provide adequate dust-control systems. Additional precautions include the incorporation of explosion-relief panels into bin tops. Similar panels are recommended in the areas of buildings surrounding the bins. Fig. 7.30 illustrates some of the key components of pneumatic systems at flour mills. Fig. 7.31 depicts a typical schematic of a flour storage and handling system.

Flour is typically conveyed via 'dilute-phase' transfer, where the air:solids ratio is >2.0, air velocities can be up to 8000 ft/min and transport capacity per pipeline can often reach up to 50 t/h. To pick up the flour particles, suspend them in the air stream and transport the flour to a desired location (e.g., unload a rail car, transfer to a bin, transfer to a process, load a rail car), an airspeed between 3600 and 4100 ft/min is required (for whole wheat the required air speed is between 5000 and 5500 ft/min).

Flour was first delivered in bulk in the 1950s, and by 1987 65% of flour delivered in the United Kingdom was in bulk (Anon., 1989). Nowadays, most flour in the United Kingdom and the United States is delivered by this method.

Bulk trucks and railcars for transport can be filled at the mill by gravity feed or blowline, or, most efficiently, by fluidized delivery

Bin fill piping & dust collectors

Bin bottom & piping

Rotary airlock valve

Blower

Active bin unloader

FIGURE 7.30   Typical components for pneumatic handling of flour.

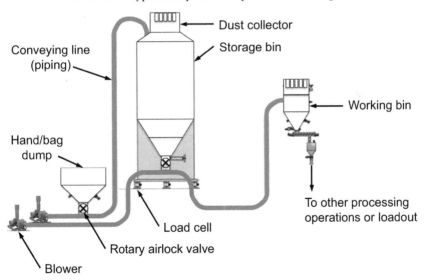

FIGURE 7.31   Schematic diagram depicting typical pneumatic handling of flour.

from a loadout bin and scale directly above the vehicle. By this method flow rates of 250–300 t/h (or more) can be achieved. Discharge of the vehicle upon arrival at its destination is then assisted by air pressure; some larger tankers have a fluidizing pad in the base of each hopper.

Blowers mounted either on the vehicle or at the customer's premises blow the delivery direct to the storage bins at the bakery or processing plant.

Mini-bulk containers, holding up to about 2 tonne, may be used in some mills for transport and delivery of products – mainly bran and germ, but in some cases flour also.

References

Aguilar, F., Crebelli, R., Di Domenico, A., Dusemund, B., Frutos, M.J., Galtier, P., Gott, D., Gundert-Remy, U., Lambré, C., Leblanc, J.-C., Lindtner, O., Moldeus, P., Mortensen, A., Mosesso, P., Parent-Massin, D., Oskarsson, A., Stankovic, I., Waalkens-Berendsen, I., Woutersen, R.A., Wright, M., and Younes, M. (2015). Scientific opinion on the re-evaluation of ascorbic acid (E 300), sodium ascorbate (E 301) and calcium ascorbate (E 302) as food additives. EFSA Journal 13 (5), 4087.

Anon., 1989. Flour Milling Correspondence Course: Product Handling, Storage and Distribution. (Module 10). Incorporated National Association of British and Irish Millers, London, UK.

Axford, D.W., Mcdermott, E.E., Redman, D.G., 1979. A note on the sodium dodecyl sulphate test for breadmaking quality: comparison with Pelshenke and Zeleny tests. Cereal Chemistry 56, 582–584.

Barnes, W.C., 1978. The rapid enzymic determination of starch damaged wheat. Staerke 30, 115–119.

Bellenger, P., Godon, B., 1972. Influence de l'aeration sur l'evolution de diverse caracteristiques biochimiques et physicochimiques. Annales de Technologie Agricole 21, 145.

Bentley, H.R., Mcdermott, E.E., Moran, T., Pace, J., Whitehead, J.K., 1950. Toxic factor from "Agenised" protein. Nature Lond. 165, 150.

Bloksma, A.H., Bushuk, W., 1988. Rheology and chemistry of dough. In: Pomeranz, Y. (Ed.). Wheat Chemistry and Technology. Wheat Chemistry and Technology Vol. II. AACCI, St. Paul, MN, USA.

Chamberlain, N., Collins, T.H., August 1977. The Chorleywood bread process: the importance of air as a dough ingredient. FMBRA Bulletin 4, 122.

Chiotelli, E., Rolee, A., le Meste, M., 2004. Rheological properties of soft wheat flour doughs: effect of salt and triglycerides. Cereal Chemistry 81, 459–468.

Cluskey, J.E., Wu, Y.V., Wall, J.S., Inglett, G.E., 1973. Oat protein concentrates from a wet-milling process: preparation. Cereal Chemistry 50, 475.

Collar, C., Bollaín, C., Rosell, C.M., 2007. Rheological behaviour of formulated bread doughs during mixing and heating. Food Science and Technology International 13 (2), 99–107.

D'Appolonia, B.L., Kunerth, W.H., 1984. The Farinograph Handbook, third ed. American Assoc of Cereal Chemists, Intl., St. Paul, MN, USA.

Donelson, J.R., Yamazaki, W.T., 1962. Note on a rapid method for the estimation of damaged starch in soft wheat flours. Cereal Chemistry 39, 460–462.

Evers, A.D., 1982. Methods for particle-size analysis of flour: a collaborative test. Laboratory Practice 31, 215–219.

Evers, A.D., 1993. On-line quantification of bran particles in white flour. Food Science and Technology Today 71 (1), 23–26.

FAO/WHO, 1973. Energy and Protein Requirements. Report of a joint FAO/WHO Ad Hoc Expert Committee. FAO Nutrition Meetings Report Series, No. 52, WHO Technical Report Series, No. 522.

Faridi, H., Faubion, J.M., 1990. Dough Rheology and Baked Product Texture. Van Norstrand Rheinhold, NY, USA.

Farrand, E.A., 1964. Modern bread processes in the United Kingdom with special reference to *alpha*-amylase and starch damage. Cereal Chemistry 41, 98–111.

Frazier, P., 1971. A Physico-chemical Investigation into the Mechanism of the Zeleny Test (PhD thesis). Leeds, UK.

Gibson, T.S., Qualla, A.L., Mccleary, B.V., 1992. An improved enzymic method for the measurement of starch damage in wheat flour. Journal of Cereal Science 15, 15–27.

Guy, R., 1993. Ingredients. In: Frame, N. (Ed.), The Technology of Extrusion Cooking. Blackie, Glasgow, UK.

Hay, R.L., Every, D., 1990. A simple glutenin turbidity test for the determination of heat damage in gluten. Journal of the Science of Food and Agriculture 53, 261–270.

Hunter, R.S., Harold, R.W., 1987. The Measurement of Appearance, second ed. John Wiley & Sons, New York, NY, USA.

Jones, C.R., 1940. The production of mechanically damaged starch in milling as a governing factor in the diastatic activity of flour. Cereal Chemistry 15, 133–169.

Kahraman, K., Sakiyan, O., Ozturk, S., Koksel, H., Sumnu, G., Dubat, A., 2008. Utilization of Mixolab to predict the suitability of flours in terms of cake quality. European Food Research and Technology 227 (2), 565–570.

Kent, N.L. (Ed.), 1985. Technical Compendium on Composite Flours. Economic Commission for Africa, Addis Ababa.

McCance, R.A., Widdowson, E.M., 1967. The Composition of Foods, Med. Res Coun., Spec. Rpt. Ser. 297. 2nd Imp. H.M.S.O, London, UK.

McCleary, B.V., Sheehan, H., 1987. Measurement of cereal alpha-amylase: a new assay procedure. Journal of Cereal Science 6, 237–251.

McDermott, E.E., 1980. The rapid, non-enzymic determination of damaged starch in flour. Journal of the Science of Food and Agriculture 31, 405–413.

Mellanby, E., 1946. Diet and canine hysteria. British Medical Journal ii, 885.

Ministry of Agriculture, Fisheries, Food, 1984. The Bread and Flour Regulations 1984 Statutory Instruments 1984, No. 1304, as Amended by the Potassium Bromate (Prohibition as a Flour Improver) Regulations 1990 (SI 1990, No. 399) H.M.S.O (London, U.K).

Pinkney, A.J., Greenaway, W.T., Zeleny, L., 1957. Further developments in the sedimentation test for wheat quality. Cereal Chemistry 34, 16.

Ranum, P., 1992. Potassium bromate in bread baking. Cereal Foods World 37, 253–258.

Rasper, V.F., Preston, K.R., 1991. The Extensograph Handbook. Amer. Assoc of Cereal Chemists Intl., St. Paul, MN, USA.

Rosell, C.M., Collar, C., Haros, M., 2007. Assessment of hydrocolloid effects on the thermo-mechanical properties of wheat using the Mixolab. Food Hydrocolloids 21, 452–462.

Saunders, J., Stauffer, S., Krishnan, P., 2007. Low-sugar bread formulations using Alice, a hard white winter wheat. South Dakota State Journal of Undergraduate Research 5, 1–9.

Shuey, W.C., Tipples, K.H., 1980. The Amylograph Handbook. Amer. Assoc. of Cereal Chemists Intl., St. Paul, MN, USA.

Further reading

American Association of Cereal Chemists International, 1962. Cereal Laboratory Methods, seventh ed. Amer. Assoc. of Cereal Chemists Intl., St. Paul, MN, USA. with ammendments to 1992.

Anon., 1989. Glossary of Baking Terms. American Institute of Baking, Manhattan, KS, USA.

Anon., 1980. Glossary of Terms for Cereals and Cereal Products. British Standards Institution, London, UK.

Anon., 1990. Flour Treatments & Flour Products. Module 12 in Workbook Series. National Association of British and Irish Millers, London, UK.

Dengate, H.N., 1984. Swelling, pasting and gelling of wheat starch. Advances in Cereal Science and Technology 6, 49–82.

Faridi, H., Rasper, V.F., 1987. The Alveograph Handbook. Amer. Assoc. of Cereal Chemists Intl., St. Paul, MN, USA.

Farrand, E.A., 1972. Controlled levels of starch damage in a commercial U.K. bread flour and effects on absorption, sedimentation value and loaf quality. Cereal Chemistry 49, 479.

Graveland, A., Bosveld, P., Lichtendonk, W.J., Moonen, J.H.E., 1984. Structure of glutenins and their breakdown during dough mixing by a complex oxidation-reduction system. In: Graveland, A., Moonen, J.H.E. (Eds.), Gluten Proteins. Inst. Cereals, Flour and Bread, TNO Wageningen, The Netherlands, pp. 59–68.

Greer, E.N., Stewart, B.A., 1959. The water absorption of wheat flour; relative effects of protein and starch. Journal of the Science of Food and Agriculture 10, 248–252.

Holland, B., Welch, A.A., Unwin, I.D., Buss, D.H., Paul, A.A., Southgate, D.A.T., 1991. McCance and Widdowson's the Composition of Foods, fifth ed. The Roy. Soc of Chem, Cambridge, UK.

Macritchie, F., 1980. Physicochemical aspects of some problems in wheat research. Advances in Cereal Science and Technology 3, 271–326.

Martin, D.J., Stewart, B.G., 1987. Dough stickiness in in rye-derived wheats. Cereal Foods World 32, 672–673.

Miskelly, D.M., Moss, H.J., 1985. Flour quality requirements for Chinese noodle manufacture. Journal of Cereal Science 3, 379–387.

Osborne, B.G., Fearn, T., Hindle, P.H., 1993. Practical Near Infrared Spectroscopy, second ed. Longmans, Harlow, Essex, UK.

Payne, P.I., Nightingale, M.A., Krattiger, A.F., Holt, L.M., 1987. The relationship between HMW glutenin subunit composition and the breadmaking quality of British-grown wheat varieties. Journal of the Science of Food and Agriculture 40, 51–65.

Pomeranz, Y., Bolling, H., Zwingelberg, H., 1984. Wheat hardness and baking properties of wheat flours. Journal of Cereal Science 2, 137–143.

Pomeranz, Y., 1983. Single, universal, bread baking test - why not? In: Holas, J., Kratochvil, J. (Eds.), Progress in Cereal Chemistry and Technology. Elsevier Science Publishers, New York, NY, USA, pp. 685–690.

Schneeweiss, R., 1982. Dictionary of Cereal Processing and Cereal Chemistry. Elsevier, Scientific Publishing Co., Amsterdam, Netherlands.

8.1 PRINCIPLES OF BAKING

Primitive humans, nomadic hunters and gatherers of fruits and nuts, started to settle down and abandon nomadic life when, in Neolithic times, they discovered how to sow the seeds of grasses and, in due time, reap crops of 'cereal grains'. With this change in way of life came the beginnings of civilization, which, in many parts of Europe and elsewhere, has historically been based on diets relying on wheat, wheaten flour and the baked products made from flour – the principal product being bread. Other civilizations developed around rice, maize and other cereal grains.

The main functions of baking are to present cereal flours in attractive, palatable, digestible and useful forms. A few examples (not comprehensive) are provided in Fig. 8.1 and Table 8.1.

While wheat is the principal cereal grain used for bread making throughout the world, other cereals, particularly rye and oat, are also used. And, combinations of various grains and/or starches from other foods with wheat are also common (Fig. 8.2).

FIGURE 8.1   Wonder Bread was one of the first mass-produced sliced breads that were commercially available in the United States. It was first released 21 May 1921 in Indianapolis, IN, by the Taggart Baking Company. It was also one of the first commercial white breads to be fortified. Scale bars represent 1 mm distance.

TABLE 8.1   Examples of breads and other bakery products

BREADS
Breads (pan)
Breads (artisan)
Breads (crisp)
Breads (flat)
ROLLS/BUNS
Bagels
Biscuits
Brown-and-serve buns
Cakes

TABLE 8.1 Examples of breads and other bakery products—cont'd

Croissants
Doughnuts
English muffins
Hamburger buns
Hot dog (wiener) buns
Muffins
Scones
OTHER BREAD PRODUCTS
Breadcrumbs
Bread stuffing/dressing
Kaiser rolls
Pizza dough
Pretzel bread/rolls
Rusks
Viennoiseries
Waffles
SWEET ROLLS
Cinnamon rolls
Dessert breads
Doughnuts
Pastries
Sticky rolls
Sweet rolls

Moreover, 'bread' comes in multiple forms. Various flatbreads are discussed in Chapter 11; some examples of leavened products (a few other baked products) are illustrated in Fig. 8.3.

The first part of this chapter will consider bread-making processes and bread (and bread products) in which wheat flour (or meal) is the sole cereal. The use of other cereals will be discussed later in this chapter.

8.1.1 Use of milled-wheat products for bread

Bread is made by baking a dough, which has for its main ingredients wheaten flour, water, yeast and salt. Other ingredients that may be added

FIGURE 8.2 Examples of breads made with various combinations of wheat flour with other starch sources. Scale bars represent 1 mm distance.

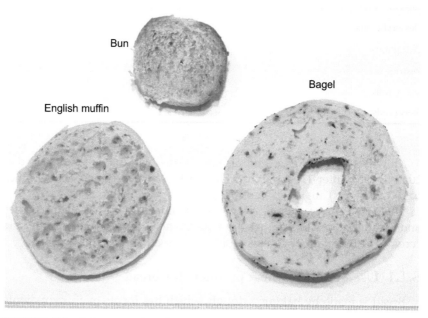

FIGURE 8.3 A few examples of yeast-leavened bakery products. Scale bars represent 1 mm distance.

include flours of other cereals, fat, malt flour, soya flour, yeast foods, emulsifiers, milk and milk products, fruit and/or gluten.

When these ingredients are mixed in correct proportions, three processes commence:

- The protein in the flour begins to hydrate, i.e., to combine with some of the water, to form gluten. Flour consists of discrete and separate particles, but the gluten is cohesive, forming a continuous three-dimensional structure that binds the flour particles together in a 'dough'. The gluten has peculiar extensible properties: it can be stretched like elastic and it possesses a degree of recoil or spring.
- Air bubbles are folded into the dough. During the subsequent handling of the dough these bubbles divide or coalesce. Eventually the dough comes to resemble a foam, with the bubbles trapped in the gluten network.
- Enzymes in the yeast start to ferment the sugars present in the flour and, later, the sugars released by diastatic action of the amylases on damaged starch in the flour, breaking them down to alcohol and carbon dioxide. The carbon dioxide gas mixes with the air in the bubbles and brings about expansion of the dough. 'Bread is fundamentally foamed gluten' (Atkins, 1971).

The requirements for making bread from wheat flour include: (1) formation of a gluten network and the creation of air bubbles within it; (2) the incorporation of carbon dioxide to turn the gluten network into a foam; (3) the development of the rheological properties of the gluten so that it retains the carbon dioxide while allowing expansion of the dough; and, finally, (4) the coagulation of the material by heating it in the oven so that the structure of the material is stabilized. The advantage of having an aerated, finely vesiculated crumb (Figs 8.4 and 8.5) in the baked product is that it is easily masticated.

Corresponding with these requirements, there are three primary stages in the manufacture of bread: mixing and dough development, dough aeration and oven baking. Fig. 8.6 provides a simple flowchart illustrating these common, primary steps to bread production. The method of dough development and aeration that has been customary since at least the time of the pharaohs in Egypt is panary fermentation by means of yeast.

8.1.2 Ingredients

8.1.2.1 Flour

Good bread-making flour is characterized as having:

- Protein that is adequate in quantity and, when hydrated, yields gluten that is satisfactory with respect to elasticity, strength and stability;
- Satisfactory gassing properties: the levels of amylase activity and of damaged starch should be adequate to yield sufficient sugars,

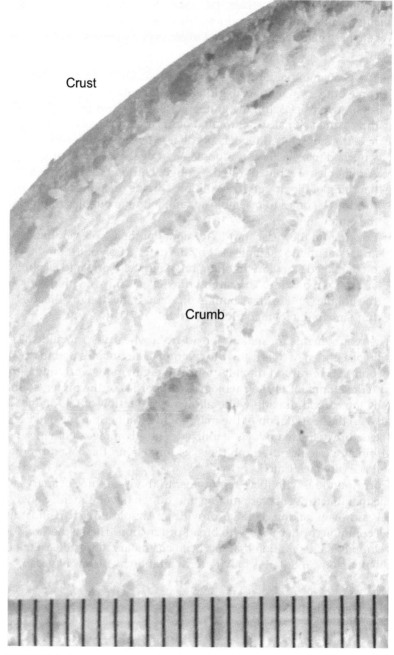

FIGURE 8.4   View of bread crust and crumb. Scale bars represent 1 mm distance.

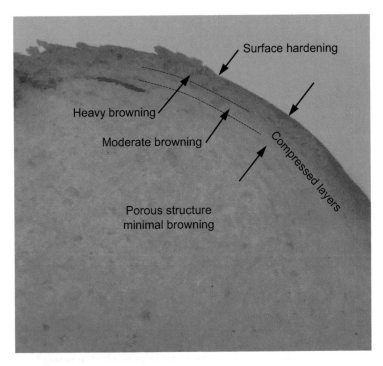

FIGURE 8.5 Top: Structure of bread varies from soft texture with loose, open pores internally, but which becomes compressed near the crust, which is hardened and brown on the surface. Bottom: Image analysis can be used to estimate porosity (black denotes pore openings; white denotes cell walls).

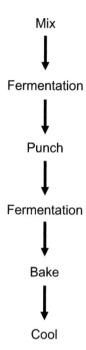

FIGURE 8.6 Simple flowchart illustrating major steps involved in baking bread. Even though there are many approaches to implementing baking bread either commercially or domestically, most use, at least to some degree, these major steps.

through diastatic action, to support the activity of the yeast enzymes during fermentation and proof;

- Satisfactory moisture content, not higher than about 14%, to permit safe storage, and satisfactory colour, and should meet specifications regarding bleach and treatment.

These requirements are met by the type of wheat called 'strong', viz. wheat having a reasonably high-protein content. Wherever possible, homegrown wheat is used for bread making, and this is the situation, for example, in Canada and in the United States, where such strong wheats, e.g., CWRS, HRS, are readily available.

In the United Kingdom, however, the homegrown wheat is, or until recently was, characteristically weak, viz. of low-protein content, and would not, by itself, yield flour from which bread, of the kind to which UK consumers are accustomed, could be made. It was therefore customary for flour millers in the United Kingdom to mill bread-making flour from a mixed grist of strong and weak wheats, the strong wheat component being imported, generally from Canada, and the weak component being homegrown UK wheat. Until the early 1960s, the average bread-making grist in the United Kingdom would consisted of 60%–70% of imported strong wheat plus 20%–30% of weak homegrown wheat (with a small proportion of 'filler' wheat of medium strength), yielding a white flour of about 12% protein content.

The imported Canadian wheat is more expensive than the homegrown UK wheat and, consequently, there was a strong urge to decrease the ratio of strong-to-weak wheat. This change was made possible in a number of ways, one of which was the advent of the Chorleywood bread process (CBP; discussed later in this chapter) because, among other advantages, the CBP permitted the use of a flour of about 1% lower protein content to produce bread of quality equivalent to that produced by the bulk fermentation process (BFP; also discussed later in this chapter).

Additional impetus to reduce still further the proportion of imported strong wheat in the bread grist followed the entry of the United Kingdom into the European Union (EU), and the imposition of heavy import levies, which have run as high as £120–130 per tonne, on the cost of wheat imported from third (i.e., non-EU) countries. Various measures have been adopted whereby the proportion of homegrown (or EU-grown) wheat in the bread-making grist could be further increased, while maintaining loaf quality. These efforts include:

- Moreover, the considerable increase in the size of the UK wheat harvest over the years has provided the flour millers with the possibility of obtaining adequate supplies of these newer varieties of good bread-making quality.
- Awarding of remunerative premiums to growers for higher protein homegrown wheats, which historically have been poorer yielders than low-protein wheats.
- Use of vital gluten as a bread ingredient (cf. later in this chapter).
- Supplementation of flours from lower protein homegrown wheats with air-classified high-protein fractions of flour.
- Use of high levels of fungal alpha-amylase (cf. later in this chapter).

The proportion of imported non-EU wheat (mostly Canadian CWRS wheat) in the UK bread-wheat grist has fallen from about 70% in 1960 to about 15% in 1990 (with a corresponding increase in the homegrown wheat proportion), with a considerable savings in the cost of the raw material. By 1992, some millers were supplying bread-making flour milled entirely from homegrown UK and EU wheats, with no non-EU component, but with the addition of 2%, or perhaps 2.5%, of vital gluten.

A similar reduction in the imported non-EU (strong) wheat content of the bread-making grist has also occurred in other countries.

One possible complication associated with the lowering of the strong-to-weak wheat ratio in the bread grist is the reduced proportion of damaged starch in the flour because of the frequent association of strength with hardness (as in the imported Canadian wheat) and, conversely, of weakness with softness (as in the EU-grown wheats). It is desirable that the content of damaged starch should be maintained at a reasonably high level, and this requirement can be met by adjustments to the milling process.

8.1.2.2 Leavening

Leavened baked goods are preferred in countries where wheat is available as a staple food. Leavening can be achieved in several ways, including the following:

1. Whisking egg into a foam with flour and other ingredients. This method is used in production of sponge and other cakes;
2. Water vapour production as in Scandinavian flatbreads and puff pastries;
3. Yeast;
4. Baking powder;
5. Starter cultures.

Yeast and baking powder are the most important leavening agents for bread. Each is appropriate for its own range of products, and in some cases, such as doughnuts, coffee cake, and pizza dough, either may be used alone or in combination.

8.1.2.3 Baking powders

Baking powders depend upon sodium bicarbonate as a source of CO_2 gas, which may be liberated by the action of sodium acid phosphate, monocalcium phosphate, sodium aluminium phosphate or glucono delta-lactone. One hundred grams of baking powder generates 15 mg (or 340 mM, or 8.2 L) of CO_2. Some is released at dough temperature and the remainder during baking.

8.1.2.4 Yeast

The quantity of yeast used is related inversely to the duration of fermentation, longer fermentation systems generally employing somewhat lower levels of yeast and also lower dough temperatures. Thus, 1% of yeast on flour wt. would be used for a 3-h straight dough system with the dough at 27°C, whereas 2%–3% of yeast on flour wt. would be required for a no-time dough at 27–30°C. Yeast activity increases rapidly with temperature, and its level of use is therefore reduced if the temperature is increased within a fixed time process.

In addition to providing CO_2 as a leavening agent, yeast also affects rheological properties of dough through the lowering of pH by CO_2 production, evolution of alcohol and the mechanical effects of bubble expansion. Further, yeast contributes significantly to the flavour and aroma of baked products.

Yeast is used in several different forms: compressed, cream (liquid), dried into pellets and instant active powders.

In recent years, attitudes to yeast production have become more enterprising. Specialized strains have been selected and bred to meet newly identified criteria. This has resulted partly from changing technologies within the baking industries and partly from new means of genetic quantification and manipulation.

Examples of these innovations are the replacement of conventional spore fusion by protoplast fusion, and genetic engineering through the use of recombinant DNA (rDNA) for introduction of an advantageous segment of the genetic materials of one strain to the genome of another. Strains that have an excellent performance in sugar-rich doughs normally show poor performance in lean doughs, but the subject of a European Patent (EP 0306107A2) is a yeast that performs well both in high-sucrose conditions and also in 'lean' conditions, where maltose is the available substrate. The technique involved was the introduction of genes coding for increased activity of the two enzymes maltose permease and maltase (α-glucosidase), allowing best use to be made of the limited quantities of maltose available in a lean dough.

Ability to ferment sugars anaerobically remains the major criterion of selection, but meeting this under different conditions has led to the introduction of specialized strains. The conditions that provide challenges include the requirements: (1) to be supplied and stored in a dry form with a longer life than the traditional compressed form; (2) to retain high activity in high-sugar formulations; and (3) to retain activity in yeast-leavened frozen doughs.

8.1.2.5 Dried yeasts

Until the early 1970s, two strains of *Saccharomyces cerevisiae* were widely used commercially. The yeast was grown to a nitrogen level of 8.2%–8.8% (dry basis), and an active dry yeast (A.D.Y.), which was grown to a nitrogen content of 7.0%. Thus, in the pelleted product, it had only 75%–80% of the gassing activity of the compressed yeasts (when compared on the same m.c. basis). New products available since that time have allowed the gap to be narrowed, although it does still exist.

Three forms of dried yeast are now available: A.D.Y., and the powdered products Instant A.D.Y. (I.A.D.Y.) and protected A.D.Y. (P.A.D.Y.).

A.D.Y. must be rehydrated in warm water (35–40°C) before adding to dough, while I.A.D.Y. and P.A.D.Y. can be added to dry ingredients before mixing. In fact, this results in more productive gassing. During storage, dried yeasts are subject to loss of activity in oxygen. The improved strains are supplied in vacuum packs or in packs with inert gas in the headspace (I.A.D.Y.), or in the presence of an antioxidant (P.A.D.Y.). P.A.D.Y. features in complete mixes containing flour and other ingredients, but the flour present must be at a very low m.c. to avoid moisture transfer and reduction in the level of production against oxidation.

8.1.2.6 High-sugar yeast

Products such as Danish pastries, doughnuts and sweet buns have a high-sugar content. The high-osmotic pressures involved are not tolerated by standard yeast strains, but good strains are available as I.A.D.Y. products. Japanese-compressed yeasts can also withstand high-osmotic conditions.

8.1.2.7 Frozen-dough yeasts

The production of breads from frozen doughs, at the point of sale, has increased dramatically and has created a requirement for cryoresistant yeasts. Most yeasts withstand freezing, but deteriorate rapidly during frozen storage. The best cryoresistant strains perform well in sweet goods but less well in lean doughs. This requirement has not yet been fully satisfied, but it has improved (Reed and Nagodawithana, 1991).

8.1.2.8 Salt

Salt is added to develop flavour. It also toughens the gluten and gives a less-sticky dough. Salt slows down the rate of fermentation, and its addition is sometimes delayed until the dough has been partly fermented. The quantity used is usually 1.8%–2.1% on flour wt., giving a concentration of 1.1%–1.4% of salt in the bread. Salt is added either as an aqueous solution (brine) or as the dry granular solid.

8.1.2.9 Fat

Fat is an essential ingredient for no-time doughs, such as the CBP. Added at the rate of about 1% on flour wt., fat improves loaf volume, reduces crust toughness and gives thinner crumb cell walls, resulting in a softer-textured loaf with improved slicing characteristics. Fat also keeps the bread soft and palatable for a longer period, which is equivalent to an antistaling effect (Hoseney, 1986).

During storage of flour, free fatty acids accumulate owing to the breakdown of the natural fats, and the gluten formed from the protein becomes less soluble and shorter in character. When flour that has been stored for a long time, e.g., a year at ambient temperature is used for the CBP, the fat level should be increased to about 1.5% on flour wt.

8.1.2.10 Sugar

Sugar is generally added to bread made in the United States, giving an acceptable sweet flavour, but it is not usually added to bread in the United Kingdom. However, sugar may be included in prover mixes.

8.1.2.11 *Vital gluten*

Vital wheat gluten, viz. gluten prepared in such a way that it retains its ability to absorb water and form a cohesive mass, is now widely used in the United Kingdom and in other EU countries as an ingredient of bread:

- At levels of 0.5%–3.0% on flour wt. to improve the texture and raise the protein content of bread, crispbread and speciality breads such as Vienna bread and hamburger rolls;
- To fortify weak flours, and to permit the use by millers of a wheat grist of lower strong-to-weak wheat ratio (particularly in the EU countries) by raising the protein content of the flour (cf. earlier in this chapter);
- In starch-reduced high-protein breads (cf. later in this chapter), in which the gluten acts both as a source of protein and as a texturing agent;
- In high-fibre breads (cf. later in this chapter) now being made in the United States, to maintain the texture and volume.

In the United States, about 70% of all vital gluten is used for breads, rolls, buns and yeast-raised goods (Magnuson, 1985). Vital gluten is also used as a binder to raise the protein level in meat products (e.g., sausages), and in ready-to-eat breakfast cereals (e.g., Kellogg's Special K), breadings, batter mixes, pasta foods, pet foods, dietary foods and textured vegetable protein products.

Ultimately, the origin of the gluten is of little importance when used to raise the flour protein content by only 1%–2%; thus, UK-grown wheat can be used to provide vital gluten, thereby further reducing the dependence on imported strong wheat. The vital gluten is generally added to the flour at the mill, particularly in the case of wholemeal (McDermott, 1985).

8.1.2.12 *Gluten flour*

Gluten flour is a blend of vital wheat gluten with wheat flour, standardized to 40% protein content in the United States.

8.1.2.13 *Fungal amylase*

Besides the use of low levels (e.g., 7–10 Farrand units) of fungal amylase to correct deficiencies in natural cereal alpha-amylase and improve gassing, fungal amylase, sold under such trade names as MYL-X and Amylozyme, has a marked effect in increasing loaf volume when used at much higher levels as a bread ingredient in rapid bread-making systems. Use of high levels is possible because the fungal amylase has a relatively low thermal inactivation temperature. The fungal amylase starts to act during the mixing stage, when it causes a softening of the dough, which must be corrected by reducing the amount of doughing water, to maintain the correct dough consistency. Use of high levels of fungal amylase in the BFP would not be desirable, as the dough-softening effect would be too severe. Hence, addition of fungal amylase at high levels is made by the baker and not at the mill.

The fungal amylase continues to act during the early part of the baking process, attacking gelatinized starch granules, improving gas retention and helping the dough to maintain a fluid condition, thus prolonging the dough expansion time and increasing loaf volume. The increase in loaf volume is directly related to the level of fungal amylase addition up to about 200 Farrand units.

The effect of the addition of about 120 Farrand units of fungal amylase is so powerful that it may permit the use of flour of up to 2% lower protein content with no loss in loaf quality.

A similar increase in loaf volume could be produced by addition of a variety of commercial carbohydrase enzyme preparations (Cauvain and Chamberlain, 1988).

8.1.2.14 Soya flour

Enzyme-active soya flour is widely used as a bread additive, at a level of about 0.7% on flour wt. Advantages claimed for its use include: beneficial oxidizing effect on the flour, bleaching effect on flour pigments (beta-carotene) due to the presence of lipoxygenase, increase in loaf volume, improvement in crumb firmness and crust appearance, and extension of shelf life (Anon., 1988a) (cf. later in this chapter).

The improving action and bleaching properties of enzyme-active soya flour are due to peroxy radicals that are released by a type-2 lipoxygenase, which has an optimum activity at pH 6.5. Enzyme-active soya flour has two effects in a flour dough: it increases mixing tolerance, and it improves dough rheology, viz. by decreasing extensibility and increasing resistance to extension. The action of the lipoxygenase is to oxidize the linoleic acid in the lipid fraction of the wheat flour, but the action only occurs in the presence of oxygen (Grosch, 1986).

Improving agents

The use and effects of improving agents – potassium bromate, ascorbic acid, azodicarbonamide, L-cysteine – have been discussed in Chapter 7.

8.1.2.15 Physical treatments

The bread-making quality of flour can be improved also by physical means, e.g., by controlled heat treatment (cf. Chapter 5) or by an aeration process, in which flour is whipped with water at high speed for a few minutes and the batter then mixed with dry flour. Improvement is brought about by oxidation with oxygen in the air, probably assisted by the lipoxidase enzymes (cf. Chapter 4) present in the flour. A similar improving effect can be obtained by overmixing normal dough (without the batter stage); cf. the Chorleywood bread process (described later in this chapter).

8.1.3 Dough making

8.1.3.1 Water absorption

The amount of water to be mixed with flour to make a dough of standard consistency is usually 55–61 pt per 100 pt of flour, increasing in proportion to the contents of protein and damaged starch (cf. Chapters 4 and 7) in the flour.

Flour contains protein, undamaged starch granules and damaged starch granules, all of which absorb water, but to differing degrees. Farrand (1964) showed that the uptake of water, per gram of component, was 2.0 g for protein, 0–0.3 g for undamaged starch and 1.0 g for damaged starch. Thus, flours from strong wheat (with higher protein content) and from hard wheat (with a higher damaged-starch content) require more water than is needed by flours from weak (lower protein) or soft (less-damaged starch) wheats to make a dough of a standard consistency.

Besides the protein and starch, the soluble part of the hemicellulose (pentosan) forming the walls of the endosperm cells also absorbs water.

The water used in dough making should have the correct temperature so that, taking account of the flour temperature and allowing for any temperature rise during mixing, the dough is made to the correct final temperature. When using a process such as the CBP, in which the temperature rise during mixing may be as much as 14°C, it may be necessary to cool the doughing water.

It is important, particularly in plant bakeries, to maintain constant dough consistency. This may be done by adjusting the level of water addition automatically or semiautomatically. Determination of water absorption of the flour by means of the Brabender Farinograph is described in Chapter 7.

A flour with high-water-absorption capacity is generally preferred for bread making. Apart from increasing the proportion of strong wheat (high protein) in the grist, which may be uneconomical, the most convenient way of increasing water absorption is to increase the degree of starch damage. The miller can bring this about by modifying the milling conditions (cf. Chapter 6).

8.1.3.2 Fermentation

The enzymes principally involved with panary fermentation are those that act upon carbohydrates: alpha-amylase and beta-amylase in flour, and maltase, invertase and the zymase complex in yeast. Zymase is the name that was formerly used for about 14 enzymes.

The starch of the flour is broken down to the disaccharide maltose by the amylase enzymes; the maltose is then split to glucose (dextrose) by maltase; glucose and fructose are fermented to carbon dioxide and alcohol by the zymase complex.

Some of the starch granules in flour become mechanically damaged during milling (cf. Chapters 4, 6, and 7), and only these damaged granules can be effectively attacked by the flour amylases. It is therefore essential that the flour should contain adequate damaged starch to supply sugar during fermentation and proof. When the amylase enzymes break down the damaged starch, water molecules bound by the starch are released, causing softening of the dough. This situation must be borne in mind when calculating the amount of doughing water required, the amount of water released being dependent not only on the level of damaged starch but also on the alpha-amylase activity, length of fermentation time and dough temperature. Excessive levels of starch damage, however, have an adverse effect on the quality of the bread (cf. Chapter 7, and later in this chapter) – loaf volume is decreased, and the bread is less attractive in appearance.

There are small quantities of sugar naturally present in flour (cf. Chapter 4) but these are soon used up by the yeast, which then depends on the sugar produced by diastatic action from the starch.

During fermentation about 0.8 kg of alcohol is produced per 100 kg of flour, but much of it is driven off during the baking process. New bread is said to contain about 0.3% of alcohol. Secondary products, e.g., acids, carbonyls and esters, may affect the gluten or impart flavour to the bread.

8.1.3.3 Amylase

Both alpha- and beta-amylases catalyze the hydrolysis of starch, but in different ways (cf. Chapter 4).

Normal flour from sound wheat contains ample beta-amylase but generally only a small amount of alpha-amylase. The amount of alpha-amylase, however, increases considerably when wheat germinates. Indeed, flour from wheat containing many sprouted grains may have too high an alpha-amylase activity, with the result that, during baking, some of the starch is changed into dextrin-like substances. Water-holding capacity is reduced, the crumb is weakened, and the dextrins make the crumb sticky (cf. Chapter 4). However, flour with too high a natural alpha-amylase activity could be used for making satisfactory bread by microwave or radio-frequency baking methods (cf. later this chapter). Another possibility would be to make use of an alpha-amylase inhibitor, e.g., one prepared from barley, as described in Canadian Patent No. 1206157 of 1987 (Zawistowska et al., 1988).

The functions of starch in the baking of bread are to dilute the gluten to a desirable consistency, to provide sugar through diastasis, to provide a strong union with gluten, and by gelatinization to become flexible and to take water from the gluten, a process that helps the gluten film to set and become rigid.

8.1.3.4 Gas production and gas retention

The creation of bubble structure in the dough is a fundamental requirement in bread making. The carbon dioxide generated by yeast activity does not create bubbles; it can only inflate gas cells already formed by the incorporation of air during mixing.

Adequate gas must be produced during fermentation, otherwise the loaf will not be inflated sufficiently. Gas production depends on the quantity of soluble sugars in the flour and on its diastatic power. Inadequate gassing (maltose value less than 1.5) may be due to an insufficiency of damaged starch or to a lack of alpha-amylase; the latter can be corrected by adding sprouted wheat to the grist, or malt flour, or fungal amylase, e.g., from *Aspergillus oryzae* or *Aspergillus awamori*, to the flour (cf. earlier this chapter). Fungal amylase is preferred to malt flour because the thermal inactivation temperature of fungal amylase is lower (75°C) than that of cereal alpha-amylase (87°C), and its use avoids the formation of gummy dextrins during baking and the consequent difficulties in slicing bread with a sticky crumb.

Gas retention is a property of the flour protein; the gluten, while being sufficiently extensible to allow the loaf to rise, must yet be strong enough to prevent gas escaping too readily, as this would lead to collapse of the loaf. The interaction of added fat with flour components also has a powerful effect on gas retention.

8.1.4 Dough development

8.1.4.1 Protein

The process of dough development, which occurs during dough ripening, concerns the hydrated protein component of the flour. It involves an uncoiling of the protein molecules and their joining together, by cross-linking, to form a vast network of protein that is collectively called 'gluten'. The coils of the protein molecules are held together by various types of bonds, including disulphide (—SS—) bonds, and it is the severing of these bonds (allowing the molecules to uncoil) and their rejoining in different positions (linking separate protein molecules together) that constitutes a major part of dough development.

Sulphydryl (—SH) groups (cf. Chapter 4) are also present in the protein molecules as side groups of the amino acid cysteine. Reactions between the —SH groups and the —SS— bonds permit new inter- and intraprotein/polypeptide relationships to be formed via —SS— bonding, one effect of this interchange being the relaxation of dough by the relief of stress induced by the mixing process.

While gluten is important in creating an extensible framework, soluble proteins in the dough liquor may also contribute to gas retention by forming an impervious lining layer within cells, effectively blocking pinholes in cell walls (Gan et al., 1990).

8.1.4.2 Dough ripening

A dough undergoing fermentation, with intermittent mechanical manipulation, is said to be 'ripening'. The dough when mixed is sticky, but as ripening proceeds, it becomes less sticky and more rubbery when moulded, and is more easily handled at the plant (or at home). The bread baked from it becomes progressively better, until an optimum condition of ripeness has been reached. If ripening is allowed to proceed beyond this point, deterioration sets in, the moulded dough gets shorter and possibly sticky again, and ultimate bread quality becomes poorer. A ripe dough has maximum elasticity after moulding and gives maximum spring in the oven; a green or underripe dough can be stretched but has insufficient elasticity and spring; an overripe dough tends to break when stretched.

If the optimum condition of ripeness persists over a reasonable period of time, the flour is said to have good fermentation tolerance. Weak flours quickly reach a relatively poor optimum, and have poor tolerance, whereas strong flours give a higher optimum, take longer to reach it and have good tolerance. Addition of improvers or oxidizing agents to the flour can speed up the rate at which dough ripens and hence shorten the time taken to achieve optimum development.

8.1.4.3 Dough stickiness

Certain agronomic advantages and improved disease resistance in wheat have been achieved by incorporating genes from rye. The short arm of the rye chromosome 1R has been substituted for the short arm of the homologous group 1 chromosome in wheat. However, the doughs made from the flour of many of the substitution lines have a major defect in that they are intensely sticky. This stickiness is not due to overmixing, excess water or excess amylolytic activity; the factor responsible for this stickiness, introduced with the rye chromosome, has not yet been identified (Martin and Stewart, 1991).

8.1.4.4 Proteolytic enzymes

Besides the enzymes that act on carbohydrates, there are many other enzymes in flour and yeast, of which those that affect proteins, the proteolytic enzymes, may be of importance in baking. Yeast contains such enzymes, but they remain within the yeast cells and hence do not influence the gluten.

The proteolytic enzymes of flour are proteases. They have both disaggregating and protein solubilizing effects, although the two phenomena may be due to distinct enzymes.

The undesirable effect on bread quality of flour milled from wheat attacked by insects is generally considered to be due to excessive proteolytic activity. Inactivation temperature is lower for proteolytic enzymes than for diastatic enzymes, and heat treatment has been recommended as a remedy for excessive proteolytic activity in buggy wheat flour. However, it is difficult to inactivate enzymes by heat treatment without damaging the gluten proteins simultaneously.

8.1.4.5 Surfactants

These substances act as dough strengtheners, to help withstand mechanical abuse during processing, and they also reduce the degree of retrogradation of starch (cf. Chapter 4, and later this chapter). They include calcium and sodium stearoyl lactylates and mono- and diacetyl tartaric esters of mono- and diglycerides of fatty acids (diacetyl tartaric acid esters; DATEM), and are used at levels of about 0.5% on flour wt. (Hoseney, 1986). The Bread and Flour Regulations 1984 permit the use of SSL, up to a maximum of 5 g/kg of bread, in all bread, and of DATEM esters, with no limit specified, in all bread.

8.1.4.6 Stearoyl-2-lactylates

As discussed in Chapter 7, calcium stearoyl-2-lactylate (CSL) and sodium stearoyl-2-lactylate (SSL) are the salts of the reaction product between lactic and stearic acids. CSL ('Verv') and SSL ('Emplex') are dough-improving and antistaling agents; they increase gas retention, shorten proving time and increase loaf volume. They increase the tolerance of dough to mixing, and widen the range over which good-quality bread can be produced. The use of CSL or SSL permits the use of a considerable proportion of nonwheat flours in 'composite flours' to make bread of good quality by ordinary procedures (cf. later this chapter). For example, a typical composite bread-making flour could contain (in parts) wheat flour 70, maize or cassava starch 25, soya flour 5, CSL 0.5–1.0, plus yeast, sugar, salt and water. The nutritive value of such bread has been shown to be superior to that of bread containing only wheat flour, salt, yeast and water. Use of CSL and SSL has been permitted in the United States since 1961; maximum use levels for a variety of food products are listed in 21 CFR 172.846 and 21 CFR 177.12 – 0.5% of flour wt. for bread.

8.1.4.7 Colour of bread crust and crumb

During oven baking, various physicochemical alterations occur due to the heating. Fig. 8.7 illustrates common temperature behaviour during the bread-making process, while Fig. 8.8 illustrates general trends for starch

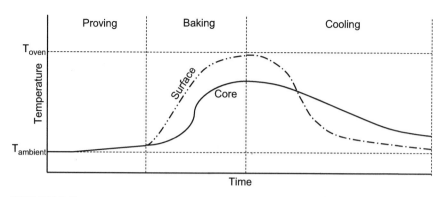

FIGURE 8.7 Typical behaviours of loaf surface temperature versus core temperature over time during final steps of the bread-making process. Note that the core lags behind the surface in both heating up and cooling down – due to conduction heat transfer through the loaf matrix. Heat transfer is affected by a number of factors, including formulation, type of cereal flour used, porosity, initial moisture content, oven temperature and temperature distribution, etc. Generally, a loaf is considered fully baked when the core reaches approximately 95°C and the starch is fully gelatinized. Additional details can be found in Cauvain and Young (1998).

gelatinization, loaf moisture content, and Hunter color score. As shown, the loaf core will have similar, yet delayed, trends compared to the surface, which develops into a crust.

The brown colour of the crust of bread is probably due to melanoidins formed by nonenzymic 'browning reactions' (Maillard type) between amino acids, dextrins and reducing carbohydrates. Addition of amino acids to flours giving pale crust colour can result in improvement of colour. The glaze on the crust of bread is due, in part, to starch gelatinization that occurs when the oven humidity is high. An underripe dough that still contains a fairly high sugar content will give a loaf of high crust colour; conversely, an overripe dough gives a loaf of pale crust colour.

The perceived colour of breadcrumb is influenced by the colour, degree of bleach and extraction rate of the flour; by the use of fat, milk powder, soya flour or malt flour in the recipe; by the degree of fermentation; by the extent to which the mixing process disperses bubbles within the dough; and by the method of panning, cross-panning and twisting to increase light reflectance.

8.1.4.8 Bread aroma and flavour

The aroma of bread results from the interaction of reducing sugars and amino compounds, accompanied by the formation of aldehydes. Aroma is also affected by the products of alcoholic and, in some cases, lactic acid fermentation: organic acids, alcohols and esters. The highest concentrations of flavour compounds in bread reside chiefly in the crust.

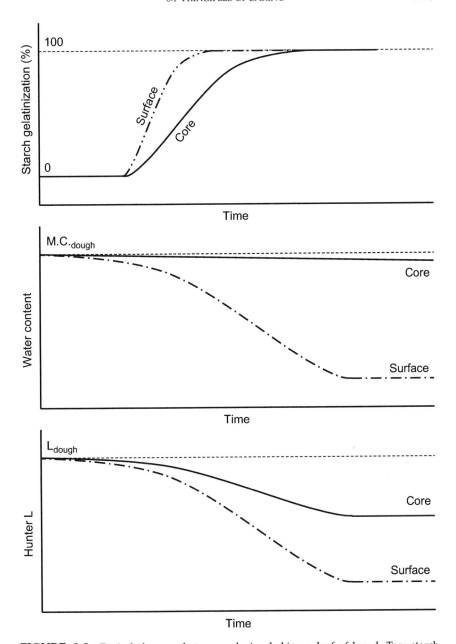

FIGURE 8.8 Typical changes that occur during baking a loaf of bread. Top: starch gelatinization – the core lags behind the surface, but eventually becomes gelatinized (due to heat transfer). Middle: moisture content – the surface loses the most moisture, while the core loses a little (due to moisture diffusion to the surface). Bottom: Hunter L colour parameter (which is reflective of Maillard reactions on the surface, while the core does not experience these to the same degree).

8.2 COMMERCIAL PROCESSES FOR MAKING WHITE BREAD

A white pan loaf of good quality is characterized by having sufficient volume, an attractive appearance regarding shape and colour, and a crumb that is finely and evenly vesiculated and soft enough for easy mastication, yet firm enough to permit thin slicing. A more open crumb structure is characteristic of other varieties, e.g., Vienna bread and French bread. The attainment of good quality in bread depends partly on the inherent characteristics of the ingredients – particularly the flour – and partly on the baking process.

In the United Kingdom, white bread comprised about 52% of the total bread eaten in the home in 1989. Methods used for commercial production of white bread differ principally according to the way in which the dough is developed, which may include:

- Biologically, by yeast fermentation. Examples: bulk (long) fermentation processes (straight dough system; sponge and dough system);
- Mechanically, by intense mixing and use of oxidizing agents. Examples: J.C. Baker's 'Do-Maker' process and AMFLOW processes (continuous); Chorleywood bread process; Spiral Mixing Method;
- Chemically, by use of reducing and oxidizing agents. Example: Activated Dough Development process.

In the bulk fermentation process, some of the starch, after breakdown to sugars, is converted to alcohol and carbon dioxide, both of which are volatile and are lost from the dough (cf. earlier this chapter). The bulk fermentation process is thus a somewhat wasteful method, and processes that utilize mechanical or chemical development of the dough offer considerable economic advantages, as there is less breakdown of the starch, as well as being much more rapid.

Other rapid methods include the Continental No-time process (or Spiral Mixing Method), the Emergency No-time process and the Aeration or Gas-injection process.

Several of the most common methods will be discussed next; the bulk fermentation method is considered the traditional approach, although the CBP has also become quite popular. Major steps of each process are illustrated in Fig. 8.9.

8.2.1 Bulk (long) fermentation processes

The bread is made by mixing a dough from flour, water, yeast, fat and salt, allowing the dough to rest at a temperature of 26–27°C while fermentation and gluten ripening take place, and then baking in the oven.

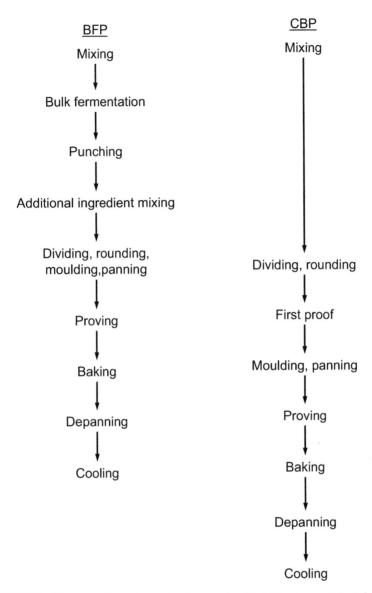

FIGURE 8.9   Flowchart illustrating major steps involved in baking bread via bulk fermentation process (BFP) and Chorleywood bread process (CBP). Detailed information regarding each process can be found in Kulp and Ponte (2000) as well as Dendy and Dobraszczyk (2001).

8.2.1.1 Straight dough system

In a representative procedure, the ingredients for a 100-kg 3-h dough would be 100 kg of flour, with probably 1 kg of yeast, 2 kg of salt, 1 kg of fat and 55–57 kg of water at a temperature that will bring the mixture to about

27°C after mixing. Until the prohibition of its use in the United Kingdom in April, 1990 (MAFF, 1990), potassium bromate would generally have been added to the flour by the miller at a rate of 15 mg/kg. The yeast is dispersed in some of the water, the salt dissolved in another portion. All these ingredients, together with the rest of the water, are then blended and mixed in a low-speed mixer during 10–20 min, during which there may be a temperature rise of 2°C. The resulting dough is set aside while fermentation proceeds.

As an alternative to treatment of the flour with potassium bromate, the miller can use ascorbic acid at a rate of 15 mg/kg, or azodicarbonamide at a level of 5 mg/kg. If the bulk fermentation time is 1 h or longer, no further treatment of the flour with oxidizing agents by the baker is required. For use in a bulk fermentation process of 30 min or less, the baker would probably add a further 50–100 mg/kg of ascorbic acid or 20–30 mg/kg of azodicarbonamide, as one of the functions of oxidizing agents is to shorten the fermentation time.

The Bread and Flour Regulations 1984 permit the use, in the United Kingdom, of up to 200 mg/kg of ascorbic acid in all bread, and of up to 45 mg/kg of azodicarbonamide in all bread except wholemeal. To avoid the accidental breaching of these regulations, a Code of Practice has been agreed to, by which millers will add not more than 50 mg/kg of ascorbic acid and/or 10 mg/kg of azodicarbonamide to flour at the mill, while improver manufacturers will ensure that their products, when added to flour at the recommended levels, will add not more than 150 mg/kg of ascorbic acid and/or 35 mg/kg of azodicarbonamide.

After about 2 h in a 3-h fermentation process, the dough is 'knocked back', i.e., manipulated to push out the gas that has been evolved in order to even out the temperature and give more thorough mixing.

After another hour's rising, the dough is divided into loaf-sized portions and these are roughly shaped. The dough pieces rest at about 27°C for 10–15 min ('first proof') and are then moulded into the final shape, during which the dough is mechanically worked to tighten it so that the gas is better distributed and retained, and then placed in tins. The final mould is very important in giving good texture in bulk-fermented bread. It is during the rounding and moulding processes that the bubble structure, resulting in a satisfactory crumb structure, is developed; bubbles that have been inflated during the fermentation are subdivided to produce a greater number of smaller bubbles.

The dough rests again in the tins for the final proof of 45–60 min at 43°C and 80%–85% r.h., and it is the carbon dioxide evolved during the final proof that inflates the dough irreversibly. The dough is then baked in the oven at a temperature of 235°C for 20–40 min, depending on loaf size, with steam injected into the oven to produce a glaze on the crust.

A number of changes take place as the temperature of the dough rises at the beginning of baking:

- The rate of gas production increases;
- At about 45°C the undamaged starch granules begin to gelatinize and are attacked by alpha-amylase, yielding fermentable sugars;
- Between 50 and 60°C the yeast is killed;
- At about 65°C the beta-amylase is thermally inactivated;
- At about 75°C the fungal amylase is inactivated;
- At about 87°C the cereal alpha-amylase is inactivated;
- Finally, the gluten is denatured and coagulates, stabilizing the shape and size of the loaf.

8.2.1.2 Sponge and dough system

When the bulk fermentation process is used in England, the straight dough system is generally employed. But in the United States, to some extent in Scotland, and occasionally in England, a sponge and dough system is used.

The sponge and dough system differs from the straight dough system in that only part of the flour is mixed at first with some or all of the yeast, some or all of the salt, and sufficient water to make a dough, which is allowed to ferment for some hours at 21°C. The sponge (as this first dough is called) is then broken down by remixing, and the remainder of the flour, water and salt, and all the fat are added to make a dough of the required consistency. Addition of oxidizing agents – ascorbic acid alone or ascorbic acid plus azodicarbonamide – would usually be made at the dough (i.e., second) stage. The dough is given a short fermentation at 27°C before proving and baking. Further details of the procedure are to be found in *The Master Bakers' Book of Breadmaking* (Brown, 1982). The sponge and dough system is said to produce bread that has a fuller flavour than that made by the straight dough system. A baker's grade flour of, say, 12% protein content, is suitable for the dough stage, but the sponge stage requires a stronger flour of, say, 13% protein content. More than 65% of all bread in the United States is made by the sponge and dough system.

8.2.1.3 Plant baking

In small bakeries, most of the processes of dividing, moulding, placing in the proving cabinets and the oven, and withdrawing therefrom, are carried out by hand – in Europe many artisan bakeries are still in operation. However, disadvantages of hand processing are lack of uniformity in the products and the excessive amount of labour involved. In large bakeries, machines carry out all these processes. Mixers are of the closed-bowl high-speed type, consuming a total quantity of energy (including no-load power) of 7.2–14.4 kJ/kg (1–4 Wh/kg) and taking 15–20 min. Dough dividers divide the dough by volume: automatic provers have built-in controls giving correct temperature and relative humidity. In the smaller bakeries, peel, reel or rack ovens have now largely replaced drawplate ovens. In larger bakeries, so-called 'travelling' ovens are used. In these, the doughs are placed on continuous bands or belts that travel through the oven. The oven is tunnel-shaped and possibly up to 18.3 m (60 ft) long.

The bulk fermentation process is now used to make probably not more than 10% of all commercially made bread in the United Kingdom; the principal users would be plant bakeries in Scotland and small bakeries, some of which prefer the sponge and dough system.

8.2.2 Mechanical development processes

8.2.2.1 *Continuous dough making*

A further stage in the mechanization of bread making is represented by the continuous bread-making process, exemplified by the Wallace & Tiernan Do-Maker Process, based on the work of J. C. Baker, which was formerly used in the United Kingdom and the United States. About 35% of US bread was made by the Baker process in 1969. In the process, first used in the United Kingdom in 1956 (see Anon., 1957), the flour, spouted from a hopper, is continuously mixed with a liquid preferment or 'brew' in electronically regulated quantities. The preferment is a mixture consisting of a sugar solution with yeast, salt, melted fat and oxidizing agents, which is fermented for 2–4 h. The dough is allowed no fermentation time, but instead is subjected to intense mechanical mixing whereby the correct degree of ripeness for proving and baking is obtained. In the absence of fermentation, it is essential to incorporate an appropriate quantity of oxidizing agent into the dough. The dough is extruded through a pipe, cut off into loaf-sized portions, proved and baked. The Do-Maker process gives bread with a characteristic and very even crumb texture. Considerable time is saved in comparison with bulk fermentation processes.

The AMFLOW process has an overall similarity to the Do-Maker process, but features a multistage preferment containing flour, and a horizontal, instead of a vertical, development chamber.

In the United States, the Do-Maker and AMFLOW processes are being replaced by the sponge and dough processes or by the CBP. The Do-Maker and AMFLOW processes are no longer used in the United Kingdom.

8.2.2.2 *Chorleywood bread process*

The CBP process is a batch or continuous process in which dough development is achieved during mixing by intense mechanical working of the dough in a short time, and bulk fermentation is eliminated. The process was devised in 1961 by cereal scientists and bakers at the

British Baking Industries Research Association, Chorleywood, Herts, England (Chamberlain et al., 1962; Axford et al., 1963). It is necessary to use a special high-speed mixer for mixing the dough. The process is characterized by:

- The expenditure of a considerable, but carefully controlled, amount of work (11 Wh/kg; 40 J/g) on the dough during a period of 2–4 min;
- Chemical oxidation with ascorbic acid (vitamin C) alone or with potassium bromate (if allowed) at a relatively high total level, viz. 100 mg/kg, or with azodicarbonamide at a level of 20–30 mg/kg or with both ascorbic acid and azodicarbonamide;
- Addition of fat (about 0.7% on flour wt.) – this is essential – of which 5% (0.035% on flour wt.) should be high–melting point fat, which will still be solid at 38°C;
- Use of extra water (3.5% more than normal, based on flour wt.);
- Absence of any preferment or liquid ferment;
- A higher level of yeast, 2% on flour wt., than is used in the BFP;
- A first proof of 2–10 min, after dividing and rounding, followed by conventional moulding and final proof.

The level of work input is critical, but is dependent on the genetic make-up or strength of the wheat. Thus, while a figure of 11 Wh/kg is applicable to dough made from an average grist, there are certain UK wheat varieties, e.g., Fresco, the flour from which, if used alone, would require 17–20 Wh/kg. If such varieties are included in the bread grist, it will be desirable for the miller to formulate the grist in such a way that the work input requirement is maintained at, or near, 11 Wh/kg. For an average grist, the quality of the bread, with respect to loaf volume and fineness of crumb structure, improves at work input levels from 7 to 11 Wh/kg, but at 13 Wh/kg or more the structure of crumb deteriorates. Work input level is monitored by a watt-hour metre and a counter unit attached to the mixer motor. The total work input required – dough wt. in kg × 11 – is set on the counter unit, and the mixer motor is automatically switched off when the determined amount of work has been performed.

One reason for the intense and rapid mixing is that it brings the molecules rapidly into contact with the oxidizing agents. During the intense mechanical development, a gas bubble structure is created in the dough, which, provided the dough is properly handled, is expanded in proof and becomes the loaf crumb structure.

Final dough temperature after mixing should be 28–30°C, but as the mixing process causes a temperature rise of 14–15°C, the doughing water may have to be cooled (a water-cooling unit is generally a part of the plant).

The use of potassium bromate is no longer permitted in the United Kingdom; if potassium bromate is not used as an oxidizing agent, the use of ascorbic acid instead, at a level of 100 mg/kg, may not be adequate to maintain loaf volume if the dough is mixed under partial vacuum. Under these conditions, the use of azodicarbonamide at a level of 20–30 mg/kg, in addition to the ascorbic acid, would be beneficial (cf. Chapter 7).

To achieve the full oxidation potential of the ascorbic acid, an adequate concentration of oxygen in the mixing machine bowl is essential; this requirement may be partially met by eliminating the vacuum, so that the dough is mixed in air, or, more effectively, by filling the headspace of the mixer with an oxygen-enriched atmosphere, e.g., 60% oxygen/40% nitrogen, equivalent to a 50/50 mixture of oxygen and air.

Further improvement in loaf volume, if potassium bromate is not being used, could be achieved by changing to a flour of slightly higher protein content or by adding vital gluten, by increasing the yeast level, by adding fungal amylase or by adding an emulsifier.

Equally satisfactory results can be obtained by replacing some or all of the fat with emulsifiers such as DATEM or sodium stearoyl-2-lactylate at a level of 0.1%–0.3% on flour wt.

The additional water required by the CBP, as compared with the BFP, is the consequence of the absence of bulk fermentation in the CBP. During fermentation, the breakdown of starch to sugars releases water, previously held by the damaged starch. The dough is softened by this released water and allowance is made for this effect, in the BFP, when calculating the amount of doughing water required. In the CBP, water is not released to the same extent because there is little or no breakdown of damaged starch, and without the released water the dough would tend to be too tight unless the amount of doughing water were increased. The extra water added in the CBP thus largely corresponds with the additional solids in the CBP that have not been lost (as in the BFP) during bulk fermentation. It leads to an increase of less than 1% in the moisture content of the bread.

Additional yeast is required in the CBP to compensate for the lower rate at which fermentation proceeds at the beginning of the final proof because, in the absence of a bulk fermentation, the yeast has not been activated as it is in the BFP. Another reason for use of extra yeast in the CBP is that the dough is relatively denser at the start of proof than it is in the BFP.

It is claimed that bread made by the CBP is indistinguishable in flavour or crumb structure from bread made by bulk fermentation, and that it stales less rapidly (Axford et al., 1968).

Advantages claimed for the CBP vis-à-vis BFP, besides the avoidance of bulk fermentation, are:

- An additional yield of about 7 pt of dough per 100 pt of flour, leading to an increase in yield of bread of 4%, and thus a net savings on raw material costs;
- A savings of 60% in processing time (see Fig. 8.10) (and even more savings vs. the sponge and dough system);

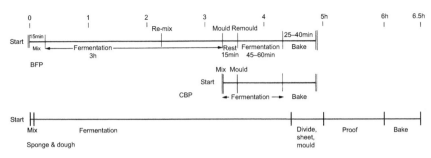

FIGURE 8.10   Approximate times required to make bread by the traditional 3-h bulk fermentation process (BFP) versus the Chorleywood Bread Process (CBP) versus the sponge and dough method. Note that times are only estimates, and may vary due to production specifics.

- A savings of 70% of space, previously occupied by fermenting dough;
- A reduction of about 70% in the amount of dough being processed at any one time and, consequently, a considerably reduced loss in case of plant breakdown;
- A lower staling rate;
- A greater amenability to control.

Moreover, by use of the CBP, flour with a protein content lower by about 1% than that required for the bulk fermentation process can be baked into bread with no loss of bread quality (cf. earlier this chapter).

About 80% of the bread baked commercially in the United Kingdom in 1986 was being made by the CBP, and that has continued to increase over the years. The process is particularly favoured for plant baking, for hot-bread shops, and for in-store bakeries, and it is adaptable to making all kinds of bread.

A mechanical dough development process, resembling the CBP, is used in New Zealand. About 90% of bread was being made in that country by this process in 1983 (Mitchell, 1983). The CBP is also used in about 30 other countries, in some of which, e.g., South Africa, it is probably the most widely used process.

8.2.3 Chemical development process

The optimum work input for mechanical dough development is lowered if a proportion of the disulphide bonds are broken chemically by the introduction of a reducing agent.

The Activated Dough Development (ADD) process achieves dough development without either bulk fermentation or mechanical development. A relatively rapid-acting reducing agent, L-cysteine, and a relatively slow-acting oxidizing agent, potassium bromate (but which is no longer allowed in the United Kingdom or Canada), or a mixture of potassium bromate and ascorbic acid, are added at the dough-mixing stage, using conventional, low-speed, mixing equipment. All the ingredients, which include 2% of yeast, 0.7%–1.0% of fat, and extra water, as for the CBP, are mixed together for 10–20 min, and the dough temperature after mixing should be 28–30°C. The reducing agent accelerates the uncoiling and reorientation of the protein molecules, and the oxidizing agent follows up by stimulating the formation of cross-links to stabilize the desired elastic three-dimensional gluten network.

During mixing, air is entrained in the dough, starting the process of cell formation, which is continued throughout the subsequent stages of dough dividing, rounding, first proof of 6–10 min and final moulding. During the final proof of 45–55 min (as for the BFP and the CBP), sufficient gas to inflate the dough is produced by activity of the yeast.

The ADD process requires the use of a baker's grade flour of about 12% protein content as used in the BFP; the lower protein content flours used in the CBP are not suitable for the ADD process. Apart from this restriction on flour quality, the ADD process offers most of the advantages over the BFP that are claimed by the CBP, and, in addition, the ADD process does not require the use of a special high-speed mixer.

The ADD process has been commonly used by hot-bread shops, in-store bakeries and family bakers, and accounted for 5%–10% of all bread made commercially in the United Kingdom.

The usual levels of addition of reducing and oxidizing agents are 35 mg/kg on flour wt. of L-cysteine hydrochloride (corresponding to about 27 mg/kg of L-cysteine) with 25 mg/kg of potassium bromate plus 50 mg/kg of ascorbic acid. These levels are based on the use of flour of 12% protein content that has already been treated at the mill with up to 20 mg/kg of potassium bromate.

The ADD process was introduced to the baking industry by the British Baking Industries Research Association in 1966, but the process could not be used commercially in the United Kingdom until 1972, when the use of L-cysteine hydrochloride was permitted by the Bread and Flour (Amendment) Regulations 1972 (MAFF, 1972). Potassium bromate and ascorbic acid were listed in the Bread and Flour Regulations 1984 (MAFF, 1984) as permitted improvers, but potassium bromate was removed from the list of improvers permitted in the United Kingdom in April 1990 (MAFF, 1990), and was banned in Canada in 1994, thereby necessitating the reformulation of additives used in the ADD process. (It has not yet been banned in the United States.) Replacement of potassium bromate by additional ascorbic acid or azodicarbonamide (within the maximum permitted limits) is not an ideal solution, as the balance between the oxidizing and reducing agents is upset, and hence, ADD is no longer a viable method in the United Kingdom.

8.2.3.1 EU policy regarding additives

Additives will only be included in a permitted list if a reasonable technological need is demonstrated, and if this need cannot be achieved by other means that are economically and technologically practicable. Furthermore, the additives must present no hazard to health at the levels of use proposed, and they must not mislead the customer. 'Need' is understood to mean preservation of nutritional quality; the meeting of special dietary requirements; enhancement of keeping quality, stability and organoleptic properties; or providing aid in manufacture, processing, preparation, treatment, packaging, transport or storage. Specified additives are to be allowed only in specified foods, and at levels not exceeding those required to achieve the desired effect (Spencer, 1989).

8.2.4 Other rapid methods

8.2.4.1 No-time continental process

This process, also known as the Spiral Mixing Method, avoids a long bulk-fermentation, the use of a high-speed mixer, and the addition of L-cysteine and potassium bromate. In this process, all the ingredients are mixed together for 8–11 min in a special open-bowl mixer equipped with a spiral-shaped beater; the action of the mixer is faster and more vigorous than that of the low-speed mixers used in the BFP, but not so intense as that which is necessary for the CBP. The mixing action incorporates more air, and hence oxygen, in the dough, thereby improving cell creation and increasing the effectiveness of ascorbic acid.

The ingredients used would include a higher level of yeast (3% on flour wt.) than used in other processes, 2% of salt, about 60 pt of water per 100 pt of flour, fat, ascorbic acid, emulsifier, fungal amylase and sugar. The dough temperature aimed at is 26–28°C. A short period of bulk fermentation (15–30 min) follows mixing. This process is widely used on the continent of Europe and is being used increasingly in the United Kingdom, particularly in small bakeries.

8.2.4.2 Emergency no-time dough

This is a short system, somewhat resembling the No-time Continental process, that is used occasionally in the United States and the United Kingdom, particularly for emergency production. The dough is made using a larger amount of yeast, e.g., 2.5% on flour wt., and a higher temperature, e.g., 30–32°C, than are usual for normal fermentation systems, and is immediately scaled off. Final moulding follows after about 15 min, and the dough pieces are proved for 1 h at 43°C before baking. The bread has a coarse, thick-walled crumb structure, and it stales rapidly.

8.2.4.3 Aeration (gas-injection) process

In 1860 Dauglish described a rapid bread-making method in which a dough was made by mixing soda water (water charged with carbon dioxide gas) and flour under pressure. When the pressure was released the dough expanded and was immediately divided and baked. The whole process, including baking, took 90 min. A modern equivalent is the Oakes Special Bread Process, a continuous system in which carbon dioxide gas is injected into the developing dough. Neither process is in commercial use for making standard bread.

8.2.4.4 Microwave and radio frequency baking

The use of microwave (MW) energy for baking bread was investigated at the Flour Milling and Baking Research Association (FMBRA), Chorleywood, England (Chamberlain, 1973). Microwave energy, generated by a magnetron, and transmitted by radiation of frequencies from about 900 MHz upwards, penetrates the dough very rapidly and cooks the loaf uniformly throughout.

Another source of energy that has been investigated is radio frequency (RF) energy, of about 27 MHz, which similarly heats the loaf rapidly throughout.

Commercial application of MW baking has not so far been possible because of the unavailability of a thermostable material for the pans that has the mechanical properties of metal but is freely permeable to microwave radiation. It is reported that the RF method can be operated with conveyors, baking pans and foil containers made of metal, and RF ovens are now commercially available for making bread, biscuits and other cereal-based products.

In both MW and RF baking, the dough is held for only a short period within the temperature range at which the activity of alpha-amylase is unwelcome, thereby permitting the use of a wheat grist of higher alpha-amylase activity than would be acceptable for conventional baking. Moreover, in conventional baking, the gases evolved are rapidly lost unless the protein content of the flour is high enough – say 10.5% or more – to confer adequate strength to the walls of the crumb cells. In both MW and RF baking, however, the rate of gas production exceeds the rate of gas loss (because the dough is heated rapidly throughout); hence, high-protein content in the flour is not obligatory. In fact, flour of 7.5% protein content was used experimentally at the FMBRA to produce bread by MW baking that compared favourably with bread conventionally baked from flour of normal protein content. Thus, a further impetus toward the commercial application of RF heating would be a substantial price differential in favour of west-European wheat (of lower protein content, and often of high alpha-amylase activity) as against imported, non-EC, strong wheat.

Baking by MW or RF alone produces crustless bread. A crust can be developed by applying thermal radiation, in the form of hot air, simultaneously during a total baking time of less than 10 min for a standard 800-g loaf. The Air Radio Frequency Assisted (ARFA) oven therefore combines radio frequency and convected hot air in a technique developed by the Electricity Council Research Centre, Capenhurst, England (Anon., 1987, 1988b). However, there is to date no known commercial application of the ARFA for bread making.

8.2.4.5 Frozen dough

The use of frozen dough, which can conveniently be stored, has recently increased in popularity, e.g., for in-store bakeries as well as home baking (e.g., breads, rolls, buns). The best results are obtained if ascorbic acid is used as the oxidant, and if the doughs are frozen before fermentation and then stored at a constant temperature to avoid problems associated with the melting of ice crystals.

8.2.5 Bread cooling

The cooling of bread is a problem in mechanical production, particularly when the bread is to be wrapped and/or sliced before sale. Bread leaves the oven with the centre of the crumb at a temperature of about 96°C and cools rapidly. During cooling, moisture moves from the interior outwards toward the crust and thence to the atmosphere. If the moisture content of the crust rises considerably during cooling, the texture of the crust becomes leathery and tough, and the attractive crispness of freshly baked bread is lost. Extensive drying during cooling results in weight loss (and possible contravention of the Weights and Measures Act in the United Kingdom) and in poor crumb characteristics. The aim in cooling is therefore to lower the temperature without much change in moisture content. This may be achieved by subjecting the loaves to a countercurrent of air conditioned to about 21°C and 80% r.h. The time taken for cooling 800-g loaves by this method is 2–3 h.

8.2.5.1 Automation

Developments that increase the efficiency of the plant baking process include the use of load cells for weighing the ingredients and controlling ingredient proportions in the mixer, and the use of the microelectronics for temperature corrective feedback and consistency corrective feedback (Baker, 1988). Another development is the introduction of computer-programmed mixers and plants.

8.2.6 Bread moisture content

There is no legal standard for the moisture content of bread in the United Kingdom. The moisture content of American (21 CFR 136.110) and of Dutch bread must not exceed 38%. In Australia the maximum permitted moisture content in any portion weighing 5 g or more is 45% for white bread, 48% for brown and wholemeal. In New Zealand, 45% is the maximum moisture content similarly permitted in any bread.

8.2.7 Bread weights

In the United Kingdom, standard bread weights were 1 and 2 lb until 6 May 1946, when weights were reduced to 14 and 28 oz. From 1 May 1978 loaves sold in the United Kingdom weighing more than 300 g (10.6 oz) were required to weigh 400 g or a multiple of 400 g.

In Belgium, loaves weighing more than 300 g must weigh 400 g or multiples of 400 g. In Germany, prescribed weights for unsliced bread were 500 g and then by multiples of 250 g up to 2000 g, and above 2000 g by multiples of 500 g. Prescribed weights for sliced bread were 125, 250 g, and then by multiples of 250–1500 g, and then by multiples of 500–3000 g.

From 1980, enforcement of bread weight regulation in the United Kingdom has taken place at the point of manufacture rather than, as formerly, at the point of sale, and is based on the average weight of a batch rather than on the weight of an individual loaf.

8.2.8 Yield of bread

Using the CBP, 100 kg of white flour at 14% m.c., produces an average 180 loaves of nominal weight 800 g (average 807 g) containing an average of 39% of moisture (total bread yield from 100 kg of flour: 145 kg of bread). Thus, a nominal 800-g loaf is made from an average of 556 g of flour at natural m.c. (478 g on dry basis) and contains on average 492 g of dry matter (i.e., 14 g of dry matter are contributed by nonflour constituents).

8.2.9 Energy consumption in bread making and environmental impacts

The CBP uses ~40 J/g of dough, or 35.7 kJ per 800-g loaf, for the mixing process. The bulk fermentation process uses about one-fifth of this amount for mixing, but some additional energy is used in heating the water for dough making.

In the baking process, the heat required comprises the heat needed to raise the temperature of the dough piece from that of the prover (about 40°C) to that at the oven exit (about 96°C); the latent heat of evaporation of the water changed to steam, and the heat required to raise the temperature of that steam to that of the oven; and the heat required to raise the temperature of the pan to the oven exit temperature.

Thus, for baking a nominal 800-g loaf (with flour at 14% m.c., water absorption 60.7% on flour wt., and oven loss 65 g), about 400 kJ (379 Btu) theoretical are required. This figure is made up of about 300 kJ (284 Btu) for the loaf itself plus about 100 kJ (95 Btu) for the pan. Oven efficiency depends on oven type and quantity of steam used for conditioning the oven atmosphere, and averages 40%, thus giving a practical requirement of 1 MJ (948 Btu) per nominal 800-g loaf plus pan: 750 kJ (711 Btu) for the 800-g loaf alone, equivalent to 937 kJ (889 Btu) per kg of bread (without pan).

Additional energy is used in conveyors, final proof, cooling, slicing, wrapping of bread, but these amounts are small in relation to the energy used for baking (Cornford, S.J., private communication, 1979).

A breakdown of the total energy requirements for making a white loaf, including energy used in growing the wheat, milling the wheat, baking the bread and selling the bread, is shown in Table 8.2. As most sectors of the food supply chain have become more efficient over the years, these distributions may have changed to some degree, but the literature is sparse in this regard.

Companies, consumers and governments have increasingly recognized that energy consumption (much of which is fossil based for many countries) generates deleterious emissions to the environment and exacerbates the potential for human-induced climate change. Life cycle assessment (LCA) is a methodology that can be used to estimate these impacts, either for specific portions of the supply chain or entirely from cradle to grave, for a specific product. Roy et al. (2009) provides an introduction and overview of LCA applied to various food products.

Unfortunately, the application of this methodology is not yet widespread, as the LCA field is still relatively new. So there is a dearth of literature regarding LCA studies for bread production. Andersson and Ohlsson (1999) assessed production of bread at the home scale, local bakery scale, and commercial industrial scale (Table 8.3). When compared on a 1 kg loaf of bread basis, even though commercial production of bread may be more efficient per kg vis-à-vis production processes at the factory, greater distribution of the finished products can offset these savings in term of energy used and environmental emissions. Thus, it appears that home production still results in the lowest environmental impacts overall. In terms of the bread supply chain, it appears that farm production and transport were the two sectors that most often had the greatest environmental impacts; production of the wheat led to the greatest eutrophication of water bodies (due to nitrogen and phosphorus runoff from fertilizers), but the processing plant led to the greatest photo-oxidation and energy use; consumer use and packaging had comparatively minimal impacts on the environment (all comparisons on a 1-kg loaf basis). Meisterling et al. (2009) examined

TABLE 8.2   Energy requirements for making one loaf of white bread[a]

	Percentage of total energy required	
GROWING WHEAT		
Tractors, etc.	5.3	
Fertilizers	11.1	
Drying, sprays	3.0	
		19.4
MILLING THE WHEAT		
Direct fuel and power	7.4	
Other	2.1	
Packaging	1.3	
Transporting	2.0	
		12.8
BAKING		
Direct fuel and power	30.2	
Other items	17.3	
Packaging	9.0	
Transporting	7.8	
		64.3
Shops	3.4	3.4
		99.9

[a] *Based upon data from Leach, G., 1975. Energy and Food Production. IPC Science & Technology Press, Guildford, Surrey, UK.*

TABLE 8.3   Environmental impacts determined by life cycle assessment of 1 kg loaf of white bread (from farm through consumer, i.e., cradle to grave)[a]

	Home baking	Local bakery	Industrial bakery (12,800–30,800 tonne/y)
Energy use (MJ/1 kg bread)	17–18	12	14–22
Global warming potential (g CO_{2eq}/1 kg bread)	520–650	660–670	630–1000
Acidification (mol H^+/1 kg bread)	0.08–0.09	0.1–0.11	0.1–0.17
Eutrophication (g O_2/1 kg bread)	88–89	120	99–160
Photo-oxidation (g NO_x/1 kg bread)	2.4–2.6	2.6–2.7	3.2–5.5

[a] *Based upon data from Andersson, K., Ohlsson, T., 1999. Life cycle assessment of bread produced on different scales. International Journal of Life Cycle Assessment 4 (1): 25–40.*

the environmental impacts of wheat production and delivery. They found that organic farm production resulted in approximately 160 g $CO_{2eq}/1$ kg bread versus conventional agricultural practices, which resulted in approximately 190 g $CO_{2eq}/1$ kg bread (thus organic production could have up to 16% lower greenhouse gas emissions). Wheat production on the farm required between 2 and 4 MJ/kg wheat. They also determined that for every 1 km that the wheat was transported, 0.07 g CO_{2eq} were emitted.

8.3 OTHER KINDS OF BREADS

In matter of fact, white bread is a modern convenience that is currently enjoyed in many developed countries, and a growing number of developing countries. Historically, and even today, around the world there is a wide variety of types of breads beyond white, cereal grains that are used, production steps used, etc. These depend upon culture and history, and in fact many have changed little for more than 1000 years; several of these are discussed in Chapter 11.

In recent years, especially in developed countries, the popularity of white breads has decreased, for a variety of reasons (see Chapter 11), and other types of breads have become increasingly more popular. Some of these will be discussed next.

8.3.1 Brown and wholemeal breads

When using the bulk fermentation process, the level of fat used in brown and wholemeal breads is generally raised to about 1.5% on flour wt. (as compared to 1% for white bread) because the fat requirement for brown flour and wholemeal is more variable than that for white flour. With the CBP, it is essential to raise the fat level in this way for brown and wholemeal breads.

A short fermentation system is generally used for wholemeal bread. For example, the dough might be allowed to ferment for 1 h before knocking back, plus 30 min to scaling and moulding, at an appropriate yeast level and temperature.

8.3.2 Wheatgerm bread

This type of bread is made from white flour with the addition of not less than 10% of processed germ, which has been heat-treated to stabilize the lipid content and to destroy glutathione, a component that has an adverse effect on bread quality. A fermentation process to inactivate glutathione, as an alternative to heat treatment, was described in Australia in 1940.

8.3.3 Gluten bread: high-protein bread

These breads are made by supplementing flour with a protein source, such as wheat vital gluten, whey extract, casein, yeast, or soya flour. Procea and Slimcea are proprietary breads in which the additional protein is provided by wheat gluten. However, most bread now made in the United Kingdom and in numerous other countries contains a small amount of added vital wheat gluten (cf. earlier this chapter) depending on the protein content of the flour.

8.3.4 High-fibre bread

This bread has both higher fibre content and fewer calories per unit than normal white bread. The high-fibre content is achieved by addition of various supplements, such as cracked or kibbled wheat, wheat bran or powdered cellulose. A type of cellulose used in the United States, called Solka-Floc, is delignified alpha-cellulose obtained from wood; usage levels are 5%–10%. The use in the United Kingdom of alpha-cellulose or of the sodium salt of carboxymethylcellulose in bread, for which a slimming claim is made, is permitted by the Bread and Flour Regulations 1984 (MAFF, 1984).

In the last few decades, inulin (Fig. 8.11) has been increasingly used as a dietary fibre supplement in a variety of food products, including breads. Inulin is a polysaccharide, which occurs in a variety of plants; chicory is the most common source of commercial inulin. It is low calorie, water

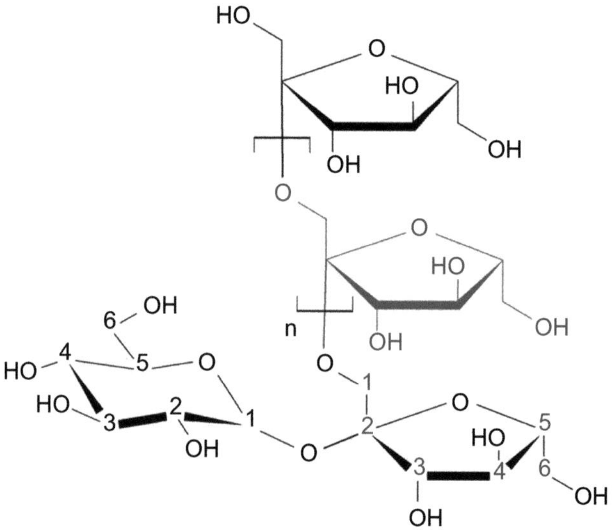

FIGURE 8.11   Chemical structure of the polysaccharide inulin ($C_{6n}H_{10n+2}O_{5n+1}$), which is classified as fructan. It is increasingly being used as a source of dietary fibre in food products.

soluble and classified as a prebiotic. According to Sirbu and Arghire (2017), a maximum level of 15% inulin can be used in bread formulations, otherwise declines in dough rheology and final bread quality will occur.

8.3.5 Granary bread

This is a proprietary bread made from a mixture of wheat and rye that has been allowed to sprout, kiln dried and rolled. To this is added barley malt.

8.3.6 Speciality breads

Other 'special' breads include pain d'épice, fruit breads, malt loaves, mixed-grain bread, bran bread, etc.

8.3.7 Baguettes

No discussion about bread would be complete without at least a brief mention of the bread that is so important to France – the baguette (Fig. 8.12).

8.4 BREAD STALING AND PRESERVATION

8.4.1 Bread staling

Staling of breadcrumb is not a drying-out process – loss of moisture is not involved in true crumb staling. The basic cause of staling is a slow change in the starch, called retrogradation, at temperatures below 55°C from an amorphous to a crystalline form, the latter binding considerably less water than the former. This change leads to a rapid hardening, a toughening of the crust and firming of the crumb, loss of flavour, increase in opaqueness of the crumb, migration of water from crumb to crust, and to shrinkage of the starch granules away from the gluten skeleton with which they are associated, with consequent development of crumbliness (Hoseney, 1986).

The rate at which staling proceeds is dependent on the temperature of storage; the rate is at a maximum at 4°C, close to the temperature inside a domestic refrigerator, decreasing at temperatures below and above 4°C. Staling can be prevented if bread is stored at temperatures above 55°C (although this leads to loss of crispness and the probability of rope development) or at −20°C, e.g., in a deep freezer.

As the amylose (straight-chain) portion of the starch is insolubilized during baking or during the first day of storage, it is considered that staling is due to heat-reversible aggregation of the amylopectin (branched-chain) portion of the starch.

FIGURE 8.12   Baguettes are produced in many places in the world, but they are of prime importance in France (where they are known as French traditional bread - baguette de tradition française). In fact, the Décret Pain law of 1993 (https://www.legifrance.gouv.fr/affich-Texte.do?cidTexte=JORFTEXT000000727617) stipulates that in France, the only ingredients may be wheat flour, water, salt and yeast (although now the wheat flour may contain up to 2% of bean flour, 0.5% of soybean flour, 0.3% malt flour); the dough cannot be frozen; and the baguettes must be kneaded, shaped and baked on the premises where they are sold. Scale bars represent 1 mm distance.

However, starch crystallization cannot account for all the crumb firming that occurs at temperatures above 21°C, and it has been suggested by Willhoft (1973) that moisture migration from protein to starch occurs, leading to rigidification of the gluten network, and contributing to crumb firming. See reviews of the subject by Radley (1968) and Elton (1969); see also Pomeranz (1971, 1980) and Pomeranz and Shellenberger (1971).

Emulsifiers like monoglycerides retard the rate of starch retrogradation. Monoglycerides, when added at the mixing stage, first react with free, soluble amylose to form an amylose-lipid complex. When the rate of addition exceeds 1%, all the free amylose is complexed and the monoglycerides begin to interact with the amylopectin, thereby retarding retrogradation (Krog et al., 1989).

8.4.2 Rope

Freshly milled flour contains bacteria and mould spores, but these normally cause no trouble in bread under ordinary conditions of baking and storage. Many mould spores and the vegetative forms of bacteria are killed at oven temperatures, but spores of some of the bacteria survive and may proliferate in the loaf if conditions are favourable, causing a disease of the bread known as 'rope'. Ropy bread is characterized by the presence in the crumb of yellow-brown spots and an objectionable odour. The organisms responsible are members of the *Bacillus subtilis* var. *mesentericus* group and *Bacillus licheniformis*. Proliferation of the bacteria is discouraged by acidic conditions in the dough, e.g., by addition of 5.4–7.1 g of ACP or 9 mL of 12% acetic acid per kg of flour.

8.4.3 Bread preservation

The expected shelf life of bread made in the United Kingdom is about 5 days for white, 3–4 days for wholemeal and brown, 2–3 days for crusty. Thereafter, bread becomes unacceptable because of staling, drying out, loss of crispness of crust or mould development.

Mould development can be delayed or prevented by addition of propionic acid or its sodium, calcium or potassium salts, all of which are permitted by the Bread and Flour Regulations 1984 at levels not exceeding 3 g/kg flour (calculated as propionic acid), or by addition of sorbic acid (not permitted in the United Kingdom or the United States). Use of propionic acid to delay mould development in bread is also permitted legally in Denmark, Germany, Italy, the Netherlands and Spain.

Other means of preservation include the use of sorbic acid–impregnated wrappers, gamma irradiation with 5×10^5 rad, or infrared irradiation.

8.4.3.1 *Packaging*

Gas packaging, with an atmosphere of carbon dioxide, nitrogen or sulphur dioxide replacing air, has been used in an attempt to extend the mould-free shelf life of baked goods. However, even in an 'anaerobic environment' of 60% CO_2 plus 40% N_2 the fungi *Aspergillus niger* and *Penicillium* spp. may appear after 16 days unless the concentration of oxygen can be kept at 0.05% or lower. The use of impermeable packaging to prevent entry of oxygen is very expensive (although B&M, in the United States, does produce a canned brown bread; http://www.bmbeans.com). A less-expensive alternative is to include an oxygen absorbent, such as active iron oxide, or a oxidizable polymer, in the packaging material. By this means, the mould-free shelf life of crusty rolls has been increased to 60 days (Smith et al., 1987). Oxygen absorbers, in the form of sachets, have become popular as well. These consist of either iron powder or ascorbic acid (Kaufman et al., 2017).

A so-called '90-day loaf' is packaged in nylon-polypropylene laminate and the interior air partly replaced by carbon dioxide. The packaged loaf is then sterilized by infrared radiation. None of these methods prevents the onset of true staling.

8.4.3.2 *Freezing*

Freezing of bread at −20°C is the most favourable method of preserving its freshness. Suitably packed, the bread remains usable almost indefinitely.

Part baking of soft rolls and French bread is a technique now widely used in hot-bread shops, in-store bakeries, bake-off units in nonbakery shops, catering establishments, and domestically. In this process, doughs for soft rolls would be proved at about 43°C and 80% r.h., those for French bread at about 32°C and 70% r.h. The proved doughs are then baked just sufficiently to kill the yeast, inactivate the enzymes and set the structure, but producing little crust colour or moisture loss. Temperature for partial baking would be about 180°C. The part-baked product can then be deep frozen at, say, −18°C, or stored at ambient temperature until required. The final bake-off at the point of use, at a temperature of up to 280°C, defrosts the frozen product, increases the crust colour to normal, and reverses all the staling that may have taken place.

8.5 USE OF CEREALS OTHER THAN WHEAT IN BREAD

The statement of Atkins (1971) that 'bread is fundamentally foamed gluten' can be correctly applied only to bread in which wheat flour or meal is the sole or dominant cereal, because the proteins in other cereals, on hydration, do not form gluten that is comparable in rheological properties with wheat gluten.

The flour of cereals other than wheat is used for bread making in two ways: either blended with wheat flour, in a form sometimes known as 'composite flour' (cf. later this chapter); or as the sole cereal component. Bread made from composite flour employs conventional baking processes, but when a nonwheat flour is used alone it is usual to make use of a gluten substitute (cf. later this chapter). Rye flour, however, is exceptional in that bread made from it as the sole cereal component does not require the addition of a gluten substitute.

8.5.1 Rye

Rye flour and meal are used for the production of numerous types of bread, both soft bread and crispbread, and rye is regarded as a bread grain in Germany and in most Eastern and Northern European countries (Drews and Seibel, 1976).

In Europe, 'rye bread' is made from all rye flour; 'rye/wheat bread' contains not less than 50% of rye flour, with wheat flour making up the remainder; 'wheat/rye bread' implies a blend of not less than 50% of wheat flour plus not less than 10% of rye flour.

Factors that influence the baking potential of rye flour include variety, environmental conditions of growth and fertilizer use, activity of amylase, protease and pentosanase enzymes, and functions of carbohydrates and proteins.

The possibility that rye may be infected, in the field, with ergot (*Claviceps purpurea*) has been discussed elsewhere in this book, and can limit its use as a food ingredient.

8.5.1.1 Soft bread

Under certain conditions rye grains germinate in the harvest field and then exhibit increased enzymatic activity that may be undesirable for bread-making purposes. Rye flour with high maltose figure (e.g., 3.5) and low amylograph value (350 or less) is of poor baking quality. Rye flours with Falling number (cf. Chapter 7) below 80 produce loaves with sticky crumb, but rye flour with FN 90-110 can be processed into acceptable bread with the aid of additives, acidifiers and emulsifiers to compensate for the effects of sprout damage. Such additives would include an acidifier to adjust the pH to 4.0–4.2, 2% of salt (on flour wt.), 0.2%–0.5% of emulsifier and 1%–3% of gelatinized flour (Gebhardt and Lehrack, 1988).

Deterioration of baking quality of rye during storage is better indicated by glutamic acid decarboxylase activity than by Falling number (Kookman and Linko, 1966).

The protein in rye flour is less important than the protein in wheat flour. The rye protein, when hydrated, does not form gluten because the proportion of the protein that is soluble is much larger in rye than in wheat (up to 80% soluble in rye sour dough as compared with 10% soluble protein in wheat dough), and because the high content of pentosans inhibits the formation of gluten (Drews and Seibel, 1976). Conversely, the pentosans and starch in rye are much more important than in wheat (Telloke, 1980). The pentosans, which comprise 4%–7% of rye flour, and the starch have an important water-binding function in forming the crumb structure of rye bread. The pentosans, in particular, play a role in raising the viscosity of rye dough. Rye flour can be fractionated according to particle size to yield fractions that vary in starch and pentosan contents. These fractions can then be blended to give an optimal ratio of pentosan to starch. A pentosan: starch ratio of 1:16 to 1:18 is considered ideal.

The starch of rye gelatinizes at a relatively low temperature, 55–70°C, at which the activity of alpha-amylase is at a maximum. In order to avoid excessive amylolytic breakdown of the starch, a normal salt level is used for making rye soft bread, and the pH of the dough is lowered by acid modification in a 'sour dough' process, preferably by lactic acid fermentation with species of *Lactobacillus*.

8.5.1.2 Straight dough process

Yeast is used for leavening, and the dough is acidified by adding lactic acid or acidic citrates. The dough is mixed slowly to prevent too much toughening that could be caused by the high viscosity of the pentosans.

8.5.1.3 Sour dough process

Sour doughs, containing lactic acid bacteria, were probably in use for making bread as long ago as 1800 BC in eastern Mediterranean regions, the process spreading to Germany between the first and sixth centuries AD, where it was used mainly by monks and guilds (Seibel and Brümmer, 1991).

The sour dough is a sponge-and-dough process. A starter dough is prepared by allowing a rye dough to stand at 24–27°C for several hours to induce a natural lactic acid fermentation caused by grain microorganisms. Alternatively, rye dough is inoculated with sour milk and rested for a few hours, after which a pure culture of organic acids (acetic, lactic, tartaric, citric, fumaric) is added to simulate the flavour of a normally soured dough. Part of the mature sour dough is retained as a subsequent starter, while the remainder is mixed with yeast and rye flour or rye wholemeal, or a blend of rye and wheat flours, and acts as a leavening agent in the making of rye sour bread (Drews and Seibel, 1976).

The flavour of San Francisco sour dough bread is due largely to lactic and acetic acids that are produced from D-glucose by *Lactobacillus*, a bacterium active in sour dough starter. The starter is fermented for 2 h at 24–26°C and then held in a retarder for 10 h at 3–6°C. The dough, which incorporates 2%–10% of starter (on flour wt. basis) besides vital gluten (1%–2%), shortening, yeast (2.5%–4.0%), salt, sugar, yeast food and water, is fermented for 30–60 min, scaled, rested for 12–15 min, proved for 5–8 h, and baked (1 lb loaves) for 30–35 min at 204–218°C. However, this long, drawn-out process can be avoided by using commercially available free-flowing sour dough bases (Ziemke and Sanders, 1988; Seibel and Brümmer, 1991).

All doughs containing rye flour have to be acidified because sour conditions improve the swelling power of the pentosans and also partly inactivate the amylase, which would otherwise have a detrimental effect on the baking process and impair normal crumb formation (Seibel and Brümmer, 1991).

Conventional dough improvers are not widely used in making rye or rye/wheat bread. Instead, pregelatinized potato flour or maize starch or rice starch may be added at a level of 3% (on flour wt.). These materials have high water-binding capacity and increase the water absorption of the dough. Other substances used with similar effect include hydrocolloids and polysaccharide gums such as locust bean and guar gums.

Staling of rye bread is less serious than that of wheat bread, and shelf life may be extended in various ways: by the addition of malt flour or pre-gelatinized potato flour or starch; by the use of sour dough; or by wrapping while still warm.

8.5.1.4 Pumpernickel

Pumpernickel is a type of soft rye bread made from very coarse rye meal by a sour dough process. A very long baking time (18–36 h) is used, with a starting temperature of about 150°C being gradually lowered to about 110°C. Pumpernickel has a long shelf life.

8.5.1.5 Crispbread (Knackerbrot)

Rye crispbread is generally made from rye wholemeal or flaked rye, using water or milk to mix the dough, and may be fermented with yeast (brown crispbread) or unfermented (white crispbread). The traditional method used in Sweden is to mix rye meal with snow or powdered ice; expansion of the small air bubbles in the ice-cold foam raises the dough when it is placed in the oven. It is desirable to use rye flour or meal of low alpha-amylase activity.

In one process, a dough made from rye wholemeal, yeast, salt and water is fermented for 2–3 h at 24–27°C. After fermentation, the dough is mixed for 5–6 min, proved for 30–60 min, sheeted, dusted with rye flour and cut to make pieces about 7.6 × 7.6 cm (3 × 3 in.) in size, which are baked for 10–12 min at 216–249°C. The baked pieces are stacked on edge and dried in a drying tunnel for 2–3 h at 93–104°C to reduce the moisture content to below 1%.

Ryvita is a crispbread made from lightly salted rye wholemeal.

8.5.1.6 Flatbreads

A type of crispbread product appeared in the 1990s under the trade name 'Cracotte', manufactured from wheat flour by a continuous extrusion cooking process. In this process the flour of about 16% m.c. is heated and sheared to form a fluid melt at 130–160°C in which starch forms the continuous phase. After extrusion, it forms a continuous strip of expanded foam (specific volume 7–10 mL/g). Individual biscuits are cut from the strip and packed like crispbreads. This type of product, which has gained popularity in many countries, may be manufactured from any cereal type, and variations have appeared that included rye, rice and maize, either as the minor or major component in blends with wheat.

8.5.2 Triticale

The bread-making characteristics of flour made from early strains of triticale were discouraging, although bread quality could be improved by addition of dough conditioners. However, bread of good quality has been made from recent triticale selections. Bread baked commercially with 65% of wheat flour blended with 35% of triticale stoneground wholemeal was first marketed (as 'Tritibread') in the United States in 1974.

Triticale flour has been tested extensively in Poland for bread making. The best results, using a blend of 90% triticale flour plus 10% of rye four, were obtained with a multiphase (preferment, sour dough) process in which the preferment was made with the rye flour (10% of the total flour) with water to a preferment yield of 400%, and a fermentation time of 24 h at 28–29°C. The sour formulation used triticale flour (50% of the total flour) with 1%–2% of yeast (on total flour basis), and water to give a sour yield of 200%. This was fermented for 3 h at 32°C. The rest of the triticale flour was then added, with salt at 1.5% on flour wt., and water, to give a dough yield of 160%–165%, and then all ingredients were fermented for 30 min at 32°C. The loaves were baked at 235–245°C (Haber and Lewczuk, 1988). Bread made from all-triticale flour has been shown to stale more rapidly than all-wheat bread.

Bread made from 50:50 or 75:25 blends of triticale flour and wheat flour had higher specific volumes (4.8; 4.9 mL/g) than the bread baked from all wheat flour (4.4 mL/g); no deleterious effect on crumb characteristics, viz. grain and texture, resulted from the admixture of triticale flour (Bakhshi et al., 1989).

8.5.3 Barley and oats

During World War II, when supplies of imported wheat were restricted in Britain, the government authorized the addition of variable quantities (up to 10% of the total grist in 1943) of barley, or of barley and oats, to the grist for making bread flour. For this purpose, the barley was generally blocked (cf. Chapter 6) to remove the husk, and the oats were used as dehusked groats (cf. Chapter 6).

Good quality bread has been made in Norway from a blend of 50% wheat flour of 78% extraction rate, 20% barley flour of 60% extraction rate, and 30% wheat wholemeal, using additional shortening (Magnus et al., 1987).

For use in bakery foods in the United States, cleaned oat grain is steam-heated to about 100°C and then held in silos to be 'ovenized' by its own heat for about 12 h. This process preserves the mineral and vitamin contents, and conditions the oats. The grain is then impact dehulled (cf. Chapter 6) without previous kilning (McKechnie, 1983).

Breads containing oats have become very popular in the United States in recent years.

8.5.4 Rice

Bread of acceptable quality has been made from a blend of 75 parts of wheat flour (12.1% protein content, 14% m.c. basis) with 25 parts of rice flour that had been partly gelatinized by extrusion. The rice flour was milled from rice grits (7.32% protein content, 14% m.c. basis), which were pregelatinized to 76.8% by extrusion, using a Creusot Loire BC-45 twin-screw extruder. The beneficial effects of extrusion treatment appeared to be due to thermal modification of the starch in the rice flour (Sharma et al., 1988). The volume of loaves baked from this blend was below that of all-wheat flour loaves, but in other respects the bread was judged acceptable.

The low contents of sodium, protein, fat and fibre, and the high content of easily digested carbohydrates favour the use of rice bread as an alternative to wheat bread for persons suffering from inflamed kidneys, hypertension and coeliac disease (cf. Chapter 4). Although a direct substitution of rice flour for wheat flour tends to result in poor quality breads, including poor load height. The volume of loaves of yeast-leavened bread made from 100% of rice flour can be improved by the addition of hydroxypropylmethylcellulose, a gum that creates a film with the flour and water that retains the leavening gases and allows expansion (Bean, 1986).

Stabilized and extracted rice bran can provide nutritional fortification, when used at levels up to 15%, for baker products such as yeast-raised goods, muffins, pancake mixes and biscuits. The rice bran contributes flavour, increases water absorption without loss of volume, adds significant amounts of essential amino acids, vitamins and minerals, but does not affect mixing tolerance or fermentation. The blood cholesterol–lowering capabilities of rice bran are a further inducement for its use (Hargrove, 1990).

8.5.5 Maize

Maize flour is used to make bread of a sort in Latin America, and also for pancake mixes, infant foods, biscuits and wafers. Pregelatinized maize starch may be used as an ingredient in rye bread (cf. earlier in this chapter). Moreover, maize is commonly used in a variety of flatbreads and other food products (cf. Chapter 11).

8.5.6 Bread made with composite flours

Flour milled from local crops can be added to wheat flour to extend the use of an imported wheat supply and thereby save the cost of foreign currency, reduce potential international trade issues, etc. This arrangement is particularly appropriate for developing countries that do not grow wheat.

Satisfactory breads can be made from such composite flours, viz. a blend of wheat flour with flour of other cereals such as maize, sorghum, millet or rice, or with flour from root crops such as cassava or potato.

The flour of the nonwheat component acts as a diluent, impairing the quality of the bread to an extent depending on the degree of substitution of the wheat flour. A higher level of substitution is possible with a strong wheat flour than with a weak one.

Possible levels of substitution, as % by wt. of the composite flour, are 15%–20% for sorghum flour and millet flour, 20%–25% for maize flour. Somewhat higher levels of substitution may be possible by the use of bread improvers or by modifying the bread-making process.

A blend of 70% wheat flour, 27% rice flour, and 3% soya flour made acceptable bread, provided that surfactant-type dough improvers were used. A more economical blend, producing acceptable bread, has been shown to be 50% wheat flour, 10% rice flour and 40% cassava flour. Rice starch can also be used, e.g., a blend of 25% rice starch with 75% wheat flour yielded acceptable bread (Bean and Nishita, 1985).

Bread of acceptable quality has been made in Senegal and Sudan from a blend of 70% of imported wheat flour of 72% extraction rate and 30% flour milled locally from white sorghum to an extraction rate of 72%–75%.

The water absorption of a blend of 15% millet flour with 85% wheat flour is about 3% higher than that of the wheat flour alone, and extra water must therefore be added. Acceptable bread can be made at an even higher rate of substitution, viz. 30%, by modifying the bread-making process in various ways, e.g., by delaying the addition of the millet flour until near the end of the mixing process; by the use of improvers, such as calcium stearoyl lactylates or tartaric esters of acetylated mono- and diglycerides of stearic acid; or by increasing the addition of sugar and fat to 4% (each) on composite flour wt.

When using a blend of maize flour and wheat flour to make bread it is desirable to increase the addition of water by about 2% for each 10% substitution of the wheat flour by maize flour, and to increase the amount of yeast to about 1.5 times that suitable for wheat flour alone. The use of hardened fat or margarine (2% on flour wt.) is recommended to achieve good bread quality. Addition of emulsifiers, such as lecithin, stearates or stearoyl lactylates, is also recommended.

8.5.6.1 Distillers grains

As discussed in the chapter on alcohol (both distilled spirits and biofuels) production, distillers grains (primarily DDG and DDGS) are coproducts that are primarily used as animal feeds, although there are many opportunities for value-added uses as well. During the fermentation of cereal grains (maize, rye, barley and wheat are the predominate grains used), the starch is converted to alcohol, carbon dioxide and other secondary products, while the nonfermentable components, e.g., proteins, fibres, fats, vitamins and minerals, remain in the residue and, in fact, are concentrated threefold due to the yeast metabolism. DDG and DDGS have potential for use in a variety of human foods, including breads and other baked products. Rosentrater (2011) provided a comprehensive review of the published literature (~60 studies) on incorporation of DDG and DDGS in a variety of food products (breads, pastas, biscuits, rolls, muffins, doughnuts, cookies, etc.). Almost universally amongst all types of foods, maximum inclusion levels were found to lie between 15% and 20%; at levels over 20% consumers began to perceive fermented flavours and odours, functionality during baking declined, and thus final product quality was reduced (i.e., lower bread expansion, denser products, reduced binding, greater brittleness, etc.); even at incorporation levels of up to 20% though, substantially increased protein and fibre levels resulted in the foods vis-à-vis the original formulations.

As discussed in Rosentrater (2011), because the biofuels production processes are mostly not designed or operated to food-grade standards, the coproducts produced from biofuels plants are considered nonfood grade. In the United States, the Food and Drug Administration (which overseas food safety) will not approve any type of treatment or process to upgrade these coproducts to food-grade status – thus they will always be relegated to animal feed or industrial uses. On the other hand, beverage distilleries may be appropriate for DDG/DDGS food use, as they already produce food-grade coproducts. In fact, some distilleries have piloted small-scale food ingredient sales in local markets, but to date none have seen nationwide marketing success. As discussed in the chapter on alcohol production, beverage distilleries are experiencing a rapid growth in the United States, Europe and many other places around the world. Perhaps use in human foods will become more widespread in the near future as a result of increasing availability.

8.5.6.2 Soya bread

A bread with good flavour, good storage properties (up to 2 weeks), and a fine-to-medium crumb structure has been made from a blend of 70%–60% wheat flour with 30%–40% soya flour. With a protein content of 19% and a reduced carbohydrate content, such bread is particularly suitable for diabetics (Anon., 1988a).

8.5.7 Bread made with gluten substitutes

It is no longer true that wheat gluten is necessary to make white bread of good quality. Acceptable bread has been made from sorghum flour – without any wheat flour and therefore without gluten – by the use of a gluten substitute, xanthan gum, a water-soluble polymer of high viscosity, which functions in the same way as gluten. Xanthan gum is made by fermenting carbohydrates with a bacterium, *Xanthomonas campestris*. The resulting viscous broth is pasteurized, precipitated with isopropyl alcohol, dried and ground. Xanthan gum is a form of bacterial cellulose, viz. a 1,4-*beta*-D-glucose polymer. One percent solutions are thixotropic, appearing gel-like at rest, but mixing, pouring and pumping easily (Anderson and Audon, 1988). Xanthan gum is already finding use as a thickening agent in foods. To achieve the best results with sorghum flour, the xanthan gum should be soaked in water before incorporation in the dough to give the bread a more open structure, and salt should be added to improve the flavour of the bread. Sorghum/xanthan bread retains its freshness for at least 6 days, which is longer than wheat bread (Satin, 1988).

Yeasted rice flour breads, using 100% rice flour or 80% of rice flour plus 20% of potato starch, but without any wheat flour, can be made if the binding function of the wheat gluten is replaced by a gum: carboxymethylcellulose (CMC) at a level of 1.6% (on flour + starch basis), or hydroxypropyl methylcellulose (HPMC) at a level of about 3% (on flour + starch basis). Bread made in this way can meet reference standards for wheat (white) bread for volume, crumb and crust colour, Instron firmness, and moisture content. The CMC or HPMC has the viscosity and film-forming characteristics to retain gas during proofing that are usually provided by gluten (Bean and Nishita, 1985; Ylimaki et al., 1988).

A mixture of galactomannans (hydrocolloids) from carob, guar and tara seeds has been used as a gluten substitute. A blend of three parts of carob seed flour (locust bean: *Ceratonia siliqua*) and one part of tara seed flour, 100–75 µm particle size, was favoured, giving good volume yield, crumb structure and flavour (Jud and Brümmer, 1990). The use of tara gum would not be permitted in the United Kingdom, while carob and guar gums would be permitted only for coeliac sufferers.

Another kind of gluten substitute, particularly for use in developing countries, can be made by boiling a 10% suspension of flour milled from tropical plants, e.g., cereal or root, in water until the starch gels, and then cooling. This material is then added to a blend of nonwheat flour, sugar, salt, yeast, vegetable oil, then mixed for 5 min, allowed to rise, and baked. The starch gel functions like gluten, trapping the gas evolved by yeast action (Anon., 1989).

8.6 BREAD QUALITY

Regardless of type of production process used, type of cereal grain used or other ingredients used, assessing the quality of the loaves is critical for both research and development efforts as well as quality control during production. Table 8.4 summarizes relevant standardized AACCI (American Association of Cereal Chemists International) methods for these purposes.

TABLE 8.4 Recommended methods for determining bread-baking performance and resulting bread properties[a]

AACCI method	Title	Primary applications
10-09	Basic straight-dough bread-baking method – long fermentation	Evaluating wheat flour and ingredients.
10-10B	Optimized straight-dough bread-making method	Evaluating wheat flour and ingredients.
10-11	Baking quality of bread flour – sponge-dough, pound-loaf method	Evaluating wheat flour and ingredients.
62-05	Preparation of sample: bread	Preparation of bread for subsequent analyses.
10-05	Guidelines for measurement of volume by rapeseed displacement	Determination of volume of breads, cakes and other baked products.
74-09	Measurement of bread firmness by universal testing machine	Determination of bread firmness of breads, cakes and other baked products for quality control.
74-10A	Measurement of bread firmness – compression test	Determination of bread firmness of breads, cakes and other baked products for quality control.
74-30	Staleness of bread – sensory perception test	Determination of bread staleness.
10-13	Guidelines for testing a variety of products	Formulation, production and testing for quality control or research and development; bagels, French hearth breads, frozen pizza doughs, hamburger buns, pita breads, soft pretzels, wheat tortillas and whole wheat breads.
10-90[b]	Baking quality of cake flour	Cell uniformity, size, thickness of cell walls, moistness, tenderness, softness and crumb colour.

[a] *Information can be found in American Association of Cereal Chemists International, 2000. Cereal Laboratory Methods. Amer. Assoc. of Cereal Chemists Inc., St. Paul, MN, USA.*
[b] *This method has been designed for assessing cake quality, but it is also useful for examining bread structures and quality.*

8.7 BREAD MACHINES FOR HOME USE

In the 1980s, small appliances were invented (in Japan) and sold for the purpose of making bread in the home (Fig. 8.13). These bread machines have been designed to replicate and simplify all processes necessary for

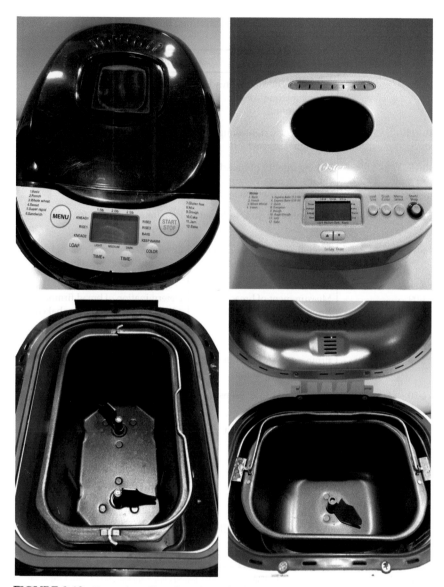

FIGURE 8.13   Examples of bread machines for home use.

the mixing, fermentation and baking (as depicted in Fig. 8.6) of bread. Most machines have pans that can accommodate loaf sizes of 1, 1.5 and 2 lb, and manual pushbutton settings can be used to adjust the settings (i.e., mixing, fermentation, baking times and temperatures) for a variety of bread types.

Because of their ease of use, these home appliances surged in popularity around the world in the 1990s, but interest has waned somewhat since the early 2000s. Although these machines are designed to mimic all necessary steps for baking bread, recipes for various breads that were originally designed for oven loaves will likely require modifications in ingredient levels and water content, as the machines do not truly replicate oven baking.

References

American Association of Cereal Chemists International, 2000. Cereal Laboratory Methods. Amer. Assoc. of Cereal Chemists Inc., St. Paul, MN, USA.

Anderson, D.M.W., Audon, S.A., 1988. Water-soluble food gums and their role in product development. Cereal Foods World 33 (10) 844, 846, 848–850.

Andersson, K., Ohlsson, T., 1999. Life cycle assessment of bread produced on different scales. International Journal of Life Cycle Assessment 4 (1), 25–40.

Anon., 1957. Breadmaking processes now available. North-West Miller 257 (13), 13.

Anon., 1987. Are you turned on to radio frequency? British Baker 184 (38) 36, 39.

Anon., July 1988a. Soya, the wonder bean. Milling 181, 13–15.

Anon., November 1988b. Radio frequency cuts baking times by half. British Baker 46, 48.

Anon., 1989. F.A.O. sees wheatless bread as boost to developing nations. Milling & Baking News 68 (3), 18–19.

Atkins, J.H.C., February 1971. Mixing requirements of baked products. Food Manufacture 47.

Axford, D.W.E., Chamberlain, N., Collins, T.H., Elton, G.A.H., 1963. The Chorleywood process. Cereal Science Today 8, 265.

Axford, D.W.E., Colwell, K.H., Cornford, S.J., Elton, G.A.H., 1968. Effect of loaf specific volume on the rate of staling in bread. Journal of the Science of Food and Agriculture 19, 95.

Baker, A.P.V., April 1988. Baking progress. Food Process U.K. 57, 37.

Bakhshi, A.K., Sehgal, K.L., Singh, R.P., Gill, K.S., 1989. Effect of bread wheat, durum wheat and triticale blends on chapati, bread and biscuit. Journal of the Science of Food and Agriculture 26 (4), 191–193.

Bean, M.M., 1986. Rice flour – its functional variations. Cereal Foods World 31 (7), 477–481.

Bean, M.M., Nishita, K.D., 1985. Rice flours for baking (Chapter 14). In: Juliano, B.O. (Ed.), Rice: Chemistry and Technology. Amer. Assoc. of Cereal Chemists Inc., St. Paul, MN, USA, pp. 539–556.

Brown, J. (Tech. Ed.), 1982. The Master Bakers' Book of Breadmaking. Nat. Assoc. of Master Bakers, Confectioners and Caterers, London, UK.

Cauvain, S.P., Chamberlain, N., 1988. The bread improving quality of fungal α-amylase. Journal of Cereal Science 8, 239–248.

Cauvain, S.P., Young, L.S., 1998. Technology of Breadmaking. Blackie Academic & Professional, Thomas Science, London, UK.

Chamberlain, N., September 8, 1973. Microwave energy in the baking of bread. Food Trade Review British Baker 167 (July 13): 20.

Chamberlain, N., Collins, T.H., Elton, G.A.H., 1962. The Chorleywood bread process. Bakers' Digest 36 (5), 52.

Dendy, D.A.V., Dobraszczyk, B.J., 2001. Cereals and Cereal Products – Chemistry and Technology. Aspen Publishers, Inc., Gaithersburg, MD, USA.

Drews, E., Seibel, W., 1976. Bread-making and other uses around the world (Chapter 6). In: Bushuk, W. (Ed.), Rye: Production, Chemistry and Technology. Amer. Assoc. of Cereal Chemists Inc., St. Paul, MN, USA.

Elton, G.A.H., 1969. Some quantitative aspects of bread staling. Bakers' Digest 43 (3), 24.

Farrand, E.A., 1964. Flour properties in relation to the modern bread processes in the United Kingdom, with special reference to *alpha*-amylase and starch. Cereal Chemistry 41, 98–111.

Gan, Z., Angold, R.E., Williams, M.R., Ellis, P.R., Vaughan, J.G., Galliard, T., 1990. The microstructure and gas retention of bread dough. Journal of Cereal Science 12, 15–24.

Gebhardt, E., Lehrack, U., 1988. The processing quality of rye in the German Democratic Republic. Bäker U. Konditor 36 (2), 55–57.

Grosch, W., 1968. Redox systems in dough (Chapter 12). In: Blanshard, J.M.V., Frazier, P.J., Galliard, T. (Eds.), Chemistry and Physics of Baking. Roy Soc. Chem., London, UK, pp. 155–169.

Haber, T., Lewczuk, J., 1988. Use of triticale in the baking industry. Acta Alimentaria Polonica 14 (3–4), 123–129.

Hargrove, K., 1990. Rice bran in bakery foods. American Institute of Baking Research Department of Technical Bulletin 12 (2), 1–6.

Hoseney, R.C., 1986. Principles of Cereal Science and Technology. Amer. Assoc. of Cereal Chemists Inc., St. Paul, MN, USA.

Jud, B., Brümmer, J.M., 1990. Production of gluten free breads using special galactomannans. Getreide Mehl und Bröt 44, 178–183.

Kaufman, J., Lacoste, A., Schulok, J., Shehady, E., Yam, K., 2017. An Overview of Oxygen Scavenging Packaging and Applications. Bakery Online: https://www.bakeryonline.com/doc/an-overview-of-oxygen-scavenging-packaging-an-0002.

Kookman, M., Linko, P., 1966. Activity of different enzymes in relation to the baking quality of rye. Cereal Science Today 11, 444.

Krog, N., Olsen, S.K., Toernaes, H., Joensson, T., 1989. Retrogradation of the starch fraction in wheat bread. Cereal Foods World 34, 281.

Kulp, K., Ponte Jr., J.G., 2000. Handbook of Cereal Science and Technology, second ed. Marcel Dekker, Inc., New York, NY, USA.

Leach, G., 1975. Energy and Food Production. IPC Science & Technology Press, Guildford, Surrey, UK.

Magnus, E.M., Fjell, K.M., Steinsholt, K., 1987. Barley flour in Norwegian bread. In: Morton, I.D. (Ed.), Cereals in a European Context. Ellis Horwood, Chichester, UK, pp. 377–384.

Magnuson, K.M., 1985. Uses and functionality of vital wheat gluten. Cereals Foods World 30, 169–171.

Martin, D.J., Stewart, B.G., 1991. Contrasting dough surface properties of selected wheats. Cereal Foods World 36, 502–504.

McDermott, E.E., 1985. The properties of commercial glutens. Cereal Foods World 30, 169–171.

McKechnie, R., 1983. Oat products in bakery foods. Cereal Foods World 28, 635–637.

Meisterling, K., Samaras, C., Schweizer, V., 2009. Decisions to reduce greenhouse gases from agriculture and product transport: LCA case study of organic and conventional wheat. Journal of Clean Production 17, 222–230.

Ministry of Agriculture, Fisheries, Food, 1972. The Bread and Flour (Amendment) Regulations 1972. Statutory Instruments 1972, No 1391. H.M.S.O., London, UK.

Ministry of Agriculture, Fisheries, Food, 1984. The Bread and Flour Regulations 1984. Statutory Instruments 1984, No. 1304. H.M.S.O., London, UK.

Ministry of Agriculture, Fisheries, Food, 1990. Potassium Bromate (Prohibition as a Flour Improver) Regulations 1990. Statutory Instruments 1990, No. 339. H.M.S.O., London, UK.

Mitchell, T.A., February 1983. Changes in raw material and manufacturing. Baker Millers' Journal 15, 18.

Pomeranz, Y. (Ed.), 1971. Wheat: Chemistry and Technology, second ed. Amer. Assoc. of Cereal Chemists Inc., St. Paul, MN, USA.

Pomeranz, Y., April 1980. Molecular approach to breadmaking – an update and new perspectives. Bakers' Digest 12.

Pomeranz, Y., Shellenberger, J.A., 1971. Bread Science and Technology. Avi Publ. Co. Inc., Westport, CT, USA.

Radley, J.A., 1968. Starch and Its Derivatives, fourth ed. Chapman & Hall, London, UK.

Reed, G., Nagodawithana, T.W., 1991. Yeast Technology, second ed. Van Norstrand Rheinhold, NY, USA.

Rosentrater, K.A., 2011. Using DDGS as a food ingredient (Chapter 18). In: Distillers Grains: Production, Properties, and Utilization. Taylor and Francis, Boca Raton, FL, USA.

Roy, P., Nei, D., Orikasa, T., Xu, Q., Okadome, H., Nakamura, N., Shiina, T., 2009. A review of life cycle assessment (LCA) on some food products. Journal of Food Engineering 90, 1–10.

Satin, M., April 28, 1988. Bread without wheat. New Scientist 56–59.

Seibel, W., Brümmer, J.-M., 1991. The sourdough process for bread in Germany. Cereals Foods World 36, 299–304.

Sharma, N.R., Rasper, V.F., Van de Voort, F.R., 1988. Pregelatinized flours in composite blends for breadmaking. Canadian Institute of Food Science and Technology Journal 21, 408–414.

Sirbu, A., Arghire, C., 2017. Functional bread: effect of inulin-type products addition on dough rheology and bread quality. Journal of Cereal Science 75, 220–227.

Smith, J.P., Coraikul, B., Koersen, W.J., 1987. Novel approach to modified atmosphere packaging of bakery products. In: Morton, I.D. (Ed.), Cereals in a European Context, pp. 332–343 (Chichester, Ellis Horwood, UK).

Spencer, B., 1989. Flour improvers in the EEC – harmony or discord? Cereal Foods World 34, 298–299.

Telloke, G.W., 1980. Private Communication.

Willhoft, E.M.A., 1973. Recent developments on the bread staling problem. Bakers' Digest 47 (6), 14.

Ylimaki, G., Hawrysh, Z.J., Hardin, R.T., Thomson, A.B.R., 1988. Applications of response surface methodology to the development of rice flour yeast breads: objective measurements. Journal of Food Science 53, 1800–1805.

Zawistowska, U., Langstaff, J., Bushuk, W., 1988. Improving effect of a natural α-amylase inhibitor on the baking quality of wheat flour containing malted barley flour. Journal of Cereal Science 8, 207.

Ziemke, W.H., Sanders, S., October 1988. Sourdough bread. American Institute of Baking, Research Department of Technical Bulletin 10, 1–4.

Further reading

Bailey, A., 1975. The blessings of bread. Paddington Press Ltd., London, UK.

Beech, G.A., 1980. Energy use in bread baking. Journal of the Science of Food and Agriculture 31, 289.

Birch, A.N., Petersen, M.A., Hansen, A.S., 2014. Aroma of wheat breadcrumb. Cereal Chemistry 91 (2), 105–114.

Chamberlain, N., April 1988. Forging ahead. Food Process U.K. 57 14, 16, 19, 20.

Dewettinck, K., Van Bockstaele, F., Kuhne, B., Van de Walle, D., Courtens, T.M., Gellynck, X., 2008. Nutritional value of bread: influence of processing, food interactions and consumer perception. Journal of Cereal Science 48, 243–257.

Fance, W.J., Wragg, B.H., 1968. Up-to-date Breadmaking. McLaren & Sons Ltd., London, UK.

Fowler, A.A., Priestley, R.J., 1980. The evolution of panary fermentation and dough development – a review. Food Chemistry 5, 283.

Goldstein, A., Nantanga, K.K.M., Seetharaman, K., 2010. Molecular interactions in starch-water systems: effect of increasing starch concentration. Cereal Chemistry 87 (4), 370–375.

Grosch, W., Schieberle, P., 1997. Flavor of cereal products – a review. Cereal Chemistry 74 (2), 91–97.

Halton, P., 1962. The development of dough by mechanical action and oxidation. Milling 138, 66.

Hutchinson, J.B., Fisher, E.A., 1937. The staling and keeping quality of bread. Bakers' National Association Review 54, 563.

Katz, J.R., 1928. Gelatinization and retrogradation of starch in relation to problems of bread staling. In: Walton, R.P. (Ed.). Walton, R.P. (Ed.), A Comprehensive Survey of Starch Chemistry, 1. Chemical Catalog Co. Inc., NY, USA, p. 100.

Kent-Jones, D.W., Mitchell, E.F., 1962. Practice and Science of Breadmaking, third ed. Northern Publ. Co. Ltd., Liverpool, UK.

Khan, K., Shewry, P.R., 2009. Wheat Chemistry and Technology, fourth ed. American Association of Cereal Chemists Inc., St. Paul, MN, USA.

Langrish, J., Gibbons, M., Evans, W.G., 1972. Wealth from Knowledge. MacMillan, London, UK.

Lazo-Velez, M.A., Chavez-Santoscoy, A., Sern-Saldivar, S.O., 2015. Selenium-enriched breads and their benefit in human nutrition and health as affected by agronomic, milling, and baking factors. Cereal Chemistry 92 (2), 134–144.

Macritchie, F., 2016. Seventy years of research into breadmaking quality. Journal of Cereal Science 70, 123–131.

Matz, S.A., 1972. Baking Technology and Engineering, second ed. Avi Publ. Co. Inc., Westport, CT, USA.

Maunder, P., 1969. The Bread Industry in the United Kingdom. University of Nottingham, Loughborough, UK.

Ministry of Agriculture, Fisheries, Food, 1962. Preservatives in Food Regulations 1962. Statutory Instruments 1962, No 1532. H.M.S.O., London, UK.

Moore, M.M., Schober, T.J., Dockery, P., Arendt, E.K., 2004. Textural comparisons of gluten-free and wheat-based doughs, batters, and breads. Cereal Chemistry 81 (5), 567–575.

Mulders, E.J., 1973. The Odour of White Bread. Agric. Res. Rep. (Versl. Landbouk, Onderz.)., p. 798 (Wageningen, Netherlands).

National Association of British, Irish Millers Ltd, 1989. Practice of Flour Milling. N.A.B.I.M., London, UK.

Pareyt, B., Finnie, S.M., Putseys, J.A., Delcour, J.A., 2011. Lipids in bread making: sources, interactions, and impact on bread quality. Journal of Cereal Science 54, 266–279.

Pyler, E.J., 1973. Baking Science and Technology, second ed. Siebel Publ. Co., Chicago, IL, USA.

Quail, K.J., 1996. Arabic Bread Production. American Association of Cereal Chemists Inc., St. Paul, MN, USA.

Ringsted, T., Siesler, H.W., Engelsen, S.B., 2017. Monitoring the staling of wheat bread using 2D MIR-NIR correlation spectroscopy. Journal of Cereal Science 75, 92–99.

Schoch, T.J., French, D., 1945. Fundamental Studies on Starch Retrogradation. (Office of the Quartermaster General, USA).

Seitz, L.M., Chung, O.K., Rengarajan, R., 1998. Volatiles in selected commercial breads. Cereal Chemistry 75 (6), 847–853.

United Nations Economic Commission for Africa, 1985. In: Kent, N.L. (Ed.), Technical Compendium on Composite Flours. Economic Commission for Africa, Addis Ababa, Ethiopia.

Van Vliet, T., 2008. Strain hardening as an indicator of bread-making performance: a review with discussion. Journal of Cereal Science 48, 1–9.

Wiener, J., Collier, D., 1975. Bread. Robert Hale, London, UK.

Williams, A. (Ed.), 1975. Breadmaking, the Modern Revolution. Hutchinsons Benham, London, UK.

Witczak, M., Ziobro, R., Luszczak, L., Korus, J., 2016. Starch and starch derivatives in gluten-free systems – a review. Journal of Cereal Science 67, 46–57.

9.1 INTRODUCTION

All cereals contain a large proportion of starch. In its natural form, the starch is insoluble, tasteless and unsuited for human consumption. To make it digestible and acceptable it must be cooked. Breakfast cereals are products that are consumed after cooking, and they fall into two main categories: those made by processes that do not include cooking and therefore have to be cooked domestically (i.e., hot cereals), and those that are cooked during processing and require no domestic cooking. The first class of products is exemplified by various types of porridge; the second by products that are described as 'ready-to-eat' cereals.

Besides the distinction regarding the need for domestic cooking vis-à-vis readiness for consumption, breakfast cereals can also be classified according to the form of the product, as well as according to the particular cereal grain used as the raw material (Holland et al., 1988).

9.2 COOKING OF CEREALS

If the cereal is cooked with excess of water and only moderate heat, as in boiling, the starch gelatinizes and becomes susceptible to starch-hydrolyzing enzymes in the digestive system. If cooked with a minimum of water, or even without water, but at a higher temperature, as in toasting, nonenzymatic browning (Maillard) reactions between proteins and reducing carbohydrates may occur, and there may be some dextrinization breakdown of the starch into dextrins. Cooking by extrusion at low moisture causes the starch granules to lose their crystallinity, but they are unable to swell as in the normal gelatinization process, which requires excess water (proteins, on the other hand, can plasticize and denature). However, when they are exposed to moisture during consumption they hydrate and swell to become susceptible to enzymatic digestion.

9.3 HOT CEREALS

9.3.1 Porridge from oats

Porridge (Figs 9.1 and 9.2) is generally made from oatmeal or oat flakes (rolled oats or 'porridge oats'), the manufacture of which is described elsewhere in this book. The milling process to make oatmeal includes no cooking (unless the oats are stabilized to inactivate the enzyme lipase, and the starch in oatmeal is ungelatinized); moreover, the particles of oatmeal are relatively coarse in size. Consequently, porridge made from coarse oatmeal requires prolonged domestic cooking, by boiling with water, to bring about gelatinization of the starch. Oatmeal of flour fineness cooks quickly, but the cooked product is devoid of the granular structure associated with the best Scotch porridge.

Rolled oats are partially cooked during manufacture; the pinhead oatmeal from which rolled oats is made is softened by treatment with steam and, in this plastic condition, is flattened on flaking rolls. Thus, porridge made from rolled oats requires only a brief domestic cooking time to complete the process of starch gelatinization (only a few minutes in a microwave oven).

The amount of domestic cooking required by rolled oats is dependent to a large extent on the processes of cutting, steaming and flaking, all of which are interrelated. The size of the pinhead oatmeal influences rate of moisture penetration in the steamer; smaller particles will be more thoroughly moistened than large particles by the steaming process, and hence the starch will be gelatinized to a greater degree, and the steamed pinhead meal will be softer. For a given roller pressure at the flaking stage, this increase in softness will result in thinner flakes being obtained from smaller-sized particles of pinhead meal. During the domestic cooking of porridge, the thinner flakes will cook more rapidly than thicker flakes because moisture penetration is more rapid.

Thin flakes would normally be more fragile than thick ones, and more likely to break during transit. However, thin flakes can be strengthened by raising the moisture content of the pinhead meal feeding the steamer, thereby increasing the degree of gelatinization of the starch. Gelatinized starch has an adhesive quality, and quite thin flakes rolled from highly gelatinized small particle-size pinhead meal can be surprisingly strong.

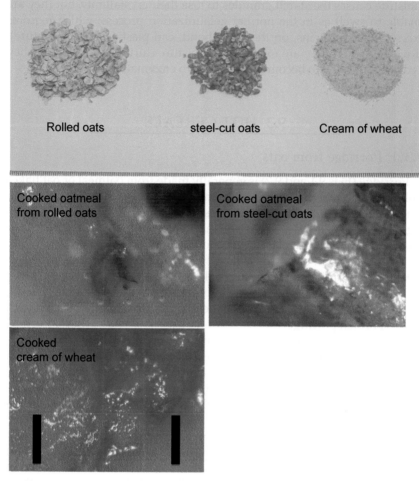

FIGURE 9.1 Hot porridge breakfast cereals. Top image shows uncooked cereals. Bottom images show cooked porridges. Note that the steel-cut oats retain much of the grain structure even after cooking, which leads to a chewy texture, compared to the creamy texture of rolled oats and cream of wheat. Scale bars represent 1 mm distance.

The average thickness of commercial rolled oats is generally 0.30–0.38 mm (0.012–0.015 in.); when tested with Congo Red stain (which colours only the gelatinized and damaged starch granules), about 30% of the starch granules in rolled oats appear to be gelatinized.

9.3.2 Ready-cooked porridge

In the search for porridge-like products that require even less cooking than rolled oats, a product called 'Porridge without the pot' has been made. Porridge can be made from this material merely by stirring with hot

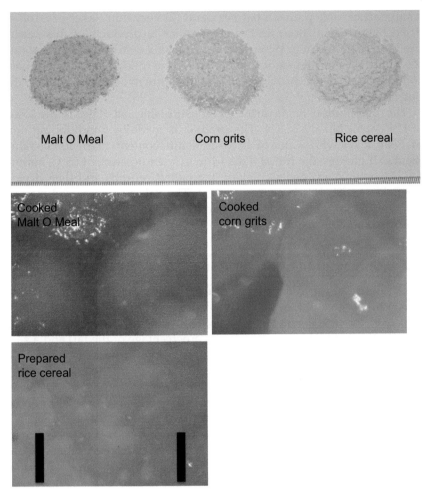

FIGURE 9.2 More porridges. Top image shows uncooked cereals. Bottom images show cooked porridges. Malt-O-Meal has similar texture before and after cooking compared to Cream of Wheat. Grits aren't technically a breakfast cereal, but are a common porridge. Rice cereal is sold precooked and is a common first food for infants. Scale bars represent 1 mm distance.

or boiling water in the bowl; it consists of oat flakes of a special type. As compared with ordinary rolled oats, these flakes are thinner, stronger and contain starch that is more completely gelatinized. They could be manufactured by steaming the pinhead oatmeal at somewhat higher moisture content than normal, rolling at a greater pressure than normal and using heated flaking rolls.

Another type of porridge mix, known as Ready Brek (manufactured by Weetabix Limited), consists of a blend of two types of flakes in approximately equal proportions: (1) ordinary rolled oats made from small particle-size pinhead oatmeal, and (2) very thin flakes of a roller-dried batter of oat flour and water, similar to products of this nature used for infant feeding. When this porridge mix is stirred with hot water, the thin flakes form a smooth paste while the rolled oats, which do not completely disperse, provide a chewy constituent and give body to the porridge.

The preparation of an instant reconstitutable oat flake is disclosed in US Patent No. 4,874,624. The product is made by conditioning normal oat flakes with water to 18.5% moisture content, extrusion cooking them at high pressure for 10–120 s to an exit temperature of 95°C, cutting the extrudates into pellets, flaking the pellets on rolls and then drying the flakes to 2%–12% moisture content. The flakes so processed may be blended (70:30) with normal oat flakes that have been steamed to inactivate enzymes.

9.3.3 Specification for oatmeal and oat flakes

Quality tests for milled oat products include determination of moisture, crude fibre and free fatty acid (FFA) contents, and of lipase activity. A recommended specification is a maximum of 5% acidity due to FFA (calculated as oleic acid, and expressed as a percentage of the fat) and a zero response for lipase activity. Other suggested tests include arsenic (which could be derived from the fuel used in the kiln), lead and copper, which catalyze oxidation of the fat (Anon., 1970).

Raw oats normally contain an active lipase enzyme, and, with the fat content of oats being some two to five times as high as that of wheat, it is desirable that the lipase should be inactivated during the processing of oats, to prevent it from catalyzing the hydrolysis of the fat, which would lead to the production of bitter-tasting free fatty acids. Lipase is inactivated by the stabilization process, as described elsewhere in this book, and is a most important safeguard for the keeping quality of oatmeal.

9.3.4 Porridge from other cereals

In Africa, maize grits or hominy grits are used to make porridge by boiling with water. In Italy, maize porridge, made from fine maize grits or coarse maize meal, and flavoured with cheese, is called 'polenta'.

Barley meal is used for making a type of porridge in many countries in the Far East, Middle East, and North Africa.

Wholemeal flour made from sorghum or millet may be cooked with water to make a porridge-like food in African countries and in India. Porridge made from parched millet grain or buckwheat in Eastern Europe (i.e., Russia, Ukraine, and Poland) is called *kasha*.

9.4 READY-TO-EAT CEREALS

While porridge-type cereals have been consumed for many years, the development of ready-to-eat (RTE) cereals is relatively recent. RTE cereals owe their origin to the Seventh Day Adventist Church, whose members, preferring an entirely vegetable-based diet, experimented with the processing of cereals in the mid-nineteenth century.

A granulated product, 'Granula', made by J.C. Jackson in 1863, may have been the first commercially available RTE breakfast cereal. A similar product, 'Granola', was made by J.H. Kellogg by grinding biscuits made from wheatmeal, oatmeal and maizemeal. Mass acceptance of ready-to-eat cereals was achieved in countries such as the United States by means of efficient advertising.

9.4.1 Processing

The stages in the processing of RTE cereals typically include the preparation of the cereal by cleaning, and possibly pearling, cutting or grinding; the addition of adjuncts such as salt, malt, sweeteners and flavouring materials; mixing with sufficient water to produce a paste or dough of the required moisture content; cooking the mixture; cooling and partially drying, and shaping the material by, e.g., rolling, puffing, shredding, etc., into the desired form, followed by toasting, which also dries the material to a safe moisture content for packaging.

9.4.2 Batch cooking

Until a few decades ago, cooking was carried out in rotating vessels, 'cookers', into which steam was injected, and the system was a batch process. The batch cooking process has now been largely superseded by continuous cooking processes in which cooking and extrusion through a die are both carried out in a single piece of equipment – a cooking extruder, or extrusion cooker (see Chapter 10). Extrusion is a high-temperature, short-time (HTST) process in which the material is plasticized at relatively high temperatures, pressures, and shear before being forced through a die and then released to ambient atmosphere (temperature and pressure) (Linko, 1989a).

9.4.3 Continuous cooking

Continuous cooking methods have many advantages over batch methods: for example, continuous methods require less floor space and less energy for operation; they permit better control of processing conditions, leading to improved quality of the final products. Moreover, batch-cooking methods were usually restricted to the use of whole grain or to relatively large grain fragments, whereas extrusion cooking can also utilize much finer materials, including flour.

9.4.4 Extrusion cookers

An extrusion cooker is a continuous processing unit based on a screw rotating within a barrel. An extensive discussion about extrusion is provided elsewhere in this book.

9.5 FLAKED PRODUCTS FROM MAIZE

Maize (for 'corn flakes'), wheat and rice are the cereals generally used for flaking. Some examples are shown in Fig. 9.3.

In the traditional batch process for making corn flakes, a blend of maize grits – chunks of about one-half to one-third of a kernel in size – plus flavouring materials, e.g., 6% (on grits wt.) of sugar, 2% of malt syrup, 2% of salt, and heat-stable vitamins and minerals, is pressure-cooked for about 2 h in rotating batch cookers at a steam pressure of about 18 lb/in.^{2} (psi) to a moisture content of about 28% after cooking. The cooking is complete when the colour of the grits has changed from chalky-white to light golden brown, the grits have become soft and translucent, and no raw starch remains (i.e., it has been gelatinized).

The cooked grits are dried by falling against a countercurrent flow of air at about 65°C under controlled humidity conditions, to ensure uniform drying, to a moisture content of about 20%, a process that takes between 2.5 and 3 h, and are then cooled and rested to allow equilibration of moisture. The resting period was formerly about 24 h, but is considerably less under controlled humidity drying conditions. The dried grits are then flaked on counterrotating rollers, which have a surface temperature of approximately 43–46°C, at a pressure of 40 tonnes at the point of contact, and the flakes thus formed are then toasted in a tunnel oven at 300°C for about 50 s. The desired blistering of the surface of the flakes is related to the roller surface temperature and to the moisture content of the grits, which should be 10%–14% moisture content when rolled. After cooling, the flakes may be sprayed with solutions of vitamins and minerals before packaging.

Extruded flakes, made from maize or wheat, are cooked in an extrusion cooker rather than in a batch pressure cooker, and can be made from fine meal or flour rather than from coarse grits alone. The dry material is fed continuously into the extrusion cooker, and is joined by a liquid solution of the flavouring materials – sugar, malt, salt, etc., and water These are

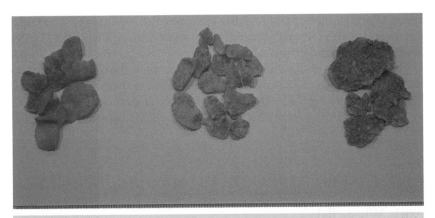

FIGURE 9.3 Flaked RTE cereals. Magnification shows blistering of the flake surface.

mixed together either prior to extrusion, or by the rotation of the extruder screw, and then conveyed through the heated barrel, thereby becoming cooked before exiting the die.

The material is extruded through the die in the form of ribbons that are cut to pellet size by a rotating knife. The pellets are then dried, tempered, flaked and toasted as described for the traditional method (Fast, 1987; Fast and Caldwell, 1990; Fast et al., 1990; Hoseney, 1986; Midden, 1989; Rooney and Serna-Saldivar, 1987).

9.6 FLAKED PRODUCTS FROM WHEAT AND RICE

9.6.1 Wheat flakes

These are traditionally made from wholewheat grain, which is conditioned with water to about 21% moisture content and then 'bumped' by passing through a pair of smooth rollers set so that the roll gap is slightly narrower than the width of the grain. Without fragmenting the grain, bumping disrupts the bran coat, assisting the penetration of water. Flavouring adjuncts – sugar, malt syrup, salt – are then added and the grain is pressure-cooked at about 15 psi for 30–35 min. The cooked wheat, at 28–30 moisture content, emerges in big lumps that have to be 'de-lumped', and then dried from about 30% moisture content to 16%–18% moisture content. After cooling to about 43°C, the grain is placed in a storage bin to temper for a short time, and then flaked (as for maize). Just before flaking, the grain is heated to about 88°C to plasticize the kernels and prevent tearing on the flaking rolls. The flakes leave the rolls at about 15%–18% moisture content and are then toasted and dried to about 3% moisture content.

9.6.2 Rice flakes

To make rice flakes by the traditional process, the preferred starting material is head rice (whole dehusked grains) or second heads (large broken kernels). Flavouring adjuncts are similar to those used with maize and wheat. The blend of rice plus adjuncts is pressure-cooked at 103.4–124.1 kPa (15–18 psi) for about 60 min. The moisture content of the cooked material should not exceed 28%, otherwise it becomes sticky and difficult to handle. De-lumping, drying (to about 17% moisture content: at lower moisture contents the particles shear; at higher moisture contents the flaking rolls become gummed up), cooling, tempering (up to 8 h), and flaking processes are similar to those used for wheat flakes.

In the toasting of the rice flakes, more heat is required than for making wheat flakes. The moisture content of the feed and the heat of the oven are adjusted so that the flakes blister and puff during toasting; accordingly, the discharge end of the oven is hotter than the feed end. The moisture content of the final product is typically 1%–3% moisture content.

The process for making rice flakes by extrusion resembles that described previously for maize and wheat, except that a colouring material is often added to offset the dull or grey appearance caused by mechanical working during extrusion. The lack of natural colour is emphasized if the formulation is low in sugar or malt syrup (which serve as sources of reducing sugars that can participate in Maillard reactions).

9.7 PUFFED PRODUCTS

Cereals may be puffed in either of two ways: by sudden application of heat at atmospheric pressure so that the water in the cereal is vaporized in situ, thereby expanding the product (i.e., oven puffing); or by the sudden transference of the cereal containing superheated steam from a high pressure to a low pressure, thereby allowing the water suddenly to vaporize and cause expansion (i.e., gun puffing). The key to the degree of puffing of the cooked grain is the suddenness of change in temperature or pressure (Hoseney, 1986), and the ability of the material's matrix to maintain a cohesive structure as the steam exits.

The preferred grains for puffing are rice, wheat, oats or pearl barley, which are prepared by cleaning, conditioning and depericarping (e.g., by a wet scouring process). Flavouring adjuncts (sugar, malt syrup, salt, etc.) are typically added, as they are for flaked products. Examples are shown in Figs 9.4–9.7.

9.7.1 Oven-puffed rice

Oven-puffed rice is made from raw or parboiled milled rice that is cooked, with the adjuncts, for 1 h at 15–18 lb/in.^{2} in a rotary cooker until uniformly translucent. The cooked grain is dried to 30% moisture content, tempered for 24 h, dried again, this time to 20% moisture content, and subjected to radiant heat to plasticize the outside of the kernels. The grain is 'bumped' through smooth rolls, just sufficiently to flatten and compress it, and then surface dried to about 15% moisture content, and tempered again for 12–15 h at room temperature. The bumped rice then passes to the toasting oven, where it remains 30–90 s. The temperature in the oven is about 300°C in the latter half of the oven cycle – as hot as possible short of scorching the grains. Due to the bumping, which has compressed the grains, and the high temperature, the grains immediately puff to 5–6 times their original size. The puffed grains are cooled, fortified with vitamins

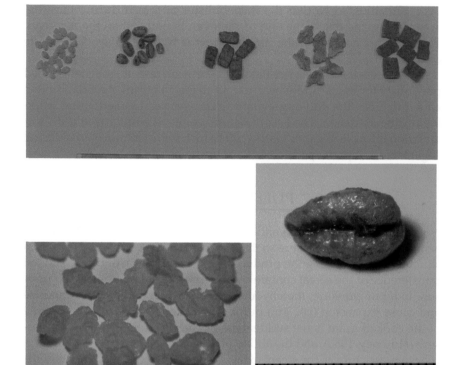

FIGURE 9.4 Puffed RTE cereals in various shapes. Magnification on puffed rice and puffed wheat. Scale bars represent 1 mm distance.

and minerals, if required, and treated with antioxidants (Hoseney, 1986; Juliano and Sakurai, 1985). Kellogg's 'Rice Krispies' is a well-known brand of oven-puffed rice. Kellogg's 'Special K', containing 20% protein, is made in a similar way to Rice Krispies up to the drying before bumping stage. The material is then wetted, coated with the enrichment, and bumped more heavily than for Krispies, then oven-puffed and toasted. The high-protein enrichment may be vital wheat gluten, defatted wheat germ, nonfat dry milk, or dried yeast, plus vitamins, minerals and antioxidants (Juliano and Sakurai, 1985).

9.7.2 Gun-puffed rice

Long-grain white rice or parboiled medium-grain rice is generally used for gun-puffing, although short-grain, low-amylose ('waxy') rice is used for gun-puffing in the United States, and parboiled waxy rice may be used in the Philippines. Puffed parboiled rice has a darker, less acceptable

FIGURE 9.5 Extruded and puffed RTE cereals in the shape of spheres. Scale bars represent 1 mm distance.

FIGURE 9.6 Extruded and puffed RTE cereals in the shape of rings/O's. Scale bars represent 1 mm distance.

colour and tends to undergo oxidative rancidity faster than puffed raw rice, but parboiled rice requires less treatment, specifically lower steam pressure and temperature, than raw rice (Juliano and Sakurai, 1985).

A batch of the prepared grain is preheated to 520–640°C and fed to the puffing gun, a pressure chamber with an internal volume of 0.5–1.0 ft³, which is heated externally and by injection of superheated steam, so that the internal pressure rapidly builds up to about 1.38 MPa (200 psi) (1.38 MN/m²) at temperatures up to around 240°C, and the starch in the material becomes gelatinized. The pressure is suddenly released by opening the chamber

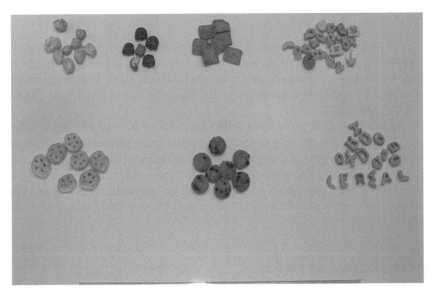

FIGURE 9.7   Examples of various extruded, puffed or sheeted RTE products. Scale bars represent 1 mm distance.

of the puffing gun. The material is 'shot' out, due to expansion of water vapour upon release of the pressure, and thus blowing up the grains or pellets to several times their original size. The puffed product is dried to 3% moisture content by toasting, then cooled and packaged (Fast, 1987; Fast and Caldwell, 1990; Juliano and Sakurai, 1985).

For satisfactory puffing, the starch should have plastic flow characteristics under pressure, and hence the temperature should be high enough to achieve this. Moreover, the material at the moment immediately before expansion requires cohesion to prevent shattering and elasticity to permit expansion. The balance between these two characteristics can be altered by adding starch, which has specific cohesive properties.

Extruded gun-puffed cereals can be made from oat flour or maize-meal, with which tapioca or rye flour can be blended as well. This material is fed into the cooking extruder and a solution of the adjuncts–sugar, malt syrup, salt – is added with more water. The dough is cooked in the cooking extruder and then transferred to a forming extruder, in which a noncooking temperature – below 71°C – is maintained. The extruded collets are subsequently dried from 20% to 24% moisture content to 9%–12% moisture content, and then gun-puffed at 260–430°C and 0.69–1.38 MPa (100–200 psi) pressure as previously described (Fast and Caldwell, 1990; Rooney and Serna-Saldivar, 1987). A 10- to 16-fold expansion will result.

9.7.3 Puffed wheat

A Plate durum wheat called Taganrog is the type of wheat preferred for puffing on account of its large grain size, which gives high yields of large puffs (this grain originated near the Russian city of Taganrog, which is located in southwest Russia near the Black Sea), but durum or CWRS wheats may also be used.

The wheat is first pretreated with about 4% of a saturated brine solution (26% salt content) to toughen the bran during preheating and make it cohesive, so that the subsequent puffing action will blow the bran away from the grain, thereby improving product appearance. Alternatively, the bran can be partially removed by pearling on carborundum stones. The puffing process is similar to that described for rice (Fast and Caldwell, 1990).

9.7.4 Continuous puffing

Using a steam-pressurized puffing chamber, the prepared grain is admitted through valves and subsequently released through an exit pore without loss of pressure in the chamber (US Patent No. 3,971,303).

9.8 SHREDDED PRODUCTS

For shredded RTE products (Fig. 9.8), wheat is the cereal generally used, often a white, starchy type, such as Australian, being preferred. The whole grain is cleaned and then cooked in boiling water with injection of steam for 30–35 min. until the centre of the kernel changes from starchy white to translucent grey, and the grain is soft and rubbery. The moisture content will be 45%–50%, and the starch will be fully gelatinized. The cooked grain is cooled to room temperature and rested for up to 24 h to allow for moisture equilibration. During this time, the kernels firm up because of retrogradation of the starch; this firming is essential for obtaining shreds of adequate strength. The conditioned grain is then fed to shredders, which consist of pairs of metal rolls – one is smooth, the other has grooves – between which the material emerges as long, parallel shreds. The shreds are detached from the grooves by the teeth of a comb and fall onto a slowly travelling band, a thick mat being built up by the superposition of several layers. The mat is then cut into tablets by a cutter that has dull cutting edges; the squeezing action of the cutter compresses the shreds and makes them adhere to one another. The tablets are then baked at 260°C in a gas-heated revolving oven or a conveyor-belt oven, for about 20 min. The major heat input is at the feed end; the biscuits increase in height as moisture is lost in the middle section of the dryer, while colour is developed in the final dryer section. The moisture content

FIGURE 9.8 Examples of shredded and granular products. Scale bars represent 1 mm distance.

of the biscuits is about 45% entering the oven but about 4% leaving the oven. The biscuits may be further dried to about 1% moisture content, passed through a metal detector, and then packaged (Fast, 1987; Fast and Caldwell, 1990).

Shredded products may also be made from the flour of wheat, maize, rice or oats, which would be cooked in batches or by continuous extrusion cooking. As with other RTE products, flavouring and nutritional adjuncts may be added. After cooking, cooling and equilibration for 4–24 h, the material is shredded and baked as described herein. When using maize or rice to make a shredded product, however, it is desirable to produce a degree of puffing to avoid product hardness. This is achieved by using a lower temperature in the first part of the baking, followed by an extremely high temperature in the last part.

9.9 GRANULAR PRODUCTS

A yeasted dough is made from either a fine wholemeal or long extraction wheat flour and malted barley flour, with added salt. The dough is fermented for about 6 h, and from it large loaves are baked. These are then broken up, dried and ground to a standard degree of fineness (typically a few mm in approximate diameter). An example is shown in Fig. 9.8.

9.10 SUGAR-COATED PRODUCTS

Flaked or puffed cereals, prepared as described before, are sometimes coated with sugar or candy. The process described in British Patent No. 754,771 uses a sucrose syrup containing 1%–8% of other sugars (e.g., honey) to provide a hard, transparent coating that does not become sticky even under humid conditions. The sugar content of corn flakes is typically raised from 7% to 43% by the coating process, that of puffed wheat from 2% to 51%. As an alternative to sugar, use of artificial sweeteners, such as aspartame, for breakfast cereals is described in US Patent No. 4,501,759 and US Patent No. 4,540,587, while the use of a dipeptide sweetener is disclosed in US Patent Nos. 4,594,252, 4,608,263 and 4,614,657.

9.11 KEEPING QUALITY OF BREAKFAST CEREALS

The keeping quality (i.e., preservation of quality and safety over time, or shelf life) of the prepared products depends to a large extent on the content and keeping quality of the oil contained within. Thus, products made from cereals having a low oil content (wheat, barley, rice, maize grits: oil contents of 1.5%–2.0%) have an advantage in keeping quality over products made from oats (oil contents of 4%–11%, average of 7%). Whole maize has high oil content too (4.4%), but most of the oil is contained in the germ, which has been removed in making grits.

The keeping quality of the oil depends on its degree of unsaturation, the presence or absence of antioxidants and prooxidants, the time and temperature of the heat treatment, the moisture content of the material when treated and the conditions of storage.

Severe heat treatment, as in toasting or puffing, may destroy antioxidants or induce formation of prooxidants, with the stability of the oil being progressively reduced as treatment temperature is raised, treatment time lengthened or moisture content of the material at the time of treatment lowered. On the other hand, momentary HTST, as at the surface of a hot roll in the roller-drying of a batter, may produce new antioxidants by interactions of proteins and sugars (nonenzymatic browning, or Maillard reactions); such a reaction is known to occur, for example, in the roller-drying of milk, and may be an explanation for the improved antioxidant activity of oat products made after steam treatment for lipase inactivation–stabilization. Similarly, enzyme-inactivated, stabilized wheat bran has a long shelf life; this material can be used for breakfast cereals, snack foods and other extruded products that need a high-fibre content and extended shelf life (Cooper, 1988).

The addition of synthetic antioxidants, such as BHA or BHT, to prepared breakfast cereals, or to packaging materials as an impregnation, as practiced in the United States, is not at present permitted in the United Kingdom. So, the use of oxygen absorbers/scavengers to restrict the onset of oxidative rancidity has gained widespread use in a variety of food products (Juliano, 1985; Cruz et al., 2012).

Another form of deterioration of breakfast cereals after processing and packaging is moisture uptake, which causes loss of the distinctive crisp texture. Moisture uptake is prevented by the use of the correct type and quality of moisture vapour-proof packaging materials as well as complete and proper sealing of the packaging (Fast, 1987).

9.12 NUTRITIVE VALUE OF BREAKFAST CEREALS

The nutritive value of breakfast cereals, as compared with that of the raw materials from which they were made, depends very much on the processing treatments involved, remembering that all heat treatment processes cause some modification or loss of nutrients. Thus, while extrusion cooking may cause the loss of essential amino acids (depending upon the length of time that the cereal received the heat treatment), it also inactivates protease inhibitors, thereby increasing the nutritional value of the proteins.

The chemical composition of some RTE breakfast cereals is shown in Table 9.1. Shredded wheat, made from low protein, soft wheat has a protein content considerably lower than that of puffed wheat, which is made from a high-protein hard wheat, such as durum or CWRS wheat.

9.12.1 Proteins and amino acids

All cereal products are deficient in the amino acid lysine, but the deficiency may be relatively greater in RTE cereals than in bread because of the changes that occur in the protein at the high-temperature treatment. The protein efficiencies of wheat-based breakfast cereals relative to casein (defined as 100), as determined by rat-growth trials, have been reported as: −15.3 for extrusion puffed; 1.8–16.3 for flaked-toasted; 2.8 for extrusion toasted; and 69.9 for extruded, lightly roasted. The differences were partly explained by the loss of available lysine as Maillard reaction products (McAuley et al., 1987). However, lysine deficiency is of less importance

TABLE 9.1   Chemical composition of ready-to-eat breakfast foods (per 100 g as sold)

Food	Water (g)	Energy (kJ)	Carbohydrates		Protein (g)
			Starch (g)	Sugars (g)	
Grape Nuts	3.5	1475	68.7	12.1	10.5
KELLOGG'S					
Corn Flakes	3	1650	76	8	8[a]
Frosties	3	1650	49	40	5[a]
Rice Krispies	3	1650	76	10	6[b]
Ricicles	3	1650	52	38	4[b]
Coco Pops	3	1600	48	39	5[b]
All-Bran	3	1150	30	15	15[c]
Bran Buds	3	1250	26	26	13[c]
Bran Flakes	3	1350	46	19	10[d]
Sultana Bran	7	1350	37	30	8[d]
Smacks	3	1650	36	48	8[d]
Fruit 'n Fibre	6	1500	43	26	9[d]
Toppas	5	1500	55	19	9[d]
Special K	3	1600	60	16	15[a,e]
Country Store	8	1500	45	24	9[a]
Start	3	1550	49	28	8[a]
NABISCO					
Shredded Wheat	9	1525	75.3	12	10.3[d]
QUAKER OATS					
Puffed Wheat	2	1360	68.8[f]	1.3[g]	13.1[d]
Oat Crunchies	2.5	1600	71.1[f]	15.9[g]	10.5[d]
Sugar Puffs	2	1554	88	51	6.0
RYVITA					
Corn Flakes		1565	71.8		8.3[a]
Morning Bran		1180	48.1	9.1	14.0[d]
WEETABIX					
Weetabix	5.6	1427	60.6	5.2	11.2[a]
Bran Fare	4.0	962	31.5	nil	17.1[a]

TABLE 9.1 Chemical composition of ready-to-eat breakfast foods (per 100 g as sold)—cont'd

| Food | Water (g) | Energy (kJ) | Carbohydrates | | Protein (g) |
			Starch (g)	Sugars (g)	
Toasted farmhouse Bran	3.0	1293	36.7	10.2	12.6[a]
Weetaflake	3.7	1490	51.7	18.9	9.7[a]

Food	Fat (g)	Ash (g)	Dietary fibre (g)	Source of data
Grape Nuts	0.5		6.2[h]	i
KELLOGG'S				
Corn Flakes	0.6	3	1	j
Frosties	0.4	2	0.6	j
Rice Krispies	0.9	3	0.7	j
Ricicles	0.5	2	0.4	j
Coco Pops	1	2	1	j
All-Bran	3	7	24	j
Bran Buds	3	5	22	j
Bran Flakes	2	5	13	j
Sultana Bran	2	4	10	j
Smacks	2	0.5	3	j
Fruit 'n Fibre	5	3	7	j
Toppas	2	1	8	j
Special K	1	3	2	j
Country Store	4	3	6	j
Start	2	2	6	j
NABISCO				
Shredded Wheat	2.3	1.5	11.4	k
QUAKER OATS				
Puffed Wheat	1.25	1.6	7.7[l]	m
Oat Crunchies	7.3	4.5	n/a	m
Sugar Puffs	1.2	0.8	3.5	m
RYVITA				
Corn Flakes	1.3	n/a	9.3	n
Morning Bran	3.8	n/a	18.5	n

Continued

TABLE 9.1   Chemical composition of ready-to-eat breakfast foods (per 100 g as sold)—cont'd

Food	Fat (g)	Ash (g)	Dietary fibre (g)	Source of data
WEETABIX				
Weetabix	2.7	2.2	12.9	o
Bran Fare	4.8	5.3	37.5	o
Roasted farmhouse Bran	3.0	4.8	20.0	o
Weetaflake	3.1	n/a	10.5	o

[a] $N \times 6.25.$
[b] $N \times 5.95.$
[c] $N \times 6.31.$
[d] $N \times 5.7.$
[e] Enriched to this level.
[f] As available monosaccharides.
[g] Total sugars as sucrose.
[h] Southgate method.
[i] McCance & Widdowson's Composition of Foods (fourth ed.; 3rd Suppl.) 1988. Reproduced with the permission of the Royal Society of Chemistry and the Controller of HMSO.
[j] Data courtesy of Kellogg Co. of Great Britain Ltd. (1990).
[k] Data courtesy of NABISCO Ltd. (1982).
[l] Nonstarch polysaccharides, soluble plus insoluble.
[m] Data courtesy of Quaker Oats Ltd. (1990).
[n] Data courtesy of Ryvita Co. Ltd. (1990).
[o] Data courtesy of Weetabix Ltd. (1990).

in RTE cereals than in bread because the former are generally consumed with milk, which is a good source of lysine. Moreover, some RTE breakfast cereals have protein supplementation.

9.12.2 Carbohydrates

The principal carbohydrate in cereal grains is starch, the complete gelatinization of which is desirable in processed foods, such as RTE cereals. Whereas ordinary cooking at atmospheric pressure requires the starch to have a moisture content of 35%–40% to achieve complete gelatinization, the same can occur at feed moisture levels less than 20% in extrusion cooking at 110–135°C, if the processing conditions are correct (Asp and Björck, 1989; Linko, 1989a). Extrusion cooking increases the depolymerization of both amylose and amylopectin by random chain splitting. The susceptibility of starch to the action of alpha-amylase has been shown to increase in the following sequence: steam cooking (least); steam flaking; popping; extrusion cooking and drum drying (most) (Asp and Björck, 1989).

9.12.3 Calorific value

The calorific value (energy content) of most RTE cereals, as eaten, is higher than that of bread (975 kJ/100 g; 233 Cal/100 g), largely on account of the relatively lower moisture content of the former. But, compared at equal moisture contents, the differences in calorific value are actually quite small. Additionally, fat and cholesterol contents may be lower than those of some other cereal foods.

The processes involved in the manufacture of RTE cereals can cause partial hydrolysis of phytic acid; the degree of destruction increases at high pressures – about 70% is destroyed in puffing, but only about 33% is lost during flaking.

9.12.4 Enzymes

Enzymes, which are proteins, are generally inactivated either partially or completely during extrusion cooking. It has been shown that peroxidase can be completely inactivated by extrusion cooking at 110–149°C with 20%–35% moisture content, although there can be some residual activity if cooked at lower moisture content. Under relatively mild extrusion cooking conditions some activity of alpha-amylase, lipase and protease can be retained, which can therefore influence product keeping quality and shelf life (Linko, 1989a,b). Conversely, wheat flour with high alpha-amylase activity that would be unsuitable for conventional bread making can be processed by extrusion cooking in which, provided the conditions are correctly chosen, the enzyme is quickly and totally inactivated, permitting the production, from such flour, of RTE breakfast cereals, flatbreads, snacks, biscuits etc. (Cheftel, 1989).

9.12.5 Minerals and vitamins

The content of some of the minerals and vitamins in RTE breakfast cereals is shown in Table 9.2. Unfortunately, information about other vitamins is meagre. About 50% of the thiamine (vitamin B_1) is destroyed during the manufacture of shredded wheat and in extrusion cooking, while nearly 100% is destroyed during puffing and flaking. These processes have little effect on riboflavin (vitamin B_2), niacin (vitamin B_3), pyridoxine (vitamin B_6) or folic acid. Extrusion cooking has been shown to cause a loss of 11%–21% of vitamin E, while in products enriched with wheat germ, extrusion cooking can cause losses of 50%–66% of vitamin E (Asp and Björck, 1989). Most of the RTE breakfast cereals manufactured in the United Kingdom are enriched with vitamins, as shown in Table 9.2; some are enriched with iron, and some with protein (viz. with the high-protein fractions of wheat and oat flour, defatted

TABLE 9.2   Mineral and vitamin content of ready-to-eat breakfast foods (per 100 g as sold)

Food	Na (g)	K (mg)	Ca (mg)	Fe (mg)	I (mg)	Zn (mg)	Mg (mg)
Grape Nuts	0.59	310	37	9.5	250	4.2	95
KELLOGG'S							
Corn Flakes	1	100	10	6.7	50		
Frosties	0.8	50	10	6.7	50		
Rice Krispies	1	150	20	6.7	150		
Ricicles	0.8	100	10	6.7	100		
Coco Pops	0.8	200	20	6.7	100		
All-Bran	1	950	70	12	700		
Bran Buds	0.5	900	60	12	650		
Bran Flakes	1	550	40	20	400		
Sultana Bran	0.7	650	50	15	350		
Smacks	0.01	200	20	6.7	150		
Fruit 'n Fibre	0.7	450	40	6.7	200		
Toppas	0.01	400	40	6.7	250		
Special K	1	250	70	13.3	150		
Country Store	0.6	500	120	6.0	300		
Start[a]	0.5	300	40	15.0	200	18.7[b]	
NABISCO							
Shredded Wheat	0.01	376	49	3.66	328	4.2	

QUAKER OATS

Food	Thiamine (mg)	Niacin (mg)	Riboflavin (mg)	Pyridoxin (mg)	Folic acid (mg)	D (μg)	B12 (μg)	C (μg)	Source of data
Puffed Wheat			26[c]	4.6[c]	350[c]				
Oat Crunchies			50.5	3.6			3.0		

RYVITA

Food	Thiamine (mg)	Niacin (mg)	Riboflavin (mg)	Pyridoxin (mg)	Folic acid (mg)	D (μg)	B12 (μg)	C (μg)	Source of data
Corn Flakes	0.65			12					
Morning Bran				12					

WEETABIX

Food	Thiamine (mg)	Niacin (mg)	Riboflavin (mg)	Pyridoxin (mg)	Folic acid (mg)	D (μg)	B12 (μg)	C (μg)	Source of data
Weetabix	0.375	375	35	7	200		2	120	
Bran Fare	0.16	1100	105	14	1100		7	140	
Roasted farmhouse Bran	0.87	725	70	44	620		4	180	
Weetaflake	0.375	375	35	7	290		2	120	

					Vitamins				
Food	Thiamine (mg)	Niacin (mg)	Riboflavin (mg)	Pyridoxin (mg)	Folic acid (mg)	D (μg)	B12 (μg)	C (μg)	Source of data
Grape Nuts	1.3	17.6	1.5	1.8	350		5		d

KELLOGG'S

Food	Thiamine (mg)	Niacin (mg)	Riboflavin (mg)	Pyridoxin (mg)	Folic acid (mg)	D (μg)	B12 (μg)	C (μg)	Source of data
Corn Flakes	1.0[b]	16[b]	1.5[b]	1.8[b]	250[b]	2.8[b]	1.7[b]		e
Frosties	1.0[b]	16[b]	1.5[b]	1.8[b]	250[b]	2.8[b]	1.7[b]		e
Rice Krispies	1.0[b]	16[b]	1.5[b]	1.8[b]	250[b]	2.8[b]	1.7[b]		e
Ricicles	1.0[b]	16[b]	1.5[b]	1.8[b]	250[b]	2.8[b]	1.7[b]		e
Coco Pops	1.0[b]	16[b]	1.5[b]	1.8[b]	250[b]	2.8[b]	1.7[b]		e

Continued

TABLE 9.2 Mineral and vitamin content of ready-to-eat breakfast foods (per 100 g as sold)—cont'd

Food	Thiamine (mg)	Niacin (mg)	Riboflavin (mg)	Pyridoxin (mg)	Folic acid (mg)	D (µg)	B$_{12}$ (µg)	C (µg)	Source of data
					Vitamins				
All-Bran	1.0[b]	16[b]	1.5[b]	1.8[b]	250[b]	2.8[b]	1.7[b]		e
Bran Buds	1.0[b]	16[b]	1.5[b]	1.8[b]	250[b]	2.8[b]	1.7[b]		e
Bran Flakes	1.0[b]	16[b]	1.5[b]	1.8[b]	250[b]	2.8[b]	1.7[b]		e
Sultana Bran	1.0[b]	16[b]	1.5[b]	1.8[b]	250[b]	2.8[b]	1.7[b]		e
Smacks	1.0[b]	15[b]	1.7[b]	1.8[b]	250[b]	2.1[b]	1.7[b]		e
Fruit 'n Fibre	1.0[b]	16[b]	1.5[b]	1.8[b]	250[b]	2.8[b]	1.7[b]		e
Toppas	1.0[b]	16[b]	1.5[b]	1.8[b]	250[b]	2.8[b]	1.7[b]		e
Special K	1.2[b]	18[b]	1.7[b]	2.2[b]	300[b]	2.8[b]	2.2[b]		e
Country Store	0.7	8	0.9						e
Start[a]	1.5[b]	24[b]	2.2[b]	2.7[b]	400[b]	4.2[b]	2.5[b]		e
NABISCO									
Shredded Wheat	0.3	4.5	0.05						f

Food	Thiamine (mg)	Niacin (mg)	Riboflavin (mg)	Pyridoxin (mg)	Folic acid (mg)	D (µg)	B_{12} (µg)	C (µg)	Source of data
QUAKER OATS									
Puffed Wheat	0.3	5	0.13	0.14					g
Oat Crunchies	0.4	4.3	0.1	0.15	57				g
RYVITA									
Corn Flakes	1.2[b]	18.0[b]	1.6[b]		300[b]	2.5[b]	2.0[b]		h
Morning Bran	1.2[b]	18.0[b]	1.6[b]		300[b]	2.5[b]	2.0[b]		h
WEETABIX									
Weetabix	0.7[b]	10.0[b]	1.0[b]	0.2				5.0	i
Bran Fare	0.7	21.0[b]	0.3	1.2					i
Roasted farmhouse Bran	1.6[b]	19.0[b]	1.8[b]						i
Weetaflake	0.7[b]	10.0[b]	1.0[b]	0.2				5.0	i

a Also contains vitamin B_1, 18.7 mg/100 g.
b Enriched to this level.
c Based upon data from Paul & Southgate.
d McCance & Widdowson's Composition of Foods (fourth ed.; 3rd Suppl.) 1988. Reproduced with the permission of the Royal Society of Chemistry and the Controller of HMSO.
e Data courtesy of Kellogg Co. of Great Britain Ltd. (1990).
f Data courtesy of NABISCO Ltd. (1982).
g Data courtesy of Quaker Oats Ltd. (1990).
h Data courtesy of Ryvita Co. Ltd. (1990).
i Data courtesy of Weetabix Ltd. (1990).

wheat germ, soya flour, nonfat dry milk, casein or vital wheat gluten) and with vitamins B_6 (pyridoxine), D_3, C and E. Some RTE breakfast cereals made in the United States are also enriched with vitamins A, B_{12} and others.

Incorporation of vitamin supplements may be accomplished in various ways: at the cooking stage, at extrusion, by surface spraying after processing or by incorporation in a sugar coating, the method chosen depending on the relative stability of the individual vitamins. Incorporation of the protein supplement may similarly be made at a number of points that are chosen to avoid subjecting the proteins to excessive heat treatment – by incorporation as a dry supplement at the extrusion stage or by coating the product with a batter of wheat gluten.

9.12.6 Oat bran

Over the last few decades a boost has been given to the use of oat bran in breakfast cereals following the discovery that this material may have hypocholesterolaemic effects in the human (de Groot et al., 1963), that is, it may lower the concentration of plasma cholesterol in the blood. As high blood cholesterol has been associated with the incidence of coronary heart disease – for each 1% fall in plasma cholesterol, coronary heart disease may fall by 2% (Nestel, 1990) – a dietary factor that may reduce it could have great societal impact. Many researchers have been investigating this aspect over the last several decades, and they continue to find evidence for this link. Several thorough reviews are available (Brown et al., 1999; Mälkki and Virtanen, 2001; Othman et al., 2014; Tang et al., 1998) and have found that, in general, at least 3 g of oat fibre per day may reduce low-density lipoprotein (LDL) levels by 5%–10%.

The content of soluble fibre is much higher in oat bran (~10.5%) than in wheat bran (~2.8%); this may be an important factor in the cholesterol-lowering activity of oat bran (which is not shown by wheat bran), and it has been suggested that a hemicellulose, β-D glucan in specific, which is the major constituent of the soluble fibre, may be the cholesterol-lowering agent, acting by increasing the faecal excretion of cholesterol (Illman and Topping, 1985; Oakenfull, 1988; Siebert, 1987).

Oat bran is obtained by milling oat flakes that have been made from stabilized oat kernels (groats), as described elsewhere in this book. It can be used as an ingredient in both hot and cold breakfast cereals. A method for making an RTE cereal from cooked oat bran is disclosed in US Patent No. 4,497,840. It can also be incorporated into bread; addition of 10%–15% of oat bran to white wheat flour yielded bread of satisfactory quality (Krishnan et al., 1987).

Rice bran, and in particular the oil in rice bran, has also been shown to have a plasma-cholesterol-lowering effect. Rice bran was not as effective as oat bran in lowering plasma total cholesterol, but rice bran favourably altered the ratio of high density lipoprotein to LDL, which may serve as an indicator for potential future coronary heart disease development (Nestel, 1990).

Preliminary work indicates that the β-glucan in the soluble fibre of a waxy, hull-less barley cultivar also may have hypocholesterolaemic effects, and the extracted β-glucans from barley have possible use as a fibre supplement in baked products (Klopfenstein and Hoseney, 1987; Newman et al., 1989). An example of RTE cereal made from oat bran is shown in Fig. 9.9.

9.13 CONSUMPTION OF BREAKFAST CEREALS

The consumption of RTE breakfast cereals in the United Kingdom has shown a steady growth from quite a small amount before World War II to about 4.2 kg/person/an. in 1972, increasing further to 5.0 kg/person/an. in 1978, and to 6.5 kg/person/an. in 1988. The consumption of oat products for hot cereals in the United Kingdom was 0.6 kg/person/an. in 1984, but increased to 0.9 kg/person/an. in 1988, possibly in response to the claim that oat bran has a blood-cholesterol-lowering effect.

The total tonnage of packaged breakfast cereals marketed in the United Kingdom in 1988 was 383,758 tonne, of which 38% was wheat based, 29% maize based, and the remainder based on other cereals or on a mixture of cereals (Business Monitor, 1989).

FIGURE 9.9 Oat bran is formulated with other ingredients, formed, baked and cut into specific shapes. Scale bars represent 1 mm distance.

In the United States in 1971 about 0.75 million tonnes of breakfast cereal were produced, of which about 35% was puffed, 35% flaked, 10% shredded, and about 20% hot cereal. Between 1980/81 and 1988/89 the total quantity of hot cereal (excluding corn grits) sold in the United States increased from 0.16 to 0.20 million tonnes, most of the increase being accounted for by oat-based products, the proportion of which increased from 71.6% in 1980/81 to 81.2% in 1988/89.

The average consumption of breakfast cereals (RTE plus hot) in the United States in 1971 was about 3.4 kg/person/an. Of the cereals used, wheat, bran or farina comprised about 37%, oatmeal or oat flour 30%, maize grits 22%, and rice 11%. However, by 1985, the consumption of RTE cereals alone had increased to 4.1 kg/person/an. (Anon., 1986), slightly lower than in the United Kingdom, with the consumption of maize-based RTE cereals in the United States increasing from 2.72 to 3.63 kg/person/an. between 1970 and 1980. Between 1974 and 1983 domestic consumption of RTE cereals in the United States grew by about 2% per annum, and growth increased further to 3.3% per annum between 1983 and 1985. In the late 1980s, an even larger growth rate of 4%–5% occurred, which was attributed to aggressive advertising campaigns (Fast, 1987).

As of 2012, the global breakfast cereal market was approximately $30 billion USD (about $10 billion from the US market alone). Although the United Kingdom, United States, Canada and Australia together accounted for 54% of all breakfast cereals manufactured globally, these markets have actually begun to stall in recent years, and have even declined somewhat as young people opt for other breakfast choices (due to convenience, time available, perceived health consequences and competition with other foods). For example, in the United States in 2001 per capita breakfast cereal consumption had risen to 5.1 kg, in 2010 that had fallen to 4.5 kg, and in 2016 it was 4.2 kg (Agri-Food Canada, 2012). The largest growth markets are no longer the United Kingdom or the United States, but instead are China, India and Brazil, although local customs and competition influence the types of products sold in the specific market locations and ultimate consumer acceptance (Schultz, 2012). Therefore, cereal companies continually strive to innovate and adapt their traditional products to meet specific expectations in various countries. Moreover, they also strive to develop new products to intrigue consumers in mature markets (i.e., addition of freeze-dried berries, using new types of extruder die inserts to produce novel RTE cereal shapes, using ancient grains, developing gluten-free products, and using novel flavour combinations, clusters (often oat), and various nuts to supplement flakes, to name a few recent innovations). Fig. 9.10 illustrates a few examples. Additionally, in order to spur interest in breakfast cereals amongst the public, Kellogg's has recently opened a cereal bar and café in New York City (www.cereality.com).

FIGURE 9.10 Consumers are now presented with many options, including combinations of fruits, nuts, clusters, textures and flavours. Healthy choices also include extruded bran. Scale bars represent 1 mm distance.

References

Agri-Food Canada, 2012. Breakfast Cereals - International Markets Bureau American Eating Trends Report. International Markets Bureau – Agriculture and Agri-Food, Ottawa, ON, Canada. Available online: http://www.agr.gc.ca/resources/prod/Internet-Internet/MISB-DGSIM/ATS-SEA/PDF/6238-eng.pdf.

Anon., October 16, 1970. Cereal specifications. Milling.

Anon., 1986. Ready-to-Eat Cereal Industry Report. Investment Report No. 609910. Kidder, Peabody & Co. Inc, Boston, USA.

Asp, N.-G., Björck, I., 1989. Nutritional properties of extruded foods. In: Mercier, C., Linko, P., Harper, J.M. (Eds.), Extrusion Cooking. Amer. Assoc. of Cereal Chemists Inc., St. Paul, MN, USA.

British Patent Specification No. 754,771 (sugar coating).

Brown, L., Rosner, B., Willett, W.W., Sacks, F.M., 1999. Cholesterol-lowering effects of dietary fiber: a meta-analysis. American Journal of Clinical Nutrition 69 (1), 30–42.

Business Monitor, 1989. PAS 4239. Miscellaneous Foods. Business Statistics Office.

Cheftel, J.C., 1989. Extrusion cooking and food safety. In: Mercier, C., Linko, P., Harper, J.M. (Eds.), Extrusion Cooking. Amer. Assoc. of Cereal Chemists Inc., St. Paul, MN, USA.

Cooper, H., April 1988. Milling moves. Food Processing 41–42.

Cruz, R.S., Camilloto, G.P., Dos Santos Pires, A.C., 2012. Chapter 2. Oxygen scavengers: an approach on food preservation. In: Eissa, A.A. (Ed.), Structure and Functioning of Food Engineering. InTech, Rijeka, Croatia.

de Groot, A.P., Luyken, R., Pikaar, N.A., 1963. Cholesterol lowering effect of rolled oats. Lancet 2, 303.

Fast, R.B., 1987. Breakfast cereals: processed grains for human consumption. Cereal Foods World 32, 241.

Fast, R.B., Caldwell, E.F. (Eds.), 1990. Breakfast Cereals and How They Are Made. Amer. Assoc. of Cereal Chemists Inc., St. Paul, MN, USA.

Fast, R.B., Lauhoff, G.H., Taylor, D.D., Getgood, S.J., 1990. Flaking ready-to-eat breakfast cereals. Cereal Foods World 35, 295.

Holland, B., Unwin, I.D., Buss, D.H., 1988. Cereals and Cereal Products, 3rd Suppl. to McCance & Widdowson's The Composition of Foods, fourth ed. (R. Soc. Chem. & Min. Agric. Fish. Food).

Hoseney, R.C., 1986. Chapter 13: Breakfast cereals. In: Principles of Cereal Science and Technology. Amer. Assoc. of Cereal Chemists Inc., St. Paul, MN, USA.

Illman, R.J., Topping, D.L., 1985. Effects of dietary oat bran on faecal steroid excretion, plasma volatile fatty acids and lipid synthesis in rats. Nutrition Research 5, 839.

Juliano, B.O. (Ed.), 1985. Rice: Chemistry and Technology, second ed. Amer. Assoc. of Cereal Chemists Inc., St. Paul, MN, USA.

Juliano, B.O., Sakurai, J., 1985. Miscellaneous rice products. In: Juliano, B.O. (Ed.), Rice: Chemistry and Technology, second ed. Amer. Assoc. of Cereal Chemists Inc., St. Paul, MN, USA.

Klopfenstein, C.F., Hoseney, R.C., 1987. Cholesterol-lowering effect of ß-glucans enriched bread. Nutrition Reports International 36, 1091.

Krishnan, P.G., Chang, K.C., Brown, G., 1987. Effect of commercial oat bran on the characteristics and composition of bread. Cereal Chemistry 64, 55.

Linko, P., 1989a. The twin-screw extrusion cooker as a versatile tool for wheat processing (Chapter 22). In: Pomeranz, Y. (Ed.), Wheat Is Unique. Amer. Assoc. of Cereal Chemists Inc., St. Paul, MN, USA.

Linko, P., 1989b. Extrusion cooking in bioconversions (Chapter 8). In: Mercier, C., Linko, P., Harper, J.R. (Eds.), Extrusion Cooking. Amer. Assoc. of Cereal Chemists Inc., St. Paul, MN, USA.

Mälkki, Y., Virtanen, E., 2001. Gastrointestinal effects of oat bran and oat gum: a review. LWT - Food Science and Technology 34 (6), 337–347.

McAuley, J.A., Hoover, J.L.B., Kunkel, M.E., Acton, J.C., 1987. Relative protein efficiency ratios for wheat-based breakfast cereals. Journal of Food Science 52, 1111.

Midden, T.M., 1989. Twin screw extrusion of corn flakes. Cereal Foods World 34, 941.

Nestel, P.J., 1990. Oat bran, rice bran. Food Australia 42, 342.

Newman, R.K., Newman, C.W., Graham, H., 1989. The hypocholesterolaemic function of barley beta-glucans. Cereal Foods World 34, 883–886.

Oakenfull, D., 1988. Oat bran. Does oat bran lower plasma cholesterol and, if so, how? CSIRO Food Research Quarterly 48, 37–39.

Othman, R.A., Moghadasian, M.H., Jones, P.J., 2014. Cholesterol-lowering effects of oat β-glucan. Nutrition Reviews 69 (6), 299–309.

Rooney, L.W., Serna-Saldivar, S.O., 1987. Corn-based ready-to-eat breakfast cereals. In: Watson, S.A., Ramstad, P.E. (Eds.), Corn: Chemistry and Technology. Amer. Assoc. of Cereal Chemists Inc., St. Paul, MN, USA.

Schultz, E.J., August 12, 2012. Cereal marketers race for global bowl domination. Advertising Age 83.

Siebert, S.E., 1987. Oat bran as a source of soluble dietary fibre. Cereal Foods World 32, 552–553.

Tang, J.L., Armitage, J.M., Lancaster, T., Silagy, C.A., Fowler, G.H., Neil, H.A.W., 1998. Systematic review of dietary intervention trials to lower blood total cholesterol in free-living subjects. BMJ 316, 1213.

US Patent Specification Nos. 3,971,303 (continuous puffing); 4,497,840 (cooked oat bran); 4,501,759 (aspartame sweetener); 4,540,587 (aspartame sweetener); 4,594,252 (dipeptide sweetener); 4,608,263 (dipeptide sweetener); 4,614,657 (dipeptide sweetener) 4,874,624 (reconstitutable oat flakes).

Further reading

Brockington, S.F., Kelly, V.J., 1972. Rice breakfast cereals and infant foods. In: Houston, D.F. (Ed.), Rice: Chemistry and Technology, first ed. Amer. Assoc. of Cereal Chemists Inc., St. Paul, MN, USA, pp. 400–418.

Guy, R.C.E., 1986. Extrusion cooking versus conventional baking. In: Chemistry and Physics of Baking, pp. 227–235 Spec. Publ. 56, R. Soc. Chem., London, UK.

Guy, R.C.E., 1989. The use of wheat flours in extrusion cooking (Chapter 21). In: Pomeranz, Y. (Ed.), Wheat Is Unique. Amer. Assoc. of Cereal Chemists Inc., St. Paul, MN, USA.

Johnson, I.T., Lund, E., 1990. Soluble fibre. Nutrition & Food Science 123, 7–9.

Luh, B.S., Bhumiratana, A., 1980. Breakfast rice cereals and baby foods. In: Luh, B.S. (Ed.), Rice: Production and Utilization. Avi Publ. Co. Inc., Westport, CT, USA, pp. 622–649.

Miller, R.C., 1988. Continuous cooking of breakfast cereals. Cereal Foods World 33, 284–291.

Morton, I.D. (Ed.), 1987. Cereals in a European Context. Ellis Horwood, Chichester, UK.

Pomeranz, Y., 1987. Extrusion products (Chapter 20). In: Pomeranz, Y. (Ed.), Modern Cereal Science and Technology. VCH Publishers Inc., New York, NY, USA.

Pomeranz, Y. (Ed.), 1989. Wheat Is Unique. Amer. Assoc. of Cereal Chemists Inc., St. Paul, MN, USA.

Watson, S.A., Ramstad, P.E. (Eds.), 1987. Corn: Chemistry and Technology. Amer. Assoc. of Cereal Chemists Inc., St. Paul, MN, USA.

Extrusion processing of pasta and other products

10.1 INTRODUCTION

Extrusion cooking, also known as extrusion, is a high-temperature, short-time process in which material is transported, mixed and plasticized at relatively high temperature, pressure and shear, and then formed upon exiting to atmospheric temperature and pressure through a die orifice. Because the process is able to bring about gelatinization, solubilization, and complex formation of starches, polymerization and unfolding of proteins, partial or complete inactivation of enzymes (according to the severity of the operating conditions), reduction of microbial load, and production of particular forms of texture, extrusion can be used to make many food and industrial products, including pastas, pet foods, aquatic feeds, texturized vegetable proteins, flatbreads, snacks, croutons, soup bases, drink bases, biscuits, confectionery products, breadings, and others. Extruders are also used in the plastics and metal processing industries. Besides these final products, extrusion cooking can also be used to make intermediate products for further processing, both for food and for nonfood use (Fichtali and van de Voort, 1989; Linko, 1989a).

10.2 PRINCIPLES OF EXTRUSION COOKING

An extrusion cooker (commonly known as an extruder) is a continuous processing unit based on a screw system (sometimes sophisticated, sometimes relatively simple) rotating within the confines of a barrel (Fig. 10.1).

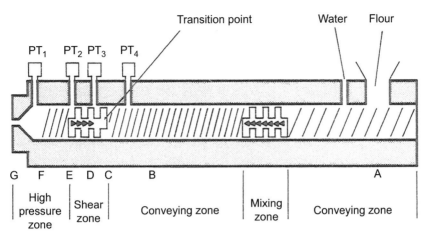

FIGURE 10.1 General schematic diagram of an extrusion cooker (extruder), showing typical components and zones. Flow through the extruder is right to left. *P* indicates a pressure sensor; *T* indicates a temperature sensor; *A* through *G* indicate various zones in the extruder barrel. *Reproduced from Guy, R.C.E., Horne, A.W., February 1989. The effects of endosperm texture on the performance of wheat flours in extrusion cooking processes. Milling 182, ix–xii, by courtesy of the A.A.C.C.*

Milled raw materials are introduced via a feed hopper, transported into a cooking zone where they are compressed and sheared at elevated temperatures and pressures to undergo a melt transition, which forms a viscous fluid. The extruder develops the fluid by heating and shearing the biopolymers, particularly the starch (Guy and Horne, 1988; Guy, 1991), and inherent water contained within the raw mix (or water is added directly into the extruder), and then shapes that fluid by pumping it through small die openings at the end of the barrel. Depending upon the temperatures achieved, proteins can be denatured as well; sometimes this is desired, other times it is not.

Extrusion equipment may consist of single- (Fig. 10.2) or twin-screws (Fig. 10.3) with spirally arranged flights for conveying, and special kneading and reversing elements for creating high-pressure shearing and kneading zones. In order to achieve the high temperatures necessary for the melt transition, the raw materials require large heat inputs. These are achieved by the dissipation of mechanical energy from the screw caused by frictional and viscous effects, by the injection of steam into the cereal mass during preconditioning or during extrusion, and sometimes by thermal conduction from heated sections of the barrel or screw, using heating systems based on electrical elements, steam or hot fluids. In these cases, the barrels are jacketed to accommodate these heating systems. In some extrusion systems, the barrel jackets can be used to either heat or cool, and thus can provide very narrow temperature control. If a jacketed system is used to heat the extruder, it has been estimated that approximately 50%

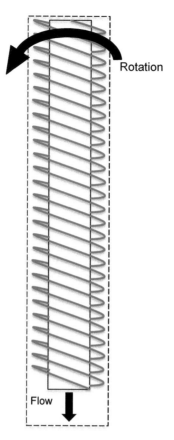

FIGURE 10.2   General schematic diagram of a single-screw extruder.

of the energy required to melt the particles into dough can arise from the jackets, whereas the other 50% of the energy will come from frictional dissipation (Figs. 10.4 and 10.5). If the extruder is not jacketed, then 100% of the energy required comes from frictional heating (Fig. 10.6).

In extruders with twin-screws, the screws may be corotating or counterrotating (Fig. 10.3). Further, there are many variations possible in screw design relating to physical dimensions, pitch, flight angles, etc. and, in the twin-screw, the extent to which the separate screws on each shaft intermesh (Fichtali and van de Voort, 1989). A key difference between single- and twin-screw extruders concerns the conveying characteristics of the screws.

Single-screw extruders were first used to manufacture ready-to-eat breakfast cereals in the 1960s, but they had problems with the transport of slippery or gummy materials because they rely on the drag flow principle

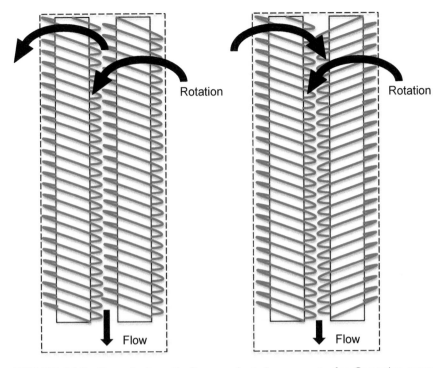

FIGURE 10.3 General schematic diagram of a twin-screw extruder. Corotating screw configuration shown on the left; counterrotating screw configuration shown on the right.

for conveying the materials within the barrel. The problems of slippage can be overcome, at least to some degree, by using grooves in the barrel walls (Hauck and Huber, 1989). The single-screw extruder has a continuous channel from the feed port to the die, and therefore its output is related to the die pressure and slippage. Sometimes the screw is designed to compress the raw materials by decreasing the flight height down the length of the barrel, thereby decreasing the volume available in the flights, which increases pressure. In some cases, the channel depth is uniform, but the screw pitch (distance between flights) decreases, which increases frictional energy dissipation. At relatively high screw speeds the screw mixes and heats the flour mass, and a melt transition is achieved, permitting the softened starch granules to be ruptured and gelatinized by the shearing action of the screw.

This transition usually occupies a fairly broad region along the screw, but the use of barrel heaters and steam injection, or use of preconditioning units, can help to induce sharper and earlier melt transitions and to increase

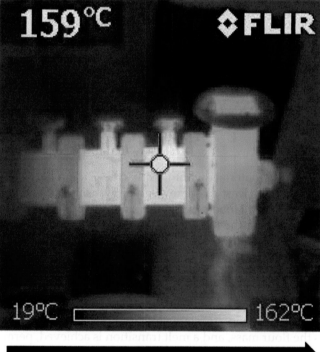

Direction of flow

FIGURE 10.4 Commercial-scale single-screw autogenous (i.e., frictional self-heating) extruder (top). Infrared imaging (bottom) indicates temperature profile down the barrel. Note how temperature rises as indicated in Figs. 10.7 or 10.8.

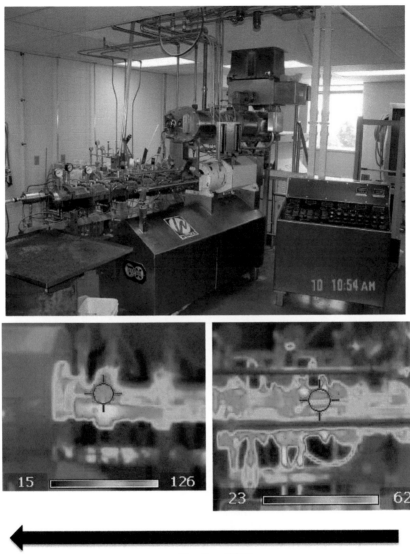

Direction of flow

FIGURE 10.5 Commercial-scale twin-screw extruder (top). Infrared imaging (bottom) indicates areas of heating. In this example, heating is accomplished via shear friction (due to screw segments used), hot water and steam injection. These extruders allow for greater temperature control than autogenous extruders, as shear effects can be countered or magnified.

extruder throughput (Harper, 1989). Considerable back-mixing may occur in the channel of the screw, giving a fairly broad residence-time distribution.

All twin-screw extruders have a positive pumping action and can convey many types of viscous materials with efficiency and narrow

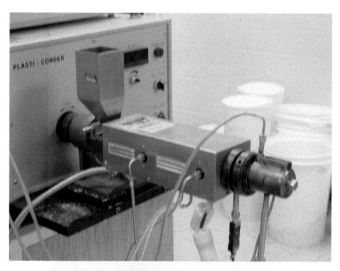

FIGURE 10.6 Laboratory-scale extruders are commonly used for product research and development. Before commercialization, however, concepts must be proven on pilot or commercial scales (such as shown in Figs. 10.4 and 10.5).

residence-time distributions. Specialized zones can be set up along the screw to improve the mixing, compression and shearing action of the screws. Corotating twin-screw extruders, which have self-wiping screws and higher operating speeds than counterrotating machines, are often the predominant choice of extruders for use in the food industry (Fichtali and van de Voort, 1989).

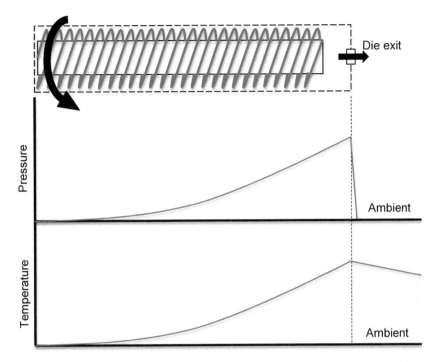

FIGURE 10.7   General pressure and temperature profiles down an extruder barrel for an autogenous extruder. Note that these increases are due only to frictional heating.

The physical changes to the raw materials occurring within the single- and twin-screw extruders are basically the same and have similar relationships to temperature, shearing forces and time. However, control of the process is simpler in the twin-screw machines because the output is not as affected by the physical nature of the melt phase being produced within the screw system, and the back-mixing can be more tightly controlled, giving better overall control and management of the process.

In general, if no external heat is provided by jackets (i.e., heating only occurs due to frictional dissipation), the temperature and pressure of the dough mass will continuously increase down the length of the barrel (Fig. 10.7). Upon exiting the die, the extrudate temperature begins slow cooling to ambient temperature, whereas pressure immediately falls to ambient conditions. Specialized screw segments (i.e., back-mixing, flight interruptions, double-flight segments, etc.) can be used, however, to increase heating in specific regions of the screw (Fig. 10.8). It is a common practice to minimize heating, and thus the transition, from occurring in the entrance of the extruder, as this could potentially result in backflow out of the hopper or potentially denature the protein to such an extent that the dough will plasticize and seize up (i.e., abruptly stop the screw rotation). Instead, it is preferred to have the transition occur past one-third or one-half way down the screw.

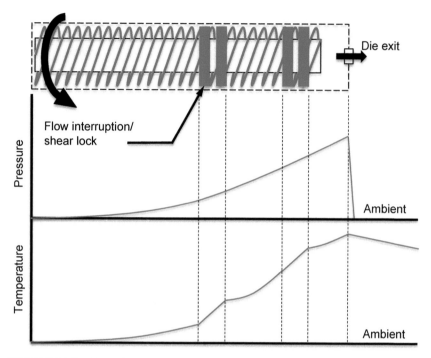

FIGURE 10.8   General pressure and temperature profiles down the extruder barrel when flight interruptions/steam locks are used. Note that use of these interruptions leads to large increases in shear stress (and thus temperature) at the locations where they are used.

Extruders are flexible machines and are used to produce a variety of food, feed and industrial products. One aspect, which allows flexibility, is the size and shape of dies that can be used. As shown in Fig. 10.9, there is an infinite variety of sizes and shapes that can be produced by using various die shapes and sizes. Sometimes extruders will use only one die insert; other times multiple die inserts will be installed on a die assembly or die plate. As the dough fluid flows through the die, maximum pressure is achieved inside the die. It is this pressure that forces the dough out of the machine. Upon exit, the extrudate mass may expand due to moisture changing state immediately from liquid to vapour (i.e., the ideal gas law determines how much potential expansion may be achieved) (Fig. 10.10). If the matrix contains a substantial portion of starch, and it is properly gelatinized, the extrudate will have enough structural integrity to expand (not explode). If the starch has not been properly gelatinized, structural integrity may be lost. Conversely, if a feed blend contains a substantial portion of protein vis-à-vis starch, expansion will likely not occur.

Extruders are one unit operation amongst many in a typical processing plant. As shown, in Fig. 10.11, prior to extrusion ingredients may be

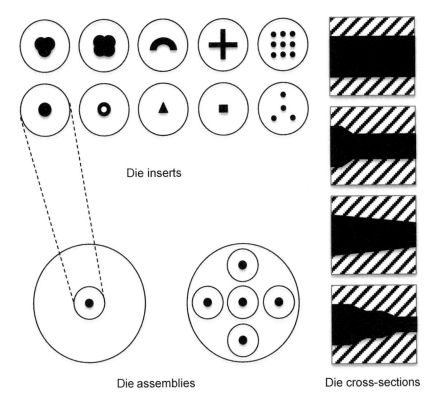

Die inserts

Die assemblies   Die cross-sections

FIGURE 10.9 Die inserts are available in a variety of shapes and sizes, depending upon the desired final product. Die inserts are then installed in die assemblies/die plates either as a single insert, or as an array of inserts.

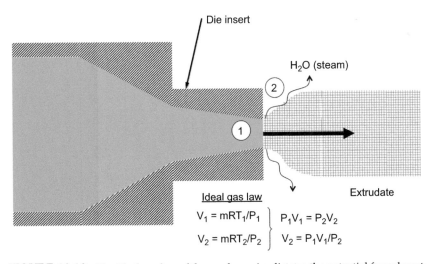

Die insert

H_2O (steam)

Ideal gas law

$$V_1 = mRT_1/P_1$$
$$V_2 = mRT_2/P_2$$

$$P_1V_1 = P_2V_2$$
$$V_2 = P_1V_1/P_2$$

Extrudate

FIGURE 10.10 The ideal gas law of thermodynamics dictates the potential for volumetric expansion of the extrudates upon die exit due to the sudden pressure drop, which results in water changing state from liquid to steam. The ability to achieve expansion depends upon ingredient composition, water content, as well as temperature and pressure achieved in the barrel and die. m, mass flow rate; P, pressure (absolute); R, gas constant; T, temperature (absolute); V, volume.

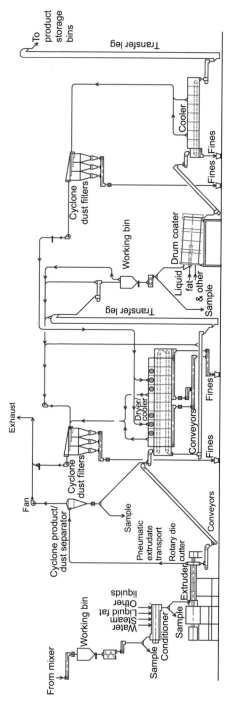

FIGURE 10.11 Extruders are one component of larger processing systems in food, feed, and industrial manufacturing plants, which include several upstream and downstream unit operations. An example of a pet food or aquaculture feed operation is shown here.

ground and mixed, then transported to a preconditioner (where water and steam may be added to moisten the particles and begin the gelatinization process). The material will be conveyed into the extruder, and upon exiting the die, the extrudates will be sent through a dryer and cooler (sometimes this will entail multiple passes), and dust and fines will be removed from the product stream. Extrudates may be coated with flavouring or fat, packaged, and then placed in warehouse storage.

10.3 PASTA

Pasta is the collective term used to describe products such as macaroni, spaghetti, vermicelli, noodles, etc., which are traditionally made from the semolina milled from hard durum wheat (*Triticum durum*) (cf. Chapter 6 for milling of semolina). Examples of long pastas are shown in Fig. 10.12; short products are shown in Figs. 10.13–10.15. Noodle products are shown

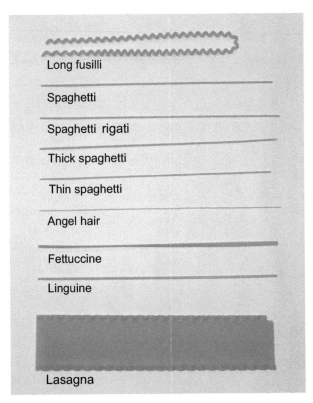

FIGURE 10.12 Examples of various long pasta products.

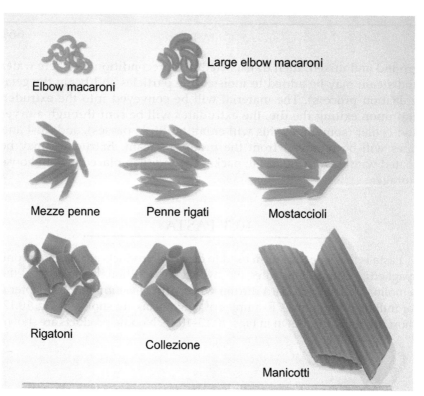

FIGURE 10.13 Examples of various short pasta products. Each scale gradation indicates 1 mm.

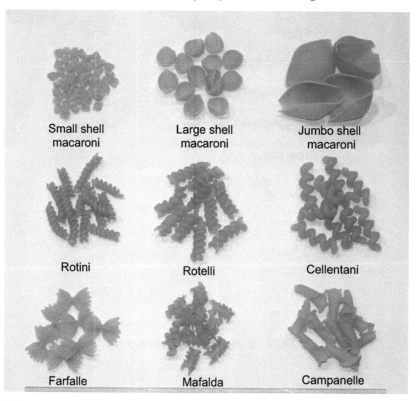

FIGURE 10.14 Examples of various short pasta products. Each scale gradation indicates 1 mm.

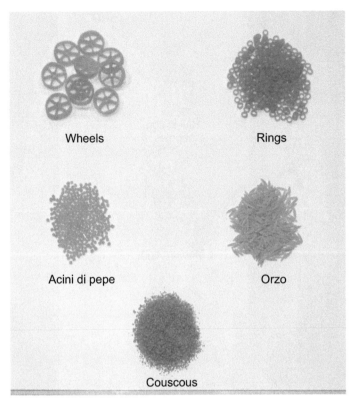

FIGURE 10.15   Examples of very small pasta products. Each scale gradation indicates 1 mm.

in Fig. 10.16. The highest quality pasta products are made from durum wheat alone, but other hard wheats, e.g., CWRS or HRS, can be substituted, but generally at the expense of quality. The Plate wheat Taganrog used for puffing (cf. Chapter 4) is not common for pasta, as it tends to yield a brownish-coloured product.

Durum wheat grown in Britain has been successfully used for pasta, although parcels with high alpha-amylase activity are unsuitable. UK-grown durum wheat is often blended in small quantities into grist in which imported durum wheat will predominate.

Historically, pasta products are believed to have been introduced into Italy from China in the 13th century, and were first produced in Europe in the 15th century in Germany (Pomeranz, 1987).

10.3.1 Traditional (kneading/sheeting) processing

In the traditional manufacturing process (in use at the beginning of the 20th century, long before modern extruders were in use), a dough is first made by mixing semolina and water to give a moisture content of about

Kluski egg
noodles

Medium egg
noodles

Extra wide egg
noodles

FIGURE 10.16 Examples of various noodle products. Each scale gradation indicates
1 mm.

30% in the dough. Depending on the pasta/noodle recipe, egg may also
be included, as it tends to improve particle binding and product texture.
This moisture content is needed to ensure that the viscosity of the dough is
low enough to avoid the generation of excessive pressure within the press.
The dough is kneaded and then forced (i.e., extruded) through either a
hydraulic press or a hand press to form a thin sheet that can then be cut
into strips, which are then carefully dried, generally at ambient conditions.
A temperature at or below 49°C is necessary to prevent cooking the product and avoid denaturing the proteins. A modification of this traditional
process, in which the dough is extruded through special dies to make
shaped products – rods, strips, tubes, etc. – was first introduced more
than 70 years ago (Pomeranz, 1987). Although most commercial pasta is
now manufactured via cooking extruders (discussed subsequently), hand
pressing is fashionable in many homes and restaurants.

10.3.2 Extrusion processing

In the extrusion cooking process, the moisture content of the dough could be about 28% (or even higher) and the extrusion temperature, i.e., the temperature of the extrusion barrel, could be around 54°C (or lower), which is low enough to avoid cooking the product. The extrusion barrel may have a jacket in which water can be circulated to maintain the required temperature, but not all companies use these for temperature control.

It is claimed that the kneading/sheeting process has the advantage of producing a springy, elastic texture in the product, leading to good eating quality, while advantages claimed for the extrusion process are that it produces a pasta with a firmer texture in a larger variety of shapes, manufacturing is more efficient and less time-consuming, and that the extruded dough has a lower moisture content, thereby requiring a shorter drying time to complete the preparation of the pasta. However, if not done correctly, pasta made by the extrusion process may lack the laminated structure and desirable eating qualities of pasta made by the kneading/sheeting process, and the colour of the pasta may be less appealing (US Patent No. 4,675,199).

10.3.3 Drying of pasta

Finally, the extruded product is dried to about 12.5% m.c. In some places in the world, drying might be done outdoors under ambient conditions using sunlight as a heat source, but in most industrialized processing operations, specialized drying equipment is used in which the temperature and relative humidity of the air are carefully controlled to produce high-quality products consistently. The rate of drying must be correct; drying at too low a rate can lead to the development of moulds, discoloration and souring, whilst drying too rapidly can cause cracking ('checking' in the United States) and curling. Until a few decades ago, drying temperatures were about 50°C, and the drying process took 14–24 h.

A three-stage drying process consisting of predrying, sweating, and drying is common (Antognelli, 1980). In the predrying stage, air at 55–90°C circulates around the product and dries it to 17%–18% m.c. in about 1 h. Moisture migrates from the centre to the periphery, where it evaporates, resulting in a marked moisture gradient in the product. In the sweating period that follows, the pasta rests to allow moisture equilibration between the inner core of the pasta and the surface to occur. In the drying stage, diminishing periods of hot-air circulation alternate with periods of sweating. Temperatures in the drying stage are 45–70°C, and the total drying time can range from 6 to 28 h.

In the late 1970s, the processing of macaroni was improved by drying at temperatures above 60°C, using shorter processing times and sterilizing the product during drying.

A rapid drying process using microwave energy has also been described (Katskee, 1978). A predrying stage uses hot air at 71–82°C to reduce the m.c. of the pasta to 17.5% in 30 min. In the microwave stage that follows, the product is fully dried to the target m.c. in 10–20 min. The microwave system operates at about 30 kW, an air temperature of 82°C, and 15%–20% r.h. The final, or cooling and equalization stage, operated at 70%–80% r.h., brings the total drying time up to 1.5 h. Advantages claimed for the microwave drying process, besides the considerable savings of time, can include equipment space reduction, improved product quality, improved cooking quality, colour enhancement, better microbiological stability, and lower installation and operating costs.

10.3.4 Processing evolutions

Throughout the last several decades, numerous processing changes have been introduced to the pasta-manufacturing industry, all of which have been directed towards achieving specific objectives, including the following:

- improvement in pasta quality, e.g., by reducing cooking loss, avoiding cracking, improving the colour of the product, eliminating dark specks, destruction of microorganisms, improving the appearance and firmness of the product (Milatovic, 1985; Manser, 1986; Abecassis et al., 1989a; European Patent Application No. 0,267,368; US Patent Nos. 4,539,214; 4,876,104);
- improvement in nutritional quality, e.g., by better retention of vitamins and other nutrients (Manser, 1986; European Patent Application No. 1,322,153);
- improvement in shelf life of the product (US Patent Nos. 4,828,852; 4,876,104);
- reduction in drying time, with consequent savings in energy expenditures (Abecassis et al., 1989b);
- faster-cooking products, e.g., a rapidly rehydratable pasta or a microwave-cookable product (European Patent Application Nos. 0,267,368; 0,272,502; 0,352,876; US Patent Nos. 4,539,214; 4,540,592);
- a ready-to-eat pasta, or an instant precooked pasta (US Patent Nos. 4,540,592; 4,828,852).

Another objective has been the diversification of the starting material, i.e., the replacement of some (or all) of the durum semolina with other cereal materials or even with noncereal materials (Molinar et al., 1975; Mestres et al., 1990).

These objectives have been addressed, and in most cases attained, by modification of the manufacturing processes in various ways, some of which include:

- modifying the moisture content of the product before drying (Abecassis et al., 1989a,b);
- modifying the temperature of the doughing water, the temperature in the cooking zone and the extrusion temperature (Milatovic, 1985; European Patent Application Nos. 0,267,368; 0,288,136);
- partially precooking the dough by preconditioning (European Patent Application No. 0,267,368);
- soaking partially cooked pasta in acidified water (US Patent No. 4,828,852);
- using higher drying temperatures (Manser, 1986; Abecassis et al., 1989a; European Patent Application Nos. 0,309,413; 0,322,053; 0,352,876);
- back-mixing the dough, combined with high drying temperatures (British Patent Specification No. 2,151,898; US Patent No. 4,540,592);
- venting the product between the cooking and the forming zones (European Patent Application No. 0,267,368);
- predrying and drying as two separate stages (Milatovic, 1985);
- using superheated steam for simultaneous cooking and drying (US Patent No. 4,539,214);
- high-temperature posttreatment to make precooked pasta (Abecassis et al., 1989a,b; Mestres et al., 1990; US Patent No. 4,830,866);
- pasteurizing (US Patent No. 4,876,104);
- extruding into N_2/CO_2 or into vacuum to improve storability (European Patent Application No. 0,146,510; US Patent No. 4,540,590);
- using extrusion cooking instead of ordinary extrusion (US Patent No. 4,540,592);
- using additives, e.g., emulsifiers, in the dough (European Patent Application Nos. 0,352,876; 0,288,136).

10.3.5 Degree of cooking during processing

Commercially available pasta products, as sold, can be described as 'uncooked', 'partially cooked', or 'fully cooked' (i.e., precooked), depending on the degree to which the protein is denatured and the starch gelatinized.

Traditionally, pasta products are manufactured and dried in an uncooked state because dry, uncooked pasta can be stored at room temperature for long periods (due to low water activity) while maintaining highly glutinous properties.

10.3.6 Quality of pasta

Particle size of the semolina is an important characteristic as it influences pasta quality: it is recommended that at least 90% of the milled particles should fall between 150 and 340 μm in size. Particles larger than 340 μm impede the activity of enzymes in the dough (Milatovic, 1985) and cause stress fractures during drying. Good pasta quality may be defined as the ability of the proteins to form an insoluble network capable of entrapping swollen and gelatinized starch granules (Feillet, 1984), and is thus related to the composition of the proteins. A strong correlation (+0.796) has been found between pasta cooking quality and acid-insoluble residue protein (Sgrulletta and De Stefanis, 1989); a highly significant correlation was also found between good cooking quality and the total sulphydryl plus disulphide (SH + SS) content of glutenins extracted by sodium tetradecanoate solution (Kobrehel and Alary, 1989; Alary and Kobrehel, 1987), and also with certain low-molecular-weight glutenin subunits, particularly the 45 band of the electrophoretic pattern. On the other hand, poor cooking quality was associated with the 42 band, and it is suggested that this association could be a useful indicator for the breeding of durum wheat for pasta-cooking qualities (du Cros and Hare, 1985; Autran et al., 1989).

In uncooked pasta, the protein is largely undenatured, and most of the starch is ungelatinized. For partially cooked and fully cooked pasta, the processing conditions of dough moisture content and extrusion temperature must be carefully controlled to prevent protein denaturation and starch gelatinization.

10.3.7 Cooking value of pasta

'Cooking value' is a measure of the texture or consistency of the pasta after cooking in the home or food service establishment. The cooked pasta should offer some resistance to chewing but should not stick to the teeth, and thus be served al dente (Figs. 10.17 and 10.18). To retain

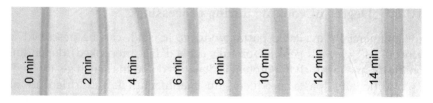

FIGURE 10.17 Pasta texture changes during cooking as the structure softens. Manufacturers recommend boiling for approximately 10 min to achieve *al dente* (i.e., firm to the bite).

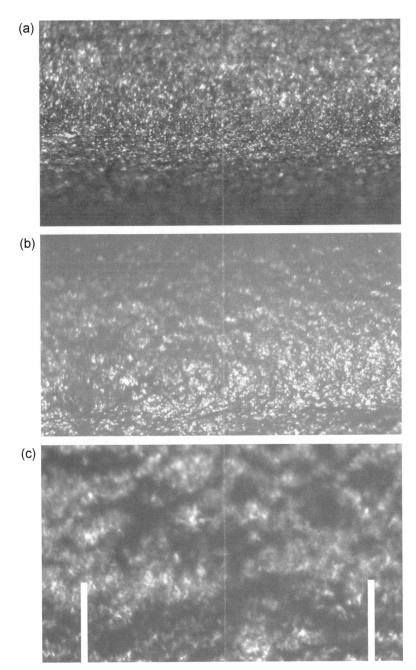

FIGURE 10.18 4× magnification of pasta surface after cooking indicates softening of the matrix. (a) Boiling for 0 min; (b) boiling for 6 min; and (c) boiling for 14 min. Scale bars indicate 1 mm distance.

these qualities and to avoid loss of nutrients, common procedures limit drying temperature to not above 60°C, but such drying takes a long time – 16–24 h. Higher drying temperatures, while still avoiding nutrient loss and adversely affecting cooking value, can be used in a stepwise process, e.g., by isothermal application of heat at temperatures between 40°C and 94°C, while maintaining the vapour pressure within the pasta (a_w) below 0.86.

In one such drying schedule, the product passes through eight areas in which the temperatures are (successively) 40, 50, 60, 70, 80, 84, 94 and 70°C, taking about 3 h to complete the entire drying cycle. The final product, with approximately 13% m.c., is then cooled to 25°C (European Patent Application No. 0,322,053).

10.3.8 Use of nonwheat materials

In some countries, e.g., France and Italy, the material used for making pasta must be durum semolina, but other countries allow the use of soft wheat flour, maize flour, or various other diluents. Countries that are nonwheat producers, e.g., many African countries, are developing pasta made from nonwheat materials, e.g., maize and sorghum. In recent years, gluten-free pasta has been gaining in popularity in many developed countries. Much of this is often made from a combination of maize and rice (white or brown), although other gluten-free grains and pseudocereals, such as quinoa, amaranth, and teff can be used.

Preparation of such pasta can be difficult because of the lack of gluten that is formed when wheat is the starting material and that contributes to dough development during mixing and extrusion, and thus prevents disaggregation of the pasta during cooking in boiling water (Feillet, 1984; Abecassis et al., 1989a). It has been suggested that the lack of gluten can be overcome by blending pregelatinized starch or corn flour before adding water and mixing, or by gelatinizing some of the starch during mixing or extruding (Molinar et al., 1975). A compromise is to use a blend of semolina and maize flour, from which good quality pasta can be obtained, provided the maize flour has fine granularity (less than 200 μm) and a low lipid content (not higher than 2% d.m.). By submitting the dried pasta to thermal treatment at 90°C for 90–180 min, a blend of 70% maize flour: 30% durum wheat semolina was found to be satisfactory (Mestres et al., 1990). Additionally, eggs, egg albumin, guar gum, and xanthan gum have been used to improve quality and particle binding, reducing leaching losses. In addition to the challenge of producing equivalent pasta quality to traditional pasta, using starting materials other than wheat often produce flavours and smells that are not acceptable to consumers – even maize or rice, which are quite mild, can be off-putting.

10.3.9 L-ascorbic acid as an additive

The use of L-ascorbic acid as an additive for improving the quality of pasta has been suggested (Milatovic, 1985). Ascorbic acid, which is oxidized to dehydroascorbic acid, a strong reducing agent, inhibits the destruction of naturally occurring pigments, leading to improvement in the colour of the product by inhibiting lipoxygenase activity. Addition of 300 mg/kg of L-ascorbic acid to the doughing water improved the colour of the pasta made from all soft wheat flour or from 50% soft wheat flour: 50% semolina, and also reduced the leaching losses of solid matter and protein from the cooked products.

10.3.10 Doughing water temperature

The temperature of the doughing water is important: water at 36–45°C is normally used for cold dough making; 45–65°C for a 'warm system' using high-temperature drying; and 75–85°C for 'very warm processing'. If egg is an additive, the doughing water temperature should not exceed 50°C, since albumen coagulates at 49° C. Similarly, when ascorbic acid is added, the water temperature should not exceed 55°C, otherwise degradation of the additive will be accelerated. In the warm system, the gluten hydrates rapidly, reducing the length of the dough preparation time. Use of the very warm system is restricted to preparation of pasta from semolina or flour having more than 32% wet gluten content. This system produces the most rapid hydration of gluten and gelatinization of the starch, but of course cannot be employed if ascorbic acid is included, because the latter decomposes at these temperatures and will thus have no beneficial effect.

10.3.11 Glyceryl monostearate as an additive

The use of glyceryl monostearate as a flour modifier can be used to permit the extrusion of the dough at a lower m.c., viz. 28%, than is customary in making uncooked pasta. The temperature of the dough at extrusion is above 54°C, but not so high as to cause gelatinization of the starch. Other flow-modifying agents suggested are whey solids and sulphydryl-reducing substances such as L-cysteine, glutathione, sodium bisulphite or calcium sulphite at levels of 0.025%–0.1% by wt (European Patent Application No. 0,288,136).

10.3.12 Prevention of starch leaching from uncooked pasta

High temperature (above 74°C) drying of pasta can cause denaturation of the proteins, thereby entrapping the starch and rendering the pasta stable to starch leaching in the presence of cold water. If dried at lower, traditional temperatures, the protein is not denatured and, as a result, starch leaches out, making a gummy, mushy product, unless the pasta is immediately placed in extremely hot water to set the protein matrix. An uncooked pasta that can withstand exposure to cold water without leaching of starch may be made by the addition of low-temperature coagulatable materials such as albumin, whole egg, whey protein concentrates, but preferably egg white, added at a level of 0.5%–3.0% by wt. Sulphydryl-reducing agents, e.g., cysteine, glutathione, which reduce disulphide ($-SS-$) bonds to sulphydryl groups ($-SH$), thereby facilitating the irreversible denaturation of the gluten, can be added at levels of 0.02%–0.04% by weight; their addition is essential at drying temperatures below 74°C. A modern drying process would use two stages: drying at 71–104°C for 2–4 h, followed by further drying at 32–71°C for 0–120 min, but with the use of a high-velocity air current (150 ft^3/min), the drying times could be shortened to 15–30 min for the first step plus 30–120 min for the second. Pasta made in this way can be rehydrated by soaking in cold water, and then cooked in about 2 min by conventional boiling or by microwaving (European Patent Application No. 0,352,876).

10.3.13 Microorganism control

Control of microorganisms in stored pasta products is essential. The moisture content of pasta is generally less than 8%, while the water activity (a_w) is approximately 0.5, both of which prevent microbial growth and ensure a long shelf life.

In addition to the control of moisture and water activity, the use of high temperature (above 60°C) drying to improve pasta product sterility has been described (Milatovic, 1985).

Another process is described in which dough is made from flour (durum wheat semolina, or whole soft wheat flour plus corn flour, rice flour or potato flour) about 75 parts, with about 25 parts of whole egg, but with no water. Additives such as wheat gluten, soya protein isolate, alginates and surfactants may be added. The dough is sheeted through a series of rollers to a thickness of 0.03 in., while keeping the m.c. about 24% (below 24% m.c. cracking may occur, while above 30% m.c. the dough becomes too soft and elastic). The sheeted dough is cut into pieces, which are partially dried in 10–60 s by, e.g., infrared heating lamps and hot air at 204°C. The product is then steamed and pasteurized to kill microorganisms, cooled to 0–10°C and packed in sterile trays with injection of CO_2/N_2 (25:75–80:20). Such packaged products should be storable at 4–10°C for at least 120 days (US Patent No. 4,876,104).

10.3.14 Quick-cooking pasta

A process described for making a quick-cooking pasta starts by making a dough from flour or semolina of which at least 90% is derived from durum wheat, with the addition of 0.5%–5.0% by wt of an edible emulsifier, such as glyceryl or sorbitan monostearate, lecithin or polysorbates. Water, at 79°C, is added to bring the m.c. to 22%–32%. The ingredients are mixed at 76–88°C, extruded through a corotating twin-screw extruder at 60–88°C. The extruder comprises three zones: a cooking zone, a venting zone and a forming zone. When passing through the venting zone, the material may be subjected to a degree of vacuum (2.5–5 psi) to draw off excess moisture. The product can be cooked by microwave energy (European Patent Application No. 0,272,502).

Another process for making rapidly rehydratable pasta simultaneously cooks and dries the extruded dough (made from semolina/flour and water) by the use of superheated steam at a temperature of 102–140°C for 7–20 min. The product is in an unexpanded condition and can be packaged without further cutting or shaping (US Patent No. 4,539,214).

A process to eliminate darkened specks in a quick-cooking pasta precooks a mixture of pasta flour and water in a preconditioner before it passes through an extruder with three zones: cooking, venting, extruding. The temperature of the mixture is kept below 101°C in the cooking zone to prevent the formation of darkened specks symptomatic of burning during cooking (European Patent Application No. 0,267,368).

10.3.15 Precooked pasta

A method of preparing instant, precooked pasta is disclosed in US Patent No. 4,540,592. The pasta dough is completely gelatinized in a corotating, twin-screw extruder, incorporating at least one high-shear back-mixing cooking zone, and using high temperature and pressure so that the final product rehydrates to a cooked pasta instantly in hot water.

Preparation of a precooked product that is shelf stable for long periods has also been described (US Patent No. 4,828,852). The starting material can be durum semolina or flour, soft wheat flour, corn flour, pregelatinized corn flour, rice flour, waxy rice flour, precooked rice flour, potato flour, precooked potato flour, lentil flour, pea flour, soya flour, kidney or pinto beans, mung bean flour, corn starch, wheat starch, potato starch, pea starch, etc. A dough of 20%–28% m.c. is extruded to make pasta of 1.0–2.0 mm thickness. The pasta is then boiled in acidified water (using acetic, malic, fumaric, tartaric, phosphoric or adipic, but preferably lactic or citric acid) and then soaked in water acidified to pH 3.8–4.3 to

61%–68% m.c. The partially cooked pasta is drained and then coated with an acidified cream at pH 4.1–4.4. The acidification gives a better shelf life and thickens the cream to improve coating. The product is then flush packaged with an inert gas or vacuum packaged. Finally, the containers are sealed and heated to 90–100°C for 20–40 min to complete the cooking of the pasta.

10.3.16 Couscous

Couscous (Fig. 10.15) is a type of pasta product made in Algeria and other areas in Northern Africa (and has been for almost 2000 years) from a paste of durum semolina and water, which is then dried and ground. The product is size graded, and is similar in size to that of very coarse semolina. Thus prepared, the couscous retains quality throughout storage. Traditionally, the couscous is placed into a steamer type of pot, with the meat and vegetables in the cooking pot underneath; this allows odours and flavours to be absorbed by the couscous above. Presteamed and dried couscous is typically sold in supermarkets, which only requires a short boiling and then it is ready to eat. Cooked couscous is typically served with meat, vegetables and sauce.

10.3.17 Codes of practice

A Code of Practice for dry pasta products in the United Kingdom, promulgated by the British Pasta Products Association, requires all pasta products (other than those containing egg, additional gluten or other additives such as tomato or spinach) to conform to the following standard: only durum wheat should be used and must not contain more than 3% common wheat; m.c. of 12.5% (max.) when packed; ash content of standard pasta of 1.3% (db); ash content for whole wheat pasta of 2.5% (db). Degree of acidity and colour are no longer specified. (The current code (2001) can be found at http://www.pasta-unafpa. org/pdf/UK.pdf.) In fact, in the European Union (EU) several codes of practice are recognized. The governing organization is the Union of Organizations of Manufactures of Pasta Products of the EU (UN A.F.P.A.), and these have been compiled (http://www.pasta-unafpa. org/ing-documents1.htm).

In the United States, pasta (known as macaroni products) is defined with standards of identity under the US Code of Federal Regulations (21 CFR 139.110 – Macaroni products). For example, macaroni products are made from semolina, durum flour, farina, or a combination of two or more of these, with water, and with or without optional ingredients (such as egg, egg white, onions, celery, garlic, salt and gluten). Macaroni is defined as pasta tubes having a diameter between 0.11 and 0.27 in.; spaghetti is defined as having a diameter between 0.06 and 0.11 in.; vermicelli must have a diameter smaller than 0.06 in.

10.3.18 Pasta composition

The composition of common pasta made from durum semolina in the United Kingdom is shown in Table 10.1, while composition of various pasta products from the United States and Italy are shown in Table 10.2. Nutrients in various novel pasta products (e.g., gluten-free, whole wheat, organic, and vegetable-based) are provided in Table 10.3. The compositions of the various products are surprisingly consistent, although some differences do arise when ingredients other than semolina are used.

10.3.19 Pasta consumption

The per capita consumption of pasta products in 1989 was (in kg) 21 in Italy, 6 in France, Greece, Portugal, Switzerland, 5 in Germany FR, 4 in Sweden, 3.5 in Austria and the United Kingdom, 3 in Spain, 2 in Finland and Ireland, 1.5 in Belgium, Denmark, Netherlands, and 0.2 in Norway (European Food Marketing Directory, 1991). By 2015, consumption had changed to 23.5 in Italy, 8 in France, 11.2 in Greece, 6.6 in Portugal, 9.2 in Switzerland, 8 in Germany, 7.7 in Sweden, 7 in Austria, 3.5 in the United Kingdom, 5 in Spain, 3.2 in Finland, 1 in Ireland, 5.4 in Belgium, 2 in Denmark, and 4.4 in the Netherlands (http://www.pasta-unafpa.org/ingstatistics4.htm, 2016).

Globally, approximately 14.3 million tonnes of pasta were produced in 2015. Of this, about 34% was manufactured in the EU (with Italy leading by producing 3.2 million tonnes), 22% in Central and South America, 17% in other European countries and 15% in North America (2 million tonnes in the United States) (http://www.pasta-unafpa.org/ingstatistics4.htm, 2016).

TABLE 10.1 Typical nutrient values of pasta[a] (g per 100 g as sold)

Carbohydrates (g)	75	Thiamine (mg)	0.09	Calcium (mg)	10
Protein (g)	12	Riboflavin (mg)	0.1	Iron (mg)	1.2
Lipoprotein (g)	1.8	Niacin (mg)	2.0	Phosphorus (mg)	144
Calories	380				

[a] Based upon data from Home Economics, December 1972, 24.

TABLE 10.2 Nutrients in various types of pasta[a] (g per 100 g as sold)

Product	Serving size (g)	Total fat	Saturated fat	Cholesterol	Sodium	Total carbohydrate	Dietary fibre	Sugars	Protein
LONG PASTA									
Angle Hair (Capellini)	56	1.8	0.0	0.0	0.0	75.0	3.6	3.6	12.5
Bucatini	56	1.8	0.0	0.0	0.0	75.0	3.6	3.6	12.5
Fettuccine	56	1.8	0.0	0.0	0.0	75.0	3.6	3.6	12.5
Linguine	56	1.8	0.0	0.0	0.0	75.0	3.6	3.6	12.5
SHORT PASTA									
Campanelle	56	1.8	0.0	0.0	0.0	75.0	3.6	3.6	12.5
Collezione Casarecce	56	1.8	0.0	0.0	0.0	75.0	3.6	3.6	12.5
Cellentani	56	1.8	0.0	0.0	0.0	75.0	3.6	3.6	12.5
Elbows	56	1.8	0.0	0.0	0.0	75.0	3.6	3.6	12.5
Farfalle	56	1.8	0.0	0.0	0.0	75.0	3.6	3.6	12.5
Gemelli	56	1.8	0.0	0.0	0.0	75.0	3.6	3.6	12.5
Mostaccioli	56	1.8	0.0	0.0	0.0	75.0	3.6	3.6	12.5
Penne	56	1.8	0.0	0.0	0.0	75.0	3.6	3.6	12.5
Pipette	56	1.8	0.0	0.0	0.0	75.0	3.6	3.6	12.5

LASAGNA

Oven-ready	51	2.9	0.0	0.0	72.5	3.9	2.0	13.7
Wavy	75	2.0	0.5	0.0	74.7	5.3	2.7	13.3

SHELLS

Jumbo	50	2.0	0.0	0.0	74.0	4.0	4.0	14.0
Large	56	1.8	0.0	0.0	75.0	3.6	3.6	12.5
Medium	56	1.8	0.0	0.0	75.0	3.6	3.6	12.5
Manicotti	51	2.0	0.0	0.0	72.5	3.9	3.9	13.7

[a]Based upon data from www.barilla.com, 2016.

TABLE 10.3   Nutrients in various novel pastas[a] (g per 100 g as sold)

Product	Serving size (g)	Total fat	Saturated fat	Cholesterol	Sodium	Total carbohydrate	Dietary fibre	Sugars	Protein
GLUTEN-FREE PASTA									
Gluten-free Spaghetti	56	1.8	0.0	0.0	0.0	78.6	1.8	0.0	7.1
Gluten-free Fettuccine	56	1.8	0.0	0.0	0.0	78.6	1.8	0.0	7.1
WHOLE GRAIN PASTA									
Whole Grain Spaghetti	56	2.7	0.0	0.0	0.0	69.6	10.7	3.6	14.3
Whole Grain Linguine	56	1.8	0.0	0.0	0.0	69.6	10.7	3.6	14.3
ORGANIC PASTA									
Organic Spaghetti	56	1.8	0.0	0.0	0.0	75.0	3.6	3.6	12.5
Organic Penne	56	1.8	0.0	0.0	0.0	75.0	3.6	3.6	12.5
VEGETABLE-BASED PASTA									
Veggie Spaghetti	56	1.8	0.0	0.0	0.0	73.2	3.6	3.6	14.3
Veggie Rotini	56	1.8	0.0	0.0	0.0	73.2	3.6	3.6	14.3
Veggie Penne	56	1.8	0.0	0.0	0.0	73.2	3.6	3.6	14.3

[a]*Based upon data from www.barilla.com, 2016.*

10.4 OTHER EXTRUSION-COOKED PRODUCTS

In 1987, about 3 million tonnes of products were made by extrusion cooking in the United States (Hauck and Huber, 1989). It is estimated that by 2019, the global snack industry may exceed $31 billion USD (Markets and Markets, 2016), with Asia consuming almost twice as much as North America, which is approximately four times larger than consumption in Europe. Snacks, however, are only one of many types of products that are manufactured by extrusion.

10.4.1 Corn snacks

Expanded corn snacks (Fig. 10.19) are manufactured by extruding corn meal, which is milled, dehulled, and degermed maize. Ingredient moisture content is adjusted (generally 20%–30%) to work in concert with the heat in the extruder to gelatinize the starch, which leads to puffed/expanded products upon die exit. If there is limited expansion/nonuniform expansion, the process produces 'curls'; if the process results in greater, consistent expansion, 'puffs' are produced; increasing the die knife speed can

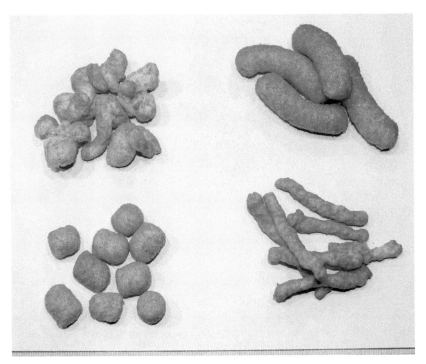

FIGURE 10.19 Examples of various expanded corn puff products. Each scale gradation indicates 1 mm.

produce 'balls', which are short puffs. Most of the time these snack products are coated in cheese powder before packaging.

10.4.2 Pet foods

Beyond human food products, one of the largest groups of products manufactured by extrusion processing is pet foods (Fig. 10.20), including both dry expanded and semimoist foods (Harper, 1986, 1989). First used by Purina in the 1950s, extruded pet foods quickly became popular with consumers and gained market share from canned moist foods. In 2015, the total market for all cat and dog food in the United States exceeded

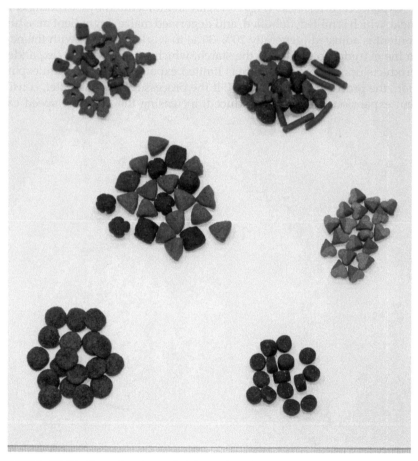

FIGURE 10.20   Examples of extruded dog and cat foods. Each scale gradation indicates 1 mm.

23 billion USD, while the next largest market was Brazil at 5.9 billion USD, followed by Japan, France, and the United Kingdom. Globally in 2015 cat and dog food sales exceeded 68 billion USD, up from approximately 58 billion USD in 2010 (http://www.euromonitor.com/pet-care, 2016).

Unlike pasta and snack products, which are high in carbohydrates, pet foods are high in protein. For example, Purina Dog Chow Adult contains 21% protein, 10% fat, 4.5% crude fibre, 12% moisture and 3532 kJ/kg (https://dogchow.com/en/dog-food/complete-and-balanced, 2016). Many pet food manufacturers produce a variety of foods, using a variety of ingredients. Formulations vary depending upon age of the animal, market segment that is targeted (low end vs high end) and costs of competing ingredients. In recent years, a multitude of premium pet foods have been developed, many of which are gluten-free, grain free, vegetable based, or even vegan in nature – many pets are considered family members, and thus their diets now parallel human diets.

Higher protein levels vis-à-vis pasta can lead to processing challenges, such as lack of expansion at the die (Fig. 10.21) and potential protein denaturing if barrel temperature is not controlled properly. Additionally, although starch is present in the ingredient matrix, and it will gelatinize during processing, because of the high level of protein, dextrinization may also occur. Screw flight configurations, screw speed, die insert configuration and size, steam and water addition rates, as well as ingredient blend moisture content all must be adjusted in order to effectively process many of the novel ingredients that have entered the marketplace in recent years. In the United States, the Food and Drug Administration regulates ingredients that are used to manufacture pet foods, in order to ensure product safety and efficacy (http://www.fda.gov/animalveterinary/products/animalfoodfeeds/petfood/, 2016).

10.4.3 Aquatic feeds

Globally, the supply of fish for human consumption has grown at a rate slightly greater than 3% annually since the 1960s, which is more than double the rate of population growth. Thus, fish protein is one of the fastest growing sources of human food. Much of this growth has occurred in the Far East – China in particular. Historically, fish has been supplied by capturing wild fish (i.e., fisheries); but in the last few decades the advent of fish farming (i.e., aquaculture) has gained momentum, and in fact, in 2016 almost half of the supply of fish was produced by aquaculture (FAO, 2016). In 2014, nearly 74 million tonnes of fish were produced via aquaculture versus approximately 93.4 million tonnes produced by capture fisheries. The leading country for aquaculture production is China, with more than 60% of all fish produced globally.

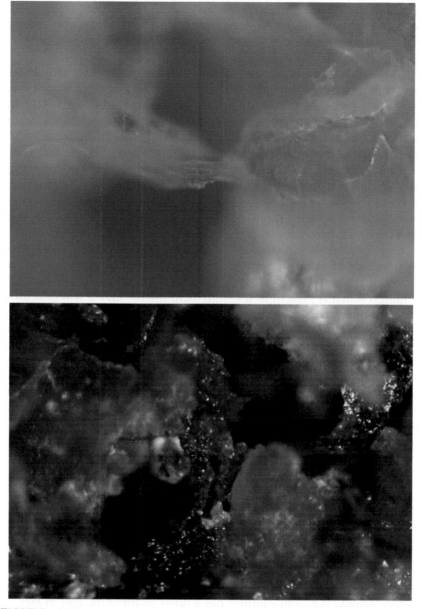

FIGURE 10.21   Left image shows the porous texture of expanded corn snack products. Right image shows nonexpanded, compacted texture of a dry pet food.

One of the largest cost components to raising fish by aquaculture is manufacturing feed (Fig. 10.22). As the industry has grown, so too has the quantity of feed required. To supply the aquaculture industry, over 24 million tonnes of aquafeeds were manufactured in 1995. But by 2014, that had grown to almost 74 million tonnes (FAO, 2016).

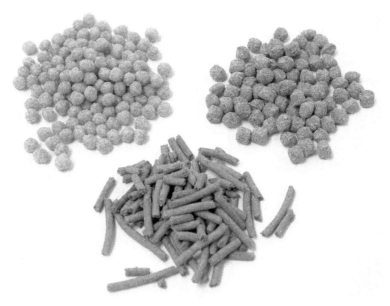

FIGURE 10.22   Some examples of extruded aquafeeds.

In many respects, aquafeeds are similar to pet foods. A variety of ingredients are used, and diets consist of high-protein and high-fat levels. Composition depends not only upon the species of fish (carnivorous fish require higher levels of protein and fat compared to omnivorous species) but also the age of the fish. For example, protein levels in catfish diets may be as high as 50% for juvenile fish, but will decrease to about 26% for adult diets; juvenile tilapia diets may be over 45% protein, but that will decrease to about 25% as the fish mature; up to 55% protein is required for juvenile trout diets, but that diminishes to about 40% for adult fish (Brown et al., 2011). Historically, the primary sources of protein and fat came from fish meal and fish oil. But, due to stagnating fishery output in recent years, and drastically higher prices for fish meal and fish oil, many companies and universities have exhaustively pursued development of fish meal–free and fish oil–free diets in the last decade – primarily based upon soybean, other oilseeds, and cereal-based grains and byproducts (Gatlin et al., 2007; Hardy, 2010).

Aquafeeds are typically spherical (or very short cylindrical) and do not come in the various shapes that either pet foods or pasta do. Pellet size depends upon the species and the age of the fish. For example, trout feeds will often start at diameters less than 0.6 mm for starter feeds, and progressively increase in size with fish age up to about 10 mm for adult fish. Moreover, some species prefer a floating feed, some a sinking feed, and some a very slow sinking feed. In order to float, each pellet must have a mass density less than $1.0 \, \text{g/cm}^3$ (the density of water). This can be obtained by adjusting the extruder setup and extrusion conditions in order to produce a slightly expanded product, and the outer surface of each pellet must be case-hardened and impermeable to water penetration. Several extruder companies have developed proprietary methods for adjusting pellet density by redesigning the die zone of the extruder so that internal pressure can be adjusted before die exit. Coating is often accomplished via a vacuum-coating machine.

10.4.4 Texturized vegetable protein

Texturized vegetable protein (TVP) (Fig. 10.23) is generally produced by extruding defatted soy protein. Because of the high-protein content in the formulation, the extrudates do not expand (due to lack of starch), and as the plasticized proteins exit the extruder, a cohesive network is formed. It was commercialized in the 1960s as a partial or complete replacement for

FIGURE 10.23   Texturized vegetable protein. Each scale gradation indicates 1 mm.

meat, it has been used as a protein source in a variety of foods, and it has been used for emergency food rations (Orcutt et al., 2005). Over the years, other plant-based proteins have also been used to produce texturized products, including cowpea and mung bean (Pham and Del Rosario, 1984).

10.4.5 Flatbreads

The extruder takes over the function of the oven, producing expansion of dough pieces, formation of structure, partial drying, and formation of flavour and colour. Thus, the process is suitable for making bread, described as 'flatbread' (Kim, 1987). The flatbreads produced by extrusion cooking are imitations of crispbreads (although the word *bread* is really a misnomer). The typical crumb-pore structure results from expansion of, and evaporation of water from, the plasticized mass on exiting from the die, when the pressure drops instantly from about $150 \, \text{kg/cm}^2$ at 160°C, resulting in drying from about 18% m.c. to about 8% m.c. The extrudate emerges as a flat strip that is cut into slices, roasted in an oven and cooled. The exactness of expansion in terms of height and weight is critical, as the slices must fit exactly into packages of prescribed volume and weight.

10.4.6 Pellets

Pellets, or half-snacks (unexpanded half-products), can be made in various shapes – flat, tube, shell, ring, screw, wheel, and other intricate shapes – and are generally made in a single-screw extruder, using a single die and with cooling of the barrel, and dried to 5%–10% m.c. The pellets are later fried, during which they expand six- to eightfold, lose water, pick up 15%–25% fat and increase in weight (Colonna et al., 1989; Meuser and Wiedmann, 1989).

10.4.7 Modified starch

Extrusion cooking can be used to treat starch at a relatively low moisture content, e.g., about 40% m.c., to bring about derivatization, plasticization and drying to make thin-cooking, pregelatinized and chemically modified starches for use in food, and also for nonfood uses, e.g., in the paper and textile industries (Meuser and Wiedmann, 1989).

10.4.8 Brewing adjuncts

Extrusion-cooked cereals have been used as brewing adjuncts. The starch granule structure is broken during extrusion cooking, and thus the starch can be more easily hydrolyzed by enzymes in the mashing process

(Smith, 1989). For example, Dale et al. (1989) discussed extrusion of sorghum for beer production; Briggs et al. (1986) examined extrusion of barley, wheat and maize; Delcour et al. (1989) examined extruded corn and sorghum. Many times the mashing, fermentation and filtration behaviour has been shown to be equivalent to conventional beer production, but in some cases, because of incomplete starch granule destruction, saccharification and filtration were reduced, and excess nitrogenous compounds led to off odours and flavours.

10.4.9 High-dextrose-equivalent syrups

These can be made by extrusion cooking of starch with the addition of thermostable alpha-amylase (Smith, 1989). Akdogan (1999) reports that the use of glucoamylase in the extrusion process could result in dextrose equivalent (DE) levels up to 94, while the use of alpha-amylase could result in DE levels as high as 98.

10.4.10 High alpha-amylase activity flour

Flour with an abnormally high alpha-amylase activity, unsuitable for making bread by conventional baking methods, can be processed by extrusion cooking – which causes rapid and complete inactivation of the enzyme – to make flatbread, snacks and biscuits. Often, indigenous raw materials can be processed by extrusion cooking in developing countries to make stable and nutritionally balanced foods, such as biscuits and precooked flours for preparation of gruels, porridges and infant foods (Cheftel, 1989; Linko, 1989b). Extruded snacks are becoming increasingly popular with the advent and distribution of very small-scale extruders.

10.4.11 Other uses for extrusion

Other suggested uses for extrusion cooking include texturizing of vital wheat gluten; pretreatment of wheat bran for the extraction of hemicellulose; and treatment of wheat flour, as an alternative to chlorination, for use in high-ratio sponge cake (Kim, 1987). Because extruders are high-temperature, short-time reactors, extrusion has been useful as a pretreatment method for producing biofuels from lignocellulosic biomass. For example, single-screw extrusion has been shown to improve hydrolyzable sugar recovery from switchgrass (Karunanithy and Muthukumarappan, 2011), prairie cord grass (Karunanithy and Muthukumarappan, 2010a) and corn stover (Karunanithy and Muthukumarappan, 2010b).

10.5 RECENT TRENDS

Recent trends indicate that pasta and other extruded food products continue to be relevant to modern diets, and in fact are growing in importance (National Restaurant Association, 2015). But, consumers are transitioning to nontraditional ingredients. Specifically, they are increasingly interested in products that contain whole grains, ancient grains and are gluten-free (e.g., contain quinoa, amaranth, chia, teff, rice, buckwheat, etc., instead of semolina). Moreover, consumers are increasingly interested in environmental sustainability, and part of that interest is eating locally sourced ingredients and products. Consumers are increasingly interested in 'healthier' pasta products, and are interested in either quick-cooking or cook-less meals. And, according to Sloan (2015), consumers are interested in novel types of pasta and pasta dishes.

References

Abecassis, J., Faure, J., Feillet, P., 1989a. Improvement of cooking quality of maize pasta products by heat treatment. Journal of the Science of Food and Agriculture 47 (4), 475–485.

Abecassis, J., Chaurand, M., Metencio, F., Feillet, P., February 1989b. Effect of moisture content of pasta during high temperature drying. Getride, Mehl und Brot 43, 58–62.

Akdogan, H., 1999. High moisture food extrusion. International Journal of Food Science and Technology 34, 195–207.

Alary, R., Kobrehel, L., 1987. The sulphydryl plus disulphide content in the proteins of durum wheat and its relationship with the cooking quality of pasta. Journal of the Science of Food and Agriculture 39, 123–136.

Antognelli, C., 1980. The manufacture and applications of pasta as a food and as a food ingredient: a review. International Journal of Food Science and Technology 15, 125.

Autran, J.C., Ait-Mouh, O., Feillet, P., 1989. Thermal modification of gluten as related to enduse properties. In: Pomeranz, Y. (Ed.), Wheat Is Unique. Amer. Assoc of Cereal Chemists Inc., St. Paul, MN, UK, pp. 563–593.

Briggs, D.E., Wadeson, A., Statham, R., Taylor, J.F., 1986. The use of extruded barley, wheat and maize as adjuncts in mashing. Journal of the Institute of Brewing 92 (5), 468–474.

British Patent Specification Nos. 2,151,898 (back-mixing of dough).

Brown, M.L., Schaeffer, T.W., Rosentrater, K.A., Bares, M.E., Muthumarappan, K., 2011. Distillers Grains: Production, Properties, and Utilization. CRC Press, Boca Raton, FL, USA.

Cheftel, J.C., 1989. Extrusion cooking and food safety. In: Mercier, C., Linko, P., Harper, J.M. (Eds.), Extrusion Cooking. Amer. Assoc. of Cereal Chemists Inc., St. Paul, MN, USA.

Colonna, P., Tayeb, J., Mercier, C., 1989. Extrusion cooking of starch and starchy products. In: Mercier, C., Limko, P., Harper, J.M. (Eds.), Extrusion Cooking. Amer. Assoc. of Cereal Chemists Inc., St. Paul, MN, USA.

Dale, C.J., Young, T.W., Makinde, A., 1989. Extruded sorghum as a brewing raw material. Journal of the Institute of Brewing 95 (3), 157–164.

Delcour, J.A., Hennebert, M.M.E., Vancraenenbroeck, R., Moerman, E., 1989. Unmalted cereal products for beer brewing. Part I. The use of high percentages of extruded or regular corn starch and sorghum. Journal of the Institute of Brewing 95 (4), 271–276.

du Cros, D.L., Hare, R.A., 1985. Inheritance of gliadin proteins associated with quality in durum wheat. Crop Science 25, 674–677.

European Food Marketing Directory, second ed., 1991. Euromonitor plc., London, UK.

European Patent Applications Nos. 0,146,510 (extrusion into N_2/CO_2); 0,267,368 (quick-cooking pasta); 0,272,502 (quick cooking pasta); 0,288,136 (shaped pasta products); 0,322,053 (drying pasta); 0,352,876 (shelf-stable, microwave-cookable pasta).

FAO, 2016. The State of World Fisheries and Aquaculture 2016: Contributing to Food Security and Nutrition for All. United Nations Food and Agriculture Organization, Rome, Italy.

Feillet, P., 1984. Present knowledge on biochemical basis of pasta cooking quality. Consequence for wheat breeders. Sciences Des Aliments 4, 551–566.

Fichtali, J., van de Voort, F.R., 1989. Fundamental and practical aspects of twin screw extrusion. Cereal Foods World 34, 921–929.

Gatlin, D.M., Barrows, F.T., Brown, P., Dabrowski, K., Gaylord, T.G., Hardy, R.W., Herman, E., Hu, G., Krogdahl, A., Nelson, R., Overturf, K., Rust, M., Sealey, W., Skonberg, D., Souza, E.J., Stone, D., Wilson, R., Wurtele, E., 2007. Expanding the utilization of sustainable plant products in aquafeeds: a review. Aquaculture Research 38 (6), 551–579.

Guy, R.C.E., January–March, 1991. Structure and formation in snack foods. Extrusion Communique 4, 8–10.

Guy, R.C.E., Horne, A.W., 1988. Cereals for extrusion cooking processes: a comparison of raw materials derived from wheat, maize and rice. In: 35th Technology Conference 1998. Biscuit, Cake, Chocolate and Confectionery Alliance, pp. 45–49.

Guy, R.C.E., Horne, A.W., February 1989. The effects of endosperm texture on the performance of wheat flours in extrusion cooking processes. Milling 182, ix–xii.

Harper, J.M., 1986. Processing characteristics of food extruders. In: Le Maguer, M., Jelen, P. (Eds.), Food Engineering and Process Applications. Unit Operations, vol. 2. Elsevier Appl. Sci. Publ., London, UK.

Hardy, R.W., 2010. Utilization of plant proteins in fish diets: effects of global demand and supplies of fishmeal. Aquaculture Research 41 (5), 770–776.

Harper, J.M., 1989. Food extruders and their applications. In: Mercier, C., Linko, P., Harper, J.M. (Eds.), Extrusion Cooking. Amer. Assoc. of Cereal Chemists Inc., St. Paul, MN, USA.

Hauck, B.W., Huber, G.R., 1989. Single screw vs. twin screw extrusion. Cereal Foods World 34, 930–939.

Karunanithy, C., Muthukumarappan, K., 2010a. Effect of extruder parameters and moisture content of switchgrass, prairie cord grass on sugar recovery from enzymatic hydrolysis. Applied Biochemistry and Biotechnology 162 (6), 1785–1803.

Karunanithy, C., Muthukumarappan, K., 2010b. Influence of extruder temperature and screw speed on pretreatment of corn stover while varying enzymes and their ratios. Applied Biochemistry and Biotechnology 162 (1), 264–279.

Karunanithy, C., Muthukumarappan, K., 2011. Optimization of switchgrass and extruder parameters for enzymatic hydrolysis using response surface methodology. Industrial Crops and Products 33 (1), 188–199.

Katskee, A.L., June 1978. Microwave macaroni drying. Macaroni Journal 12.

Kim, J.C., 1987. The potential of extrusion cooking for the utilisation of cereals. In: Morton, I.D. (Ed.), Cereals in a European Context. Ellis Horwood, United States, pp. 323–331.

Kobrehel, K., Alary, R., 1989. The role of a low molecular weight glutenin fraction in the cooking quality of durum wheat pasta. Journal of the Science of Food and Agriculture 47, 487–500.

Linko, P., 1989a. The twin-screw extrusion cooker as a versatile tool for wheat processing (Chapter 22). In: Pomeranz, Y. (Ed.), Wheat Is Unique. Amer. Assoc. of Cereal Chemists Inc., St. Paul, MN, USA.

Linko, P., 1989b. Extrusion cooking in bioconversions (Chapter 8). In: Mercier, C., Linko, P., Harper, J.R. (Eds.), Extrusion Cooking. Amer. Assoc. of Cereal Chemists Inc., St. Paul, MN, USA.

Manser, J., 1986. Einfluss von Trocknungs-Höchst-Temperaturen auf die Teigwarenqualität. Getreide, Mehl und Brot 40, 309–315.

Markets and Markets, 2016. Extruded Snacks Market by Type (Potato, Corn, Rice, Tapioca, Mixed Grain, and Others) & by Geography – Global Trends and Forecasts to 2019. Available online: http://www.marketsandmarkets.com/Market-Reports/extruded-snacks-market-139554331.html.

Mestres, C., Matencio, F., Faure, J., 1990. Optimising process for making pasta from maize in admixture with durum wheat. Journal of the Science of Food and Agriculture 51, 355–368.

Meuser, F., Wiedmann, W., 1989. Extrusion plant design (Chapter 5). In: Mercier, C., Linko, P., Harper, J.R. (Eds.), Extrusion Cooking. Amer. Assoc. of Cereal Chemists Inc., St. Paul, MN, USA.

Milatovic, L., 1985. The use of L-ascorbic acid in improving the quality of pasta. International Journal for Vitamin and Nutrition Research (Supplement) 27, 345–361.

Molinar, R., Mayorga, I., Lachance, P., Bressanti, R., 1975. Production of high protein quality pasta products using semolina-corn-soya-flour mixture. 1. Influence of thermal processing of corn flour on pasta quality. Cereal Chemistry 52, 240–247.

National Restaurant Association, 2015. 2015 Culinary Forecast. Available online: http://www.restaurant.org/Downloads/PDFs/News-Research/WhatsHot2015-Results.pdf.

Orcutt, M.W., Mcmindes, M.K., Chu, H., Mueller, I.N., Bater, B., Orcutt, A.L., 2005. Texturized soy protein utilization in meat and meat analog products. In: Riaz, M.N. (Ed.), Soy Applications in Food. CRC Press, Boca Raton, FL, USA.

Pham, C.B., Del Rosario, R.R., 1984. Studies on the development of texturized vegetable products by the extrusion process. I. Effect of processing variables on protein properties. Food Science and Technology 19 (5), 535–547.

Pomeranz, Y., 1987. Extrusion products (Chapter 20). In: Pomeranz, Y. (Ed.), Modern Cereal Science and Technology. VCH Publishers Inc., New York, NY, USA.

Sgrulletta, D., De Stefanis, E., 1989. Relationship between pasta cooking quality and acetic acid insoluble protein of semolina. Journal of Cereal Science 9, 217–220.

Sloan, A.E., 2015. The top ten food trends. Food Technology 69 (4). Available online: http://www.ift.org/food-technology/past-issues/2015/april/features/the-top-ten-food-trends/.

Smith, A., 1989. Extrusion cooking: a review. Food Science and Technology Today 3 (3), 156–161.

US Patent Specification Nos. 4,539,214 (rapidly rehydratable pasta); 4,540,590 (extrusion into vacuum); 4,540,592 (pre-cooked pasta); 4,675,199 (extrusion/compression process); 4,828,852 (shelf-stable, pre-cooked pasta); 4,830,866 (pre-cooked pasta); 4,876,104 (long shelf-life pasta).

Further reading

Autran, J.C., Galterio, G., 1989. Association between electrophoretic composition of proteins, quality characteristics and agronomic attributes of durum wheats. II. Protein-quality associations. Journal of Cereal Science 9, 195–215.

Bhattacharya, K.R., 1985. Parboiling of rice (Chapter 8). In: Juliano, B.O. (Ed.), Rice: Chemistry and Technology, second ed. Amer. Assoc. of Cereal Chemists Inc., St. Paul, MN, USA, pp. 289–348.

Certel, V.M., Mahnke, S., Gerstenkorn, P., July 6, 1989. Bulgur-nichte eine türkishe Getreide spezialität. Mühle Mischfutter Technik 126, 414–416.

Guy, R.C.E., 1986. Extrusion cooking versus conventional baking. Spec. Publ. 56, R. Soc Chem., London, UK. In: Chemistry and Physics of Baking, pp. 227–235.

Guy, R.C.E., 1989. The use of wheat flours in extrusion cooking (Chapter 21). In: Pomeranz, Y. (Ed.), Wheat Is Unique. Amer. Assoc. of Cereal Chemists Inc., St. Paul, MN, USA.

Harper, J.M., 1981. Extrusion of Foods, vol. I. CRC Press, Boca Raton, FL, USA.

Harper, J.M., 1981. Extrusion of Foods, vol. II. CRC Press, Boca Raton, FL, USA.

Houston, D.F., 1972. Rice: Chemistry and Technology. Amer. Assoc. of Cereal Chemists Inc., St. Paul, MN, USA.

Juliano, B.O. (Ed.), 1985. Rice: Chemistry and Technology, second ed. Amer. Assoc. of Cereal Chemists Inc., St. Paul, MN, USA.

Kokini, J.L., Ho, C.-T., Karwe, M.V., 1992. Food Extrusion Science and Technology. Marcel Dekker, Inc., New York, NY, USA.

Maskan, M., Altan, A., 2011. Advances in Food Extrusion Technology. CRC Press, Boca Raton, FL, USA.

Moscicki, L., 2011. Extrusion-cooking Techniques: Applications, Theory and Sustainability. Wiley-VCH Verlag GmbH & Co. KGaA, Weinheim, Germany.

Pillaiyar, P., 1990. Rice parboiling research in India. Cereal Foods World 35, 225–227.

Riaz, M.N., 2007. Extruders and Expanders in Pet Food, Aquatic and Livestock Feeds. Agrimedia GmbH, Germany.

Tolstoguzov, V.B., Muschiolik, G., Webers, V., 1989. Herstellung von Teigwaren aus Beckmehl unter Andwendung von Polysaccharid-Zusätzen. Nahrung 33, 191–201.

11.1 INTRODUCTION

Over the last century, as much of the world has industrialized, food processing technologies have continued to evolve as well, including production of an increasing variety of foods from cereal grains. Some are daily staples; others are snacks. In fact, snack foods are one of the fastest-growing sectors of the food industry, not just in the United States and Europe, but also around the world. Historically, cereals have been prepared for consumption by domestic processing on a small scale in many parts of the world, but particularly in less-industrialized countries. Industrialization has replaced domestic production to a degree, although in many parts of the world domestic production still predominates. The types of cereal grains often used are principally wheat, maize, sorghum and the millets, each of which finds greatest use in those countries in which it grows indigenously. Thus, wheat and sorghum are widely used for processing in the Indian subcontinent; maize is similarly used in Mexico and many African countries; sorghum and the millets are also used in many African countries. Geography still plays a key role, although trade agreements and international markets have made a variety of grains more available.

As many countries around the world continue to industrialize, commercial processes have increasingly become popular and make products that resemble or imitate those made by traditional domestic methods. While traditional foods continue to be available, and in many cases are the only ones available, people can be relieved of the daily tedium of domestic preparation via industrialized processing. The types of products made are many and varied, and go by many names in various countries around the world. There are, in fact, too many to describe in this work, but an overview of some will be provided in this chapter. Major categories of product include flatbreads, which may be unfermented (e.g., chapatti, roti, tortilla) or fermented (e.g., kisra, dosa, injera); porridges, which may be stiff porridges (e.g., ugali, nsima, tuwo, asida, grits) or thin porridges (e.g., ogi, ugi, nasha, madida); steam-cooked dumpling-like foods (e.g., couscous, burabusko, kenkey); boiled products (e.g., acha, kali); snack foods, which may be baked, popped, parched, puffed or fried (e.g., tortilla chips, corn chips, taco shells, crackers, biscuits, cookies); and beverages, either alcoholic (e.g., burukutu, bantu beer, kaffir beer) or nonalcoholic (e.g., mahewu). Additionally, some products now made commercially, including dry masa flour (from maize), kisra, wheat flour tortillas, etc., will be discussed in this chapter.

11.2 PRODUCTS MADE FROM WHEAT

11.2.1 Chapattis

Chapattis are one of the most widespread flatbreads in the world. Chapattis (also commonly known as chapatis or chapathis) are commonly eaten in Pakistan, India, Bangladesh, Tibet, China and other adjoining countries. Examples are shown in Fig. 11.1. In Pakistan, wheat comprised slightly more than 60% of the total cereal crop – nearly 25.5 million tonnes out of approximately 40.9 million tonnes in 2015 (FAO, 2016). About 8% was exported, and approximately 90% of the remainder was ground to make wholemeal called 'whole atta', a meal of nearly 100% extraction rate, from which chapattis – essentially wholemeal pancakes or flatbreads – are made.

To provide for the increasing demand for white flour, and with roller mills replacing stone mills, the milling of wheat in India, Pakistan and Bangladesh has been modified so as to produce white flour ('maida') and semolina (60%–65%), bran (10%–15%) and a residue called 'resultant atta' (25%–35%) from which chapattis can be made.

For making chapatti flour the wheat should have a high 1000 kernel weight, plump grains, light-coloured bran, and a protein content of 10.5%–11.0%. A strong gluten is not required, but water absorption of the flour should be high. The alpha-amylase activity need not be very low: a Falling Number (cf. Chapter 7) of 65 is satisfactory. Flour of fine granularity yields chapattis of superior quality.

Flour that is used for chapattis is typically a granular fine wheat meal of about 85% extraction rate, made by blending white flour with fine offals or bran so that the background colour is white and the large brown specks of bran are conspicuous.

(a)   (b)

(c)   (d)

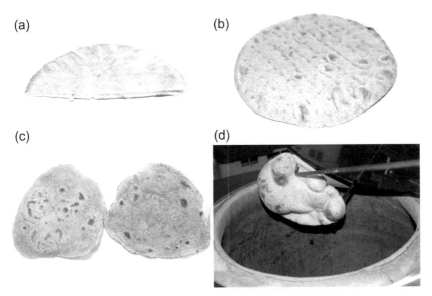

FIGURE 11.1 Examples of common flatbreads include (a) pita bread; (b) commercial (nonidentified) flatbread; (c) chapatti; and (d) naan.

Chapattis are made by mixing whole atta or resultant atta with water to form a dough, which is rested for 1 h. The dough is then divided into portions of 50–200 g, which are flattened by hand. The dough discs are baked on an iron plate over an open fire.

Types of chapattis include Tandoori Roti (baked inside a mud oven), Khameri Roti (containing yoghurt or buttermilk, sugar and salt, and the dough allowed to ferment), and Nan (or Naan) (made from white flour of 75% extraction rate by a yeasted sponge-and-dough process (cf. Chapter 8), with the addition of sugar, salt, skim milk, ghee and gram flour or eggs (Chaudri and Muller, 1970)).

The hardening or firming of chapattis may be delayed by the inclusion of shortening (3%) or of 0.5% of either glyceryl monostearate or sodium stearoyl-2-lactylate, thereby increasing the shelf life to 72 h. The best results were obtained by a combination of shortening and glyceryl monostearate (Sidhu et al., 1989).

Germination of wheat leads to an increase in reducing sugar content, diastatic activity and production of damaged starch, while decreasing the Falling Number, gluten content and chapatti water absorption. The chapattis made from sprouted wheat had better (sweetish) flavour but slightly harder texture. However, after storage for 4 days, the chapattis made from sprouted wheat had improved texture and overall quality (Leelavathi and Haridas Rao, 1988).

The inclusion of 10% full-fat soya flour made from steamed soya beans with 90% wheat flour for making chapattis almost completely eliminated the activity of trypsin inhibitor (Verma et al., 1987).

The use of triticale flour in partial replacement of wheat flour for making chapattis has was suggested, the quality of the chapattis was not impaired (Khan and Rashid, 1987).

11.2.2 Other flatbreads

Chapattis are one of a multitude of flatbreads that are consumed around the world. Many of these are fundamentally similar (i.e., the dough of each consists of water, yeast, and most often wheat flour), but they differ in terms of flour extraction level, additional ingredients, toppings, the procedure for shaping and sheeting the dough, times and temperatures for proofing, baking, cooling, as well as the type of oven in which they are baked. Table 11.1 lists some of the most common flatbreads.

Flatbreads are staples in many countries and often comprise a majority of daily caloric intake, as well as protein and fibre for many people. Often, geography and ethnicity dictate which bread is most prevalent in any given country, as well as which cereal grain is predominantly used. Some examples of flatbreads are shown in Fig. 11.1; Table 11.2 shows nutrient compositions for a few commercially available products.

TABLE 11.1 Some examples of various flatbreads around the world[a]

Primary country	Type of bread	Cereal used	Common characteristics
Afghanistan	Bolani	Wheat	Flatbread stuffed with vegetables
Common in many countries	Pita	Wheat	Round bread, typically with a pocket for stuffing
Egypt	Aish Mehahra	Fenugreek and Maize	Flat, wide loaves with diameter ~50 cm
Egypt	Baladi	Wholewheat	Round shape, with diameter ~15–20 cm
Eritrea	Injera	Teff, Wheat, Corn	Pancake-like bread
Iran	Barbari	Wheat	Oval shaped, with length of ~67–75 cm
Iran	Lavash	Wheat	Thin, round bread with diameter of ~50–60 cm
Iran	Sangak	Whole Wheat	Large bread with length of ~70–80 cm
Iran	Taftoon	Wheat	Round bread with diameter of ~40–50 cm
Israel	Matzo	Wheat and Spelt	Cracker-like flatbread, can be made into round shape with diameter of ~12 in.
Morocco	Harsha	Semolina	Pan-fried bread
Turkey	Bazlama	Wheat	Round shaped, with diameter of ~10–25 cm
Turkey	Pide	Wheat	Soft, chewy texture, similar to pita
Turkey	Yufka	Wheat	Thin, round bread with diameter of ~18 in.
Yemen	Malooga	Wheat	Yeasted flatbread, eaten with egg and buttermilk

[a] *Adapted from Pourafshar, S., Krishnan, P.G., Rosentrater, K.A., 2010. Some Middle Eastern Breads, Their Characteristics and Their Production. ASABE Paper No. 10-08667.*

11.2.3 Tortillas

A flour tortilla (Fig. 11.2; Table 11.2) is a flat, circular, light-coloured bread, about 1/16 in. thick and 6–13 in. in diameter, made from wheat flour. Wheat flour tortillas are widely consumed in Mexico, Central America and the United States.

Traditionally, tortillas were made domestically by mixing wheat flour with water, lard and salt to make dough. The dough was divided and rolled or hand-shaped to make tortilla discs, which were baked on a hot griddle (Serna-Salvidar et al., 1988). More recently, tortillas have been produced commercially, using hot-press, die-cut or hand-stretch procedures. Hot-press tortillas are baked for a relatively longer time

TABLE 11.2 Chemical composition of some commercially available flatbreads, crackers, biscuits, cookies, and tortillas (g per 100 g as sold)

Product	Serving size (g)	Total fat	Saturated fat	Cholesterol	Sodium	Total carbohydrate	Dietary fibre	Sugars	Protein	Source of data
FLAT BREADS										
Mission Wheat Flour Tortillas	70	5.7	2.1	0.0	0.8	52.9	2.9	2.9	7.1	a
Toufayan Pita Bread	52	1.0	0.0	0.0	0.4	55.8	3.8	3.8	9.6	b
Toufayan Flatbread	79	11.4	1.9	0.0	0.5	46.8	3.8	1.3	11.4	b
CRACKERS										
Cheez-It Original	30	26.7	6.7	0.0	0.8	56.7	3.3	0.0	10.0	c
Cheez-It Grooves Sharp White Cheddar	29	20.7	5.2	0.0	0.8	65.5	0.0	10.3	10.3	c
Cheez-It Reduced Fat	30	15.0	3.3	0.0	0.8	66.7	3.3	0.0	13.3	c
Cheez-It Extra Toasty	30	26.7	6.7	0.0	0.8	56.7	3.3	0.0	10.0	c
Keebler Zesta Saltine Crackers	15	10.0	0.0	0.0	1.0	73.3	6.6	0.0	6.7	d
Keebler Town House Flatbread Crisps	15	13.3	0.0	0.0	0.8	73.3	0.0	6.6	6.7	d

Melba Toast	15	0.0	0.0	0.0	0.7	80.0	0.0	0.0	13.3	e
Triscuit Rye with Caraway Seeds	28	12.5	1.8	0.0	0.6	71.4	10.7	0.0	10.7	f
Triscuit Mediterranean Style Olive	28	12.5	1.8	0.0	0.5	71.4	10.7	0.0	10.7	f
Wheat Thins	31	16.1	3.2	0.0	0.8	71.0	9.7	12.9	6.5	
BISCUITS AND COOKIES										
Keebler Sandies Shortbread	31	29.0	12.9	0.0	0.3	61.3	0.0	22.6	6.5	d
Keebler Chips Deluxe Soft 'N Chewy	31	19.4	8.1	0.0	0.4	67.7	3.2	29.0	3.2	d
Nabisco Barnum's Animals Crackers	31	12.9	1.6	0.0	0.3	77.4	3.2	25.8	6.5	g
Nilla Wafers	30	20.0	5.0	0.0	0.4	70.0	0.0	36.7	3.3	g
Oreo	34	20.6	5.9	0.0	0.3	73.5	2.9	41.2	2.9	h
Walkers Shortbread	19	31.6	21.1	0.1	0.3	57.9	5.2	15.8	5.3	i

Continued

TABLE 11.2 Chemical composition of some commercially available flatbreads, crackers, biscuits, cookies, and tortillas (g per 100g as sold)—cont'd

Product	Serving size (g)	Total fat	Saturated fat	Cholesterol	Sodium	Total carbohydrate	Dietary fibre	Sugars	Protein	Source of data
CORN TORTILLAS										
Mission White Corn	47	3.2	0.0	0.0	0.0	42.6	6.4	4.3	4.3	a
Maseca Instant Masa Flour	30	3.3	0.0	0.0	0.0	76.7	6.7	3.3	10.0	j
CORN AND TORTILLA CHIPS										
Tostitos Scoops!	28	25.0	3.6	0.0	0.4	67.9	3.6	0.0	7.1	k
Tostitos Hint of Lime	28	25.0	3.6	0.0	0.4	64.3	3.6	3.5	7.1	k
Fritos Original Corn Chips	28	35.7	5.4	0.0	0.6	57.1	3.6	0.0	7.1	k

[a] *www.missionmenus.com.*
[b] *www.toufayan.com.*
[c] *www.cheezit.com.*
[d] *www.keebler.com.*
[e] *www.bgfoods.com.*
[f] *www.triscuit.com.*
[g] *www.snackworks.com.*
[h] *www.oreo.com.*
[i] *www.walkersshortbread.com.*
[j] *www.mimaseca.com.*
[k] *www.fritolay.com.*

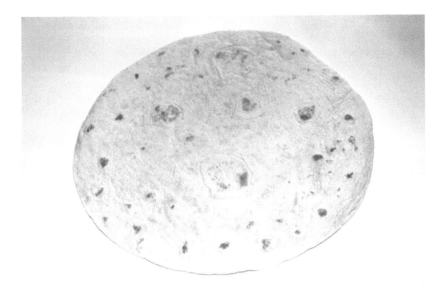

FIGURE 11.2   Example of a wheat tortilla.

at lower temperature and they puff while baking; they resist tearing and have a smooth surface. Die-cut tortillas use stronger doughs with greater water absorption, resulting in a product of lower moisture content and less resistance to cracking and breaking. Hand-stretch tortillas are irregular in shape and intermediate in quality (Serna-Salvidar et al., 1988).

The loss of flexibility of tortillas during storage may be due to retrogradation of starch, and may be prevented by the use of plasticizers, which increase flexibility and extensibility. Water is the most important plasticizer for starch and polysaccharides and should contribute 34%–45% of the dough by weight. The plasticizer components should also include glycerol or sorbitol, 5%–7% by wt, also oil and fat, 7%–9% by wt. Addition of yeast, 1%–3% on dough wt, helps the development of flavour (US Patent No. 4,735,811).

11.2.4 Pretzels

Pretzels, crisp knot-shaped biscuits, flavoured with salt, are made from wheat flour plus shortening (1.25% on flour wt), malt (1.25%), yeasts (0.25%), ammonium bicarbonate (0.04%) and water (about 42%). Typical manufacturing stages for pretzels are shown in Fig. 11.3. A dough made from these ingredients is rolled into a rope, then twisted and allowed to relax for 10 min. The rope is passed through rollers, to set

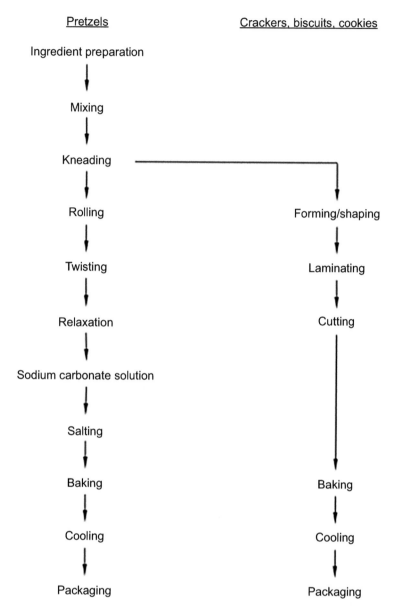

FIGURE 11.3　General stages in the preparation of pretzels, crackers, biscuits and cookies.

the knots, and allowed to ferment for 30 min. The starch on the surface is then gelatinized by passing the rope through a bath of caustic soda (1%) for 25 s at about 93°C. The dough pieces are then salted with 2% sodium chloride and baked in three stages: at 315°C for 10 min, then at 218°C to reduce the moisture content to 15%, and finally at 121°C for

(a)　　　　　　　　　　　　　　　(b)

(c)　　　　　　　　　　　　　　　(d)

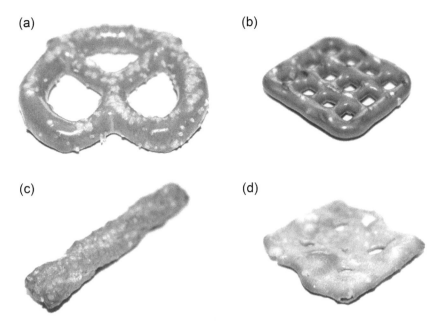

FIGURE 11.4 Pretzels are a popular snack food in their (a) traditional form, but nonconventional shapes are frequently developed for marketing, including (b) weave, (c) rod, and (d) crisp styles.

90 min (Hoseney, 1986). Examples of traditional pretzels and novel pretzel shapes that have recently been introduced into the consumer market are shown in Fig. 11.4.

11.2.5 Crackers, biscuits, cookies

Over the last several decades the snack food industries have become dominant in many developed countries, although even in developing nations the snack industry is rapidly growing. Many of these types of snack foods are cereal-based and have salty, sweet or savoury flavourings. Typical manufacturing stages for crackers, biscuits and cookies are shown in Fig. 11.3. As noted, there are many common processing steps, although differences do arise in terms of equipment used for various shaping and forming operations. Examples of various crackers are shown in Fig. 11.5, and examples of various biscuits and cookies are shown in Fig. 11.6. Not only is there great demand for snack foods in general but also novel shapes, sizes and textures in particular. As depicted in Figs 11.4–11.6, many of the newer products in the market are trying to capture consumer interest with their novelty and nontraditional formats. Table 11.2 shows nutrient compositions for a few commercially available products.

FIGURE 11.5   Styles and types of crackers are numerous. Only a few are shown here.

FIGURE 11.6 Styles and types of biscuits and cookies are numerous. Only a few are shown here.

11.3 PRODUCTS MADE FROM MAIZE

11.3.1 Tortillas

In Central America, tortillas are made from masa, which is obtained by the stone grinding of nixtamal, or lime-cooked maize. Traditionally, nixtamal, or hominy, was made by cooking maize in the leachate from wood ashes, the principal objective being to loosen and then remove the pericarp. In the modern process, whole maize is cooked in excess water containing 0.5%–2.0% of hydrated lime (on maize basis) at 83–100°C for 50–60 min, then cooled to about 68°C and allowed to steep for 8–24 h. This process is called nixtamalization, and the resulting product is called nixtamal. During this process the endosperm and germ are hydrated and softened, with partial gelatinization of the starch, and the alkali solubilizes cell walls leading to weakening of the pericarp and facilitating its removal. The nixtamal made by the traditional process is washed to remove loose pericarp and excess lime and is then stone-ground to produce masa.

A modern commercial process for making hominy (nixtamal) uses lye (caustic soda) solution, in which the maize grains remain for 25–40 min until the pericarp is free. It is then boiled and washed to remove the pericarp and the lye. The hominy is then salted and canned.

An attrition milling process for grinding the nixtamal is essential, using synthetic, lava or alumina stones that cut, knead and mash the nixtamal to form masa. The wet-ground product is obtained as a dough containing about 55% m.c.; it consists of pieces of endosperm, aleurone, germ, pericarp, free starch granules, free lipids and dissolved solids that form a 'glue-like' material that holds together the masa structure. There is no gluten in the masa dough; cohesion of the mass is due to the surface tension of the water, and is most successful when the particle size of the material is fine, and the amount of water contained is just enough to fill the spaces between the particles. The masa is sheeted, cut into triangular shapes, and baked on a griddle for 39 s at 280°C to make tortillas.

The temperature used for baking the tortillas is as high as possible, short of causing puffing (Gomez et al., 1987, 1989; Hoseney, 1986; Rooney and Serna-Salvidar, 1987; Serna-Salvidar et al., 1988).

An overview of the various stages in the production of nixtamal, masa and tortillas is shown in Fig. 11.7.

11.3.2 Nixtamalization and pellagra

The disease pellagra has been associated with a deficiency of the vitamin niacin (nicotinic acid) or niacinamide, and is prevalent among people who rely upon maize for a large proportion of their daily food. Some 50%–80% of the niacin in maize occurs in a bound form as niacytin or niacinogen, which is biologically unavailable and renders the maize deficient in niacin (Mason et al., 1971). Pellagra is not suffered by Mexicans, however, who consume maize meal in the form of tortillas. The alkaline conditions obtained during the lime cooking (nixtamalization) release the bound niacin and make it biologically available.

However, there is some evidence that the onset of black tongue in dogs (a disease similar to pellagra in humans) may be due to excessive intake of the amino acid leucine, in which maize is rich. In trials with dogs on maize diets, black tongue developed only in dogs on high leucine diets. Alkaline treatment of maize to prepare masa or hominy has been shown to result in some loss of the amino acids arginine and cystine; commercially made masa and tortillas contained lysinoalanine and lanthionine, which are breakdown products of cystine and arginine (Sanderson et al., 1978).

11.3.3 Snack products

Tortilla chips (tostadas) were traditionally produced by frying stale, leftover tortillas, and are still so prepared in Mexican restaurants.

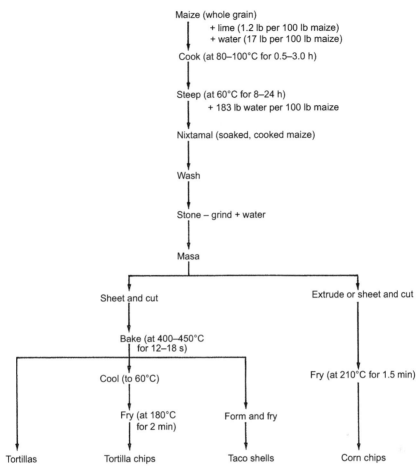

FIGURE 11.7 General stages in the preparation of tortillas and other affiliated corn products from whole maize. *Adapted from Gomez, M.H., Rooney, L.W., Waniska, R.D., Pflugfelder, R.L., 1987. Dry corn masa flours for tortilla and snack food production. Cereals Foods World 32 (5), 372–377 and Rooney, L.W., Serna-Salvidar, S.O., 1987. In: Watson, S.A., Ramstad, P.E. (Eds.), Corn: Chemistry and Technology. Amer. Assoc. of Cereal Chemists Inc., St. Paul, MN.*

Alternatively, freshly made masa, after sheeting and cutting, can be fried for 1 min at 190°C to produce tortilla chips.

Corn chips can be made from unbaked masa by extrusion through a forming extruder into hot fat, while taco shells are made by frying tempered tortillas in deep fat. The tempering allows equilibration of the moisture to occur (Hoseney, 1986).

An overview of the various production stages of nixtamal, masa, tortilla chips, taco shells and corn chips is shown in Fig. 11.7.

11.3.4 Dry masa flour

Commercial production of dry masa flour, from which tortillas and corn snacks can be made, has greatly expanded in recent years because the dry masa flour is very convenient to use. In the preparation of dry masa flour, maize is cooked and steeped in lime water by the traditional batch method, or may be cooked in a continuous process by spraying the maize with a lime solution before insertion in a steam cooker conveyor. The cooked grain is washed to remove the pericarp and then stone-ground or hammer-milled. The masa is dried, e.g., by falling against a rising stream of hot air, then hammer-milled and size-graded by sieving, the oversized particles being reground. The various particle size fractions are then blended in appropriate proportions for particular uses and packed (Gomez et al., 1987).

Tortillas can be eaten alone, like bread, or fried (to make taco shells), or eaten with fillings or toppings (nachos, tostadas, enchiladas, taquitos, burritos, tamales; eating them with salsa or guacamole has become extremely popular as a snack). In Colombia, maize grain is pounded with potash and a little water in the domestic preparation of maize meal. The alkaline effect of the potash is to loosen the bran and also to release the bound niacin. Some examples of corn tortillas, tortilla chips, and corn snack chips are shown in Fig. 11.8. Table 11.2 shows nutrient compositions for a few commercially available products.

11.3.5 Kenkey

In Ghana, traditional foods prepared from maize are known as kenkey. Cleaned maize grain is steeped in water for 3 days, the grain ground to flour, dough prepared from flour and water, spontaneous fermentation of the dough allowed for 3 days, part of the fermented dough partly cooked and then mixed with the uncooked part, the dough mass moulded into balls, the balls wrapped in maize cob sheaths and boiled to produce kenkey. Levels of niacin and lysine increase during the period of dough fermentation but fall to about the original levels during the partial cooking. There is further loss of niacin during preparation of the kenkey (Ofosu, 1971).

11.4 PRODUCTS MADE FROM SORGHUM AND THE MILLETS

Most of the products made from maize, mentioned previously, can also be made from sorghum, or from a blend of sorghum and maize, and some also from pearl millet. However, the traditional methods of preparation vary from country to country, as the following brief survey indicates.

FIGURE 11.8 Examples of a corn tortilla, tortilla chips of various shapes (including blue corn) and extruded snack chips.

The traditional method of preparation in African countries and India where sorghum and the millets form the staple food is simple pounding in a mortar to loosen the husk and to reduce the grain to wholemeal or semolina, followed by winnowing. The grain is stored as such, and from it the day's requirements of meal or flour are prepared. Preparation of larger quantities of meal is not practised because of the rapidity, in hot climates, with which the undegermed meal becomes rancid. The yield of edible products is 85%, comprised of 39% coarse semolina, 16% fine semolina and 30% flour.

Porridges made from cereal flours, such as sorghum flour, are the most important dishes consumed by the people living in Africa south of the Sahara. Both thick porridges and thin porridges are made, differing in the flour/water ratio required. Thick porridges use about 1 flour:2 water, whereas thin porridges use 1 flour:3–4 water. To make the porridge, the flour of sorghum, millet or maize, or a blend of these, is just boiled with water. Nasha is a thin porridge made from a fermented batter. Sorghum flour is mixed with a starter and water, and left to ferment for 12–16 h. The fermented batter is then diluted with water, cooked and flavoured with spices or fruit juices.

Wholemeal flour from sorghum may be cooked with water to make semisolid dumplings.

From the flour or meal, unleavened bread or chapattis are made, or the ground product may be used to make a beverage.

Millet may be consumed in the form of porridge (called kasha in the former Soviet Union) made from dry parched grain, or it may be cooked with sugar, peanuts or other foods to make desserts. Massa is made from millet flour by cooking a portion of the flour to make a thin porridge, while making a batter from the remainder of the flour, mixing it in, and leaving the mixture to ferment overnight. The mixture is flavoured with salt, pepper and onions and then fried. The product is spongy in texture, and eaten with sorghum, maize or millet porridge (Economic Commission for Africa, 1985).

In Nigeria, sorghum and pearl millet are used in four ways:

1. The dry grain is ground to make either (a) a meal or flour, from which porridge (tuwo) is prepared, or (b) grits, from which burabusko, a food resembling couscous, is prepared. The grits are agglomerated by blending with water, and then are steamed for three successive periods of 15 min. each (Galiba et al., 1987). Pancakes may be made by frying a pasta.
2. The dry grain is roasted and then ground to make roasted meal or flour, from which snacks (guguru, adun) are prepared.
3. The grain is steeped in water and a lactic fermentation allowed to proceed for 1–4 days. The moist grain is then pounded and used to prepare a fermented porridge product (ogi, akamu).
4. The grain is soaked and allowed to sprout. The sprouted grains are dried and ground to make a malt from which beverages (pito, burukutu) are prepared.

In the Sudan and Ethiopia, the flour or meal is used for making flat cakes (kisra, injera) or it may be mixed with cassava flour. Kisra, a staple food in the Sudan, is made by mixing sorghum flour, of 80%–85% extraction rate, with water and a starter, and leaving it to ferment overnight, and then baking at 160–180°C for 30 s. Kisra is now also made commercially and has a shelf life of 48 h at room temperature (Economic Commission for Africa, 1985).

The grain may be parched, popped or boiled whole.

In Ethiopia, injera are made from the flour of teff (*Eragrostis tef*), an indigenous type of millet. In the traditional domestic process, teff flour is mixed with water and allowed to ferment overnight by action of endogenous microflora to produce a sour dough, and then baked in the metad, or injera oven, to make injera, a pancake-like unleavened bread. The fermentation may be promoted by using a starter culture, called irsho, a thin paste saved from a previous fermentation. Fermented teff flour is also used for making porridge, beer (tella) and spirits (katikalla) (Umeta and Faulks, 1989).

In Uganda, sorghum grain is malted and sprouted, the radicle removed, and the remainder of the grain dried. Some of the pigment and the bitter tannins are thereby removed. The sugars produced by the malting make a sweet-tasting porridge. The grain is also used for brewing.

In India and other Asian countries, wholemeal flour from sorghum or millet may be used to make dry, unleavened pancakes (roti, chapatti, tortilla). It has been estimated that 70% of the sorghum grown in India is used for making roti, a proportion that increases to 95% in Maharashtra state (Murty and Subramanian, 1982).

If chapattis are made with cold water, the dough lacks cohesiveness because the protein in sorghum and millet is not gluten-like. The use of boiling water to make the dough results in partial gelatinization of the starch and imparts sufficient adhesiveness to permit the rolling out of thin chapattis. The water absorption of sorghum flour is higher than that of wheat flour; thus, the baking time for sorghum chapattis is longer than that for wheat chapattis. A blend of about 30% sorghum flour with 70% wheat flour produces chapattis of improved eating quality.

In India and Africa, whole sorghum grain, or dehulled and polished sorghum grain (pearl dura), may be boiled to make balila, which is used in a similar way to rice. Sorghum may also be eaten as a stiff porridge.

In India, grain sorghum and pearl millet may be popped, but whereas maize is popped in hot oil, the sorghum and millet grains are popped in hot sand (Hoseney, 1986). A more detailed account of traditional foods made from sorghum in various countries, including methods of preparation, is given by Rooney et al. (1986).

Sorghum and maize react in the same way when subjected to the alkaline cooking process of nixtamalization. This process causes the hull or pericarp to peel away from the kernels, facilitating its subsequent removal. The starch granules throughout the kernel swell, but some of the granules in the peripheral endosperm are destroyed. Tortillas made from a blend of 80% pearled or unpearled sorghum plus 20% yellow maize by the alkaline cooking process had an acceptable flavour and a soft texture. The reduced cooking and steeping times required by sorghum as compared with maize are advantageous, and the cooking time is further reduced by using pearled, rather than unpearled, sorghum (Bressani et al., 1977; Bedolla et al., 1983; Gomez et al., 1989).

Tortilla chips could be made from white sorghum by lime-cooking at boiling temperature for 20 min, using 0.5% lime, quenching to 68°C and then steeping the grains for 4–6 h to produce nixtamal, which is then stone-ground to masa (a coarse dough). The masa is sheeted, cut into pieces, baked at 280°C for 39 s and then fried in oil at 190°C for 1 min. The tortilla chips thus made from sorghum have a bland flavour; a more acceptable product, with the traditional flavour, could be made similarly from a blend of equal parts of maize and sorghum (Serna-Salvidar et al., 1988).

11.5 RICE SUBSTITUTES

11.5.1 Couscous

Couscous is a type of pasta product made in several North African countries, including Algeria, Libya, Morocco, Tunisia and other surrounding countries, from a durum semolina and water paste that is dried and ground. The product is size-graded, and is similar in size to that of very coarse semolina. Thus prepared, the couscous has excellent keeping qualities.

11.5.2 Bulgur

Bulgur consists of parboiled whole or crushed partially debranned wheat grains, and is used as a substitute for rice, e.g., in pilaf, an eastern European dish consisting of wheat, meat, oil and herbs cooked together. The ancient method of producing bulgur, which is referred to as *Arisah* in the Old Testament, consists of boiling whole wheat in open vessels until it becomes tender. The cooked wheat is spread in thin layers for drying in the sun. The outer bran layers are removed by sprinkling with water and rubbing by hand. This is followed by cracking the grains by stone or in a crude mill.

A product resembling the wheat portion of the pilaf dish was developed in the United States in 1945 as an outlet for part of the US wheat surplus. In one method of manufacture of bulgur, described by Schäfer (1962), cleaned white or red soft wheat, preferably decorticated, is cooked by a multistage process in which the moisture content is gradually increased by spraying with water and raising the temperature. Eventually, when the m.c. has reached 40%, the wheat is heated at 94°C and then steamed for 1.5 min at 206.85 kN/m^2 (30 lb/in.^{2}) pressure so that the cooked product is gummy and starchy. The starch is partially gelatinized. The m.c.

is then reduced to about 10% by drying with air at 66°C, and the dried cooked wheat is pearled or cracked. One brand of the whole grain product is called Redi wheat in the United States; the crushed is cracked bulgur (Nouri, 1968).

Other processes for making bulgur are reviewed by Shetty and Amla (1972).

In a continuous system used in the United States for the production of bulgur, the wheat is soaked in a succession of three tanks in which the m.c. is raised progressively to 25%–30% during 3.5–4h in the first tank, to 35%–40% m.c. during 2.5h in the second and to 45% m.c. during 2–2.5h in the third, giving a total soaking time of 8h. The grain is then cooked with steam at a pressure of 1.5–3bar for 70–90s and then dried to 10%–11% m.c. The outer layers are removed, the grain lightly milled and the product sieved to separate large from small bulgur (Certel et al., 1989).

During the soaking and cooking stages, a proportion of the vitamins and other nutrients present in the outer layers of the grain are mobilized, and move to the inner part of the grain, in a similar way to that described for parboiled rice (q.v.). Thus, removal of the outer layers of the grain, after soaking, cooking and drying, does not much reduce the nutritive value of the bulgur.

Bulgur is sent from the United States to countries in the Far East as part of the programme of American aid to famine areas. In 1971, 227,000tonne of bulgur were produced in the United States, of which 5% was used domestically, the remainder being used in the Foods for Peace Program. The staple food of people in these areas had always been boiled rice; the process of bread making was unknown; wheat and wheat flour were therefore unacceptable foods. Bulgur provided a cheap food that was acceptable because it could be cooked in the same way as rice and superficially resembled it. The level of bulgur exports from the United States for the food aid programme is now about 250,000 tonne/y (Certel et al., 1989).

Bulgur can be stored for 6–8months under a wide range of temperature and humidity conditions, and its hard and brittle nature discourages attack by insects and mites.

The nutritive value of bulgur is similar to that of the wheat from which it is made. Fat, ash and crude fibre levels are slightly lower, but the protein level is unchanged. Retention of thiamine and of niacin is about 98% of the original, that of riboflavin about 73%. Iron and calcium contents are slightly increased, while phosphorus content is decreased by the parboiling process. The nutrient composition of bulgur is shown in Table 11.3.

Bulgur of acceptable quality has also been prepared from triticale (Singh and Dodda, 1979) and from maize (Certel et al., 1989).

The consumption of bulgur in Turkey is estimated at 20–30 kg/head/ annum. In fact, Turkey is the world leader in bulgur production and export. Turkey is reported to export over 200,000tonne of bulgur per annum; it can produce nearly 1million tonne (Certel et al., 1989; Miller Magazine, 2016).

TABLE 11.3 Nutrients in bulgur[a] (per 100 g)

Carbohydrates (g)	75.7	Thiamine (mg)	0.28	Phosphorus (mg)	430
Protein (g)	11.2	Riboflavin (mg)	0.14	Potassium (mg)	229
Fat (g)	1.5	Niacin (mg)	5.50	Magnesium (mg)	160
Energy (kcal)	354	Vitamin B_6 (mg)	0.32	Calcium (mg)	22.6
		Pantothenic acid (mg)	0.83	Iron (mg)	7.8
		Folacin (mg)	0.038	Zinc (mg)	4.4
				Iodine (mg)	14

[a] *Based upon data from Protein Grain Products International, Washington, DC.*

11.5.3 WURLD wheat

Peeled bulgur wheat is a light-coloured, low-fibre bulgur, made to resemble rice by removal of the bran in a lye-peeling operation. The wheat grain is treated with sodium hydroxide, and the loosened bran removed by vigorous washing with water. The grain is then treated with warm dilute acetic acid to restore the surface whiteness, and then dried. This treatment leaves the aleurone layer intact. Peeled bulgur made from a red wheat is known as WURLD wheat, a US Department of Agriculture **W**estern **U**tilization **R**esearch **L**aboratory **D**evelopment (Shepherd et al., 1965). Although promising, this technology has not seen widespread commercial adoption.

11.5.4 Ricena

Ricena is a rice substitute, originating in Australia, made from wheat by a relatively inexpensive patented process, with a yield of about 65% (British Patent No. 1,199,181). The wheat is washed, cleaned, steamed under pressure and dried. The product sells at the price of low-grade rice, although the protein, iron and vitamin B_1 contents of ricena exceed those of milled rice. Again, although promising, this technology has not seen widespread commercial adoption.

11.6 PARBOILED RICE

The parboiling (viz. part-boiling) of rice is an ancient tradition in India and Pakistan, and consists of steeping the rough rice (paddy) in hot water, steaming it and then drying down to a suitable moisture content for milling. The original purpose of parboiling was to loosen the hulls, but in addition the nutritive value of the milled rice is increased by this treatment, because the water dissolves vitamins and minerals present in the hulls and bran coats and carries them into the endosperm. Thus, valuable nutrients that would otherwise be lost with the hulls and bran in the milling of raw rice are retained by the endosperm. It has been shown that rice oil migrates outwards during parboiling; the oil content of the milled kernel is lower, that of the bran higher, in parboiled than in raw rice.

By gelatinizing the starch in the outer layers of the grain, parboiling seals the aleurone layer and scutellum so that these fractions of the grain are retained in milling to a greater degree in milled parboiled than in milled raw rice (Hinton, 1948). Parboiling toughens the grain and reduces the amount of breakage in milling. Moreover, parboiled rice is less liable to insect damage than is milled raw rice, and has an improved storage life.

11.6.1 Conversion

Conversion of rice is the modern commercial development of parboiling. In the H. R. Conversion Process, wet paddy is held for about 10 min in a large vessel, which is evacuated to about 635 mm (25 in.) of mercury. The paddy is then steeped for 2–3 h in water at 75–85°C introduced under a pressure of 552–690 kN/m^2 (80–100 lb/in.^{2}). The steeping water is drained off and the paddy is heated under pressure for a short time with live steam in a steam-jacketed vessel. The steam is blown off, and the pressure in the vessel reduced to 711–737 mm (28–29 in.) of vacuum. The product is then vacuum-dried to about 15% m.c. in the steam-jacketed vessel, or it can be air-dried at temperatures not exceeding 63°C. After cooling, the converted paddy is tempered in bins for 8 h or more to permit equilibration of moisture and is then milled.

In the Malek process, paddy is soaked in water at 38°C for 4–6 h, steamed at 103.4 kN/m^2 (15 lb/in.^{2}) pressure for 15 min, dried and milled. The product is called Malekized rice.

Redistribution of thiamine (vitamin B$_1$) in rice parboiled by the Malek process was investigated by Hinton (1948). By microdissection of raw and Malekized rice grains, followed by microanalysis of the fractions, Hinton showed that the thiamine content of the endosperm adjoining the scutellum increased from 0.4 to 5.2 iu/g, while that of the scutellum decreased from 44 to 9 iu/g, and that of the pericarp/aleurone from 10 to 3 iu/g. The thiamine content of the whole endosperm increased from 0.25 to 0.5 iu/g. The redistribution of thiamine was due to the inward passage of water through the thiamine-containing layers rendered permeable by heat to the vitamin.

In a process called *double-steaming*, raw paddy is presteamed for 15–20 min and then steeped in water for 12–24 h. The water is drained off, and the product is steamed again and then dried.

Parboiling may induce discolouration of the grains or development of deteriorative flavour changes. A suggested remedy for reducing flavour changes is steeping in sodium chromate solution (0.05%). It has been shown that the additional chromium present in the milled rice is not absorbed when the rice is consumed, and thus poses no health hazard (Pillaiyar, 1990). A remedy suggested for bleaching is steeping in sodium metabisulphite solution (0.32%), but this treatment lowers the availability of thiamine.

If steeping time exceeds about 8 h, the steeping water tends to develop an off odour that is picked up by the rice. The odour is due to the activity of anaerobic bacteria in the steeping water. Various methods have been suggested to avoid this problem. If the steeping water is maintained at a temperature of 65°C, the steeping time can be limited to 4 h, and odour development is avoided.

The soaking conditions must be chosen carefully to hydrate the grain sufficiently for it to be gelatinized on subsequent heating, but to avoid splitting the hull – which leads to excessive hydration and leaching of nutrients. Rate of hydration increases with temperature of the steeping water, but above about 75°C the rate of hydration increases so rapidly that splitting of the hull occurs. Thus, the solution is to limit the moisture level in the grain to 30%–32% either by soaking at 70°C or lower, or by starting the soaking at 75°C and allowing the material to cool naturally during soaking. This method gives the fastest hydration without complications (Bhattacharya, 1985).

A process devised by F.H. Schüle GmbH does not involve steaming. Rough rice is soaked for 2–3 h in water at a medium temperature while the hydrostatic pressure is raised to 4–6 kg/cm^2 by admitting compressed air. The pressure is released, and the rice is cooked in water at about 90°C. The rice is then predried in a vibratory drier and subsequently dried by three passes through columnar driers with successively decreasing air temperature (Bhattacharya, 1985).

A high-temperature, short-time process for parboiling rice has been described (Pillaiyar, 1990). Paddy steeped to about 24% m.c. is simultaneously parboiled and dried in a sand-roaster in which the paddy is mixed with hot sand and remains in contact with the sand for about 40 s, after which the sand and the rice are separated by sieving. During this process the rice loses about 10% of moisture.

For a comprehensive survey of other processes, see Bhattacharya (1985).

11.6.2 Consumption

Parboiled or converted rice is readily consumed in India and Pakistan, while in the United States it is used as a ready-to-eat cereal, as canned rice and as a soup ingredient. Elsewhere, however, parboiled rice is not popular.

11.7 RECENT TRENDS

The food industry continues to evolve, much of which is driven by changes in demographics, taste preferences, consumption patterns, as well as new discoveries in nutritional and health sciences (e.g., impacts on cancer, heart disease, diabetes, obesity). Food companies try to increase product appeal to customers so that they remain relevant in the marketplace, to increase processing efficiencies (e.g., more efficient motors, controllers, lighting), to lower environmental impacts (e.g., reduce CO_2 and other greenhouse gas emissions, water use, energy use), and to simultaneously achieve profitability. Fig. 11.9 highlights some of the trends in the United States in recent years. Over the last decade there has been a great shift toward consumption of tortillas and wholewheat bread; this has resulted in a drastic decrease in white bread consumption. Moreover, the market for gluten-free products has seen overwhelming growth in recent years across the entire food sector (approximately 15.5 billion USD projected food sales in the United States in 2016), which has been driven by demand from consumers with gluten intolerance (such as celiac disease), as well as a shift toward perceived health benefits from consuming grains other than wheat. And, every year, the snack market continues to grow, as consumers seem to have an insatiable desire for various snack foods.

As consumers continue to demand new food products, novel shapes, textures, and flavours (especially sweet, salty, savoury) are developed and commercialized. This is true for many of the cereal products discussed in this chapter. To address these trends, food companies continue to implement new technologies. These include improvements in processing equipment, production processes, and specific ingredients, blends and seasonings that are used. A few recent developments include those for pretzels (US Patent Nos. 742,617; 8,778,428B2; D670,066S); crackers, biscuits, and cookies (US Patent Nos. 20,150,099; 9,192,168B1; D661,457S; D684,809S1; D696,837S1; D711,067S1; D731,741S1; D735,539S1; D742,617S); tortillas (US Patent Nos. 2014 0,377,424A1; D623,376S; D754,416S; 8,820,221B2); tortilla chips (US Patent Nos. D355,975S; D383,587S). Many more examples exist in the patent, scientific and trade literature – to which the reader is referred for more information.

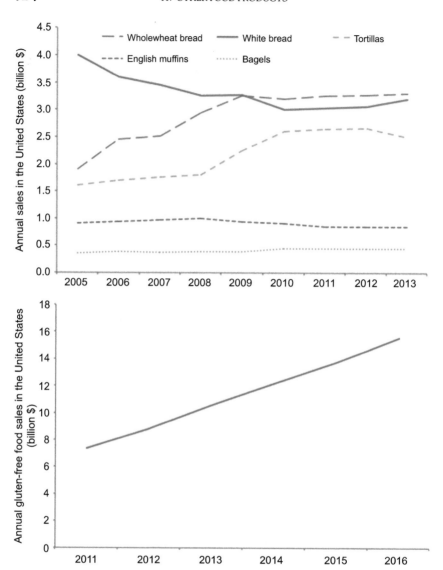

FIGURE 11.9 Sales (billion $) of various bread products in the United States indicate growing demand for wholewheat breads and tortillas, while the popularity of white bread has declined. Gluten-free foods (of which cereal products are a part) continue to grow in popularity. *Adapted from Ferdman, R.A., June 19, 2014. What America's changing bread preferences say about its politics. The Washington Post. Available online:* https://www.washingtonpost.com/news/wonk/wp/2014/06/19/what-americas-changing-bread-preferences-say-about-its-politics/ *and Statista, 2016. Retail Dollar Sales of Gluten-Free Products in the United States from 2011 to 2016 (in Million U.S. Dollars). Available online:* http://www.statista.com.

References

Bedolla, S., de Palaos, M.G., Rooney, L.W., Diehl, K.C., Khan, M.N., 1983. Cooking characteristics of sorghum and corn for tortilla preparation by several cooking methods. Cereal Chemistry 60, 263–268.

Bhattacharya, K.R., 1985. Parboiling of rice (Chapter 8). In: Juliano, B.O. (Ed.), Rice: Chemistry and Technology, second ed. Amer. Assoc. of Cereal Chemists Inc., St. Paul, MN, pp. 289–348.

Bressani, R.I., Elias, L.G., Allwood Paredas, A.E., Huezo, M.T., 1977. Processing of sorghum by lime-cooking for the preparation of tortillas. In: Dendy, D.A.V. (Ed.), Proceedings of Symposium on Sorghum and Millets for Human Food. Tropical Products Institute, London, UK.

British Patent Specification No. 1,199,181 (ricena).

Certel, V.M., Mahnke, S., Gerstenkorn, P., July 6, 1989. Bulgur-nichte eine türkishe Getreide spezialität. Mühle Mischfuttertechnik 126, 414–416.

Chaudri, A.B., Muller, H.G., 1970. Chapattis and chapati flour. Milling 152 (11), 22.

Economic Commission for Africa, 1985. In: Kent, N.L. (Ed.), Technical Compendium on Composite Flours. United Nations Economic Commission for Africa, Addis Ababa, Ethiopia.

FAO, 2016. Pakistan, GIEWS Country Briefs. Food and Agricultural Organization of the United Nations, Rome, Italy. Available online: http://www.fao.org/giews/countrybrief/country.jsp?code=PAK.

Ferdman, R.A., June 19, 2014. What America's changing bread preferences say about its politics. The Washington Post. Available online: https://www.washingtonpost.com/news/wonk/wp/2014/06/19/what-americas-changing-bread-preferences-say-about-its-politics/.

Galiba, M., Rooney, L.W., Waniska, R.D., Miller, F.R., 1987. The preparation of sorghum and millet couscous in West Africa. Cereals Foods World 32 (12), 878–884.

Gomez, M.H., Rooney, L.W., Waniska, R.D., Pflugfelder, R.L., 1987. Dry corn masa flours for tortilla and snack food production. Cereals Foods World 32 (5), 372–377.

Gomez, M.H., McDonough, C.M., Rooney, L.W., Waniska, R.D., 1989. Changes in corn and sorghum during nixtamalization and tortilla baking. Journal of Food Science 54 (2), 330–336.

Hinton, J.J.C., 1948. Parboiling treatment of rice. Nature 162, 913–915.

Hoseney, R.C., 1986. Principles of Cereal Science and Technology. Amer. Assoc. of Cereal Chemists Inc., St. Paul, MN, USA.

Khan, M.N., Rashid, J., March 1987. Nutritional quality and technological value of triticale. ASEAN Food Journal 3, 17–20.

Leelavathi, K., Haridas Rao, P., May/June 1988. Chapati from germinated wheat. Journal of Food Science and Technology 25, 162–164.

Mason, J.B., Gibson, N., Kodicek, E., 1971. The chemical nature of bound nictinic acid. The Biochemical Journal 125, 117P.

Miller Magazine, 2016. World bulgur market and Turkey. Miller Magazine. Available online: http://www.millermagazine.com/english/world-bulgur-market-and-turkey/.

Murty, D.S., Subramanian, V., 1982. Sorghum roti: I. Traditional methods of consumption and standard procedures for evaluation. In: Rooney, L.W., Murty, D.S. (Eds.), Proc. Int. Sump. On Sorghum Grain Quality. Int. Crop Res. Inst. Semi-Arid Tropics, Patancheru, A.P., India, pp. 73–78.

Nouri, N., 1968. Bulgur - ein Beitrag zur Vollwert-und vegetarisched Ernährung. Getreide, Mehl und Brot 42, 317–319.

Ofosu, A., 1971. Changes in the levels of niacin and lysine during the traditional preparation of kenkey from maize grain. Ghana Journal of Agricultural Science 4, 153.

Pillaiyar, P., 1990. Rice parboiling research in India. Cereal Foods World 35, 225–227.

Pourafshar, S., Krishnan, P.G., Rosentrater, K.A., 2010. Some Middle Eastern Breads, Their Characteristics and Their Production. ASABE Paper No. 10-08667.

Rooney, L.W., Kirleis, A.W., Murty, D.S., 1986. Traditional foods from sorghum: their production, evaluation and nutritional value (Chapter 7). In: Pomeranz, Y. (Ed.). Pomeranz, Y. (Ed.), Advances in Cereal Science and Technology, vol. VIII. Amer. Assoc. of Cereal Chemists Inc., St. Paul, MN, USA.

Rooney, L.W., Serna-Salvidar, S.O., 1987. In: Watson, S.A., Ramstad, P.E. (Eds.), Corn: Chemistry and Technology. Amer. Assoc. of Cereal Chemists Inc., St. Paul, MN, USA.

Sanderson, J., Wall, J.S., Donaldson, G.L., Cavius, J.F., 1978. Effects of alkaline processing of corn on its amino acids. Cereal Chemistry 55, 204–213.

Serna-Salvidar, S.O., Rooney, L.W., Waniska, R.D., 1988. Wheat flour tortilla production. Cereal Foods World 33 (10), 855–864.

Schäfer, W., 1962. Bulgur for underdeveloped countries. Milling 139, 688.

Shepherd, A.D., Ferrel, R.E., Bellard, N., Pence, J.W., 1965. Nutrient composition of bulgur and lye-peeled bulgur. Cereal Science Today 10, 590.

Shetty, M.S., Amla, B.L., 1972. Bulgur wheat. Journal of Food Science and Technology 9, 163.

Sidhu, J.S., Seible, W., Bruemmer, J.M., March 1989. Effect of shortening and surfactants on chapati quality. Cereal Foods World 34, 286–290.

Singh, B., Dodda, L.M., 1979. Studies on the preparation and nutrient composition of bulgur from triticale. Journal of Food Science 44, 449.

Statista, 2016. Retail Dollar Sales of Gluten-Free Products in the United States from 2011 to 2016 (in Million U.S. Dollars). Available online: http://www.statista.com.

Umeta, M., Faulks, R.M., 1989. Lactic acid and volatile (C_2–C_6) fatty acid production in the fermentation and baking of tef (*Eragrostis tef*). Journal of Cereal Science 9, 91–95.

US Patent No. 20,150,099,034A1 (Low-calorie, low-fat cracker composition containing xylose, cracker made from the composition, and method for preparing the composition).

US Patent No. 2014 0,377,424A1 (Soft shaped tortillas).

US Patent No. 4,735,811 (Plasticizers for tortillas).

US Patent No. 742617 (Pretzel-bagel and method of manufacturing thereof).

US Patent No. 8,778,428B2 (Stick-shaped snack and method for producing the same).

US Patent No. 8,820,221B2 (Compact appliance for making flat edibles).

US Patent No. 9,192,168B1 (Cracker finishing machine).

US Patent No. D355,975S (Football shaped tortilla chip).

US Patent No. D383,587S (Fish shaped tortilla chip).

US Patent No. D623,376S (Shaped tortilla).

US Patent No. D661,457S (Snack cracker food product).

US Patent No. D670,066S (Pretzel cracker).

US Patent No. D684,809S1 (Cookie pan).

US Patent No. D696,837S1 (Swirled cookie with three icing stripes).

US Patent No. D711,067S1 (Snack food).

US Patent No. D731,741S1 (Filled cracker product).

US Patent No. D735,539S1 (Cookie cutter).

US Patent No. D742,617S (Snack food product).

US Patent No. D754,416S (Shaped tortilla).

Verma, N.S., Mishra, H.N., Chauhan, G.S., September/October 1987. Preparation of full fat soy flour and its use in fortification of wheat flour. Journal of Food Science and Technology 24, 259–260.

Further reading

Adrian, J., Frangue, R., Davin, A., Gallant, D., Gast, M., 1967. The problem of the milling of millet. Nutritional interest in the traditional African process and attempts at mechanisation. Agronomie Tropicale 22 (8), 687.

Anon., 1987. Bulgur wheat production. American Miller Process 92 (10), 17–18.

Asp, N.-G., Björck, I., 1989. Nutritional properties of extruded foods (Chapter 14). In: Mercier, C., Linko, P., Harper, J. (Eds.), Extrusion Cooking. Amer. Assoc. of Cereal Chemists Inc., St. Paul, MN, USA.

Aykroyd, W.R., Gopolan, C., Balasubramanian, S.C., 1963. The Nutritive Value of Indian Foods and the Planning of Satisfactory Diets. Indian Council of Medical Research, New Delhi, India. Spec. Rpt. Ser. No. 42.

Desikachar, H.S.R., 1975. Processing of maize, sorghum and millets for food uses. Journal of Scientific and Industrial Research 34 (4), 231.

Desikachar, H.S.R., 1977. Processing of sorghum and millets for versatile food uses in India. In: Dendy, D.A.V. (Ed.), Proceedings of a Symposium on Sorghum and Millets for Human Food. Tropical Products Institute, London, UK.

Hulse, J.H., Laing, E.M., Pearson, O.B., 1980. Sorghum and the Millets. Academic Press, London, UK.

Kodicek, E., Wilson, P.W., 1960. The isolation of niacytin, the bound form of nicotinic acid. The Biochemical Journal 76, 27B.

Morgan Jr., A.I., Barta, E.J., Graham, R.P., 1966. WURLD wheat – a product of chemical peeling. Northwestern Miller 273 (6), 40.

Perten, H., 1983. Practical experience in processing and use of millet amd sorghum in Senegal and Sudan. Cereal Foods World 28 (11), 680–683.

Pomeranz, Y. (Ed.), 1988. Wheat: Chemistry and Technology. Amer. Assoc. of Cereal Chemists Inc., St. Paul, MN, USA.

Raghavendra Rao, S.N., Malleshi, N.G., Sreedharmurthy, S., Viraktamath, C.S., Desikachar, H.S.R., 1979. Characteristics of roti, dosa and vermicelli from maize, sorghum and bajra. Journal of Food Science and Technology 16, 21.

Watson, S.A., Ramstad, P.E., 1987. In: Corn: Chemistry and Technology. Amer. Assoc. of Cereal Chemists Inc., St. Paul, MN, USA.

Malting, brewing, fermentation, and distilling

Kent's Technology of Cereals, Fifth Edition
http://dx.doi.org/10.1016/B978-0-08-100529-3.00012-8

729

12.1 INTRODUCTION

The overall process involved in brewing and fermentation is the conversion of cereal starch into alcohol to make a variety of palatable intoxicating beverages, generally on a small to moderate production scale – some at a factory level, but much at a household or village level. Production, marketing and sales of alcohol-based drinks are vast on a global scale. For example, in 2015 the global sales of beer were estimated to be worth US$ 522 billion, with a total volume of 1.9×10^9 hL. Of this, 471.6×10^6 hL were produced in China, 223.5×10^6 hL were produced in the United States, 138.6×10^6 hL were produced in Brazil and 95.6×10^6 hL were produced in Germany (Statista, 2017). In terms of distilled spirits, in 2010 the global market was worth nearly US$ 91 billion, with 31% of sales being whiskey/ whisky, 22% vodka, 12% rum and 35% other (including gin, liquors, white spirits, tequila, etc.) (Ibis, 2010).

Fermentation is also used on a massive scale (compared to beverage production) to produce nonpotable industrial alcohols (e.g., biofuels) in production facilities much larger than breweries or distilleries.

Yeasts appropriate to the cereal or cereals involved actually conduct the fermentation process. Most commercial yeasts belong to the species *Saccharomyces cerevisiae* (Fig. 12.1), which now includes the 'bottom yeast' previously classified as *S. carlsbergensis* (Reed and Nagodawithana, 1991), and actually consists of several thousand specific strains.

Two key processes are involved in fermentation: first, the starch has to be converted to soluble sugars by amylolytic enzymes, and second, these sugars must be fermented to alcohol by yeast metabolism (i.e., yeasts do not ferment starch, rather they ferment glucose). In the first process the enzymes may be produced in the grains themselves (endogenously) or exogenously, i.e., in other organisms, and are then added to the fermentor or upstream of the fermentor. Alternatively they may be added as extracts.

The process by which the grain's own enzymes are employed is known as 'malting'. This involves controlled germination of the grain during which the enzymes capable of catalysing hydrolysis, not only of starch but also of other components of the grain, are produced. The most significant

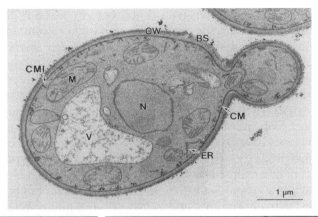

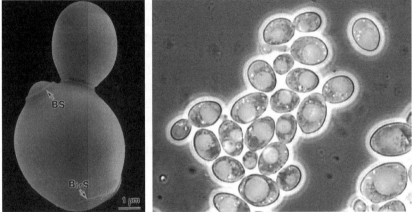

FIGURE 12.1 *Saccharomyces cerevisiae*, the most important industrial yeast, is used for the fermentation of beers, distilled spirits and biofuels. Top shows interior view of a yeast cell. *BS*, budding scar; *CM*, cell membrane; *CW*, cell wall; *ER*, endoplasmic reticulum; *M*, mitochondria; *N*, nucleus; *V*, vacuole. Bottom left image shows a budding cell and budding scar. *Images courtesy of Walker, G., Abertay University, Scotland, UK.*

are the proteases and the β-glucanases, as the products resulting from their activities ultimately affect the qualities of the beverage produced.

Other organisms are employed as a source of enzymes in the production of saké – a beer produced from rice. Enzymes are added in solution, particularly when it is required to hydrolyse the starch, etc. present in endosperm grits or flours, themselves incapable of enzyme production. Such adjuncts may provide any proportion of the total starch, depending on legislation relevant to the country of origin and the description of the alcohol product. Consequently, added enzymes may contribute different proportions of the enzyme complement.

The alcohol content of the liquor produced by fermentation is limited by the tolerance of the yeasts. Probably the most tolerant yeasts are used in saké production. They can survive alcohol contents of about 20% or more in the fermentor, although the product is often sold in a diluted form.

Distillation then concentrates the alcohol into higher levels (less than 100%, though, depending upon the azeotrope), and the resulting drinks are described as 'spirits', the special character of which depends upon flavours imparted by the processing or added to the distillate – the added flavours usually being extracts from other plant sources (e.g., cinnamon, orange, juniper, etc.). In recent years infusions of novel flavours (such as bubble gum, cotton candy, honey, etc.) have become popular, especially in vodkas and whiskies.

For alcohol production from plant materials, sugars must be present, as in fleshy fruits or other substrates from which fermentable sugars can be produced. Starch is a primary substrate (from which sugars can be produced by enzymes) in many cultures; and in fact all cereal grains can be used for alcohol production. In the West, the most commonly used cereal grains are barley, maize and wheat, but substantial quantities are derived from other grains around the world, such as wheat (for vodka), maize (for beer in central America and Africa), rye (for whiskey in the United States and *kvass* in Eastern Europe and former Soviet states), rice (for saké in Japan, *shaoshinchu*, *baijiu* and *shaojiu* in China and *soju* in Korea) and sorghum (for beer in Africa). Triticale may be used as an adjunct in beers as well. A few examples of alcohol-based drinks from around the world, and the microorganisms which are used for their fermentations, are provided in Table 12.1.

This chapter discusses the processes of malting, beer and spirits production and biofuels production.

TABLE 12.1 Some examples of cereal-based alcohol drinks from around the world and microorganisms used

Cereal grain	Drink	Microorganisms	Common locations
Barley	Beer	*Saccharomyces cerevisiae*	Worldwide
	Tella (beer)	*Lactobacillus pastorianumi*, *S. cerevisiae*	Eritrea, Ethiopia
	Whisky/whiskey	*S. cerevisiae*	China, Europe, India, Japan, United States
	Yakju (beer or wine)	*Aspergillus usamii*, *S. cerevisiae*	Korea
Maize	Busaa (beer)	*Lactobacillus plantarum*, *S. cerevisiae*	Kenya
	Chicha (beer)	*Lactobacillus* spp., *S. cerevisiae*	Central and South America
	Kaffir (bantu beer)	*Lactobacillus* spp., *S. cerevisiae*	Africa
	Munkoyo (beer)		Zambia

TABLE 12.1 Some examples of cereal-based alcohol drinks from around the world and microorganisms used—cont'd

Cereal grain	Drink	Microorganisms	Common locations
	Pito (beer)	*Candida, Lactobacillus, Leuconostoc, Saccharomyces* spp.	Ghana, Nigeria
	Tella (beer)	*L. pastorianumi, S. cerevisiae*	Eritrea, Ethiopia
	Tesguino (beer)	*Bacillus megaterium, S. cerevisiae*	Mexico
	Urwaga		Kenya
	Whiskey	*S. cerevisiae*	USA, worldwide
	Yakju (beer or wine)	*A. usamii, S. cerevisiae*	Korea
Millet	Baiju (*shaojiu*)		China
	Busaa (beer)	*L. plantarum, S. cerevisiae*	Kenya
	Ikigage		Rwanda
	Kaffir (bantu beer)	*Lactobacillus* spp., *S. cerevisiae*	Africa
	Mbege	*L. plantarum, S. cerevisiae*	Tanzania
	Munkoyo (Ibwatu beer)		Zambia
	Tella (beer)	*L. pastorianumi, S. cerevisiae*	Eritrea, Ethiopia
	Thumba	*Endomycopsis fibuliger*	Bengal
	Urwaga		Kenya
Rice	Hong-ru (wine)	*Rhizopus javanicus, Monascus purpureus, S. cerevisiae*	China, Taiwan
	Madhu (wine)		India
	Ruhi	*Lactobacillus, Mucor, Rhizopus* spp.	India
	Saké (wine)	*A. oryzae, L. sake, Pseudomomonas nitroreducens, S. cerevisiae*	China, Japan, Taiwan
	Shaoshinchu		China
	Soju/shochu	*R. javanicus, S. peka*	Canada, China, Japan, Korea, Taiwan, United States
	Tapé	*Amylomyces rouxii, Endomycopsis burtonii, R. chinensis*	China, Indonesia, Malaysia

Continued

TABLE 12.1 Some examples of cereal-based alcohol drinks from around the world and microorganisms used—cont'd

Cereal grain	Drink	Microorganisms	Common locations
	Tapuy (wine)	*Endomycopsis fibuliger*	Philippines
	Yakju (beer or wine)	*A. usamii, S. cerevisiae*	Korea
Rye	Kvass (beer)		Russia, Eastern Europe
	Whiskey	*S. cerevisiae*	Europe, United States
Sorghum	Burukutu	*Lactobacillus, S. cerevisiae*	Nigeria
	Busaa (beer)	*L. plantarum, S. cerevisiae*	Kenya
	Ikigage		Rwanda
	Kaffir (bantu beer)	*Lactobacillus* spp., *S. cerevisiae*	Africa
	Mwenge		Uganda
	Pito (beer)	*Candida, Lactobacillus, Leuconostoc, Saccharomyces* spp.	Ghana, Nigeria
	Sekete		Nigeria
	Tella (beer)	*L. pastorianumi, S. cerevisiae*	Eritrea, Ethiopia
	Urwaga		Kenya
Teff	Tella (beer)	*L. pastorianumi, S. cerevisiae*	Eritrea, Ethiopia
Wheat	Bousa (beer)	*Lactobacillus, S. cerevisiae*	Egypt
	Beer	*S. cerevisiae*	Worldwide
	Shaoxing (*sao-hsing*, rice wine)	*A. oryzae, A. rouxii, Rhizopus* sp., *S. cerevisiae*	Asia
	Tella (beer)	*L. pastorianumi, S. cerevisiae*	Eritrea, Ethiopia
	Whiskey	*S. cerevisiae*	United States, worldwide
	Yakju (beer or wine)	*A. usamii, S. cerevisiae*	Korea

Based upon data from Gelinas, P., Mckinnon, C., 2000. Fermentation and microbiological processes in cereal foods (Chapter 26). In: Kulp, K., Ponte Jr., J.G. (Eds.), Handbook of Cereal Science and Technology. Marcel Dekker, Inc., New York, NY, USA; Kubo, R., Kilasara, M., 2016. Brewing techniques of Mbege, a banana beer produced in Northeastern Tanzania. Beverages 2 (21), 1–10; Lee, M., Regu, M., Seleshe, S., 2015. Uniqueness of Ethiopian traditional alcoholic beverage of plant origin, tella. Journal of Ethnic Foods 2, 110–114; Papas, R.K., Sidle, J.E., Wamalwa, E.S., Okumu, T.O., Bryant, K.L., Goulet, J.L., Maisto, S.A., Braithwaite, R.S., Justice, A.C., 2010. Estimating alcohol content of traditional brew in Western Kenya using culturally relevant methods: the case for cost over volume. AIDS Behaviour 14 (4), 836–844; Lyumugabe, F., Uyisenga, J.P., Songa, E.B., Thonart, P., 2014. Production of traditional sorghum beer "ikigage" using Saccharomyces cerevisiae, Lactobacillus fermentum and Issatckenkia orientalis as starter cultures. Food and Nutritional Sciences 5, 507–515; and http://www.wikipedia.org.

12.2 MALTING

During malting, large molecular weight components of the endosperm cell walls, the storage proteins and the starch granules are hydrolysed by enzymes, rendering them more soluble in water – which is especially important for grain-out-style fermentations.

All cereal grains are capable of undergoing malting, but barley is particularly suitable because the adherent pales (lemma and palea – see Chapter 3) provide protection for the developing plumule, or acrospire, against damage during the necessary handling of the germinating grains. Further, the husk (pales) provides an aid to filtration when the malt liquor is being removed from the residue of insoluble grain components. A third advantage of barley lies in the firmness of the grain at high moisture content.

Both two-row and six-row barleys (see Chapter 3) are suitable: the former are generally used in Europe, the latter in North America. Distinct varieties were formerly grown for malting, but were lower yielding than varieties grown for feeding. Modern malting varieties have high yields and are thus also suitable for the less demanding alternative uses.

Four characteristics are required of a malting type of barley.

1. High germination capacity and energy, with adequate enzymatic activity.
2. Capacity of grains modified by malting to produce a maximum of extract when mashed prior to fermentation.
3. Low content of husk.
4. High starch and low protein contents.

These qualities can be affected by husbandry and handling as well as by genetic factors: loss of germination capacity can result from damage to the embryo during threshing, or overheating during drying or storage.

Provided that grains are ripe, free from fungal infestation and intact, the yield of malt extract should be directly related to starch content.

High-nitrogen barley is unsuitable for malting, however, for four reasons.

1. Starch content is lower.
2. Longer malting times are required.
3. Modification never proceeds as far as in low-nitrogen barleys.
4. The greater quantities of soluble proteins lead to haze formation, and may provide nutrients for bacteria and thus impair the keeping quality of the beer. The average nitrogen content of malting barley is 1.5%; some 38% of this appears in the beer in the form of soluble nitrogen compounds, the proportion of the total nitrogen entering the beer being somewhat larger from two-row than from six-row types.

12.2.1 Dormancy

Harvest-ripe barley may not be capable of germination immediately. While this is advantageous in the field, protecting the crop against sprouting in the ear, it is clearly a problem in relation to malting, which depends on germination occurring. The mechanism of dormancy is not fully understood, and indeed it is unlikely that a single cause is involved in all cases; in many instances it has been shown that germination is inhibited by the inability of the embryo to gain access to oxygen. A distinct phenomenon known as 'water sensitivity' can arise during steeping if a film of water is allowed to remain on the surface of the grains. The water contains too little dissolved oxygen to satisfy the needs of the developing embryo and acts as a barrier to the passage of air. Dormancy declines with time and storage is thus not just a means of holding sufficient stocks of grain, but is an essential part of the process of malting. During the storage of freshly harvested barley tests are performed to detect the time at which dormancy has declined sufficiently for malting to commence properly. Both 'dormancy' and 'water sensitivity' are defined in relation to the test performed. In one test 100 grains are germinated on filter papers with 4 and 8 mL of water; the difference between viability and the germination on 4 mL of water is called dormancy, while the difference between the levels of germination on the different volumes of water is the water sensitivity. Factors involved in controlling and breaking dormancy were reviewed by Briggs (1978).

12.2.2 Barley malting operations

The major practical steps in malting are shown schematically in Fig. 12.2. More specific details are provided in Fig. 12.3, which illustrates a typical engineering flow diagram for a malting facility. Overall plan and section views for typical malting plants are shown in Figs 12.4 and 12.5 for horizontal and vertical facility arrangements, respectively.

Selected barley is 'steeped', usually by immersion in water, for a period chosen to achieve a particular moisture level in the grain. The water is drained from the grain, which then germinates. Conditions are regulated to keep the grain cool (generally below 18°C) and minimize water losses. As the grain germinates the coleoptile (acrospire) grows beneath the husk and pericarp while the 'chit' (coleorhiza, root sheath) appears at the base of the grain, which is split by the emerging rootlets (as illustrated in Fig. 12.6).

At certain intervals the grain is mixed and turned to provide more uniform growth opportunities and prevent the roots from matting together. As the embryo grows it produces hormones, including gibberellic acid, stimulating production of hydrolytic enzymes in the scutellum and aleurone layer which lead to 'modifications' of the starchy endosperm. The malting process is regulated by the initial choice of barley, the duration of growth, the temperature, the grain moisture content, changes in

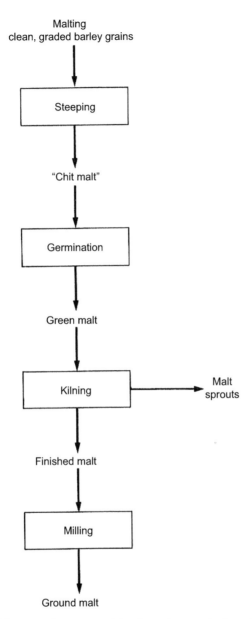

FIGURE 12.2 Flow chart summarizing the major steps in the malting process.

the steeping schedule and the use of additives. When modifications are deemed sufficient (according to intended use) the germination process is stopped by kilning the 'green malt' – that is, by drying and cooking it in a current of hot, dry air. The dry, brittle culms are then separated and the finished malt is transported to storage bins. Dry malt is stable in

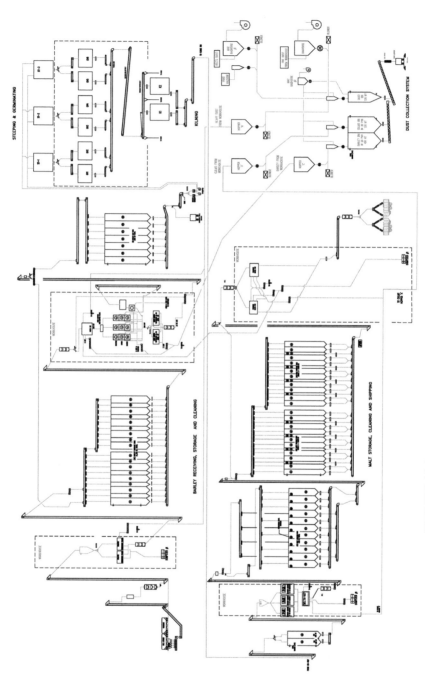

FIGURE 12.3   Process flow diagram for a typical commercial malting plant.

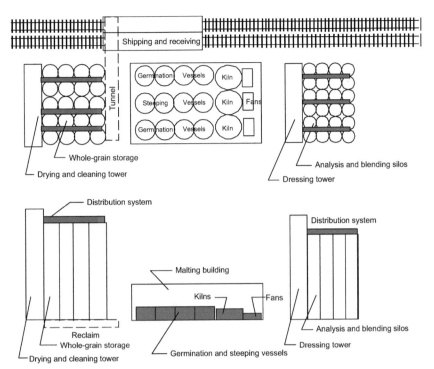

FIGURE 12.4 Plan and section views of horizontal-style malting plant layout. *Adapted from Williams, G.D., Rosentrater, K.A., 2005. Design Considerations for the Construction and Operation of Malting Facilities. Part 1: Planning, Structural, and Life Safety Considerations. Paper 054095. Annual Meeting of the American Society of Agricultural and Biological Engineers, Tampa, FL, USA.*

storage and, unlike raw barley, is readily crushed. The conditions of kilning are critical in determining the organoleptic character of the malt: it can cause a slight enhancement of the various attributes found in green malt or can completely destroy them. Malt contains relatively large quantities of soluble sugars and nitrogenous substances and, if it has been kilned at low temperatures, it contains high levels of hydrolytic enzymes. When crushed or milled malt is mixed with warm water the enzymes will then catalyse hydrolysis of the starch, other polysaccharides, proteins and nucleic acids, regardless of whether these nutrients are from the malt itself or from materials that are mixed with it. The solution of the products of hydrolysis extracted from the malt/water mixture is called the 'wort'. It forms the feedstock for fermentation for brewing or distillation (Briggs, 1978). Ultimately, malt also confers colour, aroma and flavour to the final fermented product.

One of the benefits derived from the application of engineering and technology in malting has been the reduction in resources, energy and time required to produce satisfactory malts. The amount of time saved

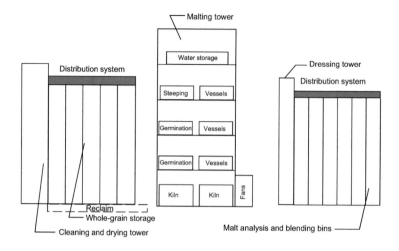

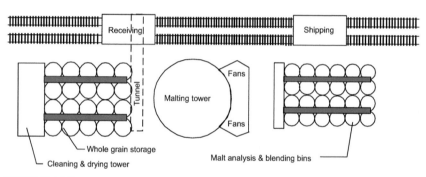

FIGURE 12.5 Plan and section views of vertical-style malting plant layout. *Adapted from Williams, G.D., Rosentrater, K.A., 2005. Design Considerations for the Construction and Operation of Malting Facilities. Part 1: Planning, Structural, and Life Safety Considerations. Paper 054095. Annual Meeting of the American Society of Agricultural and Biological Engineers, Tampa, FL, USA.*

FIGURE 12.6 Diagrammatic longitudinal sections through barley grains in the early stages of germination. 1, Imbibed grain; 2, rootlets emerged; 3, rootlets and coleoptile emerged. *From Briggs, D.E., 1978. Barley. Chapman and Hall Ltd., London, UK. Reproduced by courtesy of Chapman and Hall Ltd.*

can be inferred from the diagram in Fig. 12.7. It is clear that the greatest savings have occurred during the last century and that savings are made in all stages of malting, although the greatest benefits have been achieved in the germination stage.

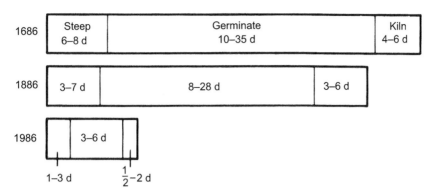

FIGURE 12.7 Diagram showing the reduction, over three centuries, in periods needed for malting and the three stages involved in the process.

12.2.3 Steeping

Traditional malting included a ditch steep followed by germination in heaps on the floor. It was a labour-intensive process, as the heaps or 'couches' required frequent manual turning, and it was very time consuming.

Current practice varies according to the size of operation and the preferences of the maltster, but self-emptying steep tanks have replaced the ditch steep. Vessels may be flat-bottomed or conical-bottomed tanks (Fig. 12.8). They have equipment for water filling and emptying, and compressed air blowers provide both aeration and 'rousing' and mixing during the steeping process. Together, these processes combine to remove carbon dioxide which accumulates as a result of respiration of grains and microorganisms associated with them. Vigorous aeration immediately after loading the barley into the steep tank also serves to raise dust, chaff and light grains to the surface for removal. These are accumulated and sold for livestock feed.

Aeration, damping and temperature must all be carefully controlled to ensure that germination occurs at the required rate and to the required degree. For poorly modified traditional pale lager, European two-rowed barley needs steeping to 41%–43% m.c., while for a pale ale malt of 43%–45% m.c. is appropriate. For a high-nitrogen barley (say 1.8% N) destined for a vinegar factory, 46%–49% m.c. may be preferred. Higher moisture levels induce faster modifications in the barley but greater losses are incurred (Briggs, 1978). Anaerobic conditions are dangerous, as they favour fermentation by the microorganisms present and the alcohols produced can harm the grain. On the other hand, excessive aeration leads to 'chitting' under water and, consequently, an unwanted rise in moisture content.

It is desirable to replace the steeping water with fresh water for a given batch of grain (between one and four times), as phosphates and organic compounds, including alcohols, accumulate and the microbial population grows. During the first steep, dissolved oxygen is depleted from steep

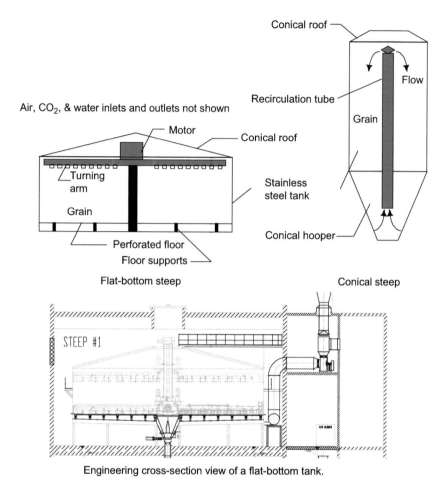

FIGURE 12.8 Typical steeping vessels. *Adapted from Williams, G.D., Rosentrater, K.A., 2005. Design Considerations for the Construction and Operation of Malting Facilities. Part 1: Planning, Structural, and Life Safety Considerations. Paper 054095. Annual Meeting of the American Society of Agricultural and Biological Engineers, Tampa, FL, USA.*

water at a rate of ~1 ppm/h, but this rate of depletion rises tenfold by the third steep (complete depletion is possible within 1 h). Temperature is controlled to a degree by the water temperature to between 10 and 16°C.

There are many variations in steeping practice, including steeping in running water ('Bavarian') or the 'flushing' regime, in which immersions are frequent but cover the grains for only a few minutes. Modern steep tanks are filled with barley to a depth of about 1.2 m. This increases to approximately 1.8 m when the grain is swollen. The uniform depth of flat-bottomed tanks provides for more uniform aeration and CO_2 removal than in conical-bottomed vessels.

Following steeping, the grain is transferred to germinating vessels. This may be by 'wet-casting', whereby it is pumped in water suspension, or by 'dry-casting' (mechanical conveyor) after draining. If additives are to be used it is convenient to add them during transfer. They may include gibberellic acid (https://en.wikipedia.org/wiki/Gibberellic_acid) or potassium bromate (https://en.wikipedia.org/wiki/Potassium_bromate). The former hastens malting, especially if bruised grains are present, while potassium bromate reduces respiration and hence the rise in temperature that accompanies it. It is also said to inhibit proteolysis and control colour development in the malting grain, as well as to reduce malting loss by reducing root growth.

12.2.4 Germination

Early mechanical maltings had rectangular germination vessels; this method was supplanted by the drum, which provided ideal control but was limited to about 100 tonnes capacity and had a high unit cost. Later, circular germination vessels were developed and capacities rose to 500 tonnes or more. Features common to all mechanical germination vessels are an automatic means of turning the germinating grains and a means of aeration. Turning may be performed by rotating spirals or augers that are moved slowly through the grain mass. In the rectangular vessels they move end to end on booms, while in the circular type vertical turners rotate on a boom, or alternatively remain stationary while the grain is transported past on a rotating floor. It is common for aeration to be provided by air passing up (usually) or down, from or into a 'plenum' beneath the floor, which is constructed of slotted steel plates.

The layout of a typical circular germination vessel is shown in Fig. 12.9. Fig. 12.10 illustrates two variants of malting towers; each has separate levels for steeping and germinating.

Temperature and humidity are controlled by humidifying and refrigerating or warming the air which passes through the grain mass. Air volumes passing are of the order of 0.15–0.2 m^3/s/tonne of barley. Temperatures of 15–19°C are common. Microprocessor control of conditions is commonplace (Gibson, 1989).

In these types of systems the danger of microbial contamination is high, as air passing with high humidity is ideal for growth of bacteria and fungi, and nutrients are plentiful. As well as introducing health hazards and off-flavours, microbes gain preferential access to oxygen, thus potentially inhibiting the germination and modification of the barley for which the system is designed. In some plants the germination and kilning are carried out in a single vessel, and this has the advantage of the microbes being killed by the heat of kilning. Cleaning the dry residue is easier than complete removal of the wet remains of germinating grain.

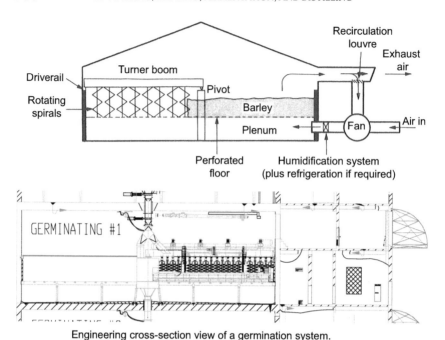

Engineering cross-section view of a germination system.

FIGURE 12.9 Typical circular germination vessel. *Based, in part, upon Williams, G.D., Rosentrater, K.A., 2005. Design Considerations for the Construction and Operation of Malting Facilities. Part 1: Planning, Structural, and Life Safety Considerations. Paper 054095. Annual Meeting of the American Society of Agricultural and Biological Engineers, Tampa, FL, USA.*

12.2.5 Kilning

The objectives of kilning are to arrest botanical growth and internal modification of the barley, to reduce moisture for grain storage and to develop colour and flavour compounds in the malt. Kilning is responsible for 90% of the energy consumption of the entire malting process unless a heat recovery unit is in use, when the proportion may be reduced to ~75%–80%.

For kilning, ambient air is heated by combustion of fuel (most often natural gas) and passed under positive or negative pressure through the bed of grains. A plenum below the floor is used, as in the earlier stages of the malting process. A recent innovation in kiln design is the multideck kiln. In this system green malt is loaded on to the upper decks and progressively transferred to the lower decks after partial drying. Warmed air is passed first through the drier, lower beds before passing, unsaturated, to the upper beds where it is capable of removing moisture from the green malt (Fig. 12.11).

The depth of green malt in a modern kiln is ~0.85–1.2 m. For maximum efficiency the bed should be level and uniformly compacted, a condition readily achieved by the automatic or semiautomatic loading machinery available today.

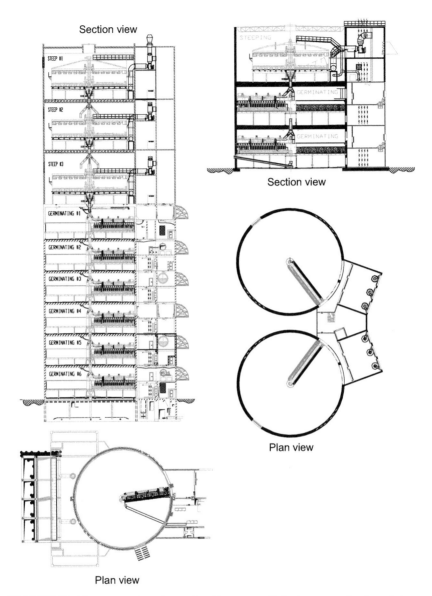

FIGURE 12.10 Engineering views of malting towers. Left: three steep levels and six germinating levels. Right: one steep level and two germinating levels.

Conditions for kilning are determined by the nature of the end product required. Variables include the extract potential, moisture content, colour, flavour profile and enzyme activity. Curing temperatures range from 80 to 100°C. In modern kilns the maltster is assisted in monitoring and controlling conditions by computerized control programs incorporated into automated systems.

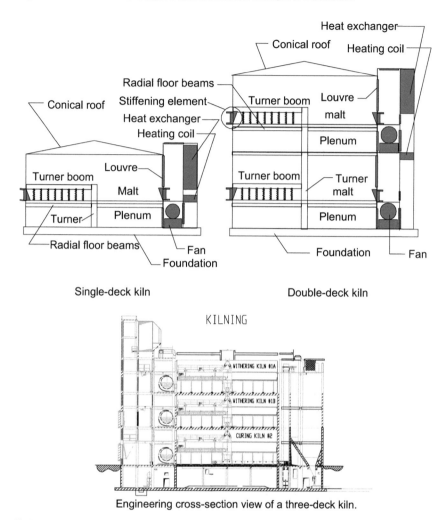

Single-deck kiln   Double-deck kiln

KILNING

Engineering cross-section view of a three-deck kiln.

FIGURE 12.11 Typical single-deck and multideck kilns. *Adapted from Williams, G.D., Rosentrater, K.A., 2005. Design Considerations for the Construction and Operation of Malting Facilities. Part 1: Planning, Structural, and Life Safety Considerations. Paper 054095. Annual Meeting of the American Society of Agricultural and Biological Engineers, Tampa, FL, USA.*

Enzymes survive high curing temperatures best if the malt is relatively dry, but under these conditions colour and flavour development are minimal. They develop mainly as a result of Maillard reactions occurring between the reducing groups of sugars and amino groups. Development of additional flavour compounds is favoured by the combination of high temperatures with wet malts (Briggs, 1978). Other factors affecting malt colour may be added caramelized sugars and oxidized polyphenols, and other contributions come from aldehydes, ketones, alcohols, amines and miscellaneous other substances, including sulphur-containing compounds and nitrogenous bases. These are discussed in detail by Briggs (1978) and Palmer (1989).

Peated distillery malts for Scotch whisky manufacture take up many substances from peat smoke and contain various alkanes, alkenes, aldehydes, alcohols, esters, fatty acids and aromatic and phenolic substances, including phenol and cresols. The use of peat as a fuel for kilning originated in cottage-industry enterprises in the crofts of the Scottish Highlands where peat was the only available fuel for domestic heating. In lowland distilleries peat heating has now been superseded by a succession of fuels, most often natural gas. Direct heating by these fuels is not permitted, though, as they can lead to introduction of nitrosamines and other deleterious chemicals into the product. Peatiness is a valued character of malt whiskies, and direct heating with peat is the rule in their production.

For production of brewing malts a major economic consideration is brewers' extract. This is measured as hot water extract available as the soluble nitrogen required to maintain fermentation and beer quality properties. An analysis of ale malt compared with that of barley is shown in Table 12.2. The characteristics of brewers' malts of various types are given in Table 12.3, and some are illustrated in Fig. 12.12.

TABLE 12.2   Analytical characteristics of barley and ale malt

Constituent	Barley	Ale malt
Moisture content (%)	15	4
Starch (%)	65	60
β-D-glucan	3.5	0.5
Pentosans (%)	9.0	10.0
Lipid (%)	3.5	3.1
Total nitrogen (%)	1.6	1.5
Total soluble nitrogen (%)	0.3	0.7
α-Amino nitrogen (%)	0.05	0.17
Sucrose (%)	1.0	2.0
Minerals (%)	approx. 2	approx. 2
Colour (°EBC)	<1.5	5.0
Hot water extract (1°/kg)	150	305
Diastatic power (*beta*-amylase °L)	20	65
Dextrinizing unit (*alpha*-amylase)	<5	30
Endo β-glucanase (IRV units)	<100	500

Based upon data from Palmer, G.H., 1989. Cereals in malting and brewing (Chapter 3). In: Palmer, G.H. (Ed.), Cereal Science and Technology. Aberdeen Univ. Press, UK. Reproduced by courtesy of Aberdeen University Press.

TABLE 12.3   Characteristics of a selection of malts

Malt type	Extract (°/kg)	Moisture (%)	Colour (°EBC)	Final kilning temperature (°C)
Ale[a]	305	4.0	5.0	100
Lager[a]	300	4.5	2.0	80
Cara Pils	265	7.0	25–35	75
Crystal malt	268	4.0	100–300	75
Amber malt	280	2.0	70–80	150
Chocolate malt	268	1.5	900–1200	220
Roasted malt	265	1.5	1250–1500	230
Roasted barley	270	1.5	1000–1550	230

[a] Of all the malts listed only ale and lager malts contain enzymes.
Based upon data from Palmer, G.H., 1989. Cereals in malting and brewing (Chapter 3). In: Palmer, G.H. (Ed.), Cereal Science and Technology. Aberdeen Univ. Press, UK. Reproduced by courtesy of Aberdeen University Press.

12.2.6   Ageing

Before use it is necessary to mill the kilned malt, but it is customary to delay this process to permit moisture equilibration. Kilning results in rapid drying of the bulk grain, but in individual grains a gradient will actually exist, from a higher inner to a lower outer (husk) moisture content. Differences in a malt of 3%–5% m.c. may actually be 4 or 5 percentage points (1%–3% m.c. outside to 5%–8% m.c. inside). Unless equilibrated, agglomeration of the damper parts can reduce extraction potential and undue fragmentation of dry husk can lead to haze in the extract (Pyler and Thomas, 1986). Storage for up to 3 months may be used. As specifications become increasingly sensitive to moisture content, the conditions of storage progressively include humidity control (Palmer, 1989).

12.2.7   Energy consumption and other costs

The Energy Technology Support Unit published a report in 1985 of a survey of British maltsters. Specific energy consumed per tonne of malt ranged from 2.48 to 6.81 GJ, with a weighted average of 3.74 GJ. Quoted costs included fuel and electricity, and power costs encompassed grain handling and process requirements. Additional estimates include 1058–1534 kWh/tonne for total energy consumption, 900–1200 kWh/tonne for heating air in the kiln, 25–75 kWh/tonne for air movement in the kiln and 27–44 kWh/tonne for steeping and germination (Briggs, 1998).

FIGURE 12.12 Examples of various malts.

Estimates of proportionate costs, including these values, are given by Gibson (1989) as:

Fuel 25%–30%
Electricity 15%–20%
Wages 15%–20%
Repairs/maintenance 10%–20%
Miscellaneous 15%–25%

As with many wet cereals processes, the costs of water treatment before discharge are increasing as environmental standards become more stringent. The biological oxygen demand (BOD) load from a 30,000 tonnes per annum malting is equivalent to a population of about 9000 persons (Gibson, 1989). It has been estimated that 44 million m³ water is used per annum for malting, which results in 30 million m³ of wastewater (http://www.ukmalt.com/water).

12.2.8 Malt production

Palmer (1989) reported that about 17 million tonnes of barley was used worldwide in 1989 to produce 12 million tonnes of malt and about 970 million hL of beer. This represented about 10% of world barley production. In 2015 global malt production was reported to be nearly 23 million tonnes, of which approximately 9.7 million tonnes came from Europe (Euromalt, 2017).

12.2.9 By-products of malting

The main by-product of malting is called 'malt sprouts' or 'culms' (Fig. 12.13). They are separated from the kilned malt by passing the malt through revolving reels or a wire screen. They account for 3%–5% of product and are incorporated into livestock feeds. Typically they contain 25%–34% N compounds, 1.6%–2.2% fat, 8.6%–11.9% fibre, 6.0%–7.1% ash and 35%–44% N-free extract (Pomeranz, 1987). More information is available from the Maltsters' Association of Great Britain (http://www.ukmalt.com/malting-co-products) and Feedipedia (www.feedipedia.org).

12.2.10 Nonbrewing uses of malt

Milled barley malt is used as a high-diastatic (i.e., amylolytic) supplement for bread flours, which are low in natural diastatic activity, and as a flavour supplement in malt loaves. Malt extracts and syrups are produced by concentrating worts by evaporation. Malt is also used in the manufacture of malt vinegar, drink products such as Ovaltine, candy and confectionery products such as Whoppers and Malted Milk

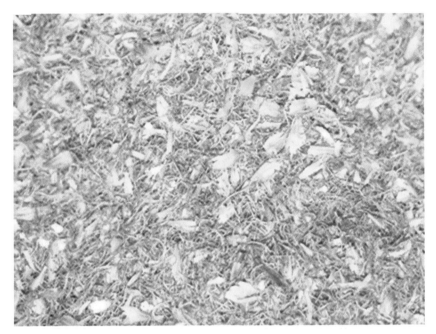

FIGURE 12.13 Malt sprouts which have been separated from the malt are used for cattle feed because of their protein and fibre levels.

Balls, and foods such as bagels, breads and breakfast cereals. Malt and malt extract have also been used as sweeteners in food products (www. maltproducts.com).

12.2.11 Adjuncts

Although malt derived from barley is generally considered to be the superior feedstock for brewing and distilling, it is common practice in many countries to supplement malt with alternative/additional sources of soluble sugars or starch capable of conversion to soluble sugars.

The principal adjuncts, as such nonmalt additives are termed, are rice, maize grits and cereal starches. Adjuncts contribute virtually no enzymes to the wort, so hydrolysis of their starch depends upon the enzymes present in the malt to which they are added. Use of adjuncts is common practice in the United States, and this is one reason for the preference there for the higher-enzyme-containing six-row barleys.

It has been estimated that in the United States 38% of total materials used in brewing (excluding hops) were adjuncts (Pyler and Thomas, 1986). Of these, 46.5% was corn grits, 31.4% rice and 0.7% barley. Sugars and syrups accounted for the remaining 21.4%.

The form in which rice is added is typically broken grains that do not meet the requirements for milled rice. As the quality of the products of fermentation is little affected by the physical nature of the adjunct (if preserved appropriately), the choice is usually made purely on economic grounds. This is not related only to the price per tonne of the adjunct, though, because the yield of extract is not the same from each. Tests for extraction carried out in the laboratory generally give higher values than those obtained in commercial practice. For example, Pyler and Thomas quote 78% for rice and 74% for maize grits in the brewhouse, but 87%–94% and 85%–90%, respectively, in the laboratory (American Society of Brewing Chemists procedure). Maize grits also contain higher levels of fat and protein than rice – constituents, which are considered undesirable.

Other adjuncts commonly used are refined maize starch, wheat and wheat starch, rye, oats, potatoes, tapioca, triticale, heat-treated (torrefied or micronized) cereals and cereal flakes. Micronization involves heating grains to nearly 200°C by infrared radiation, while torrefication achieves similar temperatures by use of hot air (Palmer, 1989). In grains treated by either method the vaporized water produced disrupts the physical structure of the endosperm, denaturing protein and partially gelatinizing starch. Digestibility is thus increased, and these products are also used in cattle feeds and whole-grain baked products. Solubility of proteins may be decreased and some flavours may be introduced through their use if adequate treatment temperatures are used in processing. Heat-treated cereals can be added to malt before grinding; their extract yield is increased if they are precooked before use. Investigations with extruded cereals gave poorer extract yields than traditional adjuncts, however (Briggs et al., 1986).

Sorghum was used considerably in the United States when maize was in short supply during World War II, while today it is used to a significant extent in Mexico and parts of Africa (examples include *bil-bil* in Cameroon, *burukutu* in Nigeria, *pombe* in East Africa and *bjala bja setso* in northern Sotho). Sorghum has lower fat and protein contents but a higher extract than maize, so it has some merit as a substrate. Moreover, it has seen a resurgence in the United States in recent years due to development of gluten-free beers (see, for example, http://www.bardsbeer.com, http://www.redbridgebeer.com).

Barley and wheat starch have lower gelatinization temperatures than maize and rice starch, hence digestion may occur at mash temperatures. It is usual, however, to premash maize, rice, wheat, barley, etc. by cooking with a small amount of malt before adding them to the mash.

Addition of barley provides a means of reducing the nitrogen content of the wort. It is disallowed, however, by the German beer law for the production of bottom-fermented beers, in which only barley malt, hops, yeast and water are allowed. Top-fermented beers follow the same regulations

TABLE 12.4 Beers made from wheat malt and their characteristics

Type	Character	Origin	Alcohol (% v/v)	Flavour features
Weizenbeer	Lager/ale	Bavaria	5–6	Full-bodied, low hops
Weisse	Lager	Berlin	2.5–3	Light flavoured
Gueuze-Lambic	Acid ale	Brussels	5+	Acidic
Hoegards wit	Ale	East of Brussels	5	Full-bodied, bitter

Based upon data from Pomeranz, Y., 1987. Barley (Chapter 18). In: Pomeranz, Y., (Ed.), Modern Cereal Science and Technology. UCH Publishers Inc., NY, USA, using information from Leach, A.A. of the Brewers' Society, UK.

but wheat malt may be included (Narziss, 1984). For special beers, pure beet-cane-invert sugar is allowed.

12.2.12 Malts from other cereals

In Africa many malts are produced from sorghum and, to a lesser extent, millets and other cereals (e.g., teff). It has been reported that in the Republic of South Africa commercial production using sorghum may be in the order of 1×10^9 L annually, and home brewing may be in the same order (Novellie, 1977).

Wheat malt is used in the production of wheat malt beers. Examples of these and their characteristics are shown in Table 12.4.

Malts made on a pilot scale from triticales grown in the United Kingdom were evaluated by Blanchflower and Briggs (1991). Viscosities of resulting worts were high due to pentosans, particularly arabinose and xylose. Hot water extracts after 5 days germination were 302–324 L°/kg. Filtered worts were turbid owing to proteinaceous materials. Malt yields were between 87% and 90%.

12.3 BREWING

12.3.1 Beer

The overall brewing process is summarized schematically in Fig. 12.14.

12.3.2 Wort production

The starting material for brewing may be pure (usually barley) malt or a mixture of malt and adjuncts. If solid adjuncts are to be included they

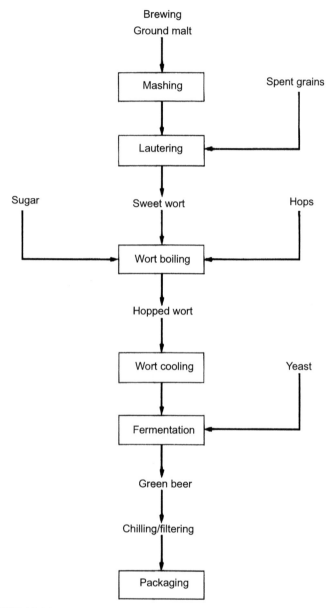

FIGURE 12.14 Flow chart summarizing the major steps in the brewing process.

may be milled with the malt. The coarsely ground material, upon hydration with brewing liquor (i.e., water, salts, sugars, syrups, etc.), produces a brewers' extract from the soluble materials in the malt and a filter bed from the husk. The quality of the filter bed depends on the size of the husk particles; they should not be too fine. The process is known as 'mashing' and is carried out in vessels called mash tuns.

After an initial rest for hydration, the temperature is raised above the gelatinization temperature of the starch. This renders the starch much more susceptible to digestion by amylase enzymes, to produce soluble sugars. The process of conversion, begun during malting, thus continues during this phase.

It is now necessary to separate the liquid wort from the solid remains of the malt and adjuncts, which is done by a process called 'lautering'. The spent grains (known as brewers grains and discussed in more depth later in this chapter) act as a filter bed when the mixture is transferred to a lauter tub (tun), which has a perforated/screened bottom. The spent grains accumulate on this and allow the liquid to pass through while retaining the solids. The sugary liquid is known as the 'sweet wort'; it may be supplemented with syrups, sugars, caramel, etc. at this stage if such adjuncts are to be used to increase the amount of fermentable sugars for fermentation. Hops are also added at this stage. As well as adding flavour they serve to sterilize the wort (hop acids have antimicrobial properties) and participate in reactions that precipitate proteins responsible for haze when the wort is boiled.

The sweet wort is placed in another tank, and the temperature is raised to boiling. Boiling may continue for 1.5–2.0 h. During boiling the humulones, or alpha-acids, are isomerized to the bitter iso-alpha-acids, enzymes are inactivated and contaminating bacteria are sterilized. Because the yield of bitter iso-alpha-acids extracted from the hops by boiling may be as low as 30%, a modern procedure is to replace part of the raw hops by a preisomerized hop extract, which is added to the beer after fermentation. It is common for half the hops to be added at the beginning of the period and half at the end. The wort is cooled (to 15.5–18°C for ales and 4–7°C for pilsners and lagers), filtered, then transferred to pitching tanks where it is 'pitched' with yeast (i.e., yeast is added). Air is also passed into the hopped wort (aeration) to provide a supply of oxygen for the yeast, which rapidly becomes active (added oxygen reduces the lag phase that would otherwise occur). Further information is provided in Briggs (1998).

12.3.3 Fermentation

Yeasts vary in their behaviour during fermentation: some strains tend to flocculate, and as a result they trap CO_2 and rise to the top. Others, which do not flocculate, sink to the bottom. Several styles of lagers are produced by bottom fermentation, while many types of ales and stouts are produced using top fermentation. Some examples of common beers are provided in Table 12.5.

TABLE 12.5   Classical beers of the world classified according to yeast types used in their brewing

Type	Character	Origin	Alcohol (% v/v)	Flavour features
BOTTOM-FERMENTED				
Münchener	Lager/ale	Munich	4–4.8	Malty, dry, moderately bitter
Vienna (Märzen)	Lager	Vienna	5.5	Full-bodied, hoppy
Pilsner	Lager	Pilsen	4.5–5	Full-bodied, hoppy
Dortmunder	Lager	Dortmund	5+	Light hops, dry, estery
Bock	Lager	Bavaria, United States, Canada	6	Full-bodied
Dopplebock	Lager/ale	Bavaria	7–13	Full-bodied, estery, winey
Light beers	Lager	U.S.	4.2–5	Light-bodied, light hops
TOP-FERMENTED				
Saissons	Ale	Belgium, France	5	Light, hoppy, estery
Trappiste	Ale	Belgian and Dutch abbeys	6–8	Full-bodied, estery
Kölsch	Ale	Cologne	4.4	Light, estery, hoppy
Alt	Ale	Düsseldorf	4	Estery, bitter
Provisie	Ale	Belgium	6	Sweet, ale-like
Ales	Ale	United Kingdom, United States, Canada, Australia	2.5–5	Hoppy, estery, bitter
Strong/old ale	Ale	United Kingdom	6–8.4	Estery, heavy, hoppy
Barley wine	Ale/wine	United Kingdom	8–12	Rich, full, estery
Stout (bitter)	Stout	Ireland	4–7	Dry, bitter
Stout (Mackeson)	Stout	United Kingdom	3.7–4	Sweet, mild, lactic, sour
Porter	Stout	London, United States, Canada	5–7.5	Very malty, rich

Based upon data from Pomeranz, Y., 1987. Barley (Chapter 18). In: Pomeranz, Y., (Ed.), Modern Cereal Science and Technology. UCH Publishers Inc., NY, USA, using information from Leach, A.A. of the Brewers' Society, UK.

A typical type of fermenter is a deep cylindrical vessel with a conical base (Fig. 12.15) into which the yeast eventually sediments (which eases cleaning and sterilization of the equipment after fermentation is complete) – flat-bottomed tanks present draining, cleaning and sanitation challenges.

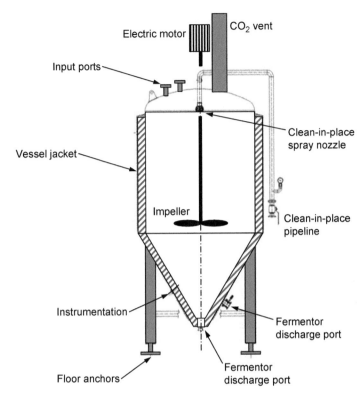

FIGURE 12.15 Conical bottom fermentation vessel.

In fermentation vessels the liberation of carbon dioxide can provide agitation, but in some cases sparging or impellers may also be used to increase agitation, which prevents the glucose substrate from becoming rate-limiting to the fermentation reactions. Accounting for the filling, fermentation, emptying and cleaning to be accomplished, a complete cycle time of approximately 5 days at 12°C or 2.5 days at 18°C is common.

As long as a substrate is available, fermentation could continue for 7–9 days, producing ethanol and carbon dioxide (in approximately equal proportions). The gas may then be collected for sale or adding back when the beer is bottled or casked, or it may be released to the atmosphere. The fermentation reaction is exothermic and the temperature would rise unduly if not controlled by heat exchangers or jackets surrounding the fermentation vessel. The yeast cell mass also increases during fermentation through asexual reproduction, and this constitutes an additional by-product. During fermentation the pH drops from ~5.2 to ~4.2 as a result of acetic and lactic acids synthesized by contaminating bacteria inevitably introduced with the yeast during the process. The green (jargon for young) beer is separated from the aggregated yeast cells and cooled to precipitate further haze-producing proteins, and then aged before filtering and carbonating. Lagers are stored at ~0°C for some weeks ('lagering') before packaging into bottles or kegs. The beer may also be pasteurized. An alternative treatment is for the green beer to be placed into casks, primed with sugars to permit secondary fermentation and 'fined' (or clarified) by including isinglass (a collagen from fish), which coagulates with the yeast for ready separation from the beer – this is then sold as 'naturally conditioned' beer, a product peculiar to the British Isles. Some common beers are listed in Table 12.5. Further information is provided in Briggs (1998).

Further technical details about malting and brewing are given in Bamforth (2006). Some of the latest developments in the brewing of beers can be found in Lewis and Young (2013).

12.3.4 Saké

Saké production was reported to be 15×10^5 kL in 1985; of this 10×10^5 kL (or ~2/3) were produced in Japan. In 2009 approximately 1.7×10^6 kL were consumed, but its popularity among 20-year-olds was only 1/2–1/4 that of older age groups (Kanauchi, 2013).

An essential difference between beer and saké is that for saké the natural enzymes present in the grain are not employed in solubilizing the starch – indeed, they are expressly deactivated before the saccharifying phase of saké brewing. Enzymes are of course needed, and these are derived from the fungus *Aspergillus oryzae*; they are provided from a culture of that mould known as 'koji', in which the microorganism is added to steamed rice and incubated. Koji contains 50 different enzymes, including the *alpha*- and *beta*-amylases present in malt, and an additional amylolytic enzyme, glucoamylase, capable of hydrolysing starch polymers to glucose. The balance of amylase to protease is influenced by cultural conditions, with higher temperatures favouring amylase production. The saké production process is summarized in Fig. 12.16 and more information about the process can be found at http://www.japansake.or.jp/sake/english/sake-basics/process.html.

The yeast used in saké production is a specialized strain of *Saccharomyces cerevisiae*.

Unlike the husk of barley in the malting process, rice husk is not valued for its filtration properties, nor indeed any other qualities. It is removed by milling. The degree of polishing is severe, typically removing 25%–30% of the brown rice weight, and in extreme cases up to 50%. Following milling the rice is washed, steeped and steamed for 30–60min. For 1 tonne of rice approximately 25kg of water is used during processing (Yoshizawa and Kishi, 1985).

Koji is added to the main mash in a seed mash: steamed rice in which an inoculum of *A. oryzae* has been cultured. Steamed rice and seed mash are added in equal proportions, and water and saké yeast are also added. Quantities involved in a mash are usually 2–7 tonnes,

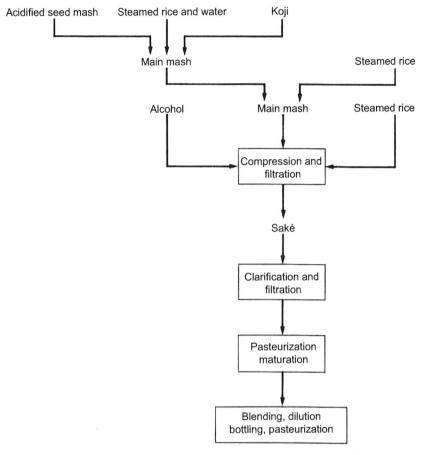

FIGURE 12.16   Flow chart summarizing the major steps in saké production.

but over 10 tonnes is possible. The seed mash is acidic as a result either of lactic acid bacteria present or of added lactic acid. The acid conditions inhibit wild microorganisms. After 2 days at 12°C the yeast population reaches the high concentration originally present in the seed mash (10^8 cells/g) and an equal amount of steamed rice and water are again added. A third addition is made the following day and fermentation increases in vigour, raising the temperature from about 9°C to 15–18°C. By 15–20 days after the final addition the alcohol content rises to 17%–19% and fermentation virtually ceases. The high alcohol content is attributable to four factors.

1. The tolerance of the yeast.
2. The low sugar content of the mass at any time.
3. The solid matrix of the mash.
4. The proteolipids in koji.

The alcohol content of the liquid is further increased to 20% before filtration and pasteurization of the saké. Maturation takes 3–8 months, after which water is added to give 15%–16% alcohol content. Bottling takes place after filtration through activated carbon to improve colour and flavour.

The by-products from the process, liquid lees and solid cake (*kasu*), are typically packaged and sold as human food products. On average, 1 tonne of rice will yield 3 kL of sake and 200 kg of *kasu* (Kanauchi, 2013).

12.4 DISTILLED SPIRITS

Distillation is a process of evaporation and recondensation used to separate liquids into various fractions according to their boiling temperature ranges. In the context of beverages it is used to produce potable spirits with an alcohol content above that of fermented drinks.

Spirits produced from grains are of two major categories, whisky (or whiskey, according to its origin – whisky is only produced in Scotland!) and neutral spirits. In whiskies care is taken to retain flavours and colours carefully introduced during production (i.e., produce character), while in neutral spirits the goal is to avoid introduction of flavours and colours during production, although flavours may be added later, for example in gin.

Whiskies of several types exist, their names denoting the carbohydrate source from which they are derived, the manner of their production and sometimes their origin. The essential characteristics of some are briefly described below. Standards of identity in the United States are provided in 27 CFR 5.22.

Scotch malt whisky is produced by traditional methods, using malt as the sole carbohydrate source (by legal definition – http://www.legislation.gov.uk/uksi/2009/2890/contents/made).

Grain whiskey is made from a mash of cooked grain and saccharified by the action of enzymes from malted barley. The grain may be maize, barley, wheat or others.

In Bourbon whiskey the grain in the mash consists of at least 51% (and usually 60%–70%) maize. Typically some rye is included to impart a spicy, estery flavour. Although included, malt contributes only 10%–15% of total carbohydrate.

Rye whiskey contains at least 51% of rye grain. Irish whiskey is made predominately from malted or unmalted barley. For both of these, wheat, rye or oats make up the remainder. Canadian whiskies are mainly blends of neutral grain whiskies (~90%) and Bourbon or rye whiskeys (Nagodawithana, 1986).

Figs 12.17(a) and (b) AB show some examples of a typical whisky/whiskey distillery.

FIGURE 12.17A Grain whiskey fermentation, showing top of the fermentor vessel (including ports for inspection, ingredient addition and CO_2 off-gassing), water cooling on the outside of the vessel for temperature control during fermentation, filling the fermentor with water and mash and CO_2 bubbles during active fermentation.

12.4.1 Traditional malt whisky (Scotch)

The traditional whisky-making process is summarized in Fig. 12.18. Well-modified peated malt is dried at relatively low temperatures (to retain enzyme activity), milled and mashed at temperatures of 64–65°C.

FIGURE 12.17B   Barrel storage warehouse.

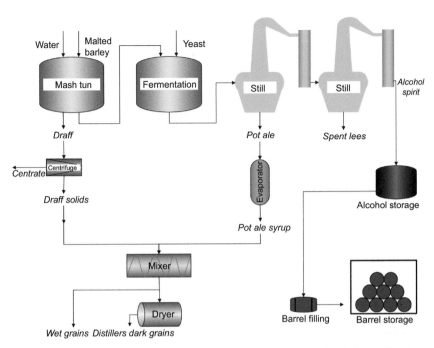

FIGURE 12.18 Flow chart summarizing the major steps in Scotch whisky production.

The first worts are run off before mashing at a higher temperature of 70°C and running off a second wort. Both worts become the liquor that is fermented. Two further worts are produced from mashes at 80 and 90°C, and are used as mashing liquor for subsequent grists. Batches of 5–10 tonnes of malt are typical.

Fermentation is preceded by cooling to 20–21°C, and is initiated by pitching with yeast. Yeasts are of the high-attenuation type (tolerating high alcohol levels) grown in molasses and ammonium salts. They are also selected to produce appropriate flavours. After 36–48 h, alcohol content reaches 8% and the temperature reaches 30°C due to heat produced during yeast metabolism.

The contents of the fermenter are transferred to the first copper pot still (Fig. 12.19; known as the 'wash' still), which produces a distillate of more than 20% alcohol known as 'low wines'. This is further distilled in a second copper pot still – the 'spirit still'. From this three fractions are collected: the first fraction (known as 'foreshots' or 'heads'), contains volatile components (such as methanol, acetaldehyde, diacetyl and sulphurous compounds) with undesirable flavour characteristics; and the last fraction, known as 'feints' (or 'tails'), contains higher alcohols/fusel oils (e.g., iso-butanol, propanol, amyl alcohol), other phenolic compounds and fatty acids, and is mixed with the foreshots and added to the low wines from the next batch

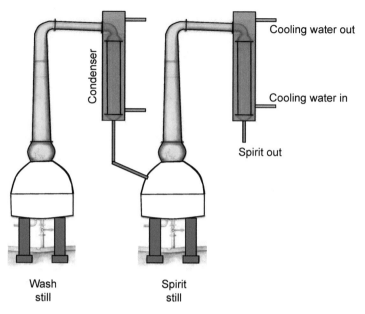

Cooling water out

Cooling water in

Spirit out

Condenser

Wash
still

Spirit
still

FIGURE 12.19 Sequential pot stills used in the distillation of Scotch whisky. Although one configuration is illustrated, a variety of designs, styles and layouts will be found in industry. Heads and tails will often be mixed with incoming streams for redistillation.

and redistilled. It is the middle fraction (the 'heart of the spirit') which is collected as the basis of the marketable product. It contains ~68% (65%–72%, depending upon the company) alcohol, but it has to mature in oak casks over a period of at least 3 years before sale (Bathgate, 1989).

The spent grains (known as draff and lees) may be further processed into distillers' grains or sent directly to livestock without further processing – this is discussed in more depth later in this chapter.

12.4.2 Grain whiskey

A summary of grain whiskey production is given in Fig. 12.20, and, as noted, it is somewhat different to the Scotch whisky process.

For many years the most-used raw material for distilleries in the United Kingdom was maize, which was imported from North or South America or South Africa. Due to import levies and changing economics during the 1980s, though, maize became largely displaced by wheat. Barley is a less attractive alternative, as β-glucans released during cooking produce undesirably high viscosity and spirit yields are low. The preferred wheat type is soft because of its lower protein content and the better-flavoured spirit that can be produced from it. More than half of the wheat used usually comes from Scotland, the remainder coming from elsewhere in the United

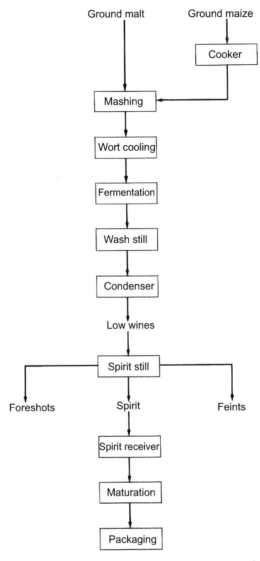

FIGURE 12.20 Flow chart summarizing the major steps in grain whiskey production.

Kingdom. In the United States and elsewhere in the world, wheat and maize are the primary substrates for whiskey production.

The mode of use of the grain varies among distilleries: in some the grain is cooked whole prior to mashing, while in others it is milled first. Mashing may be carried out in batches or as a continuous process. Individual preferences also influence the nature of the malt used, the primary variation being in the degree of kilning. The malt is milled and suspended in cold

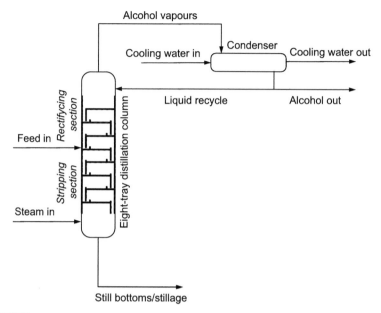

FIGURE 12.21   Eight-tray continuous distillation system showing inputs and outputs.

water before being added to the freshly cooked cereal grain. A temperature of 65°C in the combined slurries is appropriate for enzymatic conversion of the starch. Malt accounts for 8%–15% of total solids present before saccharification. Following conversion, wort is not separated and fermentation occurs in the whole cooled mash.

After fermentation, the unfermented solids, water and unwanted alcohols and other chemical compounds (as discussed previously) must be removed from the alcohol. As with malt whisky, the alcohol content is increased by multistage distillation, but in this case to a higher alcohol content of ~94%. Distillation is carried out not in pot stills but by the more economical continuous distillation method (Figs 12.21 and 12.22). Many continuous stills are based on the Coffey still, named after its purported inventor, Aeneas Coffey, who patented this process in 1830 (Bathgate, 1989). Although following the principles developed by Coffey, this method of distillation can have a variety of layouts and configurations in modern distilleries.

12.4.3 Bourbon whiskey

As defined in the United States (Code of Federal Regulations, Title 27, Part 5), bourbon whiskey must be made with at least 51% maize, be distilled to no more than 80% alcohol v/v, be aged in new, charred oak

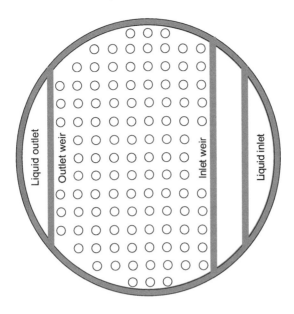

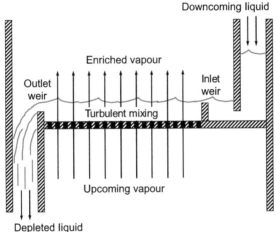

FIGURE 12.22 Circular tray for a continuous distillation column. Note that hot vapour rises through perforations, mixes with downcoming liquid and evaporates alcohol, leaving a depleted liquid to fall to the next tray. Continuous distillation columns comprise a series of stacked trays.

containers (at least 2 years to be labelled a 'straight' whiskey) and be bottled with at least 40% alcohol v/v. Blends of other cereals (often combinations of barley, wheat and rye) to be mashed with the maize are ground in a hammer mill prior to addition of malt and water. If a rye whiskey is to be made, the mash must contain at least 51% rye. Water is added at the rate of 6.0–7.4 L/kg, and some stillage (known as 'backset') may be included to adjust the pH to 5. Although a total of 10%–15% of malt is finally present, only 1%–5% is added initially, the remainder ('conversion malt') being added following cooling after the cook. Cooking consists of raising the temperature to 70°C and holding it for 30–60 min. It is cooled to 63°C for the conversion of starch to sugars to be completed. Further cooling follows before the whole mash is pumped to a fermenter, where it is pitched with 2% v/v yeast. The rise in temperature due to fermentation is not allowed to exceed 35°C. Fermentation takes about 72 h; the resulting product is known as 'drop beer', and this is then distilled in a continuous column distillation system (Fig. 12.21) (Nagodawithana, 1986).

12.4.4 Neutral spirits

The beer for neutral spirits' (e.g., vodka, grain spirits, gin, etc. – see Code of Federal Regulations, Title 27, Part 5) distillation is produced as economically as possible. Flavour is actually an undesirable attribute, so expensive means of producing flavours (as in whisky) are unnecessary. Primarily the chemistry of conversion of starch to sugars and then of sugars to alcohol must be considered. Often the cheapest source of starch is used. In countries where maize is common, that cereal is used, but elsewhere other cereals (such as wheat) may be cheaper (e.g., wheat in Russia and Canada).

For these distilled spirits, it may be more economical to use enzymes derived from microorganisms than those derived from malt. Fungal and bacterial enzymes may thus be used in this process. A further advantage is that amyloglucosidase is available from microbial sources. Adjustment of pH is achieved by use of chemicals, calcium hydroxide being added to the suspension of ground cereal introduced into the cooker to achieve pH 6.3, and dilute hydrochloric or sulphuric acid adjusting the pH to 4.5 before saccharification. Temperatures during cooking may be up to 100°C under atmospheric conditions or 160°C if high-pressure cooking is used. The pressure cooking saves time (5 min instead of 30–60 min at highest temperature).

Amylases are added at appropriate temperatures as the mash cools: bacterial at 85°C, fungal at 63°C and amyloglucosidase at 60°C. Fermentation is achieved by yeasts selected for their high rate of alcohol production and their tolerance to high alcohol levels, possibly 8%–9% or more. Continuous distillation and rectification of the resulting beer can produce an alcohol content of ~95%.

12.4.5 Craft spirits

The 20th century saw the development of large brewing and distilling companies, some on a national scale and even some on a multinational scale. A growing trend since the early 2000s, though, has been the development of small craft distilleries. Fig. 12.23 illustrates this

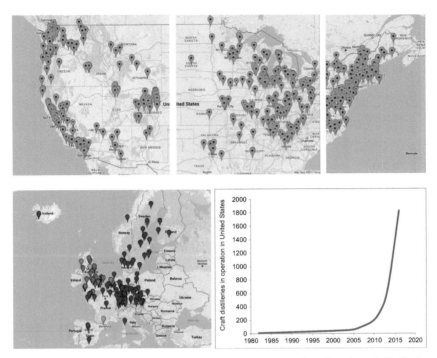

FIGURE 12.23 Google Earth maps illustrating distribution of small-scale craft distilleries that have begun operations since 1980. As of 2017, there were >2200 in the United States and >159 in Europe. *Based upon data from* www.distilling.com, *January 2017. Craft distilleries are defined as having a maximum annual sales less than 100,000 proof gallons of spirit.*

growth. This applies to all types of distilled spirits, not just whiskey. This trend has been precipitated by changes in various laws and regulations concerning the production of distilled spirits, but also a greater desire among consumers for artisan-style food and beverage products. Similar trends have been seen in the craft beer and wine industries.

Further technical details and several of the latest developments in distilled beverage production can be found in Bryce et al. (2008) and Russell (2003).

12.5 FUEL ETHANOL

During the 2000s the maize-based ethanol industry grew exponentially in the United States due to changes in government policies regarding renewable energy (Fig. 12.24) and implementation of the Renewable Fuel Standard (https://www.epa.gov/renewable-fuel-standard-program). Production and use of biofuels in other countries grew as well. Currently, the Renewable Fuel Standard in the United States mandates the use of 15 billion gal (56.8 billion L)

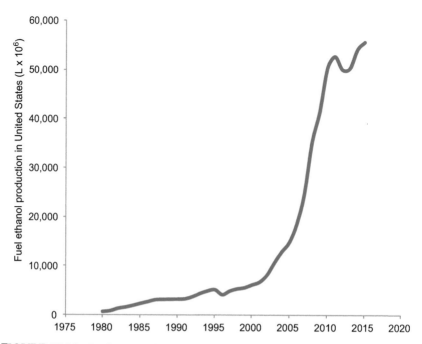

FIGURE 12.24 Production of fuel ethanol in the United States over time. *Based upon data from* www.ethanolrfa.org, *January 2017.*

annually – most of this comes from corn-based fuel ethanol. Many other countries have also developed similar mandates. Cereal grains are the most common substrate for the production of industrial alcohol.

Ethanol (ethyl alcohol) is highly combustible and has a long history of use as a motor fuel. As with beer and distilled spirits, it is produced by the metabolism of glucose by yeast (*Saccharomyces cerevisiae*); the glucose is produced by the hydrolysis of starch. In fact, since all cereal grains contain a large proportion of starch, it is technically possible to obtain ethanol from any cereal. Recall, this happens when cereal grains are malted and then brewed to make beer or distilled spirits, which are aqueous solutions of alcohol. The production of ethanol can actually be regarded as a modification of the brewing process, in which grain starch is the starting material and pure ethanol, rather than an aqueous solution, is the final product. Thus the industrial production of ethanol closely resembles the production of distilled spirits, although commercial enzymes are used, not those naturally present in the grain (i.e., the grain is not malted).

Ethanol has, in fact, a long history of use in internal combustion engines. In 1826 Samuel Morey (United States) invented a combustion engine that used ethanol and turpentine as fuel; in 1860 Nicholas Otto (Germany) invented a combustion engine that ran on pure ethanol; in 1896 Henry

Ford's (United States) quadricycle, his first automobile, was developed – it used ethanol as a fuel; and in 1908 the Hart-Parr Company (United States) manufactured farm tractors that used ethanol. The history of ethanol as a fuel is replete with many examples of setbacks and advances, drama and politics (Benton et al., 2010).

The principal reasons for making ethanol from cereals are that ethanol can be effectively used as a partial or even complete replacement for petroleum-based gasoline as a fuel for internal combustion engines; it is a domestic source of fuel; manufacturing of ethanol is a useful way of dealing with grain surpluses whenever they arise; and development of the biofuels industry can increase the number of jobs in rural localities. Historically, interest in the production of biofuels tends to increase when fuel shortages occur, when fuel prices are high and/or when prices for grains are depressed.

For example, a wheat surplus in Sweden in 1984 was dealt with by establishing a plant that separated the starch from a residue to be used for animal feed. About half the starch (the best quality) was to be used by the paper industry, and the remainder for production of ethanol.

The maximum theoretical yield of ethanol varies according to the type of cereal used: 431 L/t (2.9 gal/bu) from maize, 430 L/t from rice, 340–360 L/t from wheat and 240–250 L/t from barley and oats (Dale, 1991). Of course, due to inefficiencies in factory processing operations, achieving maximum theoretical yield is often challenging and some plants never achieve these levels. As a practical rule of thumb, 1 bu of maize (25.5 kg) will yield ~2.8 gal (10.98 L) of ethanol, and approximately 17–18 lb (7.7–8.2 kg) each of CO_2 and nonfermentable materials (e.g., proteins, lipids, fibres, minerals, etc.) – which are processed and then sold as distillers' grains.

The process of manufacture of ethanol from cereals (Fig. 12.25) closely resembles that already discussed for the production of distilled spirits, but on a much larger industrial scale. An overhead plan view for a typical ethanol facility is depicted in Fig. 12.26. One of the largest ethanol plants in the United States is shown in Fig. 12.27, while a small plant is shown in Fig. 12.28.

The production process begins by grinding the grain (Fig. 12.29) and then cooking it with water and acid or alkali (to adjust pH). Alpha-amylase and glucoamylase enzymes are added (the malting process is just too expensive and time consuming) to the cooled mash to promote the hydrolysis of starch to glucose, and the whole mash is then fermented with yeast (Fig. 12.30; fermentation vessels are nearly 1 million gal/3.8 million L), releasing carbon dioxide gas and producing alcohol. After fermentation the slurry is treated with steam in a series of continuous multitray distillation columns (Fig. 12.21), where the alcohol is separated in a rectification column, yielding 95% ethanol (5% water); the remaining water is removed from the purified ethanol by a molecular sieve technology which uses a

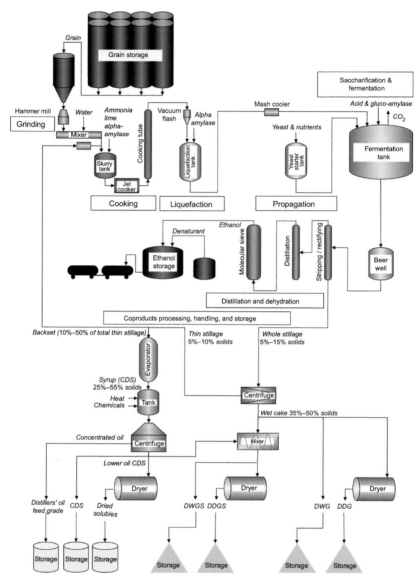

FIGURE 12.25 Flow chart summarizing the major steps in grain-based fuel ethanol processing. *Based, in part, on Rosentrater, K.A., 2011b. Manufacturing of fuel ethanol and distillers grains – current and evolving processes (Chapter 5). In: Liu, K., Rosentrater, K.A. (Eds.), Distillers Grains: Production, Processing, and Utilization. CRC Press, Boca Raton, FL, USA.*

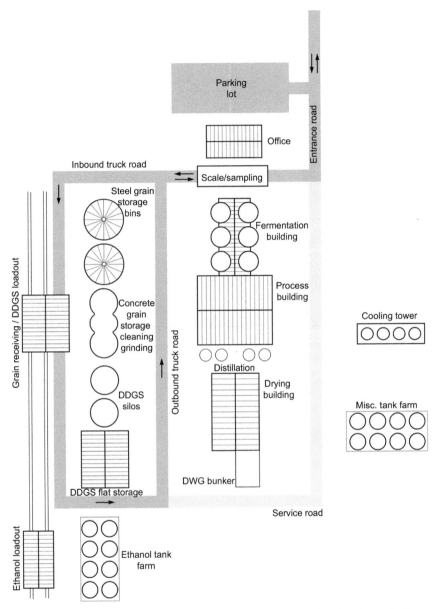

FIGURE 12.26 Overall plan view of a fuel ethanol processing facility, showing typical locations and sizes of structures. *Based, in part, on Rosentrater, K.A., 2011b. Manufacturing of fuel ethanol and distillers grains – current and evolving processes (Chapter 5). In: Liu, K., Rosentrater, K.A. (Eds.), Distillers Grains: Production, Processing, and Utilization. CRC Press, Boca Raton, FL, USA.*

FIGURE 12.27   Example of a large-scale ethanol processing facility (~450 million L/y).

FIGURE 12.28   Example of a small-scale ethanol processing facility (~79 million L/y).

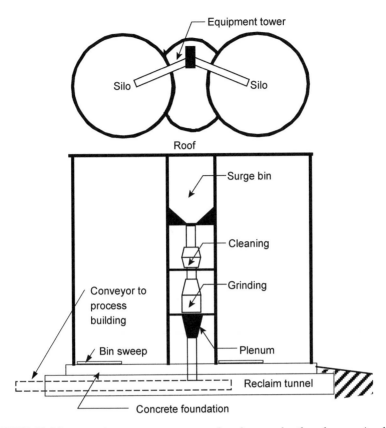

FIGURE 12.29 Typical grain storage structure for a large-scale ethanol processing facility, illustrating large-diameter concrete silos with a central surge bin, and cleaning and grinding equipment housed within. *Based, in part, on Williams, G.D. and Rosentrater, K.A., 2006. Design Considerations for the Construction and Operation of Ethanol Facilities. Part 1: Planning, Structural, and Life Safety Considerations. Paper 064115. Annual Meeting of the American Society of Agricultural and Biological Engineers, Portland, OR, USA.*

bed of zeolite to entrap the water molecules. The still bottoms are subsequently dehydrated (through centrifuges and evaporators) to produce protein-enriched distillers' grains (Figs 12.31 and 12.32), which are used for animal feed (Dale, 1991). This is discussed later.

As a motor fuel, ethanol has various advantages over gasoline: it has a very high octane number; it increases engine power; and it burns more cleanly, producing less carbon monoxide and oxides of nitrogen. On the other hand, there may be difficulties in starting the engine on ethanol alone, and accordingly a blend of ethanol with gasoline is generally used in Europe and the United States. The down side, however, is that the energy content of ethanol is lower than that of gasoline, so fuel economy may decline to a degree when using ethanol/gasoline blends.

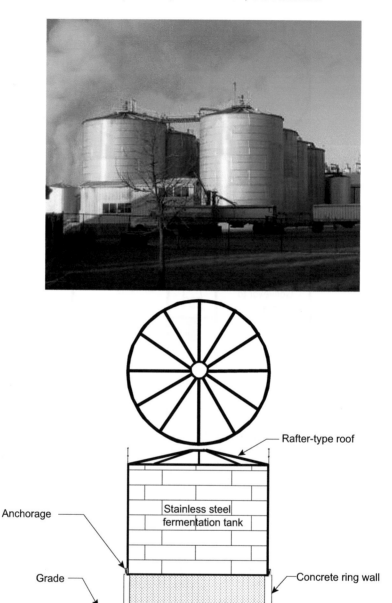

FIGURE 12.30 Typical fuel ethanol fermentation vessels, which sit upon ring foundations and are typically not enclosed in a building shell (as in beverage alcohol production). *Based, in part, on Williams, G.D., Rosentrater, K.A., 2006. Design Considerations for the Construction and Operation of Ethanol Facilities. Part 1: Planning, Structural, and Life Safety Considerations. Paper 064115. Annual Meeting of the American Society of Agricultural and Biological Engineers, Portland, OR, USA.*

FIGURE 12.31 Distillers' dried grains with solubles (DDGS) is the primary coproduct produced by fuel ethanol manufacturing plants, and consists of all nonfermented nutrients in the cereal grain (typically concentrated by a factor of 3× due to yeast metabolism). *Based, in part, on Rosentrater, K.A. 2011a. Overview of fuel ethanol production and distillers grains (Chapter 2). In: Liu, K., Rosentrater, K.A. (Eds.), Distillers Grains: Production, Processing, and Utilization. CRC Press, Boca Raton, FL, USA.*

Other industrial uses for ethanol made from cereal grains, besides motor fuel, include solvents, in antifreeze, and as a raw material for the manufacture of various industrial chemicals, e.g., acetaldehyde, ethyl acetate, acetic acid, glycols, etc. (Dale, 1991).

The carbon dioxide evolved during the fermentation stage finds uses in oilfields, for recovery of additional oil; in the manufacture of methanol; as a refrigerant; and most commonly in carbonated beverages (Dale, 1991). Capture and use of CO_2 are highly dependent upon plant location, the potential market demand and the economics involved in capturing, compressing and transporting this product to market.

The biofuels industry is dynamic, and the evolution and refinement of new processes are constantly ongoing. Some of these are described by technology providers (a few of which are http://www.katzen.com; www.icminc.com; www.poet.com). Extensive discussions about many aspects of biofuels' production using cereal grains and other cellulosic substrates can be found in Ingledew et al. (2009). Evolutions include techniques for increased energy efficiency, water efficiency, fermentation completeness

FIGURE 12.32 Distillers' dried grains with solubles (DDGS) storage typically occurs in either steel-sided flat storage buildings or concrete silos. *Based, in part, on Rosentrater, K.A., 2011b. Manufacturing of fuel ethanol and distillers grains – current and evolving processes (Chapter 5). In: Liu, K., Rosentrater, K.A. (Eds.), Distillers Grains: Production, Processing, and Utilization. CRC Press, Boca Raton, FL, USA.*

and efficiency, new strains of yeast and enzymes, and fractionation or concentration of various chemical constituents from the coproduct streams (e.g., proteins, fibres, oils, etc.) (Rosentrater, 2015).

A few examples follow. A process for the continuous production of ethanol from cereals, involving screening, filtering, saccharification, fermentation and distillation stages, has been patented (Technipetrol SPA, 1989), while a dual-purpose flour mill has been described in which the flour is air-classified to produce a high protein fraction (particle size 2–5 µm) and a residual protein-depleted fraction for use as the starting material for production of ethanol (Bonnet and Willm, 1989).

Extrusion cooking has been suggested as a method for pretreating grain to be used for the production of ethanol. The thermomechanical effects of extrusion cooking produce gelatinization and liquefaction of the starch so no liquefying enzyme is needed for the subsequent saccharification with glucoamylase. Ethanol yields from wet-extruded and steam-cooked grain were almost equal, but the extrusion method used less energy. Roller-milled whole barley, wheat or oats can be used in this process, with or without the addition of thermostable *alpha*-amylase, which appears to have little effect during extrusion cooking. The fermentation stage is carried out using either yeast (*Saccharomyces cerevisiae*) or the bacterium *Zymomonas mobilis*, the latter producing an increased initial rate of fermentation (Linko, 1989a,b). To date, however, this approach has been too expensive to implement commercially.

It is not just the starch in the grains that can be converted into ethanol: the cellulose and hemicellulose in any biological material can be transformed into biofuel. For example, corncobs can be used for the production of ethanol, and also of furfural (*vide infra*). By treating the cobs with dilute sulphuric acid, 80% of the pentosans in the cobs are converted to pentoses, from which furfural is obtained, while the residual cellulose can be hydrolysed to glucose with 65% yield (Clark and Lathrop, 1953).

The most promising raw materials for lingo-cellulosic biofuels, at least for widespread economic commercial adoption, include corn stover (e.g., leaves, cobs, stalks), residual straw from various cereal grains and forestry/wood wastes. During the tremendous growth in biofuel development in the early 2000s an enormous amount of research was conducted into developing materials and methods for production of biofuels from a variety of substrates. There have even been a few commercial factories opened. Covering these concepts is beyond the scope of this text, but the reader is referred to other sources for more information (dos Santos Bernardes, 2011; Biernat, 2015; Jacob-Lopes and Queiroz Zepka, 2017).

12.6 BY-PRODUCTS OF BREWING AND DISTILLING

This chapter focuses on techniques for commercially producing alcohol in a variety of final forms. But each conversion process discussed will also yield by-products (i.e., nonfermentable materials): malt sprouts, brewers' grains, *kasu* cake, spent lees, draff and distillers' grains. Of these, the largest quantities of by-products consist of the solids remaining after the final separation, leaving spent grains; in brewing, distilled spirits and biofuels these are termed brewers' grains, distillers' dark grains (known as DDG) and distillers' dried grains (also known as DDG), respectively. Historically these have been valuable for cattle feed, but animal scientists and nutritionists have conducted myriad studies

TABLE 12.6 Typical chemical compositions of various alcohol coproducts used as feed ingredients for animals (the majority [~99%] of use consists of beef, dairy, swine, poultry)

	Malt sprouts (barley culms), dried	Brewers' grains, fresh	Brewers' grains, dried	Draff (barley distillers' grains), fresh	Malt distillers' dark grains, dried	Pot ale syrup, fresh
Dry matter (%)	89.9	24.9	91.0	24.1	90.7	48.3
Protein (% db)	23.5	25.9	25.8	20.3	27.8	37.4
Lysine (% of protein)	4.6		3.1		4.3	
Methionine (% of protein)	1.4		1.5		1.4	
Lipid (% db)	1.7	7.0	6.7	8.2	8.5	0.2
Fibre (% db)	13.5	16.4	15.8	17.6	11.6	0.2
NDF (% db)*	45.2	49.6	56.3	65.1	39.7	0.6
Starch (% db)	14.7	5.7	7.8	1.8	3.2	1.8
Ash (% db)	6.0	4.1	4.6	3.3	5.8	9.5
Phosphorus (g/kg db)	6.1	5.8	5.7	3.3	9.7	19.0
Potassium (g/kg db)	17.5	1.6	2.9	0.3	10.5	22.3
Zinc (mg/kg db)	108.0	83.0	89.0	188.0	57.0	22.0
Copper (mg/kg db)	14.0	14.0	19.0	15.0	49.0	95.0
Iron (mg/kg db)	558.0	138.0	130.0		231.0	

	Corn, thin stillage, fresh	Corn distillers' wet grains with solubles (DWGS)	Corn distillers' dried grains with solubles (DDGS)	Rye distillers' dried grains with solubles (DDGS)	Sorghum distillers' dried grains with solubles (DDGS)	Wheat distillers' dried grains with solubles (DDGS)
Dry matter (%)	4.7	35.2	89.0	92.2	89.9	90.6
Protein (% db)	17.9	31.8	29.5	31.3	33.5	37.3
Lysine (% of protein)		3.0	3.0		2.9	2.3

TABLE 12.6 Typical chemical compositions of various alcohol coproducts used as feed ingredients for animals (the majority [~99%] of use consists of beef, dairy, swine, poultry) — cont'd

	Corn, thin stillage, fresh	Corn distillers' wet grains with solubles (DWGS)	Corn distillers' dried grains with solubles (DDGS)	Rye distillers' dried grains with solubles (DDGS)	Sorghum distillers' dried grains with solubles (DDGS)	Wheat distillers' dried grains with solubles (DDGS)
Methionine (% of protein)		1.8	2.0		1.8	1.5
Lipid (% db)	9.2	13.0	11.1	7.9	9.4	5.0
Fibre (% db)		8.2	7.9	8.2	8.1	7.7
NDF (% db)*	12.5	39.0	34.2		38.5	34.0
Starch (% db)	25.1	4.9	9.3			4.2
Ash (% db)	6.3	3.8	5.4	5.4	4.5	5.9
Phosphorus (g/kg db)		8.2	7.9	8.0	7.4	9.1
Potassium (g/kg db)		9.5	10.3	16.0	3.5	10.9
Zinc (mg/kg db)		63.0	62.0	70.0		130.0
Copper (mg/kg db)		6.0	6.0	27.0		10.0
Iron (mg/kg db)		116.0	123.0	154.0		

* NDF, neutral detergent fibre.
Based upon data from Feedipedia – Animal Feed Resources Information System, 2017. INRA CIRAD AFZ and FAO. www.feedipedia.org.

over the years, and these products are now readily used not only in ruminant (e.g., beef, dairy) diets but also in monogastric (swine, poultry, fish and other) feeds.

Although not so essential to the profitability of beverage alcohol plants, sales of these materials are critical to the profitability of ethanol plants and they truly have become coproducts (in addition to biofuel, the primary product). Table 12.6 provides typical compositions for many of these feed products. Comprehensive reviews of physical and nutritional properties, use in various species and processing evolutions are provided in Crawshaw (2001) and Liu and Rosentrater (2011), to which the reader is referred for more information.

Yeast is recoverable and saleable, and CO_2 may be worth harvesting for sale if produced in sufficient quantity and in an appropriate locality. Fusel oils, including the components furfural, ethyl acetate, ethyl lactate, ethyl decanoate, *n*-propanol, iso-butanol and amyl alcohol, are used in the perfume and other industries (Walker, 1988).

References

Bamforth, C.W., 2006. Scientific Principles of Malting and Brewing. American Association of Cereal Chemists Inc., St. Paul, MN, USA.

Bathgate, G.N., 1989. Cereals in Scotch whisky production (Chapter 4). In: Palmer, G.H. (Ed.), Cereal Science and Technology. Aberdeen University Press, UK.

Benton, H., Kovarik, W.M., Sklar, S., 2010. The Forbidden Fuel. University of Nebraska Press, Lincoln, NE, USA.

Biernat, K., 2015. Biofuels – Status and Perspective. Intech, Rijeka, Croatia.

Blanchflower, A.J., Briggs, D.E., 1991. Quality characteristics of triticale malts and worts. Journal of the Science of Food and Agriculture 56, 129–140.

Bonnet, A., Willm, C., November/December 1989. Wheat Mill for Production of Production of Proteins and Ethanol. Inds. Cer., pp. 37–46.

Briggs, D.E., 1978. Barley. Chapman and Hall Ltd., London, UK.

Briggs, D.E., 1998. Malts and Malting. Blackie Academic & Professional, London, UK.

Briggs, D.E., Wadeson, A., Statham, R., Taylor, J.F., 1986. The use of extruded barley, wheat and maize as adjuncts in mashing. Journal of the Institute of Brewing 92, 468–474.

Bryce, J.H., Piggott, J.R., Stewart, G.G., 2008. Distilled Spirits: Production, Technology and Innovation. Nottingham University Press, Nottingham, UK.

Clark, T.F., Lathrop, E.C., 1953. Corncobs – Their Composition, Availability, Agricultural and Industrial Uses. U.S. Dept. Agric., Agric. Res. Admin., Bur. Agric. Ind. Chem., AIC-177, revised 1953.

Crawshaw, R., 2001. Co-product Feeds: Animal Feeds from the Food and Drinks Industries. Nottingham University Press, Nottingham, UK.

Dale, B.E., 1991. Ethanol production from cereal grains (Chapter 24). In: Lorenz, K.J., Kulp, K. (Eds.), Handbook of Cereal Science and Technology. Marcel Dekker, Inc., NY, USA.

dos Santos Bernardes, M.A., 2011. Biofuel Production – Recent Developments and Prospects. Intech, Rijeka, Croatia.

Euromalt, 2017. Euromalt Statistics. Available online: http://www.euromalt.be.

Gelinas, P., Mckinnon, C., 2000. Fermentation and microbiological processes in cereal foods (Chapter 26). In: Kulp, K., Ponte Jr., J.G. (Eds.), Handbook of Cereal Science and Technology. Marcel Dekker, Inc., New York, NY, USA.

Gibson, G., 1989. Malting plant technology (Chapter 5). In: Palmer, G.H. (Ed.), Cereal Science and Technology. Aberdeen Univ. Press, UK.

Ibis, 2010. Global Spirits Manufacturing: C1122-gl. http://www.just-drinks.com/store/samples/2010_ibisworld%20global%20drink%20sample%20industry%20report.pdf.

Ingledew, W.M., Kelsall, D.R., Austin, G.D., Kluhsies, C., 2009. The Alcohol Textbook, fifth ed. Nottingham University Press, Nottingham, UK.

Jacob-Lopes, E., Queiroz Zepka, L., 2017. Frontiers in Bioenergy and Biofuels. Intech, Rijeka, Croatia.

Kanauchi, M., 2013. *SAKE* Alcoholic beverage production in Japanese food industry (Chapter 3). In: Muzzalupo, I. (Ed.), Food Industry. Intech, Rijeka, Croatia.

Kubo, R., Kilasara, M., 2016. Brewing techniques of *Mbege*, a banana beer produced in Northeastern Tanzania. Beverages 2 (21), 1–10.

Lee, M., Regu, M., Seleshe, S., 2015. Uniqueness of Ethiopian traditional alcoholic beverage of plant origin, tella. Journal of Ethnic Foods 2, 110–114.

Lewis, M.J., Young, T.W., 2013. Brewing. Springer, New York, NY, USA.

Linko, P., 1989a. The twin-screw extrusion cooker as a versatile tool for wheat processing (Chapter 22). In: Pomeranz, Y. (Ed.), Wheat Is Unique. Amer. Assoc. of Cereal Chemists Inc., St. Paul, MN, USA.

Linko, P., 1989b. Extrusion cooking in bioconversions (Chapter 8). In: Mercier, C., Linko, P., Harper, J. (Eds.), Extrusion Cooking. Amer. Assoc. of Cereal Chemists Inc., St. Paul, MN, USA.

Liu, K., Rosentrater, K.A., 2011. Distillers Grains: Production, Processing, and Utilization. CRC Press, Boca Raton, FL, USA.

Lyumugabe, F., Uyisenga, J.P., Songa, E.B., Thonart, P., 2014. Production of traditional sorghum beer "ikigage" using *Saccharomyces cerevisiae*, *Lactobacillus fermentum* and *Issatckenkia orientalis* as starter cultures. Food and Nutrition Sciences 5, 507–515.

Nagodawithana, T.W., 1986. Yeasts: their role in modified cereal fermentations. Advances in Cereal Science and Technology 8, 15–104.

Narziss, L.J., 1984. The German beer law. Journal of the Institute of Brewing 90, 351–358.

Novellie, L., 1977. Beverages from sorghum and millets. In: Proc Symp. Sorghum and Millets for Human Food. IACC Vienna 1976 Tropical Products Institute, London, UK.

Palmer, G.H., 1989. Cereals in malting and brewing (Chapter 3). In: Palmer, G.H. (Ed.), Cereal Science and Technology. Aberdeen Univ. Press, UK.

Papas, R.K., Sidle, J.E., Wamalwa, E.S., Okumu, T.O., Bryant, K.L., Goulet, J.L., Maisto, S.A., Braithwaite, R.S., Justice, A.C., 2010. Estimating alcohol content of traditional brew in Western Kenya using culturally relevant methods: the case for cost over volume. AIDS and Behavior 14 (4), 836–844.

Pomeranz, Y., 1987. Barley (Chapter 18). In: Pomeranz, Y. (Ed.), Modern Cereal Science and Technology. UCH Publishers Inc., NY, USA.

Pyler, R.E., Thomas, D.A., 1986. Cereal research in brewing: cereals as brewers' adjuncts. Cereal Foods World 31, 681–683.

Reed, G., Nagodawithana, T.W., 1991. Yeast Technology. Van Norstrand Rheinhold, NY, USA.

Rosentrater, K.A., 2011a. Overview of fuel ethanol production and distillers grains (Chapter 2). In: Liu, K., Rosentrater, K.A. (Eds.), Distillers Grains: Production, Processing, and Utilization. CRC Press, Boca Raton, FL, USA.

Rosentrater, K.A., 2011b. Manufacturing of fuel ethanol and distillers grains – current and evolving processes (Chapter 5). In: Liu, K., Rosentrater, K.A. (Eds.), Distillers Grains: Production, Processing, and Utilization. CRC Press, Boca Raton, FL, USA.

Rosentrater, K.A., 2015. Production and use of evolving corn-based fuel ethanol coproducts in the U.S. (Chapter 5). In: Biernat, K. (Ed.), Biofuels – Status and Perspective. Intech, Rijeka, Croatia.

Russell, I., 2003. Whisky: Technology, Production and Marketing. Academic Press, Elsevier, London, UK.

Statista, 2017. Statistics and Facts on the Beer Industry. https://www.statista.com/statistics/270269/leading-10-countries-in-worldwide-beer-production/.

Technipetrol SPA, 1989. Process and Apparatus for the Continuous Production of Ethanol from Cereals, and Method of Operating Said Apparatus. World Intellectual Property Organization Patent 89/01522.

Walker, E.W., 1988. By-Products of Distilling. Ferment. Institute of Brewing Publication, pp. 45–46 (cited by Palmer 1989).

Williams, G.D., Rosentrater, K.A., 2005. Design Considerations for the Construction and Operation of Malting Facilities Part 1: Planning, Structural, and Life Safety Considerations.

Paper 054095. Annual Meeting of the American Society of Agricultural and Biological Engineers, Tampa, FL, USA.

Williams, G.D., Rosentrater, K.A., 2006. Design Considerations for the Construction and Operation of Ethanol Facilities Part 1: Planning, Structural, and Life Safety Considerations. Paper 064115. Annual Meeting of the American Society of Agricultural and Biological Engineers, Portland, OR, USA.

Yoshizawa, K., Kishi, S., 1985. Rice in brewing. In: Juliano, B.O. (Ed.), Rice Chemistry and Technology. American Assoc. of Cereal Chemists Inc., St. Paul, MN, USA.

Further reading

Anon., 1990. HGCA Weekly Digest 15/1/90.

Aisien, A.O., Palmer, G.H., Stark, J.R., 1986. The ultrastructure of germinating sorghum and millet grains. Journal of the Institute of Brewing 92, 162–167.

Bamforth, C.W., 1985. Biochemical approaches to beer quality. Journal of the Institute of Brewing 91, 154–160.

Briggs, D.E., Hough, J.S., Stevens, R., Young, T.W., 1981. Malting and Brewing Science, second ed. Chapman and Hall, NY, USA.

Cook, A.H. (Ed.), 1962. Barley and Malt. Academic Press, London, UK.

Martin, D.T., Stewart, B.G., 1991. Contrasting dough surface properties of selected wheats. Cereal Foods World 36, 502–504.

Matz, S.A., 1991. The Chemistry and Technology of Cereals as Food and Feed, second ed. Avi. Van Norstrand Rheinhold, NY, USA.

Nagodawithana, T.L., 1986. Yeasts: their role in modified cereal fermentations. Advances in Cereal Science and Technology 8, 15–104.

Reed, G., 1981. Yeast – a microbe for all seasons. Advances in Cereal Science and Technology 4, 1–4.

Shewry, P.R., 1992. (Ed.), Barley: Genetics, Biochemistry, Molecular Biology and Biotechnology. C.A.B. International, Wallingford, Oxon, UK.

CHAPTER

13

Feed and industrial uses for cereals

The worldwide usage of all cereal grains (as well as their affiliated processing by-products and coproducts), as revealed in the Food and Agriculture Organization Food Balance Sheets, indicates that historically only a small portion is used for seed, with the remainder shared between human food, animal feed and processing and other industrial uses (including biofuels). There is, however, considerable variation among the principal cereals in terms of usage. As shown in Fig. 13.1, use of the major cereal grains as animal feed ingredients has been changing over time. Specifically, use of maize (the most commonly used cereal grain for animal feed) has grown more than use of any other cereal grain over the years; use of wheat and barley (second and third most-used cereal grains) has slowly increased and decreased, respectively, during the last few decades.

Compared to maize, wheat and barley, other cereal grains have much lower usage as animal feed and industrial products (on a mass basis); some have decreased over time, while others have slowly increased. Proportionally, very little rice or millet is used for these purposes, but over 70% of the entire crops of barley, oats, maize and sorghum are so used.

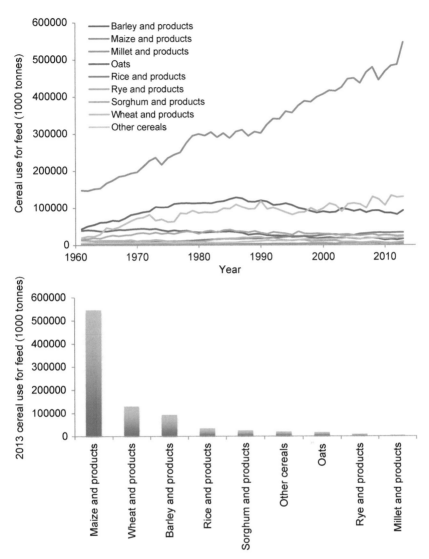

FIGURE 13.1 Top: trends in world usage of the principal cereals for animal feed production, 1960–2013. Bottom: quantity of cereal grains used in 2013 for animal feed. *Based upon data found in FAO, 2017. Food Balance Sheets. Food and Agriculture Data. Food and Agriculture Organization, Rome, Italy.* http://www.fao.org/faostat/en/#home.

13.1 RAW MATERIALS USED FOR FEED AND INDUSTRIAL PRODUCTS

The raw materials used for feed and industrial purposes (i.e., cereal grains and derivatives) generally fall into three broad categories.

1. The whole grain, as harvested, or perhaps with some minimal amount of processing.
2. Specific components of the grain provide the starting material for various chemical processing, but other components are not used, and thus are coproducts/by-products of the chemical process – these are often used as animal feed ingredients. Two examples are using starch for the production of beverage alcohol or fuel ethanol (which results in all other nonfermentable materials being used as distillers grains), and producing starch via wet milling (the resulting coproducts being corn gluten feed and corn gluten meal).
3. By-products/coproducts of the milling process, which are not usually suitable for human food but can be used for animal feed as well as a wide range of industrial purposes, including biofuels, bioplastics, fillers, adhesives, abrasives, etc.

13.2 ANIMAL FEED PROCESSING AND INGREDIENTS

Apart from human food, animal feed is one of the largest uses for cereal grains – both whole grains and milling/processing by-products/coproducts.

Over the years, maize has been the most widely used cereal for animal feeds, followed by wheat and barley (Fig. 13.1). Globally, in 2013 546,110,000 tonnes of maize were used for animal feeds, 129,669,000 tonnes of wheat and 92,660,000 tonnes of barley. All other cereal grains and affiliated products used for animal feeds accounted for 105,109,000 tonnes in 2013.

A large proportion of the cereal grains fed to animals passes through the hands of 'animal feed processors'. According to Table 13.1, global feed manufacturing surpassed 1 billion tonnes in 2016. The majority of manufactured feed around the world was produced for poultry, and the greatest rate of growth in recent years has been seen in China/Asia–Pacific and Africa. The aquaculture feed industry has been growing almost exponentially in the Far East (especially in China, Vietnam, Indonesia and Bangladesh), as the farmed-fish industry continues to grow rapidly. Even though the formulated feed industry continues to grow, there is still a considerable quantity of cereal grains fed directly to animals on the farm, not via processors.

13.2.1 Processing of cereals for animal feeds

Feed is produced for a number of animal types – livestock (including poultry, swine and ruminants), domestic animals (such as dogs, cats and horses) and fish (aquaculture). Feed mills can manufacture mash (e.g., mixed feed) or pellets (e.g., densified feed) (Fig. 13.2), and these products are shipped in bulk or bagged form. Feed-milling facilities may be dedicated to producing feed for either single or multiple species. In the United

TABLE 13.1   Global feed production in 2014–16[a]

Region	Feed production (million tonnes)	Species	% of total feed production
Asia–Pacific	368	Poultry	45
Europe	294	Swine	26
North America	191	Ruminant	20
Central and South America	158	Aquaculture	4
Africa	40	Other	5
Middle East	27		100
	1032		

Country	Feed production (million tonnes)	Country	% of total feed production
China	187	China	19
United States	170	European Union	16
Brazil	69	United States	17
Mexico	34	Brazil	7
Spain	32	Rest of world	41
India	31		100
Russia	29		
Germany	24.5		
Japan	24		
France	23.5		

[a] *Based upon data found in Donley, A., February 2017. Global feed output tops 1 billion tonnes. World Grain, 54–57; International Feed Industry Federation, 2014. Global Feed Production. http://www.ifif.org/pages/t/ Global+feed+production.*

States large feed mills tend to focus on a single species only at a given mill; in other countries it is common to have multiple-species feed mills.

The treatments/unit operations applied to cereal grains (and other noncereal-based ingredients) by animal feed processors can be both expensive and time consuming, but modern feed-milling equipment, technologies and operational efficiencies have made feed processing quite economical for livestock production – in both developed and developing countries. Obviously feed manufacturing would not be undertaken unless such treatments offered considerable advantages over the feeding of untreated whole grains and were cost-effective in providing these benefits. Both cold and hot, dry and wet and mechanical and chemical methods of treatment may be used, with the objectives of providing appropriate nutrients, optimizing animal production costs,

FIGURE 13.2   DDGS in meal (mash) form (top) and pelleted form (below).

improving palatability, densifying nutrients (in some cases), avoiding wastage, encouraging consumption (leading to a greater efficiency of food usage and faster growth) and facilitating storage and transport of the feed products. Additional objectives are to improve digestibility and/or nutritive value, prevent spoilage, detoxify poisons and deactivate antinutritional factors.

The actual operations used in a feed processing plant will depend upon the kind of cereal grains involved and the proportions of those cereals in the feed products; on the species of animal for which the feed products are intended, particularly whether for ruminants or monogastric animals; and also on the stage in the animal's life cycle, e.g., in poultry, whether for young chicks, broilers or laying hens. Further, some feed processing plants produce only mixtures of granular ingredients (known as 'mash' feed); other plants produce densified (i.e., pelleted or extruded) feed products. Many feed production plants are designed to be somewhat flexible so that they can produce a variety of feed products (Rosentrater and Williams, 2004).

Fig. 13.3 depicts a block flowchart of the general systems associated with a typical feed mill. In this type of manufacturing system, raw ingredients, such as whole grains, liquids and soft stocks

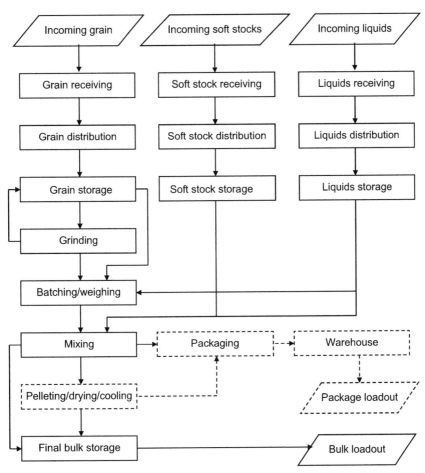

FIGURE 13.3 Block flowchart depicting major components of a typical feed mill. *Dashed lines* show additional processes required for pelleted feed production as opposed to mash feed only.

(i.e., soybean meal, cereal coproduct and by-product meals, minerals, salt and other bulk nongrain materials) are delivered to the facility by rail or truck, and then transported to appropriate storage bins via gravity flow, mechanical conveyors or pneumatic or pumping systems. When needed for various feed blends, specific levels of each of these ingredients are metered/proportioned and conveyed (i.e., batched) to the mixer, and then mixed for an appropriate length of time. This mixed feed (i.e., mash) can be pelleted or extruded, left as-is for delivery (i.e., bulk form; pellets or mash) or packaged and delivered in bag form. Fig. 13.4 shows a formalized process flow diagram for a typical facility dedicated to the production of mixed feed; Fig. 13.5 provides typical plan and section views for this type of facility. A facility that produces pelleted feeds has additional systems and equipment for the pelleting

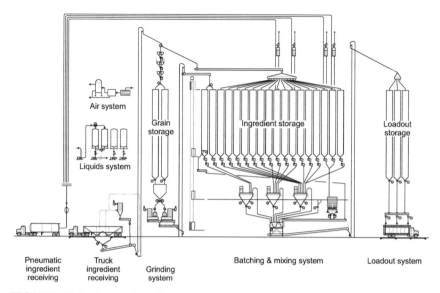

FIGURE 13.4 Process flow diagram for a typical feed mill. *Based upon Rosentrater, K.A., Williams, G.D., 2004. Design considerations for the construction and operation of feed milling facilities. Part II: process engineering considerations. In: 2004 ASAE/CSAE Annual International Meeting, Ottawa, ON, Canada, pp. 1–4.*

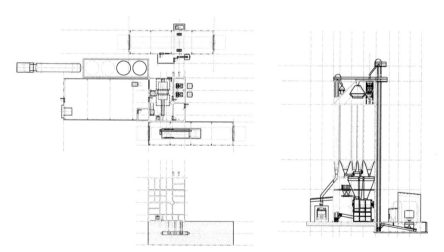

FIGURE 13.5 Plan and section views for a typical feed mill. *Based upon Rosentrater, K.A., Williams, G.D., 2004. Design considerations for the construction and operation of feed milling facilities. Part II: process engineering considerations. In: 2004 ASAE/CSAE Annual International Meeting, Ottawa, ON, Canada, pp. 1–4.*

process, such as pellet mills or extruders, dryers and/or coolers and perhaps a packaging system.

While feed processing equipment and ingredient development are continually evolving, a brief summary of some of the more salient unit operations performed by animal feed processors with cereal grains (and other ingredients), and some of the resulting benefits, are discussed below.

13.2.1.1 *Raw ingredient receipt*

Feed mills receive incoming ingredients by both rail and truck (including hopper-bottom, bulk-solids and liquids trailers). Rail receiving hoppers should be designed to provide maximum capacity, but are usually relatively shallow, which constrains carrying volume. Rail and truck hoppers between 1000 and 1200 bu in capacity are common. Because feed-mill receiving pits are generally shallow, inbound material (from either rail or truck) will pass through choke-fed flow into the receiving hopper, which is actually an effective means of controlling dust and keeping it in the hopper car. Feed mills also commonly utilize truck and rail scales (for ingredient inventories) with flow-through floors, with the hopper pit and at least one screw conveyor underneath; these systems are housed within the same receiving structure, which can be of either steel or concrete construction (Fig. 13.6). Major constituents (such as cereal grains and soybean meal) and minor ingredients (such as lime, brewers' grains, wheat middlings, etc.) are received via these systems.

Micro ingredients such as minerals, vitamins and other additives, on the other hand, are commonly delivered via bulk truck and then pneumatically conveyed to the appropriate storage bins. Pneumatic systems require blowers, delivery lines, receivers, filters and airlocks – typically one system for each ingredient to be received. It is essential to consider the terminal velocity of each ingredient to size the components of the system adequately. Generally, pneumatic transfer of ingredients requires air velocities between 4000 and 5000 ft/min. Additionally, some minor ingredients may sometimes be delivered via bulk tote bags, which will require a freight elevator (i.e., pallet hoist) in the mill structure to transport them to the batching floor.

13.2.1.2 *Raw ingredient distribution*

Once ingredients have entered the facility they are transported via multiple pieces of equipment, including bucket elevators (for vertical transfers), distributors, valves and gravity-flow spouting conveyors, belt conveyors and paddle and drag-chain conveyors. The most common type of conveyor in feed mills, however, is the screw conveyor, because it can be used not only for transporting materials but also to metre the various ingredients accurately, which is a functionality that the other conveyor types do not offer.

The two types of spouting conveyors used in feed-milling facilities are unlined round spouting, often constructed from well-casing pipes,

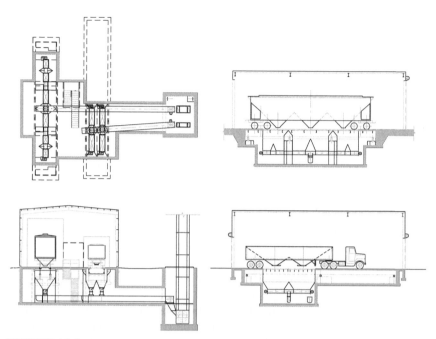

FIGURE 13.6 Plan and section views of a typical enclosed rail and truck scale hopper receiving system. *Based upon Rosentrater, K.A., Williams, G.D., 2004. Design considerations for the construction and operation of feed milling facilities. Part II: process engineering considerations. In: 2004 ASAE/CSAE Annual International Meeting, Ottawa, ON, Canada, pp. 1–4.*

and square spouting, which is constructed in a 'u-trough' shape with an attachable lid. The square-spouting type is often lined with ceramic or urethane liners to limit wear on the steel surfaces, and is primarily used for whole-grain transfer. The round-spouting type is most commonly used in feed mills for most ingredients and feeds. When designing spouting, most designers use a material flux value of 50–60 bu/h/in.^{2} for whole grains flow from receiving pits, 70–80 bu/h/in.^{2} for bin discharge and general spouting of whole grains, and 100–110 bu/h/in.^{2} for rail load-out of whole grains. Ground grain and other raw bulk ingredients, on the other hand, typically exhibit a flux of 50–60 ft^3/h/in.^{2} For material to flow properly through a given spout, the primary design consideration is the angle of installation – in other words, the flow angle. It is common practice to use an angle of 9-on-12 (approximately 37 degrees) as an absolute minimum for whole-grain flow through a spout. Preferably, though, spouts should be installed at angles between 10-on-12 and 12-on-12 (approximately 40–45 degrees), to ensure adequate whole-grain flow. Ground grain and other mill feed ingredients require flow angles between 50 and 60 degrees, while pellets require minimum angles between 45 and 50 degrees to flow properly.

13.2.1.3 *Raw ingredient storage*

The two major types of construction used for feed mills are concrete and steel. Concrete feed mills are typically slipformed, and are most prevalently used for large-scale facilities. Steel feed mills are especially common for smaller-scale facilities, and are assembled from bolted bin construction because they are sold as modular units; but they may occasionally be welded. Welded bins are usually shop-fabricated due to the economies of shop techniques compared to field fabrication. Regardless of the type of construction, no two facilities are identical; the combinations for the layout of feed-milling structures are manifold, and predominantly influenced by client/company preferences more than any other single factor. Fig. 13.7 illustrates several common bin plans.

A key consideration for any feed mill is the amount of space allocated to storage of bulk ingredients. Some owners require enough capacity to store at least 1 week's worth of raw materials; others do not. Each material's angle of repose will affect the effective storage capacity of a given bin. Moreover, compaction between the individual material particles will need to be considered when designing ingredient storage bins. This parameter is highly dependent

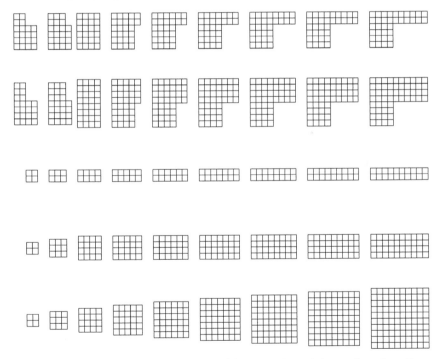

FIGURE 13.7   Feed-mill layouts (bin plans) have a variety of sizes and configurations. *Based upon Rosentrater, K.A., Williams, G.D., 2004. Design considerations for the construction and operation of feed milling facilities. Part II: process engineering considerations. In: 2004 ASAE/CSAE Annual International Meeting, Ottawa, ON, Canada, pp. 1–4.*

on a material's size, shape and moisture content, and also on the size and shape of the bin in which the material is to be stored. For whole-grain storage it is common to use a compaction factor of 7% for bin diameters up to 39 ft, 8% for bin diameters of 40–59 ft and 9% for bins with diameters greater than 60 ft. Unfortunately, little information is available for most other feed ingredients (i.e., soft stocks), so it is common to use compaction factors between 5% and 10% for these. When sizing each bin, it is essential to consider that ingredients will probably be received before the bins are completely empty, so each bin capacity must be, at a minimum, greater than the volume of one delivery truck or rail car. A common rule of thumb is to size each bin to accommodate, at a minimum, at least 150% of the delivery unit volume.

13.2.1.4 Steam-rolling and steam-flaking

Grain can be minimally treated to produce an effective feedstuff without compounding with other ingredients. For example, grain (most often corn) is treated with steam for 3–5 min (for steam-rolled) or 15–30 min (for steam-flaked) and then pressed between a pair of smooth rollers (Figs 13.8 and 13.9). These processes improve the physical texture and soften the grain. Steam-flaking makes thinner flakes than steam-rolling. The heat treatment may improve protein utilization by ruminants. In the steam-flaking process there will be some rupturing of the starch granules and partial gelatinization of the starch, resulting in more efficient use of the feed by ruminants.

13.2.1.5 Batching

To produce specific feed mixtures, appropriate quantities of ingredients must be transferred out of storage bins (or bags) and transported to the mixer. This is the function of the batching system (Fig. 13.10). For all major and minor ingredients, the equipment used to accomplish this includes screw feeders (e.g., screw conveyors), which provide excellent portioning control and are thus the conveyor of choice for this operation, and scale hoppers, which are essentially steel hoppers mounted on load cells above the mixer. These hoppers range in size from 1 tonne up to 5 tonne, and must be designed with slopes greater than 60° to prevent ingredient build-up and ensure complete clean-out. For ingredients that require precise quantities (e.g., minerals, antibiotics, etc.), microingredient systems are used. These are very small stainless or mild steel bin clusters with small feeder screws. Bulk-bag and hand dump stations are also frequently used to add minor ingredients to the feed mixture. All these components must work in concert to provide the necessary quantities of various ingredients for specific feed mixtures at the correct time. To avoid pressure differentials between the mixer and the scale hoppers during operation, which can prevent proper flow of material into the mixer, venting must be provided between them. Venting ductwork should be installed with a slope greater than 60 degrees (to prevent dust build-up), and should provide an air velocity less than

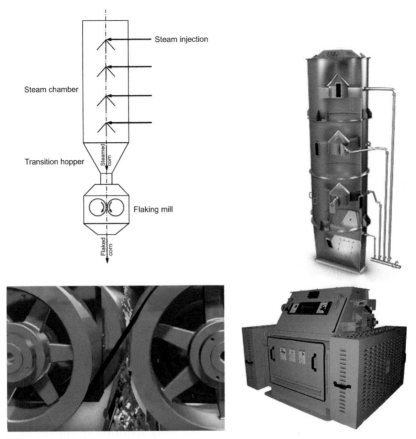

FIGURE 13.8 Steam-flaking is commonly used for corn. Schematic diagram (top left). Steam chamber (top right). Flaking mill (bottom right). Steamed corn passing between the rolls to be flaked (bottom left). *Images courtesy of Fuller, R., CPM-Roskamp.*

500 ft/min (to prevent ingredient entrainment). Furthermore, when designing a batching operation, it is essential to provide adequate access clearance and platforms so that equipment can be serviced and repaired.

13.2.1.6 Mixing

To produce specific feed mixtures, most modern feed mills utilize horizontal batch ribbon mixers (Fig. 13.11), although other types of mixers can be used too. These mixers have bottom gates that dump directly into a conveyor (typically a paddle drag) which transfers the mixed feed (e.g., mash) to a bucket elevator, where it is elevated and distributed to appropriate storage bins. Mash stays in storage until needed for

FIGURE 13.9   Steam-flaked corn.

pelleting, bagging or direct bulk load-out. Ribbon mixers vary in size, but can be constructed as large as 700 ft^3 in capacity. Most ribbons operate at a speed of approximately 40 rpm. Key to proper mixing operations is mixer cycle time, which includes the time required to fill (from the batching scales), mix, discharge and wait for another batch to begin (i.e., dead time). Most mixers can achieve a cycle time between 5 and 10 min, depending on mixer and batching efficiencies, which thus amounts to a mixing capacity of 6–12 tonne/h. Production capacity can quantitatively be determined according to:

$$Q = \frac{C \cdot E \cdot C_1}{T_f + T_m + T_e + T_d}$$

where Q is the volumetric throughput of the mixer (tonne/h), C is the effective volume of the mixer (tonne), E is the efficiency of the mixing process (%, expressed as a decimal), C_1 is a conversion factor of 60 (min/h), T_f is the time required to fill the mixer from the batching scales (min), T_m is the time required to mix the ingredients (min), T_e is the time required to empty the mixer and T_d is dead time between batches (min). When designing a mixing operation, it is essential to provide adequate access clearance so that equipment can be serviced, repaired or cleaned. This ultimately means that enough clearance must be provided for the mixer's ribbon to be removed.

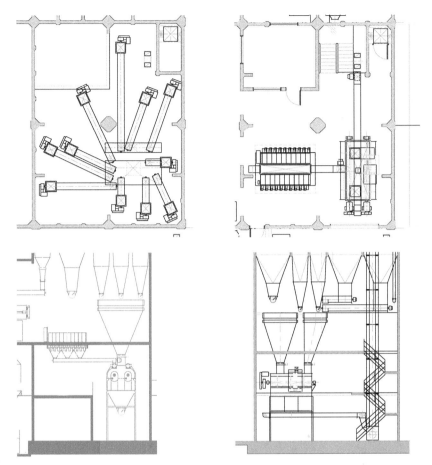

FIGURE 13.10 Plan and section views of a typical batching and mixing system. *Based upon Rosentrater, K.A., Williams, G.D., 2004. Design considerations for the construction and operation of feed milling facilities. Part II: process engineering considerations. In: 2004 ASAE/CSAE Annual International Meeting, Ottawa, ON, Canada, pp. 1–4.*

As shown in Fig. 13.11, other mixer options exist, including the vertical mixer, but these are more commonly used in smaller production facilities.

13.2.1.7 Grinding

Size reduction is one of the most common treatments in feed production. Prior to utilization in feed formulations, whole grain must be ground to reduce particle size. The objectives of grinding are to improve the digestibility of the feed products and produce a consistent particle size which can be uniformly blended with other feed ingredient particles. Size reduction is also critical to pelleting and extrusion processes. Coarsely ground

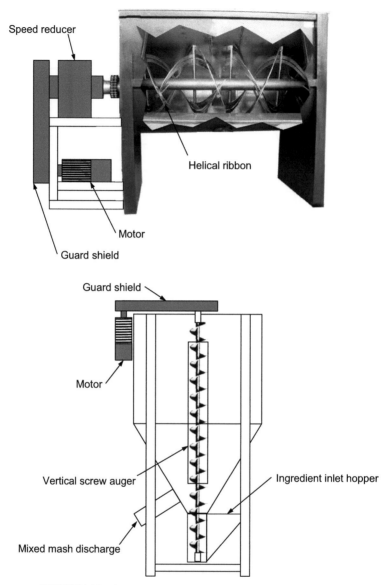

FIGURE 13.11   Ribbon mixer (top) and vertical mixer (bottom).

grain is preferred for ruminants; more finely ground grain is required for swine and poultry.

Grinding systems are generally located directly under whole-grain storage bins, in a separate room within the mill facility (Fig. 13.12), or in a separate grinding building adjacent to the mill structure. This location is driven by safety and fire codes, as discussed by Williams and Rosentrater (2004).

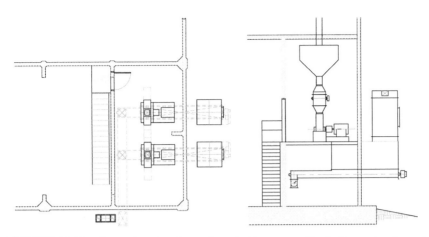

FIGURE 13.12 Plan and section views of a typical hammer-mill grinding system housed within the feed-mill structure. *Based upon Rosentrater, K.A., Williams, G.D., 2004. Design considerations for the construction and operation of feed milling facilities. Part II: process engineering considerations. In: 2004 ASAE/CSAE Annual International Meeting, Ottawa, ON, Canada, pp. 1–4.*

Hammer mills are the most common type of milling equipment, and can have diameters as large as 750 in., screen areas as large as 7000 in.[2] and operating speeds up to 3600 rpm, effectively producing hammer-tip speeds up to 21,000 ft/min, and require motors up to 600 hp. Most hammer mills are typically installed with an air system, which includes air inlets (to control dust from the process) integral to the hammer mill, a plenum under the grinder for airflow and a filter, located outside the grinder building, for dust collection. These systems require 1–2 cfm/in.[2] of screen area. Conveyors under the hammer mill (often a single-screw conveyor) must be located at least 18 in. below the hammer-mill discharge, to allow for an adequate air plenum size and prevent the ground product from becoming entrained in the air stream. It is becoming common to locate the transfer conveyor even lower and provide a cross-sectional area of up to 10 ft[2], to allow for optimal airflow through the plenum.

Roller mills or hammer mills may be used for size reduction, but hammer mills (Fig. 13.13) are generally favoured because, by choice of a screen of suitable size, the mill can yield ground material of any particular size from cracked grain to a fine powder, capital costs are generally somewhat lower than for roller mills and hammer mills can generally be operated at relatively high throughputs and efficiencies. Roller mills have been gaining in popularity over recent years, though, primarily because of their ability to produce more uniform particle sizes, with reduced noise levels and reduced power consumption. Extensive discussions about grinding processes (especially roller milling) may be found earlier in this book.

There are differences in approaches as to where the grinding operation takes place. For example, most feed mills in the United States are set up

FIGURE 13.13   Hammer mill (top), and side panel removed to show screen (bottom). *Images courtesy of Fuller, R., CPM-Roskamp.*

as premix grind (i.e., whole grains only are ground – the other ingredients are used as-is without grinding). Many feed mills around the world (outside the United States), however, are set up as postmix grind (i.e., grinding after all ingredients are mixed). Postmix grind plants produce a more consistent particle size of finished mash feed (which is important to digestibility by animals, facilitates pelleting and extrusion, and results in fewer stress cracks in the pelleted products during drying and cooling); but all ingredients are ground, which means that the grinding capacity has to be substantially larger than in a premix grind plant.

13.2.1.8 Pelleting

After grinding and mixing (or mixing and grinding, depending upon the feed-mill set-up and operation), the ingredient mixture is prepared for pelleting. Pelleting is a process intended to densify feed ingredients to improve storage, handling and shipping behaviour, and can improve the feed nutritionally by increasing its palatability and efficiency in livestock.

Mixed feed (e.g., mash) is transported from the mash storage bins to a preconditioner, where it is mixed with steam and/or water to make it more amenable to the pelleting process; this also initiates starch gelatinization, which is critical for binding the feed ingredients during the densification process. Treatment time in a conditioner of ~20s is recommended, but plants often use longer times.

After conditioning the steamed/moistened mash is conveyed into the pellet mill, where it is forced on to a die – which consists of a rotating ring with stationary rollers (Fig. 13.14) or a stationary plate with rotating rollers (Fig. 13.15) – containing a series of cylindrical die openings through which the densified mash is forced. Die openings can range in diameter from a fraction of an inch to more than 1.0 in., and die plates can be more than 1.0 in. thick. Modern pellet mills can have die ring or plate diameters up to 42 in., with effective pelleting surfaces of 1600 in.^{2}, can produce pelleted feed at a rate of up to 50 tonne/h and can consume up to 800 hp (Fig. 13.16).

Using pressure resulting from continuous feeding of mash particles between the die and a set of rollers results in pellets – a densified feed form which many domestic animals seem to prefer to meal/mash feeds. Densified/pelleted feeds lead to increased consumption by various animals, possibly because pelleting masks the flavour of unpalatable ingredients in the diet, reduces dust and wastage and facilitates transportation, logistics and storage. Pelleting also improves the utilization of amino acids by swine. The heat generated in pelleting may be effective in deactivating some heat-labile toxins and antinutritional factors; indeed, sterilization may be achieved if processing conditions provide a high enough heat/pressure treatment (i.e., within the conditioning cylinder and pellet mill).

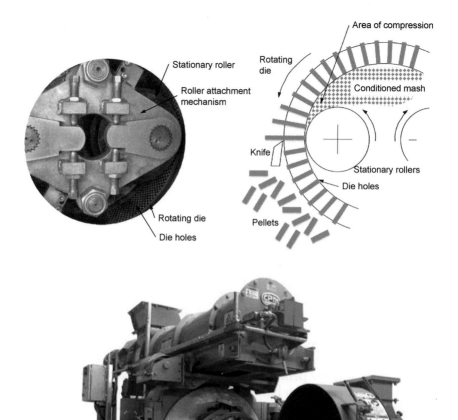

FIGURE 13.14 Ring-style pellet mill. Note the steam conditioner mounted on top of the pellet mill. Schematic diagram (top). *Image courtesy of Fuller, R., CPM-Roskamp.*

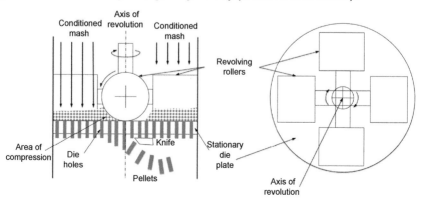

FIGURE 13.15 Plate-style pellet mill.

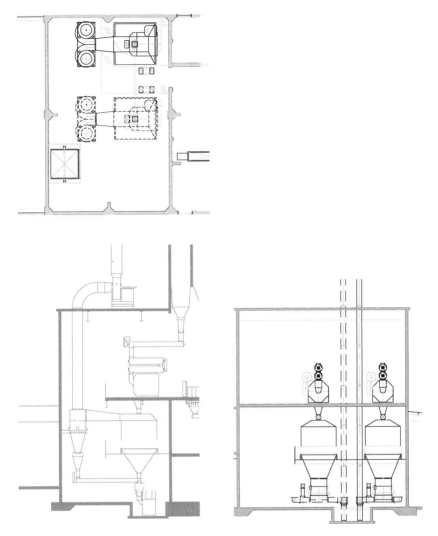

FIGURE 13.16 Plan and section views of a typical pelleting system incorporating a conditioner above the pellet mill and counterflow cooler below the pellet mill. *Based upon Rosentrater, K.A., Williams, G.D., 2004. Design considerations for the construction and operation of feed milling facilities. Part II: process engineering considerations. In: 2004 ASAE/CSAE Annual International Meeting, Ottawa, ON, Canada, pp. 1–4.*

13.2.1.9 Extrusion

Extrusion processing is discussed elsewhere in this book (for breakfast cereals, pastas and other food products). Feed extrusion is primarily limited to high-value feeds, such as aquafeeds, where pellet density must be controlled (e.g., heavier for sinking versus lighter for floating feeds), particle cohesion must be high to prevent water penetration and temperature during processing must be carefully maintained to prevent heat-labile ingredients from degrading (e.g., enzymes, vitamins, amino acids).

13.2.1.10 Drying and/or cooling

After processing the pellets are cooled (horizontal or counterflow coolers are generally used) or dried and then cooled (horizontal equipment is used) so that pellet temperature is reduced to ambient temperature (to avoid spoilage problems), screened to removed fines and broken pellets, and conveyed to storage, after which they are either bagged or loaded out in bulk. Fig. 13.17 illustrates a horizontal dryer/cooler; Fig. 13.18 depicts a counterflow cooler.

13.2.1.11 Popping and micronizing

In grain that has been popped and then rolled, rupture of the endosperm improves utilization of the starch in the digestive tract. Micronizing is a similar process to popping, but uses infrared radiation for heating the grain.

13.2.1.12 Final product storage

Bulk feed materials are stored in bins which are adjacent, and similar, to those used for whole grain and other raw bulk ingredients, so the previous discussion regarding bulk storage is germane. Load-out bins are clustered within a load-out bay (for loading trucks and/or train cars), which is located in a separate section of the mill tower. Bagged feed, on the other hand, will require warehouse storage. Commonly located adjacent to the mill structure, a warehouse is generally constructed of either concrete (precast, tilt-up or slipformed) or steel. When designing warehouse systems, it is important to provide adequate space for material storage, as well as manoeuvring room for forklift trucks. The amount of required storage will depend on many factors, including production capacity, frequency of inventory turnover, number of individual products, space required for bagged feed ingredients versus final product storage and space required for empty pallets and other materials. Pallets can range in size from 48 in. × 48 in. up to 60 in. × 60 in. As a general rule of thumb, for each 1 tonne of bagged product, which is the approximate capacity of one pallet, approximately 16 ft^2 of floor space, as a minimum, should be provided. Additionally, in practice aisles are typically 8 ft wide for forklift travel only, and 12 ft wide for forklift working space (i.e., turning, stacking, etc.).

13.2.1.13 Load out

Load-out systems for feed mills are generally different to those used in grain elevators and other facilities. The two most common options for feed mills are reversible screw conveyors and weigh lorry systems.

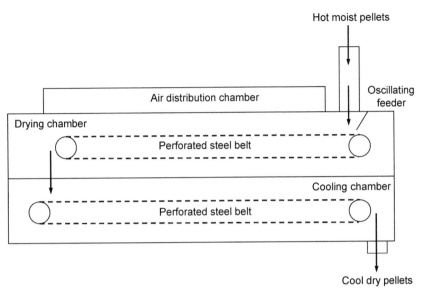

FIGURE 13.17   Horizontal-style dryer–cooler. Schematic diagram (top). *Image courtesy of Fuller, R., CPM-Roskamp.*

Collection screw conveyors (Fig. 13.19) can have either multiple discharges or an additional sliding shuttle screw conveyor, to fill multiple locations in feed delivery trucks. These systems require a truck load-out scale to achieve proper truck fill. Capacities can often exceed

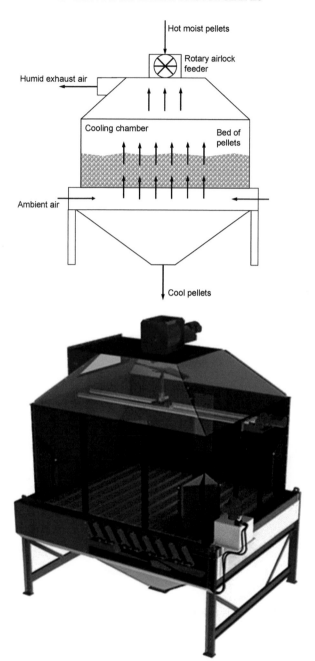

FIGURE 13.18   Counterflow (or 'vertical') cooler. Schematic diagram (top). *Image courtesy of Fuller, R., CPM-Roskamp.*

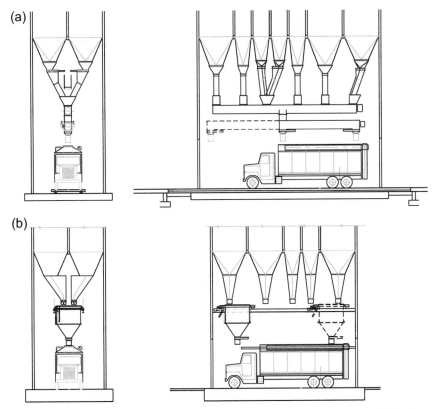

FIGURE 13.19 Section views of typical shuttle conveyor (a) and weigh lorry (b) load-out systems. *Based upon Rosentrater, K.A., Williams, G.D., 2004. Design considerations for the construction and operation of feed milling facilities. Part II: process engineering considerations. In: 2004 ASAE/CSAE Annual International Meeting, Ottawa, ON, Canada, pp. 1–4.*

300 tonne/h for large facilities. Weigh lorry systems (Fig. 13.19), on the other hand, have rail-mounted travelling scales to fill multiple feed-truck locations, often have hopper volumes between 2 and 6 tonne and can achieve load-out capacities greater than 100 tonne/h. Whichever system is implemented, however, it is essential to provide adequate clearance and access platforms so that the load-out equipment can be serviced and repaired.

13.2.1.14 Soaking

Grains may be soaked in water for 12–24 h before rolling. The soaking softens the grain kernels and causes them to swell, thereby improving rollability and palatability, thus facilitating consumption by the animals. There are, however, issues with wastewater and leaching of water-soluble nutrients. This practice is not now widely used on a commercial scale.

13.2.1.15 Reconstitution

This is a process in which grain is moistened to 25%–30% m.c. (moisture content) and then stored in an oxygen-limiting silo for 14–21 days. Although the process has been successful with maize and sorghum, and can improve the feed/growth ratio for beef cattle, it is not now widely practised on a commercial scale.

13.2.1.16 Treatment of high-moisture grain

Grain harvested at a relatively high moisture content, e.g., 20%–35% m.c., can be chemically treated (often using organic acids) to prevent the development of moulds during storage and produce stabilized feed. Recommended treatment of high-moisture grain is with acids, used at a rate of 1.0%–1.5%. Such acids could be propionic alone, or include acetic or formic acids. Maize and barley thus treated can be fed to swine, and maize and sorghum to beef cattle (Church, 1991). It is more common, however, to ensile the wet grain with leaves and stalks of the plant (i.e., corn silage), whereby anaerobic conditions are induced by covering the pile of coarsely ground materials to create silage used for feeding ruminants.

This discussion is only a brief introduction to feed-milling technology. Exhaustive information can be found in McEllhiney (1994).

In the United States maize, soybean meal and corn-based distillers dried grains with solubles (DDGS) are the feed ingredients used in largest quantity. Use of DDGS increased rapidly during the early 2000s as the biofuels industry grew. Nowadays many livestock diets have at least of portion of DDGS, and some diets even use more DDGS than all other cereal grains. In fact, DDGS may surpass the level of maize and other cereal grains in the diets of many animals in the United States today. DDGS is often used to replace maize on a 1:1 basis; the determining factor being the price of competing feed ingredients.

Beyond maize, DDGS and soybean meal, there are many cereal-based ingredients used in animal feeds. Many of these are by-products and coproducts from the various cereal processing industries (see Table 13.2 for typical compositions and photos). Some examples of the use of cereal grains and affiliated products follow.

13.2.2 Maize

For beef cattle, maize is generally fed with another cereal – the greatest benefit comes from combining slowly digested grains (maize, sorghum) with rapidly digested grains (wheat, barley, high-moisture maize) – or with roughage, because fibre and nonstarch dietary components are essential for proper rumen fermentation. A combination of 67% wheat plus 33% dry-rolled maize gave a 6% complementary effect as compared

TABLE 13.2 Some common by-products and coproducts from various cereal processing industries and typical compositions (%d.b.)[a]

Ingredient	Crude protein	Fat (ether extract)	Fibre (NDF)	Starch	Ash	
MAIZE (CORN)						
Corn bran	11.9	4.6	44.2	35.0	5.8	
Corn distillers dried grains with solubles	29.5	11.1	34.2	9.3	5.4	
Corn gluten feed	21.7	3.4	39.6	21.5	6.9	

TABLE 13.2   Some common by-products and coproducts from various cereal processing industries and typical compositions (% d.b.)[a]—cont'd

Ingredient	Crude protein	Fat (ether extract)	Fibre (NDF)	Starch	Ash	
Corn gluten meal	67.2	2.9	4.1	17.6	2.1	
Hominy feed	15.2	5.8	30.7	40.5	6.0	

BARLEY

Barley distillers grains	28.2	7.5	60.1	0.9	5.4	
Brewers' grains	25.8	6.7	56.3	7.8	4.6	

Continued

TABLE 13.2 Some common by-products and coproducts from various cereal processing industries and typical compositions (% d.b.)[a]—cont'd

Ingredient	Crude protein	Fat (ether extract)	Fibre (NDF)	Starch	Ash	
Malt culms	23.5	1.7	45.2	14.7	6.0	
OATS						
Oat hulls	5.2	2.2	75.8	9.9	4.6	
Oat mill feed	8.2	3.8	55.4	25.0	4.0	
RICE						
Rice bran (low fibre)	14.2	13.2	12.4	42.0	6.9	
Rice bran (high fibre)	8.8	10.3	48.7	14.7	13.6	

TABLE 13.2 Some common by-products and coproducts from various cereal processing industries and typical compositions (% d.b.)[a]—cont'd

Ingredient	Crude protein	Fat (ether extract)	Fibre (NDF)	Starch	Ash	
Rice bran, defatted (low fibre)	16.0	4.1	26.6	32.2	12.3	
Rice bran, defatted (high fibre)	6.7	4.8	51.7	14.3	19.1	
Rice hulls	3.7	1.5	67.8	5.3	17.5	
WHEAT						
Wheat bran	17.3	3.9	45.2	23.1	5.6	

Continued

TABLE 13.2 Some common by-products and coproducts from various cereal processing industries and typical compositions (% d.b.)[a]—cont'd

Ingredient	Crude protein	Fat (ether extract)	Fibre (NDF)	Starch	Ash	
Wheat distillers grain with solubles	37.3	5.0	34.0	4.2	5.9	
Wheat germ	29.5	9.7	16.1	19.8	4.7	
Wheat middlings	17.7	4.3	33.7	32.2	4.3	

[a] Based upon data found in Feedipedia – Animal Feed Resources Information System (www.feedipedia.com). NDF, neutral detergent fibre.

with feeding 100% dry-rolled maize (Kreikmeier, 1987). Cattle fed 75% high-moisture maize plus 25% dry-rolled grain sorghum or dry maize gained weight more rapidly and used the feed more efficiently than those fed either grain alone (Sindt, 1987). For cows, a typical diet might contain 41% high-moisture maize (along with alfalfa and soybean meal, etc.) (Schingoethe, 1991). A calf-grower feed might contain 65% maize, sorghum or barley plus 10% rolled oats and 20% soybean meal (Morrill, 1991). Grinding, cracking or rolling the grains, or steam-flaking, typically improves digestibility.

Typical feeds for early-weaned lambs could include 67% ground shelled maize plus 10% cottonseed hulls, or 74% ground ear maize (plus soybean meal and supplements) (Ely, 1991).

For pigs of all ages, maize might provide up to 85% of the grain in the rations. Use of high-lysine maize allows a reduction in the amount of soybean meal needed. The feed grains must be ground, e.g., through a hammer mill with a 3/16–3/8 in. screen. Finely ground maize is used more efficiently than coarsely ground, but very fine grinding, making a dusty meal, should be avoided. In recent years the trend for pig diets has been to use finer grinding (<500 µm) followed by pelleting to improve both the material-handling properties and the digestibility. The proportion of ground maize in the feed for swine at various stages of growth could be 80% for pregnant sows and gilts, 76% for lactating sows and gilts, 63%–71% for young pigs, 78% for growing pigs and 84% for finishing pigs (Cromwell, 1991).

For feeding poultry, maize, sorghum, wheat and barley are commonly used cereals and should be ground and perhaps pelleted. Pelleting prevents the sorting out of constituents of the diet, and is recommended for chicks and broilers. Pelleting minimizes dust and wastage and improves palatability. Maize could provide 57% of the feed for broiler starters, 62% for broiler finishers, 45% for chicks, 57% for growers (7–12 weeks) and developers (13–18 weeks) and 48% for laying hens (Nakaue and Arscott, 1991).

For horses, 25%–39% of the feed could be cracked maize, along with 30%–45% rolled oats and 7%–10% wheat bran (Ott, 1991).

By-products and coproducts of various maize-processing industries are commonly used for animal feeding. Distillers grains (DDGS and others) are discussed in the chapter on alcohol production. Corn gluten feed and corn gluten meal, coproducts of corn wet milling, are discussed in the chapter on wet milling. After the separation of the germ in the maize wet-milling process, and extraction of the oil, the residue – germ cake – is used for cattle feed.

A product known as hominy feed is a by-product of the dry milling of maize. It is a relatively inexpensive high-fibre, high-calorie material, high in carotenoids (yellow pigments, which are desirable for chicken feed) and vitamins A and D. Hominy feed is an excellent source of energy for both ruminants and monogastric animals, in this respect being equal or superior to whole maize, so it competes with other maize by-products as an animal feed ingredient. Hominy feed may partially replace grain in diets for horses, provided the feed is pelleted (Ott, 1991).

As mentioned previously, since about 2000 DDGS has increasingly replaced maize in many animal diets around the world, depending upon the price of DDGS compared to competing ingredients. It is now common to see DDGS proportions in diets of up to 50% for beef cattle, 20% for dairy cattle, 40% for swine and 15% for poultry.

13.2.2.1 Maize cobs

The maize cob (corncob in the United States) is the central rachis of the female inflorescence of the plant to which the grains are attached, and remains as agricultural waste after threshing. As about 180 kg of cobs ((dry basis) d.b.) are obtained from each tonne of maize shelled, the annual production of cobs in the United States alone was in the order of 30 million tonnes in the early 1990s, and that has grown in parallel with the increase of maize production over the years. According to maize production statistics presented in Chapter 1, in 2012 the worldwide production of cobs was nearly 180 billion kg.

Cobs consist principally of cellulose 35%, pentosans 40% and lignin 15%. Agricultural uses for maize cobs, listed by Clark and Lathrop (1953), include litter for poultry and other animals, mulch and soil conditioner and animal and poultry feeds. The feeding value of corncobs is about 62% of that of grains. Diets of up to 67% ground corncobs with 14% ground shelled maize and some soybean meal and molasses/urea can provide a suitable feed for cattle. For poultry, a feed containing corncob meal plus ground maize is preferred to one in which ground maize is the sole cereal because it results in better plumage, less feather-picking and less cannibalism. On the other hand, the corncob plus maize feed gives a reduced egg production and less bodyweight gain (Clark and Lathrop, 1953). Corncobs are typically not commonly used in modern livestock diets, though.

13.2.3 Barley

Apart from its use in malting, brewing and distilling, the next most important use for barley is as food for animals, particularly pigs, in the form of barley meal.

As whole barley typically contains about 34% crude fibre and is relatively indigestible, the preferred type of barley for animal feeding is one with a low husk content – or even hull-less barley. Low-protein barleys are favoured for malting and brewing, but barley of high protein content is more desirable for animal feeds.

The total digestible nutrients in barley are typically 79%; digestibility coefficients for constituents of ground barley are 76% for protein, 80% for fat, 92% for carbohydrate and 56% for fibre (Morrison, 1947).

The feeding value of barley is said to be equal to that of maize for ruminants (Hockett, 1991) and 85%–90% of that of maize for swine (Cromwell, 1991). In fact, for swine, barley can replace all the maize in the feed; indeed, barley is preferred to maize for certain animals, e.g., pigs. The feeding value of barley for pigs is improved by grinding, pelleting, cubing, rolling or micronizing (Hockett, 1991). It is also used extensively in compound feeds.

Barley is normally fed either crushed or as a coarse meal, thereby avoiding wastage that could result from the passage of undigested grains through the alimentary tract. The widespread use of barley by pig feeders is related to its effect on the body fat, which becomes firm and white if the ration contains a large amount of barley meal (Watson, 1953).

Swine fed barley grew faster and had a more efficient feed/gain ratio if the barley was pelleted than if fed as meal. Feed for pregnant sows and gilts can contain up to 85% of ground barley, up to 65% for lactating sows, 80% for growing pigs and 86% for finishing pigs (Cromwell, 1991).

For poultry, a feed containing barley and maize has shown improved egg production and feed efficiency as compared to either cereal fed alone (Lorenz and Kulp, 1991).

'Hiproly' (i.e., hi-pro-ly) barley is a mutant two-row barley from Ethiopia containing the 'lys' gene, which confers higher lysine content. Hiproly barley contains 20%–30% more lysine than is found in normal barley varieties. High-lysine barley has been shown to improve the growth rate of pigs (Hockett, 1991). Additionally, a high-lysine barley mutant originating in Denmark is Risø 1508, with 50% more lysine than in hiproly barley. Risø 1508 is intended to provide a feedstuff with an improved amino acid balance for the pig and dairy industries, one objective being to avoid the necessity of feeding fishmeal, which gives a taint to the product. Over the years many new varieties have been released around the world to improve yield and nutritional content of barley as an animal feed. Recently, hull-less varieties have been developed for use in the biofuels industry, which are significantly higher in starch but also useful as animal feeds. The lack of hulls facilitates milling and processing.

By-products from the dry milling of barley to make pearled barley are used for animal feed too, particularly for ruminants and horses, as constituents of compound feeds. Brewers' grains and distillers dried grains (from brewing, distilling and more recently biofuels) can be incorporated in feeds for ruminants at relatively high levels; and for pigs and poultry, but generally at lower inclusion levels. Malt culms are also used as livestock feed ingredients. Discussions of these ingredients are found elsewhere in this book.

13.2.4 Wheat

The use of wheat for animal feed is influenced by price, location and nutrient value (Mattern, 1991). The importance of wheat as an animal feedstuff is illustrated by the establishment, by the Home-Grown Cereals Authority in the United Kingdom in association with the National Farmers' Union and the UK Agricultural Supply Trade Association, of quality specifications for 'standard feed wheat' in 1978, and subsequently updated. In the United Kingdom feed wheat will typically have a moisture content of 15% and a specific weight of 72.5 kg/hL (https://cereals.ahdb.org.uk/media/658213/hgca-cereal-a5-16pp-final.pdf). Wheat fed directly to animals (not via a feed processor) could include parcels that did not meet these levels and also wheat that was unfit for milling.

When wheat has been fed to cattle, the efficiency of feed usage was greater for dry-rolled wheat than for whole wheat. Dry milling increased grain digestibility from 63% to 88%; but further processing, e.g., steam-flaking or extruding, gave no further improvement (Church, 1991). For beef cattle, wheat is best used in combination with other feed grains, e.g., maize or grain sorghum. A blend of 67% wheat plus 33% dry-milled maize improved feed efficiency as compared with either wheat or maize alone (Ward and Klopfenstein, 1991).

When fed to finishing lambs for market, wheat had 105% of the feeding value of shelled maize when the wheat comprised up to 50% of the total grain (Ely, 1991).

For pigs, wheat is an excellent food, but is often too expensive. Wheat is similar to maize on an energy basis, but has a higher content of protein, lysine and available phosphorus, and it can replace all or part of the maize in the diet for pigs. Nonmillable wheat, damaged moderately (or less) by insects or disease, can be fed to swine (Cromwell, 1991).

For feeding poultry, wheat should be ground, and preferably pelleted, to avoid the birds sorting out feed constituents. For poultry, the feed efficiency of wheat is 93%–95% of that of maize (Nakaue and Arscott, 1991).

Wheat milling by-products – bran, middlings and germ – provide palatable food for animals. Wheat middlings can replace grain in the feed, provided the diets are pelleted – otherwise they are too dusty and bulk density is too low. The energy of wheat middlings is better utilized by ruminants than by monogastric animals, due to the high fibre content. Cows fed rations containing 60% of concentrate did well if 40% of the concentrate was wheat middlings; swine did well when wheat middlings replaced up to 30% of the maize in the rations. Middlings are also fed to poultry. Wheat bran is the favoured feedstuff for horses and ruminants rather than middlings (Church, 1991). Wheat distillers grains are discussed in the chapter on alcohol production.

13.2.5 Oats

Oats have a unique nutritional value, particularly for animals which require feed with a relatively high level of good-quality protein but lower energy content. The level of protein in oat groats is higher than that in other cereals; moreover, the quality of oat protein, particularly the amino acid balance, surpasses that of the protein of other cereals, as shown by feeding tests (Webster, 1986; McMullen, 1991).

For effective feeding to animals, oats are often ground or rolled. Rolled oats can provide 10% of the feed for calves (along with 65% maize, sorghum or barley) (Klopfenstein et al., 1991).

The value of high-protein oats has been shown in diets for swine and poultry, although the nutritive value for these nonruminants can be further improved by supplementing the oats with lysine and methionine or other cereal grains (Webster, 1986). The feeding value for swine, relative to maize, is 100% for oat groats and 80% for whole oats. For poultry, oats have 93% of the value of maize for broilers and 89% for layers (Cromwell, 1991; Nakaue and Arscott, 1991). Oats are a good feed for starter pigs, although too expensive for other pigs. Ground oats can provide 25% of the feed for pregnant sows, 20% for lactating sows, 10% for young pigs and 15% for growing and finishing pigs (with maize, wheat or barley supplying most of the remainder of the feed) (Cromwell, 1991). For feeding to pigs, the oats should be ground through a hammer mill, using a 0.5 in. screen. Pelleting the ground oats gives faster growth than unpelleted meal for swine (Cromwell, 1991).

For feeding to finishing lambs for market, oats have 80% of the feed value of maize (Ely, 1991).

Historically oats have been regarded as an ideal feed for horses, and in North America this view still holds sway. For young or poor-toothed horses, whole oats are best rolled or crushed. As compared to whole oats, crushed oats give a 5% feeding advantage for working horses and a 21% advantage for weanlings and yearlings (Ott, 1973). Oats that are musty should not be used, however (Ott, 1991).

The by-products of the dry milling of oats – oat dust, hulls and oat feed meal (i.e., hulls/husks and other milling by-products) – are of reasonably good food value. Oat feed meal (or oat mill feed in the United States) is a feed ingredient suitable for ruminants and used to dilute the energy content of maize and other grains. Feed oats – the lights, doubles and thin oats removed during the cleaning of oats – are almost equal nutritionally to normal oats, and are used for livestock feeding as well (Webster, 1986).

13.2.6 Sorghum

Sorghum is a major ingredient in the feed for cattle, swine and poultry, particularly in the Western hemisphere.

For feeding to animals, sorghum is often first hammer milled and then generally steam-flaked, using high-moisture steam for 5–15 min to raise the moisture content to 18%–20%, followed by rolling to make thin flakes (Rooney and Serna-Saldivar, 1991). Steam-flaking improves the digestibility and feed efficiency of sorghum.

For beef cattle, grain sorghum has 85%–95% of the feed value of maize. The sorghum is digested slowly in the rumen and has a relatively lower total tract digestibility (Klopfenstein et al., 1991).

Ground sorghum can provide up to 80% of the feed for pregnant sows, 76% for lactating sows, 71% for young pigs, 78% for growing pigs and 84% for finishing pigs (Cromwell, 1991). For swine, low-tannin types of sorghum have a nutritive value equal to that of maize, but brown, high-tannin types, grown for their resistance to attack by birds and their decreased liability to weathering and fungal infestation, have a reduced nutritive value (Cromwell, 1991).

As compared with dry-rolled sorghum, reconstituted sorghum (moistened to 25%–30% m.c. and then stored for 14–21 days in a silo in a low-oxygen atmosphere before feeding) produced a better daily weight gain in feedlot cattle, and also a considerable improvement in feed/gain ratio (Stock, 1985). When fed to swine, reconstituted sorghum gave a slight improvement only in the case of high-tannin sorghum.

For poultry, suggested rations include 18% of sorghum for chick starters, 13% for growers (7–12 weeks), 14% for developers (13–18 weeks) and 20% for layer/breeders (fed as all-mash in a warm climate) (Nakaue and Arscott, 1991).

13.2.7 Rye

Of the annual total world usage of rye for animal feed, the majority occurs in Europe, Russia and former USSR states. Use in all of North and Central America is only a small fraction of the global total.

Rye is used in areas where it is cheaper than barley, but although it is high in energy, growth of animals is slower on rye than on other cereals, possibly because its unpalatability restricts intake. Rye contains a high level of pectin (a carbohydrate), which reduces its feeding value, so it is compounded with other cereals for animal feed. Rye also contains a resorcinol (5-alkylresorcinol) which was once thought to be toxic to animals. Attempts are being made to breed lines of rye with lower levels of resorcinol. Horses fed on rye grain show no ill effects from possible toxic constituents (Antoni, 1960), and rye can be successfully fed to swine and cattle when it contributes up to 50% in a mixed feed.

The presence of ergot in rye is a risk if the rye is fed to swine, as the ergot can cause abortions in sows and reduce the performance of growing pigs (Drews and Seibel, 1976; Cromwell, 1991; Lorenz, 1991).

13.2.8 Rice

Although rice is primarily used for human food, a small portion is used for animal feed. For swine, rice can replace up to 50% of the maize in the feed if it is pelleted, or up to 35% if fed as meal. The feeding value of pelleted broken rice for swine is 96% that of maize (Cromwell, 1991).

Considerable use for animal feeding is made of the by-products of rice milling. Rice pollards – a mixture of rice bran and rice polishings – is a high-energy, high-protein foodstuff comparing well with wheat. It contributes a useful amount of biotin, pantothenic acid, niacin, vitamin E and linoleic acid to mixed feeds, thereby reducing the requirement for premix supplementation with vitamin/minerals. The contribution of linoleic acid in rice pollards is of particular value in rations for laying hens, where it has a beneficial effect on egg size (Australian Technical Millers, 1980). For growing pigs, up to 30% of rice pollard can be fed in balanced rations without adverse effect on growth rate or carcase quality (Roese, 1978).

For young pigs, feed can contain up to 20% rice bran if pelleted (Sharp, 1991). Extracted rice bran (the residue left after oil extraction from rice bran) has an increased content of protein and a good amino acid profile for monogastric animals, and also good protein and phosphorous contents for ruminants. However, it is not a good source of fatty acids.

Rice mill feed is a mixture of rice pollards and ground rice hulls, and is used for animal feed too.

Ground rice hulls are a highly fibrous, low-energy foodstuff, suitable for diluting the energy level in rations for cattle, sheep, goats, pigs and poultry (Australian Technical Millers, 1980). The total digestible nutrients (at 14%m.c.) in rice hulls are 15% for cattle and 25% for sheep (Juliano, 1985).

Rice hulls contain 9%–20% lignin, somewhat limiting their use as animal feed. Various delignification processes have been suggested, e.g., treatment with alkali or acid (see Juliano, 1985 for details). Treatment with 12% caustic soda also reduced the high silica content of rice hulls, while treatment of the hulls with anhydrous ammonia plus monocalcium phosphate at elevated temperature and pressure increased the crude protein equivalent, broke down the harsh silica surface and softened the hulls, thus providing an acceptable feedstuff for cattle and sheep (Juliano, 1985). An even more successful treatment of rice hulls was incubation with *Bacillus* spp. for several days: this reduced the lignin and crude fibre contents to a greater degree than soaking in caustic soda alone (Juliano, 1985).

13.2.9 Millets

Of the millet used worldwide annually for animal feed over 1 million tonnes are used in the former USSR, over 1 million tonnes in Asia (the majority in China) and about 0.5Mt in Africa (FAO, 1990). There is no recorded use of millet for livestock feed in the United States, although proso millet (*Panicum miliaceum*) is grown in the United States for birdseed (Serna-Saldivar et al., 1991).

Feeding trials have shown that millets have nutritive values comparable to, or better than, those of other major cereal grains. Animals fed millet perform better than those fed sorghum; they produce better growth because the millet has a higher calorific content and better-quality protein (Serna-Saldivar et al., 1991).

Pearl millet (rolled) provided excellent protein for beef cattle, and steers gained as well on rolled pearl millet as on sorghum (Serna-Saldivar et al., 1991).

Proso millet has ~89% of the value of maize for feeding swine, and can replace 100% of the maize in rations for pigs (Cromwell, 1991). For swine, finger millet (*Elucine coracana*) was as good as maize for pig-finishing diets; proso millet had a slightly lower feed efficiency than maize, but became equal to maize when supplemented with lysine.

Poultry have been shown to produce better gains when fed millet than when fed sorghum or wheat; the efficiency of feed conversion was better for chicks fed pearl millet (*Pennisetum americanum*) than wheat, maize or sorghum; and proso millet was equivalent to sorghum or maize in respect of egg production and weight and efficiency of feed use.

13.3 PRODUCTION OF BIOCHEMICALS FROM CEREALS

As discussed elsewhere in this book, fermentation is used to convert cereal grains into ethanol. Other biobased chemicals can also be produced from grains using fermentation and other processes.

One of the first commercially viable biochemicals was furfural (Fig. 13.20). Pentosans are the starting material for the manufacture of furfural, which is a chemical with many uses. Indeed, commercial utilization of oat hulls and other pentosan-rich cereal materials lies in the manufacture of furfural (MacArthur-Grant, 1986).

Corncobs, the hulls of oats and rice and the fibrous parts of other cereals are rich in pentosans, condensation products of pentose sugars, which are associated with cellulose as constituents of cell walls, particularly of woody tissues. The pentosan content of oat hulls is approximately 29%, along with 29% cellulose and 16% lignin (McMullen, 1991).

Furfural was first produced commercially in 1922. In 1955 a furfural production facility was opened in the Dominican Republic which used sugarcane bagasse as a source of pentosans (NYT, 1955). By 1975 oat hulls were providing about 22% of the annual demand for furfural in the United

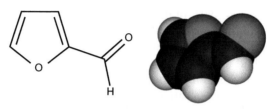

$$(C_5H_8O_4)_n + nH_2O = nC_5H_{10}O_5$$

Pentosan   Pentose

Furfural

$$C_5H_{10}O_5 = C_4H_3O{\cdot}CHO + 3H_2O$$

Pentose   Furfural

Alternate depictions of furfural molecule

FIGURE 13.20   Chemical reactions in the production of furfural from pentosans.

States, but thereafter the demand for furfural and other furan chemicals far outstripped the supply of oat hulls and increasing use was then made of other sources of pentosans, such as rice hulls, corncobs and bagasse (Shukla, 1975).

Plants for the commercial production of furfural from agricultural residues were established in the United States in the mid-20th century. The plant at Cedar Rapids, IA, used oat hulls and corncobs; the one at Memphis, TN, used rice hulls, corncobs and cottonseed hulls, while the plant at Omaha, NE, used corncobs only (Clark and Lathrop, 1953). In the late 20th century, however, these plants were closed. Almost 90% of global furfural production is now located in China (which alone accounts for approximately 74% of world production), South Africa and the Dominican Republic (Ebert, 2008).

The commercial process for manufacturing furfural involves boiling the pentosan-containing material with strong acid (sulphuric or hydro-chloric) and steam for 7–9 h at 70 psi pressure. Previous grinding of the hulls is not required. A sequence of reactions takes place. The pentosans are dissociated from the cellulose; then the pentosans are hydrolysed to pentose sugars; and finally the pentose sugars undergo cyclohydration to form furfural, a heterocyclic aldehyde, which is removed continuously by steam distillation (Dunlop, 1973; Johnson, 1991).

The theoretical yield of furfural from pentose is 64% (plus 36% water), so the theoretical yield of furfural from oat hulls containing 29% pentosans would be 22%, although a yield of only about 13% is achieved in practice. The yield from corncobs is similar, while that from rice hulls is somewhat lower, at 12% theoretical and 5% in practice (Juliano, 1985; Pomeranz, 1987).

A large proportion of the cost of the process is accounted for by the need to raise high-pressure steam: for every 1 lb of furfural produced, 15–26 lb of high-pressure steam at 188°C are required.

Furfural finds uses as a selective solvent for refining lubricating oils and petroleum spirit, and for refining animal and vegetable oils in the manufacture of margarine. It is also used for the purification of butadiene, which is needed for the manufacture of synthetic rubber.

One of the most important uses for furfural is in the manufacture of nylon. Nylon, a synthetic fibre defined chemically as a polyamide, was first produced in 1927 by the firm E I du Pont de Nemours & Co., and was introduced to the industry in 1939. 'Polyamides' are formed by the condensation of a diamide and dibasic acid, and those most often used in the manufacture of nylon are hexamethylene diamine and adipic acid. The value of furfural arises from the fact that it is an important source of hexamethylene diamine.

Other uses for furfural include production of formaldehyde furfural resins for making pipes and tanks; production of tetrahydro furfural alcohol, a solvent for dyes, paints, etc.; production of polytetramethylene ether glycol for making thermoplastics; manufacture of D-xylose, phenolic resin glues and adhesives; production of antiskid tread composition; and filter aids for breweries (MacArthur-Grant, 1986).

Besides the production of ethanol and furfural, cereal grains, especially the starch therein, can be converted (generally via pretreatments, fermentation using microorganisms other than *Saccharomyces cerevisiae*, catalysis, separations and other chemical processing operations) into a variety of high-value biobased chemicals and materials. In fact, the US Department of Energy has assessed the commercial viability of over 300 potential biochemicals and determined the top 12 with the highest potential for commercial success (Table 13.3). These 12 are deemed the most critical 'building-block' chemicals, as they are all intermediate chemicals which can be converted into a wide variety of materials and displace petroleum as building blocks. As noted in the table, while all these chemicals can serve as building blocks for other products, many are currently being manufactured (on either a pilot or a small commercial scale), and some already have direct applications. The US Department of Energy concluded that widespread scale-up and commercialization of these chemicals are essential to the development of a 'biobased' or 'green' economy.

Some fuel ethanol plants in the United States have begun making modifications to their processing operations to start producing some of these chemicals.

TABLE 13.3   Biobased building-block chemicals which can be produced from cereal grains with high potential for commercial viability[a]

Biobased chemical	Chemical structure	Potential applications
1,4-Diacids (succinic, fumaric and malic)	Succinic acid Fumaric acid Malic acid	Polymers, polyesters, resins
2,5-Furan dicarboxylic acid		Polyesters
3-Hydroxy propionic acid		Acrylic acid, sterilizing agents
Aspartic acid		Sweeteners, polymers, corrosion inhibitors
Glucaric acid		
Glutamic acid		Monosodium glutamate (MSG)
Itaconic acid		
Levulinic acid		
3-Hydroxybutyrolactone		

TABLE 13.3 Biobased building-block chemicals which can be produced from cereal grains with high potential for commercial viability[a]—cont'd

Biobased chemical	Chemical structure	Potential applications
Glycerol		Sweeteners, humectants, solvents, laxatives, lubricants, antifreeze
Sorbitol		Sweeteners, laxatives, cosmetics, humectants
Xylitol		Sweeteners

[a] *Based upon data found in Werpy, T., Petersen, G., 2004. Top Value Added Chemicals from Biomass. Results of Screening for Potential Candidates from Sugars and Synthesis Gas, vol. I. Pacific Northwest National Laboratory and National Renewable Energy Laboratory, U.S. Department of Energy. Available online: https://www.osti.gov/scitech/.*

13.4 OTHER INDUSTRIAL USES FOR CEREALS

There are many other industrial uses for cereal grains, their milled products and the by-products of milling. These uses are dependent upon the chemical characteristics and physical properties of the materials. Cereal products and/or by-products can be used as absorbents, abrasives, adhesives, binders, fillers and carriers, and for such purposes as filter aids, litter for animals, fertilizers, floor sweeping, fuels, soil conditioners and oil-well drilling aids. They are also used in the paper and mineral processing industries. A few of these uses are discussed below.

13.4.1 Wheat

Industrial uses for the milling products of wheat are listed by Pomeranz (1987). Both wheat flour and wheat starch are used in paper sizing and coating, and as adhesives in the manufacture of paper, boards, plywood, etc. Starch is also used for finishing textiles.

Gluten separated from wheat flour finds uses in paper manufacture, as an adhesive and as the starting material for the preparation of sodium glutamate and glutamic acid.

Wheatgerm is used in the production of antibiotics, pharmaceuticals and skin conditioners, while wheat bran may be used as a carrier of enzymes, antibiotics and vitamins (Pomeranz, 1987).

Starch is also used to make rigid urethane foam for insulation and paints; in plastics; and to process crude latex in the manufacture of rubber.

13.4.2 Maize and sorghum

Maize grits are used in the manufacture of wallpaper paste, and of glucose by 'direct hydrolysis'.

Coarse or granulated maize meal is used for floor wax and hand soap; fine meal or corn (maize) cones find use as a dusting agent and an abrasive in hand soap.

Corn (maize) gluten is used as a cork-binding agent, as an additive for printing dyes and in pharmaceuticals.

Acid-modified flours of maize and sorghum are used as binders for wallboard and gypsum board, providing a strong bond between the gypsum and the liner. In the building industry, maize flour is used to provide insulation of fibreboard, plywood and wafer board. In the pharmaceutical industry, maize flour is used for the production of citric acid and other chemicals by fermentation processes.

Extrusion-cooked maize flour and sorghum flour are used as core binders or foundry binders in sand–cereal–linseed-oil systems, while a thermosetting resin has been made by combining an acid-modified extruded maize flour with glyoxal or a related polyaldehyde, the mixture binding the sand particles. As adhesives, maize flour and sorghum grits are used in the production of charcoal briquettes, corrugated paper and animal feed pellets.

Both maize flour and sorghum flour find uses in ore refining, e.g., in the refining of bauxite (aluminium ore), and as binders in pelletizing iron ore. Another use for precooked maize flour or starch and sorghum grits is in oil-well drilling, where the flour or starch reduces loss of water in the drilling mud which cools and lubricates the drilling bit. For this purpose, the flours may be precooked on hot rolls or by extrusion cooking.

Maize flour has been used as an extender in polyvinyl alcohol and polyvinyl chloride films for agricultural mulches, and as extenders in rigid polyurethane resins for making furniture (Alexander, 1987).

Corncobs (maize cobs) consist chiefly of cellulose, hemicellulose (pentosan), lignin and ash. Their industrial uses are listed by Clark and Lathrop (1953), and include those discussed below.

13.4.2.1 Agricultural uses

As litter for poultry, preferably reduced to particles of 0.25–0.75 in size. Their use reportedly reduces the mortality of chickens from coccidiosis.

As mulch around plants, for retaining moisture and controlling weeds, and as a soil conditioner for improving soil texture.

As carriers and diluents for insecticides and pesticides, preferably ground to pass a US standard No. 60 sieve. Partly rolled crushed cobs have been used as material on which to grow mushrooms.

13.4.2.2 *Industrial uses*

Those based on physical properties include use for corncob pipes; in the manufacture of vinegar; as an abrasive, when finely ground, for cleaning fur and rugs; for cleaning moulds in the rubber and glass industries; and for soft-grit air-blasting, burnishing and polishing the parts of airplane engines and electric motors. Soft-grit blasting can remove or absorb rust, scale from hard water, oil, grease, wax and dirt from a variety of metals.

Finely ground cobs can replace sawdust (when mixed with sand and paraffin oil) as a floor-sweeping compound, and can be used as an abrasive in soaps.

Ground corncobs find uses in the manufacture of building materials, including asphalt shingles and roofing, brick and ceramics; and as fillers in explosives, e.g., in the manufacture of dynamite, concrete, plastics (to replace wood flour), plywood glues and adhesives (in which the corncob flour improves the spreading and binding properties) and rubber compounds and tyres (in which the corncob material adds nonskid properties).

The lignin separated from corncobs by alkaline treatment can be used as a filler for plastic moulding compounds; as a soil stabilizer in road building; in the manufacture of leather; and as an adhesive. It is more expensive than lignin obtained from wood, so economic uses for corncobs depend on those properties in which cobs show superiority over wood waste or other materials (Clark and Lathrop, 1953).

Industrial uses based on chemical properties (besides the manufacture of ethanol and furfural – see above) include the manufacture of fermentable sugars, solvents and liquid fuel; production of charcoal, gas and other chemicals by destructive distillation; use as a solid fuel (oven-dry cobs have a calorific value of about 18.6 MJ/kg, 8000 Btu/lb); and in the manufacture of pulp, paper and board (Clark and Lathrop, 1953; Klabunde, 1970).

13.4.3 Maize starch

Starch is modified in various ways for use in the food and paper-making industries. Treatments include acid modification and hydroxyethylation (with ethylene oxide) for paper coating; oxidation (with sodium hypochlorite) and phosphating (with sodium phosphate) for paper sizing and improvement of paper strength; and cationic derivatives for paper strengthening and stiffening and improved pigment retention.

Dextrinized starch, pregelatinized starch and succinate derivatives are used as adhesives, while carboxymethylated starch is used in paints, oil-drilling muds, wallpaper adhesives and detergents (Johnson, 1991).

13.4.4 Rye

Its gums, both soluble and insoluble, make rye a good substitute for other gums in wet-end additives. The starch in rye flour has a high water-binding capacity and finds use as an adhesive, for example in the glue, match and plastics industries, and as pellet binders and foundry core binders (Drews and Seibel, 1976; Lorenz, 1991).

13.4.5 Oats

Suggested uses for oat starch are as a coating agent in the pharmaceutical industry, in photocopy papers and as an adhesive; but the starch competes in these uses with rice starch and wheat starch.

Oatmeal is used in the cosmetics industry as a component of lotions, facial masks and soaps. The cleaning effect of the oatmeal may be due to the β-glucan component or to the oat oil (Webster, 1986).

13.4.6 Rice and oat hulls

As the composition of the hulls (husks) of oats and rice somewhat resembles that of corncobs, many of the industrial uses for these hulls duplicate some of those mentioned above for cobs.

Rice hulls differ, however, in yielding a large amount of ash – 22% – upon incineration, of which 95% is silica, with most of the rest being lime and potash. Thus rice hulls can be used as a source of high-grade silica in various manufacturing industries.

Oat husk is used as a filter aid in breweries, where it is mixed with the ground malt and water in the mash tun to keep the mass porous. It is also used, when finely ground, as a diluent or filler in linoleum, as an abrasive in air-blasting for removing oil and products of corrosion from machined metal components, as an antiskid tread component and as a plywood glue extender (Hutchinson, 1953; McMullen, 1991).

The cariostatic properties of oat hulls suggest possible uses as components of chewing gum and other products (McMullen, 1991).

Uses for rice hulls include chicken litter, soil amendment for potting plants, ammoniated for fertilizers, filter aid, burnt for floor sweepings, binder for pelleted feeds, insulating material, filler for building materials (e.g., wallboard: see British Patent No. 1,403,154), and binder and absorbent for pesticides and explosives (Juliano, 1985; Sharp, 1991).

Rice hulls may be used as an abrasive in soft-grit blasting of metal parts, but as the hulls contain 18%–20% silica they are too abrasive when used alone, and are preferably mixed with ground corncobs in the ratio of 40% rice hulls to 60% cob particles (Clark and Lathrop, 1953). Rice hulls have also been used successfully for mopping up oil spillages on the surface of water. After skimming off the hulls, the water was left clean enough to drink.

Fine-sieve fractions of ground rice hulls can be used as excipients (carriers) for nutrients, pharmaceuticals and biological additives in animal feed premixes and poison baits.

The high silica content of rice-hull ash is the reason for using the ash as a constituent of cement, together with slaked lime: rice-hull-ash cement is more acid resistant than Portland cement (Juliano, 1985). Rice-hull ash can be used as a silica source in glass and ceramics industries, e.g., for making Silex bricks (in Italy) and Porasil bricks (in Canada). Slaked lime reacts with the silica during firing to provide a vitreous calcium silicate bond.

Rice-hull ash can be used as a source of sodium silicate for the manufacture of water glass; for reinforcing rubber compounds; as an absorbent for oil; as an insulator for steel ingots; as an abrasive for tooth paste; as an absorbent; and as a water purifier (Australian Technical Millers Association, 1980).

Rice hulls fired at temperatures below 700°C yield amorphous silica, which has been used to make solar-grade silicon for solar cells. The silica can be chlorinated to silicon tetrachloride, which is then reacted with metallurgical-grade silicon to produce trichlorsilane for making solar-grade silicon.

13.4.7 Bioplastics

Over the last few decades considerable progress has been made in developing and commercializing biobased plastics. The majority of these utilize either starches (Zhang et al., 2014) or proteins, and thus many can be made using cereal grains as substrates – maize being foremost in the United States, as it represents the lowest-cost source of both starch and protein, which are the two primary components used to manufacture bioplastics.

Starches can be directly combined with plasticizers or other polymers to produce thermoplastics that can be made into films, sheets or injection-moulded products (Czigany et al., 2007). Likewise, concentrated cereal proteins (corn gluten meal) can be plasticized (Samarasinghe et al., 2007) and then used as thermoplastics. Zein can be separated from maize, purified and then used as coatings or films (Anderson and Lamsal, 2011). A thorough review of challenges and applications for using proteins as plastics is provided by Verbeek and Van den berg (2010).

Often various steps such as fermentation, separation, catalysis and other chemical reactions are necessary before the starch or protein is usable as a plastic. Polylactic acid (PLA) and polyhydroxyalkanoate (PHA, which is a polyester) are examples (Fig. 13.21); there are, in fact, many grades and types of PLA and PHA available on the market. Properties for a few sample PHA and PLA plastics are provided in Table 13.4. Additional information and extensive reviews can be found

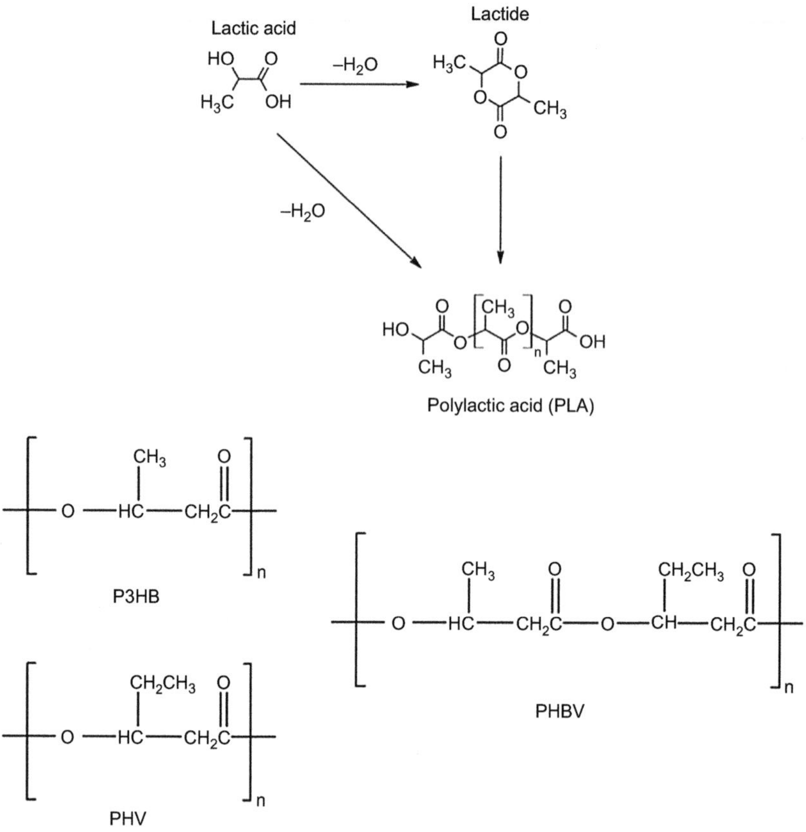

Examples of three polyhydroxyalkanoates (known as PHAs).

FIGURE 13.21 Common bioplastics include polylactic acid (PLA) and polyhydroxyalkanoate (PHA), both of which are produced by fermentation and subsequent chemical reaction engineering. Two common reaction pathways used for production of PLA are shown above.

TABLE 13.4 Example properties of polylactic acid (PLA) and polyhydroxyalkanoate (PHA) bioplastics[a]

	PHB	PHBV	MCL-PHA	PLA	PLA
Melting temperature, T_m (°C)	170–180	130–170	40–60	145–170	145–170
Glass transition temperature, T_g (°C)	−5–5	−10–0	−60 to −30	55–60	55–60
Molecular weight $\times 10^3$ (g/mol)	≤1500	≤1200	50–300		
Density (g/cm³)	1.24	1.20	1.02	1.25	1.12
Crystallinity (%)	60–80	30–80	≤30	25	25
Tensile strength (MPa)	40	30–40	≤10	62	53
Young's modulus (MPa)	3.5–4.0×10^3	0.7–3.0×10^3	≤15	3.5×10^3	3.5×10^3
Elongation to break (%)	3–8	≤100	≤450	3.5	6.0

[a] PolyFerm Canada (http://www.polyfermcanada.com/index.html) and NatureWorks (http://www.natureworksllc.com). Note: multiple grades are commercially available for each type of product, each with its own specific properties. Values in this table are for illustrative purposes only. PHA, polyhydroxyalkanoates; PHB, polyhydroxybutyrate; PHBV, Poly(3-hydroxybutyrate-co-3-hydroxyvalerate); PLA, polylactic acid.

in Siracusa et al. (2008), Snell and Peoples (2009), Luckachan and Pillai (2011) and Bernard (2014).

Many of the by-product and coproduct materials previously discussed for animal feeds can also be used in plastic products, but not as thermoplastics themselves. Rather they are most often used as fillers, which can reduce production costs of the plastic material as they replace a portion of the thermoplastic used. Sometimes they are just added directly during manufacturing, with no modifications. Often, however, they must first be chemically functionalized to improve bonding with the thermoplastic. Rosentrater and Otieno (2006) provide a review of biobased composites and the testing protocols which are necessary to ensure that these manufactured products are comparable to petroleum-based plastics in their performance. Fig. 13.22 illustrates the use of DDGS as a filler in PLA plastics.

FIGURE 13.22   Grain coproducts can be used as fillers in bioplastics to displace a portion of the plastic, which is typically much more expensive than the coproducts. For example, distillers dried grains with solubles (DDGS) can be added to polylactic acid (PLA) to displace a portion of the PLA. Scale bars represent 1 mm distance.

Indeed, the bioplastics industry, as with the biochemicals industry, holds promise for the emerging bioeconomy. The growth of both these industries depends upon the willingness of investors to build and operate processing plants, the performance of economies of countries around the world, political actions and legislation, as well as the price of petroleum –the current substrate for most chemicals and plastics.

References

Alexander, R.J., 1987. Corn dry milling: processes, products, applications (Chapter 11). In: Watson, S.A., Ramstad, P.E. (Eds.), Corn: Chemistry and Technology. Amer. Assoc. of Cereal Chemists Inc., St. Paul, MN, USA.

Anderson, T.J., Lamsal, B., 2011. Zein extraction from corn, corn products, and coproducts and modifications for various applications: a review. Cereal Chemistry 88 (2), 159–173.

Antoni, J., 1960. Rye as feedmeal. Landbauforschung 10, 69–72.

Australian Technical Millers Association, April 1980. Current use and development of rice by-products. Australasian Baker Millers' Journal 27.

Bernard, M., 2014. Industrial potential of polyhydroxyalkanoate bioplastic: a brief review. University of Saskatchewan Undergraduate Research Journal 1 (1), 1–14.

Church, D.C., 1991. Feed Preparation and processing (Chapter 11). In: Church, D.C. (Ed.), Livestock Feeds and Feeding, third ed. Prentice-Hall International, Inc., Englewood Cliffs, NJ, USA.

Clark, T.F., Lathrop, E.C., 1953. Corncobs - their composition, availability, agricultural and industrial uses. U.S. Dept. Agric., Agric. Res. Admin., Bur. Agric. Ind. Chem., AIC-177, revised 1953.

Cor Tech Research Ltd., 1972. Resin Coated Rice Hulls and the Production of Composite Articles Therefrom. British Patent Specification No. 1,403,154.

Cromwell, G.L., 1991. Feeding swine (Chapter 21). In: Church, D.C. (Ed.), Livestock Feeds and Feeding, third ed. Prentice-Hall International, Inc., Englewood Cliffs, NJ, USA.

Czigany, T., Romhany, G., Kovacs, J.G., 2007. Starch for injection molding purposes (Chapter 3). In: Handbook of Engineering Biopolymers. Hanser Publishers Verlag, Munich, Germany.

Drews, E., Seibel, W., 1976. Bread-baking and other uses around the world. In: Bushuk, W. (Ed.), Rye: Production, Chemistry and Technology. Amer. Assoc. of Cereal Chemists Inc., St. Paul, MN, USA.

Dunlop, A.P., 1973. The furfural industry. In: Pomeranz, Y. (Ed.), Industrial Uses of Cereals. Amer. Assoc. of Cereal Chemists Inc., St. Paul, MN, USA, pp. 229–236.

Ebert, J., November 2008. Furfural: future feedstock for fuels and chemicals. Ethanol Producer Magazine. BBI International Media http://biomassmagazine.com/articles/1950/furfural-future-feedstock-for-fuels-and-chemicals.

Ely, D.G., 1991. Feeding lambs for market (Chapter 19). In: Church, D.C. (Ed.), Livestock Feeds and Feeding, third ed. Prentice-Hall International, Inc., Englewood Cliffs, NJ, USA.

Food, Agriculture Organisation of the United Nations, 1990. Food Balance Sheets, 1984–86. FAO, Rome, Italy.

Hockett, E.A., 1991. Barley (Chapter 3). In: Lorenz, K.J., Kulp, K. (Eds.), Handbook of Cereal Science and Technology. Marcel Dekker, Inc., New York, NY, USA.

Home-Grown Cereals Authority, 1990. Cereal Statistics. H.G.C.A., London, UK.

Hutchinson, J.B., 1953. The quality of cereals and their industrial uses. Chemistry and Industry 578.

Johnson, L.A., 1991. Corn: production processing and their industrial utilization (Chapter 2). In: Lorenz, K.J., Kulp, K. (Eds.), Handbook of Cereal Science and Technology. Marcel Dekker, Inc., New York, NY, USA.

Juliano, B.O., 1985. Rice: Chemistry and Technology, second ed. Amer. Assoc. of Cereal Chemists Inc., St. Paul, MN, USA.

Klabunde, H., May 1970. Various methods for the industrial processing of corn. Northwestern Miller 83.

Klopfenstein, T.J., Stock, R., Ward, J.K., 1991. Feeding growing-finishing beef cattle (Chapter 14). In: Church, D.C. (Ed.), Livestock Feeds and Feeding, third ed. Prentice-Hall International, Inc., Englewood Cliffs, NJ, USA.

Kreikmeier, K., 1987. Nebr. Beef Cattle Rpt. MP 52. , p. 9.

Lorenz, K.J., 1991. Rye. In: Lorenz, K.J., Kulp, K. (Eds.), Handbook of Cereal Science and Technology. Marcel Dekker, Inc., New York, NY, USA.

Lorenz, K.J., Kulp, K., 1991. Handbook of Cereal Science and Technology. Marcel Dekker, Inc., New York, NY, USA.

Luckachan, G.E., Pillai, C.K.S., 2011. Biodegradable polymers – a review on recent trends and emerging perspectives. Journal of Polymers and the Environment 19, 637–676.

MacArthur-Grant, L.A., 1986. Sugars and non-starchy polysaccharides in oats (Chapter 4). In: Webster, F.H. (Ed.), Oats: Chemistry and Technology. Amer. Assoc. of Cereal Chemists Inc., St. Paul, MN, USA.

Mattern, P.J., 1991. Wheat (Chapter 1). In: Lorenz, K.J., Kulp, K. (Eds.), Handbook of Cereal Science and Technology. Marcel Dekker, Inc., New York, NY, USA.

McMullen, M.S., 1991. Oats (Chapter 4). In: Lorenz, K.J., Kulp, K. (Eds.), Handbook of Cereal Science and Technology. Marcel Dekker, Inc., New York, NY, USA.

McEllhiney, R.R., 1994. Feed Manufacturing Technology IV. American Feed Industry Association, Arlington, VA, USA.

Morrill, J.L., 1991. Feeding dairy calves and heifers (Chapter 16). In: Church, D.C. (Ed.), Livestock Feeds and Feeding, third ed. Prentice-Hall International, Inc., Englewood Cliffs, NJ, USA.

Morrison, F.B., 1947. Feeds and Feeding, twentieth ed. Morrison Publ. Co., Ithaca, New York, NY, USA.

Nakaue, H.S., Arscott, G.H., 1991. Feeding poultry (Chapter 22). In: Church, D.C. (Ed.), Livestock Feeds and Feeding, third ed. Prentice-Hall International, Inc., Englewood Cliffs, NJ, USA.

NYT, April 1955. Caribbean gets furfural plant. New York Times 6, 43.

Ott, E.A., 1973. Symposium on Effect of Processing on the Nutritional Value of Feeds. Nat. Acad. Sci., Washington, DC, USA. 373 pp.

Ott, E.A., 1991. Feeding horses (Chapter 23). In: Church, D.C. (Ed.), Livestock Feeds and Feeding, third ed. Prentice-Hall International, Inc., Englewood Cliffs, NJ, USA.

Pomeranz, Y., 1987. Modern Cereal Science and Technology. VCH Publishers, Inc., New York, NY, USA.

Roese, G., June 1978. Rice pollard of good value. Rice Mill News.

Rooney, L.W., Serna-Saldivar, S.O., 1991. Sorghum (Chapter 5). In: Lorenz, K.J., Kulp, K. (Eds.), Handbook of Cereal Science and Technology. Marcel Dekker, Inc., New York, NY, USA.

Rosentrater, K.A., Otieno, A., 2006. Considerations for manufacturing bio-based plastic products. Journal of Polymers and the Environment 14, 335–346.

Rosentrater, K.A., Williams, G.D., August 2004. Design considerations for the construction and operation of feed milling facilities. Part II: process engineering considerations. In: 2004 ASAE/CSAE Annual International Meeting, Ottawa, ON, Canada, pp. 1–4.

Samarasinghe, S., Easteal, A.J., Lin, A.P.-Y., 2007. Corn gluten meal-based blends and composites (Chapter 18). In: Handbook of Engineering Biopolymers. Hanser Publishers Verlag, Munich, Germany.

Schingoethe, D.J., 1991. Feeding dairy cows (Chapter 15). In: Church, D.C. (Ed.), Livestock Feeds and Feeding, third ed. Prentice-Hall International, Inc., Englewood Cliffs, NJ, USA.

Serna-Saldivar, S.O., Mcdonough, C.M., Rooney, L.W., 1991. The millets (Chapter 6). In: Lorenz, K.J., Kulp, K. (Eds.), Handbook of Cereal Science and Technology. Marcel Dekker Inc., New York, NY, USA.

Sharp, R.N., 1991. Rice: production, processing, utilization (Chapter 7). In: Lorenz, K.J., Kulp, K. (Eds.), Handbook of Cereal Science and Technology. Marcel Dekker, Inc., New York, NY, USA.

Shukla, T.P., October 1975. Chemistry of oats: protein foods and other industrial products. Critical Reviews in Food Science and Nutrition 383–431.

Sindt, M., 1987. Nebr. Beef Cattle Rpt. MP 52. , p. 9.

Siracusa, V., Rocculi, P., Romani, S., Rosa, M.D., 2008. Biodegradable polymers for food packaging: a review. Trends in Food Science & Technology 19, 634–643.

Snell, K.D., Peoples, O.P., 2009. PHA bioplastic: a value-added coproduct for biomass biorefineries. Biofpr 3, 456–467.

Stock, R., 1985. Nebr. Beef Cattle Rpt. MP 48. , p. 32.

Verbeek, C.J.R., Van den berg, L., 2010. Extrusion processing and properties of protein-based thermoplastics. Macromolecular Materials and Engineering 295 (1), 10–21.

Ward, J.K., Klopfenstein, T.J., 1991. Nutritional management of the beef cow herd (Chapter 13). In: Church, D.C. (Ed.), Livestock Feeds and Feeding, third ed. Prentice-Hall International, Inc., Englewood Cliffs, NJ, USA.

Watson, S.J., January 31, 1953. The quality of cereals and their industrial uses. Chemistry and Industry 95–97.

Webster, F.H., 1986. Oats: Chemistry and Technology. Amer. Assoc. of Cereal Chemists Inc., St. Paul, MN, USA.

Werpy, T., Petersen, G., 2004. Top Value Added Chemicals from Biomass. Results of Screening for Potential Candidates from Sugars and Synthesis Gas, vol. I. Pacific Northwest National Laboratory and National Renewable Energy Laboratory, U.S. Department of Energy. Available online: https://www.osti.gov/scitech/.

Williams, G.D., Rosentrater, K.A., 2004. Design considerations for the construction and operation of feed milling facilities. Part I: Planning, structural, and life safety considerations. ASABE Paper No. 044143.

Zhang, Y., Rempel, C., Liu, Q., 2014. Thermoplastic starch processing – a review. Critical Reviews in Food Science and Nutrition 54, 1353–1370.

Further Reading

British Patent Specification No. 585,772, 1944. Improvements in the Manufacture of Furfural.

Brown, I., Symons, E.F., Wilson, B.W., 1947. Furfural: a pilot plant investigation of its production from Australian raw materials. Journal of the Council for Scientific and Industrial Research 20, 225.

Chrch, D.C., 1991. Livestock Feeds and Feeding, third ed. Prentice-Hall International, Inc., Englewood Cliffs, NJ, USA.

Hitchcock, L.B., Duffey, H.R., 1948. Commercial production of furfural in its 25th year. Chemical Engineering Progress 44, 669.

Lathrop, A.W., Bohstedt, G., 1938. Oat Mill Feed: Its Usefulness and Value in Livestock Rations. Wis Agr. Exp. Sta. Res. Bull., vol. 135.

Martin, J.H., Macmasters, M.M., 1951. Industrial Uses for Grain Sorghum. U.S. Dept. Agric., Yearbook Agric., USA. 349 pp.

Meuser, F., Wiedmann, W., 1989. Extrusion plant design (Chapter 5). In: Mercier, M., Linko, P., Harper, J. (Eds.), Extrusion Cooking. Amer. Assoc. of Cereal Chemists Inc., St. Paul, MN, USA.

National Association of British, Irish Millers, 1991. Facts and Figures 1991. N.A.B.I.M., London, UK.

Peters, F.N., 1937. Furfural as an outlet for cellulosic waste material. Chemical and Engineering News 15, 269.

Rachie, K.O., 1975. The Millets – Importance, Utilization and Outlook. International Crop Research Institute for Semi-Arid Tropics, Begumpet, Hyderabad, India.

Smith, A., 1989. Extrusion cooking: a review. Food Science and Technology Today 3 (3), 156–161.

United States Department of Agriculture, 1987a. Agricultural Statistics 1986. U.S.D.A., Washington, DC, USA.

United States Department of Agriculture, November 1987b. World Grain Situation and Outlook. Foreign Agr. Ser. Circ. Series FG-2-87. Washington, DC.

United States Department of Agriculture, November 1987c. Feed Situation and Outlook. Econ. Res. Ser. FDS-304. Washington, DC, USA.

Wolf, M.J., Macmasters, M.M., Cannon, J.A., Rowewall, E.C., Rist, C.E., 1953. Preparation and properties of hemicelluloses from corn hulls. Cereal Chemistry 30, 451.

Wet milling: separating starch, gluten (protein) and fibre

Kent's Technology of Cereals, Fifth Edition
http://dx.doi.org/10.1016/B978-0-08-100529-3.00014-1

14.1 PURPOSE OF WET MILLING

Wet milling of cereal grains differs fundamentally from dry milling in being a maceration process in which physical and chemical changes occur in the nature of the basic constituents – starch, protein and cell-wall material – to bring about a complete dissociation of the endosperm cell contents with the release of the starch granules from the protein network in which they are enclosed. As discussed elsewhere in this book, in dry milling the endosperm is merely fragmented into cells or cell fragments with no deliberate separation of starch from protein (except in protein displacement milling by air classification, which is a special extension of dry milling).

Although all cereal grains contain starch, those most widely processed by wet milling are wheat and maize. Other cereals which are less frequently wet milled include rice, sorghum and millet, and experimental work has been carried out to separate starch and protein from triticale, rye and other grains.

14.2 WHEAT

Means for separating starch from wheat by wet-milling processes have been known from classical times. Marcus Porcius Cato (234–149 BCE) described a process in which cleaned wheat was steeped for 10 days in twice its weight of water; the excess water was then poured off, and the soaked wheat was slurried, enclosed in a cloth and the starch milk pressed out. The residue of gluten, bran and germ would have been discarded or used as animal feed.

All wet processes for the manufacture of starch and gluten include the steps of extracting the crude starch and crude protein, then purifying, concentrating and drying the two products. To obtain gluten in a relatively pure form (in addition to starch) it is necessary to separate the gluten from the bran and germ. This may readily be done by first milling the wheat by conventional dry processes, then using the white flour as the starting material for the wet process.

Until the development of recent methods, processes starting with flour were mostly variants on three long-established methods.

Martin process: dough is kneaded under water sprays. The gluten agglomerates, and the starch is washed out.

Batter process: a flour–water batter is dispersed in excess water so the gluten breaks down into small curds. The gluten is then separated from the starch milk by screening.

Alkali process: flour is suspended in an alkaline solution (e.g., 0.03 N sodium hydroxide) in which the protein disperses. Starch is removed by tabling and centrifuging, and the protein is precipitated by acidifying to pH 5.5. The protein product is in a denatured condition, i.e., non-vital.

A number of modern processes start with flour: a Canadian process (1966) resembles the alkali process, but suspends the flour in 0.2 M ammonium hydroxide; the Far-Mar-Co process (US Patent No. 3,979,375) is similar to the Martin process, but the dough moves through a tube in which it is washed and mixed; the Alfa-Laval/Raisio process (Dahlberg, 1978) resembles the batter process but uses centrifuging, decantation and hydrocyclones for separating the starch from the gluten; the Koninklijke Scholten-Honig process (British Patent No. 1,596,742) also resembles the batter process and uses hydrocyclones; a process by Walon (US Patent No. 4,217,414) uses bacterial *alpha*-amylase to solubilize the starch – the gluten is not denatured and can be separated as 'vital gluten'.

Some modern processes start with unmilled wheat, but differ from Cato's method in using a steep liquor containing 0.03%–0.70% of sulphur dioxide to inhibit development of microorganisms. After draining off the steep liquor, the wet grain is coarsely milled and slurried with water. The bran and germ are separated by screening, and the heavy starch granules are separated from the light gluten curd by sedimentation and centrifuging.

In the Pillsbury process (British Patent No. 1,357,669) grain is steeped in an acid medium with application of vacuum or carbon dioxide to remove the air pocket at the base of the kernel crease where microorganisms might develop. In the Far-Mar-Co process (US Patent No. 4,201,708) wheat is soaked in water and flaked. The flakes are disintegrated and the resulting bran-germ and endosperm particles are hydrated and form a dough-like mass which is tumbled and manipulated in water to separate and recover vital gluten, starch and bran-germ components.

In all processes, the starch and gluten are dried; in the case of gluten, this stage uses methods such as freeze drying which do not denature the gluten. 'Vital' gluten, or undenatured gluten, is gluten separated from wheat by processes which permit the retention of the characteristics of natural gluten – the ability to absorb water and form an extensible, elastic mass.

Commercial glutens are produced in the United Kingdom, Europe, Australia, Canada and several other countries (McDermott, 1985). In the United Kingdom some 260,000–270,000 tonnes of wheat were used in 1988/1989 for the manufacture of starch and vital gluten, chiefly by the dough (Martin) or batter processes, which start with flour. So far as is known, processes such as the Pillsbury and Far-Mar-Co processes that start with wheat grain are not currently being used in the United Kingdom.

With a yield of about 0.075 tonne of gluten from each tonne of wheat, the quantity of gluten produced in the United Kingdom in 1988/1989 would have been some 20,000 tonne. The demand for vital gluten in the United Kingdom considerably exceeds this figure, and has been met by imports of gluten, chiefly from the European continent. Imports of gluten in 1988/1989 amounted to 37,000 tonne, giving a total availability of about 57,000 tonne in the United Kingdom. World production of vital gluten in 1986 was reported as 253,000 tonne (Godon, 1988), of which 130,000 tonne were produced in Western Europe and 54,000 tonne in the United States, Canada, Mexico and Argentina. In 2016 gluten exports from the United Kingdom were 24,567 tonne, while imports were 50,345 tonne. During that same year, exports from the United States were 4332 tonne, while imports were 208,299 tonne (www.comtrade.un.org).

14.2.1 Vital wheat gluten

Vital gluten is used as a protein supplement and to improve bread structure:

- at levels of 0.5%–3.0% to improve the texture and raise the protein content of bread, particularly 'slimming' bread, crispbread and speciality breads such as Vienna bread and hamburger rolls;
- to fortify weak flours, and permit the use by millers of a wheat grist of lowered strong/weak wheat ratio (particularly in the European Union countries) by raising the protein content of the milled flour;
- in starch-reduced high-protein breads, in which the gluten acts as both a source of protein and a texturizing agent;
- in high-fibre breads to maintain texture and volume.

Vital gluten is also used as a binder and to raise the protein level in meat products, e.g., sausages, in breakfast foods, pet foods, dietary foods and textured vegetable protein products.

14.2.1.1 Vital gluten in bread

In the United Kingdom domestic bread consumption in 1988 averaged 30.28 oz/person/week. With a population of about 57 million, the total domestic bread consumption would have been about 2.5 million tonnes per year, or about 2.75 million tonnes total consumption allowing for about 10% consumed non-domestically. With about 45,000 tonne of vital gluten used in bread in the United Kingdom in 1988, the average level of use would have been about 1.6%. The usual rates of addition are 1.5% for white bread and 4.5% for wholemeal bread.

In the United States about 70% of all vital wheat gluten is used in the manufacture of bread, rolls, buns and other yeast-raised products. The remainder is used in breakfast cereals (e.g., Kellogg's Special K), breadings, batter mixes and pasta products (Magnuson, 1985).

Gluten flour is a blend of vital wheat gluten with wheat flour, standardized to 40% protein content in the United States.

14.3 MAIZE

As with wet milling of wheat, maize is wet milled to obtain starch and a variety of other products, including oil, cattle feed (gluten feed, gluten meal, germ cake) and the hydrolysis products of starch – liquid and solid glucose and syrup.

14.3.1 Operations

The typical sequence of operations in wet milling of maize is shown in Fig. 14.1. A typical process flow diagram illustrating these operations is shown in Fig. 14.2.

14.3.1.1 Storage and drying

For safe storage, maize must be dried because the moisture content (m.c.) at harvest is generally higher than the desirable m.c. for storage. Drying temperature should not exceed about 54°C; at higher temperatures changes occur in the protein whereby it swells less during steeping, and tends to hold the starch more tenaciously, than in grain not dried or dried at lower temperatures. In addition, if dried at temperatures above 54°C the germ becomes rubbery and tends to sink in the ground maize slurry (but the process of germ separation depends on the floating of the germ), and the starch tends to retain a high oil content.

14.3.1.2 Steeping

The cleaned maize is steeped at a temperature of about 50°C for 28–48 h in water containing 0.1%–0.2% sulphur dioxide. Steeping is carried out in a series of tanks through which the steep water is pumped counter-currently. The m.c. of the grain increases rapidly to 35%–40%, and then more slowly to 43%–45%. The steeping softens the kernel and assists subsequent

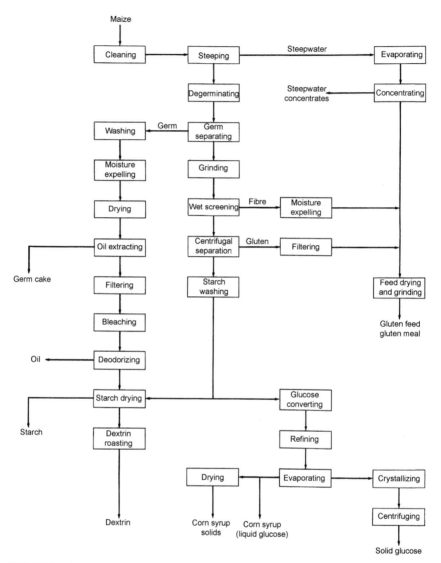

FIGURE 14.1 Maize wet-milling process. *Adapted from Anon., 1958. Food. In: Corn in Industry, fifth ed., vol. 27. Corn Industries Research Foundation Inc., New York, 291 and Matz, S.A. (Ed.), 1970. Cereal Technology, Avi Publ. Co. Inc., Westport, Conn., USA.*

separation of the hull, germ and fibre from each other. The sulphur dioxide in the steep may disrupt —SS— bonds in the matrix protein (glutelin), facilitating starch/protein separation.

After steeping, the steep water is drained off through screens (Fig. 14.3). It contains about 6% solids, of which 35%–45% is protein. The protein in the steep water is recovered by vacuum evaporation, allowed to settle out

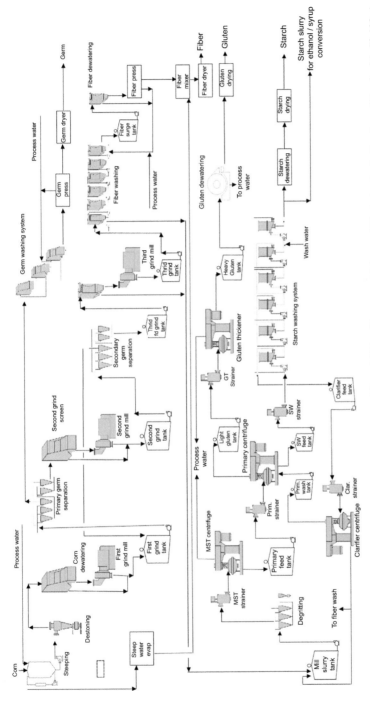

FIGURE 14.2 Typical process flow diagram for maize wet milling. *MST*, mill stream thickener; *GT*, gluten tank. *Courtesy of Franko, M., FluidQuip, Inc., Springfield, OH, USA.*

(a)   (b)

(c)   (d)

(e)   (f)

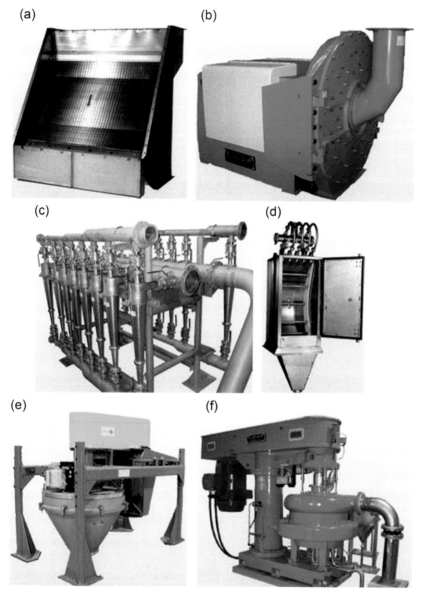

FIGURE 14.3   Equipment used in maize wet milling. (a) Gravity screen; (b) disc mill; (c) germ separation hydroclone; (d) pressure screen; (e) impact mill; and (f) disc nozzle centrifuge. *Courtesy of Franko, M., FluidQuip, Inc., Springfield, OH, USA.*

of the water in tanks and dried as 'gluten feed' for animal feeding. The water recovered is reused as steep water or, after concentration via evaporators, as a medium for the culture of organisms from which antibiotics, enzymes, etc. can be grown.

14.3.1.3 Degerming

After steeping, the maize is coarsely ground in degerming/attrition disc mills with the objective of freeing the germ from the remainder of the grain without breaking or crushing the germ (so that oil does not contaminate the resulting products). The machine generally used for this purpose is a Fuss mill, a bronze-lined chamber housing two upright metal plates studded with metal teeth (Fig. 14.3). One plate rotates at 900 rev/min, the other is stationary. Water and maize are fed into the machine, which cracks open the grain and releases the germ. By addition of starch–water suspension, the density of the ground material is adjusted to 8–10.5°Bé[1] (1.06–1.08 specific gravity; at this sp. gr. the germ floats (due to high oil content) while the grits and hulls settle).

14.3.1.4 Germ separation

The ground material flows down separating troughs in which the hulls and grits settle, while the germ overflows. Most modern plants use hydrocyclones (Fig. 14.3), which require less space and are less costly to maintain than flotation equipment. Moreover, the germ separated on hydrocyclones is cleaner than that separated by flotation. The overflow from the hydrocyclones will be rich in water and germ (often multiple passes in hydrocyclones are necessary to produce concentrated fractions), while the hydrocyclone underflow will mainly consist of water, fibre, starch and protein.

The separated germ is then washed and freed of starch on reels, dewatered in squeeze presses and dried on rotary steam driers. The dry germ is cooked by steam, and the oil is extracted by hydraulic presses or solvent extraction. The germ oil is screened, filtered and stored. The extracted germ cake is then used for cattle feed, and is termed *germ meal*.

14.3.1.5 Milling

The degermed underflow from the germ separator is strained off from the liquor and finely ground on impact mills (such as an entoleter) or attrition mills (such as the Bauer mill) (Fig. 14.3). After this process, the starch and protein of the endosperm are in a very finely divided state and remain in suspension. The hulls and fibre, which are not reduced so much in particle size, can then be separated from much of the protein and starch using pressurized screens (Fig. 14.3) or reels fitted with 18–20 mesh screens. Fine fibres, which interfere with the subsequent separation of starch from protein, are removed on gyrating shakers fitted with fine nylon cloth. Overall, the fibrous particles which are retained on the screens and filters

---

[1] Baumé Scale: a hydrometer scale on which 0° represents the sp. gr. of water at 12.5°C and 10° the sp. gr. of a 10% solution of NaCl at 12.5°C. It is also known as the Lunge Scale.

FIGURE 14.4   Corn gluten feed.

are approximately 21% protein, 15% starch, 10% fibre and 1% fat. This material is dried and sold as an animal feed ingredient known as *corn gluten feed* (Fig. 14.4).

14.3.1.6 Separation of starch from protein

In the raw grain the starch granules are embedded in a protein network which swells during the steeping stage and tends to form tiny globules of hydrated protein (Radley, 1951–52). Dispersion of the protein, which frees the starch, is accelerated by the sulphur dioxide in the steep water.

The effect of the sulphur dioxide, according to Cox et al. (1944), is due to its reducing, not to its acidic, property. The sulphur dioxide also has a sterilizing effect, preventing growth of microorganisms in the steep.

The suspension of starch and protein from the wet screening is adjusted to a density of 6°Bé (1.04 sp. gr.) by dewatering over Grinco or string filters, and the starch separated from the protein in continuous high-speed centrifuges such as the Merco centrifugal separator (Fig. 14.3).

To increase starch purity (>99.5%), the starch stream is recentrifuged in multiple-stage hydrocyclones (Fig. 14.3) to remove residual protein, and the

FIGURE 14.5 Corn gluten meal.

supernatant water is then reused in the previous hydrocyclone and eventually filtered and dried to 10%–12% m.c. in kilns or ovens, or in tunnel or flash driers. The moisture content is further reduced by vacuum drying to 5%–7% m.c. in the United States, or to 1%–2% m.c. in the United Kingdom.

The separated protein (known as gluten) is filtered and dried in rotary or flash driers; it will contain approximately 60% protein, 3% fibre and 1% fat, and is sold as another animal feed ingredient, known as *corn gluten meal* (Fig. 14.5).

Further fractionation to obtain the alcohol-soluble zein, which comprises about 50% of the maize gluten, by solvent extraction and precipitation may be carried out. Zein finds uses as a water-protective coating material for nuts and confectionery products, and as a binder for pharmaceuticals, bio-based films and bio-based plastics.

A combined dry/wet-milling process for refining maize has been described (US Patent No. 4,181,748, 1980) in which the maize is dry milled to provide endosperm, germ, hull and cleaning fractions. The endosperm fraction is then wet milled in two steps which, respectively, precede and follow an impact-milling step. The principal products are prime maize starch, corn oil and an animal feed product.

14.3.2 Wet milling in the United States

The wet milling of maize has increased greatly in the United States in recent decades. In 1960/1961 3.94 Mt of maize were made into wet-processed products; by 1984/1985 the figure had increased to 20.7 Mt, representing 10.6% of the entire maize harvest in 1984 of 195 Mt; of this, 7.9 Mt were used to produce high-fructose corn syrup (HFCS), 4.8 Mt for glucose and dextrose, 3.8 Mt for starch and 4.3 Mt for alcohol (Livesay, 1985). By 2015 approximately 40.6 Mt of maize were wet milled into 18.4 billion lb of HFCS, 5.9 billion lb of starch products, 0.9 billion lb of corn oil and 13.4 billion lb of corn gluten feed, corn gluten meal and germ meal.

The growth of the industry, especially in the 1970s, followed the development of the process to convert starch into HFCS. The sweetness of HFCS allows it to be used as a substitute for sucrose in soft drinks and other processed foods.

14.3.3 Products of wet milling

The wet milling of maize yields about 66% of starch, 4% of oil and 30% of animal feed, comprising about 24% of gluten feed of 21% protein content (made up of about 13% of fibre, 7% of steep water solubles and 4% of germ residue), plus about 5.7% of gluten meal of 60% protein content. Typical composition of coproducts from the wet milling of maize is shown in Table 14.1 (Wright, in Watson and Ramstad, 1987), and Table 14.2 provides some recent values from industrial measurements.

Most of the starch is further processed to make modified starch products, sweeteners and alcohol via fermentation (Long, 1982). The starch can also be fermented and processed into other products, such as citric acid, lactic acid, lysine, threonine and xanthan gum.

Corn gluten is also used in cork-binding agents, additives for printing dyes and pharmaceuticals. It is perhaps somewhat misleading that the protein product obtained from maize should be called 'gluten', because maize gluten in no way resembles the vital gluten that is obtained from wheat.

Ethanol can be made by yeast fermentation of maize starch, and this has a particular advantage because some of the yeast can be recycled or harvested. About 85% of the ethanol produced from maize starch is blended with gasoline, in which it acts as an octane enhancer for unleaded fuels (May, 1987). An extensive discussion about biofuel production from maize starch is provided elsewhere in this book.

An edible film has been made from maize-starch amylose obtained from high-amylose maize. Suggested uses for the film include the packaging of gravies, sauces and coffees.

TABLE 14.1 Composition of coproducts from wet milling of maize[a] (dry basis)

| Product | Moisture (%) | Protein[b] (%) | Fat (%) | Fibre (%) | | Ash (%) | NFE[d] (%) | Starch (%) |
				Crude	NDF[c]			
Maize	15.5	8.0	3.6	2.5	8.0	1.2	69.2	60.6
Corn gluten feed	9.0	22.6	2.3	7.9	25.4	7.8	50.1	low
Corn gluten meal	10.0	62.0	2.5	1.2	4.1	1.8	22.5	low
Germ meal	10.0	22.6	1.9	9.5	41.6	3.8	52.2	low
Steep liquor	50.0	23.0	0	0	0	7.3	19.2	low

[a] Based upon data from Anon., 1982. Corn Wet Milled Feed Products, second ed. Corn Refiners Assoc., Washington, DC, USA and Wright, K.N., 1987. Nutritional properties and feeding value of corn and its by-products. Based upon data in: Watson, S.A., Ramstad, P.E. (Eds.), Corn: Chemistry and Technology. Amer. Assoc. of Cereal Chemists Inc., St. Paul, MN, USA, pp. 447–478.
[b] N × 6.25.
[c] Neutral detergent fibre.
[d] Nitrogen-free extract.

TABLE 14.2　Composition of coproducts from wet milling of maize (dry basis)

Nutrient (%)[a]	Coproducts[b]						
	Dry CGF[c]	Dry CGF[d]	Wet CGF[d]	CGM[c]	CGM[d]	Germ meal[d]	Steep liquor[e]
DM	89.4	89.2	44.0	86.4	88.8	90.4	52.5
CP	23.8	24.0	25.4	65.0	67.1	25.2	44.2
NDF	35.5	36.1	38.1	11.1	9.1	41.3	2.3
ADF	12.1	11.6	12.3	8.2	4.8	14.7	0.7
Starch		14.9	12.1		15.3	20.4	
Fat	3.5	4.1	3.7	2.5	3.0	8.2	0.8
Ash	6.8	7.3	7.5	3.3	3.3	3.9	10.5
Ca	0.07	0.10	0.07	0.06	0.06	0.05	0.08
P	1.00	1.06	1.28	0.60	0.55	0.85	2.04
Mg	0.42	0.43	0.53	0.14	0.08	0.27	0.75
K	1.46	1.46	1.82	0.46	0.28	0.58	2.89
S	0.44	0.51	0.53	0.86	0.86	0.30	1.90

[a] Nutrients: ADF, acid detergent fibre; CP, crude protein; DM, dry matter; NDF, neutral detergent fibre.
[b] Coproducts: CGF, corn gluten feed; CGM, corn gluten meal.
[c] Based upon data from National Research Council (NRC), 2001. Nutrient requirements of dairy cattle, seventh revised ed. National Academies Press, Washington, DC, USA.
[d] Analysed by Dairy One Forage Lab from May 2000 to April 2014 (Number of samples > 100–400 for each nutrient analysed).
[e] Based upon data from DeFrain, J.M., Shirley, J.E., Behnke, K.C., Titgemeyer, E.C., Ethington, R.T., 2003. Development and evaluation of a pelleted feedstuff containing condensed corn steep liquor and raw soybean hulls for dairy cattle diets. Animal Feed Science and Technology 107, 75–86.

The starch obtained from the milling of waxy maize, called 'amioca', consists largely of amylopectin. Amioca paste is non-gelling and has clear, fluid-adhesive properties.

Heated and dried maize starch/water slurries yield pregelatinized starch, known as 'instant starch' as it thickens upon addition of cold water.

14.3.4 Uses for wet-milled maize products

Uses for maize starch are numerous. A few examples include paper manufacture, textiles, adhesives, various packaged foods and as the starting material for further processing by chemical treatments to make various kinds of modified starches for particular purposes, or by enzymatic hydrolysis to yield maltose, which can be further treated to make dextrose (D-glucose), regular corn syrup, HFCS and malto-dextrins.

14.3.4.1 Modified starches

When modifying starch, the objectives are to alter the physical and chemical nature to improve functional characteristics, by oxidation, esterification, etherification, hydrolysis or dextrinization. The methods commonly used are acid thinning, bleaching or oxidation, cross-linking, substitution or derivatization and instantizing.

In acid thinning, or conversion, the glucosidic linkages joining the anhydro-glucose units are broken with the addition of water. The resulting thinning reduces the viscosity of the starch paste, and allows such starches to be cooked at higher concentrations than the native starch. In a wet process of conversion, the starch is treated with 1%–3% of hydrochloric or sulphuric acids at about 50°C, then neutralized and the starch filtered off. Acid-thinned starches are used in confectionery products, particularly starch jelly candies (Moore et al., 1984). Non-food uses include paper-sizing, calendaring and coating applications (Sanford and Baird, 1983; Bramel, 1986). In a dry process of conversion, in which dry starch powder is roasted with limited moisture and a trace of hydrochloric acid, the main product is dextrins, used for adhesives and other non-food purposes.

In bleaching, or oxidation, aqueous slurries of starch are treated with hydrogen peroxide, peracetic acid, ammonium persulphate, sodium hypochlorite, sodium chlorite or potassium permanganate and then neutralized with sodium bisulphite. The xanthophyll and other pigments are bleached, thereby whitening the starch, which then becomes suitable for use as a fluidizing agent in, e.g., confectioners' sugar. Starch in aqueous slurry can be oxidized with about 5.5% (of dry weight) of chlorine as sodium hypochlorite. Hydroxyl ($-OH$) groups are oxidized, forming carboxyl ($-C=O$) or carbonyl ($OH-C=O$) groups, with cleavage of glucosidic linkages. The bulkiness of the $-C=O$ groups reduces the tendency of the starch to retrograde. Bleached starch has uses in batter and breading mixes in fried foods (Moore et al., 1984) to improve adhesion. Oxidized starch has non-food uses in paper sizing, due to its excellent film-forming and binding properties (Bramel, 1986).

Cross-linking improves the strength of swollen granules, preventing rupture. Granular starch is cross-linked by treatment with adipic acid and acetic anhydride, forming distarch adipate, or with phosphorus oxychloride, forming distarch phosphate. The slurry is neutralized, filtered, washed and dried. The viscosity of cross-linked starch is higher than that of native starch.

Derivatization or substitution consists of the introduction of substitution groups into starch by reacting the hydroxyl groups (—OH) with monofunctional reagents such as acetate, succinate, octenyl succinate, phosphate or hydroxypropyl groups. Derivatization retards the association of gelatinized amylose chains, improves clarity, reduces gelling and improves water-holding capacity (Orthoefer, 1987). Substitution may be combined with cross-linking to yield thickeners with particular processing characteristics (Moore et al., 1984). Starch esters – acetates, phosphates and octenyl succinates – have uses as thickeners in foods, e.g., fruit pies, gravies, salad dressings and filled cakes, because they withstand refrigeration and freeze/thaw cycles well. Non-food uses include warp sizing of textiles, surface sizing of paper and gummed-tape adhesives (Sanford and Baird, 1983; Bramel, 1986).

Phosphate mono-esters of starch, made by roasting starch with orthophosphates at pH 5–6.5 for 0.5–6 h at 120–160°C, have good clarity, high viscosity, cohesiveness and stability to retrogradation. They are used as emulsifiers in foods and for many non-food purposes (Orthoefer, 1987).

14.3.4.2 *Dextrose and HFCS*

Dextrose is made by enzymatic hydrolysis of starch and crystallization; **HFCS** by partial enzymatic isomerization of dextrose hydrolysates; and **regular corn syrup** and **malto-dextrins** by partial hydrolysis of corn starch with acid, acid + enzyme or enzymes only (Hebeda, 1987).

In making dextrose, thermostable bacterial *alpha*-amylase from *Bacillus subtilis* or *B. licheniformis* is used for liquefying starch to 10–15 D.E. (dextrose equivalents), followed by saccharification with glucoamylase from *Aspergillus niger* to 95%–96% dextrose (dry basis). The glucoamylase releases dextrose step-wise from the non-reducing end, cleaving both α-1,4 and α-1,6 bonds. The hydrolysate is clarified, refined and processed into crystalline dextrose, liquid dextrose, high-dextrose corn syrup or HFCS.

Glucose and dextrose are commonly used in beer, cider, soft drinks, pharmaceuticals, confectionery, baking, jams and other foods.

The refined dextrose hydrolysate can be treated with immobilized glucose isomerase of bacterial origin. This enzyme catalyses the isomerization of dextrose to D-fructose. A product containing 90% of HFCS may be obtained by chromatographic separation.

Dextrose and HFCS are important sweeteners: dextrose has 65%–76% the sweetness of sucrose, while HFCS is 1.8 times as sweet as sucrose and 2.4 times as sweet as dextrose.

The use of maize in the United States specifically to produce HFCS increased from 0.25 Mt in 1971/1972 to 8.1 Mt in 1985/1986 and 26.9 Mt in 2015. This explosive growth of production has been due to technical breakthroughs in the process for making HFCS, the high cost of sugar in the United States, taxes/subsidies/tariffs and the availability of abundant stocks of relatively low-cost maize in the country (May, 1987). HFCS and the affiliated feed coproducts are the largest product categories of the American corn wet-milling industry, accounting for more than 86% of the industrial grind in 2015.

14.4 SORGHUM

The methods used for the wet milling of sorghum (milo) closely resemble those described for maize, but the process is somewhat more difficult with sorghum than with maize. The problems are associated with the small size and spherical shape of the sorghum kernel, the large proportion of horny endosperm and the dense, high-protein peripheral endosperm layer. Varieties with dark-coloured outer layers are not satisfactory for wet milling because some of the colour leaches out and stains the starch.

14.4.1 Steeping

The cleaned sorghum is first steeped in water (1.61–1.96 L/kg) for 40–50 h in a counter-current process. Part of the water is charged with 0.10%–0.16% of sulphur dioxide, which is absorbed by the grain and weakens the protein matrix in which the starch granules are embedded.

14.4.2 Degerming

The steeped grain, in slurry form, is ground in an attrition mill. The mill has knobbed plates, one static and one rotating at about 1700 rev/min. The milling detaches the germ and liberates about half of the starch from the endosperm. The germ, which contains 40%–45% of oil, floats to the surface of the slurry and is removed from the heavier endosperm and pericarp in a continuous liquid cyclone.

The degermed endosperm and pericarp are wet screened on a nylon cloth with 70–75 μm apertures, through which the free starch and protein pass.

The residue, mostly horny endosperm, is reground in an entoleter, impact mill or other suitable mill to release more starch, and is again screened.

14.4.3 Dewatering

The washed tailings of the screen are dewatered to about 60% m.c. by continuous screw presses. The product is known as 'fibre'. The fibre, blended with concentrated steep water (containing solubles leached out during the steeping) and spent germ cake, is dried in a continuous flash drier to yield 'milo gluten feed'.

14.4.4 Starch/protein separation

The defibred starch granules and protein particles are separated in a Merco continuous centrifuge by a process of differential sedimentation; the starch granules, with a density of ~1.5, settle out of an aqueous slurry at a faster rate than the protein particles (density ~1.1). The starch slurry is dried on a Proctor & Schwartz moving-belt tunnel drier, or by flash drying.

14.4.5 Sorghum 'gluten'

The 'gluten' (protein) is concentrated, filtered and flash dried. The protein content of milo gluten is 65%–70% (dry basis). A blend of milo gluten with milo gluten feed, reducing the protein content to 45% (dry basis) is known as 'milo gluten meal'.

The yield and composition of products from the wet milling of sorghum are shown in Table 14.3.

TABLE 14.3   Yield and composition of coproducts from wet milling of sorghum[a] (dry basis)

Product	Yield[b] (%)	Protein (%)	Fat (%)	Starch (%)
Germ	6.2	11.8	38.8	18.6
Fibre	7.4	17.6	2.4	30.6
Tailings	0.8	39.2	–	25.3
Gluten	10.6	46.7	5.1	42.8
Squegee	1.2	14.0	0.6	81.6
Starch	63.2	0.4	–	67.3
Solubles	6.6	43.7	–	–

[a] Based upon data from Freeman, J.E., Bocan, B.J., 1973. Pearl millet: a potential crop for wet milling. Cereal Science Today 18, 69.
[b] % of dry substance in whole grain.

14.5 MILLET

The possibility of using pearl millet as raw material for wet-milling processing has been investigated (Freeman and Bocan, 1973). The small millet grains were much more difficult to degerm than sorghum or maize, although the potential yield of oil from millet exceeded those from the other cereals. Additionally, separation of protein from starch was also more difficult with millet than with sorghum or maize, and the products of separation were somewhat less pure than those obtained from the other cereals. The starch from millet resembled that from sorghum and maize in most respects, but the granule size was slightly smaller and the starch had a slightly lower tendency to retrograde.

The yield and composition of products from the wet milling of pearl millet are shown in Table 14.4.

14.6 RICE

14.6.1 Solvent extraction milling

Solvent extraction milling (SEM) is a process applied to rice to obtain debranned rice grain, and also rice bran and rice oil as separate products. The SEM process cannot strictly be described either as a dry-milling or as a wet-milling process, although it involves stages which could be included in each of these categories.

The customary method for milling rice uses abrasion of the rice grain in a dry condition to remove bran from the endosperm. In the SEM process (also called X-M) the bran layers are first softened and then 'wet milled' in

TABLE 14.4   Yield and composition of coproducts from wet milling of pearl millet[a] (dry basis)

Product	Yield[b] (%)	Protein (%)	Fat (%)	Starch (%)
Germ	7.5	10.4	45.6	10.4
Fibre	7.3	11.8	6.0	13.5
Tailings	1.5	34.1	1.9	34.2
Gluten	12.1	37.8	9.0	44.0
Squeegee	4.1	17.7	0.8	75.5
Starch	49.3	0.7	0.1	57.5
Solubles	16.8	46.1	–	–

[a] *Based upon data from Freeman, J.E., Bocan, B.J., 1973. Pearl millet: a potential crop for wet milling. Cereal Science Today 18, 69.*
[b] *% of dry substance in whole grain.*

the presence of a rice-oil solution. The separated bran has a higher protein content than that of the residual grain, is virtually fat-free and is thus much more stable than that separated in the conventional dry-milling process.

In the SEM process (US Patent No. 3,261,690), rice oil is applied to brown rice (dehulled rough rice) in controlled amounts, and softening of the bran is accomplished. The bran is removed by milling machines of modified conventional design in the presence of an oil solvent – rice oil/hexane miscella. The miscella acts as a washing or rinsing medium to aid in flushing bran away from the endosperm, and as a conveying medium for continuously transporting detached bran from rice. The miscella lubricates the grains, prevents rise of temperature and reduces breakage.

The debranned rice is screened, rinsed and drained, and the solvent is removed in two stages. Super-heated hexane vapour is used to flush-evaporate the bulk of the hexane remaining in the rice, and the rice is subjected to a flow of inert gas which removes the last traces of the solvent.

The bran/oil miscella slurry is pumped to vessels in which the bran settles, and is then separated centrifugally while being rinsed with hexane to remove the oil. The last traces of solvent are removed from the bran by flash desolventizing, and the bran is cooled.

The oil/hexane miscella from the bran-settling vessels is pumped to a conventional solvent recovery plant, where the hexane is stripped from the oil.

There are four main advantages of the SEM process over the conventional milling process.

1. An increase of up to 10% on head rice yield.
2. A decrease in the fat content of the rice, which improves its storage life.
3. An increase in stability of the bran product.
4. A yield of 2 kg of rice oil from each 100 kg of unmilled rice (2% yield on rice weight).

The bran product has applications in breakfast cereals, baby foods, baked goods and other foods. The oil has edible and industrial applications and is a rich source of a wax, with properties similar to those of myricyl cerotate or carnauba wax. Other applications are in margarine, cosmetics and paints (Edwards, 1967).

References

Anon., 1982. Corn Wet Milled Feed Products, second ed. Corn Refiners Assoc., Washington, DC, USA.

Bramel, G.F., 1986. Modified starches for surface coatings or paper. TAPPI 69, 54–56.

British Patent Specification No. 1,596,742 (Koninklijke Scholten-Honig NV).

British Patent Specification No. 1,357,669 (Hydroprocessing of grain; Pillsbury Co.).

Cox, M.J., Macmasters, M.M., Hilbert, G.E., 1944. Effect of the sulphurous steep in corn wet milling. Cereal Chemistry 21, 447.

Dahlberg, B.I., 1978. A new process for the industrial production of wheat starch and wheat gluten. Starke, 30 (1), 8–12.

Dairy One Forage Lab, 2014. Interactive Feed Composition Libraries. Available online at: http://dairyone.com/analytical-services/feed-and-forage/feed-composition-library/interactive-feed-composition-library/.

DeFrain, J.M., Shirley, J.E., Behnke, K.C., Titgemeyer, E.C., Ethington, R.T., 2003. Development and evaluation of a pelleted feedstuff containing condensed corn steep liquor and raw soybean hulls for dairy cattle diets. Animal Feed Science and Technology 107, 75–86.

Edwards, J.A., July 21, 1967. Solvent extraction milling. Milling 48.

Freeman, J.E., Bocan, B.J., 1973. Pearl millet: a potential crop for wet milling. Cereal Science Today 18, 69.

Godon, B., 1988. Les débouches du gluten travaux de l'IRTAC et de l'INRA. Industries Alimentaires et Agricoles 105, 819–824.

Hebeda, R.E., 1987. Corn sweeteners. In: Watson, S.A., Ramstad, R.E. (Eds.), Corn, Chemistry and Technology. Amer. Assoc. of Cereal Chemists Inc., St. Paul, MN, USA, pp. 501–534.

Livesay, J., 1985. Estimates of corn usage for major food and industrial products. In: Situation Report. U.S. Dept. Agric., Econ Res Serv., Washington, DC, USA, pp. 8–10.

Long, J.B., 1982. Food sweeteners from the maize wet milling industry. In: Swaminathan, M.R., Sprague, E.W., Singh, J. (Eds.), Processing, Utilization and Marketing of Maize. Indian Council of Agric. Technol., New Delhi, India, pp. 282–299.

Magnuson, K.M., 1985. Uses and functionality of vital wheat gluten. Cereal Foods World 30 (2), 179–181.

May, J.B., 1987. Wet milling: process and products. In: Watson, S.A., Ramstad, P.E. (Eds.), Corn: Chemistry and Technology. Amer. Assoc. of Cereal Chemists Inc., St. Paul, MN, USA, pp. 377–397.

McDermott, E.E., 1985. The properties of commercial glutens. Cereals Foods World 30 (2), 169–171.

Moore, C.O., Tuschhoff, J.V., Hastings, C.W., Schanfelt, R.V., 1984. Applications of starches in foods. In: Whistler, R.L., BeMiller, J.N., Paschall, E.F. (Eds.), Starch: Chemistry and Technology, second ed. Academic Press, Orlando, FL, USA, pp. 579–592.

National Research Council (NRC), 2001. Nutrient Requirements of Dairy Cattle, seventh revised ed. National Academies Press, Washington, DC, USA.

Orthoefer, F.T., 1987. Corn starch modification and uses. In: Watson, S.A., Ramstad, P.E. (Eds.), Corn: Chemistry and Technology. Amer. Assoc. of Cereal Chemists Inc., St. Paul, MN, USA, pp. 479–499.

Radley, J.A., 1951–52. The manufacture of maize starch. Food Manufacturing 26, 429–488 27: 20.

Sanford, P.A., Baird, J., 1983. Industrial utilization of polysaccharides. In: Aspinall, G.O. (Ed.), The Polysaccharides. vol. 2. Academic Press, New York, NY, USA, pp. 411–490.

US Patent Specification Nos. 3261690 (SEM process); 3958016 (Pillsbury Co.); 3979375 (Far-Mar-Co process); 4181748 (dry-wet milling); 4201708 (Far-mar-Co process); 4217414 (Walon).

Wright, K.N., 1987. Nutritional properties and feeding value of corn and its by-products. In: Watson, S.A., Ramstad, P.E. (Eds.), Corn: Chemistry and Technology. Amer. Assoc. of Cereal Chemists Inc., St. Paul, MN, USA, pp. 447–478.

Further reading

Andres, C., May 1980. Corn – a most versatile grain. Food Process 78.

Autran, J.-C., 1989. Soft wheat: view from France. Cereal Foods World 34 667, 668, 671, 672, 674, 676.

Finney, P.L., September 1989. Soft wheat: view from Eastern United States. Cereal Foods World 34 682, 684, 686, 687.

Leath, M.N., Hill, I.D., 1987. Economics, production, marketing and utilization of corn. In: Watson, S.A., Ramstad, P.E. (Eds.), Corn: Chemistry and Technology. Amer. Assoc. of Cereal Chemists Inc., St. Paul, MN, USA.

Linko, P., 1987. Immobilized biocatalyst systems in the context of cereals. In: Morton, I.D. (Ed.), Cereals in a European Context. Ellis Norwood, London, UK, pp. 107–118.

Meyer, P.A., March 26, 1991. High fructose starch syrup production and technology growth to continue. Milling & Baking News 70 (4), 34.

Mittleder, J.F., Anderson, D.E., Mcdonald, C.E., Fisher, N., 1978. An Analysis of the Economic Feasibility of Establishing Wheat Gluten Processing Plants in North Dakota. Bull. 508, North Dakota Agricultural Experiment Station, Fargo, ND, and U.S. Dept. Commerce.

Rao, G.V., 1979. Wet wheat milling. Cereal Foods World 24 (8), 334–335.

Watson, S.A., Ramstad, P.E., 1987. In: Corn: Chemistry and Technology. Amer. Assoc. of Cereal Chemists Inc., St. Paul, MN, USA.

Since antiquity, the growth and development of civilization has, to a large degree, depended upon the domestication, cultivation, improvement and utilization of cereal grains (Storck and Teague, 1952; Moritz, 2002; Sinclair, 2010). In fact, during ancient times grain was held in such high esteem that the cultivation, harvest and threshing of grain, and the making of food, beer and bread, was often depicted in artwork, especially in funerary sculptures and tombs (for example, see Fig. 15.1).

Although agriculture and food production allowed for the division and diversification of labour, and the growth of other disciplines as societies around the world developed, agriculture comprised the bulk of employment by most premodern societies. That began to change, however, during the Industrial Revolution, as mechanization allowed farming practices to become much more efficient (Fig. 15.2). This development was further accelerated in the 20th century, with plant hybridization techniques, development of fertilizers, herbicides, insecticides and other technologies. And computers and biotechnology increased this acceleration toward the end of the century.

During the last 200 years, scientific discoveries and engineering advances have impacted not just agriculture and food production but all aspects of societies around the world. Modern cereal grain production and processing retain some semblance to those of our ancestors, but on a much larger and advanced scale (Fig. 15.3). And technologies have been accelerating and compounding upon each other. In fact, agriculture has become

FIGURE 15.1   Painting in the tomb of Nakht (an official who lived in approximately the 14th century, BC), located near Luxor, Egypt. Note the agricultural scenes, such as harvesting (middle), threshing (top) and storing of grain (second down from top). https://commons. wikimedia.org/wiki/File:Tomb_of_Nakht_(2).jpg.

so efficient in most developed countries that less than 2% of these populations are directly employed in the agricultural production of grains, foods and animals. This has been true in developed nations for some time and now is increasingly so in various developing countries – although such a drastic reduction in proportion is quite a distant possibility.

FIGURE 15.2 Rosentrater's family farm c.1940. Note how horse-drawn equipment was used – a direct result of developments of the Industrial Revolution – and mechanization was beginning to increase the productivity of agriculture.

FIGURE 15.3 Rosentrater's family farm c.1980. Note how the grain farming operation was highly mechanized (several items from the chapter on modern grain storage are seen in this photo).

Historically, as societies develop and technologies evolve, and food production becomes more efficient, greater human populations can be supported (food is only one of a vast number of factors upon which populations depend, though). This became especially true during and after the Industrial Revolution, when populations in many countries began to rapidly increase. In fact, overall global population reached approximately 7 billion in 2011 (October 31, according to UN estimates); it is anticipated to reach nearly 10 billion by 2050 (UN, 2017); and, depending upon the growth model used, may eventually reach nearly 12 billion by 2100 (Fig. 15.4). Population growth and changes in population density are not

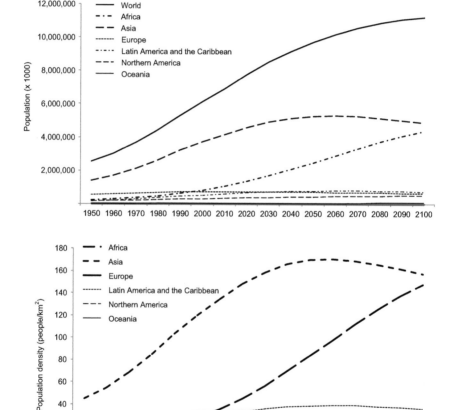

FIGURE 15.4   (Top) Global and regional populations over time. (Bottom) Regional population densities over time. *Predictions based upon data from the United Nations, 2017. 2015 Revision of World Population Prospects. United Nations, Population Division. Available online:* https://esa.un.org/unpd/wpp/.

distributed equally around the globe, though. For decades, Asia (primarily China and India) has been the fastest growing region of the planet. But Africa is quickly gaining on Asia both in terms of numbers of people and human density. As shown in Fig. 15.4, regions that are considered 'developed' appear to plateau, or even decline, over time. The greatest growth that will occur during the next 100 years is anticipated to occur in developing countries (many of which are in Africa) – and this may strain already fragile ecosystems.

Undoubtedly, as the global population continues to grow, there is an ever-growing need to increase the available food supply. Alas, even with all of the fundamentals and advances in cereal science that we have discussed in this book, this is not a simple problem to address. Regional differences in agricultural productivity, climate and technology are often complicated by economics, politics and distribution issues. In developing countries, increasing the production of cereal grains is often seen as one way to improve food security. Ironically, in many developed countries, however, increasing incidence of coeliac disease, diabetes, cancer, changing dietary attitudes and perceptions and shifting consumer choices have often led to a reduction in demand of various cereals, at least for food products (especially wheat – as discussed earlier in the book) (McGill et al., 2015; Jones and Sheats, 2016).

Indeed, cereal grains do still represent the bulk of calories in the diets of many cultures. As we discussed earlier in the book, there is growing demand for use of cereals to manufacture biofuels, biochemicals, bioplastics and other biobased products – not just in developed countries but also in many developing countries. Unfortunately, there is not an unlimited supply of arable, productive or available land with which to continue to expand grain supplies to meet so many competing demands (Foley et al., 2011; Ray et al., 2012; Ausubel et al., 2013).

Therefore, it is incumbent upon humanity (especially those of us working in this industry) to produce more grains (as well as other foods), and we must also become more efficient in growing, harvesting, storing, processing and preserving of grains and grain-based products. There are several technologies that are currently being used to help achieve these goals (each of which deserves a chapter of its own – perhaps in a subsequent edition of this book):

- Over the last few decades use of genetic engineering tools to produce genetically modified organisms (GMOs) has allowed plant scientists to implement changes faster than traditional plant breeding, and has led to tremendous increases in yields, as well as changes in field practices (such as a reduction in freshwater ecotoxicity (by ~50%) due to changes in pesticide use) (Yang and Suh, 2015). Even though many studies have shown compositional equivalence to non-GMO counterparts, and no allergenicity due to the modifications, there is still not universal acceptance of these types of grains, or consensus that these are safe and beneficial in the long run – from either the scientific community or the public at large (Domingo and Bordonaba, 2011; Séralini et al., 2011; Nicolia et al., 2014; Hilbeck et al., 2015). Although GMOs have been widely adopted in the United States (e.g., corn and soybeans) and other countries, many countries have very strict regulations, slow approval processes and limited adoption of GMO varieties (some, in fact, have outright bans). Undoubtedly, the controversies and potential benefits of GMOs will continue.

- The concepts of 'big data' and data analytics are increasingly being used for production agriculture, in which site-specific information regarding soil fertility and soil erodibility can be monitored, fertilizers and other chemicals can be dynamically applied, and crop yields can be determined for specific segments of each field (Fig. 15.5 illustrates one such example) – thus production agriculture's paradigm is shifting from a 'one-size-fits-all' approach to grain production to one that is tailored to specific portions of each field. Further, these computational and visualization techniques are increasingly being used to understand chemical analyses of grain components, distributions throughout the field,

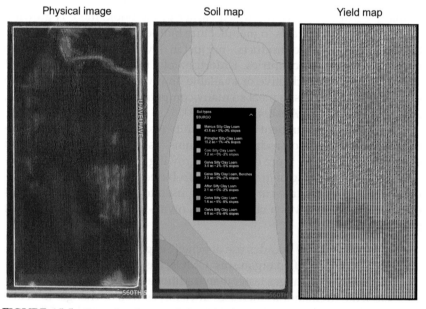

| Physical image | Soil map | Yield map |

FIGURE 15.5 Example of a map that can be developed for agricultural fields in order to understand site-specific information (e.g., fertilizer or herbicide required, precipitation received, yield achieved, etc.). Precision farming is just one approach to using big data in cereal grains production.

state or country, as well as estimations about processing yields and quality; and even societal impacts are being modelled (e.g., potential biofuel production, human health, disease probabilities, etc.). Advances in computer hardware and software have facilitated the 'age of information', but also the desire to reduce production costs and environmental impacts.

- Additional advances are occurring down the supply chain, specifically in processing plants, where sensors, instrumentation and control systems are increasingly being used to monitor processes, improve efficiencies and reduce consumption and thus operating costs. Benefits include reductions in electricity, natural gas, water and raw materials used, as well as optimizing conversion processes and processing yields, while minimizing waste materials.

What may be the greatest challenge to face humanity, however, may be climate change (i.e., global warming). The United Nations, through the Intergovernmental Panel on Climate Change (Meehl et al., 2004; IPCC, 2014, 2017), as well as scientists and researchers around the world, have determined that human-induced climate change is indeed occurring due to anthropogenic activities (including industrial, agricultural, residential, commercial and other human actions), they have been developing estimates of the eventual severity of climate change, as well as potential approaches to limit the extent of this warming. If we are to have any chance of limiting anthropogenic impacts on the climate (because it appears that we cannot now stop these impacts), it will require all aspects of society to work toward this end, including all stages in the agricultural supply chain, from producers to processors to consumers.

Truly, the challenges facing humanity during the next several decades are profound. But we have great hope that we can all rise to meet these challenges – the current generation as well as coming generations. And the production and processing of cereal grains will continue to play a pivotal role in this regard.

In this work we have covered major topics and issues related to cereal grains, including production, properties and utilization of major crops. We have discussed fundamentals as well as applications and industrial practices. We have addressed major consumer trends and how they are affecting demand for cereal products – both for existing as well as new products. Although it cannot be completely exhaustive in nature, this work should serve as a solid introduction to and resource base for cereal grains.

We feel it is of critical importance to society to provide a work that can be used to help train the next generation of professionals in the cereal grains industry, as well as to provide a comprehensive reference for those already in the field. Hopefully, we have succeeded in achieving these aims.

The cultivation and use of cereal grains have, to a great extent, both paralleled and helped humanity develop to its current state, and will likely continue to do so into the future. After all, in the words of Daniel Webster, an American statesman: 'Let us not forget that the cultivation of the earth is the most important labor of man. When tillage begins, other arts will follow'.

References

Ausubel, J.H., Wernick, I.K., Waggoner, P.E., 2013. Peak farmland and the prospect for land sparing. Population and Development Review 38 (1), 221–242.

Domingo, J.L., Bordonaba, J.G., 2011. A literature review on the safety assessment of genetically modified plants. Environment International 37 (4), 734–742.

Foley, J.A., Ramankutty, N., Brauman, K.A., Cassidy, E.S., Gerber, J.S., Johnston, M., Mueller, N.D., O'connell, C., Ray, D.K., West, P.C., Balzer, C., Bennett, E.M., Carpenter, S.R., Hill, J., Monfreda, C., Polasky, S., Rockströ, ¨M.J., Sheehan, J., Siebert, S., Tilman, D., Zaks, D.P.M., 2011. Solutions for a cultivated planet. Nature 478, 337–342.

Hilbeck, A., Binimelis, R., Defarge, N., Steinbrecher, R., Székács, A., Wickson, F., Antoniou, M., Bereano, P.L., Clark, E.A., Hansen, M., Novotny, E., Heinemann, J., Meyer, H., Shiva, V., Wynne, B., 2015. No scientific consensus on GMO safety. Environmental Sciences Europe 27, 1–6.

IPCC, 2014. Climate change 2014: synthesis report. In: Core Writing Team, Pachauri, R.K., Meyer, L.A. (Eds.), Contribution of Working Groups I, II and III to the Fifth Assessment Report of the Intergovernmental Panel on Climate Change, IPCC, Geneva, Switzerland, 151 pp.

IPCC, 2017. Intergovernmental Panel on Climate Change. Available online: http://www.ipcc.ch.

Jones, J.M., Sheats, D.B., 2016. Consumer Trends in Grain Consumption. Reference Module in Food Science. http://www.sciencedirect.com/science/article/pii/B978008100596500072X.

McGill, C.R., Fulgoni III, V.L., Devareddy, L., 2015. Ten-year trends in fiber and whole grain intakes and food sources for the United States population: National Health and Nutrition Examination Survey 2001–2010. Nutrients 7, 1119–1130.

Meehl, G.A., Washington, W.M., Ammann, C.M., Arblaster, J.M., Wigley, T.M.L., Tebaldi, C., 2004. Combinations of natural and anthropogenic forcings in twentieth-century climate. Journal of Climate 17, 3721–3727.

Moritz, L.A., 2002. Grain-Mills and Flour in Classical Antiquity. British Academy, UK.

Nicolia, A., Manzo, A., Veronesi, F., Rosellini, D., 2014. An overview of the last 10 years of genetically engineered crop safety research. Critical Reviews in Biotechnology 34 (1), 77–88.

Ray, D.K., Ramankutty, N., Mueller, N.D., West, P.C., Foley, J.A., 2012. Recent patterns of crop yield growth and stagnation. Nature Communications 3, 1293.

Séralini, G.-E., Mesnage, R., Clair, E., Gress, S., Spiroux De Vendômois, J., Cellier, D., 2011. Genetically modified crops safety assessments: present limits and possible improvements. Environmental Sciences Europe 23, 1–10.

Sinclair, T.R., 2010. Bread, Beer and the Seeds of Change: Agriculture's Imprint on World History. Centre for Agriculture and Biosciences International, Oxfordshire, UK.

Storck, J., Teague, W.D., 1952. Flour for Man's Bread: A History of Milling. University of Minnesota Press, MN, USA.

U.N., 2017, 2015. Revision of World Population Prospects. United Nations, Population Division. Available online: https://esa.un.org/unpd/wpp/.

Yang, Y., Suh, S., 2015. Changes in environmental impacts of major crops in the US. Environmental Research Letters 10 (9), 1–9. Available online: https://doi.org/10.1088/1748-9326/10/9/094016.

Abbreviations, units, equivalents

ABBREVIATIONS

AACCI	American Association of Cereal Chemists, International
ACP	acid calcium phosphate
ADA	azodicarbonamide
ADD	activated dough development
ADF	acid detergent fibre
ADY	active dried yeast
an.	annum (year)
ARFA	Air Radio Frequency Assisted
a_w	water activity
b	billion (10^9)
BC	before Christ (alternatively, B.C.E., Before Common Era)
BFP	bulk fermentation process
BHA	butylated hydroxy anisole
BHT	butylated hydroxy toluene
BOD	biological oxygen demand
b.p.	boiling point
BP	British Patent
BP	British Pharmacopoeia
BS	British Standard
BSI	British Standards Institute
Bz	benzene
cap.	head (capitum)
CBN	Commission of Biological Nomenclature
CBP	Chorleywood Bread Process

Cent.	central
cf.	compare
Ch.	chapter
CMC	carboxy methyl cellulose
COMA	Committee on Medical Aspects of Food Policy
concn	concentration
COT	Committee on Toxicology
Cs.	coarse
CSIR	Council for Scientific and Industrial Research
CSL	calcium stearoyl-2-lactylate
CTAB	cetyltrimethylammonium bromide
CWAD	Canadian Western Amber Durum wheat
CWRS	Canadian Western Red Spring wheat
CWRW	Canadian Western Red Winter wheat
CWSWS	Canadian Western Soft White Spring wheat
CWU	Canadian Western Utility wheat
D	dextrorotatory
DATEM	diacetyl tartaric esters of mono- and diglycerides of fatty acids dry basis
d.b.	dry basis
DDG	distillers dried grains
DDGS	distillers dried grains with solubles
DE	dextrose equivalents
DH	Department of Health
DHA	dehydroascorbic acid
d.m.	dry matter
DNA	deoxyribonucleic acid
DR	Democratic Republic (Germany)
DRV	dietary reference value
DSS	Department of Social Security
E	east
EAR	estimated average requirement
EC	European Community (also European Union)

edn.	edition
EP	European Patent
FAO	Food and Agriculture Organisation of the United Nations
FDA	Food and Drug Administration (of the United States)
FFA	free fatty acid
FMBRA	Flour Milling and Baking Research Association
FR	Federal Republic (Germany)
G	giga (10^9)
GATT	General Agreement on Tariffs and Trade
GC	grade colour
gg	grit gauze
GMS	glycerol monostearate
hd	head
HDL	high-density lipoprotein
HFCS	high-fructose corn syrup
HFSS	high-fructose starch syrup
HGCA	Home-Grown Cereals Authority
HPLC	high-performance liquid chromatography
HPMC	hydroxy propyl methyl cellulose
HRS	Hard Red Spring wheat
HRW	Hard Red Winter wheat
HTST	high temperature short time
IADY	instant active dried yeast
ICC	International Association of Cereal Science and Technology
ISO	International Organisation for Standardisation
IUPAC	International Union of Pure and Applied Chemistry
k	kilo (10^3)
L	laevorotatory
LDL	low-density lipoprotein
LRNI	Lower Reference Nutrient Intake
LSD	lysergic acid
m	milli (10^{-3})
μ	micro (10^{-6})

M	molar
MAFF	Ministry of Agriculture, Fisheries and Food
max.	maximum
m.c.	moisture content
Med.	medium
Midds	middlings
min.	minimum
mol.	molecular
M_r	relative molecular mass
MRC	Medical Research Council
MRL	maximum residue level
Mt	million metric tonnes
MW	microwave
n	nano (10^{-9})
N	normal
N	nitrogen
NABIM	National Association of British and Irish Millers
N.B.	nota bene
NDF	neutral detergent fibre
NIR(S)	Near-infrared reflectance (spectroscopy)
No.	number
p., pp.	page, pages
PADY	protected active dried yeast
PAGE	polyacrylamide gel electrophoresis
Pat.	patent
Propn	proportion
pt	part(s)
q.v.	quod vide; which see, reference other discussion
RDA	recommended daily amount (of nutrients)
rDNA	recombinant DNA
RF	radio frequency
r.h.	relative humidity

RNI	Reference Nutrient Intake
r.p.m.	revolutions per minute
RVA	rapid viscoanalyzer
S	south
SAP	sodium aluminium phosphate
SAPP	sodium acid pyrophosphate
SDS	sodium dodecyl sulphate
SEM	solvent extraction milling
SGP	starch granule protein
SH	disulphide
SI	Statutory Instrument
sp., spp.	species
sp.gr.	specific gravity
Sr	strontium
SRW	Soft Red Winter wheat
SS	sulphydryl
SSL	sodium stearoyl-2-lactylate
ssp.	subspecies
TD	tempering-degerming
temp.	temperature
t.v.p.	textured vegetable products/proteins
UK	United Kingdom
UN	United Nations
US	United States of America
USDA	United States Department of Agriculture
USP	United States Patent
UV	ultraviolet
vac.	vacuum
viz.	videlicet; namely, that is to say
v/v	volume for volume
W	west
w.	wire bolting cloth

w/w	weight for weight
WHO	World Health Organisation
wt	weight
yr	year
°	degree
<	less (fewer) than
>	greater (more) than
≯	not more than (alternatively, ≤)
%	percentage

UNITS

ac	acre (43,560 ft^2)
atm	atmosphere
a_w	vapour pressure, water activity
bar	steam pressure
Bé	Baume (hydrometer scale)
Btu	British thermal unit
bu	bushel (8 imperial gal)
°C	degree Celsius (centigrade scale)
cal	calorie
Cal	Calorie (kcal)
Ci	Curie
cm	centimetre (10^{-2} m)
cwt	hundredweight (112 lb)[a]
cwt (U.S.)	hundredweight (100 lb)
dL	decilitre (100 mL)
°F	degree Fahrenheit[a]
FN	Falling Number
ft	foot, feet
FU	Farrand unit

g	gramme
gal	gallon (imperial)
GJ	gigajoule (10^9 J)
h	hour
ha	hectare (10^4 m^2)
hL	hectolitre (10^2 L)
h.p.	horse power
HZ	hertz
in.	inch
i.u.	international unit
J	joule
kcal	kilocalorie (10^3 cal)
kg	kilogramme (10^3 g)
kJ	kilojoule (10^3 J)
kL	kilolitre (10^3 L)
kN	kilonewton (10^3 N)
kW	kilowatt (10^3 W)
L	litre
lb	pound (unit of force or mass)
m	metre
mM	millimolar
mg/kg	milligrammes per kilogramme (=ppm)
µg/kg	microgrammes per kilogramme (=ppb)
Mha	megahectare (10^6 ha)
MHz	megahertz (10^6 Hz)
MJ	megajoule (10^6 J)
MN	meganewton (10^6 N)
Mt	megatonne (10^6 t)
min.	minute (time)
mL	millilitre (10^{-3} L)
mm	millimetre (10^{-3} m)
µg	microgramme (10^{-6} g)

μm	micrometre (10^{-6} m)
N	Newton (unit of force)
nm	nanometre (10^{-9} m)
oz	ounce
pCi	picoCurie (10^{-12} Ci)
ppb	parts per billion (μg/kg)
ppm	parts per million (mg/kg)
psi	pounds (force) per square inch
q	quintal (10^2 kg)
rad	unit of radiation
rev	revolution
sec	second (time)
sk	sack (280 lb of flour)
t	metric tonne (10^3 kg; 2204 lb)
ton	long ton (2240 lb)[a]
ton (U.S.)	short ton (2000 lb)
W	watt
Wh	watt-hour
Yd	yard

[a]Abolished in the United Kingdom, 31 December, 1980.

EQUIVALENTS

Metric units, viz., units of the SI (Système Internationale d'Unités), have been used throughout this book, but for the convenience of readers, particularly in those countries in which metric units are not yet adopted, some conversion factors are presented below.

Further information may be obtained from a National Physical Laboratory booklet Changing to the Metric System, by Pamela Anderton and P. H. Bigg, London: H.M.S.O., 1967. See also Flour Milling by J. F. Lockwood, Stockport: Henry Simon Ltd, fourth edition, 1960, Appendix 14.

Additional information can be found at the US National Institute of Standards and Technology. 2008. NIST Special Publication 330. Available online: https://www.nist.gov/pml/special-publication-330.

Cereal crop yields are given in quintals per hectare (q/ha) in preference to the SI unit of kilogrammes per 100 square metres (kg/100 m²), which is numerically equivalent to quintals per hectare.

Length

1 m = 39.37 in.
1 m = 3.281 ft
1 m = 1.0936 yd
1 in. = 2.54 cm
1 mi = 5280 ft
1 mi = 1.609 km

Area

$1\,m^2 = 10.76\,ft^2$
$1\,ha = 2.471\,ac = 107{,}639\,ft^2 = 10{,}000\,m^2$
$1\,mi^2 = 2.59\,km^2 = 640\,ac = 259\,ha$

Mass

1 g = 0.0353 oz
1 kg = 2.205 lb
1 t = 1000 kg = 0.984 ton
1 t = 1.1023 ton (US short) = 0.98 ton (UK long) = 2204.62 lb
1 t = 10 q

Mass per unit area

1 t/ha = 0.398 ton/ac
1 q/ha = 0.79 cwt/ac

Volume

$1\,L = 1.76\,pint\,(UK) = 1.86\,pint\,(US) = 61.02\,in.^3 = 0.035\,ft^3$
1 L = 0.227 gal
1 L = 0.0275 bu (US)
$1\,bu\,(US) = 1.2445\,ft^3 = 0.046\,yd^3 = 0.035\,m^3 = 35.24\,L$
$1\,m^3 = 1000\,L = 28.38\,bu$
$1\,m^3 = 220\,gal$

Density

$1\,g/cm^3 = 62.5\,lb/ft^3$
$1\,kg/m^3 = 0.001\,g/cm^3 = 0.0625\,lb/ft^3$
$1\,g/cm^3 = 1\,kg/L = 1000\,kg/m^3$
1 kg/hL = 0.802 lb/bu

Energy/work

$1J = 0.239\,cal = 0.000278\,Wh$
$1\,kJ = 0.945\,Btu$
$1\,Btu = 1055\,J$
$1\,MJ = 0.278\,kWh$
$1\,kWh = 3.6 \times 10^6\,J$
$1\,kcal = 1\,Cal = 1.163\,Wh = 4184\,J$
$1\,therm = 100{,}000\,Btu$

Power

$1\,hp = 550\,ft\,lb/s = 745.7\,W$
$1\,W = 1\,J/s = 1\,Nm/s = 3.412\,Btu/h = 0.737\,ft\,lb/s$
$1\,kW = 1.34\,hp$

Pressure

$1\,kg/cm^2 = 14.232\,lb/in.^2$
$1\,kg/cm^2 = 0.968\,atm$
$1\,kN/m^2 = 1\,kPa = 1000\,Pa = 0.145\,lb/in.^2$
$1\,kN/m^2 = 0.0099\,atm = 0.01\,bar$

Temperature

$°F = (°C \times 1.8) + 32$
$°C = (°F - 32) \times 5/9$
$K = °C + 273.15\,K$

Concentration

$1\,mg/100g = 4.54\,mg/lb$
$1\,mg/100g = 0.0447\,oz/sk$
$1\,mg/100g = 0.3576\,oz/ton$
$1\,mg/100g = 0.3518\,oz/t$
$1\,mg/lb = 0.00985\,oz/sk$
$1\,mg/lb = 0.0788\,oz/ton$
$1\,\mu g/g = 1\,mg/kg = 1\,ppm$
$1\,\mu g/kg = 1\,ppb$
$1\,ppm = 0.0001\%$
$1\,ppm = 1000\,ppb$

Dressing surface

$1\,m^2/24h/100kg\,(wheat) = 469.5\,ft^2/sk/h\,(flour)$

Roller surface

1 cm/24 h/100 kg wheat = 16.7 in./sk/h flour (milled to 73% extraction rate)

1 in./196-lb barrel of flour (United States) = 1.43 in./sk

1 in./100-lb cwt flour (United States) = 2.8 in./sk

Index

Edwards Brothers Inc.
Ann Arbor MI. USA
May 23, 2018